ENCYCLOPEDIA OF MATHEMATICS AND ITS APPLICATIONS

Volume 44

Convex bodies: the Brunn–Minkowski theory

ENCYCLOPEDIA OF MATHEMATICS AND ITS APPLICATIONS

1 Luis A. Santaló *Integral Geometry and Geometric Probability*
2 George E. Andrews *The Theory of Partitions*
3 Robert J. McEliece *The Theory of Information and Coding: A Mathematical Framework for Communication*
4 Willard Miller, Jr *Symmetry and Separation of Variables*
5 David Ruelle *Thermodynamic Formalism: The Mathematical Structures of Classical Equilibrium Statistical Mechanics*
6 Henry Minc *Permanents*
7 Fred S. Roberts *Measurement Theory with Applications to Decisionmaking, Utility, and the Social Sciences*
8 L.C. Biedenharn and J.D. Louck *Angular Momentum in Quantum Physics: Theory and Application*
9 Lawrence C. Biedenharn and James D. Louck *The Racah–Wigner Algebra in Quantum Physics: Theory and Application*
10 John D. Dollard and Charles N. Friedman *Product Integration with Applications to Differential Equations*
11 William B. Jones and W.J. Thron *Continued Fractions: Analytic Theory and Applications*
12 Nathaniel F.G. Martin and James W. England *Mathematical Theory of Entropy*
13 George A. Baker, Jr and Peter R. Graves-Morris *Padé Approximants, Part II: Extensions and Applications*
15 E.C. Beltrametti and G. Cassinelli *The Logic of Quantum Mechanics*
16 G.D. James and A. Kerber *The Representation Theory of the Symmetric Group*
17 M. Lothaire *Combinatorics on Words*
18 H.O. Fattorini *The Cauchy Problem*
19 G.G. Lorentz, K. Jetter and S.D. Riemenschneider *Birkhoff Interpolation*
20 Rudolf Lidl and Harald Niedereitter *Finite Fields*
21 W.T. Tutte *Graph Theory*
22 Julio R. Bastida *Field Extensions and Galois Theory*
23 John Rozier Cannon *The One-Dimensional Heat Equation*
24 Stan Wagon *The Banach–Tarski Paradox*
25 Arto Salomaa *Computation and Automata*
26 N. White (ed) *Theory of Matroids*
27 N.H. Bingham, C.M. Goldie and J.L. Teugels *Regular Variation*
28 P.P. Petrushev and V.A. Popov *Rational Approximation of Real Functions*
29 N. White (ed) *Combinatorial Geometries*
30 M. Pohst and H. Zassenhaus *Algorithmic Algebraic Number Theory*
31 J. Aczel and J. Dhombres *Functional Equations in Several Variables*
32 M. Kuczma, B. Choczewski and R. Ger *Iterative Functional Equations*
33 R.V. Ambartzumian *Factorization Calculus and Geometric Probability*
34 G. Gripenberg, S.-O. Londen and O. Staffans *Volterra Integral and Functional Equations*
35 George Gasper and Mizan Rahman *Basic Hypergeometric Series*
36 Erik Torgersen *Comparison of Statistical Experiments*
37 Arnold Neumaier *Interval Methods for Systems of Equations*
38 N. Korneichuk *Exact Constants in Approximation Theory*
39 Richard A. Brualdi and Herbert J. Ryser *Combinatorial Matrix Theory*
40 N. White (ed) *Matroid Applications*
41 Shoichiro Sakai *Operator Algebras in Dynamical Systems: The Theory of Unbounded Derivations in C*-algebras*
42 W. Hodges *Model Theory*

ENCYCLOPEDIA OF MATHEMATICS AND ITS APPLICATIONS

Convex Bodies: The Brunn–Minkowski Theory

ROLF SCHNEIDER

University of Freiburg

CAMBRIDGE
UNIVERSITY PRESS

Published by the Press Syndicate of the University of Cambridge
The Pitt Building, Trumpington Street, Cambridge CB2 1RP
40 West 20th Street, New York, NY 10011-4211, USA
10 Stamford Road, Oakleigh, Victoria 3166, Australia

First published 1993

Printed in Great Britain at the University Press, Cambridge

A catalogue record for this book is available from the British Library

Library of Congress cataloguing in publication data
Schneider, Rolf, 1940–
Convex bodies: the Brunn–Minkowski theory / Rolf Schneider.
p. cm. – (Encyclopedia of mathematics and its applications; v. 44)
Includes bibliographical references and index.
ISBN 0-521-35220-7
1. Convex bodies. I. Title. II. Series.
QA649.S353 1993
516.3′74–dc20 92-11481 CIP

ISBN 0 521 35220 7 hardback

KT

CONTENTS

PREFACE

The Brunn–Minkowski theory is the classical core of the geometry of convex bodies. It originated with the thesis of Hermann Brunn in 1887 and is in its essential parts the creation of Hermann Minkowski, around the turn of the century. The well-known survey of Bonnesen and Fenchel in 1934 collected what was already an impressive body of results, though important developments were still to come, through the work of A.D. Aleksandrov and others in the thirties. In recent decades, the theory of convex bodies has expanded considerably; new topics have been developed and originally neglected branches of the subject have gained in interest. For instance, the combinatorial aspects, the theory of convex polytopes and the local theory of Banach spaces attract particular attention now. Nevertheless, the Brunn–Minkowski theory has remained of constant interest owing to its various new applications, its connections with other fields, and the challenge of some resistant open problems.

Aiming at a brief characterization of Brunn–Minkowski theory, one might say that it is the result of merging two elementary notions for point sets in Euclidean space: vector addition and volume. The vector addition of convex bodies, usually called Minkowski addition, has many facets of independent geometric interest. Combined with volume, it leads to the fundamental Brunn–Minkowski inequality and the notion of mixed volumes. The latter satisfy a series of inequalities which, due to their flexibility, solve many extremal problems and yield several uniqueness results. Looking at mixed volumes from a local point of view, one is led to mixed area measures. Quermassintegrals, or Minkowski functionals, and their local versions, surface area measures and curvature measures, are a special case of mixed volumes and mixed area measures. They are related to the differential geometry of convex hypersurfaces and to integral geometry.

Chapter 1 of the present book treats the basic properties of convex bodies and thus lays the foundations for subsequent developments. This chapter does not claim much originality; in large parts, it follows the procedures in standard books such as McMullen & Shephard [26], Roberts & Varberg [28], and Rockafellar [29]. Together with Sections 2.1, 2.2, 2.4, and 2.5, it serves as a general introduction to the metric geometry of convex bodies. Chapter 2 is devoted to the boundary structure of convex bodies. Most of its material is needed later, except for Section 2.6, on generic boundary structure, which just rounds off the picture. Minkowski addition is the subject of Chapter 3. Several different aspects are considered here such as decomposability, approximation problems with special regard to addition, additive maps and sums of segments. Quermassintegrals, which constitute a fundamental class of functionals on convex bodies, are studied in Chapter 4, where they are viewed as specializations of curvature measures, their local versions. For these, some integral-geometric formulae are established in Section 4.5. Here I try to follow the tradition set by Blaschke and Hadwiger of incorporating parts of integral geometry into the theory of convex bodies. Some of this, however, is also a necessary prerequisite for Section 4.6. The remaining part of the book is devoted to mixed volumes and their applications. Chapter 5 develops the basic properties of mixed volumes and mixed area measures and treats special formulae, extensions and analogues. Chapter 6, the heart of the book, is devoted to the inequalities satisfied by mixed volumes, with special emphasis on improvements, the equality cases (as far as they are known) and stability questions. Chapter 7 presents a small selection of applications. The classical theorems of Minkowski and the Aleksandrov–Fenchel–Jessen theorem are treated here, the latter in refined versions. Section 7.4 serves as an overview of affine extremal problems for convex bodies. In this promising field, Brunn–Minkowski theory is of some use, but it appears that for the solution of some long-standing open problems new methods still have to be invented.

Concerning the choice of topics treated in this book, I wish to point out that it is guided by Minkowski's original work also in the following sense. Some subjects that Minkowski touched only briefly have later expanded considerably, and I pay special attention to these. Examples are projection bodies (zonoids), tangential bodies, the use of spherical harmonics in convexity and strengthenings of Minkowskian inequalities in the form of stability estimates.

The necessary prerequisites for reading this book are modest: the usual geometry of Euclidean space, elementary analysis, and basic measure and integration theory for Chapter 4. Occasionally, use is made of spherical harmonics; relevant information is collected in the Appendix. My intended attitude towards the presentation of proofs cannot be summarized better than by quoting from the preface to the book on Hausdorff measures by C.A. Rogers: 'As the book is largely based on lectures, and as I like my students to follow my lectures, proofs are given in great detail; this may bore the mature mathematician, but it will I believe, be a great help to anyone trying to learn the subject *ab initio*.' On the other hand, some important results are stated as theorems but not proved, since this would lead us too far from the main theme, and no proofs are given in the survey sections 5.4, 6.8, and 7.4.

The notes at the end of nearly all sections are an essential part of the book. As a rule, this is where I have given references to original literature, considered questions of priority, made various comments and, in particular, given hints about applications, generalizations and ramifications. As an important purpose of the notes is to demonstrate the connections of convex geometry with other fields, some notes do take us further from the main theme of the book, mentioning, for example, infinite-dimensional results or non-convex sets or giving more detailed information on applications in, for instance, stochastic geometry.

The list of references does not have much overlap with the older bibliographies in the books by Bonnesen & Fenchel and by Hadwiger. Hence, a reader wishing to have a more complete picture should consult these bibliographies also, as well as those in the survey articles listed in part B of the References.

My thanks go to Sabine Linsenbold for her careful typing of the manuscript and to Daniel Hug who read the typescript and made many valuable comments and suggestions.

CONVENTIONS AND NOTATION

Here we shall fix our notation and collect some basic definitions. We shall work in n-dimensional real Euclidean vector space, $\mathbb{E}^n$, with origin o, scalar product $\langle \cdot, \cdot \rangle$ and induced norm $|\cdot|$. We shall not distinguish formally between the vector space $\mathbb{E}^n$ and its corresponding affine space, although our alternating use of the words 'vector' and 'point' is deliberate and should support the reader's intuition. As a rule, elements of $\mathbb{E}^n$ are denoted by lower-case letters, subsets by capitals and real numbers by small Greek letters. However, in later chapters the reader will notice an increasing number of exceptions to this rule.

The vector $x \in \mathbb{E}^n$ is a *linear combination* of the vectors $x_1, \ldots, x_k \in \mathbb{E}^n$ if $x = \lambda_1 x_1 + \ldots + \lambda_k x_k$ with suitable $\lambda_1, \ldots, \lambda_k \in \mathbb{R}$. If such λ_i exist with $\lambda_1 + \ldots + \lambda_k = 1$, then x is an *affine combination* of $x_1, \ldots, x_k$. For $A \subset \mathbb{E}^n$, $\operatorname{lin} A$ ($\operatorname{aff} A$) denotes the *linear hull (affine hull)* of A; this is the set of all linear (affine) combinations of elements of A and at the same time the smallest linear subspace (affine subspace) of $\mathbb{E}^n$ containing A. Points $x_1, \ldots, x_k \in \mathbb{E}^n$ are *affinely independent* if none of them is an affine combination of the others, i.e., if

$$\sum_{i=1}^{k} \lambda_i x_i = o \quad \text{with } \lambda_i \in \mathbb{R} \text{ and } \sum_{i=1}^{k} \lambda_i = 0$$

implies that $\lambda_1 = \ldots = \lambda_k = 0$. This is equivalent to the linear independence of the vectors $x_2 - x_1, \ldots, x_k - x_1$. We may also define a map $\tau: \mathbb{E}^n \to \mathbb{E}^n \times \mathbb{R}$ by $\tau(x) := (x, 1)$; then $x_1, \ldots, x_k \in \mathbb{E}^n$ are affinely independent if and only if $\tau(x_1), \ldots, \tau(x_k)$ are linearly independent.

For $x, y \in \mathbb{E}^n$ we write

$$[x, y] := \{(1-\lambda)x + \lambda y \mid 0 \leqq \lambda \leqq 1\}$$

for the *closed segment* and

$$[x, y) := \{(1-\lambda)x + \lambda y \mid 0 \leqq \lambda < 1\}$$

for a *half-open segment*, both with endpoints x, y. For A, $B \subset \mathbb{E}^n$ and $\lambda \in \mathbb{R}$ we define

$$A + B := \{a + b \mid a \in A,\ b \in B\},$$
$$\lambda A := \{\lambda a \mid a \in A\},$$

and we write $-A$ for $(-1)A$, $A - B$ for $A + (-B)$ and $A + x$ for $A + \{x\}$, where $x \in \mathbb{E}^n$. The set $A + B$ is written $A \oplus B$ and called the *direct sum* of A and B if A and B are contained in complementary affine subspaces of $\mathbb{E}^n$.

By $\operatorname{cl} A$, $\operatorname{int} A$, $\operatorname{bd} A$ we denote, respectively, the closure, interior and boundary of a subset A of a topological space. For $A \subset \mathbb{E}^n$, the sets $\operatorname{relint} A$, $\operatorname{relbd} A$ are the relative interior and relative boundary, that is, the interior and boundary of A relative to its affine hull.

The scalar product in $\mathbb{E}^n$ will often be used to describe hyperplanes and halfspaces. A *hyperplane* of $\mathbb{E}^n$ can be written in the form

$$H_{u,\alpha} = \{x \in \mathbb{E}^n \mid \langle x, u\rangle = \alpha\}$$

with $u \in \mathbb{E}^n \setminus \{o\}$ and $\alpha \in \mathbb{R}$; here $H_{u,\alpha} = H_{v,\beta}$ if and only if $(v, \beta) = (\lambda u, \lambda\alpha)$ with $\lambda \neq 0$. We say that u is a *normal vector* of $H_{u,\alpha}$. The hyperplane $H_{u,\alpha}$ bounds the two *closed halfspaces*

$$H^-_{u,\alpha} := \{x \in \mathbb{E}^n \mid \langle x, u\rangle \leqq \alpha\},$$
$$H^+_{u,\alpha} := \{x \in \mathbb{E}^n \mid \langle x, u\rangle \geqq \alpha\}.$$

Occasionally we also use $\langle \cdot\,, \cdot \rangle$ to denote the scalar product on $\mathbb{E}^n \times \mathbb{R}$ given by

$$\langle (x, \xi), (y, \eta)\rangle = \langle x, y\rangle + \xi\eta.$$

An affine subspace of $\mathbb{E}^n$ is often called a *flat*, and the intersection of a flat with a closed halfspace meeting the flat but not entirely containing it will be called a *half-flat*. A one-dimensional flat is a *line* and a one-dimensional half-flat a *ray*.

The following metric notions will be used. For $x, y \in \mathbb{E}^n$ and $\emptyset \neq A \subset \mathbb{E}^n$, $|x - y|$ is the *distance* between x and y and

$$d(A, x) := \inf\{|x - a| \mid a \in A\}$$

is the distance of x from A. For a bounded set $\emptyset \neq A \subset \mathbb{E}^n$,

$$\operatorname{diam} A := \sup\{|x - y| \mid x, y \in A\}$$

is the *diameter* of A. We write

$$B(z, \rho) := \{x \in \mathbb{E}^n \mid |x - z| \leqq \rho\}$$

and

$$B_0(z, \rho) := \{x \in \mathbb{E}^n \mid |x - z| < \rho\}$$

respectively for the closed and open balls with centre $z \in \mathbb{E}^n$ and radius

$\rho > 0$. $B^n := B(o, 1)$ is the *unit ball* and

$$S^{n-1} := \{x \in \mathbb{E}^n \big| |x| = 1\}$$

the *unit sphere* of $\mathbb{E}^n$.

By $\mathcal{H}^k$ we denote the k-dimensional Hausdorff (outer) measure on $\mathbb{E}^n$, where $0 \leqq k \leqq n$. If A is a Borel subset of a k-dimensional flat E^k or a k-dimensional sphere S^k in $\mathbb{E}^n$, then $\mathcal{H}^k(A)$ coincides respectively with the k-dimensional Lebesgue measure of A computed in E^k or with the k-dimensional spherical Lebesgue measure of A computed in S^k. Hence, all integrations with respect to these Lebesgue measures can be expressed by means of the Hausdorff measure $\mathcal{H}^k$. In integrals with respect to $\mathcal{H}^n$ we often abbreviate $d\mathcal{H}^n(x)$ by dx. The n-dimensional measure of the unit ball in $\mathbb{E}^n$ is denoted by κ_n, and its surface area by ω_n, thus

$$\kappa_n = \mathcal{H}^n(B^n) = \frac{\pi^{n/2}}{\Gamma\left(1 + \frac{n}{2}\right)}, \quad \omega_n = \mathcal{H}^{n-1}(S^{n-1}) = n\kappa_n = \frac{2\pi^{n/2}}{\Gamma\left(\frac{n}{2}\right)}.$$

Linear maps, affine maps and isometries between Euclidean spaces are defined as usual. In particular, a map $\varphi: \mathbb{E}^n \to \mathbb{E}^n$ is a *translation* if $\varphi(x) = x + t$ for $x \in \mathbb{E}^n$ with some fixed vector $t \in \mathbb{E}^n$, the *translation vector*. The set $A + t$ is called the *translate* of A by t. The map φ is a *homothety* if $\varphi(x) = \lambda x + t$ for $x \in \mathbb{E}^n$ with some $\lambda > 0$ and some $t \in \mathbb{E}^n$. The set $\lambda A + t$ with $\lambda > 0$ is called a *homothet* of A. Sets A, B are called *positively homothetic* if $A = \lambda B + t$ with $t \in \mathbb{E}^n$ and $\lambda > 0$, and *homothetic* if either they are positively homothetic or one of them is a singleton (a one-pointed set). A *rigid motion* of $\mathbb{E}^n$ is an isometry of $\mathbb{E}^n$ onto itself, and it is a *rotation* if it is an isometry fixing the origin. Each rigid motion is the composition of a rotation and a translation. A rigid motion is called *proper* if it preserves the orientation of $\mathbb{E}^n$, otherwise it is called *improper*. A rotation of $\mathbb{E}^n$ is a linear map; it preserves the scalar product and can be represented, with respect to an orthonormal basis, by an orthogonal matrix; this matrix has determinant 1 if and only if the rotation is proper. The composition of a rigid motion and a *dilatation*, by which we mean a map $x \mapsto \lambda x$ with $\lambda > 0$, is called a *similarity*.

By $SO(n)$ we denote the group of proper rotations of $\mathbb{E}^n$. With the topology induced by the usual matrix norm it is a compact topological group. The group of proper rigid motions of $\mathbb{E}^n$ is denoted by G_n and topologized as usual. Also, the Grassmannian $G(n, k)$ of k-dimensional linear subspaces of $\mathbb{E}^n$ and the set $A(n, k)$ of k-dimensional affine subspaces of $\mathbb{E}^n$ are endowed with their standard topologies.

The Haar measures on $SO(n)$, G_n, $A(n, k)$ are denoted respectively by ν, μ, μ_k. We normalize ν by $\nu(SO(n)) = 1$. The normalizations of the measures μ and μ_k will be fixed in Section 4.5 when they are needed.

For an affine subspace E of $\mathbb{E}^n$ we denote by proj_E the orthogonal projection from $\mathbb{E}^n$ onto E. We often write $\text{proj}_E A =: A|E$ for $A \subset \mathbb{E}^n$ (since A is a set, no confusion with the restriction of a function, for example $f|E$, can arise).

Some final remarks are in order. Since any k-dimensional affine subspace E of $\mathbb{E}^n$ is the image of $\mathbb{E}^k$ under some isometry, it is clear (and common practice without mention) that all notions and results that have been established for $\mathbb{E}^k$ and are invariant under isometries can be applied in E; similarly for affine-invariant notions and results.

The following notational conventions will be useful in several places. If f is a homogeneous function on $\mathbb{E}^n$, then $\bar{f}$ denotes its restriction to the unit sphere S^{n-1}. Very often, mappings of the type $f: \mathcal{K} \times \mathbb{E}^n \to M$ will occur where $\mathcal{K}$ is some class of subsets of $\mathbb{E}^n$. In this case we usually abbreviate, for fixed $K \in \mathcal{K}$, the function $f(K, \cdot): \mathbb{E}^n \to M$ by f_K.

Finally we wish to point out that in definitions the word 'if' is always understood as 'if and only if'.

1

Basic convexity

1.1. Convex sets and combinations

A set $A \subset \mathbb{E}$ is *convex* if together with any two points x, y it contains the segment $[x, y]$, thus if

$$(1-\lambda)x + \lambda y \in A \quad \text{for } x, y \in A \quad \text{and } 0 \leqq \lambda \leqq 1.$$

Examples of convex sets are obvious; but observe also that $B_0(z, \rho) \cup A$ is convex if A is an arbitrary subset of the boundary of the open ball $B_0(z, \rho)$. As immediate consequences of the definition we note that intersections of convex sets are convex, affine images and pre-images of convex sets are convex and if A, B are convex, then $A + B$ and λA ($\lambda \in \mathbb{R}$) are convex.

Remark 1.1.1. For $A \subset \mathbb{E}^n$ and λ, $\mu > 0$ one trivially has $\lambda A + \mu A \supset (\lambda + \mu)A$. Equality (for all λ, $\mu > 0$) holds precisely if A is convex. In fact, if A is convex and $x \in \lambda A + \mu A$, then $x = \lambda a + \mu b$ with a, $b \in A$ and hence

$$x = (\lambda + \mu)\left(\frac{\lambda}{\lambda + \mu} a + \frac{\mu}{\lambda + \mu} b\right) \in (\lambda + \mu)A;$$

thus $\lambda A + \mu A = (\lambda + \mu)A$. The other direction of the assertion is trivial.

A set $A \subset \mathbb{E}^n$ is called a *convex cone* if A is convex and nonempty and if $x \in A$, $\lambda \geqq 0$ implies $\lambda x \in A$. Thus a nonempty set $A \subset \mathbb{E}^n$ is a convex cone if and only if A is closed under addition and under multiplication by non-negative real numbers.

By restricting affine and linear combinations to non-negative coefficients, one obtains the following two fundamental notions. The point $x \in \mathbb{E}^n$ is a *convex combination* of the points $x_1, \ldots, x_k \in \mathbb{E}^n$ if there are numbers $\lambda_1, \ldots, \lambda_k \in \mathbb{R}$ such that

$$x = \sum_{i=1}^{k} \lambda_i x_i,\ \lambda_i \geqq 0 \quad (i = 1, \ldots, k),\ \sum_{i=1}^{k} \lambda_i = 1.$$

Similarly, the vector $x \in \mathbb{E}^n$ is a *positive combination* of the vectors $x_1, \ldots, x_k \in \mathbb{E}^n$ if

$$x = \sum_{i=1}^{k} \lambda_i x_i \quad \text{with } \lambda_i \geqq 0 \quad (i = 1, \ldots, k).$$

For $A \subset \mathbb{E}^n$ the set of all convex combinations (positive combinations) of any finitely many elements of A is called the *convex hull* (*positive hull*) of A and is denoted by $\operatorname{conv} A$ ($\operatorname{pos} A$).

Theorem 1.1.2. *If $A \subset \mathbb{E}^n$ is convex, then* $\operatorname{conv} A = A$. *For an arbitrary set $A \subset \mathbb{E}^n$,* $\operatorname{conv} A$ *is the intersection of all convex subsets of $\mathbb{E}^n$ containing A. If $A, B \subset \mathbb{E}^n$, then* $\operatorname{conv}(A + B) = \operatorname{conv} A + \operatorname{conv} B$.

Proof. Let A be convex. Trivially, $A \subset \operatorname{conv} A$. By induction we show that A contains all convex combinations of any k points of A. For $k = 2$ this holds by the definition of convexity. Suppose that it holds for $k - 1$ and that $x = \lambda_1 x_1 + \ldots + \lambda_k x_k$ with $x_1, \ldots, x_k \in A$, $\lambda_1 + \ldots + \lambda_k = 1$ and $\lambda_1, \ldots, \lambda_k > 0$, without loss of generality. Then

$$x = (1 - \lambda_k) \sum_{i=1}^{k-1} \frac{\lambda_i}{1 - \lambda_k} x_i + \lambda_k x_k \in A$$

since

$$\frac{\lambda_i}{1 - \lambda_k} > 0, \quad \sum_{i=1}^{k-1} \frac{\lambda_i}{1 - \lambda_k} = 1$$

and hence

$$\sum_{i=1}^{k-1} \frac{\lambda_i}{1 - \lambda_k} x_i \in A$$

by hypothesis. This proves $A = \operatorname{conv} A$. For arbitrary $A \subset \mathbb{E}^n$ let $D(A)$ be the intersection of all convex sets $K \subset \mathbb{E}^n$ containing A. Since $A \subset \operatorname{conv} A$ and $\operatorname{conv} A$ is evidently convex, we have $D(A) \subset \operatorname{conv} A$. Each convex K with $A \subset K$ satisfies $\operatorname{conv} A \subset \operatorname{conv} K = K$, hence $\operatorname{conv} A \subset D(A)$, which proves the equality.

Let $A, B \subset \mathbb{E}^n$. Let $x \in \operatorname{conv}(A + B)$, thus

$$x = \sum_{i=1}^{k} \lambda_i (a_i + b_i) \quad \text{with } a_i \in A,\ \ b_i \in B,\ \ \lambda_i \geqq 0,\ \ \sum_{i=1}^{k} \lambda_i = 1$$

and hence $x = \sum \lambda_i a_i + \sum \lambda_i b_i \in \operatorname{conv} A + \operatorname{conv} B$. Let $x \in \operatorname{conv} A + \operatorname{conv} B$, thus

$$x = \sum_i \lambda_i a_i + \sum_j \mu_j b_j$$

with $a_i \in A$, $b_j \in B$, $\lambda_i, \mu_j \geqq 0$, $\sum \lambda_i = \sum \mu_j = 1$. We may write

$$x = \sum_{i,j} \lambda_i \mu_j (a_i + b_j)$$

and deduce that $x \in \operatorname{conv}(A + B)$. ∎

An immediate consequence is that $\operatorname{conv}(\operatorname{conv} A) = \operatorname{conv} A$.

Theorem 1.1.3. *If* $A \subset \mathbb{E}^n$ *is a convex cone, then* $\operatorname{pos} A = A$. *For a nonempty set* $A \subset \mathbb{E}^n$, $\operatorname{pos} A$ *is the intersection of all convex cones in* $\mathbb{E}^n$ *containing* A. *If* $A, B \subset \mathbb{E}^n$, *then* $\operatorname{pos}(A + B) = \operatorname{pos} A + \operatorname{pos} B$.

Proof. As above. ∎

The following result on the generation of convex hulls is fundamental.

Theorem 1.1.4 (Carathéodory's theorem). *If* $A \subset \mathbb{E}^n$ *and* $x \in \operatorname{conv} A$, *then* x *is a convex combination of affinely independent points of* A. *In particular,* x *is a convex combination of* $n + 1$ *or fewer points of* A.

Proof. The point $x \in \operatorname{conv} A$ has a representation

$$x = \sum_{i=1}^{k} \lambda_i x_i \quad \text{with } x_i \in A, \quad \lambda_i > 0, \quad \sum_{i=1}^{k} \lambda_i = 1$$

with some $k \in \mathbb{N}$, and we may assume that k is minimal. Suppose that $x_1, \ldots, x_k$ are affinely dependent. Then there are numbers $\alpha_1, \ldots, \alpha_k \in \mathbb{R}$, not all zero, with

$$\sum_{i=1}^{k} \alpha_i x_i = o \quad \text{and} \quad \sum_{i=1}^{k} \alpha_i = 0.$$

We can choose m such that λ_m/α_m is positive and, with this restriction, as small as possible (observe that all λ_i are positive and at least one α_i is positive). In the affine representation

$$x = \sum_{i=1}^{k} \left(\lambda_i - \frac{\lambda_m}{\alpha_m} \alpha_i \right) x_i$$

all coefficients are non-negative (trivially, if $\alpha_i \leqq 0$, otherwise by the choice of m) and at least one of them is zero. This contradicts the minimality of k. Thus $x_1, \ldots, x_k$ are affinely independent, which implies $k \leqq n + 1$. ∎

The convex hull of finitely many points is called a *polytope*. A *k-simplex* is the convex hull of $k + 1$ affinely independent points, and these points are the *vertices* of the simplex. Thus Carathéodory's theorem states that conv A is the union of all simplices with vertices in A.

Another equally simple and important result on convex hulls is the following.

Theorem 1.1.5 (Radon's theorem). *Each set of affinely dependent points (in particular, each set of at least $n+2$ points) in $\mathbb{E}^n$ can be expressed as the union of two disjoint sets whose convex hulls have a common point.*

Proof. If $x_1, \ldots, x_k$ are affinely dependent, there are numbers $\alpha_1, \ldots, \alpha_k \in \mathbb{R}$, not all zero, with

$$\sum_{i=1}^{k} \alpha_i x_i = o \quad \text{and} \quad \sum_{i=1}^{k} \alpha_i = 0.$$

We may assume, after renumbering, that $\alpha_i > 0$ precisely for $i = 1, \ldots, j$; then $1 \leqq j < k$ (at least one α_i is $\neq 0$, say > 0, but not all α_i are > 0). With

$$\alpha := \alpha_1 + \ldots + \alpha_j = -(\alpha_{j+1} + \ldots + \alpha_k) > 0$$

we obtain

$$x := \sum_{i=1}^{j} \frac{\alpha_i}{\alpha} x_i = \sum_{i=j+1}^{k} \left(-\frac{\alpha_i}{\alpha}\right) x_i$$

and thus $x \in \operatorname{conv}\{x_1, \ldots, x_j) \cap \operatorname{conv}\{x_{j+1}, \ldots, x_k\}$. The assertion follows. ■

From Radon's theorem one easily deduces Helly's theorem, a fundamental and typical result of the combinatorial geometry of convex sets.

Theorem 1.1.6 (Helly's theorem). *Let $A_1, \ldots, A_k \subset \mathbb{E}^n$ be convex sets. If any $n+1$ of these sets have a common point, then all the sets have a common point.*

Proof. Suppose that $k > n+1$ (for $k < n+1$ there is nothing to prove, and for $k = n+1$ the assertion is trivial) and that the assertion is proved for $k-1$ convex sets. Then for $i \in \{1, \ldots, k\}$ there exists a point

$$x_i \in A_1 \cap \ldots \cap \check{A}_i \cap \ldots \cap A_k$$

where $\check{A}_i$ indicates that A_i has been deleted. The $k \geqq n+2$ points $x_1, \ldots, x_k$ are affinely dependent; hence from Radon's theorem we can infer that, after renumbering, there is a point

$$x \in \operatorname{conv}\{x_1, \ldots, x_j\} \cap \operatorname{conv}\{x_{j+1}, \ldots, x_k\}$$

for some $j \in \{1, \ldots, k-1\}$. Because $x_1, \ldots, x_j \in A_{j+1}, \ldots, A_k$ we have

$$x \in \operatorname{conv}\{x_1, \ldots, x_j\} \subset A_{j+1} \cap \ldots \cap A_k,$$

similarly $x \in \operatorname{conv}\{x_{j+1}, \ldots, x_k\} \subset A_1 \cap \ldots \cap A_j$. ■

Here is a little example (another one is Theorem 1.3.11) to demonstrate how Helly's theorem can be applied to obtain elegant results of a similar nature:

Theorem 1.1.7. *Let $\mathcal{M}$ be a finite family of convex sets in $\mathbb{E}^n$ and let $K \subset \mathbb{E}^n$ be convex. If any $n+1$ elements of $\mathcal{M}$ are intersected by some translate of K, then all elements of $\mathcal{M}$ are intersected by a translate of K.*

Proof. Let $\mathcal{M} = \{A_1, \ldots, A_k\}$. To any $n+1$ elements of $\{1, \ldots, k\}$, say $1, \ldots, n+1$, there are $t \in \mathbb{E}^n$ and $x_i \in A_i \cap (K+t)$, hence $-t \in K - A_i$, for $i = 1, \ldots, n+1$. Thus any $n+1$ elements of the family $\{K - A_1, \ldots, K - A_k\}$ have nonempty intersection. By Helly's theorem there is a vector $-t \in \mathbb{E}^n$ with $-t \in K - A_i$ and hence $A_i \cap (K+t) \neq \emptyset$ for $i \in \{1, \ldots, k\}$. ■

Next we look at the interplay between convexity and topological properties. We start with a simple observation.

Lemma 1.1.8. *Let $A \subset \mathbb{E}^n$ be convex. If $x \in \operatorname{int} A$ and $y \in \operatorname{cl} A$, then $[x, y) \subset \operatorname{int} A$.*

Proof. Let $z = (1-\lambda)y + \lambda x$ with $0 < \lambda < 1$. We have $B(x, \rho) \subset A$ for some $\rho > 0$; put $B(o, \rho) =: U$. First we assume $y \in A$. Let $w \in \lambda U + z$, hence $w = \lambda u + z$ with $u \in U$. Then $x + u \in A$, hence $w = (1-\lambda)y + \lambda(x+u) \in A$. This shows that $\lambda U + z \subset A$ and thus $z \in \operatorname{int} A$.

Now assume merely that $y \in \operatorname{cl} A$. Put $V := [\lambda/(1-\lambda)]U + y$. There is some $a \in A \cap V$. We have $a = [\lambda/(1-\lambda)]u + y$ with $u \in U$ and hence $z = (1-\lambda)a + \lambda(x-u) \in A$. This proves that $[x, y) \subset A$, which together with the first part yields $[x, y) \subset \operatorname{int} A$. ■

Theorem 1.1.9. *If $A \subset \mathbb{E}^n$ is convex, then* int A *and* cl A *are convex. If $A \subset \mathbb{E}^n$ is open, then* conv A *is open.*

Proof. The convexity of int A follows from Lemma 1.1.8. The convexity of cl A for convex A and the openness of conv A for open A are easy exercises. ■

The union of a line and a point not on it shows that the convex hull of a closed set need not be closed. This is different for compact sets, as a first application of Carathéodory's theorem shows.

Theorem 1.1.10. *If* $A \subset \mathbb{E}^n$, *then* $\operatorname{conv} \operatorname{cl} A \subset \operatorname{cl} \operatorname{conv} A$. *If* A *is bounded, then* $\operatorname{conv} \operatorname{cl} A = \operatorname{cl} \operatorname{conv} A$. *In particular, the convex hull of a compact set is compact.*

Proof. $\operatorname{conv} \operatorname{cl} A \subset \operatorname{cl} \operatorname{conv} A$ is easy to see. Let A be bounded; then

$$\{(\lambda_1, \ldots, \lambda_{n+1}, x_1, \ldots, x_{n+1}) | \lambda_i \geqq 0,\ x_i \in \operatorname{cl} A,\ \sum_{i=1}^{n+1} \lambda_i = 1\}$$

is a compact subset of $\mathbb{R}^{n+1} \times (\mathbb{E}^n)^{n+1}$, hence its image under the continuous map

$$(\lambda_1, \ldots, \lambda_{n+1}, x_1, \ldots, x_{n+1}) \mapsto \sum_{i=1}^{n+1} \lambda_i x_i \in \mathbb{E}^n$$

is compact. By Carathéodory's theorem, this image is equal to $\operatorname{conv} \operatorname{cl} A$. Thus $\operatorname{cl} \operatorname{conv} A \subset \operatorname{cl} \operatorname{conv} \operatorname{cl} A = \operatorname{conv} \operatorname{cl} A$. ■

The set $\operatorname{cl} \operatorname{conv} A$, which by Theorem 1.1.9 is convex, is called for short the *closed convex hull* of A. This is also the intersection of all closed convex subsets of $\mathbb{E}^n$ containing A.

To obtain information on the relative interiors of convex hulls, we first consider simplices.

Lemma 1.1.11. *Let* $x_1, \ldots, x_k \in \mathbb{E}^n$ *be affinely independent; let* $S := \operatorname{conv} \{x_1, \ldots, x_k\}$ *and* $x \in \operatorname{aff} S$. *Then* $x \in \operatorname{relint} S$ *if and only if in the unique affine representation*

$$x = \sum_{i=1}^{k} \lambda_i x_i \quad \text{with} \sum_{i=1}^{k} \lambda_i = 1$$

all coefficients λ_i *are positive.*

Proof. Clearly we may assume that $k = n + 1$. The condition is necessary since otherwise, because the representation is unique, an arbitrary neighbourhood of x would contain points not belonging to S. To prove sufficiency, let x be represented as above with all $\lambda_i > 0$. Since $x_1, \ldots, x_{n+1}$ are affinely independent, the vectors $\tau(x_1), \ldots, \tau(x_{n+1})$ (see 'Conventions and notation') form a linear basis of $\mathbb{E}^n \times \mathbb{R}$, and for $y \in \mathbb{E}^n$ the coefficients $\mu_1, \ldots, \mu_{n+1}$ in the affine representation

$$y = \sum_{i=1}^{n+1} \mu_i x_i \quad \text{with} \sum_{i=1}^{n+1} \mu_i = 1$$

(the 'barycentric coordinates' of y) are just the coordinates of $\tau(y)$ with respect to this basis. Since coordinate functions in $\mathbb{E}^{n+1}$ are continuous, the coefficients $\mu_1, \ldots, \mu_{n+1}$ depend continuously on y. Therefore, $\delta > 0$ can be chosen such that $\mu_i > 0$ $(i = 1, \ldots, n+1)$ and thus $y \in S$ for all y with $|y - x| < \delta$. This proves $x \in \operatorname{int} S$. ■

Theorem 1.1.12. *If $A \subset \mathbb{E}^n$ is convex and nonempty, then* $\operatorname{relint} A \neq \emptyset$.

Proof. Let $\dim \operatorname{aff} A = k$, then there are $k+1$ affinely independent points in A. Their convex hull S satisfies $\operatorname{relint} S \neq \emptyset$ by Lemma 1.1.11, furthermore $S \subset A$ and $\operatorname{aff} S = \operatorname{aff} A$. ■

In view of this theorem, it makes sense to define the *dimension*, $\dim A$, of a convex set A as the dimension of its affine hull. The points of $\operatorname{relint} A$ are also called *internal* points of A.

The description of $\operatorname{relint} \operatorname{conv} A$ for an affinely independent set A given by Lemma 1.1.11 can be extended to arbitrary finite sets.

Theorem 1.1.13. *Let $x_1, \ldots, x_k \in \mathbb{E}^n$; let $P := \operatorname{conv}\{x_1, \ldots, x_k\}$ and $x \in \mathbb{E}^n$. Then $x \in \operatorname{relint} P$ if and only if x can be represented in the form*

$$x = \sum_{i=1}^{k} \lambda_i x_i \quad \text{with } \lambda_i > 0 \quad (i = 1, \ldots, k), \quad \sum_{i=1}^{k} \lambda_i = 1.$$

Proof. We may clearly assume that $\dim P = n$. Suppose that $x \in \operatorname{int} P$. Put

$$y := \sum_{i=1}^{k} \frac{1}{k} x_i;$$

then $y \in P$. Since $x \in \operatorname{int} P$, we can choose $z \in P$ for which $x \in [y, z)$. There are representations

$$z = \sum_{i=1}^{k} \mu_i x_i \quad \text{with } \mu_i \geqq 0, \quad \sum_{i=1}^{k} \mu_i = 1,$$

$$x = (1 - \lambda) y + \lambda z \quad \text{with} \quad 0 \leqq \lambda < 1,$$

which gives

$$x = \sum_{i=1}^{k} \lambda_i x_i \quad \text{with } \lambda_i = (1 - \lambda) \frac{1}{k} + \lambda \mu_i > 0, \quad \sum_{i=1}^{k} \lambda_i = 1.$$

Vice versa, suppose that

$$x = \sum_{i=1}^{k} \lambda_i x_i \quad \text{with } \lambda_i > 0, \quad \sum_{i=1}^{k} \lambda_i = 1.$$

We may assume that $x_1, \ldots, x_{n+1}$ are affinely independent. Put $\lambda := \lambda_1 + \ldots + \lambda_{n+1}$ and

$$y := \sum_{i=1}^{n+1} \frac{\lambda_i}{\lambda} x_i.$$

Lemma 1.1.11 gives $y \in \operatorname{int}\operatorname{conv}\{x_1, \ldots, x_{n+1}\} \subset \operatorname{int} P$. If $k = n+1$, then $x = y \in \operatorname{int} P$. Otherwise, put

$$z := \sum_{i=n+1}^{k} \frac{\lambda_i}{1-\lambda} x_i.$$

Then $z \in P$ and $x \in [y, z) \subset \operatorname{int} P$ by Lemma 1.1.8. ■

Theorem 1.1.14. *Let $A \subset \mathbb{E}^n$ be convex. Then*

(a) $\operatorname{relint} A = \operatorname{relint}\operatorname{cl} A$,

(b) $\operatorname{cl} A = \operatorname{cl}\operatorname{relint} A$,

(c) $\operatorname{relbd} A = \operatorname{relbd}\operatorname{cl} A = \operatorname{relbd}\operatorname{relint} A$.

Proof. We may clearly assume that $\dim A = n$. Part (a): trivially, $\operatorname{int} A \subset \operatorname{int}\operatorname{cl} A$. Let $x \in \operatorname{int}\operatorname{cl} A$. Choose $y \in \operatorname{int} A$. There is $z \in \operatorname{cl} A$ with $x \in [y, z)$, and Lemma 1.1.8 shows that $x \in \operatorname{int} A$. Part (b): trivially, $\operatorname{cl} A \supset \operatorname{cl}\operatorname{int} A$. Let $x \in \operatorname{cl} A$. Choose $y \in \operatorname{int} A$. By Lemma 1.1.8 we have $[y, x) \subset \operatorname{int} A$, hence $x \in \operatorname{cl}\operatorname{int} A$. Part (c): $\operatorname{bd}\operatorname{cl} A = \operatorname{cl}(\operatorname{cl} A)\setminus\operatorname{int}(\operatorname{cl} A) = \operatorname{cl} A\setminus\operatorname{int} A = \operatorname{bd} A$, using (a). Then $\operatorname{bd}\operatorname{int} A = \operatorname{cl}(\operatorname{int} A)\setminus\operatorname{int}(\operatorname{int} A) = \operatorname{cl} A\setminus\operatorname{int} A = \operatorname{bd} A$, using (b). ■

We end this section with a definition of the central notion of this book. A nonempty, compact, convex subset of $\mathbb{E}^n$ is called a *convex body* (thus in our terminology, a convex body need not have interior points). By $\mathcal{K}^n$ we denote the set of all convex bodies in $\mathbb{E}^n$ and by $\mathcal{K}_0^n$ the subset of convex bodies with interior points. For $\emptyset \neq A \subset \mathbb{E}^n$ we write $\mathcal{K}(A)$ for the set of convex bodies contained in A and $\mathcal{K}_0(A) = \mathcal{K}(A) \cap \mathcal{K}_0^n$. Further, $\mathcal{P}^n$ denotes the set of nonempty polytopes in $\mathbb{E}^n$, and $\mathcal{P}_0^n = \mathcal{P}^n \cap \mathcal{K}_0^n$.

Notes for Section 1.1

1. The early history of the theorems of Carathéodory, Radon and Helly, and many generalizations, ramifications and analogues of these theorems forming an essential part of combinatorial convexity can be studied in the survey article of Danzer, Grünbaum & Klee [32], which is still strongly recommended. Various results related to Carathéodory's theorem can be found in Reay (1965). An important extension of Radon's theorem is Tverberg's theorem (Tverberg 1966, 1981): Each set of at least $(m-1)(n+1)+1$ points in $\mathbb{E}^n$ (where $m \geqq 2$) can be partitioned into m subsets whose convex hulls have a common point. A survey of later developments is given by Eckhoff (1979). There one also finds hints about more recent developments of the theorems of Carathéodory, Radon and Helly in the abstract setting of so-called convexity spaces.

2. It is clear how a version of Carathéodory's theorem for convex cones is to be formulated and how it can be proved. A common generalization, a version of Carathéodory's theorem for 'convex hulls of points and directions', is given by Rockafellar [29], Theorem 17.1.

1.2. The metric projection

In this section, $A \subset \mathbb{E}^n$ is a fixed nonempty closed convex set. To each $x \in \mathbb{E}^n$ there exists a unique point $p(A, x) \in A$ satisfying

$$|x - p(A, x)| \leqq |x - y| \quad \text{for all } y \in A.$$

In fact, for suitable $\rho > 0$ the set $B(x, \rho) \cap A$ is compact and nonempty, hence the continuous function $y \mapsto |x - y|$ attains a minimum on this set, say at y_0; then $|x - y_0| \leqq |x - y|$ for all $y \in A$. If, also, $y_1 \in A$ satisfies $|x - y_1| \leqq |x - y|$ for all $y \in A$, then $z =: (y_0 + y_1)/2 \in A$ and $|x - z| < |x - y_0|$, except if $y_0 = y_1$. Thus $y_0 =: p(A, x)$ is unique.

In this way a map $p(A, \cdot): \mathbb{E}^n \to A$ is defined; it is called the *metric projection* or *nearest-point map* of A. It will play an essential role in Chapter 4 when the volume of local parallel sets is investigated. It also provides a simple approach to the basic support and separation properties of convex sets (see the next section), as used by Botts (1942) and McMullen & Shephard [26].

We have $|x - p(A, x)| = d(A, x)$, and for $x \in \mathbb{E}^n \backslash A$ we denote by

$$u(A, x) := \frac{x - p(A, x)}{d(A, x)}$$

the unit vector pointing from the nearest point $p(A, x)$ to x and by

$$R(A, x) := \{p(A, x) + \lambda u(A, x) | \lambda \geqq 0\}$$

the ray through x with endpoint $p(A, x)$.

Lemma 1.2.1. *Let $x \in \mathbb{E}^n \backslash A$ and $y \in R(A, x)$; then $p(A, x) = p(A, y)$.*

Proof. Suppose that $p(A, y) \neq p(A, x)$. If $y \in [x, p(A, x))$, then

$$\begin{aligned} |x - p(A, y)| &\leqq |x - y| + |y - p(A, y)| \\ &< |x - y| + |y - p(A, x)| \\ &= |x - p(A, x)|, \end{aligned}$$

which is a contradiction. If $x \in [y, p(A, x))$, let $q \in [p(A, x), p(A, y)]$ be the point such that the segment $[x, q]$ is parallel to $[y, p(A, y)]$. Then

$$\frac{|x - q|}{|x - p(A, x)|} = \frac{|y - p(A, y)|}{|y - p(A, x)|} < 1,$$

again a contradiction. ■

Theorem 1.2.2. *The metric projection is contracting, that is,*

$$|p(A, x) - p(A, y)| \leqq |x - y| \quad \textit{for } x, y \in \mathbb{E}^n.$$

Proof. We may assume that $v := p(A, x) - p(A, y) \neq o$. We assert that

$$\langle x - p(A, x), v \rangle \geqq 0. \tag{*}$$

If this is false, then $x \notin A$ and the ray $R(A, x)$ meets the hyperplanes through $p(A, y)$ that is orthogonal to v in a point z, and we deduce from Lemma 1.2.1 that

$$|z - p(A, y)| < |z - p(A, x)| = |z - p(A, z)|,$$

which is a contradiction. Hence (*) holds. Analogously we get $\langle y - p(A, y), v \rangle \leqq 0$. Thus the segment $[x, y]$ meets the two hyperplanes that are orthogonal to v and that go through $p(A, x)$ and $p(A, y)$ respectively. Now the assertion is obvious. ■

Lemma 1.2.3. *Let S be a sphere containing A in its interior. Then $p(A, S) = \operatorname{bd} A$.*

Proof. $p(A, S) \subset \operatorname{bd} A$ is clear. Let $x \in \operatorname{bd} A$. For $i \in \mathbb{N}$ choose x_i in the interior of S (that is, of the ball bounded by S) such that $x_i \notin A$ and $|x_i - x| < 1/i$. From Theorem 1.2.2 we have

$$|x - p(A, x_i)| = |p(A, x) - p(A, x_i)| \leqq |x - x_i| < 1/i.$$

The ray $R(A, x_i)$ meets S in a point y_i and we have $p(A, y_i) = p(A, x_i)$, hence $|x - p(A, y_i)| < 1/i$. A subsequence $(y_{i_j})_{j \in \mathbb{N}}$ converges to a point $y \in S$. From $\lim p(A, y_i) = x$ and the continuity of the metric projection we see that $x = p(A, y)$. Thus $\operatorname{bd} A \subset p(A, S)$. ■

The existence of a unique nearest-point map is characteristic of convex sets. We prove this result here to complete the picture, although no use will be made of it.

Theorem 1.2.4. *Let $A \subset \mathbb{E}^n$ be a closed set with the property that to each point of $\mathbb{E}^n$ there is a unique nearest point in A. Then A is convex.*

Proof. Suppose A satisfies the assumption but is not convex. Then there are points x, y with $[x, y] \cap A = \{x, y\}$, and one can choose $p > 0$ such that the ball $B = B((x + y)/2, \rho)$ satisfies $B \cap A = \emptyset$. By an elementary compactness argument, the family $\mathcal{B}$ of all closed balls B' containing B and satisfying $(\operatorname{int} B') \cap A = \emptyset$ contains a ball C with maximal radius. By this maximality, there is a point $p \in C \cap A$, and by the assumed uniqueness of nearest points in A it is unique. If $\operatorname{bd} B$ and bd C have a common point, let this (unique) point be q, otherwise let q be the centre of B. For sufficiently small $\varepsilon > 0$, the ball $C + \varepsilon(q - p)$ includes B and does not meet A. Hence, the family $\mathcal{B}$ contains an element with greater radius than that of C, a contradiction. ∎

Note for Section 1.2

Theorem 1.2.4 was found independently (in a more general form) by Bunt (1934) and Motzkin (1935); it is usually associated with the name of Motzkin. In general, a subset A of a metric space is called a Chebyshev set if for each point of the space there is a unique nearest point in A. There are several results and interesting open problems concerning the convexity of Chebyshev sets in normed linear spaces. For more information, see Valentine [30], Chapter VII, Marti [25], Chapter IX, Vlasov (1973) and §6 of the survey article by Burago & Zalgaller (1978).

1.3. Support and separation

The simplest support and separation properties of convex sets seem intuitively obvious, and they are easy to prove. Nevertheless, their many applications make them a basic tool in convexity.

Let $A \subset \mathbb{E}^n$ be a subset and $H \subset \mathbb{E}^n$ a hyperplane and let H^+, H^- denote the two closed halfspaces bounded by H. We say that H *supports A at x* if $x \in A \cap H$ and either $A \subset H^+$ or $A \subset H^-$. H is a *support plane* of A or *supports* A if H supports A at some point x, which is necessarily a boundary point of A. If $H = H_{u,\alpha}$ supports A and $A \subset H^-_{u,\alpha} = \{y \in \mathbb{E}^n | \langle y, u \rangle \leqq \alpha\}$, then $H^-_{u,\alpha}$ is called a *supporting halfspace* of A and u is called an *exterior* or *outer normal vector* of both $H_{u,\alpha}$ and $H^-_{u,\alpha}$. If, moveover, $H_{u,\alpha}$ supports A at x, then u is an *exterior normal vector of A at x*. A flat E *supports A at x* if $x \in A \cap E$ and E lies in some support plane of A.

Lemma 1.3.1. *Let $A \subset \mathbb{E}^n$ be nonempty, convex and closed and let $x \in \mathbb{E}^n \setminus A$. The hyperplane H through $p(A, x)$ orthogonal to $u(A, x)$ supports A.*

Proof. Clearly $H \cap A \neq \emptyset$. Let H^- be the closed halfspace bounded by H that does not contain x. Suppose there exists some $y \in A$ with $y \notin H^-$. Let z be the point in $[p(A, x), y]$ nearest to x. Then $|x - z| < |x - p(A, x)|$, which contradicts the definition of $p(A, x)$ since $z \in A$. This shows that $A \subset H^-$. ∎

Theorem 1.3.2. *Let $A \subset \mathbb{E}^n$ be convex and closed. Then through each boundary point of A there is a support plane of A. If $A \neq \emptyset$ is bounded,* *then to each vector* $u \in \mathbb{E}^n \setminus \{o\}$ *there is a support plane to A with exterior normal vector u.*

Proof. Let $x \in \operatorname{bd} A$. First let A be bounded. By Lemma 1.2.3 there is a point $y \in \mathbb{E}^n \setminus A$ such that $x = p(A, y)$. By Lemma 1.3.1 the hyperplane through $p(A, y) = x$ orthogonal to $y - x$ supports K at x.

If A is unbounded, there exists through x a support plane H of $A \cap B(x, 1)$; let H^- be the corresponding supporting halfspace of $A \cap B(x, 1)$. If there is a point $z \in A \setminus H^-$, then $[z, x] \subset A$, but $[z, x) \cap B(x, 1) \not\subset H^-$, a contradiction. Hence H supports A.

Let A be bounded and $u \in \mathbb{E}^n \setminus \{o\}$. Since A is compact, there is a point $x \in K$ satisfying $\langle x, u \rangle = \sup \{\langle y, u \rangle | y \in K\}$. Evidently $\{y \in \mathbb{E}^n | \langle y, u \rangle = \langle x, u \rangle\}$ is a support plane to A with exterior normal vector u. ∎

The existence of support planes through arbitrary boundary points is characteristic for convex sets, in the following precise sense:

Theorem 1.3.3. *Let* $A \subset \mathbb{E}^n$ *be a closed set such that* $\operatorname{int} A \neq \emptyset$ *and such that through each boundary point of A there is a support plane to A. Then A is convex.*

Proof. Suppose that A satisfies the assumptions but is not convex. Then there are points $x, y \in A$ and $z \in [x, y]$ with $z \notin A$. Since $\operatorname{int} A \neq \emptyset$ (and $n \geqq 2$, as we may clearly assume), we can choose $a \in \operatorname{int} A$ such that x, y, a are affinely independent. There is a point $b \in \operatorname{bd} A \cap [a, z)$. By assumption, through b there exists a support plane H to A, and $a \notin H$ because $a \in \operatorname{int} A$. Hence H intersects the plane $\operatorname{aff}\{x, y, a\}$ in a line. The points x, y, a must lie on the same side of this line, which is obviously a contradiction. ∎

We turn to separation. Let $A, B \subset \mathbb{E}^n$ be sets and $H_{u,\alpha} \subset \mathbb{E}^n$ a hyperplane. The hyperplane $H_{u,\alpha}$ *separates* A and B if $A \subset H^-_{u,\alpha}$ and $B \subset H^+_{u,\alpha}$, or vice versa. This separation is said to be *proper* if A and B do not both lie in $H_{u,\alpha}$. The sets A and B are *strictly separated* by $H_{u,\alpha}$ if $A \subset \operatorname{int} H^-_{u,\alpha}$ and $B \subset \operatorname{int} H^+_{u,\alpha}$, or vice versa, and they are *strongly separated* by $H_{u,\alpha}$ if there is an $\varepsilon > 0$ such that $H_{u,\alpha-\varepsilon}$ and $H_{u,\alpha+\varepsilon}$ both separate A and B. Separation of A and a point x means separation of A and $\{x\}$. We first consider this special case:

Theorem 1.3.4. *Let* $A \subset \mathbb{E}^n$ *be convex and let* $x \in \mathbb{E}^n \setminus A$. *Then A and x can be separated. If A is closed, then A and x can be strongly separated.*

Proof. If A is closed, the hyperplane through $p(A, x)$ orthogonal to $u(A, x)$ supports A and hence separates A and x. The parallel hyperplane through $(p(A, x) + x)/2$ strongly separates A and x. If A is not closed and $x \notin \mathrm{cl}\, A$, then a hyperplane separating $\mathrm{cl}\, A$ and x a fortiori separates A and x. If $x \in \mathrm{cl}\, A$, then $x \in \mathrm{bd}\,\mathrm{cl}\, A$ by Theorem 1.1.14, and by Theorem 1.3.2 there is a support plane to $\mathrm{cl}\, A$ through x; it separates A and x. ■

Corollary 1.3.5. *Each nonempty closed convex set in $\mathbb{E}^n$ is the intersection of its supporting halfspaces.*

Separation of pairs of sets can be reduced to separation of a set and a point:

Lemma 1.3.6. *Let $A, B \subset \mathbb{E}^n$ be nonempty subsets. A and B can be separated (strongly separated) if and only if $A - B$ and o can be separated (strongly separated).*

Proof. We consider only strong separation; the other case is analogous (or put $\varepsilon = 0$). Suppose that $H_{u,\alpha}$ strongly separates A and B, say $A \subset H^-_{u,\alpha-\varepsilon}$ and $B \subset H^+_{u,\alpha+\varepsilon}$ for some $\varepsilon > 0$. Let $x \in A - B$; thus $x = a - b$ with $a \in A$, $b \in B$. From $\langle a, u\rangle \leqq \alpha - \varepsilon$ and $\langle b, u\rangle \geqq \alpha + \varepsilon$ we get $\langle x, u\rangle \leqq -2\varepsilon$, so that $A - B$ and o are strongly separated by $H_{u,-\varepsilon}$.

Suppose that $A - B$ and o can be strongly separated. Then there are $u \in \mathbb{E}^n \setminus \{o\}$ and $\varepsilon > 0$ such that $\langle x, u\rangle \leqq -2\varepsilon$ for all $x \in A - B$. Let

$$\alpha := \sup\{\langle a, u\rangle \mid a \in A\},$$
$$\beta := \inf\{\langle b, u\rangle \mid b \in B\}.$$

For $a \in A$, $b \in B$ we have $\langle a, u\rangle - \langle b, u\rangle \leqq -2\varepsilon$, hence $\beta - \alpha \geqq 2\varepsilon$. Thus $H_{u,(\alpha+\beta)/2}$ strongly separates A and B. ■

If $A, B \subset \mathbb{E}^n$ are convex, then $A - B$ is convex. If A is compact and B is closed, then $A - B$ is easily seen to be closed. The condition $o \notin A - B$ is equivalent to $A \cap B = \emptyset$. Hence from Lemma 1.3.6 and Theorem 1.3.4 we deduce:

Theorem 1.3.7. *Let $A, B \subset \mathbb{E}^n$ be nonempty convex sets with $A \cap B = \emptyset$. Then A and B can be separated. If A is compact and B is closed, then A and B can be strongly separated.*

The following examples should be kept in mind. Let

$$A := \{(\xi, \eta) \in \mathbb{E}^2 \mid \xi > 0,\ \eta \geqq 1/\xi\},$$

$$B := \{(\xi, \eta) \in \mathbb{E}^2 | \xi > 0, \eta \leqq -1/\xi\},$$
$$G := \{(\xi, \eta) \in \mathbb{E}^2 | \eta = 0\}.$$

These are pairwise disjoint, closed, convex subsets of $\mathbb{E}^2$. A and B can be strictly separated (by G), but not strongly. $A - B$ and o cannot be strictly separated. A and G can be separated, but not strictly.

On the other hand, convex sets may be separable even if they are not disjoint. The exact condition is given by the following theorem.

Theorem 1.3.8. *Let $A, B \subset \mathbb{E}^n$ be nonempty convex sets. Then A and B can be properly separated if and only if*

$$\operatorname{relint} A \cap \operatorname{relint} B = \emptyset. \qquad (*)$$

Proof. Suppose that (*) holds. Put $C := \operatorname{relint} A - \operatorname{relint} B$. Then $o \notin C$, and C is convex. By Theorem 1.3.4 there exists a hyperplane $H_{u,0}$ with $C \subset H^-_{u,0}$. Let

$$\beta := \inf\{\langle b, u\rangle | b \in B\},$$

then $B \subset H^+_{u,\beta}$. Suppose there exists a point $a \in A$ with $\langle a, u\rangle > \beta$. By Theorem 1.1.12 there exists a point $z \in \operatorname{relint} A$, and Lemma 1.1.8 states that $[z, a) \subset \operatorname{relint} A$. Hence, there is a point $\bar{a} \in \operatorname{relint} A$ with $\langle \bar{a}, u\rangle > \beta$. There is a point $b \in B$ with $\langle b, u\rangle < \langle \bar{a}, u\rangle$ and then, by a similar argument as before, a point $\bar{b} \in \operatorname{relint} B$ with $\langle \bar{b}, u\rangle < \langle \bar{a}, u\rangle$. Thus $\bar{a} - \bar{b} \in C$ and $\langle \bar{a} - \bar{b}, u\rangle > 0$, a contradiction. This shows that $A \subset H^-_{u,\beta}$. Thus A and B are separated by $H_{u,\beta}$. If $A \cup B$ lies in some hyperplane, then this argument yields a hyperplane relative to $\operatorname{aff}(A \cup B)$ separating A and B, and this can be extended to a hyperplane in $\mathbb{E}^n$ that properly separates A and B.

Vice versa, let H be a hyperplane properly separating A and B, say with $A \subset H^-$ and $B \subset H^+$. Suppose there exists $x \in \operatorname{relint} A \cap \operatorname{relint} B$. Then $x \in H$. Since $A \subset H^-$ and $x \in \operatorname{relint} A$, we must have $A \subset H$, similarly $B \subset H$, a contradiction. Thus (*) holds. ∎

Occasionally we shall have to use support and separation of convex cones. For these we have:

Theorem 1.3.9. *Let $C \subset \mathbb{E}^n$ be a closed convex cone. Each support plane of C contains* o. *If $x \in \mathbb{E}^n \setminus C$, then there exists a vector $u \in \mathbb{E}^n$ such that*

$$\langle c, u\rangle \geqq 0 \text{ for all } c \in C \text{ and } \langle x, u\rangle < 0.$$

Proof. Let H be a support plane to C. There is a point $y \in H \cap C$. Then $\lambda y \in C$ for all $\lambda > 0$, which is impossible if $o \notin H$. Hence $o \in H$. The rest is clear from Lemma 1.3.1. ∎

We shall now prove two more results in the spirit of the theorems of Carathéodory and Helly. They are treated in this section since the first of them needs support planes in its proof and the second one deals with separation.

Theorem 1.3.10 (Steinitz's theorem). *Let $A \subset \mathbb{E}^n$ and $x \in \operatorname{int} \operatorname{conv} A$. Then $x \in \operatorname{int} \operatorname{conv} A'$ for some subset $A' \subset A$ with at most $2n$ points.*

Proof. The point x lies in the interior of a simplex with vertices in $\operatorname{conv} A$; hence, by Carathéodory's theorem applied to each vertex, $x \in \operatorname{int} \operatorname{conv} B$ for some subset B of A with at most $(n+1)^2$ points. We can choose a line G through x that does not meet the affine hull of any $n-1$ points of B. Let x_1, x_2 be the endpoints of the segment $G \cap \operatorname{conv} B$. By Theorem 1.3.2, through x_j there is a support plane H_j to $\operatorname{conv} B$ $(j = 1, 2)$. Clearly $x_j \in \operatorname{conv}(B \cap H_j)$; hence by Carathéodory's theorem there is a representation

$$x_j = \sum_{i=1}^{n} \lambda_{ji} y_{ji} \quad \text{with } y_{ji} \in B, \ \ \lambda_{ji} \geqq 0, \ \ \sum_{i=1}^{n} \lambda_{ji} = 1$$

$(j = 1, 2)$, and here necessarily $\lambda_{ji} > 0$ by the choice of G. With suitable $\lambda \in (0, 1)$ we have

$$x = (1 - \lambda)x_1 + \lambda x_2 = \sum_{i=1}^{n} [(1 - \lambda)\lambda_{1i} y_{1i} + \lambda \lambda_{2i} y_{2i}]$$

$$\in \operatorname{relint} \operatorname{conv} \{y_{11}, \ldots, y_{1n}, y_{21}, \ldots, y_{2n}\}$$

by Theorem 1.1.13. Here relint can be replaced by int, since by the choice of G the points $y_{11}, \ldots, y_{1n}$ are affinely independent and for at least one k also $y_{11}, \ldots, y_{1n}, y_{2k}$ are affinely independent. ■

Theorem 1.3.11 (Kirchberger's theorem). *Let $A, B \subset \mathbb{E}^n$ be compact sets. If for any subset $M \subset A \cup B$ with at most $n + 2$ points the sets $M \cap A$ and $M \cap B$ can be strongly separated, then A and B can be strongly separated.*

Proof. First we assume that A and B are finite sets. For $x \in \mathbb{E}^n$ define (with $\tau(x) := (x, 1)$)

$$H_x^{\pm} := \{v \in \mathbb{E}^n \times \mathbb{R} \mid \pm \langle v, \tau(x) \rangle > 0\}.$$

Let $M \subset A \cup B$ and $\operatorname{card} M \leqq n + 2$. By the assumption there exist $u \in \mathbb{E}^n$ and $\alpha \in \mathbb{R}$ such that $\langle u, a \rangle > \alpha$ for $a \in M \cap A$ and $\langle u, b \rangle < \alpha$ for $b \in M \cap B$. Writing $v := (u, -\alpha)$, we see that $\langle v, \tau(a) \rangle = \langle u, a \rangle - \alpha > 0$; thus $v \in H_a^+$ for $a \in M \cap A$. Similarly $v \in H_b^-$ for $b \in M \cap B$. Thus the family $\{H_a^+ | a \in A\} \cup \{H_b^- | b \in B\}$ of finitely many convex sets in $\mathbb{E}^n \times \mathbb{R}$ has the property that any $n+2$ or fewer of the sets have nonempty intersection. By Helly's theorem, the intersection of all sets in the family is not empty. Since this intersection is open, it contains an element of the form $v = (u, -\alpha)$ with $u \neq o$. For $a \in A$ we have $v \in H_a^+$, hence $\langle u, a \rangle > \alpha$, and for $b \in B$ similarly $\langle u, b \rangle < \alpha$. Hence A and B, being finite sets, are strongly separated by $H_{u,\alpha}$.

Now let A, B be compact sets satisfying the assumption. Suppose that A and B cannot be strongly separated. Then $\operatorname{conv} A$ and $\operatorname{conv} B$ cannot be strongly separated. By Theorem 1.1.10 these sets are compact and hence by Theorem 1.3.7 they cannot be disjoint. Let $x \in \operatorname{conv} A \cap \operatorname{conv} B$. Then $x \in \operatorname{conv} A' \cap \operatorname{conv} B'$ with finite subsets $A' \subset A$ and $B' \subset B$, which hence cannot be strongly separated. This contradicts the result shown above. ■

We conclude this section with another application of a separation theorem, which will be useful in the study of Minkowski addition.

Lemma 1.3.12. *Let $A, B \subset \mathbb{E}^n$ be nonempty convex sets. If*

$$x \in \operatorname{relint}(A + B),$$

then x can be represented in the form $x = a + b$ with $a \in \operatorname{relint} A$ and $b \in \operatorname{relint} B$.

Proof. There is a representation $x = y + z$ with $y \in A$ and $z \in B$. We may assume that $x = y = z = o$ and also that $\dim(A + B) = n$. Since $o \in \operatorname{int}(A + B)$, $A + B$ and o cannot be separated. Hence, by Lemma 1.3.6 and Theorem 1.3.8, there is a point

$$a \in \operatorname{relint} A \cap \operatorname{relint}(-B).$$

Then $-a \in \operatorname{relint} B$ and $o = a - a$. ■

Notes for Section 1.3

1. Separation and support properties of convex sets in finite and infinite dimensions are of fundamental importance in various fields such as functional analysis, optimization, control theory, mathematical economy, and others. For infinite-dimensional separation and support theorems we refer only to Bourbaki [9], Marti [25], Holmes (1975) and Bair & Fourneau [3]; see also the survey article by Klee (1969b).

   A thorough study of several types of separation in $\mathbb{E}^n$ was made by Klee (1968).
2. A stronger version of Theorem 1.3.3 (existence of local support planes) is associated with the name of Tietze. A survey of results of this type is given in the article by Burago & Zalgaller (1978).
3. Historical information on the theorems of Steinitz and Kirchberger can be found in the survey article by Danzer, Grünbaum & Klee [32].

4. *Positive bases*. Let $B \subset \mathbb{E}^n$. Using Theorem 1.1.13 one easily sees that $\operatorname{pos} B = \mathbb{E}^n$ holds if and only if $o \in \operatorname{int} \operatorname{conv} B$. The set B is called a *positive basis* of $\mathbb{E}^n$ if $\operatorname{pos} B = \mathbb{E}^n$ but $\operatorname{pos} B' \neq \mathbb{E}^n$ for each proper subset $B' \subset B$. Thus Steinitz's theorem implies that a positive basis of $\mathbb{E}^n$ contains at most $2n$ vectors. If B is a linear basis of $\mathbb{E}^n$, then $B \cup (-B)$ is a positive basis, and up to multiplication by positive numbers it is only in this way that the maximal number $2n$ can be achieved. Positive bases have been investigated by Davis (1954), McKinney (1962), Bonnice & Klee (1963), Reay (1965) and Shephard (1971).

1.4. Extremal representations

The purpose of this section is to represent a closed convex set as the convex hull of a smaller set, and here the smallest possible sets will be of particular interest. A first candidate for a smaller set with the same convex hull is the relative boundary. Only the obvious trivial cases must be excluded:

Lemma 1.4.1. *If $A \subset \mathbb{E}^n$ is a closed convex set with $A \neq \operatorname{conv} \operatorname{relbd} A$, then A is either a flat or a half-flat.*

Proof. Clearly we may assume that $\dim A = n$. There is a point $x \in \operatorname{int} A$ with $x \notin \operatorname{conv} \operatorname{bd} A$ (since otherwise $A = \operatorname{int} A \cup \operatorname{bd} A = \operatorname{conv} \operatorname{bd} A$). By the separation theorem, 1.3.4, there is a closed halfspace H^- such that $x \in H^-$ and $\operatorname{conv} \operatorname{bd} A \subset H^+$. Each point $y \in \operatorname{int} H^-$ satisfies $[x, y] \cap \operatorname{bd} A = \emptyset$ and hence $y \in \operatorname{int} A$; thus $H^- \subset A$. By the convexity and closedness of A, each translate of H^- with a point of A in its boundary is contained in A. Thus A is either equal to $\mathbb{E}^n$ or is a halfspace. ■

We will exclude the exceptional cases that are the subject of Lemma 1.4.1 by demanding that A be *line-free*, meaning that A does not contain a line. Owing to Lemma 1.4.2 below, this is not a severe restriction. First let $A \subset \mathbb{E}^n$ be a closed convex set. Suppose that A contains a ray $G_{x,u} := \{x + \lambda u \mid \lambda \geqq 0\}$ with $x \in \mathbb{E}^n$ and $u \in \mathbb{E}^n \setminus \{o\}$. Let $y \in A$. Let $z \in G_{y,u}$ and $w \in [x, z)$. The ray through w with endpoint y meets $G_{x,u}$, hence $w \in A$. Thus $[x, z) \subset A$ and hence $z \in A$. This shows that also $G_{y,u} \subset A$. For this reason, it makes sense to define

$$\operatorname{rec} A := \{u \in \mathbb{E}^n \setminus \{o\} \mid G_{x,u} \subset A\} \cup \{o\}$$

where $x \in A$; this set does then not depend upon the choice of x. It is evidently a closed convex cone, called the *recession cone* of A. One may also write

$$\operatorname{rec} A = \{u \in \mathbb{E}^n \mid A + u \subset A\}.$$

Lemma 1.4.2. *Each closed convex set $A \subset \mathbb{E}^n$ can be represented in the form $A = \bar{A} \oplus V$, where V is a linear subspace of $\mathbb{E}^n$ and $\bar{A}$ is a line-free closed convex set in a subspace complementary to V.*

Proof. Assume that A is not line-free. Then

$$V := \operatorname{rec} A \cap (-\operatorname{rec} A)$$

(the *lineality space* of A) is the linear subspace consisting of all vectors that are parallel to some line contained in A. Let U be a linear subspace complementary to V and put $\bar{A} := A \cap U$; then $\bar{A} + V \subset A$. Let $x \in A$. Through x there exists a line $G \subset A$; since it is parallel to V, it meets U in a point y. Then $x = y + (x - y)$ with $y \in \bar{A}$ and $x - y \in V$; hence $x \in \bar{A} + V$. This proves $A = \bar{A} \oplus V$. Clearly $\bar{A}$ is closed, convex and line-free. ■

The representation by convex hulls of minimal sets requires some definitions. Let $A \subset \mathbb{E}^n$ be a convex set. A *face* of A is a convex subset $F \subset A$ such that each segment $[x, y] \subset A$ with $F \cap \operatorname{relint}[x, y] \neq \emptyset$ is contained in F or, equivalently, such that $x, y \in A$ and $(x + y)/2 \in F$ implies $x, y \in F$. If $\{z\}$ is a face of A, then z is called an *extreme point* of A. In other words, z is an extreme point of A if and only if it cannot be written in the form $z = (1 - \lambda)x + \lambda y$ with $x, y \in A$ and $\lambda \in (0, 1)$. The set of all extreme points of A is denoted by $\operatorname{ext} A$. An *extreme ray* of A is a ray that is a face of A. By $\operatorname{extr} A$ we denote the union of the extreme rays of A.

Theorem 1.4.3. *Each line-free closed convex set $A \subset \mathbb{E}^n$ is the convex hull of its extreme points and extreme rays*;

$$A = \operatorname{conv}(\operatorname{ext} A \cup \operatorname{extr} A).$$

Proof. The assertion is clear for $n \leqq 1$. Suppose that $n \geqq 2$, $\dim A = n$ (w.l.o.g.) and the assertion has been proved for convex sets of smaller dimension. By Lemma 1.4.1, $A = \operatorname{conv} \operatorname{bd} A$. By the support theorem, 1.3.2, each point $x \in \operatorname{bd} A$ lies in some support plane H of A. By the induction hypothesis, x lies in the convex hull of the extreme points and extreme rays of $H \cap A$, and it is an immediate consequence of the definition of a face that these are respectively extreme points and extreme rays of A itself. The assertion follows. ■

Corollary 1.4.4. *If $A \subset \mathbb{E}^n$ is a line-free closed convex set then*

$$A = \operatorname{conv} \operatorname{ext} A + \operatorname{rec} A.$$

Proof. By Theorem 1.4.3, a point $x \in A$ can be written as

$$x = \sum_{i=1}^{k} \lambda_i x_i + \sum_{i=k+1}^{m} \lambda_i v_i, \quad \lambda_i \geqq 0, \quad \sum_{i=1}^{m} \lambda_i = 1$$

with $x_i \in \operatorname{ext} A$ and $v_i \in \operatorname{extr} A$. Then v_i, lying in some extreme ray, can be written as $v_i = y_i + u_i$ where y_i, the endpoint of the ray, is an extreme point of A, and $u_i \in \operatorname{rec} A$. The assertion follows. ■

Corollary 1.4.5 (Minkowski's theorem). *Each convex body $K \in \mathcal{K}^n$ is the convex hull of its extreme points.*

Here the set of extreme points cannot be replaced by a smaller set, since a point $x \in K$ is an extreme point of K if and only if $K \setminus \{x\}$ is convex.

Extreme points can also be characterized in a slightly different way that is sometimes useful. Let $A \subset \mathbb{E}^n$ be closed and convex and let $x \in A$. A *cap of A around x* is a set of the form $A \cap H^+$, where H^+ is a closed halfspace with $x \in \operatorname{int} H^+$.

Lemma 1.4.6. *Let $A \subset \mathbb{E}^n$ be closed and convex and let $x \in A$. Then x is an extreme point of A if and only if each neighbourhood of x contains a cap of A around x.*

Proof. Let $x \in \operatorname{ext} A$. A given neighbourhood of x contains an open ball B_0 with centre x. Let $\bar{A} := \operatorname{conv}(A \setminus B_0)$. Then $x \notin \bar{A}$, hence there is a hyperplane H strongly separating $\bar{A}$ and x. If H^+ is the closed halfspace bounded by H and containing x, then $A \cap H^+$ is the required cap.

If $x \notin \operatorname{ext} A$, then $x \in \operatorname{relint}[y, z]$ for suitable $y, z \in A$. Each cap of A around x contains y or z, hence sufficiently small neighbourhoods of x cannot contain a cap around x. ■

The notion of an extreme point of a convex set A involves convex combinations of points in A and is thus related to an 'intrinsic' description of A. Looking at a convex set from outside, one is led to a related but different class of special boundary points. The point $x \in A$ is called an *exposed point* of A if there is a support plane H to A with $H \cap A = \{x\}$. The set of exposed points of A is denoted by $\exp A$. Clearly $\exp A \subset \operatorname{ext} A$, but even for $A \in \mathcal{K}^2$ this inclusion is in general strict. This is shown by the example of a rectangle with a semicircle attached to one of its edges. However, each extreme point of a closed convex set is a limit of exposed points. We formulate the corresponding result for convex bodies only.

Theorem 1.4.7 (Straszewicz's theorem). *For* $K \in \mathcal{K}^n$

$$\operatorname{ext} K \subset \operatorname{cl} \exp K,$$

hence

$$K = \operatorname{cl} \operatorname{conv} \exp K.$$

Proof. Let $x \in \operatorname{ext} K$ and let U be a neighbourhood of x. By Lemma 1.4.6, U contains a cap $K \cap H^+$ of K around x. Let G be the ray orthogonal to $H = \operatorname{bd} H^+$ with endpoint x and meeting H. To any point $z \in G$ there is a point $y_z \in K$ with maximal distance from z. Obviously, $y_z \in \exp K$. If $|z - x|$ is sufficiently large, then $y_z \in H^+$ by elementary geometry. Thus $y_z \in K \cap H^+ \subset U$ and hence $\operatorname{ext} K \subset \operatorname{cl} \exp K$.

Using Minkowski's theorem, we get

$$K = \operatorname{conv} \operatorname{ext} K \subset \operatorname{conv} \operatorname{cl} \exp K$$
$$\subset \operatorname{cl} \operatorname{conv} \exp K \subset \operatorname{cl} \operatorname{conv} \operatorname{ext} K = K,$$

hence $K = \operatorname{cl} \operatorname{conv} \exp K$. ■

Notes for Section 1.4

1. *Minkowski's theorem and its aftermath.* Minkowski's theorem was first proved by Minkowski (1911, §12). Of particular importance is its extension to infinite-dimensional spaces: Each compact convex subset of a locally convex Hausdorff linear space is the closed convex hull of its extreme points. This was proved, in a more special case, by Krein & Milman (1940); see, e.g., Bourbaki [9]. A certain converse is due to Milman (1948); this states that in the Krein–Milman theorem the set of extreme points cannot be replaced by a set whose closure does not contain the extreme points. We refer the reader to Jacobs (1971) for an elementary introduction to extreme points, including applications, and to N. M. Roy (1987) for a survey article on extreme points of convex sets in infinite-dimensional spaces. An extension of the Krein–Milman theorem to locally compact sets, of which Theorem 1.4.3 is the finite-dimensional version, is due to Klee (1957b); see Jongmans (1968) for another extension.

   The Choquet theory of integral representations can be considered as a further extension of the Krein–Milman theorem; see, e.g., Bauer (1964), Phelps (1966, 1980), Choquet (1969a,b), Alfsen (1971), and the survey article by Saškin (1973) which contains many references up to 1972.
2. Straszewicz's theorem was first proved by Straszewicz (1935); see also Wets (1974) and, for the infinite-dimensional case, Klee (1958) and Bair (1976a).
3. A good source for the theorems of Minkowski and Straszewicz and their extensions to unbounded closed convex sets is Rockafellar [29].

   Klee (1959a) made an extensive study of the closed sets that are the convex hull of a finite system of points and rays (polyhedral sets).

   If $A \subset \mathbb{E}^n$ is an unbounded line-free closed convex set, then Corollary 1.4.4 states that $A = \operatorname{conv} \operatorname{ext} A + \operatorname{rec} A$. Starting from this representation, Batson (1988) found characterizations of the sets A for which $\operatorname{ext} A$ is bounded.

4. The following result is due to Dubins (1962); see also Pranger (1973). If $K \in \mathcal{K}^n$ and $M \subset \mathbb{E}^n$ is a flat of codimension k, then each extreme point of $K \cap M$ is a convex combination of $k+1$ (or fewer) extreme points of K. In fact, Dubins proved an infinite-dimensional version. A simplified proof and an extension are due to Klee (1963a).
5. *Continuous barycentre functions*. By Minkowski's theorem, each point $x \in K$, where $K \in \mathcal{K}^n$, is a convex combination of extreme points of K. For a polytope P, such a convex combination may involve all the extreme points of P. Let P have the extreme points $x_1, \ldots, x_k$. A k-tuple $(\varphi_1, \ldots, \varphi_k)$ of real-valued continuous functions $\varphi_1, \ldots, \varphi_k$ on P such that

$$x = \sum_{i=1}^{k} \varphi_i(x) x_i, \quad \varphi_i(x) \geqq 0 \quad (i = 1, \ldots, k), \quad \sum_{i=1}^{k} \varphi_i(x) = 1,$$

is called a *continuous barycentre function*. It is called *extreme* if for each $x \in P$, the set $\{x_i | \varphi_i(x) > 0\}$ is affinely independent. The existence of continuous barycentre functions was proved by Kalman (1961) and the existence of extreme continuous barycentre functions by Fuglede (1986a). Brøndsted (1986) showed that the extreme continuous barycentre functions coincide with the extreme points of the convex set of all continuous barycentre functions on P. Related investigations for general compact convex sets were done by Fuglede (1986a). Answering a question of Kalman (1961), Wiesler (1964) showed that not all the functions φ_i can be convex.

1.5. Convex functions

The investigation of convex sets is closely tied up with convex functions. We treat convex functions in a slightly more general fashion than would be necessary for our later applications.

For convex functions it is convenient to admit as the range the extended system $\bar{\mathbb{R}} = \mathbb{R} \cup \{-\infty, \infty\}$ of real numbers with the usual rules. These are the following conventions. For $\lambda \in \mathbb{R}$, $-\infty < \lambda < \infty$, $\infty + \infty = \lambda + \infty = \infty + \lambda = \infty$, $-\infty - \infty = -\infty + (-\infty) = \lambda - \infty = -\infty + \lambda = -\infty$, and finally $\lambda\infty = \infty$, 0 or $-\infty$ according to whether $\lambda > 0$, $\lambda = 0$ or $\lambda < 0$. For a given function $f\colon \mathbb{E}^n \to \bar{\mathbb{R}}$ and for $\alpha \in \bar{\mathbb{R}}$ we use the abbreviation

$$\{f = \alpha\} := \{x \in \mathbb{E}^n | f(x) = \alpha\},$$

and $\{f \leqq \alpha\}$, $\{f < \alpha\}$ etc. are defined similarly.

A function $f\colon \mathbb{E}^n \to \bar{\mathbb{R}}$ is called *convex* if f is *proper*, which means that $\{f = -\infty\} = \emptyset$ and $\{f = \infty\} \neq \mathbb{E}^n$, and if

$$f((1-\lambda)x + \lambda y) \leqq (1-\lambda)f(x) + \lambda f(y)$$

for all $x, y \in \mathbb{E}^n$ and for $0 \leqq \lambda \leqq 1$. A function $f\colon D \to \bar{\mathbb{R}}$ with $D \subset \mathbb{E}^n$ is called *convex* if its extension $\tilde{f}$ defined by

$$\tilde{f}(x) := \begin{cases} f(x) & \text{for } x \in D \\ \infty & \text{for } x \in \mathbb{E}^n \setminus D \end{cases}$$

is convex. A function f is *concave* if $-f$ is convex.

Trivial examples of convex functions are the affine functions; these are the functions $f: \mathbb{E}^n \to \mathbb{R}$ of the form $f(x) = \langle u, x \rangle + \alpha$ with $u \in \mathbb{E}^n$ and $\alpha \in \mathbb{R}$. A real-valued function on $\mathbb{E}^n$ is affine if and only if it is convex and concave.

The following assertions are immediate consequences of the definition. The supremum of (arbitrarily many) convex functions is convex if it is proper. If f, g are convex functions, then $f + g$ and αf for $\alpha \geqq 0$ are convex if they are proper.

Remark. If f is convex, then

$$f(\lambda_1 x_1 + \ldots + \lambda_k x_k) \leqq \lambda_1 f(x_1) + \ldots + \lambda_k f(x_k)$$

for all $x_1, \ldots, x_k \in \mathbb{E}^n$ and all $\lambda_1, \ldots, \lambda_k \in [0, 1]$ with $\lambda_1 + \ldots + \lambda_k = 1$. This is called *Jensen's inequality*; it follows by induction.

Convex functions have the important property (important for optimization etc.) that each local minimum is a global minimum. In fact, let $f: \mathbb{E}^n \to \bar{\mathbb{R}}$ be convex and suppose that $x_0 \in \mathbb{E}^n$ and $\rho > 0$ are such that $f(x_0) < \infty$ and $f(x_0) \leqq f(x)$ for $|x - x_0| \leqq \rho$. For $x \in \mathbb{E}^n$ with $|x - x_0| > \rho$ let

$$y := \frac{\rho}{|x - x_0|} x + \left(1 - \frac{\rho}{|x - x_0|}\right) x_0,$$

then $|y - x_0| = \rho$ and hence

$$f(x_0) \leqq f(y) \leqq \frac{\rho}{|x - x_0|} f(x) + \left(1 - \frac{\rho}{|x - x_0|}\right) f(x_0),$$

which gives $f(x_0) \leqq f(x)$.

A convex function determines in a natural way several convex sets. Let $f: \mathbb{E}^n \to \bar{\mathbb{R}}$ be convex. Then the sets

$$\operatorname{dom} f := \{f < \infty\},$$

the *effective domain* of f, and for $\alpha \in \mathbb{R}$ the *sublevel sets* $\{f < \alpha\}$, $\{f \leqq \alpha\}$, are convex. The *epigraph* of f,

$$\operatorname{epi} f := \{(x, \zeta) \in \mathbb{E}^n \times \mathbb{R} \mid f(x) \leqq \zeta\}$$

is a convex subset of $\mathbb{E}^n \times \mathbb{R}$. The asserted convexity is in each case easy to see. Vice versa, a nonempty convex set $A \subset \mathbb{E}^n$ determines a convex function by

$$I_A(x) := \begin{cases} 0 & \text{for } x \in A \\ \infty & \text{for } x \in \mathbb{E}^n \setminus A, \end{cases}$$

the *indicator function* of A.

We investigate the general analytic properties of convex functions, starting with continuity.

Theorem 1.5.1. *Each convex function* $f: \mathbb{E}^n \to \bar{\mathbb{R}}$ *is continuous on* int dom f *and Lipschitzian on any compact subset of* int dom f.

Proof. First we show the continuity. Let $x_0 \in \operatorname{int}\operatorname{dom} f$. We can choose a simplex S with $x_0 \in \operatorname{int} S \subset S \subset \operatorname{int}\operatorname{dom} f$ and a number $\rho > 0$ with $B(x_0, \rho) \subset S$. For $x \in S$ there is a representation

$$x = \sum_{i=1}^{n+1} \lambda_i x_i \quad \text{with } \lambda_i \geqq 0, \ \sum_{i=1}^{n+1} \lambda_i = 1,$$

where $x_1, \ldots, x_{n+1}$ are the vertices of S, and we deduce that

$$f(x) \leqq \sum_{i=1}^{n+1} \lambda_i f(x_i) \leqq c := \max\{f(x_1), \ldots, f(x_{n+1})\}.$$

Now let $y = x_0 + \alpha u$ with $\alpha \in [0, 1]$ and $|u| = \rho$. From $y = (1-\alpha)x_0 + \alpha(x_0 + u)$ we get

$$f(y) \leqq (1-\alpha)f(x_0) + \alpha f(x_0 + u),$$

hence

$$f(y) - f(x_0) \leqq \alpha(c - f(x_0))$$

since $x_0 + u \in S$. On the other hand,

$$x_0 = \frac{1}{1+\alpha}\, y + \frac{\alpha}{1+\alpha}(x_0 - u)$$

and hence

$$f(x_0) \leqq \frac{1}{1+\alpha} f(y) + \frac{\alpha}{1+\alpha} f(x_0 - u),$$

which gives

$$f(x_0) - f(y) \leqq \alpha(c - f(x_0)).$$

Thus

$$|f(y) - f(x_0)| \leqq \frac{1}{\rho}[c - f(x_0)]|y - x_0|$$

for $y \in B(x_0, \rho)$, which shows the continuity of f at x_0 in a sharpened version. Thus f is continuous on int dom f.

Now let $C \subset \operatorname{int}\operatorname{dom} f$ be a compact subset. By compactness, there exists a number $\rho > 0$ such that $C_\rho := C + B(o, \rho) \subset \operatorname{int}\operatorname{dom} f$. On the compact set C_ρ, the continuous function $|f|$ attains a maximum a. Let x, $y \in C$ be given. Then

$$z := y + \frac{\rho}{|y - x|}(y - x) \in C_\rho$$

and

$$y = (1-\lambda)x + \lambda z \quad \text{with } \lambda = \frac{|y - x|}{\rho + |y - x|},$$

hence $f(y) \leqq (1-\lambda)f(x) + \lambda f(z)$ yields

$$f(y) - f(x) \leqq \lambda[f(z) - f(x)] \leqq \frac{2a}{\rho}|y - x|.$$

Interchanging x and y we get $|f(y) - f(x)| \leqq b|y - x|$ with b independent of x and y. Thus f is Lipschitzian on C. ■

The differentiability of convex functions will be considered first for the case $n = 1$.

Theorem 1.5.2. *Let $f: \mathbb{R} \to \bar{\mathbb{R}}$ be convex. Then on* int dom f *the following holds. The right derivative f'_r and the left derivative f'_l exist and are monotonically increasing functions. The inequality $f'_l \leqq f'_r$ is valid, and with the exception of at most countably many points, $f'_l = f'_r$ holds and hence f is differentiable. Further, f'_r is continuous from the right and f'_l is continuous from the left (in particular, if f is differentiable on* int dom f, *then it is continuously differentiable).*

Proof. In the following, all arguments of f are taken from int dom f. Let $0 < \lambda < \mu$. Then

$$f(x + \lambda) = f\left(\frac{\mu - \lambda}{\mu} x + \frac{\lambda}{\mu}(x + \mu)\right) \leqq \frac{\mu - \lambda}{\mu} f(x) + \frac{\lambda}{\mu} f(x + \mu),$$

hence

$$\frac{f(x + \lambda) - f(x)}{\lambda} \leqq \frac{f(x + \mu) - f(x)}{\mu}.$$

Analogously,

$$f(x - \lambda) = f\left(\frac{\mu - \lambda}{\mu} x + \frac{\lambda}{\mu}(x - \mu)\right) \leqq \frac{\mu - \lambda}{\mu} f(x) + \frac{\lambda}{\mu} f(x - \mu),$$

hence

$$\frac{f(x) - f(x - \mu)}{\mu} \leqq \frac{f(x) - f(x - \lambda)}{\lambda}.$$

For arbitrary $\lambda, \mu > 0$,

$$f(x) = f\left(\frac{\lambda}{\lambda + \mu}(x - \mu) + \frac{\mu}{\lambda + \mu}(x + \lambda)\right)$$

$$\leqq \frac{\lambda}{\lambda + \mu} f(x - \mu) + \frac{\mu}{\lambda + \mu} f(x + \lambda),$$

hence

$$\frac{f(x) - f(x - \mu)}{\mu} \leqq \frac{f(x + \lambda) - f(x)}{\lambda}.$$

From the monotoneity and boundedness properties thus established one deduces the existence of

$$f'_{\mathrm{r}}(x) = \lim_{\lambda \downarrow 0} \frac{f(x+\lambda) - f(x)}{\lambda},$$

$$f'_{\mathrm{l}}(x) = \lim_{\lambda \downarrow 0} \frac{f(x-\lambda) - f(x)}{-\lambda},$$

as well as the inequality $f'_{\mathrm{l}}(x) \leqq f'_{\mathrm{r}}(x)$.

Thus, for $x < y$,

$$f'_{\mathrm{l}}(x) \leqq f'_{\mathrm{r}}(x) \leqq \frac{f(y) - f(x)}{y - x} \leqq f'_{\mathrm{l}}(y) \leqq f'_{\mathrm{r}}(y).$$

Hence f'_{l} and f'_{r} are increasing and thus have only countably many discontinuities. At each continuity point of f'_{l} the above inequalities yield $f'_{\mathrm{l}} = f'_{\mathrm{r}}$ and hence the existence of the derivative f'.

Let $x < y$. We have

$$\frac{f(y) - f(x)}{y - x} = \lim_{z \downarrow x} \frac{f(y) - f(z)}{y - z} \geqq \lim_{z \downarrow x} f'_{\mathrm{r}}(z),$$

hence

$$\lim_{z \downarrow x} f'_{\mathrm{r}}(z) \leqq f'_{\mathrm{r}}(x) \leqq \lim_{z \downarrow x} f'_{\mathrm{r}}(z)$$

by the monotoneity of f'_{r}. Thus f'_{r} is continuous from the right. Analogously one obtains that f'_{l} is continuous from the left. ■

Remark 1.5.3. Let $f: \mathbb{R} \to \mathbb{R}$ be convex; let $x_0 \in \mathbb{R}$ and m be a number with $f'_{\mathrm{l}}(x_0) \leqq m \leqq f'_{\mathrm{r}}(x_0)$. As noted in the above proof, one has

$$\frac{f(x) - f(x_0)}{x - x_0} \begin{cases} \geqq f'_{\mathrm{r}}(x_0) \geqq m, & \text{if } x > x_0, \\ \leqq f'_{\mathrm{l}}(x_0) \leqq m, & \text{if } x < x_0, \end{cases}$$

thus $f(x) \geqq f(x_0) + m(x - x_0)$ for all $x \in \mathbb{R}$. This shows that the line

$$\{(x, y) \in \mathbb{R} \times \mathbb{R} \mid y = f(x_0) + m(x - x_0)\}$$

supports the epigraph of f at the point $(x_0, f(x_0))$.

Now let $f: \mathbb{E}^n \to \bar{\mathbb{R}}$ be a convex function. Theorem 1.5.2 implies, in particular, for each point $x \in \operatorname{int}\operatorname{dom} f$, the existence of all the one-sided directional derivatives

$$f'(x; u) := \lim_{\lambda \downarrow 0} \frac{f(x + \lambda u) - f(x)}{\lambda},$$

where $u \in \mathbb{E}^n \setminus \{o\}$. We first collect some remarks on these directional derivatives in connection with sublinearity.

A function $f: \mathbb{E}^n \to \mathbb{R}$ is called *positively homogeneous* if

$$f(\lambda x) = \lambda f(x) \quad \text{for all } \lambda \geqq 0 \text{ and all } x \in \mathbb{E}^n$$

and f is called *subadditive* if

$$f(x + y) \leqq f(x) + f(y) \quad \text{for all } x, y \in \mathbb{E}^n.$$

A function that is positively homogeneous and subadditive is called *sublinear*. Every sublinear function is convex.

Lemma 1.5.4. *Let $f: \mathbb{E}^n \to \bar{\mathbb{R}}$ be convex and $x \in \operatorname{int} \operatorname{dom} f$. Then the directional derivative function*

$$f'(x; \cdot): \mathbb{E}^n \to \mathbb{R}$$

is sublinear.

Proof. Let $u \in \mathbb{E}^n \setminus \{o\}$. For $\lambda, \tau > 0$ we may write

$$\frac{f(x + \tau\lambda u) - f(x)}{\tau} = \lambda \frac{f(x + \tau\lambda u) - f(x)}{\tau\lambda},$$

and $\tau \downarrow 0$ gives $f'(x; \lambda u) = \lambda f'(x; u)$. For $u, v \in \mathbb{E}^n$ the convexity of f yields

$$\begin{aligned} f(x + \tau(u + v)) &= f(\tfrac{1}{2}(x + 2\tau u) + \tfrac{1}{2}(x + 2\tau v)) \\ &\leqq \tfrac{1}{2} f(x + 2\tau u) + \tfrac{1}{2} f(x + 2\tau v), \end{aligned}$$

hence

$$\frac{f(x + \tau(u + v)) - f(x)}{\tau} \leqq \frac{f(x + 2\tau u) - f(x)}{2\tau} + \frac{f(x + 2\tau v) - f(x)}{2\tau}$$

for $\tau > 0$, and $\tau \downarrow 0$ gives $f'(x; u + v) \leqq f'(x; u) + f'(x; v)$. ∎

Let $f: \mathbb{E}^n \to \mathbb{R}$ be a sublinear function. The condition $f(-u) = -f(u)$ is necessary and sufficient in order that f be linear on the subspace $\operatorname{lin}\{u\}$, and in this case we say that u is a *linearity direction* of f.

Lemma 1.5.5. *Let $f: \mathbb{E}^n \to \mathbb{R}$ be a sublinear function. The set $L[f]$ of all linearity directions of f is a linear subspace. Let $x \in \mathbb{E}^n$. Then $f'(x; \cdot) \leqq f$ and*

$$\operatorname{lin}(L[f] \cup \{x\}) \subset L[f'(x; \cdot)].$$

Proof. Let $u, v \in L[f]$. Then $f(u + v) \leqq f(u) + f(v) = -f(-u) - f(-v) \leqq -f(-u - v) \leqq f(u + v)$, the latter because $0 = f(o) = f(w - w) \leqq f(w) + f(-w)$. It follows that $u + v \in L[f]$ and hence that $L[f]$ is a linear subspace.

For $u \in \mathbb{E}^n$ and $\tau > 0$, the sublinearity of f yields

$$\frac{f(x + \tau u) - f(x)}{\tau} \leqq f(u),$$

and $\tau \downarrow 0$ gives $f'(x; u) \leqq f(u)$.

Let $u \in L[f]$. From

$$\begin{aligned} 2f(x) &\leqq f(x + \tau u) + f(x - \tau u) \\ &\leqq 2f(x) + \tau[f(u) + f(-u)] \\ &= 2f(x) \end{aligned}$$

it follows that

$$\frac{f(x + \tau u) - f(x)}{\tau} + \frac{f(x - \tau u) - f(x)}{\tau} = 0, \qquad (*)$$

and $\tau \downarrow 0$ gives $f'(x; u) + f'(x; -u) = 0$, hence $u \in L[f'(x; \cdot)]$. By positive homogeneity, equality $(*)$ holds also for $u = x$ if $0 < \tau < 1$, hence $\tau \downarrow 0$ gives $f'(x; x) + f'(x; -x) = 0$ and thus $x \in L[f'(x; \cdot)]$. ■

We return to general convex functions and consider stronger differentiability properties. For a convex function, the existence of all two-sided directional derivatives (that is, Gâteaux differentiability) implies (total or Fréchet) differentiability. In fact, this is already implied by the existence of the partial derivatives. The ith partial derivative of f at x, denoted by $\partial_i f(x)$, is the two-sided directional derivative $f'(x; e_i)$, where $e_1, \ldots, e_n$ is a fixed orthonormal basis of $\mathbb{E}^n$.

Theorem 1.5.6. *Let $f: \mathbb{E}^n \to \bar{\mathbb{R}}$ be convex and $x \in \operatorname{int} \operatorname{dom} f$. If f has partial derivatives (of first order) at x, then f is differentiable at x.*

Proof. If f has partial derivatives at x, we can define the gradient of f at x by

$$\operatorname{grad} f(x) := \sum_{i=1}^{n} \partial_i f(x) e_i$$

and we then have to show that

$$\lim_{|h| \to 0} \frac{g(h)}{|h|} = 0, \qquad (*)$$

where

$$g(h) := f(x + h) - f(x) - \langle \operatorname{grad} f(x), h \rangle \quad \text{for } h \in \mathbb{E}^n.$$

The function g is convex and finite in a neighbourhood of o. For $h = \sum_{i=1}^{n} \eta_i e_i$ we have

$$g(h) = g\left(\frac{1}{n} \sum_{i=1}^{n} n\eta_i e_i\right) \leqq \frac{1}{n} \sum_{i=1}^{n} g(n\eta_i e_i).$$

Using the inequality

$$\langle h, v\rangle \leqq |h||v| \leqq |h| \sum_{i=1}^{n} |v_i| \quad \text{for } v = \sum_{i=1}^{n} v_i e_i,$$

we obtain

$$g(h) \leqq \sum_{i=1}^{n} \eta_i \frac{g(n\eta_i e_i)}{n\eta_i} \leqq |h| \sum_{i=1}^{n} \left| \frac{g(n\eta_i e_i)}{n\eta_i} \right|$$

(where the summands with $\eta_i = 0$ have to be replaced by 0) and a similar inequality with h replaced by $-h$. Since $g(h) + g(-h) \geqq g(\text{o}) = 0$ by the convexity of g, we obtain for $h \neq \text{o}$

$$-\sum_{i=1}^{n} \left| \frac{g(-n\eta_i e_i)}{n\eta_i} \right| \leqq \frac{-g(-h)}{|h|} \leqq \frac{g(h)}{|h|} \leqq \sum_{i=1}^{n} \left| \frac{g(n\eta_i e_i)}{n\eta_i} \right|.$$

Since the partial derivatives of g at o exist and are zero, the left and right sides of this chain of inequalities tend to 0 for $h \to \text{o}$, hence (*) follows. ∎

For differentiable functions one has some useful criteria for convexity, which we now collect.

Theorem 1.5.7. *Let $D \subset \mathbb{R}$ be an open interval and $f: D \to \mathbb{R}$ a differentiable function. Then f is convex if and only if f' is increasing.*

Proof. If f is convex, then f' is increasing, by Theorem 1.5.2. Let f' be increasing, and let $x, y \in D$, $x < y$, and $\lambda \in (0, 1)$. By the mean value theorem, there exist numbers $\theta_1 \in [x, (1-\lambda)x + \lambda y]$ and $\theta_2 \in [(1-\lambda)x + \lambda y, y]$ with

$$f'(\theta_1) = \frac{f((1-\lambda)x + \lambda y) - f(x)}{\lambda(y - x)},$$

$$f'(\theta_2) = \frac{f(y) - f((1-\lambda)x + \lambda y)}{(1-\lambda)(y - x)}.$$

From $f'(\theta_1) \leqq f'(\theta_2)$ we obtain

$$f((1-\lambda)x + \lambda y) \leqq (1-\lambda)f(x) + \lambda f(y).$$ ∎

Corollary 1.5.8. *Let $D \subset \mathbb{R}$ be an open interval and $f: D \to \mathbb{R}$ a twice differentiable function. Then f is convex if and only if $f'' \geqq 0$.*

The extension of these criteria to differentiable functions on $\mathbb{E}^n$ relies on the simple fact, following from the definition of convexity, that a function on $\mathbb{E}^n$ is convex if and only if its restriction to an arbitrary line is convex.

Theorem 1.5.9. *Let $D \subset \mathbb{E}^n$ be convex and open and let $f: D \to \mathbb{R}$ be a differentiable function. Then the following assertions are equivalent:*

(a) f is convex;

(b) $f(y) - f(x) \geqq \langle \operatorname{grad} f(x), y - x \rangle$ for $x, y \in D$;

(c) $\langle \operatorname{grad} f(y) - \operatorname{grad} f(x), y - x \rangle \geqq 0$ for $x, y \in D$.

Proof. Suppose (a) holds. Let $x, y \in D$ and $0 < \lambda < 1$. From

$$f(x + \lambda(y - x)) = f((1 - \lambda)x + \lambda y) \leqq (1 - \lambda)f(x) + \lambda f(y)$$

we get

$$\frac{f(x + \lambda(y - x)) - f(x)}{\lambda} - \langle \operatorname{grad} f(x), y - x \rangle$$

$$\leqq f(y) - f(x) - \langle \operatorname{grad} f(x), y - x \rangle.$$

For $\lambda \to 0$ the left side converges to 0, hence (b) follows.

Suppose (b) holds. Interchanging the roles of x and y and adding both inequalities, we get (c).

Suppose (c) holds. Let $x, y \in D$. Let

$$g(\lambda) := f(x + \lambda(y - x)) \quad \text{for } 0 \leqq \lambda \leqq 1.$$

For $0 \leqq \lambda_0 < \lambda_1 \leqq 1$ the chain rule gives

$$\begin{aligned} &g'(\lambda_1) - g'(\lambda_0) \\ &= \langle \operatorname{grad} f(x + \lambda_1(y - x)), y - x \rangle - \langle \operatorname{grad} f(x + \lambda_0(y - x)), y - x \rangle \\ &\geqq 0. \end{aligned}$$

Thus g' is an increasing function and, therefore, g is a convex function. Since $x, y \in D$ were arbitrary, this implies that f is convex. ■

Theorem 1.5.10. *Let $D \subset \mathbb{E}^n$ be convex and open and let $f: D \to \mathbb{R}$ be twice differentiable. Then f is convex if and only if its Hessian matrix*

$$\operatorname{Hess} f(x) := (\partial_i \partial_j f(x))_{i,j=1}^n$$

is positive semidefinite for all $x \in D$.

Proof. For $x, y \in D$ define

$$g(\lambda) := f(x + \lambda(y - x)) \quad \text{for } x + \lambda(y - x) \in D.$$

Then f is convex if and only if all such functions g are convex, which is equivalent to $g'' \geqq 0$. We need only check whether $g''(0) \geqslant 0$ holds generally. In fact, let $\lambda_1 \in (0, 1)$. Writing $x_1 := x + \lambda_1(y - x)$ and $h(\tau) := f(x_1 + \tau(y - x_1))$, we have $g(\lambda_1 + \lambda) = h(\lambda/(1 - \lambda_1))$, hence $g''(\lambda_1) \geqq 0$ if and only if $h''(0) \geqq 0$.

Now for $x = \sum_{i=1}^n \xi_i e_i$, $y = \sum_{i=1}^n \eta_i e_i$ we have

$$\begin{aligned} g''(0) &= \lim_{\lambda\to 0} \frac{g'(\lambda) - g'(0)}{\lambda} \\ &= \lim_{\lambda\to 0} \frac{1}{\lambda} \langle \operatorname{grad} f(x + \lambda(y - x)) - \operatorname{grad} f(x), y - x\rangle \\ &= \lim_{\lambda\to 0} \sum_{j=1}^n \frac{1}{\lambda} [\partial_j f(x + \lambda(y - x)) - \partial_j f(x)](\eta_j - \xi_j) \\ &= \sum_{i=1}^n \sum_{j=1}^n \partial_i \partial_j f(x)(\eta_i - \xi_i)(\eta_j - \xi_j), \end{aligned}$$

from which the assertion follows. ■

For convex functions that are not necessarily differentiable, the notions of gradient and differential have natural extensions that are particularly important in optimization. In our context they are of interest due to their simple geometric interpretation. If $f: \mathbb{E}^n \to \bar{\mathbb{R}}$ is convex and differentiable at x, then Theorem 1.5.9 states that

$$f(y) \geqq f(x) + \langle \operatorname{grad} f(x), y - x\rangle \quad \text{for all } y \in \mathbb{E}^n.$$

This gives rise to the following definition. Let $f: \mathbb{E}^n \to \bar{\mathbb{R}}$ be convex and let $x \in \operatorname{dom} f$. The set

$$\partial f(x) := \{v \in \mathbb{E}^n \mid f(y) \geqq f(x) + \langle v, y - x\rangle \quad \text{for all } y \in \mathbb{E}^n\}$$

is called the *subdifferential* of f at x, and each element of $\partial f(x)$ is called a *subgradient* of f at x. Clearly the subdifferential of f at x is a closed convex subset of $\mathbb{E}^n$. It parametrizes the set of support planes to the epigraph of f above x, in a sense made precise in the following theorem.

Theorem 1.5.11. *Let $f: \mathbb{E}^n \to \bar{\mathbb{R}}$ be convex and let $x \in \operatorname{dom} f$. The vector v is a subgradient of f at x if and only if the vector $(v, -1) \in \mathbb{E}^n \times \mathbb{R}$ is an exterior normal vector to* epi f *at $(x, f(x))$.*

Proof. $(v, -1)$ is an exterior normal vector to epi f at $(x, f(x))$ if and only if

$$\langle (y, \eta) - (x, f(x)), (v, -1)\rangle \leqq 0$$

for all $(y, \eta) \in \operatorname{epi} f$, and this is equivalent to

$$\langle (y, f(y)) - (x, f(x)), (v, -1)\rangle \leqq 0$$

for all $y \in \mathbb{E}^n$, hence to

$$f(y) \geqq f(x) + \langle v, y - x\rangle$$

for all $y \in \mathbb{E}^n$. ■

Differentiability of a convex function is equivalent to the uniqueness of subgradients.

Theorem 1.5.12. *Let $f: \mathbb{E}^n \to \bar{\mathbb{R}}$ be convex and let $x \in \operatorname{int} \operatorname{dom} f$. Then f is differentiable at x if and only if f has only one subgradient at x.*

Proof. Suppose that f is differentiable at x. Let v be a subgradient of f at x, then

$$\frac{f(x + \lambda u) - f(x)}{\lambda} \geqq \langle v, u \rangle$$

for $\lambda > 0$, hence $f'(x; u) \geqq \langle v, u \rangle$ for all $u \in \mathbb{E}^n$. Differentiability of f at x implies $f'(x; u) = -f'(x; -u) \leqq -\langle v, -u \rangle = \langle v, u \rangle$, hence $f'(x; u) = \langle v, u \rangle$. Since this holds for all $u \in \mathbb{E}^n$, v is unique.

Vice versa, suppose that f has at most one subgradient at x. Let $u \in \mathbb{E}^n \setminus \{o\}$. Put

$$g(\lambda) := f(x + \lambda u) \quad \text{for } \lambda \in \mathbb{R}$$

and choose a number m with $g_l'(0) \leqq m \leqq g_r'(0)$. By Remark 1.5.3, we have $g(\lambda) \geqq g(0) + m\lambda$ and thus $f(x + \lambda u) \geqq f(x) + m\lambda$. Hence the line

$$G := \{(x + \lambda u, f(x) + m\lambda) | \lambda \in \mathbb{R}\} \subset \mathbb{E}^n \times \mathbb{R}$$

satisfies $G \cap \operatorname{int} \operatorname{epi} f = \emptyset$. By Theorem 1.3.7 there exists in $\mathbb{E}^n \times \mathbb{R}$ a hyperplane $H_{(v,-1),\alpha}$ separating G and $\operatorname{int} \operatorname{epi} f$ (the normal vector of this hyperplane can in fact be assumed to be of the form $(v, -1)$ since, due to $x \in \operatorname{int} \operatorname{dom} f$, the hyperplane cannot be 'vertical'). Since $H_{(v,-1),\alpha}$ supports $\operatorname{epi} f$ at $(x, f(x))$, v is a subgradient of f at x and hence unique. The inclusion $G \subset H_{(v,-1),\alpha}$ implies that

$$\langle (v, -1), (x + \lambda u, f(x) + m\lambda) \rangle = \alpha$$

for all $\lambda \in \mathbb{R}$ and hence that $m = \langle v, u \rangle$. Since v is unique, m is also unique and thus g is differentiable at 0. Since $u \in \mathbb{E}^n \setminus \{o\}$ was arbitrary, f has two-sided directional derivatives at x and hence, by Theorem 1.5.6, is differentiable at x. ■

Notes for Section 1.5

1. Standard references for convex functions are Rockafellar [29] and Roberts & Varberg [28], which we have followed in many respects; see also Marti [25].
2. *Differentiability almost everywhere of convex functions.* Let f be a finite convex function on an open convex subset $D \subset \mathbb{E}^n$. By Theorem 1.5.1, f is locally Lipschitzian and hence, by Rademacher's theorem (e.g., Federer (1969), p. 216), differentiable almost everywhere on D. This follows also from Theorem 1.5.12 and Theorem 2.2.4, the latter applied to the epigraph

of f, or from the one-dimensional case (Theorem 1.5.2) and Fubini's theorem (Roberts & Varberg [28], p. 116). Rather precise information on the set of non-differentiability is available from work of Zajíček (1979), who also has a related result for continuous convex functions on a separable Banach space.

The convex function f is in fact twice differentiable almost everywhere. Since the first differential need not exist in a full neighbourhood of the point considered, this assertion must be interpreted carefully. One possible way of defining the twice differentiability of f at x is by the existence of a second-order Taylor expansion. This means that f is differentiable at x and that there exists a symmetric linear map $Af(x)\colon \mathbb{E}^n \to \mathbb{E}^n$ such that

$$f(y) = f(x) + \langle \operatorname{grad} f(x), y - x\rangle + \tfrac{1}{2}\langle Af(x)(y - x), y - x\rangle + o(|y - x|^2)$$

for all $y \in D$. Another possibility is to say that a map $\vartheta\colon D \to \mathbb{E}^n$ is a subgradient choice for f if $\vartheta(x) \in \partial f(x)$ for each $x \in D$. In this case the function f is called twice differentiable at x if the family of subgradient choices for f is uniformly differentiable at x, which means that there exist a neighbourhood V of x, a linear map $Af(x)\colon \mathbb{E}^n \to \mathbb{E}^n$ and a function $\psi\colon \mathbb{R}^+ \to \mathbb{R}^+$ with $\psi(\tau) \to 0$ for $\tau \to 0$ such that

$$|\vartheta(y) - \vartheta(x) - Af(x)(y - x)| \leqq \psi(|y - x|)|y - x|$$

for $y \in V \cap D$ and each subgradient choice ϑ for f. It follows from the work of Aleksandrov (1939c) (see also Bangert (1979)) that for a convex function twice differentiability in this sense and the existence of a second-order Taylor expansion are equivalent.

With this definition, it is true that the convex function f is twice differentiable almost everywhere. For $n = 1$, this is a consequence of the differentiability almost everywhere of a monotone function, as was first pointed out by Jessen (1929). For $n = 2$, it was proved by Busemann & Feller (1936a), and by using their result and an induction argument, Aleksandrov (1939c) obtained the general case. Other proofs, more analytic in character, are due to Rešetnjak (1968b), who used techniques from the theory of distributions, and to Bangert (1977, 1979). Essential tools of Bangert's proof are the Lipschitz property of the metric projection (Theorem 1.2.2), Rademacher's theorem on the almost everywhere differentiability of Lipschitz maps and a version of Sard's lemma for Lipschitz maps.

Asplund (1973) applied Aleksandrov's theorem to show that a metric projection on any closed (not necessarily convex) subset of $\mathbb{R}^n$ is almost everywhere differentiable.

3. Dudley (1977) showed that a Schwartz distribution on $\mathbb{E}^n$ is a convex function if and only if its second derivative is a non-negative $n \times n$ matrix-valued Radon measure. He also showed the absolute continuity of such a measure with respect to $(n - 1)$-dimensional Hausdorff measure, and other related results.
4. *The second-order subdifferential.* Given a finite convex function f on $\mathbb{E}^n$, Hiriart-Urruty (1986) introduced for all $x \in \mathbb{E}^n$ and all $v \in \partial f(x)$ a closed convex set containing o, which he (later) called the second-order subdifferential of f at (x, v), denoting it by $\partial^2 f(x, v)$. Then

$$\partial^2 f(x) := \bigcap_{v \in \partial f(x)} \partial^2 f(x, v)$$

is called the second-order subdifferential of f at x. $\partial^2 f(x)$ is a compact convex set. If f is twice differentiable at x, then $\partial^2 f(x)$ is the ellipsoid associated with the ordinary second differential of f at x. The second-order subdifferential was further investigated by Hiriart-Urruty & Seeger (1989), who also studied it in relation to a notion of generalized Dupin indicatrices for non-smooth convex surfaces.

1.6. Duality

Associated with convex sets, cones and functions, which satisfy some weak conditions, are dual objects of the same kind. This duality permits us, for example, to translate certain results on boundary points of convex sets into results on support planes (or normal vectors), and vice versa. But there are numerous other applications.

First, let $K \in \mathcal{K}_0^n$ (the set of convex bodies with interior points) be a body with $o \in \operatorname{int} K$. We define

$$K^* := \{x \in \mathbb{E}^n | \langle x, y\rangle \leqq 1 \quad \text{for all } y \in K\}.$$

(More properly, K^* should be defined as the subset

$$\{x^* \in (\mathbb{E}^n)^* | x^*(y) \leqq 1 \quad \text{for all } y \in K\}$$

of the dual space $(\mathbb{E}^n)^*$ of $\mathbb{E}^n$. As, however, we are working with a fixed scalar product in $\mathbb{E}^n$, we shall always use the canonical isomorphism between $(\mathbb{E}^n)^*$ and $\mathbb{E}^n$ induced by this scalar product to identify both spaces.)

The next theorem shows that K^*, which is called the *polar body* of K, satisfies the same assumptions as K and that this correspondence is a true duality.

Theorem 1.6.1. *Let* $K \in \mathcal{K}_0^n$ *and* $o \in \operatorname{int} K$. *Then* $K^* \in \mathcal{K}_0^n$, $o \in \operatorname{int} K^*$, *and* $K^{**} = K$.

Proof. If $x_1, x_2 \in K^*$ and $\lambda \in [0, 1]$, then $\langle (1-\lambda)x_1 + \lambda x_2, y\rangle \leqq 1$ for all $y \in K$, hence $(1-\lambda)x_1 + \lambda x_2 \in K^*$. Thus K^* is convex, and it is equally easy to see that K^* is closed. For a ball $B(o, \varepsilon)$ one has $B(o, \varepsilon)^* = B(o, 1/\varepsilon)$ as an immediate consequence of the definition. Further, $K_1 \subset K_2$ implies $K_1^* \supset K_2^*$. We can choose $\varepsilon, \rho > 0$ with $B(o, \varepsilon) \subset K \subset B(o, \rho)$; hence $B(o, 1/\rho) \subset K^* \subset B(o, 1/\varepsilon)$, showing that $o \in \operatorname{int} K^*$ and that K^* is compact.

Let $y \in K$. For arbitrary $x \in K^*$ one has $\langle x, y\rangle \leqq 1$, hence $y \in K^{**}$. Thus $K \subset K^{**}$ (observe that up to now neither convexity nor closedness of K have been needed).

Let $z \in \mathbb{E}^n \setminus K$. Since K is convex and closed then by Theorem 1.3.4 there is a hyperplane $H_{u,\alpha}$ with $K \subset H_{u,\alpha}^-$ and $\langle z, u\rangle > \alpha$; here $\alpha > 0$ since $o \in \operatorname{int} K$. For all $y \in K$ we have $\langle y, u/\alpha\rangle \leqq 1$, hence $u/\alpha \in K^*$. Now $\langle z, u/\alpha\rangle > 1$ shows that $z \notin K^{**}$. This finishes the proof of $K^{**} = K$. ■

The following theorem describes the extent to which the forming of polar bodies interchanges the operations of intersection and union.

Theorem 1.6.2. *Let* $K_1, K_2 \in \mathcal{K}_0^n$ *and* $o \in \operatorname{int} K_i$ $(i = 1, 2)$. *Then*

$$(K_1 \cap K_2)^* = \operatorname{conv}(K_1^* \cup K_2^*),$$
$$[\operatorname{conv}(K_1 \cup K_2)]^* = K_1^* \cap K_2^*.$$

If $K_1 \cup K_2$ *is convex, then* $K_1^* \cup K_2^*$ *is convex, hence*

$$(K_1 \cap K_2)^* = K_1^* \cup K_2^*,$$
$$(K_1 \cup K_2)^* = K_1^* \cap K_2^*.$$

Proof. From $K_1 \cap K_2 \subset K_i$ we get $K_i^* \subset (K_1 \cap K_2)^*$, hence $\operatorname{conv}(K_1^* \cup K_2^*) \subset (K_1 \cap K_2)^*$. From $K_i \subset \operatorname{conv}(K_1 \cup K_2)$ we get $[\operatorname{conv}(K_1 \cup K_2)]^* \subset K_i^*$, hence $[\operatorname{conv}(K_1 \cup K_2)]^* \subset K_1^* \cap K_2^*$. Applying both inclusions to K_i^* instead of K_i and using $K^{**} = K$, we arrive at the first two equalities of the theorem.

Suppose that $K_1 \cup K_2$ is convex. Let $x \in \mathbb{E}^n \setminus (K_1^* \cup K_2^*)$. By Theorem 1.3.4 there are $u_i \in \mathbb{E}^n \setminus \{o\}$ and $\alpha_i \in \mathbb{R}$ with $\langle x, u_i \rangle > \alpha_i$ and $\langle y, u_i \rangle \leqq \alpha_i$ for $y \in K_i^*$ (hence $\alpha_i > 0$), which implies $u_i/\alpha_i \in K_i^{**} = K_i$ $(i = 1, 2)$. Since $K_1 \cup K_2$ is convex, there is a point $z \in [u_1/\alpha_1, u_2/\alpha_2] \cap K_1 \cap K_2$. Since z is a convex combination of u_1/α_1 and u_2/α_2, we get $\langle x, z \rangle > 1$ and hence $x \notin (K_1 \cap K_2)^*$. This proves $(K_1 \cap K_2)^* \subset K_1^* \cup K_2^*$. Since the opposite inclusion is trivial, we get $(K_1 \cap K_2)^* = K_1^* \cup K_2^*$, which shows that $K_1^* \cup K_2^*$ is convex. ■

A similar duality exists for convex cones. Let $C \subset \mathbb{E}^n$ be a closed convex cone. We define

$$C^* := \{x \in \mathbb{E}^n \mid \langle x, y \rangle \leqq 0 \quad \text{for all } y \in C\}$$

and call this the *dual cone* of C. This is again a closed convex cone, and one has $C^{**} = C$. The proof is analogous to the corresponding one for polar bodies, except that instead of Theorem 1.3.4 the separation theorem, 1.3.9, has to be applied. Also, the proof of the following theorem is so similar to that of Theorem 1.6.2 that it can be omitted.

Theorem 1.6.3. *Let* $C_1, C_2 \subset \mathbb{E}^n$ *be closed convex cones. Then*

$$(C_1 \cap C_2)^* = C_1^* + C_2^*,$$
$$(C_1 + C_2)^* = C_1^* \cap C_2^*.$$

If $C_1 \cup C_2$ *is convex, then* $C_1^* \cup C_2^*$ *is convex, hence*

$$(C_1 \cap C_2)^* = C_1^* \cup C_2^*,$$
$$(C_1 \cup C_2)^* = C_1^* \cap C_2^*.$$

To treat duality for convex functions, a few preliminaries are necessary. To establish the dualities above, we had to use the fact that a closed convex set is the intersection of its supporting halfspaces. In the case of a convex function, this fact will be applied to the epigraph. One considers, therefore, only functions whose epigraph is closed; such a function is called, in brief, *closed*. Closedness of a function is evidently equivalent to lower semicontinuity. The supporting halfspaces to the epigraph of a convex function are of two kinds. Generally, a closed halfspace in $\mathbb{E}^n \times \mathbb{R}$ containing the epigraph of a proper function is of the form

$$\{(x, \zeta) \in \mathbb{E}^n \times \mathbb{R} \mid \langle (x, \zeta), (u, \eta)\rangle \leqq \alpha\}$$

with $\eta \leqq 0$. If $\eta = 0$, the halfspace is called *vertical*. If it is not vertical, we can assume that $\eta = -1$, so that the halfspace is given by

$$\begin{aligned} H^-_{(u,-1),\alpha} &= \{(x, \zeta) \mid \langle x, u\rangle - \alpha \leqq \zeta\} \\ &= \operatorname{epi} h \end{aligned}$$

with $h(x) := \langle x, u\rangle - \alpha$. Observing that $\operatorname{epi} f \subset \operatorname{epi} h$ for functions h, f is equivalent to $h \leqq f$, we are led to the following lemma.

Lemma 1.6.4. *Let $f: \mathbb{E}^n \to \bar{\mathbb{R}}$ be convex and closed. Then*

$$f = \sup\{h \mid h: \mathbb{E}^n \to \mathbb{R} \text{ affine}, \quad h \leqq f\}.$$

Proof. The closed convex set $\operatorname{epi} f$ is the intersection of its supporting halfspaces. We have to show that it is already the intersection of its nonvertical supporting halfspaces. Let

$$H_1^- := \{(x, \zeta) \in \mathbb{E}^n \times \mathbb{R} \mid h_1(x) := \langle x, u_1\rangle - \alpha_1 \leqq 0\}$$

be a vertical supporting halfspace of $\operatorname{epi} f$, and let $(x_0, \zeta_0) \in \mathbb{E}^n \times \mathbb{R}$ be a point such that $(x_0, \zeta_0) \notin H_1^-$. There is some non-vertical supporting halfspace to $\operatorname{epi} f$ (for instance, at a point $(x, f(x))$ with $x \in \operatorname{relint} \operatorname{dom} f$), say

$$H_2^- := \{(x, \zeta) \in \mathbb{E}^n \times \mathbb{R} \mid h_2(x) := \langle x, u_2\rangle - \alpha_2 \leqq \zeta\}.$$

Define $h := \lambda h_1 + h_2$ with $\lambda > 0$ chosen below. For $x \in \operatorname{dom} f$ we have $h_1(x) \leqq 0$ and $h_2(x) \leqq f(x)$, hence $h(x) \leqq f(x)$. If $x \in \mathbb{E}^n \setminus \operatorname{dom} f$, then $f(x) = \infty$ and hence $h(x) \leqq f(x)$ again. Thus

$$\operatorname{epi} f \subset H^- := \{(x, \zeta) \in \mathbb{E}^n \times \mathbb{R} \mid h(x) \leqq \zeta\}.$$

Since $(x_0, \zeta_0) \notin H_1^-$, we have $h_1(x_0) > 0$ and hence $\lambda > 0$ can be chosen such that $h(x_0) > \zeta_0$ and thus $(x_0, \zeta_0) \notin H^-$. We deduce the existence of a non-vertical supporting halfspace (a translate of H^-) to $\operatorname{epi} f$ that does not contain (x_0, ζ_0). Thus $\operatorname{epi} f$ is the intersection of all its non-vertical supporting halfspaces; equivalently f is the supremum of all affine functions $\leqq f$. ■

We are now in a position to treat the announced duality for convex functions. For a closed convex function $f: \mathbb{E}^n \to \bar{\mathbb{R}}$ the *conjugate function* is defined by

$$f^*(u) := \sup\{\langle u, x\rangle - f(x) | x \in \mathbb{E}^n\} \quad \text{for } u \in \mathbb{E}^n.$$

The following reformulation of the definition sheds some light on the intuitive meaning of the conjugate function. For given $u \in \mathbb{E}^n \setminus \{o\}$, each real number α determines a non-vertical halfspace $H^-_{(u,-1),\alpha}$ in $\mathbb{E}^n \times \mathbb{R}$, and we have

$$(x, f(x)) \in H^-_{(u,-1),\alpha} \Leftrightarrow \langle u, x\rangle - f(x) \leqq \alpha,$$

where the latter inequality is satisfied for all $x \in \mathbb{E}^n$ if and only if $\alpha \geqq f^*(u)$. Hence we may also write

$$f^*(u) = \inf\{\alpha \in \mathbb{R} | \operatorname{epi} f \subset H^-_{(u,-1),\alpha}\},$$

which can be interpreted as saying that the conjugate function f^* provides a parametrization of the set of supporting halfspaces of the epigraph of f.

Theorem 1.6.5. *Let $f: \mathbb{E}^n \to \bar{\mathbb{R}}$ be a closed convex function. Then f^* is a closed convex function, and $f^{**} = f$.*

Proof. From $\{f = \infty\} \neq \mathbb{E}^n$ it follows that $\{f^* = -\infty\} = \emptyset$. As the epigraph $\operatorname{epi} f$ has some non-vertical supporting halfspace $\{(x, \zeta) | \langle u, x\rangle - \alpha \leqq \zeta\}$, we have $\langle u, x\rangle - f(x) \leqq \alpha$ for $x \in \mathbb{E}^n$ and therefore $f^*(u) \leqq \alpha$, hence $\{f^* = \infty\} \neq \mathbb{E}^n$. Thus f^* is proper. Being the supremum of a family of affine functions, which are convex and continuous, f^* is convex and lower semicontinuous, hence closed. The definition of f^* can be rewritten as

$$f^*(u) = \sup\{\langle u, x\rangle - \zeta | (x, \zeta) \in \operatorname{epi} f\},$$

thus

$$f^{**}(x) = \sup\{\langle x, u\rangle - \alpha | (u, \alpha) \in \operatorname{epi} f^*\}.$$

Now $(u, \alpha) \in \operatorname{epi} f^*$ is equivalent to $\langle \cdot, u\rangle - \alpha \leqq f$, hence Lemma 1.6.4 gives

$$\begin{aligned} f(x) &= \sup\{h(x) | h: \mathbb{E}^n \to \mathbb{R} \text{ affine}, h \leqq f\} \\ &= \sup\{\langle x, u\rangle - \alpha | (u, \alpha) \in \operatorname{epi} f^*\} \\ &= f^{**}(x). \end{aligned}$$

■

Note for Section 1.6

Infimal convolution. Under conjugation, addition of closed convex functions is transformed into a process called *infimal convolution* (see Rockafellar [29]). For functions $f_1, \ldots, f_m: \mathbb{E}^n \to \bar{\mathbb{R}}$ one defines

$$(f_1 \square \ldots \square f_m)(x) := \inf\{f_1(x_1) + \ldots + f_m(x_m) | x_i \in \mathbb{E}^n, x_1 + \ldots + x_m = x\}.$$

If $f_1, \ldots, f_m$ are closed convex functions, then

$$(f_1 + \ldots + f_m)^* = f_1^* \square \ldots \square f_m^*.$$

1.7. The support function

Since a closed convex set is the intersection of its supporting halfspaces, such a set can conveniently be described by specifying the position of its support planes, given their exterior normal vectors. This description is provided by the support function, which is of fundamental importance in the Brunn–Minkowski theory of convex bodies.

For a nonempty closed convex set $K \subset \mathbb{E}^n$ the *support function* $h(K, \cdot) = h_K$ is defined by

$$h(K, u) := \sup\{\langle x, u\rangle | x \in K\} \quad \text{for } u \in \mathbb{E}^n.$$

For $u \in \operatorname{dom} h(K, \cdot)\backslash\{o\}$ we put

$$H(K, u) := \{x \in \mathbb{E}^n | \langle x, u\rangle = h(K, u)\},$$
$$H^-(K, u) := \{x \in \mathbb{E}^n | \langle x, u\rangle \leqq h(K, u)\},$$
$$F(K, u) := H(K, u) \cap K.$$

$H(K, u)$, $H^-(K, u)$ and $F(K, u)$ are respectively the *support plane, supporting halfspace* and *support set* of K, each with exterior normal vector u. Note that these definitions of support plane and supporting halfspace extend the notions of Section 1.3. They obviously coincide with the latter if $F(K, u) \neq \emptyset$, but for an unbounded set K it may happen that $F(K, u) = \emptyset$.

The intuitive meaning of the support function is simple. For a unit vector $u \in S^{n-1} \cap \operatorname{dom} h(K, \cdot)$, $h(K, u)$ is the signed distance of the support plane to K with exterior normal vector u from the origin; the distance is negative if and only if u points into the open halfspace containing the origin.

From the definition of the support function we see immediately that $h(K, \cdot) = \langle z, \cdot\rangle$ if and only if $K = \{z\}$, that $h(K + t, u) = h(K, u) + \langle t, u\rangle$ for $t \in \mathbb{E}^n$ and further that $h(K, \lambda u) = \lambda h(K, u)$ for $\lambda \geqq 0$ and $h(K, u + v) \leqq h(K, u) + h(K, v)$. Hence, $h(K, \cdot)$ is a convex function if $K \neq \mathbb{E}^n$. It is also clear from the definition that $h_K \leqq h_L$ if and only if $K \subset L$ and that, for any linear subspace E of $\mathbb{E}^n$, one has

$h(K|E, u) = h(K, u)$ for $u \in E$. Further, we note that $h(\lambda K, \cdot) = \lambda h(K, \cdot)$ for $\lambda \geqq 0$ and $h(-K, u) = h(K, -u)$.

For $K \neq \mathbb{E}^n$, the fact that $x \in K$ if and only if $\langle x, u \rangle \leqq h(K, u)$ for all $u \in \mathbb{E}^n$ can also be expressed by saying that

$$\partial h_K(o) = K,$$

that is, K is the subdifferential of its support function at o.

The support function of a compact convex set $K \subset \mathbb{E}^n$ can also be described as follows. Let

$$C_K := \{\lambda(x, -1) \in \mathbb{E}^n \times \mathbb{R} | x \in K, \lambda \geqq 0\}.$$

Then C_K is a closed convex cone in $\mathbb{E}^n \times \mathbb{R}$, and its dual cone is the epigraph of the support function of K,

$$C_K^* = \operatorname{epi} h_K.$$

In fact, $(y, \eta) \in \operatorname{epi} h_K$ if and only if $\eta \geqq \langle x, y \rangle$ for all $x \in K$, and this is equivalent to $\langle (y, \eta), (\lambda x, -\lambda) \rangle \leqq 0$ for all $x \in K$ and $\lambda \geqq 0$ and hence to $(y, \eta) \in C_K^*$.

To simplify the exposition, the following considerations will be restricted to compact convex sets, since in later chapters the support function will be used only for these. For a convex body K, the supremum in the definition of $h(K, u)$ is attained and finite for each u, and $h(K, \cdot)$ is a sublinear function.

Theorem 1.7.1. *If $f: \mathbb{E}^n \to \mathbb{R}$ is a sublinear function, then there is a unique convex body $K \in \mathcal{K}^n$ with support function f.*

The uniqueness is clear. For the fundamental existence result we give three proofs, each of which has its merits. The first proof is very short (provided the duality theorem for convex functions is taken for granted), the second one is perhaps the simplest and the third is of historical interest.

First proof. By Theorem 1.5.1, the function f is continuous and hence closed. For each real $\lambda > 0$ its conjugate function f^* satisfies

$$\begin{aligned} f^*(x) &= \sup\{\langle x, u \rangle - f(u) | u \in \mathbb{E}^n\} \\ &= \sup\{\langle x, \lambda u \rangle - f(\lambda u) | u \in \mathbb{E}^n\} \\ &= \lambda f^*(x), \end{aligned}$$

which is only possible if $f^*(x) \in \{0, \infty\}$ for $x \in \mathbb{E}^n$. By Theorem 1.6.5, f^* is closed and convex, hence it is the indicator function I_K of a closed convex set K, which is not empty because $\{f^* = \infty\} \neq \mathbb{E}^n$. Theorem 1.6.5 further implies

$$\begin{aligned} f(u) = f^{**}(u) &= \sup\{\langle x, u\rangle - f^*(x) | x \in \mathbb{E}^n\} \\ &= \sup\{\langle x, u\rangle | x \in K\} \\ &= h(K, u). \end{aligned}$$

Clearly K is bounded. ■

Thus the support function of a convex body (or, more generally, of a closed convex set) is the conjugate of its indicator function.

For the second and third proofs we define

$$K := \{x \in \mathbb{E}^n | \langle x, v\rangle \leqq f(v) \quad \text{for all } v \in \mathbb{E}^n\}.$$

Then K, being the intersection of closed halfspaces, one for each normal direction, is convex and compact, and if $K \neq \emptyset$, we evidently have $h(K, u) \leqq f(u)$ for $u \in \mathbb{E}^n$. Therefore, it remains to show that $K \neq \emptyset$ and $h(K, u) \geqq f(u)$ for $u \in \mathbb{E}^n$.

Second proof. Since f is sublinear and continuous, its epigraph is a closed convex cone in $\mathbb{E}^n \times \mathbb{R}$. Let $u \in \mathbb{E}^n \setminus \{o\}$ be given. Since $(u, f(u)) \in \operatorname{bd} \operatorname{epi} f$, Theorem 1.3.2 implies the existence of a support plane $H_{(y,\eta),\alpha}$ to $\operatorname{epi} f$ through $(u, f(u))$ such that $\operatorname{epi} f \subset H^-_{(y,\eta),\alpha}$. From Theorem 1.3.9 it follows that $\alpha = 0$. Clearly $\eta \neq 0$ since $\operatorname{dom} f = \mathbb{E}^n$, hence we may assume that $\eta = -1$. Then we have

$$\operatorname{epi} f \subset H^-_{(y,-1),0} = \{(v, \nu) \in \mathbb{E}^n \times \mathbb{R} | \langle (y, -1), (v, \nu)\rangle \leqq 0\}$$

and thus $\langle y, v\rangle \leqq f(v)$ for all $v \in \mathbb{E}^n$. From $(u, f(u)) \in H_{(y,-1),0}$ we get $\langle y, u\rangle = f(u)$. This shows that $y \in K$, hence $K \neq \emptyset$, and $h(K, u) \geqq f(u)$. ■

Third proof. The assertion $K \neq \emptyset$ will be proved by induction with respect to the codimension of the linearity space $L[f]$ (cf. Lemma 1.5.5). If $\dim L[f] = n$, then f is linear, say $f(u) = \langle z, u\rangle$ with $z \in \mathbb{E}^n$, and then $K = \{z\}$. Suppose that $\dim L[f] = k < n$ and that in the case of greater dimension of the linearity space the assertion is true. Let $x \in \mathbb{E}^n \setminus L[f]$. By Lemma 1.5.5, $\dim L[f'(x; \cdot)] \geqq k + 1$, hence the induction hypothesis applied to the function $f'(x; \cdot)$, which is sublinear by Lemma 1.5.4, yields the existence of a point $y \in \mathbb{E}^n$ satisfying

$$\langle y, v\rangle \leqq f'(x; v) \quad \text{for all } v \in \mathbb{E}^n.$$

By Lemma 1.5.5, $f'(x; \cdot) \leqq f$, hence $y \in K$. The assertion $K \neq \emptyset$ is proved.

Let $u \in \mathbb{E}^n \setminus \{o\}$. Since $f'(u; \cdot)$ is sublinear, the result already proved yields the existence of a point $y \in \mathbb{E}^n$ with

$$\langle y, v\rangle \leqq f'(u; v) \leqq f(v) \quad \text{for } v \in \mathbb{E}^n.$$

This shows that $y \in K$, and from $\langle y, -u\rangle \leqq f'(u; -u) = -f(u)$ we get $h(K, u) \geqq \langle y, u\rangle \geqq f(u)$. ■

The directional derivatives of support functions appearing implicitly in the last proof have a simple intuitive meaning: they yield the support functions of the corresponding support sets.

Theorem 1.7.2. *For $K \in \mathcal{K}^n$ and $u \in \mathbb{E}^n \setminus \{o\}$,*

$$h'_K(u; x) = h(F(K, u), x) \quad \text{for } x \in \mathbb{E}^n.$$

Proof. By Lemma 1.5.4 and Theorem 1.7.1, $h'_K(u; \cdot)$ is the support function of a convex body K' that satisfies $K' \subset K$ because $h'_K(u; \cdot) \leqq h_K$ (Lemma 1.5.5). Let $y \in K'$. Then $\langle y, u\rangle \leqq h(K, u)$; from $\langle y, -u\rangle \leqq h'_K(u; -u) = -h(K, u)$ we get $\langle y, u\rangle = h(K, u)$ and hence $y \in F(K, u)$. Thus $K' \subset F(K, u)$.

Let $y \in F(K, u)$. Then $\langle y, u\rangle = h(K, u)$ and $\langle y, v\rangle \leqq h(K, v)$ for $v \in \mathbb{E}^n$. Choosing $v = u + \lambda x$ ($\lambda > 0$, $x \in \mathbb{E}^n$) we get

$$\langle y, x\rangle \leqq \frac{h(K, u + \lambda x) - h(K, u)}{\lambda}$$

and then $\langle y, x\rangle \leqq h'_K(u; x)$. This proves $F(K, u) \subset K'$ and thus the theorem. ■

Corollary 1.7.3. *Let $K \in \mathcal{K}^n$ and $u \in \mathbb{E}^n \setminus \{o\}$. The support function h_K is differentiable at u if and only if the support set $F(K, u)$ contains only one point x. In this case,*

$$x = \operatorname{grad} h_K(u).$$

Proof. By Theorem 1.5.6, differentiability of h_K at u is equivalent to partial differentiability at u, hence to $h'_K(u; e_i) = -h'_K(u; -e_i)$ for the vectors $e_1, \ldots, e_n$ of an orthonormal basis. By Theorem 1.7.2 this is equivalent to the fact that $F(K, u)$ is contained in the intersection of n pairwise orthogonal hyperplanes, hence to $F(K, u) = \{x\}$ for some point x.

If $F(K, u) = \{x\}$, then $\langle x, v\rangle = h(F(K, u), v) = h'_K(u; v)$ for $v \in \mathbb{E}^n$, in particular $\langle x, e_i\rangle = h'_K(u; e_i) = \partial_i h_K(u)$, thus $x = \operatorname{grad} h_K(u)$. ■

Corollary 1.7.3 also follows from Theorem 1.5.12, but we have preferred to give a more direct proof. The connection to Theorem 1.5.12 is established by the following observation.

Theorem 1.7.4. *If $K \in \mathcal{K}^n$ and $u \in \mathbb{E}^n \setminus \{o\}$, then the subdifferential of the support function h_K at u is precisely the support set $F(K, u)$.*

Proof. By the definition of the subdifferential, $x \in \partial h_K(u)$ if and only if

$$h(K, v) \geqq h(K, u) + \langle x, v - u \rangle \quad \text{for } v \in \mathbb{E}^n. \tag{$*$}$$

If $x \in F(K, u)$, then $\langle x, u \rangle = h(K, u)$ and $\langle x, v \rangle \leqq h(K, v)$ for $v \in \mathbb{E}^n$, hence $(*)$ holds. If $(*)$ holds, then

$$h(K - x, v) \geqq h(K - x, u) \quad \text{for } v \in \mathbb{E}^n.$$

Applying this to $v = o$ and to $v = 2u$, we get $h(K - x, u) = 0$, thus $h(K, u) = \langle x, u \rangle$; then $h(K - x, v) \geqq 0$ for $v \in \mathbb{E}^n$ shows that $o \in K - x$, hence $x \in K$. Thus $x \in F(K, u)$. ■

We now come to the property of the support function that, for subsequent applications, is of prime importance, namely its additive behaviour (in the first argument) under addition. If K and L are convex bodies, then $K + L$ is also a convex body. The convexity is clear, and compactness follows from the compactness of $K \times L$ and the continuity of addition as a map from $\mathbb{E}^n \times \mathbb{E}^n$ to $\mathbb{E}^n$. The addition of convex bodies is usually called *Minkowski addition*. A function f on $\mathcal{K}^n$ with values in some abelian semigroup is called *Minkowski additive* if

$$f(K + L) = f(K) + f(L) \quad \text{for } K, L \in \mathcal{K}^n.$$

Important examples are obtained as follows.

Theorem 1.7.5. *For $K, L \in \mathcal{K}^n$ one has*

(a) $h(K + L, \cdot) = h(K, \cdot) + h(L, \cdot)$,
(b) $H(K + L, \cdot) = H(K, \cdot) + H(L, \cdot)$,
(c) $F(K + L, \cdot) = F(K, \cdot) + F(L, \cdot)$,

Proof. Let $u \in \mathbb{E}^n \setminus \{o\}$. There are points $x \in K$ with $h(K, u) = \langle x, u \rangle$ and $y \in L$ with $h(L, u) = \langle y, u \rangle$. This implies $h(K, u) + h(L, u) = \langle x + y, u \rangle \leqq h(K + L, u)$. Each point $z \in K + L$ has a representation $z = x + y$ with $x \in K$, $y \in L$, and it follows that $\langle z, u \rangle = \langle x, u \rangle + \langle y, u \rangle \leqq h(K, u) + h(L, u)$. Since $z \in K + L$ was arbitrary, one has $h(K + L, u) \leqq h(K, u) + h(L, u)$. This proves (a), and (b) is an immediate consequence. Equality (c) follows from Theorem 1.7.2 and (a) or by an easy direct argument. ■

A first important consequence of Theorem 1.7.5(a) is that the equality $K + M = L + M$ for convex bodies $K, L, M \in \mathcal{K}^n$ implies $K = L$. Hence, $(\mathcal{K}^n, +)$ is a commutative semigroup with cancellation law. It also has a natural multiplication with non-negative real numbers, $(\lambda, K) \mapsto \lambda K$, satisfying the rules $\lambda(K + L) = \lambda K + \lambda L$, $(\lambda + \mu)K = \lambda K + \mu K$, $\lambda(\mu K) = (\lambda\mu)K$, $1K = K$. Thus $\mathcal{K}^n$ together with Minkowski addition and this multiplication is an *abstract convex cone*. By the map

$$\begin{aligned} \varphi\colon \mathcal{K}^n &\to C(S^{n-1}) \\ K &\mapsto \bar{h}_K \end{aligned} \tag{1.7.1}$$

(where $\bar{h}_K = h_K|S^{n-1}$) this abstract cone is isomorphically embedded into the real vector space $C(S^{n-1})$ of continuous real functions on S^{n-1}.

The additivity of the support function (in the first argument) carries over to some important functionals derived from it. Let $K \in \mathcal{K}^n$. The number

$$w(K, u) := h(K, u) + h(K, -u), \quad u \in S^{n-1},$$

is the *width* of K *in the direction* u; this is the distance between the two support planes of K orthogonal to u. The maximum of the width function,

$$D(K) := \max\{w(K, u) | u \in S^{n-1}\}$$

is at the same time the diameter of K, thus $D(K) = \operatorname{diam} K$. In fact, if $x, y \in K$ are points with maximal distance, then the hyperplanes through x and y orthogonal to $x - y$ are support planes to K. On the other hand, to each $u \in S^{n-1}$ there are points $x, y \in K$ with $w(K, u) \leqq |x - y|$.

The *minimal width*

$$\Delta(K) := \min\{w(K, u) | u \in S^{n-1}\}$$

is called, in brief, the *width* of K. (Of course, diameter and width are not Minkowski additive; one merely has $D(K + L) \leqq D(K) + D(L)$ and $\Delta(K + L) \geqq \Delta(K) + \Delta(L)$ for $K, L \in \mathcal{K}^n$.)

The mean value of the width function over S^{n-1} is called the *mean width* b, thus

$$b(K) := \frac{2}{\omega_n} \int_{S^{n-1}} h(K, u) \, \mathrm{d}\mathcal{H}^{n-1}(u). \tag{1.7.2}$$

The vector-valued integral

$$s(K) := \frac{1}{\kappa_n} \int_{S^{n-1}} h(K, u) u \, \mathrm{d}\mathcal{H}^{n-1}(u) \tag{1.7.3}$$

defines the *Steiner point* $s(K)$ of K.

It is clear that mean width b and Steiner point s are Minkowski-additive functions. They also have important invariance properties. The mean width is rigid motion invariant, that is, it satisfies $b(gK) = b(K)$ for $K \in \mathcal{K}^n$ and each rigid motion $g\colon \mathbb{E}^n \to \mathbb{E}^n$. This is obvious from (1.7.2), since for a rotation g we have

$$h(gK, u) = h(K, g^{-1}u) \tag{1.7.4}$$

and the spherical Lebesgue measure is rotation invariant; further, for a translation by a vector $t \in \mathbb{E}^n$ we have

$$h(K + t, u) = h(K, u) + \langle t, u \rangle.$$

The invariance of b then follows from

$$\int_{S^{n-1}} \langle t, u \rangle \, \mathrm{d}\mathcal{H}^{n-1}(u) = 0,$$

which holds since the integrand is an odd function on S^{n-1}.

The Steiner point is equivariant under rigid motions, that is to say, $s(gK) = gs(K)$ for each rigid motion g. For a rotation this follows from (1.7.3) and (1.7.4), and for translations one has to use

$$\frac{1}{\kappa_n} \int_{S^{n-1}} \langle t, u \rangle u \, \mathrm{d}\mathcal{H}^{n-1}(u) = t \quad \text{for } t \in \mathbb{E}^n. \tag{1.7.5}$$

For the proof, note that

$$\int_{S^{n-1}} \langle t, u \rangle u \, \mathrm{d}\mathcal{H}^{n-1}(u) = \alpha t$$

with a real number α independent of t, since the integral is linear in t and invariant under rotations and reflections fixing t. Choosing $|t| = 1$ we obtain, if $(e_1, \ldots, e_n)$ is an orthonormal basis of $\mathbb{E}^n$,

$$\begin{aligned} \alpha &= \int_{S^{n-1}} \langle t, u \rangle^2 \, \mathrm{d}\mathcal{H}^{n-1}(u) = \frac{1}{n} \sum_{i=1}^{n} \int_{S^{n-1}} \langle e_i, u \rangle^2 \, \mathrm{d}\mathcal{H}^{n-1}(u) \\ &= \frac{1}{n} \int_{S^{n-1}} |u|^2 \, \mathrm{d}\mathcal{H}^{n-1}(u) = \kappa_n. \end{aligned}$$

We mention one further property of the Steiner point, namely

$$s(K) \in \operatorname{relint} K \quad \text{for } K \in \mathcal{K}^n. \tag{1.7.6}$$

This will be clear when in Section 5.4 the Steiner point $s(K)$ is interpreted as the centroid of a certain measure concentrated on the relative boundary of K (formula (5.4.11) for $r = n$).

Besides the support function, other functions can be used to describe a convex body analytically, but in our context these will be of less importance.

Let $K \in \mathcal{K}_0^n$ be a convex body such that $o \in \operatorname{int} K$. The function $g(K, \cdot)$ defined by

$$g(K, x) := \min \{\lambda \geqq 0 \,|\, x \in \lambda K\} \quad \text{for } x \in \mathbb{E}^n$$

is called the *gauge function* of K, and

$$\begin{aligned} \rho(K, x) &:= \max \{\lambda \geqq 0 \,|\, \lambda x \in K\} \qquad \text{for } x \in \mathbb{E}^n \setminus \{o\} \\ &= \frac{1}{g(K, x)} \end{aligned}$$

defines the *radial function* $\rho(K, \cdot)$ of K. It is immediately clear from the definitions that

$$K = \{x \in \mathbb{E}^n | g(K, x) \leqq 1\},$$

$$g(K, x) = \frac{|x|}{|\lambda x|} \quad \text{if } x \in \mathbb{E}^n \setminus \{o\}, \quad \lambda > 0, \quad \lambda x \in \operatorname{bd} K$$

and

$$\rho(K, x)x \in \operatorname{bd} K \quad \text{for } x \in \mathbb{E}^n \setminus \{o\}. \tag{1.7.7}$$

The gauge function $g(K, \cdot)$ is sublinear; this follows easily from the definition, but also from the following result.

Theorem 1.7.6. *For $K \in \mathcal{K}^n$ with $o \in \operatorname{int} K$,*

$$g(K, \cdot) = h(K^*, \cdot).$$

Proof. Writing $\widetilde{K} := \{x \in \mathbb{E}^n | h(K^*, x) \leqq 1\}$, we have $g(\widetilde{K}, \cdot) = h(K^*, \cdot)$ by the definition of the gauge function. Let $x \in \widetilde{K}$. For $u \in K^*$ we have $\langle u, x\rangle \leqq h(K^*, x) \leqq 1$, hence $x \in K^{**} = K$. Thus $\widetilde{K} \subset K$. Let $x \in K$. There is some $v \in K^*$ with $h(K^*, x) = \langle v, x\rangle \leqq 1$, hence $x \in \widetilde{K}$. Thus $K = \widetilde{K}$. ■

Remark 1.7.7. By Theorem 1.7.6 we have

$$\rho(K^*, u) = \frac{1}{h(K, u)} \quad \text{for } u \in S^{n-1},$$

which expresses a simple way to find the boundary points of the polar body K^* from the support planes of the body K. Further, let $x \in \operatorname{bd} K$, let u be an outer unit normal vector of K at x, and let $y = |y|u \in \operatorname{bd} K^*$. Then

$$\begin{aligned}\langle y, x\rangle &= \langle u, x\rangle / g(K^*, u) = \langle u, x\rangle / h(K, u) = 1\\ &= g(K, x) = h(K^*, x),\end{aligned}$$

hence x is an outer normal vector to K^* at y. (A general version of this remark is contained in Lemma 2.2.3.)

Remark 1.7.8. If $K \in \mathcal{K}^n$ with $o \in \operatorname{int} K$ is centrally symmetric with respect to o (i.e., $K = -K$), then

$$\|x\|_K := g(K, x) \quad \text{for } x \in \mathbb{E}^n$$

defines a norm $\|\cdot\|_K$ on $\mathbb{E}^n$ for which K is the unit ball. If we identify (as we always do) the dual space of $\mathbb{E}^n$ with $\mathbb{E}^n$ itself via the scalar product $\langle \cdot, \cdot \rangle$, then the dual Banach space $(\mathbb{E}^n, \|\cdot\|_K^*)$ of the Banach space $(\mathbb{E}^n, \|\cdot\|_K)$ has its norm given by

$$\|y\|_K^* = \sup\{\langle y, x\rangle | x \in \mathbb{E}^n, \|x\|_K \leqq 1\}$$
$$= \sup\{\langle y, x\rangle | x \in K\}$$
$$= h(K, y) = g(K^*, y),$$

and its unit ball is K^*.

We return to support functions and conclude this section with a useful remark on differences of support functions.

Lemma 1.7.9. *If $f: S^{n-1} \to \mathbb{R}$ is a twice continuously differentiable function, there exists a convex body $K \in \mathcal{K}^n$ and a number $r > 0$ such that*

$$f = \bar{h}_K - \bar{h}_{rB^n}. \tag{1.7.8}$$

Hence, the real vector space spanned by the differences of support functions (restricted to S^{n-1}) is dense in the space $C(S^{n-1})$ (with the maximum norm).

Proof. We extend f to $\mathbb{E}^n \setminus \{o\}$ by putting $f(x) := |x| f(x/|x|)$, and we define $g(x) := |x| = h(B^n, x)$ for $x \in \mathbb{E}^n$. Then the function $f + rg$ is positively homogeneous of degree one, for any $r \in \mathbb{R}$. Let $\mathrm{d}^2(f + rg)_x$ denote its second differential at x, considered as a bilinear form on $\mathbb{E}^n$. By homogeneity, $\mathrm{d}^2(f + rg)_x(x, \cdot) = 0$, and for unit vectors x, y with $y \perp x$ we obtain $\mathrm{d}^2(f + rg)_x(y, y) = \mathrm{d}^2 f_x(y, y) + r$. By continuity and compactness, $\mathrm{d}^2 f_x(y, y)$ attains a minimum, hence we can choose $r > 0$ so that $\mathrm{d}^2(f + rg)_x(y, y) \geqq 0$ for $x \in S^{n-1}$ and arbitrary y. By homogeneity, $\mathrm{d}^2(f + rg)_x$ is positive semidefinite for all $x \in \mathbb{E}^n \setminus \{o\}$. Theorem 1.5.10 now shows that $f + rg$ is convex and thus, by Theorem 1.7.1, that it is the support function of a convex body K that satisfies (1.7.8). The rest of the assertion is clear. ■

Notes for Section 1.7

1. *Determination of a convex body by its support function.* The second proof given for Theorem 1.7.1 essentially goes back to Rademacher (1921). It also appears in Hille & Phillips (1957), p. 253, where it is attributed to Fenchel. Separation or support properties are used in several versions of the proof appearing in the literature, at least implicitly, as in our first proof, which can be found in Rockafellar [29]. These methods have the advantage of being extendable to infinite dimensions. Our third proof is the one of Bonnesen & Fenchel [8]. Yet other proofs are due to Aleksandrov (1939a), who used the topological theorem on the invariance of domain, and to McMullen (1976a), who found a particularly elementary approach involving Helly's theorem.
2. *Continuous convex sets.* The nonempty closed convex sets K in $\mathbb{E}^n$ for which the support function

$$h(K, u) = \sup\{\langle x, u\rangle | x \in K\} \quad \text{for } u \in S^{n-1}$$

(restricted to the unit sphere) is continuous (with the usual definition of continuity for functions with values in $\mathbb{R} \cup \{\infty\}$) have been called *continuous convex sets* by Gale & Klee (1959). These authors have characterized those sets in several ways and have shown that the system $\mathcal{S}^n$ of continuous convex sets shares some properties with $\mathcal{K}^n$. For instance, $A, B \in \mathcal{S}^n$ implies that $A + B \in \mathcal{S}^n$, $\operatorname{conv}(A \cup B) \in \mathcal{S}^n$ and that A and B can be strongly separated if they are disjoint.

3. *The support function of an intersection*. Let $(K_i)_{i \in I}$ (where I is an arbitrary index set) be a family of convex bodies in $\mathcal{K}^n$ and suppose that their intersection K is not empty. Then the support function of K can be represented in the form
$$h(K, u) = \inf \left\{ \sum_{i \in I} h(K_i, u_i) \,\middle|\, \sum_{i \in I} u_i = u \right\},$$
where the infimum is taken over all representations $u = \sum u_i$ with $u_i = o$ for all but finitely many $i \in I$. This was proved by Sandgren (1954), where the result is attributed to F. Riesz, and was rediscovered by Kneser (1970). It can also be deduced from the more general Theorem 5.6 in Rockafellar [29].
4. *The semiaxis function*. Let $K \in \mathcal{K}_0^n$ be a convex body with $o \in \operatorname{int} K$. For $u \in S^{n-1}$, the value $h(K, u)$ of the support function is the distance of the support plane to K with exterior normal vector u from o, while $u/g(K, u)$ is the boundary point of K on the ray with endpoint o and direction u. The polarity $K \mapsto K^*$ interchanges hyperplanes and boundary points, hence support function and gauge function interchange their roles under this polarity (Theorem 1.7.6). Leichtweiß (1965) (see also Leichtweiß [23], §13) suggested a way of describing convex bodies by a class of functions that serve equally well for a body and its polar. Instead of support planes or boundary points one has to consider support elements. A *support element* of K is a pair (x, u) where x is a boundary point of K and u is an exterior unit normal vector to K at x. Let Nor K denote the set of all support elements of K with its natural topology. The map $\sigma\colon \operatorname{Nor} K \to S^{n-1}$ defined by
$$\sigma((x, u)) := \frac{x_0 + u}{|x_0 + u|}, \quad x_0 = \frac{x}{|x|},$$
is a homeomorphism. To each $v \in S^{n-1}$ there exists a unique two-sheeted hyperboloid of revolution, with axis of revolution through o and of direction v, that touches K at $\sigma^{-1}(v)$. Let $a(K, v)$ denote the length of the semiaxis of this hyperboloid. The *semiaxis function* $a(K, \cdot)$ determines K uniquely, and one has $a(K^*, \cdot) = a(K, \cdot)^{-1}$. The semiaxis function was further investigated by Baum (1973a, b), who also showed how to express the boundary of K explicitly in terms of the semiaxis function and who obtained necessary and sufficient conditions for a function to be a semiaxis function. Later Baum (1987) defined, for $K \in \mathcal{K}^n$ with $o \in \operatorname{int} K$, a function $F_K\colon \mathbb{E}^n \to \mathbb{R}$ by
$$F_K(u) := \sup \{2\langle u, y\rangle^2 - |y|^2 \,|\, y \in K, \langle u, y\rangle \geqq 0\};$$
then the semiaxis function $a(K, \cdot)$ is the restriction of $\sqrt{F_K}$ to S^{n-1}. The supremum in the definition of F_K is attained at a unique point $y \in K$. The function F_K is convex and differentiable. Baum also considered infinite-dimensional generalizations of these concepts. For boundary representations, see Baum (1989).
5. *Support flats to convex bodies*. The support function of a convex body $K \in \mathcal{K}^n$ determines the position of its support planes, and the gauge function of K (if $o \in \operatorname{int} K$) fixes its boundary points on rays through the origin. The following common generalization was investigated by Firey (1973). Let $q \in \{0, \dots, n-1\}$. To each q-flat E_q tangent to S^{n-1}, take a translate E'_q in

the subspace spanned by E_q and o. What conditions are necessary and sufficient for the family $\{E'_q\}$ to be the full set of supporting q-flats to some convex body K? (E'_q supports K if $E'_q \cap K \neq \emptyset$ and E'_q is contained in a support plane of K.) Firey defined the q-support function H_q of a convex body by means of the distance of the flats E'_q from o, and he answered the question completely by characterizing the q-support functions of convex bodies containing o. He also treated some applications.

6. *A convexity criterion*. In the paper mentioned in the preceding note, Firey (1973) also extended Rademacher's plane polar coordinate test for convexity (Bonnesen & Fenchel [8], p. 28) to higher dimensions: If a given function $h: S^{n-1} \to \mathbb{R}$ is extended to $\mathbb{E}^n$ by positive homogeneity, that is, by $h(\lambda u) = \lambda h(u)$ for $\lambda \geqq 0$, when is the extension subadditive and hence a support function? Firey proved that this holds if and only if

$$\det \begin{pmatrix} \langle u_1, e_1\rangle & \dots & \langle u_1, e_n\rangle & h(u_1) \\ \vdots & & \vdots & \vdots \\ \langle u_n, e_1\rangle & \dots & \langle u_n, e_n\rangle & h(u_n) \\ \langle v, e_1\rangle & \dots & \langle v, e_n\rangle & h(v) \end{pmatrix}$$

$$\times \det \begin{pmatrix} \langle u_1, e_1\rangle & \dots & \langle u_1, e_n\rangle \\ \vdots & & \vdots \\ \langle u_n, e_1\rangle & \dots & \langle u_n, e_n\rangle \end{pmatrix} \leqq 0$$

for all choices of $u_1, \dots, u_n, v \in S^{n-1}$ for which $v \in \operatorname{pos}\{u_1, \dots, u_n\}$; here $e_1, \dots, e_n$ is an orthonormal basis of $\mathbb{E}^n$.

In the plane, one has another simple criterion. A function $f: \mathbb{R} \to \mathbb{R}$ of period 2π is said to be a sub-sine function if the following holds: If $f(\alpha)$ agrees with $A\cos\alpha + B\sin\alpha$ at $\alpha = \alpha_1, \alpha_2$, where $\alpha_1 < \alpha_2 < \alpha_1 + \pi$, then $f(\alpha) \leqq A\cos\alpha + B\sin\alpha$ for $\alpha_1 < \alpha < \alpha_2$. Now there is a support function h on $\mathbb{E}^2$ for which $f(\alpha) = h(e_1\cos\alpha + e_2\sin\alpha)$ if and only if f is a sub-sine function (see Green 1950). Another formulation is found in Kallay (1974).

7. Lemma 1.7.9 appears, more or less explicitly, at several places in the literature, e.g., Ewald (1965), Valette (1971), Schneider (1971a). The last part of the lemma was proved in a different way by Fenchel & Jessen (1938).

   Differences of support functions were investigated, by, e.g., Weil (1974a).

8. *Power means of support functions*. For $K \in \mathcal{K}_0^n$ and $x \in K$, let $h_k(x)$ be the $(-k)$th power mean of $h(K - x, \cdot)$ over S^{n-1}. Aleksandrov (1967) obtained several estimates for the maximum of h_k.

1.8. The Hausdorff metric

The set $\mathcal{K}^n$ of convex bodies can be made into a metric space in several geometrically reasonable ways. The Hausdorff metric is particularly convenient and applicable. The natural domain for this metric is the set $\mathcal{C}^n$ of nonempty compact subsets of $\mathbb{E}^n$. We shall first treat the

Hausdorff metric on this more general class of sets; this will involve no extra effort and is useful in several respects.

For $K, L \in \mathcal{C}^n$ the *Hausdorff distance* is defined by

$$\delta(K, L) := \max\{\sup_{x\in K} \inf_{y\in L} |x - y|, \sup_{x\in L} \inf_{y\in K} |x - y|\} \tag{1.8.1}$$

or, equivalently, by

$$\delta(K, L) = \min\{\lambda \geqq 0 | K \subset L + \lambda B^n, L \subset K + \lambda B^n\}. \tag{1.8.2}$$

Then δ is a metric on $\mathcal{C}^n$, the *Hausdorff metric*.

Since we are considering compact sets, sup and inf in (1.8.1) may be replaced by max and min. Of the two definitions, (1.8.1) can be used in arbitrary metric spaces if $|x - y|$ is replaced by the distance between x and y in terms of the metric. The equivalence of (1.8.1) and (1.8.2) in $\mathbb{E}^n$ is seen as follows. Denote the right-hand side of (1.8.2) by α. For $x \in K$ one then has $x \in L + \alpha B^n$, hence $x = y + \alpha b$ with suitable $y \in L$ and $b \in B^n$; hence $|x - y| \leqq \alpha$ and, therefore, $\inf_{y\in L} |x - y| \leqq \alpha$. Since $x \in K$ was arbitrary, $\sup_{x\in K} \inf_{y\in L} |x - y| \leqq \alpha$. Interchanging K and L, we get $\delta(K, L) \leqq \alpha$. Let $0 < \lambda < \alpha$ and, say, $K \not\subset L + \lambda B^n$. Then $x \notin L + \lambda B^n$ for some $x \in K$, hence $|x - y| \geqq \lambda$ for all $y \in L$. This yields $\delta(K, L) \geqq \lambda$. Since $\lambda < \alpha$ was arbitrary, we get $\delta(K, L) \geqq \alpha$ and thus $\delta(K, L) = \alpha$, as asserted.

That δ is in fact a metric is easy to see. To prove, for instance, the triangle inequality, let $K, L, M \in \mathcal{C}^n$ and put $\delta(K, L) = \alpha$, $\delta(L, M) = \beta$. Then $K \subset L + \alpha B^n$ and $L \subset M + \beta B^n$, hence $K \subset M + \beta B^n + \alpha B^n = M + (\alpha + \beta)B^n$; analogously, $M \subset K + (\alpha + \beta)B^n$, hence $\delta(K, M) \leqq \alpha + \beta$.

In the following, all metrical and topological notions occurring in connection with $\mathcal{C}^n$ or $\mathcal{K}^n$ are tacitly understood to refer to the Hausdorff metric and the topology induced by it.

We want to show (in a stronger form) that the metric space $(\mathcal{C}^n, \delta)$ is locally compact. A few preparations are needed.

Lemma 1.8.1. *If $(K_i)_{i\in\mathbb{N}}$ is a decreasing sequence in $\mathcal{C}^n$, that is, if $K_{i+1} \subset K_i$ for $i \in \mathbb{N}$, then*

$$\lim_{i\to\infty} K_i = \bigcap_{j=1}^{\infty} K_j.$$

Proof. $K := \bigcap_{j=1}^{\infty} K_j$ is compact and not empty. If the assertion were false, then $K_m \not\subset K + \varepsilon B^n$ for all $m \in \mathbb{N}$ with some fixed $\varepsilon > 0$. Let $A_m := K_m \setminus \operatorname{int}(K + \varepsilon B^n)$, then $(A_m)_{m\in\mathbb{N}}$ is a decreasing sequence of nonempty, compact sets and hence has nonempty intersection A. Clearly $A \cap K = \emptyset$, but $A_m \subset K_m$ implies $A \subset K$, a contradiction. ■

Theorem 1.8.2. *The metric space* $(\mathcal{C}^n, \delta)$ *is complete.*

Proof. Let $(K_i)_{i\in\mathbb{N}}$ be a Cauchy sequence in $\mathcal{C}^n$. Put $A_m := \operatorname{cl}\bigcup_{i=m}^{\infty} K_i$. Then $(A_m)_{m\in\mathbb{N}}$ is a decreasing sequence of nonempty compact sets (the boundedness of $\bigcup K_i$ is an immediate consequence of the Cauchy property), hence Lemma 1.8.1 yields $A_m \to A := \bigcap_{i\in\mathbb{N}} A_i$ for $m \to \infty$. Hence, for given $\varepsilon > 0$, there exists $n_0 \in \mathbb{N}$ such that $A_m \subset A + \varepsilon B^n$ for $m \geqq n_0$, hence $K_i \subset A + \varepsilon B^n$ for $i \geqq n_0$. Since $(K_i)_{i\in\mathbb{N}}$ is a Cauchy sequence, there exists $n_1 \geqq n_0$ with $K_j \subset K_i + \varepsilon B^n$ for $i, j \geqq n_1$. Thus for $i, m \geqq n_1$ one has $\bigcup_{j=m}^{\infty} K_j \subset K_i + \varepsilon B^n$ and hence $A_m \subset K_i + \varepsilon B^n$, which implies $A \subset K_i + \varepsilon B^n$. We have proved that $\delta(K_i, A) \leqq \varepsilon$ for $i \geqq n_1$, from which the assertion follows. ■

Theorem 1.8.3. *In* $(\mathcal{C}^n, \delta)$ *each closed, bounded subset is compact.*

In particular, the space $(\mathcal{C}^n, \delta)$ is locally compact. By a known compactness criterion in metric spaces, Theorem 1.8.3 is a consequence of the following result.

Theorem 1.8.4. *From each bounded sequence in* $\mathcal{C}^n$ *one can select a convergent subsequence.*

Proof. Let $(K_i^0)_{i\in\mathbb{N}}$ be a sequence in $\mathcal{C}^n$ whose elements are contained in some cube C of edge length γ. For each $m \in \mathbb{N}$, the cube C can be written as a union of 2^{mn} cubes of edge length $2^{-m}\gamma$. For $K \in \mathcal{C}^n$ let $A_m(K)$ denote the union of all such cubes that meet K. Since (for each m) the number of subcubes is finite, the sequence $(K_i^0)_{i\in\mathbb{N}}$ has a subsequence $(K_i^1)_{i\in\mathbb{N}}$ such that $A_1(K_i^1) =: T_1$ is independent of i. Similarly, there is a union T_2 of subcubes of edge length $2^{-2}\gamma$ and a subsequence $(K_i^2)_{i\in\mathbb{N}}$ of $(K_i^1)_{i\in\mathbb{N}}$ such that $A_2(K_i^2) = T_2$. Continuing in this way, we obtain a sequence $(T_m)_{m\in\mathbb{N}}$ of unions of subcubes (of edge length $2^{-m}\gamma$ for given m) and to each m a sequence $(K_i^m)_{i\in\mathbb{N}}$ such that

$$A_m(K_i^m) = T_m \tag{$*$}$$

and

$$(K_i^m)_{i\in\mathbb{N}} \text{ is a subsequence of } (K_i^k)_{i\in\mathbb{N}} \text{ for } k < m. \tag{$**$}$$

By $(*)$ we have $K_i^m \subset K_j^m + \lambda B^n$ with $\lambda = 2^{-m}\sqrt{n}\gamma$, hence $\delta(K_i^m, K_j^m) \leqq 2^{-m}\sqrt{n}\gamma$ $(i, j, m \in \mathbb{N})$ and thus by $(**)$

$$\delta(K_i^m, K_j^k) \leqq 2^{-m}\sqrt{n}\gamma \text{ for } i, j \in \mathbb{N} \text{ and } k \geqq m.$$

For $K_m := K_m^m$ it follows that

$$\delta(K_m, K_k) \leqq 2^{-m}\sqrt{n}\gamma \quad \text{for } k \geqq m.$$

Thus $(K_m)_{m\in\mathbb{N}}$ is a Cauchy sequence and hence convergent by Theorem 1.8.2. This is the subsequence that proves the assertion. ■

From now on we again restrict our considerations to the space $\mathcal{K}^n$ of convex bodies.

Theorem 1.8.5. *$\mathcal{K}^n$ is a closed subset of $\mathcal{C}^n$.*

Proof. Let $K \in \mathcal{C}^n \setminus \mathcal{K}^n$. Then there are points $x, y \in K$ and numbers $\lambda \in (0,1)$, $\varepsilon > 0$ such that $B(z, \varepsilon) \cap K = \emptyset$ for $z = (1-\lambda)x + \lambda y$. Let $K' \in \mathcal{C}^n$ satisfy $\delta(K, K') < \varepsilon/2$. There are points $x', y' \in K'$ with $|x' - x| < \varepsilon/2$ and $|y' - y| < \varepsilon/2$, hence the point $z' := (1-\lambda)x' + \lambda y'$ satisfies $|z' - z| < \varepsilon/2$. If $z' \in K'$, then there is a point $w \in K$ with $|w - z'| < \varepsilon/2$, hence with $|w - z| < \varepsilon$, a contradiction. Thus K' is not convex. We have proved that $\mathcal{C}^n \setminus \mathcal{K}^n$ is open. ■

Theorems 1.8.4 and 1.8.5 together yield the famous *Blaschke selection theorem*:

Theorem 1.8.6. *From each bounded sequence of convex bodies one can select a subsequence converging to a convex body.*

This theorem is a very useful tool in proving the existence of convex bodies with various specific properties.

It is sometimes convenient to have a description of the convergence of convex bodies in terms of convergent sequences of points.

Theorem 1.8.7. *The convergence $\lim_{i\to\infty} K_i = K$ in $\mathcal{K}^n$ is equivalent to the following conditions taken together:*

(a) *each point in K is the limit of a sequence $(x_i)_{i\in\mathbb{N}}$ with $x_i \in K_i$ for $i \in \mathbb{N}$;*

(b) *the limit of any convergent sequence $(x_{i_j})_{j\in\mathbb{N}}$ with $x_{i_j} \in K_{i_j}$ for $j \in \mathbb{N}$ belongs to K.*

Proof. Assume that $K_i \to K$ for $i \to \infty$. Let $x \in K$. Put $x_i := p(K_i, x)$. Then $x_i \in K_i$ and $|x - x_i| \leqq \delta(K, K_i) \to 0$ for $i \to \infty$. Thus (a) holds. Let $(i_j)_{j\in\mathbb{N}}$ be an increasing sequence in $\mathbb{N}$ and suppose that $x_{i_j} \in K_{i_j}$ $(j \in \mathbb{N})$ and $x_{i_j} \to x$ for $j \to \infty$, but $x \notin K$. Then $B(x, \rho) \cap (K + \rho B^n) = \emptyset$ for some $\rho > 0$. But for j sufficiently large we have $|x_{i_j} - x| < \rho$ and $x_{i_j} \in K_{i_j} \subset K + \rho B^n$, a contradiction. Thus (b) holds.

Now assume that (a) and (b) are satisfied. Let $\varepsilon > 0$ be given. We have to show that

$$K \subset K_i + \varepsilon B^n \qquad \text{for all large } i, \tag{$*$}$$

$$K_i \subset K + \varepsilon B^n \qquad \text{for all large } i. \tag{$**$}$$

If ($*$) is false, there exists a sequence $(y_{i_j})_{j \in \mathbb{N}}$ in K, converging to some $y \in K$, with $d(K_{i_j}, y_{i_j}) \geqq \varepsilon$ ($j \in \mathbb{N}$). By (a), there exist $x_i \in K_i$ with $x_i \to y$ for $i \to \infty$. Then $|x_{i_j} - y_{i_j}| \to 0$ for $j \to \infty$, a contradiction. Hence ($*$) holds. If ($**$) is false, there exists a sequence $(y_{i_j})_{j \in \mathbb{N}}$ with $y_{i_j} \in K_{i_j}$ and $y_{i_j} \notin K + \varepsilon B^n$ for $j \in \mathbb{N}$. By (a) there exist points $x_{i_j} \in K_{i_j}$ with $x_{i_j} \in K + \varepsilon B^n$ for sufficiently large j. By the convexity of K_{i_j} (only the connectedness is needed) there are points $z_{i_j} \in K_{i_j} \cap \operatorname{bd}(K + \varepsilon B^n)$. The sequence $(z_{i_j})_{j \in \mathbb{N}}$ has a convergent subsequence whose limit must belong to K by (b), a contradiction. Hence ($**$) holds. ■

Having endowed $\mathcal{C}^n$ and $\mathcal{K}^n$ with a topology, we should check the continuity of the mappings that we have encountered in previous sections. Some of them are even Lipschitz mappings, for trivial reasons. This holds for the convex hull operator as a map from $\mathcal{C}^n$ to $\mathcal{K}^n$, in fact

$$\delta(\operatorname{conv} K, \operatorname{conv} L) \leqq \delta(K, L),$$

and for addition and union as maps from $\mathcal{C}^n \times \mathcal{C}^n$ to $\mathcal{C}^n$, since

$$\delta(K + K', L + L') \leqq \delta(K, L) + \delta(K', L'),$$

$$\delta(K \cup K', L \cup L') \leqq \max\{\delta(K, L), \delta(K', L')\}.$$

The intersection as a map from $\mathcal{C}^n \times \mathcal{C}^n$ to $\mathcal{C}^n$ is not continuous. For convex bodies, however, the following can be shown.

Theorem 1.8.8. *Let $K, L \in \mathcal{K}^n$ be convex bodies that cannot be separated by a hyperplane. If $K_i, L_i \in \mathcal{K}^n$ ($i \in \mathbb{N}$) are convex bodies with $K_i \to K$ and $L_i \to L$ for $i \to \infty$, then $K_i \cap L_i \neq \emptyset$ for almost all i and $K_i \cap L_i \to K \cap L$ for $i \to \infty$.*

Proof. We use the criterion 1.8.7. Let $x \in K \cap L$ ($K \cap L \neq \emptyset$ since otherwise K and L could be separated). Define $x_i := p(K_i \cap L_i, x)$ for those i for which $K_i \cap L_i \neq \emptyset$. We assert that x_i exists for almost all i and that $x_i \to x$ for $i \to \infty$. If this is false, there exists a ball B with centre x (and positive radius) such that $K_i \cap L_i \cap B = \emptyset$ for infinitely many i. For sufficiently large i we have $K_i \cap B \neq \emptyset$, since $x \in K$ and $K_i \to K$; similarly $L_i \cap B \neq \emptyset$ for large i. Thus there is a sequence $(H_{i_j})_{j \in \mathbb{N}}$ of hyperplanes H_{i_j} separating $K_{i_j} \cap B$ and $L_{i_j} \cap B$. A subsequence converges to a hyperplane H, and H separates $K \cap B$ and $L \cap B$ (as one easily sees using Theorem 1.8.7). Since $x \in K \cap L \cap B$, necessarily $x \in H$. From the convexity of K and L it now follows that H separates K and L, a contradiction. Hence $x_i \to x$ for $i \to \infty$.

On the other hand, if $(i_j)_{j \in \mathbb{N}}$ is an increasing sequence in $\mathbb{N}$, if $x_{i_j} \in K_{i_j} \cap L_{i_j}$ $(j \in \mathbb{N})$ and $x_{i_j} \to y$ for $j \to \infty$, then clearly $y \in K \cap L$ by Theorem 1.8.7. Now it follows from this theorem that $K_i \cap L_i \to K \cap L$ for $i \to \infty$. ■

The following example shows that the non-separability condition in Theorem 1.8.8 cannot be omitted. Let $K = L = L_i = [x, y]$, where $x \neq y$, and let $K_i = [x, y_i]$ be a segment with $[x, y_i] \cap [x, y] = \{x\}$ and such that $y_i \to y$ for $i \to \infty$. Then it follows that $K_i \to K$ and $L_i \to L$, but it does not follow that $K_i \cap L_i \to K \cap L$.

Next we show that the metric projection is continuous in both variables simultaneously, that is, the map

$$p: \mathcal{K}^n \times \mathbb{E}^n \to \mathbb{E}^n$$
$$(K, x) \mapsto p(K, x)$$

is continuous. This is a consequence of the following more precise result.

Lemma 1.8.9. *Let* $K, L \in \mathcal{K}^n$, $x, y \in \mathbb{E}^n$, *and put* $D := \operatorname{diam}(K \cup L \cup \{x, y\})$. *Then*

$$|p(K, x) - p(L, y)| \leqq |x - y| + \sqrt{5D\delta(K, L)}.$$

Proof. Taking Theorem 1.2.2 into account (and the triangle inequality), we only have to show that

$$|p(K, x) - p(L, x)| \leqq \sqrt{5D\delta(K, L)}.$$

If $x \in K \cap L$, this is trivial. We assume therefore, say, $x \notin K$. Put $d(K, x) = d$ and $\delta(K, L) = \delta$. The ball $B(p(K, x), \delta)$ contains a point of L, hence $d(L, x) \leqq d + \delta$ and thus $p(L, x) \in B(x, d + \delta)$. Let H^- be the supporting halfspace of K with exterior normal vector $u(K, x)$. Then L is contained in the translate $H^- + \delta u(K, x)$, hence

$$p(L, x) \in B(x, d + \delta) \cap [H^- + \delta u(K, x)].$$

Elementary geometry now yields

$$|p(K, x) - p(L, x)| \leqq \sqrt{4d\delta + \delta^2} \leqq \sqrt{5D\delta},$$

since $d \leqq D$ and $\delta \leqq D$. ■

Also the support function $h: \mathcal{K}^n \times \mathbb{E}^n \to \mathbb{R}$ is continuous as a function of two variables; it is in fact locally Lipschitz continuous, as a consequence of the following result.

Lemma 1.8.10. *For* $K, L \in \mathcal{K}^n$ *with* $K, L \subset RB^n$ $(R > 0)$ *and for* u, $v \in \mathbb{E}^n$,

$$|h(K, u) - h(L, v)| \leqq R|u - v| + \max\{|u|, |v|\}\delta(K, L).$$

Proof. Put $\delta(K, L) = \delta$ and choose a point $x \in K$ with $h(K, u) = \langle x, u\rangle$. From $x \in K \subset L + \delta B^n$ we get $\langle x, v\rangle \leqq h(L + \delta B^n, v) = h(L,v) + \delta|v|$, hence $h(K, u) - h(L, v) \leqq \langle x, u - v\rangle + \delta|v| \leqq |x||u - v| + \delta|v| \leqq R|u - v| + \delta|v|$. The assertion follows after interchanging (K, u) and (L, v). ■

In terms of the support function, the convergence of convex bodies can be treated very conveniently, as a result of the following simple observation. Here $\|\cdot\|$ denotes the maximum norm for real functions on S^{n-1}.

Theorem 1.8.11. *For* $K, L \in \mathcal{K}^n$,

$$\delta(K, L) = \sup_{u \in S^{n-1}} |h(K, u) - h(L, u)| =: \|\bar{h}_K - \bar{h}_L\|.$$

Proof. Let $\delta(K, L) \leqq \alpha$. Then $K \subset L + \alpha B^n$, hence $h(K, u) \leqq h(L + \alpha B^n, u) = h(L, u) + \alpha$ for $u \in S^{n-1}$. Interchanging K and L, we get $|h(K, u) - h(L, u)| \leqq \alpha$ for $u \in S^{n-1}$ and thus $\|\bar{h}_K - \bar{h}_L\| \leqq \alpha$. The argument can be reversed. ■

From Theorem 1.8.11 and the definitions (1.7.2) and (1.7.3) it follows immediately that the mean width b and the Steiner point s are Lipschitz maps. In both cases, the optimal Lipschitz constants are easily determined. From (1.7.2) one obtains

$$|b(K) - b(L)| \leqq 2\delta(K, L)$$

for $K, L \in \mathcal{K}^n$, with equality if and only if $K = L + \alpha B^n$ or $L = K + \alpha B^n$ with $\alpha \geqq 0$. For given $K, L \in \mathcal{K}^n$, take $v \in S^{n-1}$ so that $|s(K) - s(L)| = \langle s(K) - s(L), v\rangle$; then

$$|s(K) - s(L)| = \left|\frac{1}{\kappa_n}\int_{S^{n-1}}[h(K, u) - h(L, u)]\langle u, v\rangle \,\mathrm{d}\mathcal{H}^{n-1}(u)\right|$$

$$\leqq \frac{1}{\kappa_n}\,\delta(K, L)\int_{S^{n-1}}|\langle u, v\rangle|\,\mathrm{d}\mathcal{H}^{n-1}(u).$$

The last integral is twice the $(n-1)$-dimensional measure of the orthogonal projection of S^{n-1} on to a hyperplane orthogonal to v (this is easy to see and is a special case of (5.3.32)). Thus

$$|s(K) - s(L)| \leqq \frac{2\kappa_{n-1}}{\kappa_n}\,\delta(K, L). \tag{1.8.3}$$

It is clear that for $K \neq L$ equality in (1.8.3) cannot be attained, and it is easy to see that the constant cannot be replaced by a smaller one.

Theorem 1.8.11 shows that the map φ defined by (1.7.1) maps $\mathcal{K}^n$ not only isomorphically as a convex cone into the vector space $C(S^{n-1})$, but also isometrically with respect to the Hausdorff metric on $\mathcal{K}^n$ and the metric induced on $C(S^{n-1})$ by the maximum norm. In particular, convergence of convex bodies is equivalent to uniform convergence on S^{n-1} of the corresponding support functions. This uniform convergence follows in any case from pointwise convergence.

Theorem 1.8.12. *If a sequence of support functions converges pointwise (on $\mathbb{E}^n$ or, equivalently, on S^{n-1}), then it converges uniformly on S^{n-1} to a support function.*

Proof. Let $(h(K_i, \cdot))_{i \in \mathbb{N}}$ be a pointwise converging sequence of support functions. Clearly the limit function, say h, is sublinear and, hence, a support function by Theorem 1.7.1. For each $u \in S^{n-1}$, $\alpha(u) := \sup_i h(K_i, u)$ is finite. We can choose a number $R > 0$ with

$$\bigcap_{u \in S^{n-1}} H^-_{u,\alpha(u)} \subset RB^n,$$

then $K_i \subset RB^n$ for $i \in \mathbb{N}$. Let h_i be the restriction of $h(K_i, \cdot)$ to S^{n-1}; then $|h_i(u) - h_i(v)| \leqq R|u - v|$ by Lemma 1.8.10. Choose a finite set $S \subset S^{n-1}$ such that to each $u \in S^{n-1}$ there is some $v \in S$ with $|u - v| < \varepsilon/3R$. Since S is finite, there exists $n_0 \in \mathbb{N}$ such that $|h_i(v) - h_j(v)| < \varepsilon/3$ for $i, j \geqq n_0$ and all $v \in S$. Now let $u \in S^{n-1}$. Choose $v \in S$ such that $|u - v| < \varepsilon/3R$. Then

$$\begin{aligned} |h_i(u) - h_j(u)| &\leqq |h_i(u) - h_i(v)| + |h_i(v) - h_j(v)| + |h_j(v) - h_j(u)| \\ &\leqq R|u - v| + \varepsilon/3 + R|u - v| < \varepsilon \end{aligned}$$

for $i, j \geqq n_0$ and hence $|h_i(u) - h(u)| \leqq \varepsilon$ for $i \geqq n_0$. Since $u \in S^{n-1}$ was arbitrary, the uniform convergence on S^{n-1} of the sequence $(h_i)_{i \in \mathbb{N}}$ to h is proved. ∎

Having introduced a metric on the set $\mathcal{K}^n$ of convex bodies, we can now consider approximation. The approximation of general convex bodies by simpler ones such as polytopes or bodies with differentiable boundaries is a useful tool for many investigations. Here we note only the most basic facts on approximation by polytopes. Further approximation results will be treated in Sections 2.4 and 3.3.

Theorem 1.8.13. *Let $K \in \mathcal{K}^n$ and $\varepsilon > 0$. Then there is a polytope $P \in \mathcal{K}^n$ with $P \subset K \subset P + \varepsilon B^n$, hence with $\delta(K, P) \leqq \varepsilon$.*

Proof. Cover K by finitely many balls with radius ε and centres in K, and let P be the convex hull of their centres. Evidently P satisfies the requirements. ■

Remark. In the preceding proof we may impose the condition that the centres of the balls have rational coordinates. Hence, the space $(\mathcal{K}^n, \delta)$ is separable, that is, it has a countable dense subset.

Lemma 1.8.14. *Let $K_1, K_2 \in \mathcal{K}^n$ and $K_2 \subset \operatorname{int} K_1$. Then there is a number $\eta > 0$ such that each convex body $K \in \mathcal{K}^n$ with $\delta(K_1, K) < \eta$ satisfies $K_2 \subset K$.*

Proof. Since $K_2 \subset \operatorname{int} K_1$, the function $h(K_1, \cdot) - h(K_2, \cdot)$ is positive on $\mathbb{E}^n \setminus \{o\}$ and hence, being continuous, attains a positive minimum η on S^{n-1}. Let $K \in \mathcal{K}^n$ be a convex body with $\delta(K_1, K) < \eta$. Then $|h(K_1, u) - h(K, u)| < \eta$ and hence $h(K_2, u) \leqq h(K_1, u) - \eta < h(K, u)$ for $u \in S^{n-1}$; thus $K_2 \subset K$. ■

Theorem 1.8.15. *Let $K \in \mathcal{K}^n$ and $o \in \operatorname{int} K$. For each $\lambda > 1$ there exists a polytope $P \in \mathcal{K}^n$ with $P \subset K \subset \lambda P$.*

Proof. Choose a number $\rho > 0$ with $B(o, \rho) \subset \operatorname{int} K$ and a number ε with $0 < \varepsilon \leqq (\lambda - 1)\rho$. By Theorem 1.8.13 and Lemma 1.8.14 we can choose ε small enough that there is a polytope P with

$$B(o, \rho) \subset P \subset K \subset P + \varepsilon B^n.$$

For $u \in S^{n-1}$ we deduce

$$\begin{aligned} h(\lambda P, u) &= h(P, u) + (\lambda - 1)h(P, u) \\ &\geqq h(P, u) + \varepsilon = h(P + \varepsilon B^n, u) \\ &\geqq h(K, u), \end{aligned}$$

thus $K \subset \lambda P$. ■

We prove a further continuity result. The *volume functional* V_n on $\mathcal{K}^n$ is defined as the restriction of the n-dimensional measure $\mathcal{H}^n$ to $\mathcal{K}^n$.

Theorem 1.8.16. *The volume functional V_n is continuous on $\mathcal{K}^n$.*

Proof. Let $K \in \mathcal{K}^n$ be given. If $V_n(K) = 0$ and $\bar{K} \in \mathcal{K}^n$ satisfies $\delta(K, \bar{K}) = \alpha \leqq 1$ (without loss of generality), then K lies in a hyperplane and $\bar{K} \subset K + \alpha B^n$. Hence, $V_n(\bar{K}) \leqq V_n(K + \alpha B^n) \leqq c(K)\alpha$ with

$c(K)$ independent of α, by an easy estimation using Fubini's theorem. Suppose, therefore, that $V_n(K) > 0$. Then we may assume that $o \in \operatorname{int} K$. Let $\varepsilon > 0$ be given. Choose $\lambda > 1$ with $(\lambda^n - 1)\lambda^n V_n(K) < \varepsilon$ and $\rho > 0$ with $\rho B^n \subset \operatorname{int} K$. By Lemma 1.8.14, there is a number $\alpha > 0$ with $\alpha \leqq (\lambda - 1)\rho$ and such that $\rho B^n \subset \bar{K}$ for all $\bar{K} \in \mathcal{K}^n$ satisfying $\delta(K, \bar{K}) < \alpha$. Suppose that the latter holds. Then

$$K \subset \bar{K} + \alpha B^n \subset \bar{K} + (\lambda - 1)\rho B^n \subset \bar{K} + (\lambda - 1)\bar{K} = \lambda \bar{K}$$

and similarly $\bar{K} \subset \lambda K$. It follows that

$$V_n(K) \leqq V_n(\lambda \bar{K}) + \lambda^n V_n(\bar{K}),$$

hence

$$V_n(K) - V_n(\bar{K}) \leqq (\lambda^n - 1)V_n(\bar{K}) \leqq (\lambda^n - 1)\lambda^n V_n(K),$$
$$V_n(\bar{K}) - V_n(K) \leqq (\lambda^n - 1)V_n(K) \leqq (\lambda^n - 1)\lambda^n V_n(K),$$

thus

$$|V_n(K) - V_n(\bar{K})| \leqq (\lambda^n - 1)\lambda^n V_n(K) < \varepsilon. \qquad \blacksquare$$

Notes for Section 1.8

1. *Hausdorff metric and topologies on spaces of subsets*. The Hausdorff metric is usually considered in a more general context. Let (X, d) be a metric space. Then

$$\delta(C, D) := \max\{\sup_{x \in C} \inf_{y \in D} d(x, y), \sup_{x \in D} \inf_{y \in C} d(x, y)\}$$

defines the Hausdorff metric δ on the set of nonempty closed and bounded subsets of X. It was introduced by Hausdorff (1914, pp. 293ff), who also described the convergence with respect to this metric in terms of closed limits (see also Hausdorff 1927, p. 149).

Let $\mathcal{C}(X)$ denote the set of nonempty compact subsets of X. Different common topologies introduced on spaces of subsets of a topological space coincide, on $\mathcal{C}(X)$, with the topology induced by the Hausdorff metric. This is true for the Vietoris topology and for the topology of closed convergence. The latter, which is usually considered on the set of closed subsets of a locally compact, second countable Hausdorff space, has applications in mathematical economics (Hildenbrand 1974) and stochastic geometry (Matheron 1975). For properties of the metric space $(\mathcal{C}(X), \delta)$ we refer the reader to Kuratowski (1968), §42, and for general investigations on topologies for spaces of subsets to Michael (1951) and the monograph of Klein & Thompson (1984).

2. *Selection theorems*. The Blaschke selection theorem for convex bodies was first stated and used by Blaschke [5] (pp. 62ff). A general theorem, which implies the compactness of $(\mathcal{C}(X), \delta)$ for a compact metric space X, appears in Hausdorff (1927), p. 147; see also Kuratowski (1968), §42 and Rogers (1970), p. 91. An elegant proof for compact subsets of $\mathbb{E}^n$ was given by Hadwiger (1949a), [18], Section 4.3.2; we have followed his approach. A very general form appears in Dierolf (1970).

By Lemma 1.8.10, one has $|h(K, u) - h(K, v)| \leqq R|u - v|$ for $K \in \mathcal{K}^n$ with $K \subset RB^n$ (for given $R > 0$) and for $u, v \in \mathbb{E}^n$. Therefore, the family

$$\{\bar{h}_K | K \in \mathcal{K}^n, K \subset RB^n\}$$

is equicontinuous; further, it is uniformly bounded and closed in $C(S^{n-1})$ (with the maximum norm). By the Arzelà–Ascoli theorem, it is compact. This is another proof of the Blaschke selection theorem. Two different methods of deducing the Blaschke selection theorem from the Arzelà–Ascoli theorem were described by Heil (1967a); he also deduced the selection theorem for compact sets in this way.

Selection theorems for star-shaped sets and for more general sets were treated by Hirose (1965), Drešević (1970), Beer (1975) and Spiegel (1975).

3. *Convergence in sense of the Hausdorff metric*. Theorem 1.8.7 can be generalized and refined in the following way. We consider the space $\mathcal{C}^n$ of nonempty compact sets in $\mathbb{E}^n$. Define

$$\lambda(K, L) := \max_{x \in K} \min_{y \in L} |x - y| \quad \text{for } K, L \in \mathcal{C}^n,$$

so that $\delta(K, L) = \max\{\lambda(K, L), \lambda(L, K)\}$. For $K \in \mathcal{C}^n$ and a sequence $(K_i)_{i \in \mathbb{N}}$ in $\mathcal{C}^n$, the following assertions may hold.

(1) $\lambda(K, K_i) \to 0$ for $i \to \infty$.

(2) Each point $x \in K$ is the limit of a sequence $(x_i)_{i \in \mathbb{N}}$ with $x_i \in K_i (i \in \mathbb{N})$.

(3) Each open set intersecting K intersects K_i for almost all i.

(1′) $\lambda(K_i, K) \to 0$ for $i \to \infty$.

(2′) The limit of any convergent sequence $(x_{i_j})_{j \in \mathbb{N}}$ with $x_{i_j} \in K_{i_j}$ $(j \in \mathbb{N})$ belongs to K, and the sequence $(K_i)_{i \in \mathbb{N}}$ is bounded.

(3′) Each closed set having empty intersection with K has empty intersection with almost all K_i.

Then the following can be shown: The conditions (1), (2), (3) are equivalent; the conditions (1′), (2′), (3′) are equivalent. Hence

$$\lim_{i \to \infty} K_i = K \Leftrightarrow (2) \text{ and } (2') \Leftrightarrow (3) \text{ and } (3').$$

The proof is left to the reader.

4. *Convergence of closed convex sets*. On the set of (not necessarily bounded) closed convex subsets of $\mathbb{E}^n$ several different types of convergence have been considered, which coincide on $\mathcal{K}^n$. A comparison of four such types of convergence is made in Salinetti & Wets (1979). See also Wijsman (1964, 1966) and Beer (1989).

5. *Topology of hyperspaces*. For a nonempty subset X of $\mathbb{E}^n$, the space $\mathcal{K}(X)$ (of convex bodies contained in X), metrized by the Hausdorff metric, is also called the cc-*hyperspace* of X. The topology of $\mathcal{K}(X)$, in dependence on the geometry of X, has received some attention. The following results are due to Nadler, Quinn & Stavrakas (1975, 1977, 1979); see also Chapter XVIII of Nadler (1978). If $X \in \mathcal{K}^n$ and $\dim X \geqq 2$, then $\mathcal{K}(X)$ is homeomorphic to the Hilbert cube $I_\infty = \prod_{i=1}^{\infty}[-1/2^i, 1/2^i]$. For $n \geqq 2$, $\mathcal{K}(\text{int } B^n)$ and $\mathcal{K}(\mathbb{E}^n)$ are homeomorphic to I_∞ with a point removed. If $X \subset \mathbb{E}^2$ is compact, connected and such that $\mathcal{K}(X)$ is homeomorphic to I_∞, then X is topologically a 2-cell. Curtis, Quinn & Schori (1977) found a complete characterization of the polyhedral 2-cells whose cc-hyperspaces are homemorphic to I_∞.

6. *Embedding theorems for spaces of convex sets*. The map

$$\varphi: \mathcal{K}^n \to C(S^{n-1})$$
$$K \mapsto \bar{h}_K$$

given by (1.7.1) maps the cone of convex bodies in $\mathbb{E}^n$ isomorphically, and by Theorem 1.8.11 also isometrically, onto a closed convex cone in a Banach space. For certain spaces of convex subsets of infinite-dimensional vector spaces, similar embedding theorems are known. Rådström (1952) proved the following result.

Let L be a real normed linear space and M a set of closed, bounded convex sets in L with the following properties.

(1) M is closed under addition and multiplication by non-negative scalars.
(2) If $A \in M$ and S is the unit ball of L, then $A + S$ is closed.
(3) M is metrized by the Hausdorff metric.

Then M can be embedded as a convex cone in a real normed linear space N in such a way that

(a) the embedding is isometric,
(b) addition in N induces addition in M,
(c) multiplication by non-negative scalars in N induces the corresponding operation in M.

Rådström showed that the conditions on M are satisfied, for instance, by the set of all finite-dimensional compact convex sets and by the set of all compact convex sets.

Rådström proof was by means of an abstract construction, but it was pointed out later that one can also use support functionals in a similar way to their use in the finite-dimensional case. For locally convex spaces this was done by Hörmander (1955), who proved also a generalization of Theorem 1.7.1, characterizing the support functionals of closed convex sets in such a space (essentially by an argument that can be viewed as extending Rademacher's method, which is given as the second proof of Theorem 1.7.1). More general embedding theorems of a similar kind were obtained by Godet-Thobie and The Lai (1970), Urbański (1976), Tolstonogov (1977) and Schmidt (1986).

Essential for the existence of such embeddings is the cancellation law demanding that $A + C = B + C$ implies $A = B$. An algebraic study of the validity of this cancellation law in arbitrary real vector spaces under suitable convexity and other assumptions on the subsets A, B, C was carried out by Jongmans (1973).

7. *Representation of semigroups as systems of compact convex sets.* If the multiplication $(\lambda, K) \mapsto \lambda K = \{\lambda x | x \in K\}$ $(K \in \mathcal{K}^n)$ is considered for all real $\lambda \in \mathbb{R}$, then the rules listed before (1.7.1) remain the same except that $(\lambda + \mu)K = \lambda K + \mu K$ holds only for $\lambda\mu \geqq 0$. The structure $(\mathcal{K}^n, +)$ with this multiplication is then an example of an abelian $\mathbb{R}$-semigroup with cancellation law. Ratschek & Schröder (1977) have characterized the abstract $\mathbb{R}$-semigroups that can be represented in this way by systems of compact convex sets, with $\mathbb{E}^n$ replaced by a suitable locally convex space.
8. *Different metrics for convex bodies.* In the theory of convex bodies, the Hausdorff metric was first used by Blaschke [5] (§§17, 18). It is, therefore, sometimes called the Blaschke–Hausdorff metric. In the geometry of convex bodies, other metrics have been proposed also. The following are defined on $\mathcal{K}_0^n$ only. The *symmetric difference metric* δ^S is given by

$$\delta^S(K, L) := \mathcal{H}^n(K \,\Delta\, L) \quad \text{for } K, L \in \mathcal{K}_0^n,$$

where $K \,\Delta\, L = (K \setminus L) \cup (L \setminus K)$ denotes the symmetric difference of K and L. This metric is invariant under volume-preserving affinities of $\mathbb{E}^n$. Two affine-invariant metrics are defined on $\mathcal{K}_0^n$ as follows:

$$\delta^D(K, L) := \log(1 + 2\inf\{\lambda \geqq 0 | K \subset L + \lambda DL,\ L \subset K + \lambda DK\}),$$

where $DK := K - K$ is the difference body of K, and

$$\delta^Q(K, L) := \mathcal{H}^n(K \,\Delta\, L)/\mathcal{H}^n(K \cup L),$$

A detailed investigation and comparison of the four metrics δ, δ^S, δ^D, δ^Q on $\mathcal{K}_0^n$ was carried out by Shephard & Webster (1965). In particular, they showed that all these metrics induce the same topology on $\mathcal{K}_0^n$, but not the same uniform structures. Further, they studied the completeness and uniform continuity of some functions with respect to these metrics.

Dinghas (1957a) showed that the symmetric difference metric (for not necessarily convex sets) is decreased under simultaneous symmetrization at an affine subspace.

Another class of metrics is obtained if in Theorem 1.8.11 the maximum norm is replaced by an L_p norm: For $1 \leqq p < \infty$ and $K, L \in \mathcal{K}^n$ let

$$\delta_p(K, L) := \left(\int_{S^{n-1}} |h_K - h_L|^p \, d\mathcal{H}^{n-1} \right)^{1/p};$$

then δ_p is a metric on $\mathcal{K}^n$, called the L_p metric. These metrics were investigated by Vitale (1985a). One has $\delta_p(K, L) \geqq \omega_n^{1/p} \delta(K, L)$ trivially, and Vitale established estimates in the other direction. As a consequence of a sharp, but more complicated, inequality he showed, for instance, that

$$c_p(K, L)[\delta(K, L)]^{(p+n-1)/p} \geqq \delta_p(K, L)$$

for $K, L \in \mathcal{K}^n$, where the factor $c_p(K, L)$ depends (for fixed n and p) only on $\operatorname{diam}(K \cup L)$. He deduced that all of the δ_p metrics ($1 \leqq p \leqq \infty$, with $\delta_\infty = \delta$), induce the same topology on $\mathcal{K}^n$ and yield complete metric spaces in which closed, bounded sets are compact. For δ_1, see also Florian (1989).

9. *Isometries*. Because the Hausdorff metric on $\mathcal{K}^n$ has a simple geometric meaning, the study of the metric space $(\mathcal{K}^n, \delta)$ deserves some interest of its own. A natural first question asks for the isometries that this space permits. Suppose that the map $i: \mathcal{K}^n \to \mathcal{K}^n$ is an isometry with respect to the Hausdorff metric. Under the assumption that i is surjective, it was proved by Schneider (1975a) that i is induced by an isometry of $\mathbb{E}^n$, in the sense that there is a rigid motion $g: \mathbb{E}^n \to \mathbb{E}^n$ so that $i(K) = gK$ for all $K \in \mathcal{K}^n$. Without assuming surjectivity, Gruber & Lettl (1980) have shown that one can find a rigid motion g of $\mathbb{E}^n$ and a convex body L such that $i(K) = gK + L$ for all $K \in \mathcal{K}^n$. Every isometry of $\mathcal{K}_0^n$ with respect to the symmetric difference metric is induced by an equi-affinity of $\mathbb{E}^n$ (Gruber 1978). For the subspace $\mathcal{K}_0^n(B^n)$ of n-dimensional convex bodies contained in the unit ball, every isometry with respect to the symmetric difference metric is induced by a rotation of the ball (Gruber 1982). The corresponding result for the Hausdorff metric on $\mathcal{K}^n(B^n)$ follows from Bandt (1986). Similar investigations for non-convex sets and for subsets of other spaces were carried out by Gruber (1980a, b), Gruber & Lettl (1979), Lettl (1980), Gruber & Tichy (1982) and Bandt (1986).

10. *Metric convexity and metric segments*. For a metric space it is, furthermore, of interest to study its properties in terms of the geometric notions introduced by Menger (1928) (see also Blumenthal 1953). The space $(\mathcal{K}^n, \delta)$ is a metric segment space, since it follows from Theorem 1.8.11 that

$$\delta(K, (1-\alpha)K + \alpha L) = \alpha\delta(K, L)$$

for $K, L \in \mathcal{K}^n$ and $\alpha \in [0, 1]$. (Some generalizations to infinite-dimensional spaces were investigated by Bantegnie 1975). Shephard & Webster (1965) showed that $(\mathcal{K}_0^n, \delta^S)$ is metrically convex (in Menger's sense), but not a metric segment space, while $(\mathcal{K}_0^n, \delta^D)$ and $(\mathcal{K}_0^n, \delta^Q)$ are not metrically convex.

In the space $(\mathcal{K}^n, \delta)$ there is in general a great variety of metric segments joining two given bodies K and L. A study of their totality was begun by Jongmans (1979). Answering a question left open by him, Schneider (1981c) showed the following. Let $K, L \in \mathcal{K}^n$ be convex bodies joined by only one metric segment (equivalently, $A = \frac{1}{2}(K + L)$ is the only convex body satisfying $\delta(K, A) = \delta(A, L) = \frac{1}{2}\delta(K, L)$). Then either $K = L + \rho B^n$ or

$L = K + \rho B^n$ with some $\rho \geqq 0$, or else $\dim K < n$ and $L = K + t$ where the vector t is orthogonal to aff K.

11. *Metric entropy*. A characteristic of a totally bounded metric space (X, d) that is designed to describe its 'massivity', is the *metric entropy*. This is, by definition, the function H given by
$$H(\varepsilon) := \log N(\varepsilon) \quad \text{for } \varepsilon > 0,$$
where $N(\varepsilon)$ is the minimal number of closed d-balls of radius ε covering X. For the spaces $(\mathcal{K}^n(A), \delta)$ and $(\mathcal{K}_0^n(A), \delta^S)$, where $\emptyset \neq A \subset \mathbb{E}^n$ is a bounded open set, the asymptotic behaviour of the metric entropy for $\varepsilon \to 0$ was investigated by Dudley (1974). For $(\mathcal{K}^n(B^n), \delta)$, Bronshtein (1976) showed more precisely that
$$c_1(n)\varepsilon^{(1-n)/2} \leqq H(\varepsilon) \leqq c_2(n)\varepsilon^{(1-n)/2} \quad \text{for } \varepsilon \leqq c_3(n),$$
where $c_i(n)$ is a positive constant depending only on n. In particular, the *exponent of entropy*, defined by
$$\liminf_{\varepsilon \downarrow 0} (\log\log N(\varepsilon)/|\log \varepsilon|),$$
is equal to $(n-1)/2$ for the space $(\mathcal{K}^n(B^n), \delta)$ (see also Bandt 1988).

12. *Diameters of sets of convex bodies*. It is an interesting and, in general, difficult task to determine the diameter, with respect to the Hausdorff metric, of a given bounded subset of $\mathcal{K}^n$ that is defined by some geometric conditions. As an example, McMullen (1984) proved that the space of all convex bodies in $\mathcal{K}^n$ with mean width b equal to that of a line segment of length 2 and with Steiner point s at the origin, has diameter 1. As a corollary, McMullen obtained the sharp estimate
$$\delta(K, L) \leqq \frac{\omega_n}{4\kappa_{n-1}} \max\{b(K), b(L)\} + |s(K) - s(L)|$$
for $K, L \in \mathcal{K}^n$.

13. *Affine equivalence classes of convex bodies*. Let $\mathcal{AK}_0$ denote the set of all affine equivalence classes of n-dimensional convex bodies in $\mathbb{E}^n$. Thus L belongs to the equivalence class $[K] \in \mathcal{AK}_0$ if and only if $K = \alpha L$ for some non-degenerate affine transformation α of $\mathbb{E}^n$. If $\mathcal{AK}_0$ is equipped with the quotient topology (induced by the standard topology on $\mathcal{K}_0$), then $\mathcal{AK}_0$ is compact. This was proved by Macbeath (1951). In the course of the proof, Macbeath introduced the function
$$\rho(K, L) := \inf_{\alpha} \{\mathcal{H}^n(\alpha L)/\mathcal{H}^n(K) \mid K \subset \alpha L\},$$
where α ranges over the affinities of $\mathbb{E}^n$, and showed that
$$\Delta([K], [L]) = \log \rho(K, L) + \log \rho(L, K), \quad [K], [L] \in \mathcal{AK}_0,$$
defines a metric Δ on $\mathcal{AK}_0$ that induces the quotient topology.

The exponent of entropy for the space $(\mathcal{AK}_0, \Delta)$ was determined by Bronshtein (1978a); it is equal to $(n-1)/2$.

14. *Affine equivalence classes of compact sets*. Let $\mathcal{AC}^n$ denote the set of affine equivalence classes of nonempty compact subsets of $\mathbb{E}^n$. Webster (1965) showed how the Hausdorff metric can be used to define a metric on $\mathcal{AC}^n$ that makes $\mathcal{AC}^n$ into a compact space. This metric, however, does not induce the quotient topology on $\mathcal{AC}^n$ (obtained from $(\mathcal{C}^n, \delta)$ and the affine-equivalence relation), since the latter fails to be metrizable. Let $\mathcal{AC}_n^n$ $(\mathcal{AC}_+^n)$ be the subset of all affine equivalence classes containing an element C with $\dim \operatorname{conv} C = n$ (respectively, $\mathcal{H}^n(C) > 0$). Then Webster's metric and the quotient topology on $\mathcal{AC}^n$ induce the same topology on $\mathcal{AC}_n^n$, which is compact. On the other hand, there does not exist a metric on $\mathcal{AC}_+^n$ that

makes $\mathcal{AC}_+^n$ compact and induces the quotient topology. Webster's results essentially answered a question posed by Grünbaum [37] (p. 263).

15. *Applications of the Steiner point*. The exact Lipschitz constant in (1.8.3) was determined by Przesławski (1985), Saint Pierre (1985) and Vitale (1985a).

    The properties of the Steiner point, in particular Minkowski additivity, the inclusion property (1.7.6), and the Lipschitz continuity (1.8.3), make it a useful tool for some applications. For instance, the Steiner point has been used for the construction of continuous selections of multi-valued mappings (Linke 1980, Przesławski 1985) and in the theory of random sets (Vitale 1984, 1990, Giné & Hahn 1985a, b). Motivated by this, Vitale (1985b) showed the non-existence of a continuous extension of the Steiner point to all convex bodies in infinite dimensional Hilbert spaces.

16. *Minkowski additive selections*. Generalizing the Steiner point map, Zivaljević (1989) studied Minkowski additive functions $f: \mathcal{K}^n \to \mathbb{E}^n$ satisfying $f(K) \in K$ for all $K \in \mathcal{K}^n$. In particular, he characterized all such Minkowski additive selections f for which $f(K) \in \text{ext}\, K$ for all $K \in \mathcal{K}^n$.

17. *Isometry invariant measures on* $\mathcal{K}^n$. The following question was posed by McMullen (see Gruber & Schneider [36], p. 268): Is there a natural and useful (isometry) invariant measure on the space $\mathcal{K}^n$ with the Hausdorff metric? It appears (depending on the interpretation of 'useful') that the answer is no. Bandt & Baraki (1986) have shown: For $n > 1$ there is no σ-finite Borel measure on $(\mathcal{K}^n, \delta)$ that is invariant with respect to all isometries from the whole space into itself and is not identically zero.

2

Boundary structure

2.1. Facial structure

The notions of face, extreme point and exposed point of a convex set were defined in Section 1.4. In the present section we shall study the boundary structure of closed convex sets in relation to these and similar or more specialized notions. We shall assume in the following that $K \subset \mathbb{E}^n$ is a nonempty closed convex set.

An i-dimensional face of K is referred to as an *i-face*. $\mathcal{F}(K)$ is the set of all faces and $\mathcal{F}_i(K)$ the set of all i-faces of K. A face of dimension $(\dim K) - 1$ is usually called a *facet*. The empty set $\emptyset$ and K itself are faces of K; the other faces are called *proper*. It follows from the definition of a face and from Lemma 1.1.8 that the faces of K are closed. If $F \neq K$ is a face of K, then $F \cap \operatorname{relint} K = \emptyset$. (If $z \in F \cap \operatorname{relint} K$, we choose $y \in K \setminus F$. There is some $x \in K$ with $z \in \operatorname{relint} [x, y]$. Then $[x, y] \subset F$, a contradiction.) In particular, $F \subset \operatorname{relbd} K$ and $\dim F < n$.

Theorem 2.1.1. *If F_i is a face of K for $i \in I$* (a nonempty index set), *then $\bigcap_{i \in I} F_i$ is a face of K. If F is a face of K and G is a face of F, then G is a face of K.*

Proof. This follows immediately from the definition of a face. ■

Theorem 2.1.2. *If $F_1, F_2 \in \mathcal{F}(K)$ are distinct faces, then* $\operatorname{relint} F_1 \cap \operatorname{relint} F_2 = \emptyset$. *To each nonempty relatively open convex subset A of K there is a unique face $F \in \mathcal{F}(K)$ with $A \subset \operatorname{relint} F$. The system*

$$\{\operatorname{relint} F \mid F \in \mathcal{F}(K)\}$$

is a disjoint decomposition of K.

Proof. Assume that $F_1, F_2 \in \mathcal{F}(K)$, $F_1 \neq F_2$ and $z \in \operatorname{relint} F_1 \cap \operatorname{relint} F_2$. There is some $x \in F_1 \setminus F_2$, say. We can choose $y \in F_1$ such that $z \in \operatorname{relint}[x, y]$. From $[x, y] \subset F_1 \subset K$, $z \in F_2$ and $F_2 \in \mathcal{F}(K)$ it follows that $[x, y] \subset F_2$, a contradiction. Thus $\operatorname{relint} F_1 \cap \operatorname{relint} F_2 = \emptyset$.

Let $A \neq \emptyset$ be a relatively open convex subset of K. Let F be the intersection of all faces of K containing A; then F is a face of K. Suppose there is some point $x \in A \setminus \operatorname{relint} F$. Then there is a support plane H to F with $x \in H$ and $H \cap F \neq F$. Since $x \in A \subset F$ and A is relatively open, it follows that $A \subset H$. Now $H \cap F$ is clearly a face of F and hence a face of K. But then $F \subset H \cap F$ by the definition of F, a contradiction. Thus $A \subset \operatorname{relint} F$. It follows from the first part that F is the only face with this property.

In particular, the set A may be one-pointed. Thus each point $x \in K$ is contained in relint F for a unique face $F \in \mathcal{F}(K)$. ■

In the proof above we have used the obvious fact that each support set $H \cap K$, where H is a support plane of K, is a face of K. The support sets are also called *exposed faces* of K. The singleton $\{x\}$ is an exposed face if and only if x is an exposed point. We denote by $\mathcal{S}(K)$ the set of all exposed faces and by $\mathcal{S}_i(K)$ the set of all i-dimensional exposed faces of K. Clearly not every face is an exposed face: not every extreme point is an exposed point, and taking direct sums yields examples with higher-dimensional faces. However, each proper face F of K is contained in some exposed face. In fact, since $F \cap \operatorname{relint} K = \emptyset$, there exists a hyperplane H separating F and K. From $F \subset K$ it follows that $F \subset H$, hence F is contained in the exposed face $H \cap K$.

Theorem 2.1.3. *If F_i is an exposed face of K for $i \in I \neq \emptyset$, then $\bigcap_{i \in I} F_i$ is either empty or an exposed face of K.*

Proof. Let $F := \bigcap_{i \in I} F_i$ and assume that $F \neq \emptyset$. After a translation, we may assume that $o \in F$. For $i \in I$ there exists a vector $u_i \in \mathbb{E}^n \setminus \{o\}$ such that $F_i = K \cap H_{u_i,0}$ and $K \subset H^-_{u_i,0}$. We may assume that $u_1, \ldots, u_r$ are linearly independent and that any other u_i linearly depends on these vectors. Then $F = \bigcap_{i=1}^r F_i$. Let $u := \sum_{i=1}^r u_i$. Then $K \subset H^-_{u,0}$ and $o \in K \cap H_{u,0}$, hence $H_{u,0}$ supports K. For $x \in F$ we have $\langle x, u_i \rangle = 0$ for $i \in I$ and hence $\langle x, u \rangle = 0$; thus $F \subset H_{u,0} \cap K$. For $y \in K \setminus F$, $\langle y, u_j \rangle < 0$ for some $j \in \{1, \ldots, r\}$, hence $\langle y, u \rangle < 0$ and thus $y \notin H_{u,0}$. This proves that $H_{u,0} \cap K = F$ and therefore that $F \in \mathcal{S}(K)$. ■

One should observe that the second part of Theorem 2.1.1 has no counterpart for exposed faces: An exposed face of an exposed face of K

need not itself be an exposed face of K, but it is, of course, a face of K.

From now on in this section we assume that $K \in \mathcal{K}^n$.

Exposed faces behave well under polarity, as we shall now see. Suppose that $K \in \mathcal{K}_0^n$ and $o \in \operatorname{int} K$. In Section 1.6 the polar body of K was defined as

$$K^* = \{x \in \mathbb{E}^n \mid \langle x, y \rangle \leqq 1 \quad \text{for all } y \in K\}.$$

For a subset $F \subset K$ we now define the *conjugate face* of F by

$$\hat{F} := \{x \in K^* \mid \langle x, y \rangle = 1 \quad \text{for all } y \in F\}.$$

One should keep in mind that that $\hat{F}$ depends also on K and not only on F; but this slightly incomplete notation is quite common. If we write $\hat{\hat{F}}$, this means $(\hat{F})\hat{}$, where the right-hand circumflex refers to K^*. Instead of $\{y\}\hat{}$ we write $\hat{y}$.

Theorem 2.1.4. *Let* $o \in \operatorname{int} K$ *and* $\emptyset \neq F \subset K$. *Then the following assertions* (a), (b), (c) *are equivalent.*

(a) *F lies in an exposed face of K,*

(b) $\hat{F} \neq \emptyset$,

(c) *$\hat{F}$ is an exposed face of K^*.*

If (a)–(c) *hold, then $\hat{\hat{F}}$ is the smallest exposed face of K containing F.*

Proof. For $y \in \operatorname{bd} K$, the hyperplane $\{x \in \mathbb{E}^n \mid \langle x, y \rangle = 1\}$ supports K^* (e.g., by Theorem 1.7.6); hence $\hat{y} = F(K^*, y)$ is an exposed face of K^*. For $y \in \operatorname{int} K$, $\langle x, y \rangle < 1$ for all $x \in K^*$, hence $\hat{y} = \emptyset$. By definition,

$$\hat{F} = \bigcap_{y \in F} \{x \in K^* \mid \langle x, y \rangle = 1\} = \bigcap_{y \in F} \hat{y}.$$

Thus $\hat{F}$ is either empty or (by 2.1.3) an exposed face of K^*.

Suppose that F lies in the exposed face $S := K \cap H_{u,1}$ (without loss of generality), where $K \subset H_{u,1}^-$. Then $u \in \hat{F}$ and hence $\hat{F} \neq \emptyset$. Vice versa, if $u \in \hat{F}$, then $u \in K^*$ and $\langle u, y \rangle = 1$ for all $y \in F$, hence F lies in an exposed face of K. Thus (a), (b) and (c) are equivalent.

Suppose again that F lies in the exposed face $S = K \cap H_{u,1}$. Let $z \in \hat{\hat{F}}$. Then $z \in K^{**} = K$ and $\langle z, x \rangle = 1$ for all $x \in \hat{F}$, and in particular $\langle z, u \rangle = 1$, thus $z \in S$. This shows that $\hat{\hat{F}} \subset S$ for each exposed face S of K containing F. Since $\hat{\hat{F}}$ is itself an exposed face of K containing F, it follows that $\hat{\hat{F}}$ is the smallest exposed face containing F. ■

Remark 2.1.5. For a proper face F of K that is not an exposed face it follows from Theorem 2.1.4 that $\hat{\hat{F}} \neq F$.

By Theorem 2.1.2, each point $x \in K$ belongs to the relative interior of a uniquely determined face F_x of K. This leads to a classification of the points of K according to the dimension of the containing face. The point x is called an *r-extreme* point of K if $\dim F_x \leqq r$. Equivalently, x is r-extreme if and only if there is no $(r+1)$-dimensional convex set $A \subset K$ with $x \in \operatorname{relint} A$. Further, the point x is called an *r-exposed* point of K if x is contained in an exposed face $F \in \mathcal{S}(K)$ with $\dim F \leqq r$. The set of all the r-extreme (r-exposed) points of K is called its *r-skeleton* (*exposed r-skeleton*) and denoted by $\operatorname{ext}_r K$ ($\exp_r K$). We have $\operatorname{ext}_0 K = \operatorname{ext} K$, $\exp_0 K = \exp K$; evidently $\exp_r K \subset \operatorname{ext}_r K$ and, trivially, $\operatorname{ext}_r K \subset \operatorname{ext}_s K$ and $\exp_r K \subset \exp_s K$ for $r < s$.

Theorem 1.4.7, which is due to Straszewicz and which states that each extreme point is a limit of exposed points, can be extended to r-extreme and r-exposed points. This extension was obtained by Asplund (1963). His proof, however, is difficult to understand, and we prefer to give a proof due to McMullen (1983b). It requires the following lemma.

Lemma 2.1.6. *Let $G \in \mathcal{F}_r(K)$ (for some $r \in \{0, \ldots, n-1\}$) and $q \in$* relint *G. If there is a $(n-r-1)$-flat supporting K precisely at q, then $G \in \mathcal{S}_r(K)$.*

Proof. The assumption demands the existence of an $(n-r-1)$-flat L and of a support plane H to K such that $L \subset H$ and $L \cap K = \{q\}$. Let $F := H \cap K$. From $g \in \operatorname{relint} G$ we have $G \subset F$. Suppose that $F \neq G$. Let $x \in F \setminus G$. We may assume that $q = o$. By $L \cap G = \{o\}$ and $o \in \operatorname{relint} G$ we have $L \cap \operatorname{lin} G = \{o\}$; from $\dim G = r$ and $\dim L = n - r - 1$ we get $L + \operatorname{lin} G = H$. Hence, there is a representation $x = y + z$ with $y \in \operatorname{lin} G$ and $o \neq z \in L$. Since $o \in \operatorname{relint} G$, there is a number $\lambda > 0$ with $-\lambda y \in G$. Then

$$\frac{\lambda}{1+\lambda} z = \frac{\lambda}{1+\lambda} x + \frac{1}{1+\lambda}(-\lambda y) \in F,$$

thus $o \neq [\lambda/(1+\lambda)]z \in L \cap K$, which contradicts the assumption. We conclude that $F = G$ and thus $G \in \mathcal{S}_r(K)$. ∎

Theorem 2.1.7. *For $r = 0, \ldots, n-1$,*

$$\operatorname{ext}_r K \subset \operatorname{cl} \exp_r K.$$

Proof. For $r = 0$ this is true by Theorem 1.4.7. Assume that $r \geqq 1$ and that the assertion is true for $r-1$ instead of r. Let $p \in \operatorname{ext}_r K$ and a neighbourhood V of p be given. By Theorem 2.1.2 there is a unique face $F \in \mathcal{F}(K)$ such that $p \in \operatorname{relint} F$, and $p \in \operatorname{ext}_r K$ implies $\dim F \leqq r$.

If $\dim F < r$, then $p \in \text{ext}_{r-1} K$ and by the induction hypothesis we have $p \in \text{cl}\,\text{exp}_{r-1} K \subset \text{cl}\,\text{exp}_r K$. Assume, therefore, that $\dim F = r$. Let M be an $(n-r)$-flat through p that is complementary to aff F. Then $p \in \text{ext}(M \cap K)$, hence by Theorem 1.4.7 there is a point $q \in V \cap \text{exp}(M \cap K)$. Let $G \in \mathcal{F}_s(K)$ be the face with $q \in \text{relint}\, G$. Since $q \in \text{exp}(M \cap K)$ and $q \in \text{relint}\, G$, we must have $s \leqq r$. If $s < r$, then $q \in \text{ext}_{r-1} K$, and the induction hypothesis yields $q \in \text{cl}\,\text{exp}_{r-1} K \subset \text{cl}\,\text{exp}_r K$. If $s = r$, let L be an $(n-r-1)$-flat in M that supports $M \cap K$ precisely at q. Lemma 2.1.6 now shows that $G \in \mathcal{S}_r(K)$, hence $q \in \text{exp}_r K$. Since V was an arbitrary neighbourhood of p, the assertion is proved. ■

We conclude this section with a simple remark on the topology of the r-skeleton. If $\dim K \leqq 2$, it is easy to see that ext K is closed, but for $\dim K \geqq 3$ this is not true in general. For example, let K be the convex hull of a circular disc C and of an orthogonal segment S such that there is a point $x \in \text{relbd}\, C \cap \text{relint}\, S$. Then x is a limit of extreme points of K but is itself not an extreme point.

It is, however, easy to see that $\text{ext}_r K$ is a G_δ set (an intersection of countably many open sets): For $k \in \mathbb{N}$ let A_k be the set of the centres of all $(r+1)$-dimensional closed balls of radius $1/k$ contained in K. It is not difficult to show that A_k is closed, and obviously

$$\text{ext}_r K = \bigcap_{k \in \mathbb{N}} ([K + B_0(o, 1/k)] \setminus A_k).$$

Notes for Section 2.1

1. *Characterization of sets of extreme points.* If $S = \text{ext}\, K$ for some $K \in \mathcal{K}^n$, then clearly (a) $\text{cl}\, S \subset \text{conv}\, S$, (b) $\text{cl}\, S$ is compact, (c) $S \cap \text{conv}\, A \subset A$ for all $A \subset S$. Björck (1958) proved that for a nonempty set $S \subset \mathbb{E}^n$ these three conditions are also sufficient for the existence of a convex body $K \in \mathcal{K}^n$ such that $S = \text{ext}\, K$, and he considered an infinite-dimensional extension of this result.
2. *A semicontinuity property of skeletons.* Let $(K_i)_{i \in \mathbb{N}}$ be a sequence in $\mathcal{K}^n$ converging to $K \in \mathcal{K}^n$. Then, for $r = 0, \ldots, n-1$,
$$\text{cl}\,\text{ext}_r K \subset \liminf_{i \to \infty} \text{ext}_r K_i.$$
(By definition, $x \in \liminf A_i$ if every neighbourhood of x meets A_i for almost all i.) For a proof, see Schneider [43] (Theorem (2.13) and §8).
3. *The face-function.* Let $K \in \mathcal{K}_0^n$. By Theorem 2.1.2, each point $x \in K$ is contained in the relative interior of a unique face of K; denote this face by F_x. The function $F: x \mapsto F_x$ is called the *face-function* of K; its restriction to bd K is the face-function of bd K. Klee & Martin (1971) have investigated the semicontinuity properties of the face-function. For $X = K$ or bd K, let X_l (X_u) be the set of all points $x \in X$ at which F is lower (upper) semicontinuous. Klee and Martin studied these sets and showed the following:

Theorem. The face-function of bd K is lower semicontinuous almost everywhere in the sense of category and upper semicontinuous almost everywhere in the sense of measure. However, when $n \geqq 3$ it may be lower semicontinuous almost nowhere in the sense of measure and upper semicontinuous almost nowhere in the sense of category.

Actually, upper semicontinuity almost everywhere in the sense of measure for $n > 3$ was only conjectured, but the assertion was completed by Larman (1971) who proved the following:

Theorem. The union of the relative boundaries of the proper faces of K of dimension at least 1 has zero $(n-1)$-dimensional Hausdorff measure.

Lower semicontinuity of face-functions is related to the continuity of so-called envelopes of functions. For a continuous function $f: K \to \mathbb{R}$ one defines the *envelope* f_e by

$$f_e := \sup \{g \mid g: \mathbb{E}^n \to \mathbb{R} \text{ affine}, \ g \leqq f \text{ on } K\}.$$

Let K_e denote the set of all points of K at which every envelope is continuous. Kruskal (1969) gave a three-dimensional example with $K_e \neq K$. Klee & Martin (1971) proved $K_e = K$ for $n = 2$, and Eifler (1977) showed that $K_e = K_l$ for $K \in \mathcal{K}_0^n$.

4. *Stable convex bodies.* The convex body $K \in \mathcal{K}_0^n$ is called *stable* if the map $(x, y) \mapsto \frac{1}{2}(x + y)$ from $K \times K$ onto K is open. The following assertions are equivalent:
   (a) K is stable;
   (b) the face-function of K is lower semicontinuous;
   (c) all skeletons $\text{ext}_r K$, $r = 0, \ldots, n$, of K are closed.

   For the proof, see Papadopoulou (1977) or Debs (1978); the survey by Papadopoulou (1982) explains why stable convex sets are of interest and contains further results. For $n > 2$, the convex bodies K with properties (a), (b), (c) are also characterized by the fact that the metric space consisting of the proper faces of K with the Hausdorff metric is compact. Similarly, the metric space of all exposed faces is compact if and only if all the exposed r-skeletons of K are closed, $0 \leqq r \leqq n-2$. These results are due to Reiter & Stavrakas (1977).
5. Let $K \in \mathcal{K}^n$, and for $x \in \text{relbd}\, K$ let S_x be the smallest exposed face of K containing x. In his investigation of the convexity of Chebyshev sets, Brown (1980) conjectured the following. If F is an exposed face of K, then there exists an $x \in \text{relbd}\, F$ such that either $S_x = \{x\}$ or $S_x = F$. He proved this for $n \leqq 5$.
6. *Topology of skeletons.* In general, the topological structure of ext K and exp K for $K \in \mathcal{K}_0^n$ is not easy to describe. (For the situation in the infinite-dimensional case, see Klee (1955), Bensimon (1987).) The following is quoted from Klee (1958) (where $n = 3$): 'Find a useful and simple characterization of the class $\mathcal{X}_n$ of all subsets X of the unit sphere S^{n-1} such that there is a homeomorphism of S^{n-1} onto the boundary of a convex body K mapping X into ext K'. The following partial answer was given by Collier (1975): Let X be a subset of a compact 0-dimensional metric space Z, and let $n \geqq 3$ be an integer. There is a homeomorphism of Z into the boundary of a convex body $K \in \mathcal{K}_0^n$ mapping X onto ext K if and only if X is a G_δ set with at least $n + 1$ points. On the other hand, non-trivial examples of subsets of S^2 that do not belong to $\mathcal{X}_3$ can be constructed by means of the following result of Collier (1976): If $K \in \mathcal{K}_0^3$, then each component of $(\text{cl}\,\text{ext}\, K) \setminus \text{ext}\, K$ is a subset of a 1-dimensional face of K.

Bronshtein (1981) considered the collection $\mathcal{M}_{n,N}$ of sets M in $\mathbb{E}^n$ such that there exist $K \in \mathcal{K}^N$ and a homeomorphism of cl M into K mapping M onto ext K. He showed that $\mathcal{M}_{n,n+1}$ contains all compacta of $\mathbb{E}^n$, further that each locally compact, bounded set is in $\mathcal{M}_{n,n+2}$ but not necessarily in $\mathcal{M}_{n,n+1}$, and similar results.

The following is known about exp K. For dim $K = 2$, exp K is a G_δ set; but Corson (1965) constructed a three-dimensional convex body K for which exp K is not the union of a G_δ set with an F_σ set and contains no dense G_δ set. He also constructed a body $K \in \mathcal{K}_0^3$ for which exp K is the union of a countable number of closed sets each of which has no interior with respect to exp K. Choquet, Corson & Klee (1966) proved for $K \in \mathcal{K}_0^n$ that exp K is the union of a G_δ set, an F_σ set and $n-2$ sets each of which is the intersection of a G_δ set with an F_σ set. For $K \in \mathcal{K}_0^3$ they showed that exp K is the union of a G_δ set and a set that is the intersection of a G_δ and an F_σ set.

Klee (1958) proved for smooth $K \in \mathcal{K}_0^n$ that exp K is a G_δ set and ext $K \setminus$exp K is a first category F_σ set in ext K. For $n \geqq 2$ he showed the existence of a smooth body $K \in \mathcal{K}_0^n$ such that ext K is closed and ext K $\setminus$exp K is dense in ext K, and proved the existence of a body $K \in \mathcal{K}_0^3$ such that bd $K\setminus$ext K, ext $K\setminus$exp K and exp K are all dense in bd K; see also Edelstein (1965).

7. *Paths in the 1-skeleton*. The 1-skeleton of a polytope has well-known connectivity properties, which are important, for instance, for the edge-following procedures of the simplex algorithm. Larman & Rogers (1970, 1971) initiated a programme of extending these properties to general convex bodies. In their 1970 paper they proved the following theorem: Let a and b be two distinct exposed points of a convex body $K \in \mathcal{K}_0^n$. Then there are n simple arcs $P_1, \ldots, P_n$ in the 1-skeleton of K, each joining a to b, such that $P_i \cap P_j = \{a, b\}$ for $1 \leqq i < j \leqq n$. In their 1971 paper, Larman and Rogers obtained, in two refined versions, the following result on the existence of increasing paths: Let L be a non-constant linear function on $\mathbb{E}^n$ and let $K \in \mathcal{K}_0^n$. Then there is a continuous map s of the closed interval $[0, 1]$ to the exposed 1-skeleton of K with
$$L(s(0)) = \inf_{x \in K} L(x), \quad L(s(1)) = \sup_{x \in K} L(x),$$
$$L(s(t_1)) < L(s(t_2)), \quad \text{when } 0 \leqq t_1 < t_2 \leqq 1.$$
This line of research was continued by Gallivan (1979), Gallivan & Larman (1981) and Gallivan & Gardner (1981). Similar investigations for infinite-dimensional compact convex sets were carried out by Larman & Rogers (1973), Larman (1977) and Dalla (1986).

8. It was pointed out by Fedotov (1979) that a convex body K with $o \in \text{int}\, K$ can have a face F with $\hat{\hat{F}} \neq F$ (Remark 2.1.5); this corrects an erroneous statement in Bourbaki [10] (Chapter IV, §1, Exercise 2a).

9. *Hausdorff measures of skeletons*. For a polytope $P \subset \mathbb{E}^n$ and for $r \in \{0, 1, \ldots, n\}$ it is natural to consider the total r-dimensional volume, $\eta_r(P)$, of the r-faces of P. Inequalities for these functionals were investigated by Eggleston, Grünbaum & Klee (1964) and by Larman & Mani (1970). The definition
$$\eta_r(K) := \mathcal{H}^r(\text{ext}_r\, K),$$
where $\mathcal{H}^r$ is the $r$dimensional Hausdorff measure, extends the function η_r to arbitrary convex bodies $K \in \mathcal{K}^n$. A different way of extending η_r from polytopes to general convex bodies was proposed by Eggleston, Grünbaum & Klee (1964). They showed that η_r is lower semicontinuous on the set of n-dimensional polytopes and defined

$$\zeta_r(K) := \liminf_{P \to K} \eta_r(P)$$

for $K \in \mathcal{K}^n$, where lim inf is taken over all sequences of polytopes converging to K. Then ζ_r is lower semicontinuous on $\mathcal{K}^n$. The comparison of η_r and ζ_r is non-trivial for $1 \leqq r \leqq n-2$. Burton (1979) showed that η_r is lower semicontinuous and that $\zeta_r \geqq \eta_r$. It has been asked whether $\zeta_r = \eta_r$ in general (Problem 76 posed by Schneider in [36]). For $r = n-2$ this was proved by Larman (1980), but the other cases remain open.

Eggleston, Grünbaum & Klee (1964) also asked for a characterization of those convex bodies $K \in \mathcal{K}^n$ for which $\zeta_r(K) < \infty$. It was observed by Schneider [43] (p. 19) that a convex body K for which $\zeta_r(K) < \infty$ or $\eta_r(K) < \infty$ has the property that (in the sense of Haar measure) almost every $(n-r)$-flat intersects K in a polytope. This property, however, is not so restrictive as one might expect. Dalla & Larman (1980) constructed a convex body $K \in \mathcal{K}^3$ almost all of whose two-dimensional plane sections are polygons but for which ext K has Hausdorff dimension one. (This answered in the negative a question posed by Schneider; see Problem 77 in [36]). These authors also showed that a convex body $K \in \mathcal{K}^n$ has $\operatorname{ext}(K \cap E)$ of (Hausdorff) dimension at most r for almost all k-flats E if and only if the dimension of $\operatorname{ext}_{n-k} K$ is at most $n-k+r$. Burton (1979b) proved that $\eta_r(K) < \infty$ (where $1 \leqq r \leqq n-2$) implies $\mathcal{H}^r(\operatorname{ext}_{r-1} K) = 0$. He also obtained results on the facial structure of a convex body K having $\eta_r(K) < \infty$, exhibiting some resemblance to polytopal behaviour.

Burton (1980a) has found integral-geometric formulae for the functions η_r. The simplest of these says that $\eta_r(K)$ is, up to a factor depending only on n and r, the integral of $\eta_0(K \cap E)$ over all $(n-r)$-flats E, where the integration is with respect to the rigid motion invariant measure on the space of $(n-r)$-flats.

A number of inequalities for the functions η_r are known. Schneider (1978b) proved

$$\eta_r(K) > (n-r+1)V_r(K)$$

for $K \in \mathcal{K}_0^n$ and $r \in \{1, \ldots, n-2\}$, where $V_r(K)$ is the intrinsic volume, defined in Section 4.2. This inequality is sharp, that is, the quotient $\eta_r(K)/V_r(K)$ comes arbitrarily close to $n-r+1$ by proper choice of K. A weaker form of the inequality was obtained earlier by Firey & Schneider (1979). For non-negative integers r and s with $r+s \leqq n$, Burton (1979b) showed

$$(r+1)(s+1)\eta_{r+s}(K) \leqq \eta_r(K)\eta_s(K),$$

which strengthens an inequality obtained by Eggleston, Grünbaum & Klee (1964) for polytopes.

A general problem of isoperimetric type can be formulated as follows: For $1 \leqq s < r \leqq n$ determine the least number $q(n, r, s)$ such that

$$\eta_r(K) \leqq q(n, r, s)(\eta_s(K))^{r/s}$$

for all $K \in \mathcal{K}^n$. For polytopes this problem was first studied by Eggleston, Grünbaum & Klee (1964). It is not known whether $q(n, r, s) < \infty$ in general. The known results, coming from various sources, are listed in Burton & Larman (1981): The isoperimetric inequality gives the value of $q(n, n, n-1)$. Explicit upper bounds are known for $q(n, r, s)$ when s divides r, and furthermore for $q(n, n-1, s)$, with a better value for $q(3, 2, 1)$, and for $q(3, 3, 1)$. Burton & Larman (1981) found an upper bound for $q(n, n-2, n-3)$ if $n \geqq 4$.

The following inequality of the Brunn–Minkowski type was established by

Dalla (1987): For $K, L \in \mathcal{K}^n$ and $\lambda \in [0, 1]$,

$$\eta_1((1-\lambda)K + \lambda L) \geqq (1-\lambda)\eta_1(K) + \lambda\eta_1(L).$$

One may speculate whether $\eta_r((1-\lambda)K + \lambda L)^{1/r}$ is in general a concave function of λ (it is for $r = n$ and $r = n-1$; see Sections 6.1 and 6.5).

Hausdorff measures of finite-dimensional skeletons of infinite-dimensional convex sets were investigated by Dalla (1985).

2.2. Singularities

Through each boundary point of a convex body there is a supporting hyperplane, but not necessarily a unique one. Non-uniqueness gives rise to singularities, and these will be studied in the present section.

First we introduce some convex cones that describe the behaviour of a convex body at one of its points. Let $K \in \mathcal{K}^n$ be a convex body and let $x \in K$. Define

$$P(K, x) := \{\lambda(y - x) \mid y \in K, \lambda > 0\} = \bigcup_{\lambda>0} \lambda(K - x).$$

Then $P(K, x)$ is a convex cone, and $P(K, x) = \operatorname{aff} K - x$ if and only if $x \in \operatorname{relint} K$. The closed convex cone $S(K, x) := \operatorname{cl} P(K, x)$ can evidently be represented as

$$S(K, x) = \bigcap_{x \in H(K,u)} H^-(K, u) - x,$$

where the intersection extends over all $u \in \mathbb{E}^n \setminus \{o\}$ (or $u \in S^{n-1}$) for which x lies in the supporting hyperplane $H(K, u)$. (For $x \in \operatorname{int} K$, we adopt the usual convention that the intersection of an empty family of subsets of $\mathbb{E}^n$ is equal to $\mathbb{E}^n$.) $S(K, x)$ is called the *support cone* of K at x. (In Bonnesen & Fenchel [8], $S(K, x) + x$ is called the *projection cone* of K at x.) Further, we define the *normal cone* of K at x by

$$N(K, x) := \{u \in \mathbb{E}^n \setminus \{o\} \mid x \in H(K, u)\} \cup \{o\}.$$

If $x \in K \cap H(K, u)$, then u is called an *outward* (or *outer*, or *exterior*) *normal vector* of K at x. Thus $N(K, x)$ consists of all outward normal vectors of K at x together with the zero vector. Clearly $N(K, x)$ is a closed convex cone and

$$N(K, x) = S(K, x)^*, \tag{2.2.1}$$

that is, $N(K, x)$ and $S(K, x)$ are a pair of dual convex cones. The normal cone is closely related to the metric projection, in fact, it follows from Lemma 1.3.1 that

$$N(K, x) = p_K^{-1}(x) - x \tag{2.2.2}$$

for $x \in K$.

If K lies in a proper linear subspace E of $\mathbb{E}^n$ and if $S_E(K, x)$ and

$N_E(K, x)$ are, respectively, the support cone and normal cone of K at x, taken with respect to the subspace E, then obviously

$$S(K, x) = S_E(K, x),$$

while

$$N(K, x) = N_E(K, x) + E^{\perp}, \tag{2.2.3}$$

where $E^{\perp}$ is the orthogonal complement of E.

The following theorem describes the behaviour of the support cone and normal cone under addition and intersection.

Theorem 2.2.1. *Let* $K, L \in \mathcal{K}^n$.

(a) *If* $x \in K$ *and* $y \in L$, *then*

$$S(K + L, x + y) = S(K, x) + S(L, y),$$
$$N(K + L, x + y) = N(K, x) \cap N(L, y).$$

(b) *If* $x \in K \cap L$ *and* relint $K \cap$ relint $L \neq \emptyset$, *then*

$$S(K \cap L, x) = S(K, x) \cap S(L, x),$$
$$N(K \cap L, x) = N(K, x) + N(L, x).$$

Proof of (a). Performing translations if necessary, we may clearly assume that $x = y = o$. Let $u \in N(K + L, o)\setminus\{o\}$. Then $K + L \subset H^-_{u,0}$. From $K = K + \{o\} \subset H^-_{u,0}$ and $o \in K$ it follows that $u \in N(K, o)$. Similarly we get $u \in N(L, o)$.

Let $u \in N(K, o) \cap N(L, o)\setminus\{o\}$. From $u \in N(K, o)$ we have $K \subset H^-_{u,0}$; similarly $L \subset H^-_{u,0}$ and thus $K + L \subset H^-_{u,0}$. Since $o \in K + L$, we get $u \in N(K + L, o)$. This proves $N(K + L, o) = N(K, o) \cap N(L, o)$.

From (2.2.1) and Theorem 1.6.3 we now get $S(K + L, o) = S(K, o) + S(L, o)$.

Proof of (b). Let $x \in K \cap L$ and $y \in \text{relint}\, K \cap \text{relint}\, L$. The first equality of (b) is clear if $x = y$; hence we assume that $x \neq y$. Let $z \in S(K, x) \cap S(L, x)$. Since $y - x \in \text{relint}\, K - x \subset \text{relint}\, P(K, x)$ and $z \in \text{cl}\, P(K, x)$, it follows from Lemma 1.1.8 that $[y - x, z) \subset \text{relint}\, P(K, x)$. Hence, for $w \in [y - x, z)$ there is a number $\lambda > 0$ such that $[x, x + \lambda w] \subset K$. Similarly, there is a number $\mu > 0$ such that $[x, x + \mu w] \subset L$. It follows that $w \in P(K \cap L, x)$. Since $w \in [y - x, z)$ was arbitrary, we deduce that $z \in \text{cl}\, P(K \cap L, x) = S(K \cap L, x)$. Thus $S(K, x) \cap S(L, x) \subset S(K \cap L, x)$. The opposite inclusion is trivial. This proves the first equality of (b). The second equality is again obtained by applying (2.2.1) and Theorem 1.6.3. ∎

For a point $x \in \operatorname{bd} K$ we denote by S_x the smallest exposed face of K containing x.

Lemma 2.2.2. *Let $K \in \mathcal{K}^n$ and $x, y \in \operatorname{bd} K$. Then $N(K, x) = N(K, y)$ if and only if $S_x = S_y$ (in particular, if $y \in \operatorname{relint} S_x$).*

Proof. If $N(K, x) = N(K, y)$, then each exposed face containing x contains y also, and each exposed face containing y contains x. It follows that $S_x = S_y$. Suppose that $S_x = S_y$ (which is clearly satisfied if $y \in \operatorname{relint} S_x$). Let $u \in N(K, x)$. Then $S_x \subset K \cap H(K, u)$, hence $y \in H(K, u)$ and thus $u \in N(K, y)$. The assertion follows. ■

We can use this to extend the definition of the normal cone. Let F be a nonempty convex subset of K. Then we define $N(K, F) := N(K, x)$, where $x \in \operatorname{relint} F$. This does not depend on the choice of x: if $F \cap \operatorname{int} K \neq \emptyset$, then $N(K, x) = \{o\}$. Otherwise, let $x, y \in \operatorname{relint} F$. Then S_x is the smallest exposed face of K containing F, hence $S_x = S_y$ and Lemma 2.2.2 gives $N(K, x) = N(K, y)$.

The normal cone $N(K, F)$ is closely related to the conjugate face $\hat{F}$, as shown in the following lemma.

Lemma 2.2.3. *Suppose that $o \in \operatorname{int} K$ and let F be a nonempty convex subset of K. Then*

$$N(K, F) = \operatorname{pos} \hat{F} \cup \{o\}.$$

Proof. Let $u \in N(K, F) \setminus \{o\}$. Then $H_{v,1}$ with $v = h(K, u)^{-1} u$ is a support plane of K satisfying $x \in K \cap H_{v,1}$ for some point $x \in \operatorname{relint} F$. It follows that $F \subset H_{v,1}$; hence $\langle y, v \rangle \leqq 1$ for $y \in K$ and $\langle y, v \rangle = 1$ for $y \in F$. This shows that $v \in \hat{F}$ and hence $u \in \operatorname{pos} \hat{F}$. Thus $N(K, F) \subset \operatorname{pos} \hat{F} \cup \{o\}$. The argument can be reversed. ■

It follows from Lemma 2.2.3, (2.2.3) and Theorem 2.1.2 that distinct normal cones of a convex body have disjoint relative interiors. In particular, we have

$$\operatorname{int} N(K, x) \cap \operatorname{int} N(K, y) = \emptyset \quad \text{for } N(K, x) \neq N(K, y). \tag{2.2.4}$$

We note that

$$\bigcup_{x \in \operatorname{ext} K} N(K, x) = \mathbb{E}^n, \tag{2.2.5}$$

since each support set of K contains an extreme point of K.

The normal cones lead, in a natural way, to a classification of the singular boundary points of a convex body. Let $K \in \mathcal{K}_0^n$ be a convex body with interior points and let $x \in \operatorname{bd} K$. We say that x is an *r-singular* point of K if $\dim N(K, x) \geqq n - r$. Bonnesen & Fenchel [8] (p. 14) call x a p-Kantenpunkt (ridge point of order p) if $\dim N(K, x) = n - p$. Thus x is r-singular if and only if x is a p-Kantenpunkt for some $p \leqq r$. An $(n-2)$-singular point is briefly called *singular*. If the supporting hyperplane to K at x is unique, i.e., if x is not singular, then x is called a *regular* or *smooth* point of K. The convex body K (of dimension n) is called *regular* or *smooth* if all its boundary points are regular. We denote the set of regular boundary points of K by $\operatorname{reg} K$.

The following theorem shows that a convex body cannot have large sets of r-singular boundary points.

Theorem 2.2.4. *Let $K \in \mathcal{K}_0^n$ and $r \in \{0, \ldots, n-1\}$. The set of r-singular points of K can be covered by countably many compact sets of finite r-dimensional Hausdorff measure.*

Proof. We use the metric projection p_K. Let M_r denote the intersection of a fixed closed ball containing K in its interior with the union of all r-dimensional affine subspaces of $\mathbb{E}^n$ that are spanned by points with rational coordinates. Let x be an r-singular point of K. Then the translated normal cone

$$N(K, x) + x = p_K^{-1}(x)$$

is of dimension at least $n - r$. It is, therefore, easy to see that this set meets M_r; hence $x \in p_K(M_r)$. As M_r is a union of countably many compact sets of finite r-dimensional Hausdorff measure and as p_K is a Lipschitz map by Theorem 1.2.2, the assertion follows. ■

Theorem 2.2.4 implies, in particular, that the set of r-singular points has $(r+1)$-dimensional Hausdorff measure zero, and furthermore that the set of singular points has $(n-1)$-dimensional Hausdorff measure zero. It also implies that K has at most countably many 0-singular points. This assertion can be generalized as follows. By a *perfect* face of $K \in \mathcal{K}_0^n$ one understands a face F of K for which

$$\dim F + \dim N(K, F) = n.$$

Theorem 2.2.5. *A convex body $K \in \mathcal{K}_0^n$ has at most countably many perfect faces.*

Proof. If F is a perfect face of K, then the set $p_K^{-1}(\operatorname{relint} F)$ is of dimension n and hence contains a point z with rational coordinates in its interior. Since $p_K(z) \in \operatorname{relint} F$ and different faces have disjoint relative interiors by Theorem 2.1.2, the assertion follows. ∎

The classification of support planes of a convex body, which we now describe, will in later chapters turn out to be of greater importance than the classification of boundary points. Let $K \in \mathcal{K}^n$ be a convex body. Let $u \in \mathbb{E}^n \setminus \{o\}$. Then u is an outer normal vector of K at each point of the support set $F(K, u) = K \cap H(K, u)$. The normal cone $N(K, F(K, u))$ (which is equal to $N(K, x)$ for any point $x \in \operatorname{relint} F(K, u)$) has a unique face, denoted by $T(K, u)$, that contains u in its relative interior (cf. Theorem 2.1.2). We call $T(K, u)$ the *touching cone* of K at u, since for every vector $v \in \operatorname{relint} T(K, u)$ the support plane $H(K, v)$ touches K at the same set: choose $x \in \operatorname{relint} F(K, u)$ and let D be the flat through x totally orthogonal to $\operatorname{lin} T(K, u)$; then

$$F(K, v) = D \cap K$$

for each $v \in \operatorname{relint} T(K, u)$. In fact, $D \cap K \subset F(K, v)$ is clear. On the other hand, let $y \in F(K, v)$; then $\langle y - x, v \rangle = 0$ and $\langle y - x, t \rangle \leqq 0$ for each $t \in T(K, u)$. Let $t \in T(K, u)$. Since $v \in \operatorname{relint} T(K, u)$, there is a vector $w \in T(K, u)$ and $\lambda > 0$ such that $\lambda v = t + w$. We get $0 = \langle y - x, \lambda v \rangle = \langle y - x, t + w \rangle$ and hence $\langle y - x, t \rangle = 0$. This shows that $y \in D$.

The vector u will be called an *r-extreme normal vector* of K, and $H(K, u)$ an *r-extreme support plane* of K, if $\dim T(K, u) \leqq r + 1$ $(r = 0, \ldots, n - 1)$. Thus u is an r-extreme normal vector of K if and only if there do not exist $r + 2$ linearly independent normal vectors $u_1, \ldots, u_{r+2}$ at one and the same boundary point of K such that $u = u_1 + \ldots + u_{r+2}$. For 0-extreme we say *extreme*.

***Remark* 2.2.6.** The set D is the affine hull of the set which in Bonnesen & Fenchel [8] (p. 16) is denoted by $\mathcal{D}_m$ and is found in a different way. (The construction described there can be interpreted as a dual way of finding the minimal face $T(K, u)$ of $N(K, F(K, u))$ containing u by successively determining the minimal exposed face containing u.) Bonnesen and Fenchel call $H(K, u)$ a p-Kantenstützebene if $\dim D = p$. Thus $H(K, u)$ is a p-Kantenstützebene if and only if it is an $(n - p - 1)$-extreme, but not an $(n - p - 2)$-extreme, support plane of K.

Further, the vector u is called an *r-exposed normal vector* of K, and $H(K, u)$ an *r-exposed support plane* of K, if $\dim N(K, F(K, u)) \leqq r + 1$.

The choice of notation suggests an underlying duality. In fact, let $K \in \mathcal{K}_0^n$ and $o \in \operatorname{int} K$. A given support plane H of K can be written as $H = H_{u,1}$ with $u \in \mathbb{E}^n \setminus \{o\}$. Writing $F(K, u) = F$, we have $N(K, F) = \operatorname{pos} \hat{F}$ by Lemma 2.2.3 and $u \in \hat{F}$. Hence, the touching cone $T(K, u)$ is the smallest face of $\operatorname{pos} \hat{F}$ containing u, and this is the positive hull of the smallest face of $\hat{F}$ containing u. It follows that u is an r-extreme point of K^* if and only if u is an r-extreme normal vector of K. If u is an r-exposed normal vector of K, then $\dim \operatorname{pos} \hat{F} = \dim N(K, F) \leqq r + 1$, hence $\dim \hat{F} \leqq r$. Thus u is contained in the exposed face $\hat{F}$ of dimension at most r and is, therefore, an r-exposed point of K^*. Vice versa, let u be an r-exposed point of K^*. Since $\hat{u} = F(K, u) = F$ (where the circumflex refers to K^*), $\hat{F} = \hat{\hat{u}}$ is, by Theorem 2.1.4, the smallest exposed face of K^* containing u. Hence, $\dim \hat{F} \leqq r$ and thus $\dim N(K, F) \leqq r + 1$, that is, u is an r-exposed normal vector of K.

The preceding observation leads to a dual version of Theorem 2.1.7:

***Theorem* 2.2.7.** *Let $K \in \mathcal{K}^n$ and $r \in \{0, \ldots, n-1\}$. Then each r-extreme support plane of K is a limit of r-exposed support planes of K.*

Proof. If $\dim K = n$, we may assume that $o \in \operatorname{int} K$ and then apply Theorem 2.1.7 to K^* to get the assertion. Suppose that $\dim K = k < n$. Let $u \in \mathbb{E}^n \setminus \{o\}$ be a given vector. We may assume that $o \in K$; let $E = \operatorname{lin} K$ and write $u = u_0 + u_1$ with $u_0 \in E$ and $u_1 \in E^\perp$. If u is an r-extreme normal vector of K and $u_0 = o$, then u is an r-exposed normal vector of K, since $T(K, u) = N(K, F(K, u)) = E^\perp$ in this case. Suppose that $u_0 \neq o$. Then $F(K, u) = F(K, u_0) = F$, say. Let $N_E(K, F) \subset E$ be the normal cone of K at F and $T_E(K, u_0) \subset E$ the touching cone of K at u_0, both with respect to E. Then

$$N(K, F) = N_E(K, F) + E^\perp,$$
$$T(K, u) = T_E(K, u_0) + E^\perp.$$

It is now clear that the vector u is an r-extreme (r-exposed) normal vector of K if and only if u_0 is an $(r - n + k)$-extreme ($(r - n + k)$-exposed) normal vector of K. Thus the assertion of the theorem is obtained in an obvious way if the result already proved is applied in E. ■

The classification of support planes plays a role since certain problems in later chapters lead to pairs of convex bodies that have some, but not all, support planes in common. In particular, the convex body $K \in \mathcal{K}^n$ containing the convex body $L \in \mathcal{K}^n$ is called a *p-tangential body* of L if each $(n-p-1)$-extreme support plane of K is a support plane of L ($p \in \{0, \ldots, n-1\}$). Thus a 0-tangential body of L is the body L itself, and each p-tangential body of L is also a q-tangential body for $p < q \leqq n-1$. An $(n-1)$-tangential body is briefly called a *tangential body*. The following theorem characterizes p-tangential bodies in a slightly different way.

Theorem 2.2.8. *Let $K, L \in \mathcal{K}^n$, $L \subset K$ and $p \in \{0, \ldots, n-1\}$. Then K is a p-tangential body of L if and only if the following condition holds.*

(*) *Each support plane of K that is not a support plane of L contains only $(p-1)$-singular points of K.*

Proof. Suppose that K is a p-tangential body of L. Let H be a support plane of K and assume that H contains a point x of K that is not $(p-1)$-singular, which means that $\dim N(K, x) \leqq n-p$. If u is the outer normal vector of H, then $N(K, F(K, u)) \subset N(K, x)$, hence $\dim T(K, u) \leqq n-p$, so that u is an $(n-p-1)$-extreme normal vector of K. Therefore, H is a support plane of L. Thus condition (*) holds.

Assume that (*) holds. Let H be an $(n-p-1)$-extreme support plane of K. By Theorem 2.2.7, H is the limit of a sequence $(H_i)_{i\in\mathbb{N}}$ of $(n-p-1)$-exposed support planes of K. Each H_i contains a point x_i of K with $\dim N(K, x_i) \leqq n-p$, that is, x_i is not $(p-1)$-singular. By condition (*), H_i is a support plane of L. Since this is true for all i, H is a support plane of L. This proves that K is a p-tangential body of L. ∎

If K is a 1-tangential body of L, then each point of K in a support plane of K that does not support L is a 0-singular point of K. There are at most countably many such points, and K is clearly the convex hull of L and these points. If x and y are any two of these 0-singular points, then $[x, y]$ meets L, since otherwise one finds a support plane of K not supporting L but containing a point of K that is not 0-singular. For intuitive reasons that are now evident, 1-tangential bodies are also called *cap-bodies*.

The notion of the p-extreme normal vector of a convex body admits a generalization that will be useful in Section 6.6. Let $K_1, \ldots, K_{n-1} \in \mathcal{K}^n$ be $n-1$ convex bodies. We say that the vector $u \in \mathbb{E}^n \setminus \{o\}$ is $(K_1, \ldots, K_{n-1})$-*extreme* if there exist $(n-1)$-dimensional linear subspaces $E_1, \ldots, E_{n-1}$ of $\mathbb{E}^n$ such that

$$T(K_i, u) \subset E_i \quad \text{for } i = 1, \ldots, n-1$$

and

$$\dim E_1 \cap \ldots \cap E_{n-1} = 1.$$

In particular, u is a p-extreme normal vector of the convex body K if and only if u is

$$(\underbrace{K, \ldots, K}_{n-1-p}, \underbrace{B^n, \ldots, B^n}_{p})\text{-extreme.}$$

Here, of course, the p-tuple of unit balls B^n may be replaced by any other p-tuple of regular convex bodies.

We could also dualize the notion of r-singular boundary points in a similar way to that in which we dualized the notions of p-extreme and p-exposed boundary points. We need only the following case. Let $K \in \mathcal{K}^n$ and $u \in S^{n-1}$. The vector u is called a *regular normal vector* of K, and $H(K, u)$ is a *regular support plane* of K, if $\dim F(K, u) = 0$; otherwise u is called *singular*. Thus $u \in S^{n-1}$ is a singular normal vector of K if and only if $H(K, u) \cap K$ contains a segment. A convex body is called *strictly convex* if its boundary does not contain a segment. We denote the set of regular normal vectors of K by $\operatorname{regn} K$.

Theorem 2.2.9. *The set of singular unit normal vectors of a convex body $K \in \mathcal{K}^n$ has $(n-1)$-dimensional Hausdorff measure zero.*

Proof. Obviously we may assume that $\dim K = n$ and $o \in \operatorname{int} K$. Let u be a singular unit normal vector of K. Then $H(K, u)$ contains a segment F. The point $x := h(K, u)^{-1}u$ is a boundary point of the polar body K^*, and $F \subset \hat{x}$ (where the circumflex refers to K^*). From Lemma 2.2.3 it follows that $N(K^*, x) \supset F$, hence x is a singular boundary point of K^*. Thus u is contained in the image of M of the set of singular boundary points of K^* under the radial projection from $\operatorname{bd} K^*$ to S^{n-1}. Since this radial projection is obviously a Lipschitz map, it follows from Theorem 2.2.4 that M has $(n-1)$-dimensional Hausdorff measure zero. ∎

It is now in order to define spherical images and reverse spherical images. Let $K \in \mathcal{K}^n$. For a subset $\beta \subset \mathbb{E}^n$ we denote by $\sigma(K, \beta)$ the set of all outer unit normal vectors of K at points of β, thus

$$\sigma(K, \beta) := \bigcup_{x \in K \cap \beta} N(K, x) \cap S^{n-1},$$

and call this the *spherical image of* K at β. Similarly, for a subset $\omega \subset S^{n-1}$ we denote by $\tau(K, \omega)$ the set of all boundary points of K at which there exists a normal vector of K belonging to ω, thus

$$\tau(K, \omega) := \bigcup_{u \in \omega} F(K, u).$$

We may call this the *reverse spherical image of* K *at* ω. The *spherical image map* of K is the map

$$\sigma_K\colon \operatorname{reg} K \to S^{n-1}$$

defined by letting $\sigma_K(x)$, for $x \in \operatorname{reg} K$, be the unique outer unit normal vector of K at x. The *reverse spherical image map* of K (not the inverse of σ_K, which does not exist in general) is the map

$$\tau_K\colon \operatorname{regn} K \to \operatorname{bd} K$$

for which $\tau_K(u)$, for $u \in \operatorname{regn} K$, is the unique point in $F(K, u)$. Let x_j, $x \in \operatorname{reg} K$ and suppose that $x_j \to x$ for $j \to \infty$. Every convergent subsequence of the sequence $(\sigma_K(x_j))_{j\in\mathbb{N}}$ converges to a normal vector of K at x, hence to $\sigma_K(x)$. Therefore the map σ_K is continuous. Similarly, the map τ_K is continuous. In particular, if K is smooth and strictly convex, then σ_K and τ_K are homeomorphisms and are inverse to each other.

Since we want to be able to speak of the Lebesgue measure of the spherical image $\sigma(K, \beta)$, we show that it exists for a sufficiently large family of sets β.

Lemma 2.2.10. *Let $K \in \mathcal{K}^n$ and let $\beta \subset \mathbb{E}^n$ be a Borel set. Then the spherical image $\sigma(K, \beta)$ of K at β is Lebesgue measurable on S^{n-1}.*

Proof. Let $\mathcal{A}$ be the family of all subsets $\beta \subset \mathbb{E}^n$ for which $\sigma(K, \beta)$ is Lebesgue measurable. First let β be closed. Let $(u_j)_{j\in\mathbb{N}}$ be a sequence in $\sigma(K, \beta)$ for which $u_j \to u$ for $j \to \infty$. For $j \in \mathbb{N}$, we can choose $x_j \in K \cap \beta$ such that $u_j \in N(K, x_j)$. Since $K \cap \beta$ is compact, a subsequence of $(x_j)_{j\in\mathbb{N}}$ converges to a point $x \in K \cap \beta$, and from $u_j \to u$ it follows that $u \in N(K, x)$, hence $u \in \sigma(K, \beta)$. Thus $\sigma(K, \beta)$ is closed and hence $\beta \in \mathcal{A}$. Next let $\beta \in \mathcal{A}$ and $\beta^c := \mathbb{E}^n \setminus \beta$. Suppose that $u \in \sigma(K, \beta) \cap \sigma(K, \beta^c)$. Then $F(K, u)$ contains a point of β and a point of β^c, hence u is a singular normal vector of K. From Theorem 2.2.9 we deduce that

$$\mathcal{H}^{n-1}(\sigma(K, \beta) \cap \sigma(K, \beta^c)) = 0.$$

Hence the set $\sigma(K, \beta^c)$ differs from the set $S^{n-1} \setminus \sigma(K, \beta)$, which is Lebesgue measurable, only by a set of (spherical Lebesgue) measure zero. It follows that $\sigma(K, \beta^c)$ is Lebesgue measurable and thus $\beta^c \in \mathcal{A}$. Finally, it is clear that $\sigma(K, \bigcup_i \beta_i) = \bigcup_i \sigma(K, \beta_i)$ for any family (β_i). We have shown that $\mathcal{A}$ is a σ-algebra containing the closed sets. Hence, $\mathcal{A}$ contains all Borel subsets of $\mathbb{E}^n$, as we set out to prove. ■

To formulate an analogous result for the reverse spherical image $\tau(K, \omega)$ we define the *radial map*

$$f\colon \mathbb{E}^n \backslash \{o\} \to S^{n-1} \quad \text{by} \quad f(x) := \frac{x}{|x|}.$$

Lemma 2.2.11. *Let $K \in \mathcal{K}_0^n$ with $o \in \operatorname{int} K$ and let $\omega \subset S^{n-1}$ be a Borel set. Then $f(\tau(K, \omega))$ is Lebesgue measurable on S^{n-1}.*

The proof, which uses Theorem 2.2.4, is so similar to that of Lemma 2.2.10 that we can omit it.

Notes for Section 2.2

1. Some considerations using the first part of Theorem 2.2.1 can be found in Jongmans (1981a); for instance, the remark that the sum of a polytope and a non-smooth convex body cannot be smooth, and some related results.
2. Theorem 2.2.4 is essentially due to Anderson & Klee (1952). For $n = 3$, a different proof was given by Besicovitch (1963a). A generalization to certain non-convex sets can be found in Federer (1959, 4.15(3) on p. 447). A weaker form of Theorem 2.2.4 was claimed without proof by Favard (1933a, p. 228). That the set of singular boundary points of a convex body $K \in \mathcal{K}_0^n$ is of $(n-1)$-dimensional Hausdorff measure zero was first proved by Reidemeister (1921); see also Aleksandrov [1], (Chapter V, §2) for an equivalent result.

   A stronger, and in a sense optimal, version of Theorem 2.2.4 was obtained by Zajíček (1979), who showed that the countably many sets covering the r-singular points can be chosen to be of a more special type.

   Theorem 2.2.4 implies, in particular, that the set of regular points of a closed convex set with interior points is dense in the boundary of the set. This is also true in arbitrary separable Banach spaces (Mazur 1933; see also Marti [25], p. 112).
3. The special case of Theorem 2.2.5 for one-dimensional faces was first proved by Fujiwara (1916). The general case appears (with different proofs) in Bourbaki [10] (Chapter IV, §1, Exercise 2e) and in Brøndsted (1966). The latter author introduced the term 'perfect face', while Bourbaki speaks of 'regular' faces.

   Fedotov (1979a) called a face F of $K \in \mathcal{K}^n$ *isolated at* a point $x \in F$ if there exists a neighbourhood U of x for which

   $$U \cap \operatorname{ext}_{\dim F} K = U \cap F.$$

   If this holds, then he showed that x is an internal point of F, that F is isolated at each of its internal points and that F is a perfect face.
4. Fedotov (1978a) made a study of faces, and of extreme and exposed points in relation to normal vectors, and obtained, for example, the following results. Let $K \in \mathcal{K}_0^n$ and $\beta \subset \mathbb{E}^n$. If $x \in \exp_i K$ and β is a neighbourhood of x, then the spherical image $\sigma(K, \beta)$ of K at β contains an $(n-i-1)$-dimensional submanifold. Vice versa, if β is open and the spherical image of β contains an $(n-i-1)$-dimensional Lipschitz submanifold, then $\beta \cap \exp_i K \neq \emptyset$.
5. Lemma 2.2.10 appears (essentially) in Aleksandrov [1], (Chapter V, §2) and Lemma 2.2.11 in Aleksandrov (1937a), §2. A more detailed study of spherical images can be found in Aleksandrov (1939c), §5.

6. *Minkowski's characterization of extreme support planes.* Let $K \in \mathcal{K}_0^n$ and $u \in \mathbb{E}^n \setminus \{o\}$. For small $\delta > 0$, let $r(\delta)$ be the radius of the largest $(n-1)$-dimensional ball contained in $K \cap H_{u,h(K,u)-\delta}$. Then u is an extreme normal vector of K if and only if $r(\delta)/\delta \to \infty$ for $\delta \to 0$. This was proved by Minkowski (1911) for $n = 3$ and extended to higher dimensions by Favard (1933a).
7. *Tangential bodies.* The definition of a p-tangential body given above is essentially the one in Bonnesen & Fenchel [8], p. 14. Favard (1933), p. 273, used property (*) of Theorem 2.2.8 as a definition (and talked of a tangential body of order $n - p$). The equivalence expressed by Theorem 2.2.8 was proved in Schneider (1978c).
8. The convex cone $P(K, x)$ defined at the beginning of Section 2.2 is often denoted by cone (x, K) and called the *cone of K at x*. Above we were only interested in its closure, but of course the cone $P(K, x)$ itself carries more information about the behaviour of K at its point x. Let $C \subset \mathbb{E}^n$ be a convex cone. There exists a convex body $K \in \mathcal{K}^n$ and a point $x \in K$ such that $P(K, x) = C$ if and only if C is an F_σ-set. This was proved by Sung & Tam (1987). Their work is also related to earlier work by Larman (1974) and Brøndsted (1977) on certain cones (inner aperture, barrier cone) associated with an unbounded convex set and to a certain classification of boundary points of convex bodies by Waksman & Epelman (1976).

2.3. Segments in the boundary

We have called a convex body *strictly convex* if its boundary does not contain a segment. As the example of a polytope shows, the boundary of a convex body can contain many segments. The directions of these segments, however, make up a relatively small set; in the example it is confined to finitely many great spheres (where directions are represented by elements of the unit sphere S^{n-1}). It is easy to construct convex bodies where the set of directions of segments in the boundary is dense in S^{n-1}, but still of measure zero. The latter holds true in general, but no easy proof seems to be known. With some effort, however, the following stronger result can be shown.

Theorem 2.3.1 (Ewald–Larman–Rogers). *Let $K \in \mathcal{K}^n$ be a convex body and let $U(K) \subset S^{n-1}$ be the set of all unit vectors that are parallel to a segment in the boundary of K. Then $U(K)$ has σ-finite $(n-2)$-dimensional Hausdorff measure.*

Since the set of unit vectors parallel to a segment in a facet of K has positive $(n-2)$-dimensional Hausdorff measure and since a convex body can have infinitely many facets, the result of Theorem 2.3.1 is best possible.

Theorem 2.3.1 can be considered as one of the most remarkable results on the boundary structure of general convex bodies. For this reason, and since a generalization to be treated below will be applied in Chapter 4, we shall give the full proof, essentially following Ewald, Larman & Rogers (1970), with a variant at the end of the proof that was proposed by Zalgaller (1972).

An important tool, which is useful in other contexts, too, is the following cap-covering theorem due to Ewald, Larman and Rogers. Recall that a cap of a convex body $K \in \mathcal{K}_0^n$ is an n-dimensional set of the form $K \cap H^+$ where H^+ is a closed halfspace. For given $r, R > 0$ we denote by $\mathcal{K}_0^n(r, R)$ the set of all convex bodies $K \in \mathcal{K}^n$ satisfying

$$rB^n + x \subset K \subset RB^n + x$$

for some $x \in K$.

***Theorem* 2.3.2** (Ewald–Larman–Rogers). *For $K \in \mathcal{K}_0^n(r, R)$ there exist positive constants $c_1, \ldots, c_4$ depending only on n, r, R such that the following holds. For $0 < \varepsilon < c_1$ there exist caps $K_1, \ldots, K_m$ of K whose union covers the boundary of K and whose widths $\Delta(K_i)$ and volumes $V_n(K_i)$ satisfy*

$$c_2\varepsilon < \Delta(K_i) < c_3\varepsilon \quad \text{for } i = 1, \ldots, m,$$

$$\sum_{i=1}^{m} V_n(K_i) < c_4\varepsilon.$$

The proof will be split into a sequence of lemmas. The first of these is well known (Bonnesen & Fenchel [8], p. 53) and quoted here without proof.

***Lemma* 2.3.3.** *If $K \in \mathcal{K}_0^n$ and o is the centroid of K, then $-K \subset nK$.*

For $K \in \mathcal{K}^n$, $x \in K$ and $\lambda > 0$ the convex body

$$M(x, \lambda) := x + \lambda[(K - x) \cap (x - K)]$$

is called a *Macbeath region*.

We assume in the following that a convex body $K \in \mathcal{K}_0^n(r, R)$ is given and, without loss of generality, that

$$rB^n \subset K \subset RB^n.$$

In particular, the origin o is a point of K.

***Lemma* 2.3.4.** *If $x, y \in K$ and $M(x, \frac{1}{2}) \cap M(y, \frac{1}{2}) \neq \emptyset$, then $M(y, 1) \subset M(x, 5)$.*

Proof. A point $z \in M(x, \frac{1}{2}) \cap M(y, \frac{1}{2})$ is of the form

$$z = x + \tfrac{1}{2}(x - k_1) = y + \tfrac{1}{2}(k_2 - y) \tag{2.3.1}$$

with $k_1, k_2 \in K$. Let $w \in M(y, 1)$. Then

$$w = y + (k_3 - y) = y + (y - k_4)$$

with suitable $k_3, k_4 \in K$; hence

$$w = k_3 = x + 5[(\tfrac{4}{5}x + \tfrac{1}{5}k_3) - x] \in x + 5(K - x)$$

and, using (2.3.1),

$$\begin{aligned} w = y + (y - k_4) &= 6x - 2k_1 - 2k_2 - k_4 \\ &= x + 5[x - (\tfrac{2}{5}k_1 + \tfrac{2}{5}k_2 + \tfrac{1}{5}k_4)] \\ &\in x + 5(x - K), \end{aligned}$$

hence $w \in M(x, 5)$. ■

***Lemma* 2.3.5.** *Let $H^+ = H^+_{u,\alpha}$ be a closed halfspace meeting K but not rB^n and such that the distance between its bounding hyperplane H and the parallel support plane $H(K, u)$ is at most $\frac{1}{2}r$. Let c be the centre of gravity* (w.r.t. H) *of $K \cap H$. Then $K \cap H^+ \subset M(c, 3n)$.*

Proof. We suppose that the assertion is false. Since $K \subset c + 3n(K - c)$, there must exist a point $k \in K \cap H^+$ with $k \notin c + 3n(c - K)$. Writing $\eta = 1/(3n)$, we have

$$c - \eta(k - c) \notin K. \tag{2.3.2}$$

We may assume that $|u| = 1$. The point $k \in K \cap H^+$ can be written in the form $k = h + tu$ with $h \in H$ and $0 \leqq t \leqq \frac{1}{2}r$. We can also write $c = su + z$ with $\langle z, u \rangle = 0$. Since $rB^n \subset \mathbb{E}^n \setminus H^+$, we have $s \geqq r \geqq 2t$. Define

$$v := \frac{s}{s + t}(h - c) = \frac{t}{s + t}z;$$

then $\langle v, u \rangle = 0$ and

$$c + v = \frac{s}{s + t}k \in K \cap H. \tag{2.3.3}$$

On the other hand, using (2.3.3),

$$c - \frac{\eta(s + t)}{s - \eta t}v = \frac{s}{s - \eta t}[c - \eta(k - c)]$$

belongs to H but not to K, by (2.3.2) and $\eta t < \frac{1}{2}r < s$. Since c is the centre of gravity of $K \cap H$, Lemma 2.3.3 now implies

$$\frac{\eta(s + t)}{s - \eta t} > \frac{1}{n - 1} > \frac{1}{n}.$$

Using $\eta t \leqq t \leqq \frac{1}{2}r \leqq \frac{1}{2}s$, we arrive at

$$\frac{1}{3n} = \eta > \frac{1}{n}\frac{s - \eta t}{s + t} \geqq \frac{1}{n}\frac{s - \frac{1}{2}s}{s + \frac{1}{2}s} = \frac{1}{3n},$$

a contradiction. ■

For $0<\varepsilon<r$ we put

$$K_{-\varepsilon} := \bigcap_{u\in S^{n-1}} [H^-(K, u) - \varepsilon u]$$

(this is the inner parallel body of K at distance ε, cf. Section 3.1).

Lemma 2.3.6. *If* $0<\varepsilon<\frac{1}{4}r^2R^{-1}$, *then*

$$(1 - 4r^{-2}R\varepsilon)K \subset K_{-\varepsilon} \subset K$$

and

$$V_n(K\backslash K_{-\varepsilon}) < 4nr^{-2}RV_n(K)\varepsilon.$$

Proof. Let $x \in \operatorname{bd} K_{-\varepsilon}$. The non-negative function $h(K-x,\cdot)-\varepsilon$ attains a minimum on S^{n-1}, say at u. This minimum is equal to 0, since otherwise $x \in \operatorname{int} K_{-\varepsilon}$. Hence $x \in H(K,u)-\varepsilon u$ and, therefore, $|x| \geqq \langle x, u\rangle = h(K,u)-\varepsilon \geqq r-\varepsilon > \frac{1}{2}r$ (observe that $\varepsilon<\frac{1}{4}r$ since $r<R$). Since $|x|<R$, the angle θ between u and the vector x satisfies $\cos\theta \geqq r/(2R)$. Let $x'=\xi x$ with $\xi>0$ and $x' \in \operatorname{bd} K$. Then $|x'-x|\cos\theta \leqq \varepsilon$ and hence $|x'-x|/|x| \leqq 4r^{-2}R\varepsilon$. Thus

$$(1 - 4r^{-2}R\varepsilon)K \subset K_{-\varepsilon} \subset K$$

and

$$\begin{aligned} V_n(K\backslash K_{-\varepsilon}) &\leqq V_n(K\backslash(1 - 4r^{-2}R\varepsilon)K) \\ &= [1-(1-4r^{-2}R\varepsilon)^n]V_n(K) \\ &< 4nr^{-2}R\varepsilon V_n(K). \end{aligned}$$

■

Lemma 2.3.7. *Let* $0<\varepsilon<r$ *and* $x \in \operatorname{bd} K_{-\varepsilon}$. *Then*

$$B(x, \varepsilon) \subset M(x, 1) \quad \textit{and} \quad \Delta(M(x, 1)) = 2\varepsilon.$$

If $\varepsilon<(R/n)[r/(2R)]^{2n}$, *there is a cap* K° *of* K, *bounded by a hyperplane through* x, *so that*

$$K^\circ \subset M(x, 3n).$$

Proof. From the definition of $K_{-\varepsilon}$ it is clear that $B(x,\varepsilon)\subset K$ and thus $B(x,\varepsilon)\subset M(x,1)$. This implies $\Delta(M(x,1)) \geqq 2\varepsilon$. With u chosen as in the proof of the preceding lemma, we have

$$M(x, 1) \subset H^-(K, u) \cap [x - (H^-(K, u) - x)].$$

The right-hand side is a strip bounded by two parallel hyperplanes at a distance 2ε apart, hence $\Delta(M(x,1))=2\varepsilon$.

Among all caps of K bounded by hyperplanes through x, there is one with minimal volume, say K°. The special cap $K\cap[H^+(K,u)-\varepsilon u]$, with u as above, can be covered by a cylinder of height ε whose base is an $(n-1)$-ball of radius R, hence

$$V_n(K^\circ) \leqq \kappa_{n-1} R^{n-1} \varepsilon. \tag{2.3.4}$$

There is a point $y \in K^\circ$ lying in a support hyperplane $\bar{H}$ of K parallel to the bounding hyperplane H of K° through x. Let t be the distance of x from $\bar{H}$ and τ the distance of o from $\bar{H}$. The convex body K contains the convex hull of y and the intersection of rB^n with the hyperplane through o parallel to $\bar{H}$. The cap K° contains a set similar to this convex hull, with similarity ratio t/τ. Hence,

$$V_n(K^\circ) \geqq \left(\frac{t}{\tau}\right)^n \frac{1}{n} \kappa_{n-1} r^{n-1} \tau,$$

which together with (2.3.4) and $\tau \leqq R$ yields

$$t^n \leqq \frac{nR^{2n-2}}{r^{n-1}} \varepsilon.$$

If now $\varepsilon < (R/n)[r/(2R)]^{2n}$, this yields $t < \frac{1}{4}r$. In particular, the hyperplane H does not meet the ball $\frac{1}{2}rB^n$. From the minimum property of K° it follows that x is the centre of gravity of $K \cap H$; otherwise a suitable small rotation of H around x would yield a cap with smaller volume. We can now apply Lemma 2.3.5, with r replaced by $\frac{1}{2}r$, to conclude that $K^\circ \subset M(x, 3n)$. ■

We are now in a position to prove the cap covering theorem.

Proof of Theorem 2.3.2. Put $c_1 = (R/n)[r/(2R)]^{2n}$, the constant appearing in Lemma 2.3.7, and let $0 < \varepsilon < c_1$. Choose points $x_1, \dots, x_m \in \operatorname{bd} K_{-\varepsilon}$ with

$$M(x_i, \tfrac{1}{2}) \cap M(x_j, \tfrac{1}{2}) = \emptyset \quad \text{for } i \neq j \tag{2.3.5}$$

and such that m is maximal (clearly a finite m exists, since $V_n(M(x_i, \frac{1}{2})) \geqq V_n(B(x_i, \frac{1}{2}\varepsilon))$ by Lemma 2.3.7).

Let $z \in \operatorname{bd} K$, and let $y \in [\mathrm{o}, z] \cap \operatorname{bd} K_{-\varepsilon}$. As in the proof of Lemma 2.3.6 and by the choice of ε,

$$|z - y| < 4r^{-2}R\varepsilon|y| < 4r^{-2}R^2\varepsilon < \tfrac{1}{4}r,$$

which implies that $y - (z - y) \in K$ and hence $z \in M(y, 1)$. By the maximality property of $x_1, \dots, x_m$, there is an index $i \in \{1, \dots, m\}$ for which

$$M(x_i, \tfrac{1}{2}) \cap M(y, \tfrac{1}{2}) \neq \emptyset.$$

Lemma 2.3.4 now shows that $z \in M(y, 1) \subset M(x_i, 5)$. Since $z \in \operatorname{bd} K$ was arbitrary, we have proved that

$$\operatorname{bd} K \subset \bigcup_{i=1}^{m} M(x_i, 5). \tag{2.3.6}$$

Let $i \in \{1, \ldots, m\}$. The point x_i lies in the boundary of a supporting halfspace H_i^- of $K_{-\varepsilon}$. Clearly

$$M(x_i, \tfrac{1}{2}) \setminus H_i^- \subset K \setminus K_{-\varepsilon}$$

and hence, by (2.3.5),

$$\sum_{i=1}^{m} V_n(M(x_i, \tfrac{1}{2}) \setminus H_i^-) \leqq V_n(K \setminus K_{-\varepsilon}).$$

Since $M(x_i, \frac{1}{2})$ is symmetric with respect to the point x_i, we have $V_n(M(x_i, \frac{1}{2}) \setminus H_i^-) = \frac{1}{2} V_n(M(x_i, \frac{1}{2}))$ and conclude from Lemma 2.3.6 that

$$\sum_{i=1}^{m} V_n(M(x_i, \tfrac{1}{2})) < 8nr^{-2} R V_n(K)\varepsilon. \qquad (2.3.7)$$

By Lemma 2.3.7, for each $i \in \{1, \ldots, m\}$ a cap K_i° of K, bounded by a hyperplane through x_i, can be found such that

$$K_i^\circ \subset M(x_i, 3n). \qquad (2.3.8)$$

There is a supporting halfspace H_{u_i, α_i}^- of K so that

$$K_i^\circ = K \cap H_{u_i, \alpha_i - t_i}^+$$

with suitable $t_i > 0$. We define the cap K_i of K by

$$K_i := K \cap H_{u_i, \alpha_i - 6t_i}^+.$$

From

$$x_i - 5(K - x_i) \subset x_i - 5(H_{u_i, \alpha_i}^- - x_i) = H_{u_i, \alpha_i - 6t_i}^+$$

it follows that $K \cap M(x_i, 5) \subset K_i$ and thus

$$\mathrm{bd}\, K \subset \bigcup_{i=1}^{m} K_i$$

by (2.3.6).

Choose a point $y_i \in K \cap H_{u_i, \alpha_i}$. From the convexity of K and from (2.3.8) we infer that

$$\begin{aligned} K_i &\subset y_i + 6(K_i^\circ - y_i) \subset y_i + 6[M(x_i, 3n) - y_i] \\ &= M(x_i, 18n) + 5x_i - 5y_i. \end{aligned}$$

From Lemma 2.3.7 it follows that $\Delta(K_i) \leqq 36n\varepsilon$, and since $B(x_i, \varepsilon) \subset M(x_i, 1) \subset K_i$, we have $\Delta(K_i) \geqq 2\varepsilon$. Finally,

$$\begin{aligned} \sum_{i=1}^{m} V_n(K_i) &\leqq \sum_{i=1}^{m} V_n(M(x_i, 18n)) = (36n)^n \sum_{i=1}^{m} V_n(M(x_i, \tfrac{1}{2})) \\ &< (36n)^n 8nr^{-2} R V_n(K)\varepsilon \end{aligned}$$

by (2.3.7), which completes the proof of Theorem 2.3.2. ■

The proof of Theorem 2.3.1 requires two more lemmas. For the first one, a rectangular parallelepiped is referred to as a *box*.

***Lemma* 2.3.8.** *A convex body $K \in \mathcal{K}_0^n$ can be covered by a box P with smallest edge length $\Delta(K)$ and with volume $V_n(P) \leqq n!\, V_n(K)$.*

Proof. We use induction on n. The case $n = 1$ being trivial, assume that $n \geqq 2$ and the assertion has been proved in dimension $n - 1$. The convex body of width $\Delta(K)$ has two parallel supporting hyperplanes H_0, H_1 of mutual distance $\Delta(K)$. Let π denote the orthogonal projection from $\mathbb{E}^n$ onto H_0. There must be a point $x_0 \in K \cap H_0 \cap \pi(K \cap H_1)$, since otherwise $K \cap H_0$ and $\pi(K \cap H_1)$ could be strongly separated by an $(n - 2)$-plane E in H_0, and small rotations of H_0 and H_1 around E, respectively $H_1 \cap \pi^{-1}(E)$, would yield a parallel strip containing K, of width strictly less than $\Delta(K)$. Let $x_1 \in K \cap H_1$ be the point for which $\pi x_1 = x_0$ and let C be the convex hull of πK and x_1. For $y \in \pi K$, let $y_1 \in \operatorname{relbd} \pi K$ be such that $y \in [x_0, y_1]$. If $z \in \pi^{-1}(y_1) \cap K$, then $\operatorname{conv}\{z, x_0, x_1\} \subset K$, and the intersections of the line $\pi^{-1}(y)$ with this triangle and with C have the same length. By Cavalieri's principle, $V_n(K) \geqq V_n(C) = 1/n\Delta(K)V_{n-1}(\pi K)$, where V_{n-1} denotes $(n - 1)$-dimensional volume. By the induction hypothesis, πK can be covered by an $(n - 1)$-dimensional box P' of smallest edge length $\Delta(\pi K_1) \geqq \Delta(K)$ and $(n - 1)$-volume $V_{n-1}(P') \leqq (n - 1)!\, V_{n-1}(\pi K)$. Hence, K can be covered by a box P of smallest edge length $\Delta(K)$ and volume $V_n(P) = \Delta(K)V_{n-1}(P') \leqq \Delta(K)(n - 1)!\, V_{n-1}(\pi K) \leqq n!\, V_n(K)$. ■

***Lemma* 2.3.9.** *Let $K_0, K_1 \in \mathcal{K}_0^n$. Let $\Lambda(K_0, K_1)$ be the set of all differences $x_1 - x_0$ of boundary points x_0 of K_0 and x_1 of K_1 lying in parallel support planes, thus*

$$\Lambda(K_0, K_1) = \bigcup_{u \in S^{n-1}} [F(K_1, u) - F(K_0, u)].$$

If $C_1, \ldots, C_m$ are caps of $K_0 + K_1$ covering the boundary of $K_0 + K_1$, then

$$\Lambda(K_0, K_1) \subset \bigcup_{i=1}^{m} (\mathrm{D}C_i + a_i),$$

where $\mathrm{D}C_i$ is the difference body of C_i and a_i is a suitable translation vector.

Proof. For $i \in \{1, \ldots, m\}$ we can write $C_i = (K_0 + K_1) \cap H^+_{u_i, \alpha_i - \tau_i}$, where H_{u_i, α_i} is a support plane of $K_0 + K_1$, with exterior unit normal vector u_i, and $\tau_i > 0$. Let $\alpha_i^\nu := h(K_\nu, u_i)$ and $C_i^\nu := K_\nu \cap H^+_{u_i, \alpha_i^\nu - \tau_i}$ for $\nu = 0, 1$. A given element of $\Lambda(K_0, K_1)$ is of the form $x_1 - x_0$ with $x_\nu \in F(K_\nu, v)$ for suitable $v \in S^{n-1}$. Then $x_0 + x_1 \in$

$F(K_0 + K_1, v) \subset \mathrm{bd}(K_0 + K_1)$, hence there is an index $j \in \{1, \dots, m\}$ for which $x_0 + x_1 \in C_j$. We have

$$\alpha_j - \tau_j \leqq \langle x_0 + x_1, u_j \rangle \leqq \alpha_j,$$
$$\langle x_\nu, u_j \rangle \leqq \alpha_j^\nu \quad \text{for } \nu = 0, 1.$$

From $\langle x_\nu, u_j \rangle < \alpha_j^\nu - \tau_j$ for $\nu = 0$ or 1 it would follow that

$$\langle x_0 + x_1, u_j \rangle < \alpha_j^0 + \alpha_j^1 - \tau_j = \alpha_j - \tau_j,$$

a contradiction. Thus $\alpha_j^\nu - \tau_j \leqq \langle x_\nu, u_j \rangle$ and, therefore, $x_\nu \in C_j^\nu$ for $\nu = 0, 1$. This shows that

$$x_1 - x_0 \in \bigcup_{i=1}^{m} (C_i^1 - C_i^0). \tag{2.3.9}$$

If $b_i^\nu \in F(K_\nu, u_i)$, then $C_i^0 + b_i^1 \subset C_i$ and $C_i^1 + b_i^0 \subset C_i$, hence

$$C_i^1 - C_i^0 \subset DC_i + b_i^1 - b_i^0,$$

which together with (2.3.9) proves the assertion. ■

Proof of Theorem 2.3.1. Let $K \in \mathcal{K}^n$ be given. Since each boundary segment of K is contained in a support set $F(K, u)$ and since $F(K + B^n, u) = F(K, u) + u$, we may assume from the beginning that $K = K' + B^n$ with a convex body K'; thus K is 1-smooth. (A convex body $L \in \mathcal{K}^n$ is called *η-smooth* if to each $x \in \mathrm{bd}\, L$ there is a vector t such that $x \in \eta B^n + t \subset L$.)

Let Π_0, Π_1 be a pair of parallel hyperplanes in $\mathbb{E}^n$. We say that a vector $u \in S^{n-1}$ is *associated with* (K, Π_0, Π_1) if the boundary of K contains a segment parallel to u that intersects Π_0 and Π_1. Let $\mathcal{R}$ denote the set of all hyperplanes in $\mathbb{E}^n$ that are spanned by points with rational coordinates. We fix a pair Π_0, Π_1 of parallel hyperplanes in $\mathcal{R}$ meeting the interior of K. Let U^* be the set of all $u \in S^{n-1}$ that are associated with (K, Π_0, Π_1). Since there are only countably many possible choices for Π_0, Π_1 and since the union of the sets U^* corresponding to these different choices covers the set $U(K)$ of Theorem 2.3.1, it suffices to show that U^* has finite $(n-2)$-dimensional Hausdorff measure.

Since directions of boundary segments remain unchanged under homotheties, we may assume that $\Pi_0 = H_{w,-1}$ and $\Pi_1 = H_{w,1}$ with $w \in S^{n-1}$. Let $K_0 = K \cap \Pi_0$ and $K_1 = K \cap \Pi_1$. In the following, c_1, $c_2, \dots$ denote constants depending only on n, K_0, K_1, but not on ε. By Λ we denote the set of all differences $x_1 - x_0$ where $x_0 \in K_0$, $x_1 \in K_1$ and x_0 and x_1 lie in a support plane of K. Then $U^* = \{y/|y| \,|\, y \in \Lambda \cup (-\Lambda)\}$. Since U^* is the image of $\Lambda \cup (-\Lambda)$ under a Lipschitz map, it suffices to show that Λ has finite $(n-2)$-dimensional Hausdorff measure.

We apply Theorem 2.3.2 to the $(n-1)$-dimensional convex body $K_0 + K_1$, with $\mathbb{E}^n$ replaced by $\operatorname{aff}(K_0 + K_1) = H_{w,0}$. Let $c_1, \ldots, c_4$ be the constants appearing in that theorem (they now depend only on n, K_0, K_1). Let $0 < \varepsilon < \sqrt{c_1}$ be given. There exist $(n-1)$-dimensional caps $C_1, \ldots, C_m$ of $K_0 + K_1$ covering the boundary of $K_0 + K_1$ and such that the widths $\Delta'(K_i)$ and volumes $V_{n-1}(K_i)$, computed in $H_{w,0}$, satisfy

$$c_2\varepsilon^2 < \Delta'(C_i) < c_3\varepsilon^2, \tag{2.3.10}$$

$$\sum_{i=1}^{m} V_{n-1}(C_i) < c_4\varepsilon^2. \tag{2.3.11}$$

Let $x_1 - x_0 \in \Lambda$, with $x_0 \in K_0$ and $x_1 \in K_1$ lying in a support plane of K. Then $x_1 - w$ and $x_0 + w$ lie in parallel support planes (with respect to $H_{w,0}$) of $K_1 - w$ and $K_0 + w$, respectively. From Lemma 2.3.9 we now infer that

$$\Lambda \subset \bigcup_{i=1}^{m} (\mathrm{D}C_i + t_i) \tag{2.3.12}$$

with suitable translation vectors t_i.

Let $i \in \{1, \ldots, m\}$. The difference body $\mathrm{D}C_i$ has width $\Delta'(\mathrm{D}C_i) = 2\Delta'(C_i)$ and $(n-1)$-volume

$$V_{n-1}(\mathrm{D}C_i) \leqq n^{n-1} V_{n-1}(C_i),$$

since $\mathrm{D}C_i = C_i - C_i$ is contained in a translate of $C_i + (n-1)C_i$, according to Lemma 2.3.3. By Lemma 2.3.8, $\mathrm{D}C_i$ can be covered by an $(n-1)$-dimensional box P_i with smallest edge length $w_1 = \Delta'(\mathrm{D}C_i)$ and with volume $V_{n-1}(P_i) \leqq (n-1)!\, V_{n-1}(\mathrm{D}C_i)$.

We have assumed that K is 1-smooth. Since Π_0, Π_1 meet the interior of K, it follows that K_0 and K_1, and thus also $K_0 + K_1$, are η-smooth for some $\eta > 0$. Hence, each cap C_i contains a cap, of the same width, cut from an $(n-1)$-ball of radius η. Now it follows from (2.3.10) that the edge lengths $w_1 \leqq w_2 \leqq \ldots \leqq w_{n-1}$ of the box P_i satisfy $w_2 > c_5\varepsilon$. The box P_i can be covered by N_i $(n-1)$-dimensional cubes of edge length ε, where

$$\begin{aligned} N_i &\leqq \prod_{k=1}^{n-1}\left(\frac{w_k}{\varepsilon} + 1\right) \leqq \varepsilon^{-n} w_1 \cdots w_{n-1}\left(\varepsilon + \frac{\varepsilon^2}{\Delta'(\mathrm{D}C_i)}\right)\left(1 + \frac{1}{c_5}\right)^{n-2} \\ &\leqq c_6\varepsilon^{-n} V_{n-1}(P_i) \leqq c_7\varepsilon^{-n} V_{n-1}(\mathrm{D}C_i) \leqq c_8\varepsilon^{-n} V_{n-1}(C_i). \end{aligned}$$

By (2.3.12), the set Λ can be covered by

$$N \leqq c_8\varepsilon^{-n} \sum_{i=1}^{m} V_{n-1}(C_i) \leqq c_9\varepsilon^{2-n}$$

cubes of edge length ε, using (2.3.11). Since this holds for all sufficiently small numbers $\varepsilon > 0$, the set Λ has finite $(n-2)$-dimensional Hausdorff measure. This completes the proof of Theorem 2.3.1. ■

Theorem 2.3.1 can be considered as a special case of a more general theorem referring to corresponding segments in the boundaries of a pair of convex bodies. We denote by $SO(n)$ the group of proper rotations of $\mathbb{E}^n$, equipped with its usual topology and differentiable structure. The normalized Haar measure on $SO(n)$ is denoted by ν. On the compact Lie group $SO(n)$ we can choose a bi-invariant Riemannian metric. This induces, in the usual way, a metric (distance function), which we denote by d_1. Hausdorff measures on $SO(n)$ refer to this distance function. (The σ-finiteness of Hausdorff measures, in which we are interested, does not depend on the choice of the Riemannian metric, since any two continuous Riemannian metrics on $SO(n)$ induce equivalent distance functions.)

Theorem 2.3.10. *Let $K, K' \in \mathcal{K}^n$ be convex bodies. The set $U(K, K')$ of all rotations $\rho \in SO(n)$ for which K and $\rho K'$ contain parallel segments lying in parallel supporting hyperplanes has σ-finite $[\frac{1}{2}n(n-1)-1]$-dimensional Hausdorff measure.*

It is not difficult to obtain Theorem 2.3.1 from this result by taking K' equal to a segment. Theorem 2.3.1 is interesting in itself as a strong result on the boundary structure of convex bodies. The following corollary of Theorem 2.3.10 will be needed in Section 4.5.

Corollary 2.3.11. *Let $K \in \mathcal{K}^n$. For $K' \in \mathcal{K}^n$, the set $U(K, K')$ of Theorem* 2.3.10 *has ν-measure zero.*

Let E be a k-dimensional linear subspace of $\mathbb{E}^n (k \in \{1, \dots, n-1\})$. The set of all rotations $\rho \in SO(n)$ for which ρE is parallel to some supporting k-flat of K that contains more than one point of K is of ν-measure zero.

Recall that a supporting flat of K is an affine subspace of $\mathbb{E}^n$ contained in a supporting hyperplane of K and having nonempty intersection with K. The first part of the corollary is true since Theorem 2.3.10 implies that $U(K, K')$ has $\frac{1}{2}n(n-1)$-dimensional Hausdorff measure zero and hence ν-measure zero, because $\frac{1}{2}n(n-1)$ is the dimension of $SO(n)$. The second part is obtained by taking $K' = E \cap B^n$.

Although Corollary 2.3.11 is much weaker than Theorem 2.3.10, no simpler proof of the former seems to be known. The proof of Theorem 2.3.10 requires some preliminaries.

A pair of orthogonal unit vectors in $\mathbb{E}^n$ will be called a *2-frame*. We denote by $V_{n,2}$ the set of all 2-frames in $\mathbb{E}^n$. For $(u, v) \in V_{n,2}$ and $\rho \in SO(n)$ we write $\rho(u, v) = (\rho u, \rho v)$. On $V_{n,2}$ a metric τ is introduced by

$$\tau((u_1, v_1), (u_2, v_2)) := |u_1 - u_2| + |v_1 - v_2|.$$

In the following, $c_{10}, c_{11}, \ldots$ denote positive constants independent of ε (but c_k may depend on those c_j with $j < k$). The assertions below are understood to hold for all sufficiently small $\varepsilon > 0$.

Lemma 2.3.12. *Let $A, B \subset V_{n,2}$ be sets of diameter $c_{10}\varepsilon$. The set $X(A, B)$ of all rotations $\rho \in SO(n)$ for which $A \cap \rho B \neq \emptyset$ can be covered by less than $c_{14}\varepsilon^{-(n-2)(n-3)/2}$ balls, in $(SO(n), d_1)$, of diameter ε.*

Proof. First we assume that $a, b \in V_{n,2}$ are 2-frames such that $\tau(a, b) \leqq c_{10}\varepsilon$, and we assert that there exists a rotation $\rho \in SO(n)$ in a $c_{12}\varepsilon$-neighbourhood of the neutral element e such that $b = \rho a$.

We embed $SO(n)$ into $\mathbb{R}^{n^2}$ by associating with each $\rho \in SO(n)$ the entries, in the usual order, of its matrix with respect to a given orthonormal basis $(b_1, \ldots, b_n)$ of $\mathbb{E}^n$. The standard Euclidean metric on $\mathbb{R}^{n^2}$ induces a second metric d_2 on $SO(n)$. It also induces a Riemannian metric on the smooth embedded submanifold $SO(n)$, and this induces another metric d_3 on $SO(n)$. Since the submanifold $SO(n)$ is compact, the metrics d_2 and d_3 are equivalent, that is, there exist constants γ_1, γ_2 such that $\gamma_1 d_2(\rho, \sigma) \leqq d_3(\rho, \sigma) \leqq \gamma_2 d_2(\rho, \sigma)$ for ρ, $\sigma \in SO(n)$. Also, the metrics d_1 and d_3 are equivalent, since both are induced by continuous Riemannian metrics.

Since the metric τ is invariant under the operation of $SO(n)$ and since d_1 is bi-invariant, we may, without loss of generality, assume that $b = (b_1, b_2)$. Let $a = (a_1, a_2)$. Applying the Gram–Schmidt orthonormalization process to the n-tuple $(a_1, a_2, b_3, \ldots, b_n)$ (which is linearly independent if ε is sufficiently small) and using $|a_1 - b_1| + |a_2 - b_2| \leqq c_{10}\varepsilon$, we get an orthonormalized n-tuple $(a_1, \ldots, a_n)$ satisfying $|a_i - b_i| \leqq c_{11}\varepsilon$ for $i = 1, \ldots, n$. If we take the coordinate vectors (with respect to the basis $b_1, \ldots, b_n$) of $a_1, \ldots, a_n$ as the row vectors of a matrix, then this matrix represents a rotation $\rho \in SO(n)$ for which $\rho a = b$. By definition of the metric d_2 we have

$$d_2(\rho, e) = (|a_1 - b_1|^2 + \ldots + |a_n - b_n|^2)^{1/2} \leqq \sqrt{n}\, c_{11}\varepsilon$$

and hence $d_1(\rho, e) \leqq c_{12}\varepsilon$.

To prove the lemma, now let $a \in A, b \in B$. The set M of all $\sigma \in SO(n)$ for which $a = \sigma b$ is a left coset of the isotropy subgroup of b, which can be identified with $SO(n-2)$, hence M is a submanifold of dimension $\frac{1}{2}(n-2)(n-3)$. Let $\rho \in X(A, B)$. Then there are elements

$a' \in A$ and $b' \in B$ for which $a' = \rho b'$. By assumption we have $\tau(a, a') \leqq c_{10}\varepsilon$ and $\tau(b, b') \leqq c_{10}\varepsilon$. As shown above, there exist rotations ρ_1, ρ_2 in a $c_{12}\varepsilon$-neighbourhood of e such that $a' = \rho_1 a$ and $b' = \rho_2 b$. It follows that $a = \rho_1^{-1}\rho\rho_2 b$, hence $\mu := \rho_1^{-1}\rho\rho_2 \in M$. Since the metric d_1 is bi-invariant, we get

$$\begin{aligned} d_1(\rho, \mu) &= d_1(\rho_1\mu\rho_2^{-1}, \mu) = d_1(\rho_1\mu, \mu\rho_2) \\ &\leqq d_1(\rho_1\mu, \mu) + d_1(\mu, \mu\rho_2) = d_1(\rho_1, e) + d_1(e, \rho_2) \\ &\leqq 2c_{12}\varepsilon. \end{aligned}$$

Thus $X(A, B)$ is contained in a $2c_{12}\varepsilon$-neighbourhood of M. The smooth, compact submanifold M of dimension $\frac{1}{2}(n-2)(n-3)$ can be covered by less than $c_{13}\varepsilon^{-(n-2)(n-3)/2}$ balls of diameter ε (for a proof, see Flatto & Newman (1977), Theorem 2.1). Then $X(A, B)$ can be covered by less than $c_{14}\varepsilon^{-(n-2)(n-3)/2}$ balls of diameter ε. ∎

The proof of Theorem 2.3.10 is an extension of the one given for Theorem 2.3.1, and we use the same notation, as well as some results established in the course of that proof.

Proof of Theorem 2.3.10. Let $K, K' \in \mathcal{K}^n$ be given. Both bodies can be assumed to be 1-smooth. Let Π_0, Π_1 be a pair of parallel hyperplanes in $\mathbb{E}^n$. The 2-frame (u, v) is said to be *associated with* (K, Π_0, Π_1) if the support set $F(K, v)$ contains a segment parallel to u that intersects Π_0 and Π_1 and if $\langle u, w\rangle \geqq \frac{1}{2}$, where w is the unit vector orthogonal to Π_0 and pointing from Π_0 to Π_1. Let $\rho \in U(K, K')$. Then there exists a 2-frame (u, v) such that $F(K, v)$ and $F(\rho K', v)$ each contain a segment parallel to u. Hence, there exist four hyperplanes Π_0, Π_1, Π_0', Π_1' in $\mathcal{R}$ such that Π_0 and Π_1 are parallel, say at distance a, Π_0' and Π_1' are parallel and at the same distance a, (u, v) is associated with (K, Π_0, Π_1) and $(\rho^{-1}u, \rho^{-1}v)$ is associated with (K', Π_0', Π_1').

We fix a pair Π_0, Π_1 of parallel hyperplanes in $\mathcal{R}$ that meet the interior of K, and a pair Π_0', Π_1' of parallel hyperplanes in $\mathcal{R}$ at the same distance that meet the interior of K'. Let $U^*(K, K')$ be the set of all rotations $\rho \in SO(n)$ for which there exists a 2-frame (u, v) associated with (K, Π_0, Π_1) and such that $(\rho^{-1}u, \rho^{-1}v)$ is associated with (K', Π_0', Π_1'). Clearly it suffices to show that $U^*(K, K')$ has finite $[\frac{1}{2}n(n-1) - 1]$-dimensional Hausdorff measure.

We may assume that $\Pi_0 = H_{w,-1}$ and $\Pi_1 = H_{w,1}$. In the following, $c_{15}, c_{16}, \ldots$ denote positive constants depending only on n, on $K_\nu = K \cap \Pi_\nu$ and on $K_\nu' = K' \cap \Pi_\nu'$ $(\nu = 0, 1)$. Let V^* be the set of all 2-frames associated with (K, Π_0, Π_1). If $(u, v) \in V^*$, then there are points x_0, x_1 satisfying

$$\begin{aligned} x_0 \in K_0,\ x_1 &\in K_1, \\ x_0,\ x_1 &\in F(K, v), \\ u &= (x_1 - x_0)/|x_1 - x_0| \end{aligned} \tag{2.3.13}$$

The point $x_0 + x_1$ lies in one of the caps $C_1, \ldots, C_m$ used in the proof of Theorem 2.3.1, say $x_0 + x_1 \in C_i$. Then $x_1 - x_0 \in \mathrm{D}C_i + t_i$ with some vector t_i (compare the proof of Lemma 2.3.9). The set $\mathrm{D}C_i$ was covered by N_i $(n-1)$-dimensional cubes of edge length ε, say by $W_1^{(i)}, \ldots, W_{N_i}^{(i)}$. We denote by V_{ij}^* the set of all $(u, v) \in V^*$ for which points x_0, x_1 satisfying (2.3.13) exist such that $x_0 + x_1 \in C_i$ and $x_1 - x_0 \in W_j^{(i)} + t_i$. Then V^* is the union of all these V_{ij}^*.

For $x \in \mathbb{E}^n$, we denote by $\bar{x}$ the orthogonal projection of x onto the linear subspace $H_{w,0}$. Let $u_1, u_2 \in S^{n-1}$, $\langle u_\nu, w\rangle \geqq \frac{1}{2}$ and $x_\nu \in H_{u_\nu,0}$ for $\nu = 1, 2$. Then

$$x_\nu = \bar{x}_\nu - \frac{\langle \bar{x}_\nu, u_\nu\rangle}{\langle w, u_\nu\rangle} w$$

and hence

$$\begin{aligned} |x_1 - x_2| &\leqq |\bar{x}_1 - \bar{x}_2| + \left|\frac{\langle \bar{x}_1, u_1 - u_2\rangle}{\langle w, u_1\rangle}\right| \\ &\quad + \left|\frac{\langle \bar{x}_1, u_2\rangle}{\langle w, u_1\rangle} - \frac{\langle \bar{x}_1, u_2\rangle}{\langle w, u_2\rangle}\right| + \left|\frac{\langle \bar{x}_1 - \bar{x}_2, u_2\rangle}{\langle w, u_2\rangle}\right| \\ &\leqq 3|\bar{x}_1 - \bar{x}_2| + 6|\bar{x}_1||u_1 - u_2|. \end{aligned} \tag{2.3.14}$$

Now let $(u_1, v_1), (u_2, v_2) \in V_{ij}^*$. Then

$$|u_1 - u_2| \leqq \operatorname{diam} W_j^{(i)} \leqq \sqrt{n-1}\,\varepsilon.$$

The vectors $\bar{v}_1$, $\bar{v}_2$ are exterior normal vectors of $K_0 + K_1$ at points belonging to the same cap C_i. As in the proof of Theorem 2.3.1, this cap contains a cap, of the same width $\Delta'(C_i) < c_3\varepsilon^2$, that is cut, by the same $(n-2)$-plane, from an $(n-1)$-ball of some positive radius η. It follows that the angle between $\bar{v}_1$ and $\bar{v}_2$ is less than $c_{15}\varepsilon$. Let $\lambda_1, \lambda_2 > 0$ be such that $\lambda_1\bar{v}_1$, $\lambda_2\bar{v}_2$ are unit vectors, then

$$|\lambda_1\bar{v}_1 - \lambda_2\bar{v}_2| \leqq c_{15}\varepsilon.$$

Since $\lambda_1, \lambda_2 \geqq 1$ and v_1, v_2 are unit vectors, we have

$$|v_1 - v_2| \leqq |\lambda_1 v_1 - \lambda_2 v_2|.$$

From (2.3.14) we obtain

$$|\lambda_1 v_1 - \lambda_2 v_2| \leqq 3|\lambda_1\bar{v}_1 - \lambda_2\bar{v}_2| + 6|u_1 - u_2|$$

and thus

$$|v_1 - v_2| \leqq c_{16}\varepsilon.$$

It follows that, in terms of the metric τ introduced on $V_{n,2}$,

$$\operatorname{diam} V_{ij}^* \leqq c_{17}\varepsilon.$$

By the proof of Theorem 2.3.1, the total number of the sets V_{ij}^* is $N \leqq c_{18}\varepsilon^{2-n}$. Thus V^* can be covered by $N \leqq c_{18}\varepsilon^{2-n}$ subsets of $A_1, \ldots, A_N$ of $V_{n,2}$ of diameter $c_{17}\varepsilon$.

Similarly, the set of all 2-frames associated with (K', Π_0', Π_1') can be covered by N subsets $A_1', \ldots, A_N'$ of $V_{n,2}$ of diameter $c_{17}\varepsilon$.

Now let $\rho \in U^*(K, K')$. Then there exists a 2-frame (u, v) associated with (K, Π_0, Π_1) such that $\rho^{-1}(u, v)$ is associated with (K', Π_0', Π_1'). For suitable i, j we have $(u, v) \in A_i$ and $\rho^{-1}(u, v) \in A_j'$, hence $(u, v) \in A_i \cap \rho A_j'$ and therefore $\rho \in X(A_i, A_j')$ (as defined in Lemma 2.3.12). Thus

$$U^*(K, K') \subset \bigcup_{i,j=1}^{N} X(A_i, A_j').$$

By Lemma 2.3.12, each set $X(A_i, A_j')$ can be covered by less than $c_{19}\varepsilon^{-(n-2)(n-3)/2}$ balls of diameter ε, hence $U^*(K, K')$ can be covered by less than

$$(c_{18}\varepsilon^{2-n})^2 c_{19}\varepsilon^{-(n-2)(n-3)/2} = c_{20}\varepsilon^{-[n(n-1)/2-1]}$$

balls of diameter ε. Since this is true for all sufficiently small $\varepsilon > 0$, the set $U^*(K, K')$ has finite $[n(n-1)/2-1]$-dimensional Hausdorff measure. This completes the proof of Theorem 2.3.10. ■

Notes for Section 2.3

1. The question of whether the set $U(K)$ in Theorem 2.3.1 has $(n-1)$-dimensional measure zero was first asked by Klee (1957a). (Ewald, Larman & Rogers (1970) mention an example due to Bing and Klee of a topological 2-sphere in $\mathbb{E}^3$ containing segments of all directions; thus some assumption such as convexity is necessary.) For $n = 3$ (the two-dimensional case is trivial), McMinn (1960) showed that $U(K)$ can be covered by the ranges of countably many Lipschitz mappings from $[-1, 1]$ to S^2. This implies that $U(K)$ is of σ-finite one-dimensional Hausdorff measure. A shorter proof was given by Besicovitch (1963b). For $n \geqq 4$ Klee (1969a) posed the problem of showing that $U(K) \neq S^{n-1}$. This was proved in the stronger form given in Theorem 2.3.1 by Ewald, Larman & Rogers (1970). In fact, these authors obtained a more general result. To formulate it, let $G(n, r)$ be the Grassmann manifold of all r-dimensional linear subspaces of $\mathbb{E}^n$. It is a compact differentiable manifold of dimension $r(n-r)$ and can be equipped with a rotation invariant Riemannian metric. This induces a distance function and thus Hausdorff measures. The following general result can be stated.

Theorem. Let $K \in \mathcal{K}^n$ and $1 \leqq s \leqq r \leqq n-1$. The set of all r-dimensional linear subspaces of $\mathbb{E}^n$ parallel to some supporting r-flat of K that contains an s-dimensional convex subset of K has σ-finite $[r(n-r)-s]$-dimensional Hausdorff measure.

The case $s = r$ of this theorem was proved by Ewald, Larman & Rogers (1970). By extending their methods, Zalgaller (1972) treated the case $s = 1$, and briefly sketched a proof of the general theorem above.

Some of these results can be interpreted in terms of the existence of r-dimensional Chebyshev subspaces of n-dimensional Banach spaces; see Klee (1969a) and Zalgaller (1972). The case $s = 1$ implies, in particular, that for almost all r-dimensional linear subspaces E, the shadow boundary of K in direction E is sharp. This was stated, without a complete proof, by Ewald (1964).

By further extending the methods of Zalgaller, Schneider (1978d) obtained Theorem 2.3.10.

As mentioned after Theorem 2.3.1, the possible presence of facets is an obstruction to improving the result, and a similar remark concerns the generalizations mentioned above. However, Ewald, Larman & Rogers (1970) showed the following. If $2 \leqq r \leqq n - 2$ and $K \in \mathcal{K}_0^n$ has no facets, then the set of all r-dimensional linear subspaces of $\mathbb{E}^n$ parallel to some r-dimensional convex subset in the boundary of K has $r(n - r - 1)$-dimensional Hausdorff measure zero. Further, the following was proved by Larman & Rogers (1971). Let H be a hyperplane in $\mathbb{E}^n$. The set of unit vectors parallel to segments in the boundary of a convex body K that are parallel to H but not lying in the support planes of K parallel to H has $(n - 2)$-dimensional Hausdorff measure zero.

2. Using the result of Ewald, Larman & Rogers, Ivanov (1973) deduced the following result on segments in the boundary: The union of all lines in $\mathbb{E}^n$ that meet the boundary of the convex body $K \in \mathcal{K}^n$ in a segment has σ-finite $(n - 1)$-dimensional Hausdorff measure. Consequently, almost all points $x \in \mathbb{E}^n \setminus K$ have the property that the shadow boundary of K under central projection from x is sharp.
3. For convex subsets in the boundary of a convex body cut out by concurrent sections, Ivanov (1976) showed the following. Let $K \in \mathcal{K}_0^n$ be a convex body with $o \in \operatorname{int} K$ and let $2 \leqq r \leqq n - 1$. The set of all r-dimensional linear subspaces of $\mathbb{E}^n$ containing an $(r - 1)$-dimensional convex subset of $\operatorname{bd} K$ that does not meet the relative interior of a facet of K has Haar measure zero in $G(n, r)$.
4. Methods related to the cap covering theorem of Ewald, Larman & Rogers, which is Theorem 2.3.2 above, have proved useful in other contexts, too. See, for instance, Bárány & Larman (1988) and Bárány (1989).

2.4. Polytopes

The boundary structure of polytopes is particularly simple as far as the facial structure and the classification of boundary points are concerned (not, of course, from the combinatorial or metrical viewpoint). In this book, polytopes are mainly considered as tools for obtaining results on general convex bodies. We treat, therefore, only the most basic properties of polytopes.

A polytope in $\mathbb{E}^n$ is, by definition, the convex hull of a finite (possibly empty) subset of $\mathbb{E}^n$. If $P = \operatorname{conv}\{x_1, \ldots, x_k\}$, then clearly the extreme points of P are among the $x_1, \ldots, x_k$; hence, by Minkowski's theorem, the polytopes are precisely the convex bodies with finitely many extreme points. If H is a support plane of P, then

$$H \cap P = \operatorname{conv}(H \cap \{x_1, \ldots, x_k\}),$$

hence each support set of a polytope is itself a polytope. Since each extreme point of a Minkowski sum $P_1 + P_2$ is the sum of an extreme point of P_1 and an extreme point of P_2, the sum of two polytopes is a polytope.

First we show that for polytopes the distinction between faces and exposed faces (support sets) is unnecessary.

Theorem 2.4.1. *Let $P \subset \mathbb{E}^n$ be a polytope, F_1 a support set of P and F a support set of F_1. Then F is a support set of P.*

Proof. We may assume that $o \in F$. There is a support plane $H_{u,0}$ to P with $H_{u,0} \cap P = F_1$ and $P \subset H^-_{u,0}$. In $H_{u,0}$ there is a support plane H to F_1 with $H \cap F_1 = F$, say

$$H = \{x \in H_{u,0} | \langle x, v \rangle = 0\},$$
$$F_1 \subset \{x \in H_{u,0} | \langle x, v \rangle \leqq 0\}.$$

Define

$$\eta_0 := \max\{-\langle x, v \rangle / \langle x, u \rangle | x \in \operatorname{ext} P \backslash \operatorname{ext} F_1\}$$

and $H(\eta) := H_{\eta u + v, 0}$ with $\eta > \eta_0$. We have $\langle x, u \rangle < 0$ for $x \in \operatorname{ext} P \backslash \operatorname{ext} F_1$, hence

$$\langle x, \eta u + v \rangle < \eta_0 \langle x, u \rangle + \langle x, v \rangle \leqq 0$$

by the definition of η_0. For $x \in \operatorname{ext} F_1 \backslash \operatorname{ext} F$ we get

$$\langle x, \eta u + v \rangle = \langle x, v \rangle < 0,$$

and for $x \in \operatorname{ext} F$, $\langle x, \eta u + v \rangle = 0$. Thus $\operatorname{ext} F \subset H(\eta)$, whereas $\operatorname{ext} P \backslash \operatorname{ext} F \subset \operatorname{int} H^-_{\eta u + v, 0}$. We see that $H(\eta)$ is a support plane to P with $H(\eta) \cap P = F$ and therefore F is a support set of P. ■

Corollary 2.4.2. *Each proper face of a polytope is a support set.*

Proof. Assume that P is a polytope of dimension n and that the assertion has been proved for polytopes of smaller dimension (for zero-dimensional polytopes there is nothing to prove). Let F be a proper face of P. As remarked earlier (before Theorem 2.1.3), F is contained in a support set F_1 of P and it is a face of F_1. Now either $F = F_1$ and then F is a support set of P, or F is a proper face of F_1 and then a support set of F_1, by the induction hypothesis. By Theorem 2.4.1, F is also a support set of P. ■

In particular, each extreme point of a polytope P is an exposed point of P. The extreme points of a polytope are called its *vertices*; the set of vertices of P is often denoted by vert P. The one-dimensional faces of a polytope P are its *edges*, and the ($\dim P - 1$)-dimensional faces are its *facets*. If $x_1, \ldots, x_k$ are the vertices of the polytope P and F is a proper face of P, then there is a hyperplane H with

$$F = H \cap P = \operatorname{conv}(H \cap \{x_1, \ldots, x_k\}).$$

Thus each face is the convex hull of a subset of the vertices, and $\mathcal{F}(P)$ is finite.

A polytope was defined as the convex hull of finitely many points. Alternatively, it can be represented as the intersection of finitely many halfspaces.

***Theorem* 2.4.3.** *Each polytope is the intersection of finitely many closed halfspaces.*

Proof. Let $P \subset \mathbb{E}^n$ be a polytope. Since flats and half-flats can be represented as intersections of finitely many closed halfspaces, it suffices to assume that $\dim P = n$. Let $F_1, \ldots, F_k$ be the facets of P; then $F_i = H_i \cap P$ where H_i is a unique support plane of P. Let H_i^- be the closed halfspace bounded by H_i and containing P ($i = 1, \ldots, k$). We assert that

$$P = H_1^- \cap \ldots \cap H_k^-. \tag{*}$$

The inclusion $P \subset H_1^- \cap \ldots \cap H_k^-$ is trivial. Let $x \in \mathbb{E}^n \setminus P$. Let A be the union of the affine hulls of x and any $n-1$ vertices of P. We can choose a point $y \in (\operatorname{int} P) \setminus A$. There is a point $z \in \operatorname{bd} P \cap [x, y]$; it lies in some support plane to P and hence in some face F of P. Suppose that $\dim F := j \leqq n-2$. By Carathéodory's theorem, z belongs to the convex hull of some $j+1 \leqq n-1$ vertices of P and hence to A. But then $y \in A$, a contradiction. This shows that F is a facet, hence $F = F_i$ for suitable $i \in \{1, \ldots, k\}$. From $y \in \operatorname{int} P \subset \operatorname{int} H_i^-$ we deduce that $x \notin H_i^-$. This proves (*). ■

Let $P \in \mathcal{K}_0^n$ be a polytope and F a facet of P. Then $F = P \cap H_{u,\alpha}$ and $P \subset H_{u,\alpha}^-$ for suitable u and α. We may assume that $u \in S^{n-1}$; then u is called the *outer unit normal vector* of the facet F. The outer unit normal vectors of the facets of P are called, in brief, the *normal vectors* of P.

***Corollary* 2.4.4.** *If $P \in \mathcal{K}_0^n$ is a polytope and $u_1, \ldots, u_k$ are its normal vectors, then*

$$P = \bigcap_{i=1}^{k} H^-_{u_i, h(P,u_i)}.$$

In particular, the numbers $h(P, u_1), \ldots, h(P, u_k)$ determine P uniquely.

Proof. If F_i is the facet of P with normal vector u_i, then $F_i = P \cap H_{u_i, h(P,u_i)}$. The assertion is now clear from the proof of Theorem 2.4.3. ■

At this point, it is convenient to use polarity, and this requires the following lemma. It shows, in particular, that the polar body of a polytope (with respect to an interior point) is again a polytope.

Lemma 2.4.5. *Let $u_1, \ldots, u_k \in \mathbb{E}^n \setminus \{o\}$. If*

$$P := \bigcap_{i=1}^{k} H^-_{u_i, 1}$$

is bounded, then

$$P^* = \operatorname{conv}\{u_1, \ldots, u_k\}.$$

Proof. (We remark that $o \in \operatorname{int} P$, hence P^* is well defined.) Let $Q := \operatorname{conv}\{u_1, \ldots, u_k\}$. For $x \in P$ we have $\langle x, u_i \rangle \leqq 1$, hence $u_i \in P^*$. Thus $Q \subset P^*$.

Let $v \in \mathbb{E}^n \setminus Q$. Then there is a hyperplane $H_{z,\alpha}$ strongly separating Q and v, and by an appropriate choice of signs we may assume that $\langle z, u_i \rangle < \alpha$ for $i = 1, \ldots, k$ and $\langle z, v \rangle > \alpha$. Since P is bounded, clearly $\operatorname{pos}\{u_1, \ldots, u_k\} = \mathbb{E}^n$ (otherwise the closed convex cone $\operatorname{pos}\{u_1, \ldots, u_k\}$ has a boundary point and hence a supporting hyperplane, necessarily through o, each outer normal vector of which is an element of P); hence there is a representation

$$o = \sum_{i=1}^{k} \lambda_i u_i \quad \text{with } \lambda_i \geqq 0 \text{ and } \sum_{i=1}^{k} \lambda_i > 0.$$

This yields $0 < \sum \lambda_i \alpha$ and thus $\alpha > 0$; without loss of generality, $\alpha = 1$. But then $z \in P$, hence $v \notin P^*$. This proves the equality $Q = P^*$. ■

Theorem 2.4.6. *Each bounded intersection of finitely many closed halfspaces is a polytope.*

Proof. Let $P = \bigcap_{i=1}^{k} H_i^-$ be bounded, where $H_i^- \subset \mathbb{E}^n$ is a closed halfspace. We may assume that $\dim P = n$ (otherwise we work in aff P) and, after a translation, that $o \in \operatorname{int} P$. Then $o \in \operatorname{int} H_i^-$, hence H_i^- can be represented in the form $H_i^- = H^-_{u_i,1}$. Lemma 2.4.5 shows that P^* is a polytope. By Theorem 2.4.3, P^* is the intersection of finitely many closed halfspaces (and $o \in \text{int}\, P^*$, as always). Hence $P^{**} = P$ is a polytope. ■

Theorem 2.4.6 implies, in particular, that the intersection of finitely many polytopes is a polytope and that the intersection of a polytope with a flat is a polytope.

More information on the faces of a polytope is now easy to obtain.

Theorem 2.4.7. *Each proper face of a polytope P is contained in some facet of P.*

Proof. We may assume that $\dim P = n$. In the proof of Theorem 2.4.3 it has been shown that P has a representation

$$P = \bigcap_{i=1}^{k} H_i^-,$$

where H_i^- is a closed halfspace and $F_i := P \cap H_i$ is a facet of P $(i = 1, \dots, k)$.

Since $\text{bd}\, P \subset \bigcup_{i=1}^{k}(P \cap H_i)$, each boundary point of P is contained in some facet F_j.

Now let F be a proper face of P and choose $x \in \text{relint}\, F$. Then $x \in F_j = P \cap H_j$ for some $j \in \{1, \dots, k\}$. Since $x \in \text{relint}\, F$ and H_j is a support plane of P, we deduce $F \subset H_j$, hence $F \subset F_j$. ■

Corollary 2.4.8. *Let P be a polytope, let $F^j \in \mathcal{F}_j(P)$ and $F^k \in \mathcal{F}_k(P)$ be faces such that $F^j \subset F^k$. Then there are faces $F^i \in \mathcal{F}_i(P)$ $(i = j+1, \dots, k-1)$ such that $F^j \subset F^{j+1} \subset \dots \subset F^{k-1} \subset F^k$.*

Proof. Clearly F^j is a face of F^k (being a support set of P and hence of F^k). If F^j is a proper face of F^k, then F^j is contained in a facet F^{k-1} of F^k, and by Theorem 2.1.1, F^{k-1} is a face of P. By repeating this argument, the assertion is obtained. ■

By Lemma 2.2.3, the normal cones of an n-dimensional polytope can be related to the faces of its polar polytope (with respect to an interior point). This yields:

Theorem 2.4.9. *Let $P \in \mathcal{K}_0^n$ be a polytope, let $F_1, \dots, F_k$ be its facets, and let u_i be the outer unit normal vector of F_i $(i = 1, \dots, k)$. If F is a proper face of P, then*

$$N(P, F) = \text{pos}\, \{u_i | F \subset F_i\}.$$

Proof. We may assume that $o \in \operatorname{int} P$. From Lemma 2.2.3 we know that $N(P, F) = \operatorname{pos} \hat{F}$. Writing $v_i = u_i/h(P, u_i)$, we have

$$P = \bigcap_{i=1}^{k} H^-_{v_i,1}$$

by Corollary 2.4.4, hence Lemma 2.4.5 yields

$$P^* = \operatorname{conv}\{v_1, \ldots, v_k\}.$$

The face $\hat{F}$ of P^* is the convex hull of those v_i that it contains, and $\operatorname{pos} \hat{F}$ is the positive hull of these v_i and hence of the corresponding u_i. Now $v_i \in \hat{F}$ is equivalent to $F \subset H_{v_i,1}$, which is equivalent to $F \subset F_i$. ■

We note some consequences of Theorem 2.4.9. Let $P \in \mathcal{K}^n_0$ be a polytope and F a proper face of P. Then F is the intersection of the facets of P containing F; this follows from Theorem 2.4.7 and the fact that the intersection of all facets containing F is a face of the same dimension as F. Let G be another proper face of P. From Theorem 2.4.9 it follows that

$$F \subset G \Leftrightarrow N(P, F) \supset N(P, G) \tag{2.4.1}$$

and that

$$\dim N(P, F) = n - \dim F. \tag{2.4.2}$$

As a consequence, a point x of P is r-singular if and only if it belongs to some face of dimension $\leqq r$. Thus for n-polytopes, the notions of r-extreme, r-exposed and r-singular boundary points all coincide. In particular, the extreme points (vertices) of P are precisely the points x with $\dim N(P, x) = n$.

From (2.4.1) one further deduces for $v \in \mathbb{E}^n \setminus \{o\}$ that

$$F = F(P, v) \Leftrightarrow v \in \operatorname{relint} N(P, F) \tag{2.4.3}$$

and that

$$N(P, F) = \bigcap_{x \in \operatorname{ext} F} N(P, x). \tag{2.4.4}$$

If $o \in \operatorname{int} P$, so that the polar polytope P^* is defined, then $F \subset G$ for proper faces F and G of P implies $\hat{F} \supset \hat{G}$ by (2.4.1) and Lemma 2.2.3. Thus the map $F \mapsto \hat{F}$ from the face lattice $\mathcal{F}(P)$ to the face lattice $\mathcal{F}(P^*)$ (observe that $\hat{\emptyset} = P^*$ and $\hat{P} = \emptyset$) is an antimorphism, that is, a bijection that reverses the inclusion relation.

A polytope is called *simplicial* if all its proper faces (equivalently, all its facets) are simplices. An n-polytope P is called *simple* if each of its vertices is contained in exactly n facets. Clearly this holds if and only if P^* (taken with respect to some interior point of P as origin) is simplicial.

The normal cones of a polytope give rise to a series of important metric invariants: Let $P \in \mathcal{K}^n$ be a polytope and $F \in \mathcal{F}_k(P)$ a k-face of P. Then the number

$$\gamma(F, P) := \mathcal{H}^{n-k-1}(N(P, F) \cap S^{n-1})/\omega_{n-k}$$

(with the convention $\gamma(F, P) = 1$ if $k = n$) is called the *external angle* of P at its face F. Thus $\gamma(F, P)$ is the $(n-k-1)$-dimensional measure of the spherical image $\sigma(P, \operatorname{relint} F)$ of P at $\operatorname{relint} F$, divided by the total measure of the $(n-k-1)$-dimensional unit sphere. If $\dim P < n$, the number $\gamma(F, P)$ remains the same if computed in any affine subspace containing P; this follows from (2.2.3) and Fubini's theorem. From (2.2.4) and (2.2.5) one deduces that

$$\sum_{F \in \mathcal{F}_0(P)} \gamma(F, P) = 1. \tag{2.4.5}$$

We shall now establish a special result dealing with approximation by polytopes, which will be useful in later chapters. Some preparations are required.

Two polytopes P_1, P_2 are called *strongly isomorphic* if

$$\dim F(P_1, u) = \dim F(P_2, u)$$

for all $u \in S^{n-1}$. Obviously this is an equivalence relation; the corresponding equivalence class of a polytope P will be called its *a-type*.

Lemma 2.4.10. *If P_1, P_2 are strongly isomorphic polytopes then, for each $u \in S^{n-1}$, the support sets $F(P_1, u)$, $F(P_2, u)$ are strongly isomorphic.*

Proof. Let $u \in S^{n-1}$ be given. Let $v \in S^{n-1}$ and put $F(F(P_i, u), v) := F_i'$ $(i = 1, 2)$. We may assume that $o \in F_1' \cap F_2'$. As shown in the proof of Theorem 2.4.1, for η sufficiently large the hyperplane $H_{\eta u+v,0}$ supports P_i and $H_{\eta u+v,0} \cap P_i = F_i'$ for $i = 1, 2$. Since P_1, P_2 are strongly isomorphic, we get $\dim F_1' = \dim F_2'$ and thus the assertion. ∎

Lemma 2.4.11. *The polytopes P_1, $P_2 \in \mathcal{K}^n$ are strongly isomorphic if and only if*

$$\{N(P_1, x) | x \in \operatorname{ext} P_1\} = \{N(P_2, y) | y \in \operatorname{ext} P_2\}. \tag{*}$$

Proof. Suppose that P_1, P_2 are strongly isomorphic. Let $x \in \operatorname{ext} P_1$. Choose $u \in \operatorname{int} N(P_1, x)$ and $y \in F(P_2, u)$, then $y \in \operatorname{ext} P_2$. Let $v \in N(P_1, x)$ and $\lambda \in [0, 1)$; then $v_\lambda := (1-\lambda)u + \lambda v \in \operatorname{int} N(P_1, x)$, hence $\dim F(P_2, v_\lambda) = \dim F(P_1, v_\lambda) = \dim \{x\} = 0$ and, therefore, $F(P_2, v_\lambda) = \{y_\lambda\}$ for some $y_\lambda \in \operatorname{ext} P_2$, which implies $v_\lambda \in \operatorname{int} N(P_2, y_\lambda)$.

Since this holds for all $\lambda \in [0,1)$, we conclude that $y_\lambda = y$. It follows that $v \in N(P_2, y)$, thus $N(P_1, x) \subset N(P_2, y)$. Since $x \in \operatorname{ext} P_1$ was arbitrary and the roles of P_1 and P_2 can be interchanged, it follows that $N(P_1, x) = N(P_2, y)$. Thus (*) holds.

Suppose, conversely, that (*) is true. Let $u \in S^{n-1}$ and choose a vertex x_1 of $F(P_1, u)$. There is a vertex x_2 of P_2 for which $N(P_2, x_2) = N(P_1, x_1)$. The vector u lies in the relative interior of a face of $N(P_i, x_i)$ of dimension $n - \dim F(P_i, u)$ (as follows from (2.4.2) and (2.4.3)), hence $\dim F(P_1, u) = \dim F(P_2, u)$. Since u was arbitrary, P_1 and P_2 are strongly isomorphic. ■

Let P_1, P_2 be strongly isomorphic polytopes. Let F be a proper face of P_1. The normal cone $N(P_1, F)$ is a face of $N(P_1, x)$ for some vertex x of F, hence by Lemma 2.4.11 a face of $N(P_2, y)$ for some vertex y of P_2 and thus the normal cone of a unique face $\varphi(F)$ of P_2. It is clear (using (2.4.1)) that φ (extended to improper faces) is a bijective map from $\mathcal{F}(P_1)$ to $\mathcal{F}(P_2)$ that preserves inclusions, that is, a combinatorial isomorphism. This explains the term *strongly isomorphic*. In the literature, strongly isomorphic polytopes have also been called *analogous* (for this reason, we speak of the a-type of a polytope), *strongly combinatorially isomorphic* (*strongly combinatorially equivalent*), or *locally similar*.

Lemma 2.4.12. *Let P be a simple n-polytope with normal vectors $u_1, \ldots, u_k$. Then there is a number $\beta > 0$ such that each polytope of the form*

$$P' := \bigcap_{i=1}^{k} H^-_{u_i, h(P,u_i)+\alpha_i} \tag{*}$$

with $|\alpha_i| \leqq \beta$ is simple and strongly isomorphic to P.

Proof. We can choose pairwise disjoint balls $B(x, \rho)$ for $x \in \operatorname{ext} P$ such that each ball $B(x, \rho)$ meets only those facets of P that contain x. Next, we can choose a number $\beta > 0$ with the following property: If x is a vertex of P and if, say, $u_1, \ldots, u_n$ are the outer normal vectors of the facets containing x, then for $|\alpha_i| \leqq \beta$ the intersection point of the hyperplanes $H_{u_i, h(P,u_i)+\alpha_i}$ $(i = 1, \ldots, n)$ lies in $B(x, \rho)$, and $H_{u_j, h(P,u_j)+\alpha_j}$ does not meet $B(x, \rho)$ for $j \notin \{1, \ldots, n\}$. It is then clear that P' defined by (*) with $|\alpha_i| \leqq \beta$ is a polytope having a vertex x' in each of the balls $B(x, \rho)$ $(x \in \operatorname{ext} P)$ such that $N(P', x') = N(P, x)$. We deduce that P' has no further vertices and that it is simple and strongly isomorphic to P, by Lemma 2.4.11. ■

Lemma 2.4.13. *Let P be an n-polytope with normal vectors $u_1, \ldots, u_k$. Then to each $\beta > 0$ numbers $\alpha_1, \ldots, \alpha_k$ with $|\alpha_i| \leqq \beta$ can be chosen such that*

$$P' := \bigcap_{i=1}^{k} H^-_{u_i, h(P,u_i)+\alpha_i}$$

is a simple n-polytope with normal vectors $u_1, \ldots, u_k$ and the property that each normal cone of a vertex of P' is contained in some normal cone of P.

Proof. We choose balls $B(x, \rho)$ $(x \in \operatorname{ext} P)$ as in the proof of Lemma 2.4.12. If $|\alpha_1|, \ldots, |\alpha_k|$ are sufficiently small, then P' is an n-polytope with normal vectors $u_1, \ldots, u_k$ and such that any vertex x' of P' is contained in some ball $B(x, \rho)$. By Theorem 2.4.9, $N(P', x')$ is then the positive hull of only those normal vectors that belong to facets of P containing x; hence $N(P', x') \subset N(P, x)$. By choosing the α_i successively in an appropriate way, we can clearly achieve that the resulting polytope P' is simple. ■

We are now in a position to prove the following result on simultaneous approximation of convex bodies by strongly isomorphic polytopes.

Theorem 2.4.14. *Let $K_1, \ldots, K_m \in \mathcal{K}^n$ be convex bodies. To each $\varepsilon > 0$ there exist simple strongly isomorphic polytopes $P_1, \ldots, P_m \in \mathcal{K}^n$ satisfying $\delta(K_i, P_i) < \varepsilon$ for $i = 1, \ldots, m$.*

Proof. Let $\varepsilon > 0$ be given. By Theorem 1.8.13 there are polytopes $Q_1, \ldots, Q_m \in \mathcal{K}^n$ with $\delta(K_i, Q_i) < \varepsilon/2$ for $i = 1, \ldots, m$. Let $P := Q_1 + \ldots + Q_m$ (which is also a polytope) and let $u_1, \ldots, u_k$ be the normal vectors of P. Starting with P, we construct P' as in Lemma 2.4.13 and put $P_i := Q_i + \alpha P'$, where $\alpha > 0$ is sufficiently small that $\delta(P_i, Q_i) < \varepsilon/2$ and hence $\delta(K_i, P_i) < \varepsilon$ for $i = 1, \ldots, m$. Let x be a vertex of some P_i. Then $x = q + \alpha p$ with suitable $q \in \operatorname{ext} Q_i$ and $p \in \operatorname{ext} P'$. From Theorem 2.2.1 we have

$$N(P_i, x) = N(Q_i, q) \cap N(P', p).$$

By the construction of P', the normal cone $N(P', p)$ is contained in some normal cone of P, and by Theorem 2.2.1 the latter is contained in some normal cone of Q_i, necessarily $N(Q_i, q)$, because $N(P', p) \cap N(Q_i, q)$ is n-dimensional. Thus $N(P_i, x) = N(P', p)$. Since x was an arbitrary vertex of P_i, we deduce from Lemma 2.4.11 that P_i is strongly isomorphic to P'. This implies also that P_i is simple. ■

Notes for Section 2.4

1. Standard references on convex polytopes, from which we have taken some of the basic arguments, are the books by Grünbaum [15], McMullen & Shephard [26] and Brøndsted [11].
2. *Strongly isomorphic polytopes.* The approximation theorem, 2.4.14, is due to Aleksandrov (1937b). He defined strongly isomorphic polytopes in a different, but equivalent way (and called them *analogous*). Our definition appears in Shephard (1963), where the term 'locally similar' was used. Still another term was introduced by Grünbaum [15], p 50.
   Lemma 2.4.11 can also be found, in essence, in W. Meyer (1969, Corollary 2.7), and in Kallay (1982).
3. *Monotypic polytopes.* McMullen, Schneider & Shephard (1974) studied the n-polytopes P, called *monotypic*, with the property that the a-type of P is uniquely determined by the system of normal vectors of P (see also Section 3.2, Note 3).
4. *Approximation by polytopes.* While in this book the approximation of convex bodies by polytopes is only used as a tool, it is an interesting topic in its own right and can be studied under various aspects. We refer the reader to the survey article by Gruber [33], which contains many references. Later contributions to this subject are by Bokowski & Mani-Levitska (1984, 1987), Florian (1986), Gruber (1988, 1990+, 1991b) and Schneider (1986a, 1987b). Schneider (1988c) gives a survey on the approximation of convex bodies by random polytopes.

2.5. Higher regularity and curvature

Another special class of convex bodies, besides the polytopes, with an accessible boundary structure are the convex bodies whose boundaries are regular submanifolds of $\mathbb{E}^n$, in the sense of differential geometry, and which satisfy suitable differentiability properties. The present section is mainly concerned with the second-order boundary structure of such convex bodies. The central notions are those of curvature, principal curvatures and radii of curvature. The last part is devoted to the curvature properties of general convex bodies.

First we define curvatures for a general convex body $K \in \mathcal{K}_0^n$ with interior points. We assume that x is a smooth boundary point of K. Let u be the unique outward unit normal vector to K at x. The linear subspace $H(K, u) - x$ of $\mathbb{E}^n$ is called the *tangent space* of K at x and is denoted by T_xK. We can choose a number $\varepsilon > 0$ and a neighbourhood U of x such that $U \cap \operatorname{bd} K$ can be described in the form

$$U \cap \operatorname{bd} K = \{x + t - f(t)u \mid t \in T_xK \cap B(o, \varepsilon)\}, \qquad (2.5.1)$$

where $f\colon T_xK \cap B(o, \varepsilon) \to \mathbb{R}$ is a convex function satisfying $f \geqq 0$ and $f(o) = 0$.

Let $t \in T_xK \cap S^{n-1}$ be a given unit tangent vector of K at x. For $0 < \tau < \varepsilon$, let $c(\tau)$ be the centre of the circle lying in the plane through x spanned by u and t, going through the point $x + \tau t - f(\tau t)u$ and touching $H(K, u)$ at x. If $(\tau_j)_{j\in\mathbb{N}}$ is a sequence such that $\tau_j \to 0$ for $j \to \infty$ and $(c(\tau_j))_{j\in\mathbb{N}}$ converges to some point $x - \rho u$, and thus

$$\rho = \lim_{j\to\infty} \frac{\tau_j^2}{2f(\tau_j t)},$$

then ρ is called a *radius of curvature* of K at x in direction t. The set of all numbers ρ (∞ admitted) arising in this way is an interval $[\rho_i(x, t), \rho_s(x, t)]$ with $0 \leqq \rho_i(x, t) \leqq \rho_s(x, t) \leqq \infty$. We call

$$\rho_i(x, t) = \liminf_{\tau\downarrow 0} \frac{\tau^2}{2f(\tau t)}$$

the *lower radius of curvature* and

$$\rho_s(x, t) = \limsup_{\tau\downarrow 0} \frac{\tau^2}{2f(\tau t)}$$

the *upper radius of curvature* of K at x in direction t. The reciprocal values (possibly ∞)

$$\kappa_i(x, t) := 1/\rho_s(x, t), \quad \kappa_s(x, t) := 1/\rho_i(x, t)$$

are, respectively, the *lower curvature* and *upper curvature* of K at x in direction t. If both values coincide and are finite, their common value $\kappa(x, t)$ is the *curvature* of K at x in direction t.

Now we introduce differentiability assumptions. A convex body $K \in \mathcal{K}_0^n$ is said to be of class C^k, for some $k \in \mathbb{N}$, if its boundary hypersurface is a regular submanifold of $\mathbb{E}^n$, in the sense of differential geometry, that is k-times continuously differentiable. K is of class C^∞ if it is of class C^k for each $k \in \mathbb{N}$. Since locally bd K can be represented as the graph of a convex function, as in (2.5.1), it follows from Theorems 1.5.12 and 1.5.11 that K is of class C^1 if and only if it is smooth (i.e., it has a unique support plane at each boundary point). Here one has to use the property that a differentiable convex function is continuously differentiable, as follows from Theorem 1.5.2, applied to the partial derivatives.

We assume that the reader is familiar with the elementary differential geometry of hypersurfaces of class C^2. We recall some of the basic definitions and facts and then concentrate on some special aspects of the convex case.

Let $K \in \mathcal{K}_0^n$ be of class C^2. We identify, in the canonical way, the abstract tangent spaces of $\mathbb{E}^n$ with the vector space $\mathbb{E}^n$ itself and thus also the abstract tangent space of the submanifold bd K at x with the linear subspace $T_x K$, as introduced above. For $x \in$ bd K, let $\nu(x)$ be the outward unit normal vector of K at x. The map ν: bd $K \to S^{n-1} \subset \mathbb{E}^n$ thus defined (also denoted by σ_K in Section 2.2) is called the *spherical image map* (or *Gauss map*) of K; it is of class C^1. Its differential

$dv_x := W_x$ maps T_xK into itself. The linear map $W_x: T_xK \to T_xK$ is called the *Weingarten map*. The bilinear form defined on T_xK by

$$II_x(v, w) := \langle W_x v, w \rangle \quad \text{for } v, w \in T_xK$$

is the *second fundamental form* of bd K at x. The *first fundamental form* is just the restriction of the scalar product to T_xK,

$$I_x(v, w) := \langle v, w \rangle \quad \text{for } v, w \in T_xK.$$

In local coordinates, the foregoing can be expressed as follows. Let $X: M \to \mathbb{E}^n$, where $M \subset \mathbb{R}^{n-1}$ is open, be a parametrization of class C^2 of a neighbourhood of x on bd K where, say, $x = X(y)$ with $y \in M$. Put $N := \nu \circ X$, so that $N: M \to S^{n-1}$ is a map of class C^1. We have $dN_y = dv_x \circ dX_y$ and hence $W_x = dN_y \circ dX_y^{-1}$. For a differentiable function φ on M we denote by φ_i the partial derivative with respect to the ith (Cartesian) coordinate. Then $dX(e_i) = X_i$, $dN(e_i) = N_i$ (where e_i is the ith basis vector of $\mathbb{R}^{n-1}$) and hence $W_x X_i = N_i$ (as usual, we omit the argument y). The matrix $(l_{ij})_{i,j=1}^{n-1}$ of the bilinear form II_x with respect to the standard basis $(X_1, \dots, X_{n-1})$ of T_xK has entries

$$\begin{aligned} l_{ij} &= II_x(X_i, X_j) = \langle W_x X_i, X_j \rangle \\ &= \langle N_i, X_j \rangle = -\langle N, X_{ij} \rangle, \end{aligned} \tag{2.5.2}$$

the latter because $\langle N, X_j \rangle = 0$. Thus II_x is symmetric, and the Weingarten map is self-adjoint (with respect to $\langle \cdot, \cdot \rangle$). The matrix $(g_{ij})_{i,j=1}^{n-1}$ of the first fundamental form is given by $g_{ij} = \langle X_i, X_j \rangle$. For the matrix $(l^j_i)_{i,j=1}^{n-1}$ of the Weingarten map, which is defined by

$$N_i = \sum_{j=1}^{n-1} l^j_i X_j, \quad i = 1, \dots, n-1, \tag{2.5.3}$$

one obtains $l_{ik} = \langle N_i, X_k \rangle = \sum_j l^j_i \langle X_j, X_k \rangle = \sum_j l^j_i g_{jk}$, hence

$$l^j_i = \sum_{k=1}^{n-1} l_{ik} g^{kj}, \tag{2.5.4}$$

where $(g^{ij})_{i,j}$ denotes the inverse matrix of $(g_{ij})_{i,j}$.

The eigenvalues of the Weingarten map W_x are, by definition, the *principal curvatures* of K at x. The equality $W_x v = \lambda v$ for $v \in T_xK \setminus \{o\}$ and $\lambda \in \mathbb{R}$ is equivalent to $II_x(v, w) = \lambda I_x(v, w)$ for all $w \in T_xK$. Thus the principal curvatures are also the eigenvalues of the second fundamental form relative to the first fundamental form. In local coordinates as above, $v = \sum_{i=1}^{n-1} v^i X_i$ is an eigenvector of the Weingarten map and λ is a corresponding principal curvature if and only if

$$\sum_{i=1}^{n-1} (l^j_i - \lambda \delta^j_i) v^i = 0,$$

or equivalently

$$\sum_{i=1}^{n-1}(l_{ij} - \lambda g_{ij})v^i = 0,$$

for $j = 1, \ldots, n-1$.

For the special representation given by (2.5.1), that is, $X(t) = x + t - f(t)u$ for $t \in T_xK \cap B_0(o, \varepsilon)$, one obtains (at the point x) $l_{ij} = f_{ij}$ and $g_{ij} = \delta_{ij}$ and hence that the principal curvatures at x are the eigenvalues of the Hessian matrix of f. This provides a simple interpretation of the principal curvatures and shows that they are non-negative, by Theorem 1.5.10.

By H_j we denote that jth normalized elementary symmetric function of the principal curvatures; thus $H_0 = 1$ and

$$H_j = \binom{n-1}{j}^{-1} \sum_{1 \leqq i_1 < \ldots < i_j \leqq n-1} k_{i_1} \cdots k_{i_j} \tag{2.5.5}$$

for $j = 1, \ldots, n-1$. We refer to H_1 as the *mean curvature* and to H_{n-1} as the *Gauss–Kronecker curvature*.

As is well known from differential geometry, under our present assumptions the curvature $\kappa(x, t)$ defined initially exists and is given by

$$\kappa(x, t) = \frac{II_x(t, t)}{\langle t, t \rangle} \quad \text{for } t \in T_xK \setminus \{o\}. \tag{2.5.6}$$

Let $(v_1, \ldots, v_{n-1})$ be an orthonormal basis of T_xK consisting of eigenvectors of the Weingarten map W_x, and let k_i be the principal curvature corresponding to v_i. Then

$$\kappa(x, t) = k_1\langle t, v_1\rangle^2 + \ldots + k_{n-1}\langle t, v_{n-1}\rangle^2 \tag{2.5.7}$$

for each unit vector $t \in T_xK$. This result is known as *Euler's theorem*.

The following assumption, stronger than C^k, will often be important. We say that K is of class C_+^k (for $k \geqq 2$) if K is of class C^k and the spherical image map ν: bd $K \to S^{n-1}$ is a diffeomorphism (of class C^1, and hence C^{k-1}). This is equivalent to the assumption that its differential, the Weingarten map, is everywhere of maximal rank, and thus to the assumption that all principal curvatures are non-zero, or equivalently that $H_{n-1} \neq 0$.

Let K be of class C_+^2. Then the map ν has an inverse ν^{-1} of class C^1, and for the support function h_K of K we get

$$h_K(u) = \langle \nu^{-1}(u), u \rangle;$$

hence h_K is differentiable on S^{n-1} and thus on $\mathbb{E}^n \setminus \{o\}$. From Corollary 1.7.3 we have

$$\operatorname{grad} h_K(u) = \nu^{-1}\left(\frac{u}{|u|}\right)$$

for $u \in \mathbb{E}^n \setminus \{o\}$. It follows that the support function h_K is, in fact, of class C^2.

For several investigations on convex bodies, particularly in connection with Minkowski addition, it appears more natural to impose differentiability assumptions on the support function than on the boundary. Let $K \in \mathcal{K}_0^n$. By Corollary 1.7.3, the support function h_K is differentiable on $\mathbb{E}^n \setminus \{o\}$ if and only if K is strictly convex, and if this is satisfied, h_K is continuously differentiable. Let h_K be of class C^1 (note that K may well have singularities, so that $\operatorname{bd} K$ need not be of class C^1). For $u \in \mathbb{E}^n \setminus \{o\}$, let $\xi(u)$ be the unique point of $\operatorname{bd} K$ at which u is attained as an outward normal vector. By Corollary 1.7.3,

$$\xi(u) = \operatorname{grad} h_K(u). \tag{2.5.8}$$

The map ξ thus defined on $\mathbb{E}^n \setminus \{o\}$ is positively homogeneous of degree zero.

Now we assume that h_K is of class C^2. Then ξ is of class C^1. Its restriction to S^{n-1}, the map $\bar{\xi}: S^{n-1} \to \operatorname{bd} K$, will be called the *reverse spherical image map*. For $u \in S^{n-1}$, let T_u be the $(n-1)$-dimensional linear subspace of $\mathbb{E}^n$ that is orthogonal to u; thus T_u is the tangent space of S^{n-1} at u. The differential $\mathrm{d}\bar{\xi}_u := \overline{W}_u$ maps T_u into T_u. We call the linear map $\overline{W}_u: T_u \to T_u$ the *reverse Weingarten map* and the bilinear form

$$\overline{II}_u(v, w) := \langle \overline{W}_u v, w \rangle, \quad v, w \in T_u,$$

the *reverse second fundamental form* of K at u.

Let $N: M \to \mathbb{E}^n$, where $M \subset \mathbb{R}^{n-1}$ is open, be a parametrization of class C^2 of a neighbourhood of u on S^{n-1} and let $u = N(y)$. Put $X := \bar{\xi} \circ N$. We have $\mathrm{d}X_y = \mathrm{d}\bar{\xi}_u \circ \mathrm{d}N_y$, hence $\overline{W}_u = \mathrm{d}X_y \circ \mathrm{d}N_y^{-1}$ and thus $\overline{W}_u N_i = X_i$. The matrix $(b_{ij})_{i,j=1}^{n-1}$ of $\overline{II}_u$ with respect to the standard basis $(N_1, \ldots, N_{n-1})$ of T_u is given by

$$\begin{aligned} b_{ij} &= \overline{II}_u(N_i, N_j) = \langle \overline{W}_u N_i, N_j \rangle = \langle X_i, N_j \rangle \\ &= \sum_{k=1}^{n} \sum_{m=1}^{n} (\partial_k \partial_m h_K) \circ N n_i^k n_j^m \end{aligned} \tag{2.5.9}$$

by (2.5.8), where $N = (n^1, \ldots, n^n)$. Thus $\overline{II}_u$ is symmetric and $\overline{W}_u$ is self-adjoint. The first fundamental form $\overline{I}_u$ of the sphere S^{n-1} at u is given by $\overline{I}_u(v, w) := \langle v, w \rangle$ for $v, w \in T_u$. Its matrix with respect to $(N_1, \ldots, N_{n-1})$ is denoted by $(e_{ij})_{i,j=1}^{n-1}$; thus $e_{ij} = \overline{I}_u(N_i, N_j) = \langle N_i, N_j \rangle$. Introducing the matrix $(b_i^j)_{i,j=1}^{n-1}$ of the reverse Weingarten map by

$$X_i = \sum_{j=1}^{n-1} b_i^j N_j, \quad i = 1, \ldots, n-1, \tag{2.5.10}$$

we find $b_{ik} = \langle X_i, N_k \rangle = \sum_j b_i^j \langle N_j, N_k \rangle = \sum_j b_i^j e_{jk}$ and hence

$$b_i^j = \sum_{k=1}^{n-1} b_{ik} e^{kj}, \tag{2.5.11}$$

where $(e^{ij})_{i,j}$ is the matrix inverse to $(e_{ij})_{i,j}$.

The eigenvalues $r_1, \ldots, r_{n-1}$ of the reverse Weingarten map are called the *principal radii of curvature* of K at u. It should be kept in mind that, while the principal curvatures are functions on the boundary of K, the principal radii of curvature are considered as functions of the outer unit normal vector, in other words, as functions on the spherical image. We denote by s_j the jth normalized elementary symmetric function of the principal radii of curvature:

$$s_j = \binom{n-1}{j}^{-1} \sum_{1 \leqq i_1 < \ldots < i_j \leqq n-1} r_{i_1} \cdots r_{i_j}. \tag{2.5.12}$$

The vector $v \in T_u \backslash \{o\}$ is an eigenvector of the reverse Weingarten map and λ is a corresponding radius of curvature if $\overline{W}_u v = \lambda v$, or, equivalently, $\overline{II}_u(v, w) = \lambda \overline{I}_u(v, w)$ for all $w \in T_u$. In terms of the local parametrization N of S^{n-1} used above and with $v = \sum_{i=1}^{n-1} v^i N_i$, this is equivalent to

$$\sum_{i=1}^{n-1} (b_i^j - \lambda \delta_i^j) v^i = 0, \quad j = 1, \ldots, n-1 \tag{2.5.13}$$

and to

$$\sum_{i=1}^{n-1} (b_{ij} - \lambda e_{ij}) v^i = 0, \quad j = 1, \ldots, n-1. \tag{2.5.14}$$

In particular, the product of the principal radii of curvature is given by

$$s_{n-1} \circ N = \det (b_i^j) = \frac{\det (b_{ij})}{\det (e_{ij})}. \tag{2.5.15}$$

Since the principal radii of curvature, $r_1, \ldots, r_{n-1}$, are the eigenvalues of the matrix $(b_i^j)_{i,j}$, the matrix $(b_i^j + \rho \delta_i^j)_{i,j}$ with $\rho \in \mathbb{R}$ has the eigenvalues $r_1 + \rho, \ldots, r_{n-1} + \rho$, hence

$$\det (b_i^j + \rho \delta_i^j) = \sum_{k=0}^{n-1} \rho^k \binom{n-1}{k} s_{n-1-k} \circ N. \tag{2.5.16}$$

It may sometimes be desirable to express the foregoing notions more directly in terms of the support function. This can be achieved by means of the following lemma. Let $\mathrm{d}^2 h_u$ be the second differential of $h = h_K$ at u, considered as a bilinear form on $\mathbb{E}^n$. For $u \in \mathbb{E}^n \backslash \{o\}$, let π_u be the orthogonal projection onto the orthogonal subspace T_u.

Lemma 2.5.1. *For* $u \in \mathbb{E}^n \backslash \{o\}$,

$$\mathrm{d}^2 h_u(a, b) = |u|^{-1} \langle \overline{W}_{u/|u|} \pi_u a, \pi_u b \rangle \tag{2.5.17}$$

for all $a, b \in \mathbb{E}^n$.

Proof. Let $u \in \mathbb{E}^n \backslash \{o\}$ be given. By homogeneity, we may assume that $|u| = 1$. Let $(e_1, \ldots, e_n)$ be an orthonormal basis of $\mathbb{E}^n$ with $e_n = u$. We write $(y^1, \ldots, y^n)$ for the coordinates of $y \in \mathbb{E}^n$ and h_{ij} for the second partial derivatives of h with respect to this basis. The homogeneity relation

$$\sum_{j=1}^{n} h_{ij}(y)y^j = 0 \quad \text{for } i = 1, \ldots, n \tag{2.5.18}$$

yields $h_{in}(u) = 0$ and hence

$$d^2 h_u(a, b) = \sum_{i,j=1}^{n-1} h_{ij}(u)a^i b^j \quad \text{for } a, b \in \mathbb{E}^n.$$

From (2.5.8) we get for $i, j \leqq n - 1$,

$$\begin{aligned} h_{ij}(u) &= \langle \xi_i(u), e_j \rangle = \langle d\xi_u(e_i), e_j \rangle \\ &= \langle d\bar{\xi}_u(e_i), e_j \rangle = \langle \overline{W}_u e_i, e_j \rangle, \end{aligned}$$

hence

$$\langle \overline{W}_u \pi_u a, \pi_u b \rangle = \sum_{i,j=1}^{n-1} h_{ij}(u)a^i b^j,$$

which completes the proof of the lemma. ■

Immediately from (2.5.18) and (2.5.17) we see the following.

Corollary 2.5.2. *The eigenvectors (with respect to $\langle \cdot, \cdot \rangle$) of the second differential of h_K at $u \in S^{n-1}$ are u, with corresponding eigenvalue 0, and the eigenvectors of the reverse Weingarten map at u, with corresponding eigenvalues $r_1, \ldots, r_{n-1}$, the principal radii of curvature.*

It now follows from Theorem 1.5.10 that the principal radii of curvature are non-negative.

Corollary 2.5.3. *For $j \in \{1, \ldots, n-1\}$, $\binom{n-1}{j}s_j(u)$, the jth elementary symmetric function of the principal radii of curvature at $u \in S^{n-1}$, is equal to the sum of the principal minors of order j of the Hessian matrix (with respect to any orthonormal basis of $\mathbb{E}^n$) of h_K at u.*

In particular

$$(n-1)s_1 = r_1 + \ldots + r_{n-1} = \Delta h_K \quad \text{on } S^{n-1}, \tag{2.5.19}$$

where Δ denotes the Laplace operator on $\mathbb{E}^n$.

For $n = 2$, there is only one principal radius of curvature, and we call it *the* radius of curvature, denoted by r.

For K as before, let $u \in S^{n-1}$ and a unit vector $t \in T_u$ be given. We denote by $r(u, t)$ the radius of curvature, at u, of the image $\text{proj}_E K$ of

K under orthogonal projection onto the linear subspace E spanned by u and t. To compute $r(u, t)$, we can choose an orthonormal basis $(e_1, \ldots, e_n)$ of $\mathbb{E}^n$ such that $e_1 = t$ and $e_n = u$. The support function of $\operatorname{proj}_E K$ is just the restriction of h_K to E, hence (2.5.19) gives

$$r(u, t) = h_{11}(u) + h_{nn}(u) = h_{11}(u),$$

where the notation is as in the proof of Lemma 2.5.1. By that lemma,

$$\begin{aligned}\overline{II}_u(t, t) &= \langle \overline{W}_u t, t\rangle = \mathrm{d}^2 h_u(t, t)\\ &= \sum_{i,j=1}^{n-1} h_{ij}(u)\delta_1^i\delta_1^j = h_{11}(u).\end{aligned}$$

Thus

$$r(u, t) = \frac{\overline{II}_u(t, t)}{\langle t, t\rangle} \quad \text{for } t \in T_u\backslash\{o\}, \tag{2.5.20}$$

in analogy to (2.5.6), and in analogy to Euler's theorem (2.5.7) we have

$$r(u, t) = r_1\langle t, v_1\rangle^2 + \ldots + r_{n-1}\langle t, v_{n-1}\rangle^2 \tag{2.5.21}$$

for each unit vector $t \in T_u$, where v_i is the unit eigenvector of $\overline{W}_u$ corresponding to r_i.

In the planar case, $n = 2$, it is often more convenient not to use the homogeneous support function h_K, but the function $h\colon [0, 2\pi] \to \mathbb{R}$ defined by $h(\alpha) = h_K(u_\alpha)$, where (e_1, e_2) is an orthonormal basis of $\mathbb{E}^2$ and $u_\alpha = e_1 \cos\alpha + e_2 \sin\alpha$. Using the homogeneity relations, one then computes $\Delta h_K(u_\alpha) = (h'' + h)(\alpha)$, hence the radius of curvature is given by

$$r(u_\alpha) = (h'' + h)(\alpha). \tag{2.5.22}$$

To extend (2.5.22) to $n > 2$, we introduce the spherical Laplace operator Δ_S on S^{n-1}. A simple way to do this is as follows (see, e.g., Seeley 1966). If $f\colon S^{n-1} \to \mathbb{R}$ is a function of class C^2, one extends it to $\mathbb{E}^n\backslash\{o\}$ as a positively homogeneous function of degree zero; then $\Delta_S f$ is the restriction of Δf to S^{n-1}. If $f\colon \mathbb{E}^n\backslash\{o\} \to \mathbb{R}$ is positively homogeneous of degree k, then

$$\Delta f = \Delta_S f + k(k + n - 2)f \quad \text{on } S^{n-1}.$$

Hence, (2.5.19) gives

$$s_1 = h_K + \frac{1}{n-1}\Delta_S h_K. \tag{2.5.23}$$

All this can be applied to convex bodies $K \in \mathcal{K}_0^n$ that are of class C_+^2. As seen before, the support function h_K of such a body is of class C^2, and we have $\xi(u) = \nu^{-1}(u)$ for $u \in S^{n-1}$. This yields

$$\overline{W}_u = \mathrm{d}\overline{\xi}_u = (\mathrm{d}\nu_{\xi(u)})^{-1} = W_{\xi(u)}^{-1}, \tag{2.5.24}$$

so that the reverse Weingarten map at u coincides with the inverse Weingarten map taken at $\xi(u)$. In particular, the eigenvectors of $W_{\xi(u)}$ and W_u are the same, all principal radii of curvature are positive and

$$r_i(u) = \frac{1}{k_i(\xi(u))} \tag{2.5.26}$$

if the ordering is chosen properly. This implies

$$s_j(u) = \frac{H_{n-1-j}}{H_{n-1}}(\xi(u)), \tag{2.5.27}$$

$$H_j(x) = \frac{s_{n-1-j}}{s_{n-1}}(\nu(x)), \tag{2.5.28}$$

for $j = 1, \ldots, n-1$.

To have complete symmetry, it remains to show that a convex body $K \in \mathcal{K}_0^n$ with a support function of class C^2 and with everywhere positive radii of curvature is necessarily of class C_+^2. For the proof, we may assume that $o \in \operatorname{int} K$. By Remark 1.7.7, the radius function of the polar body K^* is given by $\rho(K^*, \cdot) = h(K, \cdot)^{-1}$ on S^{n-1}, hence K^* is of class C^2. As before, we use a local parametrization $N: M \to \mathbb{E}^n$ of class C^2 of the sphere S^{n-1} around u and put $\overline{X} = \overline{\xi} \circ N$. From $\overline{W}_u N_i = \overline{X}_i$ and the assumption that all eigenvalues of $\overline{W}_u$ are non-zero, it follows that $\overline{X}_1, \ldots, \overline{X}_{n-1}$ are linearly independent. Hence, $\overline{X}$ is a local parametrization of class C^1 of bd K. In particular, it follows that K is of class C^1 and that $\overline{\xi}$ is a diffeomorphism of class C^1 from S^{n-1} onto bd K. The spherical image map ν^* of K^* can be described by

$$\nu^*: \rho(K^*, u)u \mapsto \frac{\overline{\xi}(u)}{|\overline{\xi}(u)|}, \quad u \in S^{n-1},$$

(see Remark 1.7.7) and thus is a composition of C^1-diffeomorphisms. Since ν^* is a diffeomorphism, K^* is of class C_+^2 and hence has a support function of class C^2 and nowhere vanishing radii of curvature. Repetition of the argument, with the roles of K and K^* interchanged, shows that K is of class C_+^2.

The elementary symmetric functions s_j and H_j appear in some formulae related to surface area and volume computations. Let $K \in \mathcal{K}_0^n$ be a convex body of class C_+^2. Let $X: M \to \mathbb{E}^n$, where $M \subset \mathbb{R}^{n-1}$ is open, be a local C^2-parametrization of bd K. As is well known from differential geometry, the $\mathcal{H}^{n-1}$-measure of $X(M)$ can be computed from

$$\mathcal{H}^{n-1}(X(M)) = \int_M \mathrm{d}A,$$

with the differential geometric surface area element given by

$$\mathrm{d}A = \sqrt{\det(g_{ij})}\,\mathrm{d}y^1 \cdots \mathrm{d}y^{n-1},$$

where $y^1, \ldots, y^{n-1}$ are Cartesian coordinates in M. Since K is of class C^2_+, the map $N = \nu \circ X$ is a local C^1-parametrization of S^{n-1}, hence

$$\mathcal{H}^{n-1}(N(M)) = \int_M \mathrm{d}\sigma$$

with

$$\mathrm{d}\sigma = \sqrt{\det(e_{ij})}\,\mathrm{d}y^1 \cdots \mathrm{d}y^{n-1}.$$

Now (2.5.10) gives

$$g_{ij} = \langle X_i, X_j \rangle = \left\langle \sum_r b_i^r N_r, \sum_s b_j^s N_s \right\rangle = \sum_{r,s} b_i^r b_j^s e_{rs},$$

hence $\det(g_{ij}) = [\det(b_i^j)]^2 \det(e_{ij})$ and thus

$$\mathrm{d}A = |\det(b_i^j)|\mathrm{d}\sigma = s_{n-1} \circ N \mathrm{d}\sigma$$

by (2.5.15). We deduce that for any integrable real function f on bd K we have

$$\int_{\mathrm{bd}\,K} f(x)\,\mathrm{d}\mathcal{H}^{n-1}(x) = \int_{S^{n-1}} f(\xi(u)) s_{n-1}(u)\,\mathrm{d}\mathcal{H}^{n-1}(u). \quad (2.5.29)$$

Similarly, for any integrable real function g on S^{n-1} we have

$$\int_{S^{n-1}} g(u)\,\mathrm{d}\mathcal{H}^{n-1}(u) = \int_{\mathrm{bd}\,K} g(\nu(x)) H_{n-1}(x)\,\mathrm{d}\mathcal{H}^{n-1}(x). \quad (2.5.30)$$

Next, we compute a 'local parallel volume', which will play an essential role in Chapter 4. For a relatively open subset β of bd K, let

$$A_\rho(K, \beta) := \{x \in \mathbb{E}^n | 0 < \mathrm{d}(K, x) \leqq \rho \text{ and } p(K, x) \in \beta\}$$

for $\rho > 0$. Thus $A_\rho(K, \beta)$ is the set of all points in $\mathbb{E}^n$ at positive distance at most ρ from K and with the nearest point in K falling in β. Similarly, for an open subset ω of S^{n-1} we define

$$B_\rho(K, \omega) := \{x \in \mathbb{E}^n | 0 < d(K, x) \leqq \rho \text{ and } u(K, x) \in \omega\}$$

for $\rho > 0$. If K is smooth and strictly convex, we clearly have $A_\rho(K, \beta) = B_\rho(K, \nu(\beta))$. For K of class C^2_+, the measure of this set can be computed as follows. It suffices to assume that $\beta = X(M)$ for some local parametrization X as above. Then, obviously,

$$A_\rho(K, \beta) = \{X(y) + \lambda N(y) | y \in M,\ 0 < \lambda \leqq \rho\}.$$

Thus $A_\rho(K, \beta)$ is the image of the region $M \times (0, \rho] \subset \mathbb{R}^{n-1} \times \mathbb{R}$ under the injective map $(y, \lambda) \mapsto X(y) + \lambda N(y)$, which has Jacobian

$$|X_1 + \lambda N_1, \ldots, X_{n-1} + \lambda N_{n-1}, N|,$$

where $|\cdot, \ldots, \cdot|$ denotes the determinant. For this Jacobian, we obtain from (2.5.10) and (2.5.16) that it is equal to

$$\left|\sum_j (b_1^j + \lambda\delta_1^j)N_j, \ldots, \sum_j (b_{n-1}^j + \lambda\delta_{n-1}^j)N_j, N\right|$$
$$= \det(b_i^j + \lambda\delta_i^j)|N_1, \ldots, N_{n-1}, N|$$
$$= \pm \sum_{k=0}^{n-1} \lambda^k \binom{n-1}{k} s_{n-1-k} \circ N\sqrt{\det(e_{ij})}.$$

Here we have used the fact that $|N_1, \ldots, N_{n-1}, N|^2$ is equal to the square of the $(n-1)$-volume of the parallelepiped spanned by $N_1, \ldots, N_{n-1}$, which is equal to the Gram determinant $\det(\langle N_i, N_j\rangle)$. Integration of the absolute value now yields, with $\omega = \nu(\beta)$, or $\beta = \xi(\omega)$,

$$\begin{aligned}\mathcal{H}^n(A_\rho(K, \beta)) &= \mathcal{H}^n(B_\rho(K, \omega))\\ &= \frac{1}{n}\sum_{m=0}^{n-1} \rho^{n-m}\binom{n}{m}\int_\omega s_m \, d\mathcal{H}^{n-1}\\ &= \frac{1}{n}\sum_{m=0}^{n-1} \rho^{n-m}\binom{n}{m}\int_\beta H_{n-1-m} \, d\mathcal{H}^{n-1},\end{aligned} \tag{2.5.31}$$

where (2.5.30) and (2.5.27) have been used.

So far, we have considered only one convex body K and have, therefore, not indicated that the notation introduced depends on K. In the following, we shall write $\kappa(K, x, t)$, $II_x(K, v, w)$, $II_x(K)$, $r(K, u, t)$, respectively, for $\kappa(x, t)$, $II_x(v, w)$, II_x, $r(u, t)$, and so on, where necessary.

The following theorem, on pairs of convex bodies, allows us to deduce a global inclusion result (up to translations) from local curvature comparisons. Here $II_x \geqq II'_y$ means $II_x(t, t) \geqq II'_y(t, t)$ for all t.

Theorem 2.5.4. *Let $K, L \in \mathcal{K}^n$ be convex bodies of class C_+^2. Then the following conditions are equivalent.*

(a) *$II_x(L) \geqq II_y(K)$ for all pairs of points $x \in \operatorname{bd} L$ and $y \in \operatorname{bd} K$ at which the outer unit normal vectors are the same;*

(b) *$r(L, u, t) \leqq r(K, u, t)$ for each orthonormal pair of vectors u, t;*

(c) *$h_K - h_L$ is a support function.*

Proof. Of course, (a) is also equivalent to

(a′) $\kappa(L, x, t) \leqq \kappa(K, y, t)$ for x, y as in (a) and all $t \in T_xL = T_yK$ and (b) is equivalent to

(b′) $\bar{II}_u(L) \leqq \bar{II}_u(K)$ for $u \in S^{n-1}$.

By Theorems 1.5.10 and 1.7.1, $h_k - h_L$ is a support function if and only if $d^2(h_K - h_L)$ is positive semidefinite. By Lemma 2.5.1, this is equivalent to (b′) and thus, by (2.5.20), to (b). Using (2.5.24), we have, for $t \in T_u \cap S^{n-1}$,

$$\bar{II}_u(t, t) = \langle \overline{W}_u t, t \rangle = \langle W_{\xi(u)}^{-1} t, t \rangle,$$
$$II_{\xi(u)}(t, t) = \langle W_{\xi(u)} t, t \rangle.$$

The equivalence of (a) and (b′) now follows, since $A \geqq B$ for self-adjoint positive endomorphisms A, B of T_u implies $B^{-1} \geqq A^{-1}$. ($A \geqq B$ means $\langle At, t \rangle \geqq \langle Bt, t \rangle$ for all $t \in T_u$ and hence is equivalent to the fact that each eigenvalue of A relative to B is $\geqq 1$. But $Ae_i = \lambda_i Be_i$ is equivalent to $B^{-1}(Be_i) = \lambda_i A^{-1}(Be_i)$; thus the eigenvalues of A relative to B coincide with the eigenvalues of B^{-1} relative to A^{-1}.) ■

The assertion (c) in Theorem 2.5.4 implies that, for suitable $z \in \mathbb{E}^n$, $h_K - h_{L+z} \geqq 0$ and thus $L + z \subset K$ (see also Theorem 3.2.9).

The reverse Weingarten map and the notions derived from it have the advantage that they show simple behaviour under Minkowski addition. Let $K_1, \ldots, K_m \in \mathcal{K}_0^n$ be convex bodies with support functions of class C^2, and let

$$K = \lambda_1 K_1 + \ldots + \lambda_m K_m$$

with $\lambda_1, \ldots, \lambda_m \geqq 0$. Then (2.5.8) implies that

$$\xi(K, u) = \sum_{r=1}^{m} \lambda_r \xi(K_r, u),$$

which in turn yields

$$\overline{W}_u(K) = \sum_{r=1}^{m} \lambda_r \overline{W}_u(K_r),$$

$$\bar{II}_u(K) = \sum_{r=1}^{m} \lambda_r \bar{II}_u(K_r).$$

If $N\colon M \to \mathbb{E}^n$ is a local C^2-parametrization of S^{n-1} and if $X^{(r)} := \xi(K_r, N(\cdot))$, then the matrix of $\bar{II}_u(K_r)$ with respect to the basis $(N_1, \ldots, N_{n-1})$ is given by $b_{ij}(K_r) = \langle X_i^{(r)}, N_j \rangle$, hence

$$b_{ij}(K) = \sum_{r=1}^{m} \lambda_r b_{ij}(K_r).$$

In particular, for the product of the principal radii of curvature of the Minkowski linear combination K we get, by (2.5.15),

$$s_{n-1}(K, N(\cdot)) = [\det(e_{ij})]^{-1} \det(\lambda_1 b_{ij}(K_1) + \ldots + \lambda_m b_{ij}(K_m)).$$

Generally, if $A_1, \ldots, A_m$ are symmetric real $k \times k$ matrices, the determinant of $\lambda_1 A_1 + \ldots + \lambda_m A_m$ is a homogeneous polynomial of degree k in $\lambda_1, \ldots, \lambda_m$. It can be written as

$$\det(\lambda_1 A_1 + \ldots + \lambda_m A_m) = \sum_{i_1,\ldots,i_k=1}^{m} \lambda_{i_1} \cdots \lambda_{i_k} D(A_{i_1}, \ldots, A_{i_k}), \tag{2.5.32}$$

since the coefficient of $\lambda_{i_1} \cdots \lambda_{i_k}$ does, in fact, depend only on $A_{i_1}, \ldots, A_{i_k}$. One may assume that the coefficients are symmetric in their arguments, and then they are uniquely determined. $D(A_1, \ldots, A_k)$ is called the *mixed discriminant* of $A_1, \ldots, A_k$.

We deduce that we have an identity

$$s_{n-1}(\lambda_1 K_1 + \ldots + \lambda_m K_m, u) = \sum_{i_1,\ldots,i_{n-1}=1}^{m} \lambda_{i_1} \cdots \lambda_{i_{n-1}} s(K_{i_1}, \ldots, K_{i_{n-1}}, u) \tag{2.5.33}$$

for $u \in S^{n-1}$, where s is symmetric in its first $n-1$ arguments. We call $s(K_1, \ldots, K_{n-1}, \cdot)$, which is defined by (2.5.33), the *mixed curvature function* of $K_1, \ldots, K_{n-1}$. In local coordinates as above,

$$s(K_1, \ldots, K_{n-1}, N(\cdot)) = [\det(e_{ij})]^{-1} D((b_{ij}(K_1)), \ldots, (b_{ij}(K_{n-1}))). \tag{2.5.34}$$

In particular, we see from this and (2.5.16) that

$$s_j(K, \cdot) = s(\underbrace{K, \ldots, K}_{j}, \underbrace{B^n, \ldots, B^n}_{n-1-j}, \cdot) \tag{2.5.35}$$

for $j = 0, \ldots, n-1$ and any convex body K with a support function of class C^2.

If we want to express $s(K_1, \ldots, K_{n-1}, u)$ by means of support functions, we may, for fixed $u \in S^{n-1}$, choose an orthonormal basis $(e_1, \ldots, e_n)$ of $\mathbb{E}^n$ with $e_n = u$. If $K \in \mathcal{K}^n$ has a support function h of class C^2, then $h_{in}(u) = 0$ by (2.5.18), hence Corollary 2.5.3 gives

$$s_{n-1}(K, u) = \det(h_{ij}(u))_{i,j=1}^{n-1}.$$

We deduce that

$$s(K_1, \ldots, K_{n-1}, u) = D((h_{ij}(K_1, u))_{i,j=1}^{n-1}, \ldots, (h_{ij}(K_{n-1}, u))_{i,j=1}^{n-1}). \tag{2.5.36}$$

Finally, in this section, we come back to general convex bodies and describe, without proofs, their curvature behaviour. Let $K \in \mathcal{K}_0^n$, let x be a smooth boundary point of K and let u be the outer unit normal vector of K at x. We may assume that bd K is represented around x by a convex function f, as in (2.5.1). For $h > 0$, define

$$D(h) := (2h)^{-1/2}\{t \in T_x K \cap B(o, \varepsilon) \mid f(t) \leqq h\}.$$

If the Hausdorff closed limit $\lim_{h \downarrow 0} D(h) := D$ exists, this set is called the (Dupin) *indicatrix* of K at x. If K is of class C^2, then differential geometry tells us that D exists and is given by

$$D = \left\{ \sum_{i=1}^{n-1} y^i e_i \middle| k_1(y^1)^2 + \ldots + k_{n-1}(y^{n-1})^2 \leqq 1 \right\} \tag{2.5.37}$$

if $(e_1, \ldots, e_{n-1})$ is an orthonormal basis of eigenvectors of the Weingarten map at x and k_i is the principal curvature corresponding to e_i. For a general convex body K, the (smooth) point x is called a *normal* point or *Euler point* of K if at x the indicatrix exists and its boundary (if any) is a quadric in T_xK with centre x. If x is a normal point, then the indicatrix at x can be represented in the form (2.5.37), with a suitable orthonormal basis $(e_1, \ldots, e_{n-1})$, and this defines the principal curvatures $k_1, \ldots, k_{n-1}$. They are non-negative, since the indicatrix is always a convex set. Furthermore, it follows from the definition of D that the curvature $\kappa(x, t)$ exists for each unit vector $t \in T_x$ and is given by

$$\kappa(x, t) = k_1\langle t, e_1\rangle^2 + \ldots + k_{n-1}\langle t, e_{n-1}\rangle^2;$$

thus Euler's theorem holds.

It follows from work of Aleksandrov (1939c) that x is a normal point of K if and only if the representing function f is twice differentiable at o, in the sense explained in Section 1.5, Note 2. The almost everywhere twice differentiability of convex functions then yields the following theorem due to Busemann & Feller (1936a) for $n \leqq 3$ and to Aleksandrov (1939c) for arbitrary n.

Theorem 2.5.5. *For each convex body $K \in \mathcal{K}_0^n$, $\mathcal{H}^{n-1}$-almost all boundary points are normal.*

Although this is a strong theorem, it should be kept in mind that essential information on the shape of a general convex body may be carried by those points that are not normal. This is shown clearly by the example of a polytope.

Notes for Section 2.5

1. *Jessen's radii of curvature.* The definitions of lower and upper radii of curvature given at the beginning of Section 2.5 are not the only natural ones. To describe a different possibility, we assume that $n = 2$ and that $K \in \mathcal{K}_0^n$ is represented in a neighbourhood of the smooth boundary point x in the way given by (2.5.1). Let $t \in T_xK \cap S^{n-1}$. For $0 < \tau_j < \varepsilon$ let v_j be an outer unit normal vector of K at $x(\tau_j) := x + \tau_j t - f(\tau_j t)u$, and let $z(\tau_j, v_j)$ be the intersection point of the normals $\{x - \lambda u | \lambda \geqq 0\}$ and

$\{x(\tau_j) - \lambda v_j | \lambda \geqq 0\}$. If $\tau_j \to 0$ for $j \to \infty$ and $z(\tau_j, v_j)$ converges to some point $x - ru$, thus

$$r = \lim_{j\to\infty} \frac{\tau_j}{\sqrt{1 - \langle u, v_j\rangle^2}},$$

then r is called a *Jessen radius of curvature* of K at x in the direction t. The set of all numbers r (∞ admitted) obtainable in this way is an interval $[r_i, r_s]$ with $0 \leqq r_i \leqq r_s \leqq \infty$. It contains the interval $[\rho_i, \rho_s]$. Jessen (1929) showed that each Jessen radius of curvature, say r, satisfies

$$\rho_s - \sqrt{\rho_s(\rho_s - \rho_i)} \leqq r \leqq \rho_s + \sqrt{\rho_s(\rho_s - \rho_i)},$$

and that these inequalities comprise the only general restrictions for ρ_i, ρ_s, r_i, r_s, In particular, $\rho_i = \rho_s$ implies $r_i = r_s$. Connections with the second-order differentiability properties of f at o are described in Busemann [12], Section 2.

2. *Complete Riemannian submanifolds with non-negative sectional curvatures*. If $K \in \mathcal{K}_0^n$ is of class C^∞ (say), then bd K with the induced Riemannian metric is a Riemannian manifold with non-negative sectional curvatures. This assertion has an inverse that is not restricted to the compact case:

   Theorem (Sacksteder). If M is a C^∞ $(n-1)$-dimensional complete orientable Riemannian manifold of non-negative sectional curvature that is not identically zero, and if $X: M \to \mathbb{E}^n$ is an isometric immersion, then X is an embedding and $X(M)$ is the boundary of a convex set.

   This theorem is due to Sacksteder (1960), with special cases proved before by Hadamard (1897), Stoker (1936), van Heijenoort (1952) and Chern & Lashof (1958).

   The Gauss map of such submanifolds was investigated by Wu (1974). More generally, let $K \subset \mathbb{E}^n$ be a closed convex set with interior points, $K \neq \mathbb{E}^n$, and let $\sigma(K)$ be the set of all outer unit normal vectors at boundary points of K. Then Wu showed that there exists a k-dimensional great subsphere S^k of S^{n-1} $(0 \leqq k \leqq n-1)$ and a unique subset σ_0 of S^k that is relatively open and (geodesically) convex such that $\sigma_0 \subset \sigma(K) \subset \operatorname{cl} \sigma_0$. Wu also showed by an example that $\sigma(K)$ itself need not be convex.

3. *Boundaries and support functions of class C^2*. That the support function of a convex surface of class C^2 with positive curvatures is of class C^2 was noted by Wintner (1952, Appendix). Here the assumption of positive curvatures cannot be omitted. Hartman & Wintner (1953) gave an example showing that the support function of a convex body need not be of class C^2 even if the boundary is real-analytic and also treated some related questions.

4. *Differentiability properties of projections and Minkowski sums*. Let $K \in \mathcal{K}_0^n$ be a convex body of class C_+^2, and let K' be its projection onto a k-dimensional linear subspace E. The support function h_K of K is of class C^2, and the support function of K' is obtained from h_K by restriction; hence it is also of class C^2. Since all radii of curvature of K are positive, the same holds for K', as can be deduced from (2.5.20) in connection with Lemma 2.5.1. Thus K' is of class C_+^2.

   Without the assumption of positive curvatures, no such argument is possible. The case $n = 3$, $k = 2$ was investigated by Kiselman (1986). He showed that K' need not be of class C^2 even if K is of class C^∞. Further, he proved: If the boundary of K is of class C^2 with Lipschitz continuous second derivatives, then K' has a twice differentiable boundary. If bd K is real-analytic, then the boundary of K' is of Hölder class $C^{2+\varepsilon}$ for some $\varepsilon > 0$ (but it may be exactly of class $C^{2+2/q}$ for any odd integer $q \geqq 3$).

Even more unexpected are the results that Kiselman (1987) obtained for Minkowski sums. He posed the question of when it is true that $K + L$ is of class C^k if $K, L \in \mathcal{K}_0^n$ are of class C^k? He showed that for $n = 2$ the answer is in the affirmative for $k = 1, 2, 3, 4$ but not for $k = 7$. Further, the boundary of $K + L$, where $K, L \in \mathcal{K}_0^2$ have real-analytic boundaries, is of Hölder class $C^{6+2/3}$, and this is best possible. He further mentions an example of convex bodies $K, L \in \mathcal{K}_0^3$ with C^∞ boundaries such that $K + L$ is not of class C^2.

5. A weaker form of Theorem 2.5.4 is due to Rauch (1974). He proved, in a different way, that condition (a) of Theorem 2.5.4 implies that some translate of L is contained in K. For the proof given above, see Schneider (1988b); compare also Note 6 in Section 3.2.
6. *Differential-geometric properties of general convex surfaces*. Theorem 2.5.5 is, of course, closely related to second-order differentiability properties of convex functions, and we refer to Section 1.5, Note 2, for relevant references. A detailed investigation of the differential-geometric properties of convex surfaces without differentiability assumptions was undertaken by Busemann & Feller (1936a, 1935a, b, 1936b). Among the topics that they treated, for convex surfaces for which no regularity was assumed, were versions of the theorems of Meusnier, Olinde Rodrigues and Euler, umbilics, shortest curves, upper and lower indicatrices, Gauss curvature and spherical images; see also Busemann [12]. The extension of Theorem 2.5.5 from $n = 3$ (Busemann & Feller 1936a) to general n by Aleksandrov (1939c) is not straightforward. Among several other related results, Aleksandrov also investigated consequences of the twice differentiability almost everywhere of support functions.
7. *ε-smooth points.* If $K \in \mathcal{K}_0^n$ and $x \in \operatorname{bd} K$ is a normal point of K, then x is an ε-smooth point of K for some $\varepsilon > 0$, that is, there exists a ball B of radius ε such that $x \in B \subset K$. Hence, it follows from Theorem 2.5.5 that the set of all boundary points of K that are not ε-smooth for any $\varepsilon > 0$ is of $\mathcal{H}^{n-1}$-measure zero. For this result, McMullen (1974b) has given a simple direct proof, thus answering a question posed by Sallee (1972).
8. *Limit sections*. The concept of the Dupin indicatrix can be generalized as follows, say, for $n = 3$. Let $K \in \mathcal{K}_0^3$, $x \in \operatorname{bd} K$, H a support plane to K at x, and $(H_i)_{i\in\mathbb{N}}$ a sequence of planes meeting K and converging to H. If there is a sequence $(\sigma_i)_{i\in\mathbb{N}}$ of similarities such that the sequence $(\sigma_i(H_i \cap K))_{i\in\mathbb{N}}$ converges to some two-dimensional convex body C, then C is called a *limit section* of K at x. The point x is called *universal* if every plane convex body is a limit section of K at x, for suitable choices of H_i and σ_i; it is called *p-universal* if an analogous statement is true where the planes H_i have to be parallel to H. Melzak (1958) proved the existence of a convex body with a p-universal point (and otherwise of class C^∞), the existence of a convex body for which every boundary point is universal and related results.
9. *Approximation with converging radii of curvature*. From the counterpart of Theorem 2.5.5 for support functions (Aleksandrov 1939c) it follows that a convex body $K \in \mathcal{K}_0^n$ has principal radii of curvature $r_i(u)$, where

$$0 \leqq r_1(u) \leqq r_2(u) \leqq \ldots \leqq r_{n-1}(u) < \infty$$

for σ-almost all $u \in S^{n-1}$; here σ denotes the spherical Lebesgue measure on S^{n-1}. The $r_i(u)$ can be defined as the eigenvalues, corresponding to eigenvectors orthogonal to u, of the second differential of the support function h_K at $u \in S^{n-1}$ (cf. Corollary 2.5.2); they are defined σ-almost everywhere on S^{n-1} and are measurable and σ-integrable. The following approximation result was proved and applied by Weil (1973); for an extension, see Weil (1974a).

Theorem. For any convex body $K \in \mathcal{K}^n$ there exists a sequence $(K_i)_{i \in \mathbb{N}}$ of convex bodies of class C^2_+ converging to K such that the radii of curvature of K_i and K, denoted by $r_1^{(i)}, \ldots, r_{n-1}^{(i)}$ and $r_1, \ldots, r_{n-1}$, respectively, satisfy the following assertions:

(a) $r_j^{(i)} \to r_j$ σ-almost everywhere for $i \to \infty$, $j = 1, \ldots, n-1$;

(b) $r_j^{(i)} \to r_j$ in the $L^1(S^{n-1})$ norm for $i \to \infty$, $j = 1, \ldots, n-1$;

(c) the measures $\int r_j^{(i)} \, d\sigma$ converge weakly to the measure $\int r_j \, d\sigma$ for $i \to \infty$, $j = 1, \ldots, n-2$.

10. *Tangential radii of curvature.* The number $r(u, t)$ defined by (2.5.20) is sometimes called the *tangential radius of curvature* of K at $\xi(u)$ in the direction t. Formula (2.5.21) was proved by Blaschke [5], p. 117, for $n = 3$; see also Firey (1967b), p. 12, for the general case.

2.6. Generic boundary structure

Due to their simple boundary structure, polytopes as well as C^2_+-regular convex bodies are favourite objects of investigation. On the other hand, one may argue that these are very special convex bodies, and one may ask whether there exists something like the boundary behaviour of a 'general' convex body. This question can be made precise if the notion of Baire category is used, and then several affirmative answers, often surprising ones, are possible.

The space $\mathcal{K}^n$ of convex bodies with the Hausdorff metric is a complete metric space, by Theorems 1.8.2 and 1.8.5, and thus a Baire space. This means that the complement of every first category subset of $\mathcal{K}^n$ is dense in $\mathcal{K}^n$. A subset is called *of first category* or *meagre* if it is a countable union of nowhere dense subsets. A subset that is not of first category is said to be *of second category*. The complement of a meagre set is called *residual*. In particular, a dense open subset is residual, and an intersection of countably many residual sets is residual. Meagre subsets of a Baire space can be considered as 'small'. We say, therefore, that a certain property E for elements of a Baire space is *generic* if the subset of all elements having property E is residual. For the space $\mathcal{K}^n$ it has become customary to say instead that *most* convex bodies have the property E, or that a *typical* convex body has this property.

Category arguments can be used to give (non-constructive) existence proofs for objects with certain 'irregularity' properties, which otherwise may be hard to obtain. When such an argument works it shows that, in fact, most elements of the space have the property in question. For the boundary structure of convex bodies, which is our concern in the present chapter, there are several results showing that a typical convex body has properties that, at first sight, one would be inclined to consider as 'pathological'. This holds, at least, for second-order properties, whereas the typical first-order boundary behaviour is not very exciting.

The following theorem shows, in a slightly more precise form, that most convex bodies are smooth and strictly convex.

Theorem 2.6.1. *In $\mathcal{K}^n$, the set of smooth convex bodies and the set of strictly convex bodies are dense G_δ-sets; hence, most convex bodies are smooth and strictly convex.*

Proof. For $k \in \mathbb{N}$, let $\mathcal{A}_k$ be the set of all $K \in \mathcal{K}^n$ having a boundary point at which there exist two outer normal vectors forming an angle $\geqq 1/k$, and let $\mathcal{B}_k$ be the set of all $K \in \mathcal{K}^n$ for which $\operatorname{bd} K$ contains a segment of length $\geqq 1/k$. Since $\mathcal{K}^n \setminus \bigcup_{k\in\mathbb{N}} \mathcal{A}_k$ is the set of all smooth convex bodies in $\mathcal{K}^n$ and $\mathcal{K}^n \setminus \bigcup_{k\in\mathbb{N}} \mathcal{B}_k$ is the set of all strictly convex bodies, it suffices to show that $\mathcal{A}_k$ and $\mathcal{B}_k$ are closed and nowhere dense. Let $(K_i)_{i\in\mathbb{N}}$ be a sequence in $\mathcal{A}_k$ converging to $K \in \mathcal{K}^n$. For each $i \in \mathbb{N}$, we choose a point $x_i \in \operatorname{bd} K_i$ at which there exist two outward unit normal vectors u_i, v_i of K_i, forming an angle $\geqq 1/k$. After selecting subsequences and changing the notation, we may assume that $x_i \to x$, $u_i \to u$ and $v_i \to v$ for $i \to \infty$. Then $x \in K$ (by Theorem 1.8.7), and from $\langle x_i, u_i \rangle = h(K_i, u_i)$, $\langle x_i, v_i \rangle = h(K_i, v_i)$ and Lemma 1.8.10 it follows that $\langle x, u \rangle = h(K, u)$ and $\langle x, v \rangle = h(K, v)$; thus $x \in \operatorname{bd} K$ and $u, v \in N(K, x)$. Since u and v form an angle $\geqq 1/k$, we conclude that $K \in \mathcal{A}_k$. Thus $\mathcal{A}_k$ is closed. In a similar way, one easily shows that $\mathcal{B}_k$ is closed. Since the set of smooth and strictly convex bodies is dense in $\mathcal{K}^n$ (this follows, e.g., from Section 3.3), the closed sets $\mathcal{A}_k$ and $\mathcal{B}_k$ have empty interior and thus are nowhere dense. ∎

While Theorem 2.6.1 shows that the boundary and the support function of a typical convex body are differentiable, in general no stronger differentiability properties are satisfied. This is shown by the behaviour of the curvatures described in the theorems below. These results were discovered by Zamfirescu (1980a, b).

By $\mathcal{K}_r^n \subset \mathcal{K}^n$ we denote the subset of smooth (= regular) convex bodies. For these bodies, the lower and upper radii of curvature were defined in Section 2.5. By Theorem 2.6.1, $\mathcal{K}_r^n$ is a dense G_δ-set in $\mathcal{K}^n$. In particular, it is itself a Baire space, and every residual subset of $\mathcal{K}_r^n$ is residual in $\mathcal{K}^n$. Hence, it causes no loss of generality if we restrict the following considerations to the space $\mathcal{K}_r^n$.

Theorem 2.6.2. *For most convex bodies K in $\mathcal{K}_r^n$, at every boundary point $x \in \operatorname{bd} K$ and for every tangent direction t at x,*

$$\rho_i(x, t) = 0 \quad \textit{or} \quad \rho_s(x, t) = \infty.$$

Thus the typical convex body is smooth and strictly convex, and its curvatures $\kappa(x, t)$ are zero wherever they exist and are finite. By Aleksandrov's theorem, 2.5.5, the latter holds for almost all boundary points x of a convex body and for every tangent direction t at x.

***Corollary* 2.6.3.** *For most convex bodies K in $\mathcal{K}_r^n$, $\kappa(x, t) = 0$ for $\mathcal{H}^{n-1}$-almost all $x \in \operatorname{bd} K$ and every tangent direction t at x.*

The preceding results do not exclude the possibility that simultaneously $\rho_i(x, t) = 0$ and $\rho_s(x, t) = \infty$ for the points x in a boundary set of K of $\mathcal{H}^{n-1}$-measure zero. Typically, in fact, this holds in a residual set:

***Theorem* 2.6.4.** *For most convex bodies K in $\mathcal{K}_r^n$, at most boundary points x of K and for every tangent direction t at x,*

$$\rho_i(x, t) = 0 \quad \textit{and} \quad \rho_s(x, t) = \infty.$$

The ideas involved in the proofs given below for Theorems 2.6.2 and 2.6.4 are taken from Zamfirescu (1980a, 1985). We start with some preparations. Let $K \in \mathcal{K}_r^n$, $x \in \operatorname{bd} K$ and u be the unique outer unit normal vector of K at x; let $t \in T_x K$ be a unit tangent vector. We define a two-dimensional half-plane by

$$D(x, u, t) := \{x + \lambda u + \mu t \mid \lambda \in \mathbb{R}, \mu \geqq 0\}$$

and a convex arc by

$$S(K, x, t) := \operatorname{bd} K \cap D(x, u, t).$$

For $\alpha > 0$, we say that K is *α-flat at* (x, t) if

$$S(K, x, t) \cap \operatorname{int} B(x - \alpha u, \alpha) = \emptyset,$$

and K is *α-curved at* (x, t) if

$$S(K, x, t) \subset B(x - \alpha u, \alpha).$$

It follows from the definitions of the lower and upper radii of curvature that $\rho_i(x, t) > 0$ if and only if K is α-flat at (x, t) for some $\alpha > 0$, and that $\rho_s(x, t) < \infty$ if and only if K is α-curved at (x, t) for some $\alpha > 0$.

For $k \in \mathbb{N}$ we define the following sets.

$\mathcal{A}_k := \{K \in \mathcal{K}_r^n \mid$ There exist $x \in \operatorname{bd} K$ and $t \in T_x K$ such that K is k^{-1}-flat at (x, t) and k-curved at $(x, t)\}$.

For given $K \in \mathcal{K}_r^n$, let

$A_k^1(K) := \{x \in \operatorname{bd} K \mid$ There exists $t \in T_x K$ such that K is k^{-1}-flat at $(x, t)\}$,

$A_k^2(K) := \{x \in \operatorname{bd} K \mid$ There exists $t \in T_x K$ such that K is k-curved at $(x, t)\}$.

For $m \in \mathbb{N}$ and $i = 1, 2$ we put

$$\mathscr{C}^i_{k,m} := \{K \in \mathcal{K}^n_r \mid \text{There exists } x \in \operatorname{bd} K \text{ such that } B_0(x, m^{-1}) \cap \operatorname{bd} K \subset A^i_k(K)\}.$$

Lemma 2.6.5. *The sets* $A^1_k(K)$, $A^2_k(K)$ *are closed in* $\operatorname{bd} K$. *The sets* $\mathscr{A}_k$, $\mathscr{C}^1_{k,m}$, $\mathscr{C}^2_{k,m}$ *are closed in* $\mathcal{K}^n_r$ $(k, m \in \mathbb{N})$.

Proof. We start with $\mathscr{A}_k$. Let $(K_j)_{j\in\mathbb{N}}$ be a sequence in $\mathscr{A}_k$ converging to a convex body $K \in \mathcal{K}^n_r$. For each $j \in \mathbb{N}$, we choose a point $x_j \in \operatorname{bd} K_j$ and a unit tangent vector $t_j \in T_{x_j}K_j$ such that

$$S(K_j, x_j, t_j) \cap \operatorname{int} B(x_j - k^{-1}u_j, k^{-1}) = \emptyset,$$
$$S(K_j, x_j, t_j) \subset B(x_j - ku_j, k),$$

where u_j is the outer unit normal vector of K_j at x_j. After selecting suitable subsequences and changing the notation, we may assume that $x_j \to x$ and $t_j \to t$ for $j \to \infty$. Then it is clear that $x \in \operatorname{bd} K$, the sequence $(u_j)_{j\in\mathbb{N}}$ converges to the unique outer unit normal vector u of K at x, and t is a unit tangent vector of K at x. We assert that

$$S(K, x, t) \cap \operatorname{int} B(x - k^{-1}u, k^{-1}) = \emptyset, \tag{2.6.1}$$
$$S(K, x, t) \subset B(x - ku, k). \tag{2.6.2}$$

Suppose that (2.6.1) were false. Then there is a point $z \in S(K, x, t)$ and a ball B with centre z such that $B \subset \operatorname{int} B(x - k^{-1}u, k^{-1})$. Now

$$K_j \cap D(x_j, u_j, t_j) \to K \cap D(x, u, t) \quad \text{for} \quad j \to \infty.$$

This follows from Theorem 1.8.8, applied to K_j and the intersection L_j of $D(x_j, u_j, t_j)$ with some ball containing all the bodies $K_1, K_2, \dots,$ and from the fact that K and $D(x, u, t)$ cannot be separated by a hyperplane since K is smooth. It follows that

$$S(K_j, x_j, t_j) \cap \operatorname{int} B(x_j - k^{-1}u_j, k^{-1}) \neq \emptyset$$

for sufficiently large j, a contradiction. Thus (2.6.1) holds. Suppose that (2.6.2) were false. Then there is a point $z' \in S(K, x, t)$ and a ball B' with centre z' such that $B' \cap B(x - ku, k) = \emptyset$. As above, we conclude that

$$S(K_j, x_j, t_j) \not\subset B(x_j - ku_j, k)$$

for sufficiently large j, again a contradiction. Thus (2.6.2) holds. Hence, K is k^{-1}-flat and k-curved at (x, t). This proves that $\mathscr{A}_k$ is closed.

Let $(x_j)_{j\in\mathbb{N}}$ be a sequence in $A^1_k(K)$ converging to a point x. For each j, we choose a unit tangent vector $t_j \in T_{x_j}K$ such that

$$S(K, x_j, t_j) \cap \operatorname{int} B(x_j - k^{-1}u_j, k^{-1}) = \emptyset,$$

where u_j is the outer unit normal vector of K at x_j. Precisely as above, with K_j replaced by K, we find that K is k^{-1}-flat at (x, t) for suitable $t \in T_x(K)$. Thus $A_k^1(K)$ is closed. In a similar way, we may show that $A_k^2(K)$ is closed.

Let $(K_j)_{j\in\mathbb{N}}$ be a sequence in $\mathscr{C}_{k,m}^1$ converging to some $K \in \mathscr{K}_r^n$. For each $j \in \mathbb{N}$, we choose $x_j \in \operatorname{bd} K_j$ such that

$$B_0(x_j, m^{-1}) \cap \operatorname{bd} K_j \subset A_k^1(K_j).$$

After selecting a subsequence and changing the notation, we may assume that $x_j \to x \in \operatorname{bd} K$ for $j \to \infty$. Let $y \in B_0(x, m^{-1}) \cap \operatorname{bd} K$. For each $j \in \mathbb{N}$, we can choose $y_j \in \operatorname{bd} K_j$ in such a way that $y_j \to y$ for $j \to \infty$. For j sufficiently large, we have $|x_j - y_j| < m^{-1}$ and thus $y_j \in A_k^1(K_j)$. Hence, there is a unit tangent vector $t_j \in T_{y_j}K_j$ such that K_j is k^{-1}-flat at (y_j, t_j). The argument used before shows that K is k^{-1}-flat at (y, t) for suitable $t \in T_yK$. Thus $y \in A_k^1(K)$ and hence $B_0(x, m^{-1}) \cap \operatorname{bd} K \subset A_k^1(K)$, which shows that $\mathscr{C}_{k,m}^1$ is closed. Similarly one shows that $\mathscr{C}_{k,m}^2$ is closed. ■

Proof of Theorem 2.6.2. Let $\mathscr{A} \subset \mathscr{K}_r^n$ be the subset of smooth convex bodies having a boundary point x and a tangent direction t at x such that $\rho_i(x, t) \neq 0$ and $\rho_s(x, t) \neq \infty$. Then

$$\mathscr{A} = \bigcup_{k\in\mathbb{N}} \mathscr{A}_k.$$

Let $k \in \mathbb{N}$ be given. If $P \in \mathscr{K}^n$ is a polytope and $0 < \varepsilon < k^{-1}$, then $P + \varepsilon B^n \in \mathscr{K}_r^n \setminus \mathscr{A}_k$. Since the set of bodies of this type is dense in $\mathscr{K}_r^n$ and $\mathscr{A}_k$ is closed in $\mathscr{K}_r^n$ by Lemma 2.6.5, $\mathscr{A}_k$ is nowhere dense in $\mathscr{K}_r^n$. Thus $\mathscr{A}$ is of first category in $\mathscr{K}_r^n$. ■

Proof of Theorem 2.6.4. Let $\mathscr{K}^* \subset \mathscr{K}_r^n$ be the set of smooth convex bodies K for which the set of all boundary points x with some tangent direction t at x satisfying

$$\rho_i(x, t) > 0 \quad \text{or} \quad \rho_s(x, t) < \infty$$

is of second category in $\operatorname{bd} K$. We have to show that $\mathscr{K}^*$ is meagre in $\mathscr{K}_r^n$. Clearly

$$\mathscr{K}^* = \left\{K \in \mathscr{K}_r^n \,\middle|\, \bigcup_{k\in\mathbb{N}} A_k^1(K) \cup \bigcup_{k\in\mathbb{N}} A_k^2(K) \text{ is of second category}\right\}.$$

Let $K \in \mathscr{K}^*$. Then there are a number $i \in \{1, 2\}$ and a number $k \in \mathbb{N}$ such that $A_k^i(K)$ is not nowhere dense. Since the set $A_k^i(K)$ is closed by Lemma 2.6.5, it has nonempty interior relative to $\operatorname{bd} K$. Hence $K \in \mathscr{C}_{k,m}^i$ for some $m \in \mathbb{N}$, and we conclude that

$$\mathscr{K}^* \subset \bigcup_{k,m\in\mathbb{N}} (\mathscr{C}_{k,m}^1 \cup \mathscr{C}_{k,m}^2).$$

Let $k, m \in \mathbb{N}$ and $i \in \{1, 2\}$ be given. Let $P \in \mathcal{K}^n$ be a polytope with faces of diameter less than m^{-1} and choose $0 < \varepsilon < (2km)^{-1}$. Then $P + \varepsilon B^n \in \mathcal{K}_r^n \backslash \mathcal{C}_{k,m}^i$. Since the bodies of this type are dense in $\mathcal{K}_r^n$ and $\mathcal{C}_{k,m}^i$ is closed in $\mathcal{K}_r^n$ by Lemma 2.6.5, the set $\mathcal{C}_{k,m}^i$ is nowhere dense in $\mathcal{K}_r^n$. Thus $\mathcal{K}^*$ is meagre in $\mathcal{K}_r^n$. ■

Notes for Section 2.6

1. Theorem 2.6.1 was first proved, in an infinite-dimensional version, by Klee (1959b); it was rediscovered by Gruber (1977). Klee also proved that in each infinite-dimensional Banach space most compact convex sets are the closures of their sets of extreme points.

   Theorem 2.6.1 can be strengthened as follows. In a metric space (X, ρ), a set M is called *porous* if there is a number $\alpha > 0$ such that, for every $x \in X$ and any ball $B(x, \varepsilon)$ with centre x and radius ε, there exists a point $y \in B(x, \varepsilon)$ such that

   $$B(y, \alpha\rho(x, y)) \cap M = \varnothing.$$

   A countable union of porous sets is *σ-porous*. One says that *nearly all* elements of a metric Baire space have a property E if the set of elements not having property E is a σ-porous set. Zamfirescu (1987a) proved that nearly all convex bodies in $\mathcal{K}^n$ are smooth and strictly convex.
2. *Typical curvature properties*. Gruber (1977) proved that the set of convex bodies in $\mathcal{K}^n$ that have a boundary of class C^2 or which are, in a certain sense, uniformly strictly convex, is of first category. Theorem 2.6.2 and Corollary 2.6.3 (as well as some other corollaries) are due to Zamfirescu (1980a). In connection with Corollary 2.6.3, the following simple example in $\mathbb{E}^2$ is of interest: Take a convex polygon $P \in \mathcal{K}_0^2$ and let $\mathrm{T}P$ be the convex hull of all the points dividing each side of P into three equal parts. Then $\lim_{k\to\infty} \mathrm{T}^k P$ is a smooth, strictly convex body with curvature vanishing almost everywhere on the boundary. This example is due to de Rham (1957).

   A weaker form of Theorem 2.6.4, in which the set of boundary points x of K at which $\rho_i(x, t) = 0$ and $\rho_s(x, t) = \infty$ for every tangent direction t is merely dense in bd K, was proved by Schneider (1979b). The general form of Theorem 2.6.4 was obtained by Zamfirescu (1980b), and the simpler proof given above appears in Zamfirescu (1985). The curvature properties of typical convex bodies were further investigated by Zamfirescu (1988a). He proved, among other results, that on the boundary of a typical convex body K, the set of points x with $\kappa(x, t) = \infty$ for some $t \in T_x(K)$ and that of points in which the lower curvature in some direction equals the upper curvature in the opposite direction are dense. Further results on the boundary structure of typical convex bodies can be found in Zamfirescu (1985, 1988a).
3. *Typical properties of normals.* The curvature behaviour of a convex body is closely related to the behaviour of its normals (that is, the lines through boundary points spanned by corresponding normal vectors). The behaviour of the normals of typical convex bodies was investigated by Zamfirescu (1982a, 1984b). For instance, he proved that for most convex bodies K in $\mathcal{K}^n$, most points of $\mathbb{E}^n$ lie on infinitely many normals of K.
4. Other topics treated under category aspects include geodesics (Zamfirescu 1982b, 1989+, Gruber 1991a), convex curves on convex surfaces (Zamfirescu 1987b), diameters (Zamfirescu 1984a, Bárány & Zamfirescu 1990), approximation (Schneider & Wieacker 1981, Gruber & Kenderov 1982, Gruber

1983), shadow boundaries (Zamfirescu 1988b, 1991, Gruber & Sorger (1989), non-differentiability properties of the metric projection (Zamfirescu 1990); see also Theorem 3.2.14.

There are several further category results in convexity, concerning properties that are farther away from the main themes of this book. One can find references in the survey articles by Zamfirescu (1985), Gruber (1985) and Zamfirescu [45].

5. *Convex functions*. Smoothness properties of typical convex functions were studied by Gruber (1977) and Klima & Netuka (1981); see also Howe (1982).
6. Schwarz & Zamfirescu (1987) considered typical convex sets of convex sets.
7. *Typical compact sets*. By Theorem 1.8.2, the space $\mathcal{C}^n$ of nonempty compact subsets of $\mathbb{E}^n$ with the Hausdorff metric is also a Baire space. Some category results are known for this space. Most compact sets in $\mathcal{C}^n$ have Hausdorff measure zero, if any Hausdorff measure is given (see Gruber (1983, 1989) for more general results), and they are homeomorphic to Cantor's ternary set (see Wieacker 1988b for references). Wieacker found some surprising results on the (first-order) boundary structure of the convex hull of a typical compact set. In the following theorem, which collects the main results of Wieacker (1988b), $\exp^* K$ denotes the set of farthest points of the convex body K; thus $x \in \exp^* K$ if and only if there is a ball B such that $K \subset B$ and $x \in K \cap \operatorname{bd} B$.

Theorem. For most-compact sets $C \in \mathcal{C}^n$, the convex hull $K = \operatorname{conv} C$ has the following properties:

(a) $\operatorname{bd} K$ is of class C^1 but not of class C^2;
(b) $\exp^* K$ and $\operatorname{ext} K \setminus \exp K$ are dense in $\operatorname{ext} K$;
(c) $\operatorname{ext} K$ is homeomorphic to Cantor's ternary set;
(d) $\exp K$ is homeomorphic to the space of irrational numbers;
(e) $\exp^* K$ is homemorphic to the topological product of the space of rational numbers and Cantor's ternary set,
(f) The k-skeleton $\operatorname{ext}_k K$ of K is of (Hausdorff and topological) dimension k $(k = 0, \ldots, n-1)$.

3

Minkowski addition

3.1. Minkowski addition and subtraction

The aim of the third chapter is a more systematic study of Minkowski addition. In the present section we shall first collect some elementary formal properties of this addition and then investigate its convexifying effect. Finally, we shall consider a certain counterpart to Minkowski addition, a useful operation that has been called Minkowski subtraction.

The sum of two subsets $A, B \subset \mathbb{E}^n$ was defined by

$$A + B := \{a + b | a \in A, b \in B\},$$

and the scalar multiple by

$$\lambda A := \{\lambda a | a \in A\}$$

for real numbers λ. One may rewrite the definition of the sum in the form

$$A + B = \bigcup_{b \in B} (A + b).$$

This gives a 'kinematic' interpretation that may be helpful to the intuition: $A + B$ is the set that is covered if A undergoes all translations by vectors of B.

We recall some of the information on Minkowski sums that is scattered in former sections. If A and B are convex, compact or polytopes, then $A + B$ is, respectively, convex, compact or a polytope. As a map from $\mathcal{C}^n \times \mathcal{C}^n$ into $\mathcal{C}^n$, Minkowski addition is continuous. $\mathcal{C}^n$ and $\mathcal{K}^n$ with Minkowski addition are commutative semigroups with unit element $\{o\}$. $\mathcal{K}^n$ satisfies the cancellation law (Section 1.7), that is, if $K, L, M \in \mathcal{K}^n$ and $K + M = L + M$, then $K = L$.

There are some trivial rules valid for arbitrary subsets A, B, C of $\mathbb{E}^n$ and non-negative real numbers λ, μ, namely

$$(A \cup B) + C = (A + C) \cup (B + C), \tag{3.1.1}$$

$$(A \cap B) + C \subset (A + C) \cap (B + C), \tag{3.1.2}$$

$$\lambda A + \lambda B = \lambda(A + B),$$

$$\lambda A + \mu A \supset (\lambda + \mu)A.$$

If A is convex, then $\lambda A + \mu A = (\lambda + \mu)A$ (Remark 1.1.1). Special relations hold for convex bodies with convex union:

Lemma 3.1.1. *Let $K, L \in \mathcal{K}^n$ be convex bodies such that $K \cup L$ is convex. Then*

$$(K \cap L) + C = (K + C) \cap (L + C) \tag{3.1.3}$$

for $C \in \mathcal{K}^n$, and

$$(K \cup L) + (K \cap L) = K + L. \tag{3.1.4}$$

Proof. Let $x \in (K + C) \cap (L + C)$; then $x = y + c = z + d$ with $y \in K$, $z \in L$ and $c, d \in C$. Since $K \cup L$ is convex, there is a number $\lambda \in [0, 1]$ such that $(1 - \lambda)y + \lambda z \in K \cap L$ and hence

$$\begin{aligned} x &= (1 - \lambda)(y + c) + \lambda(z + d) \\ &= (1 - \lambda)y + \lambda z + (1 - \lambda)c + \lambda d \\ &\in (K \cap L) + C. \end{aligned}$$

Thus $(K + C) \cap (L + C) \subset (K \cap L) + C$, which together with (3.1.2) yields (3.1.3).

For the proof of (3.1.4), we note that $(K \cup L) + (K \cap L) \subset K + L$ holds trivially. Let $x \in K$, $y \in L$. Since $K \cup L$ is convex, the segment $[x, y]$ contains a point $x' = (1 - \lambda)x + \lambda y \in K \cap L$, where $\lambda \in [0, 1]$. Then $y' := \lambda x + (1 - \lambda)y \in K \cup L$ and $x + y = x' + y'$. This shows that $K + L \subset (K \cap L) + (K \cup L)$ and thus proves (3.1.4). ∎

We remind the reader that $A - B$ for $A, B \subset \mathbb{E}^n$ is the set

$$A - B = \{a - b \mid a \in A, b \in B\},$$

and $-A = (-1)A = \{-a \mid a \in A\}$ is the image of A under reflection in the origin.

If $K \in \mathcal{K}^n$ is a convex body, the set

$$\mathrm{D}K := K - K = \{x - y \mid x, y \in K\}$$

is a convex body centrally symmetric with respect to the origin; it is called the *difference body* of K. Since $h(-K, u) = h(K, -u)$, we have

$$h(\mathrm{D}K, u) = h(K, u) + h(K, -u) = w(K, u)$$

for $u \in S^{n-1}$; thus the support function of $\mathrm{D}K$ at the unit vector u is the width of K in direction u. If the difference body of K is a ball, then $w(K, \cdot)$ is constant, and K is called a *body of constant width.*

The Minkowski addition of non-convex sets is, in more than one sense, a convexifying operation. Roughly speaking, vector sums of many bounded sets are approximately convex. This observation can be made quantitatively precise and can be connected with several possible ways of measuring the non-convexity of a compact set. The following results of this kind have proved useful in mathematical economics and in stochastic geometry.

Theorem 3.1.2 (Shapley–Folkman). *Let $A_1, \ldots, A_k \subset \mathbb{E}^n$ and let*

$$x \in \sum_{i=1}^{k} \operatorname{conv} A_i.$$

Then there is an index set $I \subset \{1, \ldots, k\}$ with card $I \leqq n$ *such that*

$$x \in \sum_{i \in I} \operatorname{conv} A_i + \sum_{i \notin I} A_i.$$

Proof. If

$$x \in \sum_{i=1}^{k} \operatorname{conv} A_i = \operatorname{conv}(A_1 + \ldots + A_k)$$

(by Theorem 1.1.2), then $x \in \operatorname{conv}(A_1' + \ldots + A_k')$ with suitable finite sets $A_i' \subset A_i$. Hence, it suffices to prove the result under the assumption that the sets $A_1, \ldots, A_k$ are finite.

Define $f: (\mathbb{E}^n)^k \to \mathbb{E}^n$ by $f(y_1, \ldots, y_k) := y_1 + \ldots + y_k$ and let $P \subset (\mathbb{E}^n)^k$ be the polytope defined by

$$P := \operatorname{conv} A_1 \times \ldots \times \operatorname{conv} A_k;$$

then $x \in f(P)$. Let z be an extreme point of $f^{-1}(x) \cap P$, and let F be the face of P with $z \in \operatorname{relint} F$. Since $f^{-1}(x)$ is of codimension n and z is extreme, it follows that $\dim F \leqq n$. The face F is of the form $F = F_1 \times \ldots \times F_k$, where F_i is a face of $\operatorname{conv} A_i$. Writing $I := \{i | \dim F_i > 0\}$, we deduce that card $I \leqq n$. For $i \in \{1, \ldots, k\} \setminus I$ we have $F_i = \{a_i\}$ with an extreme point a_i of $\operatorname{conv} A_i$, hence with $a_i \in A_i$. This yields

$$x = f(z) \in f(F_1 \times \ldots \times F_k) = F_1 + \ldots + F_k$$

$$\subset \sum_{i \in I} \operatorname{conv} A_i + \sum_{i \notin I} A_i. \qquad \blacksquare$$

For convenience, we formulate the subsequent results only for compact subsets (although some of them could be generalized).

Corollary 3.1.3. *If $A_1, \ldots, A_k \in \mathcal{C}^n$ and $A_i \subset RB^n$ for $i = 1, \ldots, k$ with some real number R, then*

$$\delta\left(\sum_{i=1}^{k} A_i, \operatorname{conv} \sum_{i=1}^{k} A_i\right) \leqq nR.$$

Proof. Let $x \in \operatorname{conv}(A_1 + \ldots + A_k)$. By Theorem 3.1.2 we can assume, after renumbering, that $x = a + a_{m+1} + \ldots + a_k$ with $m \leqq n$, $a \in \sum_{i=1}^{m} \operatorname{conv} A_i$ and $a_i \in A_i$ for $i = m+1, \ldots, k$. By the definition of the Hausdorff distance, there is a point $a' \in A_1 + \ldots + A_m$ such that

$$|a - a'| \leqq \delta(A_1 + \ldots + A_m, \operatorname{conv}(A_1 + \ldots + A_m)) \leqq mR,$$

the latter because $A_1 + \ldots + A_m \subset mRB^n$. Hence, the point $\bar{a} := a' + a_{m+1} + \ldots + a_k \in A_1 + \ldots + A_k$ satisfies $|x - \bar{a}| = |a - a'| \leqq mR \leqq nR$. The assertion follows. ■

Remark 3.1.4. The essential feature of the estimate of Corollary 3.1.3 is that the right-hand side is independent of k. In particular, if $(A_i)_{i \in \mathbb{N}}$ is a sequence of uniformly bounded compact sets in $\mathbb{E}^n$, it follows that

$$\delta\left(\frac{1}{k}(A_1 + \ldots A_k), \frac{1}{k} \operatorname{conv}(A_1 + \ldots + A_k)\right) \to 0 \text{ for } k \to \infty.$$

Thus averaging of compact sets in the Minkowski sense is 'asymptotically convexifying'.

There are sharper estimates that demonstrate the convexifying effect of Minkowski addition. They can, for instance, be deduced from the following lemma. For a compact set $A \in \mathcal{C}^n$ there is a unique ball of smallest radius containing A, the *circumball* of A. Its radius, $R(A)$, is called the *circumradius* of A.

Lemma 3.1.5. *Let $A_1, \ldots, A_m \in \mathcal{C}^n$ and $x \in \operatorname{conv}(A_1 + \ldots + A_m)$. Then there is a point $a \in A_1 + \ldots + A_m$ such that*

$$|x - a|^2 \leqq \sum_{i=1}^{m} R(A_i)^2.$$

Proof. We use induction with respect to m. If $m = 1$, let z be the centre of the circumball of A_1. Since $x \in \operatorname{conv} A_1$, we can choose $a \in A_1$ with $\langle z - x, a - x \rangle \leqq 0$ and then obtain

$$\begin{aligned} |x - a|^2 &\leqq |x - a|^2 + |z - x|^2 + 2\langle z - x, x - a \rangle \\ &= |x - a|^2 \leqq R(A_1)^2. \end{aligned}$$

Thus the assumption holds for $m = 1$. Suppose it is true for some $m \geqq 1$. Let $x \in \operatorname{conv}(A_1 + \ldots + A_{m+1})$. Then $x = y + z$ with $y \in$

$\operatorname{conv}(A_1 + \ldots + A_m)$ and $z \in \operatorname{conv} A_{m+1}$. By the inductive assumption, there exists a point $a \in A_1 + \ldots + A_m$ such that

$$|y - a|^2 \leqq \sum_{i=1}^{m} R(A_i)^2.$$

Let $w := p(\operatorname{conv} A_{m+1}, x - a)$; then

$$|x - a - w|^2 \leqq |x - a - z|^2 = |y - a|^2 \leqq \sum_{i=1}^{m} R(A_i)^2.$$

Suppose for a moment that $x - a - w \neq o$. The hyperplane H through w orthogonal to $x - a - w$ separates $x - a$ and $\operatorname{conv} A_{m+1}$ (Section 1.2), and $w \in H \cap \operatorname{conv} A_{m+1} = \operatorname{conv}(H \cap A_{m+1})$. By the case $m = 1$ of the lemma, there is $a_{m+1} \in H \cap A_{m+1}$ such that

$$|w - a_{m+1}|^2 \leqq R(H \cap A_{m+1})^2 \leqq R(A_{m+1})^2.$$

We have

$$\langle x - a - w, a_{m+1} - w \rangle = 0. \tag{3.1.5}$$

If $x - a - w = o$, we choose $a_{m+1} \in A_{m+1}$ with $|w - a_{m+1}|^2 \leqq R(A_{m+1})^2$, and (3.1.5) holds trivially. Now $a + a_{m+1} \in A_1 + \ldots + A_{m+1}$ and

$$\begin{aligned}|x - (a + a_{m+1})|^2 &= |(x - a - w) - (a_{m+1} - w)|^2 \\ &= |x - a - w|^2 + |a_{m+1} - w|^2 \leqq \sum_{i=1}^{m+1} R(A_i)^2,\end{aligned}$$

which completes the proof. ■

To formulate a consequence, we define, for a compact set $A \in \mathcal{C}^n$, the *inner radius* of A by

$$\rho(A) := \sup_{x \in \operatorname{conv} A} \inf \{\lambda \leqq 0 | x \in \operatorname{conv}(A \cap B(x, \lambda))\}.$$

Then $\rho(A) = 0$ if and only if A is convex, and trivially $\rho(A) \leqq \operatorname{diam} A$, with equality if A is affinely independent.

Theorem 3.1.6 (Shapley–Folkman–Starr). *Let* $A_1, \ldots, A_k \in \mathcal{C}^n$. *If* $x \in \operatorname{conv}(A_1 + \ldots + A_k)$, *then there exists a point* $a \in A_1 + \ldots + A_k$ *such that*

$$|x - a| \leqq \sqrt{n} \max_{1 \leqq i \leqq k} \rho(A_i), \tag{3.1.6}$$

hence

$$\delta\left(\sum_{i=1}^{k} A_i, \operatorname{conv} \sum_{i=1}^{k} A_i\right) \leqq \sqrt{n} \max_{1 \leqq i \leqq k} \rho(A_i). \tag{3.1.7}$$

Proof. Let $x \in \operatorname{conv}(A_1 + \ldots + A_k)$. By Theorem 3.1.2, there is an index set $I \subset \{1, \ldots, k\}$ with $\operatorname{card} I \leqq n$ such that $x = \sum_{i=1}^{k} a_i$ with $a_i \in \operatorname{conv} A_i$ for $i \in I$ and $a_i \in A_i$ for $i \notin I$. By Lemma 3.1.5, there exists a point $b \in \sum_{i\in I} A_i$ such that

$$\left|\sum_{i\in I} a_i - b\right|^2 \leqq \sum_{i\in I} R(A_i)^2.$$

Hence, the point $a := b + \sum_{i\notin I} a_i$ satisfies $a \in \sum_{i=1}^k A_i$ and

$$|x - a|^2 \leqq \sum_{i\in I} R(A_i)^2 \leqq n \max_{1\leqq i\leqq k} R(A_i)^2. \tag{3.1.8}$$

We can write

$$x = \sum_{i=1}^k x_i \quad \text{with } x_i \in \operatorname{conv} A_i.$$

Put $A_i' := A_i \cap B(x_i, \rho(A_i))$; then $x_i \in \operatorname{conv} A_i'$ by the definition of $\rho(A_i)$ (and by compactness), hence $x \in \operatorname{conv}(A_1' + \ldots + A_k')$. By the result proved above, there is a point $a \in \sum_{i=1}^k A_i' \subset \sum_{i=1}^k A_i$ such that

$$|x - a| \leqq \sqrt{n} \max_{1\leqq i\leqq k} R(A_i') \leqq \sqrt{n} \max_{1\leqq i\leqq k} \rho(A_i),$$

which proves (3.1.6). The estimate (3.1.7) is an immediate consequence. ■

Both functions appearing in Theorem 3.1.6, $\delta(A, \operatorname{conv} A)$ and $\rho(A)$, are measures for the deviation of a compact set A from its convex hull. As such, they also measure the non-convexity of the set, but in a way that depends on the Euclidean metric and that is not invariant under similarities. Since convexity is an affine notion, a geometrically significant measure for the non-convexity of sets should be invariant under affine transformations. Such a function can also be derived from the theorem of Shapley and Folkman. Applying Theorem 3.1.2 with $k = n + 1$ and $A_i = A \subset \mathbb{E}^n$, we get

$$A + n \operatorname{conv} A = \operatorname{conv} A + n \operatorname{conv} A.$$

Hence, if we define

$$c(A) := \inf\{\lambda \geqq 0 \mid A + \lambda \operatorname{conv} A \text{ is convex}\},$$

then $c(A) \leqq n$. It is clear that the function c is invariant under non-singular affine maps and that $c(A) = 0$ for convex sets A. The following theorem (Schneider 1975d) shows that for compact sets c can serve as a measure of non-convexity and that it describes the sets that are least convex, in this sense.

Theorem 3.1.7. *For every set $A \subset \mathbb{E}^n$,*

$$0 \leqq c(A) \leqq n.$$

For compact sets A, equality holds on the left if and only if A is convex, and on the right if and only if A consists of $n + 1$ affinely independent points.

Proof. It is convenient to write

$$A_\lambda := (1+\lambda)^{-1}(A + \lambda \operatorname{conv} A)$$

for $\lambda \geqq 0$; then $A_\lambda \subset \operatorname{conv} A \subset \operatorname{conv} A_\lambda$, hence

$$c(A) = \inf\{\lambda \geqq 0 \mid A_\lambda = \operatorname{conv} A\}.$$

The inequalities $0 \leqq c(A) \leqq n$ are clear. Now let $A \in \mathcal{C}^n$ be compact. Assume that $c(A) = 0$. Let $x \in \operatorname{conv} A$, and let $\lambda > 0$. Then $x \in A_\lambda$, hence $x = (1+\lambda)^{-1}(a_\lambda + \lambda b_\lambda)$ with $a_\lambda \in A$ and $b_\lambda \in \operatorname{conv} A$. This implies $x - a_\lambda = \lambda(b_\lambda - x)$ and thus $\delta(A, x) \leqq 2\lambda \operatorname{diam} A$. Since $\lambda > 0$ was arbitrary, we see that $x \in A$, hence A is convex.

Now suppose that $c(A) = n$. Then A cannot be contained in a hyperplane (since otherwise $c(A) \leqq n-1$), hence $\operatorname{conv} A$ has interior points.

Let $x \in \operatorname{bd} \operatorname{conv} A$. Then x lies in a supporting hyperplane H of $\operatorname{conv} A$, hence

$$x \in H \cap \operatorname{conv} A = \operatorname{conv}(H \cap A) = (H \cap A)_{n-1} \subset A_{n-1}.$$

Let $\lambda > n-1$ and put $\alpha := (\lambda - n + 1)(1+\lambda)^{-1}$; then

$$x + \alpha(\operatorname{conv} A - x) \subset (1-\alpha)A_{n-1} + \alpha \operatorname{conv} A = A_\lambda.$$

Since $\alpha > 0$ and this holds for all $x \in \operatorname{bd} \operatorname{conv} A$, we see that the set

$$R_\lambda := (\operatorname{conv} A) \backslash A_\lambda$$

satisfies

$$\operatorname{cl} R_\lambda \subset \operatorname{int} \operatorname{conv} A \quad \text{for } \lambda > n-1. \tag{3.1.9}$$

If $\mu > \lambda$, then

$$\begin{aligned} A_\mu &= (1+\mu)^{-1}A + (\mu-\lambda)(1+\mu)^{-1}(1+\lambda)^{-1} \operatorname{conv} A \\ &\quad + \lambda(1+\lambda)^{-1} \operatorname{conv} A \\ &\supset (1+\mu)^{-1}A + (\mu-\lambda)(1+\mu)^{-1}(1+\lambda)^{-1}A + \lambda(1+\lambda)^{-1} \operatorname{conv} A \\ &= A_\lambda; \end{aligned}$$

hence $\mu > \lambda$ implies $R_\mu \subset R_\lambda$. Since the sets $\operatorname{cl} R_\lambda$ are compact and nonempty for $\lambda < n$, there exists a point

$$z \in \bigcap_{0<\lambda<n} \operatorname{cl} R_\lambda.$$

By (3.1.9), $z \in \operatorname{int} \operatorname{conv} A$. On the other hand,

$$z \notin \operatorname{int}(1+n)^{-1}(a + n \operatorname{conv} A) \text{ for } a \in A, \tag{3.1.10}$$

since otherwise, for sufficiently large $\lambda < n$,

$$z \in \operatorname{int}(1+\lambda)^{-1}(a + \lambda \operatorname{conv} A) \subset \operatorname{int} A_\lambda$$

and thus $z \notin \operatorname{cl} R_\lambda$, a contradiction.

By Carathéodory's theorem, there exists an affinely independent set $Y \subset A$ such that $z \in \operatorname{conv} Y$, and some subset $X \subset Y$ satisfies $z \in \operatorname{relint} \operatorname{conv} X$. Suppose that $\dim \operatorname{aff} X < n$. Then

$$z \in \operatorname{conv} X = (1 + \lambda)^{-1}(X + \lambda \operatorname{conv} X) \quad \text{for } \lambda \geqq n - 1$$

and hence $z \in \operatorname{relint}(1 + n)^{-1}(x + n \operatorname{conv} X)$ for suitable $x \in X$. Since $z \in \operatorname{int} \operatorname{conv} A$, we must have $\operatorname{relint} \operatorname{conv} X \subset \operatorname{int} \operatorname{conv} A$ and hence $z \in \operatorname{int}(1 + n)^{-1}(x + n \operatorname{conv} A)$. This contradicts (3.1.10). Hence $\dim \operatorname{aff} X = n$, and Y is the set of vertices of an n-simplex S. From (3.1.10) it follows that z is the centroid of S, since this is the only point not contained in $\bigcup_{x \in Y} \operatorname{int}(1 + n)^{-1}(x + nS)$. If A contains a point $a \notin Y$, we can replace an appropriate point of Y by a and still obtain an affinely independent set $\bar{Y} \subset A$ satisfying $z \in \operatorname{conv} \bar{Y}$. By the same argument, $\bar{Y}$ is the set of vertices of an n-simplex of which z is the centroid. But as Y and $\bar{Y}$ differ in precisely one point, this is impossible. Hence $A = Y$, that is, A consists of $n + 1$ affinely independent points. Vice versa, if A is of this kind and if $A + \lambda \operatorname{conv} A$ is convex for some $\lambda \geqq 0$, then the barycentre of A belongs to A_λ, which implies $\lambda \geqq n$. This completes the proof. ∎

Having studied some properties of Minkowski addition, we now introduce a complementary operation called *Minkowski subtraction* (although it was not introduced by Minkowski). While the sum of two sets $A, B \subset \mathbb{E}^n$ can be defined by

$$A + B = \bigcup_{b \in B} (A + b),$$

the *Minkowski difference* of A and B is, by definition, the set

$$A \sim B := \bigcap_{b \in B} (A - b).$$

(If B is empty, $A \sim B$ is, by convention, equal to $\mathbb{E}^n$.) Evidently we may also write

$$A \sim B = \{x \in \mathbb{E}^n | B + x \subset A\}.$$

There are some trivial rules connecting addition and subtraction, namely

$$(A + B) \sim B \supset A, \tag{3.1.11}$$

$$(A \sim B) + B \subset A, \ (B \neq \emptyset) \tag{3.1.12}$$

$$(A \sim B) + C \subset (A + C) \sim B, \tag{3.1.13}$$

$$(A \sim B) \sim C = A \sim (B + C), \tag{3.1.14}$$

$$A + B \subset C \Leftrightarrow A \subset C \sim B. \tag{3.1.15}$$

The verification is immediate. Under convexity assumptions, a little more is true. If A is convex, then $A \sim B$ is an intersection of convex sets and hence convex. For convex bodies $K, L \in \mathcal{K}^n$ we say that L is a *summand* of K if there exists a convex body M such that $K = L + M$. A closer look at summands is the aim of the next section. Here we remark:

Lemma 3.1.8. *Let $K, L \in \mathcal{K}^n$ be convex bodies. Then*

$$(K + L) \sim L = K.$$

The relation

$$(K \sim L) + L = K$$

holds if and only if L is a summand of K.

Proof. Let $x \in (K + L) \sim L$; hence $L + x \subset K + L$ and, therefore, $h_L + h_{\{x\}} \leqq h_K + h_L$. Subtracting h_L, we get $x \in K$. Thus $(K + L) \sim L \subset K$, which together with (3.1.11) proves the first assertion. If $(K \sim L) + L = K$, then L is a summand of K. Vice versa, suppose that $K = M + L$ for some $M \in \mathcal{K}^n$. Then $K \sim L = (M + L) \sim L = M$, which proves the second assertion. ∎

For convex bodies $K, L \in \mathcal{K}^n$, the Minkowski difference can also be represented as an intersection of halfspaces, in the following form:

$$\begin{aligned} K \sim L &= \bigcap_{u \in S^{n-1}} [H^-(K, u) - h(L, u)u] \\ &= \bigcap_{u \in S^{n-1}} H^-_{u,h(K,u)-h(L,u)}. \end{aligned} \tag{3.1.16}$$

In fact, $x \in K \sim L$ is equivalent to $L + x \subset K$, which is equivalent to $h(L + x, u) \leqq h(K, u)$ and hence to $\langle x, u \rangle \leqq h(K, u) - h(L, u)$ for all $u \in S^{n-1}$.

In particular, (3.1.16) implies

$$h(K \sim L, u) \leqq h(K, u) - h(L, u) \quad \text{for } u \in S^{n-1}.$$

Let $x \in \operatorname{bd}(K \sim L)$. Then $L + x$ must contain a boundary point y of K (otherwise, $L + x + v \subset K$ for all v in some neighbourhood of o and thus $x \in \operatorname{int}(K \sim L)$). If u is an outer unit normal vector of K at y, then $h(K, u) = h(L + x, u)$, hence $h(K, u) - h(L, u) = \langle x, u \rangle \leqq h(K \sim L, u)$. Thus, to each boundary point x of $K \sim L$ there is a normal vector u of $K \sim L$ at x such that $h(K \sim L, u) = h(K, u) - h(L, u)$.

The following notion of parallel bodies is useful in several respects. Let $K, B \in \mathcal{K}^n$ be convex bodies and let $\lambda \geqq 0$ be a real number. Then the convex body $K + \lambda B$ is called an (*outer*) *parallel body* of K, and $K \sim \lambda B$ is an *inner parallel body*, both *relative to* B. The greatest number λ for which $K \sim \lambda B$ is not empty is called the *inradius of K relative to B* and denoted by $r(K, B)$; thus

$$r(K, B) := \max\{\lambda \geqq 0 | \lambda B + t \subset K \text{ for some } t \in \mathbb{E}^n\}.$$

For fixed B, one defines the full system of relative parallel bodies of K by

$$K_\rho := \begin{cases} K + \rho B & \text{for } 0 \leqq \rho < \infty, \\ K \sim -\rho B & \text{for } -r(K, B) \leqq \rho \leqq 0. \end{cases}$$

Then for $K, L \in \mathcal{K}^n$ and arbitrary $\rho, \sigma \in \mathbb{R}$ the rule

$$K_\rho + L_\sigma \subset (K + L)_{\rho+\sigma} \tag{3.1.17}$$

is valid.

Proof. Let $\rho, \sigma \geqq 0$. We have

$$\begin{aligned} K_\rho + L_\sigma &= K + \rho B + L + \sigma B = K + L + (\rho + \sigma)B \\ &= (K + L)_{\rho+\sigma}. \end{aligned}$$

Next,

$$\begin{aligned} &K_{-\rho} + L_{-\sigma} + (\rho + \sigma)B \\ &\qquad = (K \sim \rho B) + \rho B + (L \sim \sigma B) + \sigma B \subset K + L, \end{aligned}$$

hence $K_{-\rho} + L_{-\sigma} \subset (K + L) \sim (\rho + \sigma)B = (K + L)_{-\rho-\sigma}$.

If $\rho \geqq \sigma$, then

$$\begin{aligned} K_\rho + L_{-\sigma} &= K + (\rho - \sigma)B + (L \sim \sigma B) + \sigma B \subset K + (\rho - \sigma)B + L \\ &= (K + L)_{\rho-\sigma}. \end{aligned}$$

If $\rho \leqq \sigma$, then

$$\begin{aligned} K_\rho + L_{-\sigma} + (\sigma - \rho)B &= K + \rho B + (\sigma - \rho)B + (L \sim \sigma B) \\ &= K + (L \sim \sigma B) + \sigma B \subset K + L, \end{aligned}$$

hence $K_\rho + L_{-\sigma} \subset (K + L) \sim (\sigma - \rho)B = (K + L)_{\rho-\sigma}$. ■

From (3.1.17) we deduce the following lemma which later will turn out to be very useful.

Lemma 3.1.9. *If $K, B \in \mathcal{K}^n$, then the full system $\rho \mapsto K_\rho$ of parallel sets of K relative to B is concave with respect to inclusion, that is,*

$$(1 - \lambda)K_\rho + \lambda K_\sigma \subset K_{(1-\lambda)\rho+\lambda\sigma}$$

for $\lambda \in [0, 1]$ and $\rho, \sigma \in [-r(K, B), \infty)$.

Proof. $(1 - \lambda)K_\rho + \lambda K_\sigma = [(1 - \lambda)K]_{(1-\lambda)\rho} + [\lambda K]_{\lambda\sigma} \subset K_{(1-\lambda)\rho+\lambda\sigma}$ by (3.1.17). ■

When the preceding lemma is applied, it is important to know in which cases the inner parallel bodies of K are homothetic to K. The following lemma is a slightly improved version of a result of Bol (1943b). Recall (Section 2.2) that the convex body K is a tangential body of the body $B \in \mathcal{K}^n$ if each extreme support plane of K is a support plane of B. For the following, an equivalent description is convenient. Let K be a tangential body of B. If x is a boundary point of K, then through x there is an extreme support plane of K, which is hence a support plane of B. Vice versa, suppose that through each boundary point of K there is a support plane of B. Then a support plane of K that is not a support plane of B contains only singular points of K. By Theorem 2.2.8, K is a tangential body of B. Thus a convex body K is a tangential body of B if and only if through each boundary point of K there exists a support plane to K that also supports B.

Lemma 3.1.10. *Let $K, B \in \mathcal{K}_0^n$ be convex bodies, and let $\rho \in (0, r(K, B))$. Then $K \sim \rho B$ is homothetic to K if and only if K is homothetic to a tangential body of B.*

Proof. Suppose that $K \sim \rho B$ is homothetic to K. After a suitable choice of the origin we may assume that $\lambda K = K \sim \rho B$, where $0 < \lambda < 1$. Then

$$\lambda^2 K = \lambda(\lambda K) = \lambda(K \sim \rho B) = \lambda K \sim \lambda\rho B = (K \sim \rho B) \sim \lambda\rho B$$
$$= K \sim (1 + \lambda)\rho B$$

by (3.1.14). By induction we get

$$\lambda^k K = K \sim (1 + \lambda + \ldots + \lambda^{k-1})\rho B$$

for $k \in \mathbb{N}$, hence

$$\{o\} = K \sim \tau B \quad \text{with } \tau = \frac{\rho}{1 - \lambda},$$

because $\rho_k \uparrow \tau$ obviously implies $K \sim \rho_k B \to K \sim \tau B$. Let $x \in \operatorname{bd} K$. Then $\lambda x \in \lambda K = K \sim \rho B$, hence $\rho B + \lambda x \subset K$. Since λx is a boundary point of $K \sim \rho B$, the body $\rho B + \lambda x$ contains a boundary point y of K, say $y = \rho b + \lambda x$ with $b \in B$. Then

$$v := \frac{y - \lambda x}{1 - \lambda} = \frac{\rho b}{1 - \lambda} \in \tau B \subset K.$$

The point $y \in \operatorname{bd} K$ is a convex combination, with positive coefficients, of the points $x \in \operatorname{bd} K$ and $v \in K$, hence any support plane to K through y contains x and v and thus supports K at x and τB at v. Thus

K is a tangential body of τB (and hence is homothetic to a tangential body of B).

Conversely, suppose that K is a tangential body of τB, for some $\tau > 0$. Then

$$K = \bigcap_{u \in \omega} H^-(\tau B, u) = \bigcap_{u \in \omega} H^-(K, u),$$

where $\omega \subset S^{n-1}$ is the set of those normal vectors u for which the support plane $H(K, u)$ also supports τB. (To see this, observe that K is the intersection of those of its supporting halfspaces that contain a regular point of K in their boundaries.) Now

$$K \sim \rho B = \bigcap_{u \in S^{n-1}} H^-_{u,h(K,u)-\rho h(B,u)} \subset \bigcap_{u \in \omega} H^-_{u,h(K,u)-\rho h(B,u)}.$$

For $u \in \omega$ we have

$$h(K, u) - \rho h(B, u) = \frac{\tau - \rho}{\tau} h(K, u),$$

hence $K \sim \rho B \subset (1 - \rho/\tau)K$. From Lemma 3.1.9 we get

$$\frac{\rho}{\tau} K_{-\tau} + \left(1 - \frac{\rho}{\tau}\right) K_0 \subset K_{-\rho},$$

and since $K_{-\tau} = \{o\}$, this gives $K \sim \rho B = (1 - \rho/\tau)K$. ∎

Notes for Section 3.1

1. *Minkowski addition and subtraction.* All the formal rules for Minkowski addition and subtraction that we have mentioned, and many more, can be found in Hadwiger [18] (who uses $A \times B$ and A/B instead of $A + B$ and $A \sim B$), with one exception. The exception is equation (3.1.4), which presumably first appeared in Sallee (1966).

   More recently, addition and subtraction (for not necessarily convex sets) have found applications in mathematical morphology and image processing; see Matheron (1975), Serra (1982, 1988) and Stoyan, Kendall & Mecke (1987); in the latter, note in particular the remarks in Section 1.4 on quantitative image processing. A useful survey is given by Haralick, Sternberg & Zhuang (1987). In these and related references, different notation is used, namely

$$A \oplus B := A + B,$$
$$A \ominus B := A \sim -B.$$

   These operations are complementary to each other, in the sense that

$$A \ominus B = (A^c \oplus B)^c,$$
$$A \oplus B = (A^c \ominus B)^c,$$

   where A^c is the complement of the set A. In applications, B is usually a fixed convex body, for instance in the plane a disc or a hexagon, which is then called the *structuring element*. It is used to transform a given set A in various ways, in particular by means of the following operations:

$A \oplus -B$, the *dilatation* of A by B;

$A \ominus -B$, the *erosion* of A by B;

$A_B := (A \ominus -B) \oplus B$, the *opening* of A by B,

$A^B := (A \oplus -B) \ominus B$, the *closing* of A by B.

2. *Analysis of $\mathcal{K}^n$-valued functions*. It is not surprising that some parts of the elementary analysis of real-valued functions can be carried over to $\mathcal{K}^n$-valued functions, with Minkowski addition replacing the addition of real numbers. Integrals of functions from an interval into $\mathcal{K}^n$ were constructed by Dinghas (1956, 1961). For later developments see, e.g., Artstein (1974) and the references given there. Artstein emphasizes the usefulness of the support function. It appears that a consequent use of the support function, together with Theorem 1.8.12 and some elementary estimates, would also permit one to deduce many of the results in Ahrens (1970, 1972) and in Chapter IV of Meschkowski & Ahrens (1974) directly from the corresponding results for real-valued functions.

   Vitale (1979) has studied a Bernstein type of approximation for functions from $[0, 1]$ into $\mathcal{K}^n$, where Minkowski addition is used.
3. *Vector spaces consisting of classes of convex bodies*. The set $\mathcal{K}^n$ of convex bodies, equipped with Minkowski addition and multiplication by non-negative real numbers, forms a commutative semigroup, satisfying the cancellation law, with scalar operator. More algebraic structure can be introduced if suitable equivalence classes of convex bodies are considered. Vector spaces of classes of convex bodies, in addition endowed with norms or scalar products, have been investigated by Ewald & Shephard (1966), Ewald (1965), Shephard (1966c), Schmitt (1967), Lewis (1975) and Geivaerts (1974b, 1976); see also Sorokin (1968) and Zamfirescu (1975).

   In the background of part of these investigations is, roughly speaking, the idea of decomposing a support function into its even and odd parts, and of concentrating attention on one of these parts. Valette (1971) treats a related topic in this sense also.
4. *Prolongation of linear series*. Let $K_0, K_1 \in \mathcal{K}^n$ and $K_\lambda := (1-\lambda)K_0 + \lambda K_1$ for $\lambda \in [0, 1]$. We say that the linear series $(K_\lambda)_{\lambda\in[0,1]}$ can be prolonged via K_1 if, for some $\lambda > 1$, the function $(1-\lambda)h(K_0, \cdot) + \lambda h(K_1, \cdot)$ is still a support function; in other words, if there exists a convex body K_λ such that $K_1 = (1-\lambda^{-1})K_0 + \lambda^{-1}K_\lambda$. Evidently this is equivalent to the condition that K_0 be homothetic to a summand of K_1. Known criteria for summands (see Section 3.2) can therefore be used to study this prolongation. It was investigated by Vincensini (1935, 1936). He applied it to the 'construction' of convex bodies with given difference body (domaine vectoriel). With his notation, but using the support function with advantage, we can describe this procedure as follows. Let $D \in \mathcal{K}^n$ be a given convex body, centrally symmetric with respect to the origin. Let Γ be an arbitrary convex body with the property that $C := \mathrm{D}\Gamma = \Gamma - \Gamma$ is homothetic to a summand of D (this is always the case if D and Γ are of class C^2_+); say, αC is a summand of D where $\alpha \in (0, 1)$. Then

$$h(\Delta, u) := \frac{\alpha}{2}\,[h(\Gamma, u) - h(\Gamma, -u)] + \frac{1}{2}\,h(D, u)$$

   defines a support function, and the body Δ evidently satisfies $\mathrm{D}\Delta = D$. In a series of papers, Vincensini (1937a, b, c, 1938, 1956, 1957, 1965) studied related questions and applications; see also Vincensini & Zamfirescu (1967).
5. *Extremal and facial structure of sums of convex sets*. The following was observed by Klee (1959b, Proposition 6.5). Let $A, B \subset \mathbb{E}^n$ be convex sets, A

closed and B compact. Then for each point $z \in \text{ext}(A+B)$ there are unique points $x_z \in A$ and $y_z \in B$ (necessarily extreme points) such that $z = x_z + y_z$. Further, for each point $x \in \text{ext}\, A$ there is a point $y \in B$ such that $x + y \in \text{ext}(A+B)$. However, there may be points $y \in \text{ext}\, B$ such that $x + y \notin \text{ext}(A+B)$ for arbitrary $x \in A$. This was further studied by Bair (1979).

The extremal and facial structure of sums of closed convex sets in certain infinite dimensional spaces was investigated by Husain & Tweddle (1970), A. K. Roy (1972), Edelstein & Fesmire (1975), Jongmans (1976) and Bair, Fourneau & Jongmans (1977).

Some observations on smoothness properties of Minkowski sums are found in Jongmans (1979, 1981a, b).

6. The behaviour of certain cones associated with convex sets in real vector spaces under Minkowski addition of the convex sets was studied by Dubois & Jongmans (1982).
7. *Bodies of constant width.* A comprehensive survey of bodies of constant width was given by Chakerian & Groemer [31].
8. *Pairs of constant width.* As a generalization of bodies of constant width, Maehara (1984) introduced pairs of constant width. A pair (K, L) of convex bodies $K, L \in \mathcal{K}^n$ is called a *pair of constant width* r if

$$h(K, u) + h(L, -u) = r \quad \text{for } u \in S^{n-1}.$$

In particular, K is of constant width r if and only if (K, K) is a pair of constant width. If (K, L) is a pair of constant width r, then $K + L$ is of constant width $2r$. For $A \subset \mathbb{E}^n$ and $r > 0$, let $\Omega_r(A)$ be the intersection of all closed balls of radius r whose centres belong to A. Maehara proves the following: A pair (K, L) is of constant width r if and only if $\Omega_r(K) = L$ and $\Omega_r(L) = K$. If $A \neq \emptyset$ is a set having circumradius at most r, then $(\Omega_r(A), \Omega_r^2(A))$ is a pair of constant width.

Sallee (1987) extended this to pairs of sets of constant relative width with respect to a centrally symmetric convex body S. This led to interesting additional results and also to some open problems. (See also Note 4 in Section 3.2.)

9. *The theorem of Shapley, Folkman and Starr.* Theorem 3.1.2, Lemma 3.1.5 and Theorem 3.1.6 are all from Starr (1969), where credit for most of the results (namely, up to (3.1.8)) is given to unpublished work of Folkman and Shapley; see also Arrow & Hahn (1971), Appendix B, Ekeland & Temam (1976), Appendix I, and Starr (1981). Our proof of Theorem 3.1.2 essentially follows Artstein (1980), Theorem 5.1. Simple proofs of Theorems 3.1.2 and 3.1.6 were also given by Cassels (1975). He introduced, for compact sets $A \in \mathcal{C}^n$, the function

$$c_3(A) := \sup_{x \in \text{conv}\, A} \inf \left(\sum_{i=1}^{m} \alpha_i |x - a_i|^2 \right)^{1/2},$$

where the infimum is taken over all representations $x = \sum_{i=1}^{m} \alpha_i a_i$ with $m \in \mathbb{N}$, $a_i \in A$, $\alpha_i \geqq 0$ and $\sum \alpha_i = 1$, and he showed that

$$c_3\left(\sum_{i=1}^{k} A_i\right) \leqq \sqrt{n} \max_{1 \leqq i \leqq k} c_3(A_i)$$

for $A_1, \ldots, A_k \in \mathcal{C}^n$. Like the functions defined by

$$c_1(A) := \delta(A, \text{conv}\, A),$$

$$c_4(A) := \rho(A),$$

c_3 measures the deviation of A from its convex hull. Since

$$c_1(A) \leqq c_3(A) \leqq c_4(A),$$

the result of Cassels implies the inequality (3.1.7), which can be written as

$$c_1\left(\sum_i A_i\right) \leqq \sqrt{n} \max_i c_4(A_i).$$

W. Weil (private communication) asked whether also

$$c_1\left(\sum_i A_i\right) \leqq \sqrt{n} \max_i c_1(A_i).$$

Wegmann (1980) investigated relations between c_1, c_3, c_4 and a further function c_2 defined in a similar way. In particular, he showed that $c_3 = c_4$, while simple examples $A \in \mathcal{C}^n$ exist with $c_1(A) < c_3(A)$.

The theorem of Shapley, Folkman and Starr has been applied in mathematical economics; see Starr (1969), Arrow & Hahn (1971) and Hildenbrand (1974).

10. *Minkowski addition in the theory of random sets*. The theorem of Shapley, Folkman and Starr has also found applications in stochastic geometry. For these applications, the crude estimate of Corollary 3.1.3 is sufficient, since one only needs Remark 3.1.4. The latter is used to deduce results on sequences of independent random compact sets and their partial Minkowski sums from results on the corresponding convex hulls, and for convex sets the embedding into a Banach space via the suppport function can be utilized. For such applications, see Artstein & Vitale (1975), Ljašenko (1979) (who gives independent proofs for the required estimates), Weil (1982b), Giné, Hahn & Zinn (1983) and Artstein (1987+).

    Further applications of Minkowski addition in the theory of random sets are found in Cressie (1978), Mase (1979), Schürger (1983) and Giné & Hahn (1985a, b); for infinite dimensions see Giné & Hahn (1985c) and the references given there.

11. The main part of Theorem 3.1.7 was formulated and proved above for compact sets only. Some more general and additional results were proved in Schneider (1975d). Let $A \subset \mathbb{E}^n$ be an arbitrary set. If A is bounded and $c(A) = 0$, then $\operatorname{cl} A$ is convex; if $c(A) = n$, then A consists of $n + 1$ affinely independent points. If A is either unbounded or connected, then $c(A) \leqq n - 1$; in both cases the bound $n - 1$ is sharp.

    The relation

$$A + n \operatorname{conv} A = (n + 1) \operatorname{conv} A$$

for a compact set $A \subset \mathbb{E}^n$, which was noted before Theorem 3.1.7, as a consequence of Theorem 3.1.2, appears also in Borwein & O'Brien (1978). These authors prove: A Banach space X is finite dimensional if and only if for any compact set $A \subset X$ there exists $\alpha > 0$ with

$$A + \alpha \operatorname{cl} \operatorname{conv} A = (\alpha + 1) \operatorname{cl} \operatorname{conv} A.$$

The least such α that works for all A is the dimension of the space.

12. *Computational aspects of Minkowski addition.* Gritzmann & Sturmfels (1990) determine the computational complexity of computing the Minkowski sum of k convex polytopes in $\mathbb{E}^n$, which are presented either in terms of vertices or in terms of facets. For fixed n, they obtain a polynomial time algorithm for adding k polytopes with up to N vertices. Among other combinatorial results, they have bounds for the number of vertices of the sum in terms of the vertex numbers of the given polytopes.

13. *Hammer's associated bodies and reducibility*. For $K \in \mathcal{K}^n$, Hammer (1951) defined and studied the associated bodies

$$K(\rho) := \begin{cases} \bigcap_{b \in \operatorname{bd} K} [\rho(K - b) + b] & \text{for } \rho \in (0, 1), \\ \bigcup_{b \in \operatorname{bd} K} [\rho(K - b) + b] & \text{for } \rho \in [1, \infty). \end{cases}$$

This construction can also be described in terms of addition and subtraction. In fact, it is not difficult to show (see also Voiculescu 1966) that

$$K(\rho) = K + (\rho - 1)\mathrm{D}K = \rho K + (\rho - 1)(-K) \quad \text{for } \rho \geqq 1,$$
$$K(\rho) = K \sim (1 - \rho)\mathrm{D}K = \rho K \sim (1 - \rho)(-K) \quad \text{for } \rho \leqq 1.$$

Hammer further proposed a notion of reducibility, as follows. There is a number $r_K \leqq 1$ such that

$$K = K(\rho)(\rho/(2\rho - 1)) \quad \text{for } \rho \geqq r_K,$$
$$K \neq K(\rho)(\rho/(2\rho - 1)) \quad \text{for } \rho < r_K.$$

Then K is called *reducible* if $r_K < 1$, otherwise *irreducible*. This notion of reducibility was studied in a series of papers by Zamfirescu (1966a, b, c, 1967a, b, c, d, 1973, 1975). It appears that part of the results can be obtained more easily using the following reformulation. Write

$$\alpha_K := \sup \{\alpha \geqq 0 \mid -\alpha K \text{ is a summand of } K\}$$

or, equivalently, using support functions,

$$\alpha_K = \sup \{\alpha \geqq 0 \mid h_K - \alpha h_{(-K)} \text{ is convex}\}.$$

Then it follows from the representation of $K(\rho)$ above that $r_K = (1 + \alpha_K)^{-1}$. Denoting by h_K^+ the even part and by h_K^- the odd part of h_K, we may also write

$$r_K = \inf \{\rho \geqq 0 \mid (2\rho - 1)h_K^+ + h_K^- \text{ is convex}\}.$$

In particular, K is reducible if and only if it has a summand positively homothetic to $-K$ (Zamfirescu 1973). Now, elementary properties of Minkowski addition and criteria for convex functions (Sections 1.5, 2.5) and for summands (Section 3.2) can be used to obtain results on reducibility.

14. *The Minkowski measure of symmetry*. The number $r(K, -K)$, the inradius of K relative to $-K$, is an affine invariant of K which, under different names, has often been discussed in the literature. By definition, $r(K, -K) = \max \{\lambda \geqq 0 \mid -\lambda K + t \subset K \text{ for some } t \in \mathbb{E}^n\}$. Grünbaum [37], Section 6.1, calls this *Minkowski's measure of symmetry* and denotes it by $F_1(K)$. Clearly $F_1(K) = 1$ if and only if K is centrally symmetric. If $K \in \mathcal{K}_0^n$ has its centroid at o, then

$$h(K, u) \geqq \frac{1}{n} h(K, -u) = \frac{1}{n} h(-K, u) \quad \text{for } u \in S^{n-1},$$

which for special cases goes back to Minkowski (1897) (see also Bonnesen & Fenchel [8], §34); hence $-K \subset nK$ and thus $F_1(K) \geqq 1/n$. The equality $F_1(K) = 1/n$ holds if and only if K is a simplex (see the references in Grünbaum [37], 6.1). The set

$$C(K) := \{a \in \mathbb{E}^n \mid -F_1(K)(K - a) + a \subset K\}$$
$$= \frac{K \sim F_1(K)(-K)}{1 + F_1(K)} = K \sim \frac{F_1(K)}{1 + F_1(K)} \mathrm{D}K$$

is called the *critical set* of K. Klee (1953) proved

$$\frac{1}{F_1(K)} + \dim C(K) \leqq n,$$

which implies that $\dim C(K) \leqq n-2$ for $K \in \mathcal{K}^n$. Dziechcińska-Halamoda & Szwiec (1985) for polytopes and Laget (1987) in general proved that every convex body in $\mathbb{E}^n$ of dimension at most $n-2$ is the critical set of some convex body $K \in \mathcal{K}^n$.

15. *Asymmetry in mathematical morphology*. Inner parallel bodies of a convex body K with respect to its difference body $\mathrm{D}K$ and the Minkowski measure of symmetry have also been discussed in mathematical morphology; see Jourlin & Laget (1988). Jourlin & Laget (1987) asked for a characterization of convex bodies K with the property that the inner parallel body $K \sim \tau \mathrm{D}K$, for some $\tau > 0$, is positively homothetic to K. For the case of the plane, an answer was given by Schneider (1989a): If $K \in \mathcal{K}_0^2$ has this property and is not centrally symmetric, then K is a polygon with the property that some homothet Q of $-K$ is properly contained in K and is such that each edge of K contains a vertex of Q. Examples of such polygons are the affine images of regular polygons with an odd number of edges, but there are many others. The corresponding problem in higher dimensions is open.

3.2. Summands and decomposition

Let $K, L \in \mathcal{K}^n$ be convex bodies. As defined in Section 3.1, the body L is a *summand* of K if there exists a convex body $M \in \mathcal{K}^n$ such that $K = L + M$. In this section, we shall study different aspects of the summand relation. After a remark on direct summands, we collect some criteria for summands; then we investigate convex bodies having only trivial summands.

We start with a result concerning direct summands, to which we shall not come back. We write $K = K_1 \oplus \ldots \oplus K_m$ if $K = K_1 + \ldots + K_m$ for suitable K_i lying in linear subspaces E_i of $\mathbb{E}^n$ such that $\mathbb{E}^n = E_1 \oplus \ldots \oplus E_m$. The convex body L is called a *direct summand* of K if there exists a convex body M such that $K = L \oplus M$. If a representation $K = L \oplus M$ is only possible with $\dim L = 0$ or $\dim M = 0$, then K is called *directly indecomposable*.

Theorem 3.2.1. *Every convex body* $K \in \mathcal{K}^n$ *with* $\dim K \geqq 1$ *has a representation*

$$K = K_1 \oplus \ldots \oplus K_m$$

with $m \in \{1, \ldots, n\}$ *and directly indecomposable bodies* $K_1, \ldots, K_m$ *with* $\dim K_i \geqq 1$. *The representation is unique up to the order of the summands.*

Proof. The existence of the representation is obvious. To show the essential uniqueness, suppose that

$$K = K_1 \oplus \ldots \oplus K_m = L_1 \oplus \ldots \oplus L_r,$$

where the bodies K_i, L_j are directly indecomposable. Then $m \geqq 2$, $r \geqq 2$. Put $\bar{K} := K_2 \oplus \ldots \oplus K_m$. We can choose a vector $u \neq o$ orthogonal to K_1 such that the support set $F(\bar{K}, u)$ contains only one point, say x. Then $F(K, u) = F(K_1, u) + F(\bar{K}, u) = K_1 + x$ by Theorem 1.7.5(c); thus

$$K_1 + x = F(K, u) = F(L_1, u) \oplus \ldots \oplus F(L_r, u).$$

Since K_1 is directly indecomposable, the decomposition on the right-hand side must be trivial, that is, at most one of the summands has positive dimension. Thus K_1 is a translate of a subset of some L_j. By a similar argument, L_j is a translate of a subset of some K_i. Since $\dim K_1 \geqq 1$ and K_1, K_i for $1 \neq i$ are in complementary subspaces, we must have $i = 1$ and thus $K_1 = L_j$. Repeating the argument for $K_2 \oplus \ldots \oplus K_m$, and so on, we arrive at the assertion. ■

We turn to summands in the general case and first prove a simple intuitive criterion. For convex bodies $K, L \in \mathcal{K}^n$ one says that L *slides freely inside* K if to each boundary point x of K there exists a translation vector $t \in \mathbb{E}^n$ such that $x \in L + t \subset K$.

***Theorem* 3.2.2.** *Let $K, L \in \mathcal{K}^n$. Then L is a summand of K if and only if L slides freely inside K.*

Proof. If $K = L + M$ and $x \in K$, there exist $y \in L$ and $t \in M$ such that $x = y + t$, hence $x \in L + t \subset M = K$.

Vice versa, suppose that L slides freely inside K. Let $x \in \operatorname{bd} K$ be given. By assumption, there exists $t \in \mathbb{E}^n$ such that $x \in L + t \subset K$. Then $t \in K \sim L$ and $x \in L + (K \sim L)$. We deduce that $\operatorname{bd} K \subset (K \sim L) + L$ and hence $K = (K \sim L) + L$. By Lemma 3.1.8, L is a summand of K. ■

As a consequence, we can show that every summand L of a polytope P is itself a polytope: To a given vertex x of P, there is a unique vector t_x such that $x \in L + t_x \subset P$. Clearly $N(L, x - t_x) \supset N(P, x)$. Thus L has finitely many boundary points $y_1, \ldots, y_k$ such that $\bigcup_{i=1}^k N(L, y_i) = \mathbb{E}^n$. If u is a normal vector of L at some exposed point a, then $u \in N(L, y_i)$ for some i and hence $a = y_i$. Consequently, L has only finitely many exposed points and thus is a polytope.

The possible summands of a polytope have a structure that is strongly related to the polytope itself. Let P, Q_1, $Q_2 \in \mathcal{P}^n$ and $P = Q_1 + Q_2$. Let F be a face of P. Choose $u \in S^{n-1}$ such that $F(P, u) = F$ and put $G_i := F(Q_i, u)$ for $i = 1, 2$. Then $F = G_1 + G_2$, hence G_1 is a summand of F. In particular, the normal vectors (of the facets) of Q_1 are among the normal vectors of P. Choose $x \in \operatorname{relint} F$. Then x has a representation $x = y_1 + y_2$ with $y_i \in \operatorname{relint} G_i$ (Lemma 1.3.12). By Theorem 2.2.1(a) we have

$$\begin{aligned} N(P, F) &= N(P, x) = N(Q_1, y_1) \cap N(Q_2, y_2) \\ &= N(Q_1, G_1) \cap N(Q_2, G_2), \end{aligned}$$

hence $N(P, F) \subset N(Q_1, G_1)$. Since $u \in \operatorname{relint} N(Q_1, G_1)$, the face G_1 does not depend on the choice of u. We say that G_1 is the face of Q_1 *corresponding to* F. If F' is a face of P containing F and G_1' is the face of Q_1 corresponding to F', then a vector v with $F' = F(P, v)$ satisfies $v \in N(P, F) \subset N(Q_1, G_1)$ and $v \in \operatorname{relint} N(Q_1, G_1')$, hence $N(Q_1, G_1') \subset N(Q_1, G_1)$. Thus G_1 is contained in G_1'.

Particularly easy to describe are the summands of a polygon $P \in \mathcal{P}^2$. Choose an orthonormal basis e_1, e_2 of $\mathbb{E}^2$ and let $x_0 = x_0(P)$ be the vertex of P with smallest e_2-coordinate and, if there are two vertices with this property, also with smallest e_1-coordinate. Let $x_0, x_1, \ldots, x_k$, $x_{k+1} = x_0$ be the vertices of P in cyclic order, in such a way that the vectors

$$v_i := x_i - x_{i-1}, \quad i = 1, \ldots, k+1$$

satisfy

$$v_i = L_i(e_1 \cos \alpha_i + e_2 \sin \alpha_i) \tag{3.2.1}$$

with $L_i > 0$ and

$$0 \leqq \alpha_1 < \alpha_2 < \ldots < \alpha_{k+1} < 2\pi. \tag{3.2.2}$$

Then

$$\sum_{i=1}^{k+1} v_i = o. \tag{3.2.3}$$

Vice versa, if vectors $v_1, \ldots, v_{k+1}$ satisfying (3.2.1), (3.2.2) and (3.2.3) are given, then

$$x_0 \text{ (arbitrary)}, \quad x_j := x_0 + \sum_{i=1}^{j} v_i, \; j = 1, \ldots, k$$

are the vertices of a convex polygon P. In fact, if $j \in \{1, \ldots, k+1\}$ is given, we can choose $u \in S^1$ with

$$\begin{aligned} \langle v_i, u \rangle &> 0 \quad \text{for } i = j, j-1, \ldots, m, \\ \langle v_i, u \rangle &< 0 \quad \text{for } i = j+1, \ldots, m-1, \end{aligned}$$

where the indices i have to be understood as residue classes modulo $(k+1)$. For $r \in \{j-1, j-2, \ldots, m-1\}$ we then have

$$\langle x_j - x_r, u \rangle = \langle v_{r+1} + v_{r+2} + \ldots + v_j, u \rangle > 0$$

and for $r \in \{j+1, \ldots, m-1\}$,

$$\langle x_j - x_r, u\rangle = -\langle v_{j+1} + v_{j+2} + \ldots + v_r, u\rangle > 0.$$

Thus each x_j is a vertex of $P := \operatorname{conv}\{x_0, x_1, \ldots, x_k\}$.

Now suppose that $P = Q + Q'$. Since each edge of Q is a summand of an edge of P, the polygon Q has the vertices

$$x_0(Q) \text{ and } y_j = \sum_{i=1}^{j} \lambda_i v_i, \quad j = 1, \ldots, k \tag{3.2.4}$$

(not necessarily distinct), where $0 \leqq \lambda_i \leqq 1$. Clearly, Q' then has the vertices

$$x_0(P) - x_0(Q) \text{ and } y_j' = \sum_{i=1}^{j} (1-\lambda_i) v_i, \quad j = 1, \ldots, k. \tag{3.2.5}$$

Vice versa, if Q is a polygon whose vertices can be given by (3.2.4), then Q is a summand of P, since (3.2.5) can be used to define the vertices of the other summand. Thus we see that Q is a summand of the polygon P if and only if for each edge $F(Q,u)$ of Q the support set $F(P,u)$ is an edge of P of at least the same length. This remark will later be extended (Theorem 3.2.8).

We return to summands of general convex bodies and first observe that there is a close relation to intersections of translates. Let $K, L \in \mathcal{K}^n$ be convex bodies such that L is a summand of K. Then $K = L + M$ for some $M \in \mathcal{K}^n$, and since $K = (K \sim M) + M$ by Lemma 3.1.8, we get

$$L = K \sim M = \bigcap_{t \in M} (K - t).$$

Thus each summand of K is an intersection of a family of translates of K. For $n = 2$, this necessary condition for a summand is also sufficient:

Theorem 3.2.3. *For $K \in \mathcal{K}^2$, each nonempty intersection of translates of K is a summand of K.*

Proof. Suppose that $\emptyset \neq L = \bigcap_{t \in T}(K + t)$ with some set $\emptyset \neq T \subset \mathbb{E}^2$. First let K be a polygon, say

$$K = \bigcap_{i=1}^{k} H^-_{u_i, \alpha_i}.$$

Then

$$L = \bigcap_{t \in T}(K+t) = \bigcap_{t \in T} \bigcap_{i=1}^{k} H^-_{u_i, \alpha_i + \langle u_i, t\rangle}$$

$$= \bigcap_{i=1}^{k} \bigcap_{t \in T} H^-_{u_i, \alpha_i + \langle u_i, t\rangle} = \bigcap_{i=1}^{k} H^-_{u_i, \alpha_i + \beta_i},$$

where $\beta_i := \inf_{t \in T} \langle u_i, t \rangle$. It is now easy to see that the polygon L satisfies the sufficient assumption stated above for summands of a polygon K and hence that L is a summand of K.

Now let $K \in \mathcal{K}^2$ be arbitrary. We can choose a sequence $(P_i)_{i \in \mathbb{N}}$ of polygons such that

$$P_1 \supset P_2 \supset P_3 \dots \quad \text{and} \quad \bigcap_{i \in \mathbb{N}} P_i = K,$$

hence $\lim_{i \to \infty} P_i = K$ (cf. Lemma 1.8.1). We have

$$L = \bigcap_{t \in T} (K + t) = \bigcap_{t \in T} \bigcap_{i \in \mathbb{N}} (P_i + t) = \bigcap_{i \in \mathbb{N}} P_i'$$

with $P_i' := \bigcap_{t \in T} (P_i + t)$. By the first part, $P_i = P_i' + Q_i$ with $Q_i \in \mathcal{K}^2$. From $P_1' \supset P_2' \supset \dots$ we infer that $\lim P_i' = L$, and since $\lim P_i = K$, there exists $\lim Q_i := M \in \mathcal{K}^2$, and $K = L + M$. Thus L is a summand of K. ■

In $\mathbb{E}^n$ with $n \geqq 3$, an intersection of two translates of a convex body $K \in \mathcal{K}^n$ is in general not a summand of K. This is shown by the example of a square pyramid in $\mathbb{E}^3$.

For further investigation of the intersections of translates, the following may be remarked.

Lemma 3.2.4. *Let $K, L \in \mathcal{K}^n$. Then L is an intersection of translates of K if and only if*

$$K \sim (K \sim L) = L.$$

Proof. If $L = \bigcap_{t \in T} (K - t)$ with some family $T \subset \mathbb{E}^n$, then

$$L = \bigcap_{t \in T} (K - t) \supset \bigcap_{L \subset K - t} (K - t) \supset L,$$

hence

$$L = \bigcap_{L \subset K - t} (K - t) = \bigcap_{t \in K \sim L} (K - t) = K \sim (K \sim L).$$

The latter equality also shows that $K \sim (K \sim L)$ is an intersection of translates of K. ■

For $K, L \in \mathcal{K}^2$ we deduce from Theorem 3.2.3 that $K \sim L$, if not empty, is always a summand of K, and that L is a summand of K if and only if $K \sim (K \sim L) = L$.

Criteria for summands involving intersections of translates can also be formulated in higher dimensions. Lemma 3.2.5 and Theorem 3.2.7 below are due, in more general forms, to Wieacker (1988a).

***Lemma* 3.2.5.** *Let $K, L \in \mathcal{K}^n$. Then L is a summand of K if and only if, for all $x \in K$ and $y_1, \ldots, y_{n+1} \in L$,*

$$(x - L) \cap (K - y_1) \cap \ldots \cap (K - y_{n+1}) \neq \emptyset.$$

Proof. By Lemma 3.1.8, L is a summand of K if and only if $(K \sim L) + L = K$, hence if and only if, for each $x \in K$, there is $z \in L$ such that

$$x - z \in K \sim L = \bigcap_{y \in L} (K - y)$$

or, equivalently,

$$(x - L) \cap \bigcap_{y \in L} (K - y) \neq \emptyset.$$

Now Helly's theorem, 1.1.6, applied to the finite subfamilies of $\{(x - L) \cap (K - y) | y \in L\}$, together with the compactness of L and K, yields the assertion. ■

Lemma 3.2.5 is improved by Theorem 3.2.7. Its proof uses the following lemma.

***Lemma* 3.2.6.** *Let $K, L \in \mathcal{K}^n$. If* $\mathrm{proj}_E L$ *is a summand of* $\mathrm{proj}_E K$ *for all two-dimensional linear subspaces E in some dense subset of $G(n, 2)$, then L is a summand of K.*

Proof. Define $g(u) := h(K, u) - h(L, u)$ for $u \in \mathbb{E}^n$. Let $E \in G(n, 2)$ be a two-dimensional linear subspace for which $\pi K = \pi L + M_E$ with a convex body $M_E \subset E$; here $\pi = \mathrm{proj}_E$. For $u, v \in E$ we have $u + v \in E$ and hence

$$\begin{aligned} g(u + v) &= h(K, u + v) - h(L, u + v) \\ &= h(\pi K, u + v) - h(\pi L, u + v) \\ &= h(M_E, u + v) \leqq h(M_E, u) + h(M_E, v) \\ &= g(u) + g(v). \end{aligned}$$

The set of all pairs (u, v) for which the inequality $g(u + v) \leqq g(u) + g(v)$ has thus been established is dense in $\mathbb{E}^n \times \mathbb{E}^n$; therefore the continuous, positively homogeneous function g is sublinear and hence, by Theorem 1.7.1, is the support function of a convex body $M \in \mathcal{K}^n$. Then $K = L + M$. ■

***Theorem* 3.2.7.** *Let $K, L \in \mathcal{K}^n$. Then L is a summand of K if and only if, for all $x \in K$ and $y_1, y_2 \in L$,*

$$(x - L) \cap (K - y_1) \cap (K - y_2) \neq \emptyset. \tag{3.2.6}$$

Proof. Only the sufficiency of condition (3.2.6) has to be proved. Suppose first that $n = 2$. We assert that

$$(K - y_1) \cap (K - y_2) \cap (K - y_3) \neq \emptyset \tag{3.2.7}$$

for $y_1, y_2, y_3 \in L$. Suppose this were false for some $y_1, y_2, y_3 \in L$. Then there is a line H that strongly separates $(K - y_1) \cap (K - y_2)$ (which is not empty, by (3.2.6)) and $K - y_3$. For $x \in K$, the set $x - L$ meets $(K - y_1) \cap (K - y_2)$, $(K - y_2) \cap (K - y_3)$ and $(K - y_1) \cap (K - y_3)$ and, hence, also the disjoint segments $H \cap (K - y_1)$ and $H \cap (K - y_2)$. Thus $x - L$ contains the open segment S between them. From

$$S \subset \bigcap_{x \in K} (L - x) = L \sim K$$

we deduce that K is strictly contained in a translate of L, say in L. Then $h(K, \cdot) \leqq h(L, \cdot)$, while $h(K, u) < h(L, u)$ for some $u \in S^1$. For $y, z \in L$ with $\langle y, u \rangle = h(L, u)$ and $\langle z, u \rangle = h(L, -u)$ we get

$$h(K - y, u) = h(K, u) - \langle y, u \rangle < 0,$$
$$h(K - z, -u) = h(K, -u) - \langle z, -u \rangle \leqq 0$$

and therefore $(K - y) \cap (K - z) = \emptyset$. This contradicts (3.2.6); thus (3.2.7) must hold.

From (3.2.6), (3.2.7) and Helly's theorem we now get

$$(x - L) \cap (K - y_1) \cap (K - y_2) \cap (K - y_3) \neq \emptyset$$

whenever $x \in K$ and $y_1, y_2, y_3 \in L$. Lemma 3.2.5 then shows that L is a summand of K. Thus the theorem is proved for $n = 2$.

Now let $n \geqq 3$. Let $E \in G(n, 2)$ and $\pi = \mathrm{proj}_E$. For $x \in \pi K$ and $y_1, y_2 \in \pi L$ we choose $\bar{x} \in K$ and $\bar{y}_1, \bar{y}_2 \in L$ with $x = \pi \bar{x}$, $y_i = \pi \bar{y}_i$ and find

$$(\bar{x} - L) \cap (K - \bar{y}_1) \cap (K - \bar{y}_2) \neq \emptyset;$$

hence

$$(x - \pi L) \cap (\pi K - y_1) \cap (\pi K - y_2) \neq \emptyset.$$

By the first part of the proof, πL is a summand of πK. By Lemma 3.2.6, L is a summand of K. ∎

For special convex bodies in $\mathcal{K}^n$, such as polytopes or sufficiently smooth bodies, more effective criteria for summands can be proved. The case of polytopal summands is easy to check by means of the following result.

***Theorem* 3.2.8.** *Let $P, K \in \mathcal{K}^n$, where P is a polytope. Then P is a summand of K if and only if the support set $F(K, u)$ contains a translate of $F(P, u)$ whenever $F(P, u)$ is an edge of P $(u \in S^{n-1})$.*

Proof. If $K = P + M$ with $M \in \mathcal{K}^n$, then $F(K, u) = F(P, u) + F(M, u) \supset F(P, u) + t$ for some $t \in M$, so that the condition is necessary. Suppose now that the condition is satisfied. First let $n = 2$. We can easily construct a sequence $(P_i)_{i \in \mathbb{N}}$ of polygons such that $P_i \to K$ for $i \to \infty$ and $F(K, u_j) \subset F(P_i, u_j)$ for each normal vector u_j of an edge of P. By the criterion for summands of polygons described earlier, we have $P_i = P + Q_i$ with some polygon Q_i. From $P_i \to K$ we get $Q_i \to Q$ for some $Q \in \mathcal{K}^2$ and $K = P + Q$; thus P is a summand of K.

Now let $n \geqq 3$. Let $E \in G(n, 2)$ be a two-dimensional linear subspace with the property that $\dim F(P, w) \leqq 1$ for each $w \in E$. Let $\pi = \operatorname{proj}_E$. If $w \in E \setminus \{o\}$ is such that $F(\pi P, w)$ is an edge of the polygon πP, then, by the choice of E, $\pi^{-1} F(\pi P, w) \cap P = F(P, w)$ is an edge of P. By the assumption of the theorem the support set $F(K, w)$ contains a translate of the edge $F(P, w)$, hence $F(\pi K, w)$ contains a translate of the edge $F(\pi P, w)$. We have proved that πP and πK satisfy the assumptions of the theorem. By the first part of the proof, πP is a summand of πK. Since the set of all 2-subspaces E that could have been chosen above is dense in $G(n, 2)$, it follows from Lemma 3.2.6 that P is a summand of K. ■

For a convex body $K \in \mathcal{K}^n$ we denote by $\mathcal{S}(K)$ the set of all convex bodies that are homothetic to a summand of K. Thus $\mathcal{S}(K)$ is a convex cone in $\mathcal{K}^n$, in the sense that $L, M \in \mathcal{S}(K)$ and $\lambda \geqq 0$ imply $L + M \in \mathcal{S}(K)$ and $\lambda L \in \mathcal{S}(K)$. Further, for convex bodies $L, K \in \mathcal{K}^n$ we write $L \leqq K$ if $\dim F(L, u) \leqq \dim F(K, u)$ for all $u \in S^{n-1}$. Then the special case of Theorem 3.2.8 where K is also a polytope (this is due to Shephard 1963), evidently implies: If $P, K \in \mathcal{P}^n$ are polytopes, then $P \in \mathcal{S}(K)$ if and only if $P \leqq K$.

Another class of convex bodies for which criteria for summands are easily formulated are the bodies of class C^2_+. The following result is just a reformulation of Theorem 2.5.4.

Theorem 3.2.9. *Let K, $L \in \mathcal{K}^n$ be convex bodies of class C^2_+. Then L is a summand of K if and only if the radii of curvature satisfy*

$$r(L, u, v) \leqq r(K, u, v)$$

for each orthogonal pair of vectors $u, v \in S^{n-1}$; equivalently, if the curvatures satisfy

$$\kappa(L, x, t) \geqq \kappa(K, y, t)$$

for $x \in F(L, u)$, $y \in F(K, u)$, $u \in S^{n-1}$ and $t \perp u$.

Consequently, if K is a convex body of class C^2_+, then there are positive numbers r, R such that rK is a summand of the unit ball B^n and B^n is a summand of RK.

Let $K, L \in \mathcal{K}^n$. One says that the convex body L *rolls freely inside* K if for each rotation $\rho \in SO(n)$ and each boundary point x of K there is a translation vector $t \in \mathbb{E}^n$ such that $x \in \rho L + t \subset K$. By Theorem 3.2.2, this is equivalent to the condition that each congruent image of L is a summand of K. Hence, Theorem 3.2.9 yields the following.

Corollary 3.2.10. *Let $K, L \in \mathcal{K}^n$ be convex bodies of class C^2_+. Then L rolls freely inside K if and only if*

$$\max_{u,v} r(L, u, v) \leqq \min_{u,v} r(K, u, v).$$

This is a generalization of Blaschke's 'rolling theorem'. Blaschke [5], §24, considered the case where $n \leqq 3$ and L is a ball.

We turn now to the investigation of convex bodies having only trivial summands. The convex body $K \in \mathcal{K}^n$ of positive dimension is called *indecomposable* if a representation $K = K_1 + K_2$ with $K_1, K_2 \in \mathcal{K}^n$ is only possible with K_1, K_2 homothetic to K; otherwise K is called *decomposable*. Thus K is indecomposable precisely if $\mathcal{S}(K) = \{\lambda K + t | \lambda \geqq 0,\ t \in \mathbb{E}^n\}$.

In dealing with indecomposable convex bodies (and elsewhere too) it is convenient to introduce the class

$$\mathcal{K}^n_s := \{K \in \mathcal{K}^n | s(K) = o,\ b(K) = 1\},$$

where s is the Steiner point and b is the mean width (see Section 1.7). We call $\mathcal{K}^n_s$ the class of *normalized convex bodies*. Evidently it contains precisely one element from each homothety class of convex bodies with more than one point. From the linearity properties of Steiner point and mean width it follows that $\mathcal{K}^n_s$ is a convex subset of $\mathcal{K}^n$, by which we mean, of course, that $K, L \in \mathcal{K}^n_s$ and $\lambda \in [0, 1]$ implies $(1 - \lambda)K + \lambda L \in \mathcal{K}^n_s$. By the continuity of s and b, $\mathcal{K}^n_s$ is closed. If $K \in \mathcal{K}^n_s$ and $x \in K$, then $[o, x] \subset K$ and hence $(2\kappa_{n-1}/n\kappa_n)|x| = b([o, x]) \leqq b(K) = 1$. Thus $\mathcal{K}^n_s$ is bounded, and by the Blaschke selection theorem it is compact. We see that the map φ (see (1.7.1) and Theorem 1.8.11) maps $\mathcal{K}^n_s$ bijectively onto a compact convex set in the Banach space $C(S^{n-1})$. Under this map, the indecomposable bodies in $\mathcal{K}^n_s$ correspond precisely to the extreme points of $\varphi(\mathcal{K}^n_s)$. (The definition of an extreme point is the same as in Section 1.4: $x \in \varphi(\mathcal{K}^n_s)$ is an extreme point of $\varphi(\mathcal{K}^n_s)$ if and only if it cannot be represented in the form $x = (1 - \lambda)y + \lambda z$ with $y, z \in \varphi(\mathcal{K}^n_s)$ and $\lambda \in (0, 1)$.) In fact, let $K \in \mathcal{K}_s^n$ be decomposable. Then $K = L + M$ with suitable $L, M \in \mathcal{K}^n$ that are not homothetic to K. We have $b(L) + b(M) = b(L + M) = b(K) = 1$ and similarly $s(L) + s(M) = o$. Writing $\bar{L} := [L - s(L)]/b(L)$, $\bar{M} := [M - s(M)]/b(M)$ and $\lambda = b(M)$, we get $\bar{L}, \bar{M} \in \mathcal{K}_s^n$ and $(1 - \lambda)\bar{L} + \lambda\bar{M} = K$, hence $\varphi(K) = (1 - \lambda)\varphi(\bar{L}) + \lambda\varphi(\bar{M})$ with $\varphi(\bar{L}) \neq \varphi(K)$, so that $\varphi(K)$ is not an extreme point of $\varphi(\mathcal{K}_s^n)$. The converse conclusion is obvious.

In the plane, the indecomposable convex bodies are easily determined.

Theorem 3.2.11. *The indecomposable bodies in $\mathcal{K}^2$ are the segments and the triangles.*

Proof. That segments and triangles are indecomposable is clear (a summand of a triangle must be a polygon with the same normal vectors). Vice versa, let $K \in \mathcal{K}^2$ be indecomposable and not a segment. By Theorem 3.2.3, each nonempty intersection $K \cap (K + t)$ must be homothetic to K. Let U denote the set of all exterior unit normal vectors at regular boundary points of K. Choose $u, v \in U$, $u \neq \pm v$, and let x, y be regular points of $\operatorname{bd} K$ where these normal vectors are attained. Then $\bar{K} := K \cap (K + x - y)$ has interior points. Since $\bar{K}$ is homothetic to K, the body K must have a boundary point where both u and v, are normal vectors. This implies that no vector $w \in \operatorname{int} \operatorname{pos} \{u, v\}$ can be a normal vector at a regular point of K. Thus, U must either consist of three vectors or of four vectors in opposite pairs. Since the support lines at regular points (which are dense in $\operatorname{bd} K$) determine K uniquely, K is either a triangle or a parallelogram, but the latter is a sum of two segments and thus decomposable. ■

Remark. A regular hexagon in the plane can be written as the Minkowski sum of two triangles and also as the sum of three segments. Thus a representation of a convex body as a finite sum of indecomposable bodies, if possible, is in general not unique.

The assertion of Theorem 3.2.11 is typical for the plane; in higher dimensions the situation changes drastically. In fact, most convex bodies in $\mathbb{E}^n$, for $n \geqq 3$, are indecomposable. We first show that there are many indecomposable polytopes.

There are several criteria from which the indecomposability of a polytope can be deduced; we choose one due to McMullen (1987). Let $P \in \mathcal{P}^n$ be a polytope. A *strong chain* of faces of P is a sequence $F_0, F_1, \ldots, F_k$ of faces of P such that $\dim(F_{j-1} \cap F_j) \geqq 1$ for $j = 1$,

$\ldots, k$. A family $\mathcal{F}$ of faces of P is called *strongly connected* if for each $F, G \in \mathcal{F}$ there exists a strong chain $F = F_0, F_1, \ldots, F_k = G$ with $F_j \in \mathcal{F}$ for $j = 1, \ldots, k$. The family $\mathcal{F}$ *touches* the face F if $F \cap G \neq \emptyset$ for some $G \in \mathcal{F}$.

Theorem 3.2.12. *Let $P \in \mathcal{P}^n$ be a polytope having a strongly connected family $\mathcal{F}$ of indecomposable faces that touches each facet of P. Then P is indecomposable.*

Proof. We may assume that $\dim P = n$. Let Q be a summand of P. Let $F_0, F_1, \ldots, F_k \in \mathcal{F}$ form a strong chain. Let G_j be the face of Q corresponding to F_j (in the sense explained after Theorem 3.2.2). Then G_j is a summand of F_j and F_j is indecomposable, hence $G_j = \lambda_j F_j + t_j$ with suitable $\lambda_j \geqq 0$ and $t_j \in \mathbb{E}^n$. Because $\dim(F_{j-1} \cap F_j) \geqq 1$ for each j, we conclude that $\lambda_{j-1} = \lambda_j$ and $t_{j-1} = t_j$ for $j = 1, \ldots, k$. Since the family $\mathcal{F}$ is strongly connected, there are a number $\lambda \geqq 0$ and a vector $t \in \mathbb{E}^n$ such that $F(Q, v) = \lambda F(P, v) + t$ whenever $F(P, v) \in \mathcal{F}$.

Let $u_1, \ldots, u_m$ be the outer unit normal vectors of the facets of P; let $i \in \{1, \ldots, m\}$. By assumption, $F(P, u_i)$ has a vertex x lying in some face $F(P, v) \in \mathcal{F}$. The corresponding vertex y of Q lies in $F(Q, v)$, hence $y = \lambda x + t$. This yields $h(Q, u_i) = \langle y, u_i \rangle = \langle \lambda x + t, u_i \rangle = \lambda h(P, u_i) + \langle t, u_i \rangle = h(\lambda P + t, u_i)$ for $i = 1, \ldots, m$ and hence $Q = \lambda P + t$ by Corollary 2.4.4. Since Q was an arbitrary summand of P, the polytope P is indecomposable. ■

Corollary 3.2.13. *If all two-dimensional faces of the polytope P are triangles, then P is indecomposable.*

This can be used to show that, for $n \geqq 3$, most convex bodies in $\mathcal{K}^n$ are indecomposable.

Theorem 3.2.14. *For $n \geqq 3$, the set of indecomposable bodies in $\mathcal{K}^n$ is a dense G_δ-set.*

Proof. Since the simplicial polytopes are dense in $\mathcal{K}^n$ if $n \geqq 3$, it follows from Corollary 3.2.13 that the set $\mathcal{A}$ of indecomposable polytopes is a dense subset of $\mathcal{K}^n$. Let $K \in \mathcal{K}^n$ be decomposable. Then one easily checks that K can be written in the form $K = (L + M)/2$ with convex bodies $L, M \in \mathcal{K}^n$ such that $s(K) = s(L) = s(M)$, $b(K) = b(L) = b(M)$ and $L \neq M$. For $k \in \mathbb{N}$, let $\mathcal{B}_k$ be the set of all $K \in \mathcal{K}^n$ having such a representation with $\delta(L, M) \geqq 1/k$. Using the Blaschke selection theorem, one shows that $\mathcal{B}_k$ is closed. Since $\mathcal{A} = \mathcal{K}^n \setminus \bigcup_{k \in \mathbb{N}} \mathcal{B}_k$, the assertion follows. ■

Since in a Baire space the intersection of two dense G_δ-sets is a dense G_δ-set, the last result together with Theorem 2.6.1 shows that most convex bodies in $\mathcal{K}^n$, $n \geqq 3$, are smooth, strictly convex, and indecomposable. It appears that no concrete example of such a body is explicitly known. This is not too surprising, since it is hard to imagine how such a body should be described. By a result of Sallee (1972), to be indecomposable it cannot be too smooth. More precisely, a neighbourhood U in the boundary of a convex body $K \in \mathcal{K}^n$ is called *ε-smooth*, for some $\varepsilon > 0$, if for each $x \in U$ there exists $t \in \mathbb{E}^n$ such that $x \in \varepsilon B^n + t \subset K$. Sallee proved: If the boundary of a strictly convex body K contains a neighbourhood U that is ε-smooth for some $\varepsilon > 0$, then K is decomposable.

Notes for Section 3.2

1. *Direct decompositions*. The proof of Theorem 3.2.1 can be extended to the case of line-free closed convex sets. In this form, the theorem was proved by Gruber (1970a), although in a less direct way. For an approach involving matroids, see Kinczes (1987), Theorem 2.5. From a general and abstract viewpoint, such decompositions were studied by Gale & Klee (1975).
2. Theorem 3.2.2 appears at several places in the literature (sometimes described as 'obvious' or 'folklore'); see Geivaerts (1972), Sallee (1974), Weil (1974b) and Firey (1979).
3. *Intersections of translates*. Theorem 3.2.3 was proved in a special case (the intersections of two translates) by W. Meyer (1972). His proof can be extended to the general case, which appears in Geivaerts (1972). The theorem does not extend verbally to higher dimensions; in fact, higher-dimensional analogues lead to interesting problems. In [36] (Problem 71) it was conjectured that a smooth convex body $K \in \mathcal{K}^n$, $n \geqq 3$, with the property that each nonempty intersection of K with a translate of K is a summand of K, is necessarily an ellipsoid. This is still unknown. More information is available in the case of polytopes. McMullen, Schneider & Shephard (1974) called a polytope $P \in \mathcal{P}^n$ *monotypic* if every polytope Q with the same system of unit normal vectors (of facets) as P is strongly isomorphic to P or, equivalently, if every polytope with the same system of normal vectors as P is simple. They showed, among other equivalent conditions, that a polytope is monotypic if and only if, for all $t \in \mathbb{E}^n$, $P \cap (P + t)$ is either empty or homothetic to a summand of P. They further proved that a centrally symmetric monotypic polytope is the direct sum of centrally symmetric polygons and line segments, and they classified the three-dimensional monotypic polytopes as well as the monotypic n-polytopes with at most $n + 3$ facets.

   Let $\mathcal{S}^n$ be the class of all convex bodies $K \in \mathcal{K}^n$ for which each nonempty intersection of translates of K is a summand of K. By Lemma 3.2.4, $K \in \mathcal{S}^n$ if and only if $K \sim (K \sim L) = L$ implies that L is a summand of K, thus if and only if $K \sim L = M$ and $K \sim M = L$ implies that $L + M = K$. From a slightly different viewpoint, this problem (for centrally symmetric K) was also studied by Sallee (1987). A result of Maehara (1984) implies that n-balls belong to $\mathcal{S}^n$.

A general study of relations between summands and intersections of translates was presented by Wieacker (1988a), from which Theorem 3.2.7 is taken. He has other, more general, results in this spirit, also for unbounded convex sets. Restricted to convex bodies, one of his results reads as follows.

Theorem. Let $K, L_i \in \mathcal{K}^n$ for $i \in I$ (an index set) and suppose that $\bigcap_{i\in I} L_i \neq \emptyset$. If for each subset J of I with $n-1$ or fewer elements there is a subset $J' \subset I$ such that $\bigcap_{j\in J\cup J'} L_j$ is a summand of K, then $\bigcap_{i\in I} L_i$ is a summand of K.

This result can be considered as a higher-dimensional extension of Theorem 3.2.3, to which it reduces if $n = 2$ and each L_i is a translate of K.

4. Geivaerts (1972, 1974a) has studied the class $\mathcal{S}(K)$ of all convex bodies homothetic to a summand of the body $K \in \mathcal{K}^n$. For $n = 2$ he proved that $\mathcal{S}(K)$ is closed if and only if K is a polygon, and that $\mathcal{S}(K)$ is dense in $\mathcal{K}^n$ if and only if K is smooth. He also investigated the class $\mathcal{S}'(K) \subset \mathcal{K}^n$ of all bodies L with the property that for each $x \in \operatorname{bd} K$ there exist $\lambda_x > 0$ and $t_x \in \mathbb{E}^n$ such that $x \in \lambda_x L + t_x \subset K$.
5. *Criteria for summands.* Theorem 3.2.8 was proved by Shephard (1963) in the case where K is a polytope and by Weil (1974b) in the general case, though in a more indirect way using deeper results. The proof given here is taken from Schneider (1974b).

Weil (1980b) proved the following theorem and explained that it can be considered as a generalization of Theorem 3.2.8 to general convex bodies.

Theorem. Let $K, L \in \mathcal{K}^n$. Then L is a summand of K if and only if, for each translate L' of L not contained in K, the set of all points of L' farthest from K is a support set of L'.

If L is strictly convex, this simplifies as follows: L is a summand of K if and only if each translate L' of L not contained in K has a unique point farthest from K.

Wieacker (1988c) found a way of generalizing Weil's theorem to unbounded closed convex sets. Since for such sets the vector sum need not be closed, and sets of farthest points may be empty, this requires a careful reformulation.

Weil (1980b) and Firey (oral communication, Oberwolfach 1980) proposed to characterize summands L of convex bodies K by imposing topological conditions on $(L + t) \cap \operatorname{bd} K$. Such a result was proved by Goodey (1982):

Theorem. Let $K, L \in \mathcal{K}_0^n$. If $\operatorname{bd} K \cap \operatorname{int} L'$ is acyclic (in particular, if it is a topological ball) for every translate L' of L, then L is a summand of K.

In fact, Goodey did not speak of summands but of freely sliding bodies, but this is equivalent by Theorem 3.2.2.

Burton (1979a) obtained the following sufficient criterion for a summand: Let K, $L \in \mathcal{K}^n$ be convex bodies with $L \subset K$ whose nearest-point maps commute, that is, $p_L \circ p_K = p_L$ $(= p_K \circ p_L)$. Then L is a summand of K. He also proved: Suppose that L is smooth. Then $p_L \circ p_K = p_L$ if and only if K is an outer parallel body of L.

Criteria for summands involving mixed volumes or mixed area measures were proved by Weil (1974b) and McMullen (1990); see also Matheron (1978b) (compare Section 4.2, Note 2, and Section 5.1, Note 9).

6. *Rolling theorems*. Extensions of Blaschke's rolling theorem, also to non-compact convex sets, were treated by Koutroufiotis (1972), Rauch (1974), Delgado (1979), and very thoroughly and in great generality by Brooks & Strantzen (1989). Analogues for convex curves in spherical or hyperbolic two-dimensional space are due to Karcher (1968). In a question involving geometric probabilities (see Section 4.5, Note 10), freely rolling convex bodies were used, and some characterizations proved, by Firey (1979) and Weil (1982a). In particular, these authors (Firey, §5, for the plane case; Weil, Theorem 1, in $\mathbb{E}^n$) give versions of Theorem 3.2.9 that hold for general convex bodies, without differentiability assumptions. Brooks & Strantzen (1989, Theorem 4.3.2) have a very general result of this kind where, however, the conclusion is not in terms of the summand relation but in terms of the inclusion relation.

7. Apparently the first to consider indecomposable convex bodies was Gale (1954). Without proof he gave some examples of decomposable and indecomposable convex bodies, and stated the assertion of Theorem 3.2.11. For this theorem, an analytic proof was given by Silverman (1973a) and a geometric one by W. Meyer (1969, 1972). Meyer proved a weaker form of Theorem 3.2.3 and then used a characterization due to Rogers & Shephard (1957) of the simplex, which holds in n-dimensional space (see Section 7.3, Note 2, for further results related to this theorem).

   Simple examples of indecomposable convex bodies in $\mathbb{E}^n$ ($n \geqq 3$) that are not polytopes can be constructed in the following way. A *general frustum* over A and B in $\mathbb{E}^n$ is, by definition, the convex hull of two convex bodies A and B that lie in different parallel hyperplanes. Sallee (1974) proved: Suppose that K is a general frustum over A and B where A is indecomposable and no homothet of A is a summand of B. Then K is indecomposable.

8. *Summands and decomposability of polytopes*. The first results on summands of polytopes and on indecomposability are due to Shephard (1963) (see also Grünbaum [15], Chapter 15). Shephard proved the polytope case of Theorem 3.2.8 and deduced the following sufficient condition for indecomposability.

Theorem. A polytope $P \in \mathcal{P}^n$ is indecomposable if it has an edge to which each vertex is connected by some strong chain of indecomposable faces.

Among the consequences that Shephard deduced are Corollary 3.2.13 and the fact (in a more general version) that every simple n-polytope, except the simplex, is decomposable.

W. Meyer (1969, 1970b, 1974) has some weaker sufficient indecomposability conditions for 3-polytopes. He further studied the cone $\mathcal{S}(P)$ of all polytopes homothetic to a summand of the polytope P and deduced a necessary and sufficient condition for indecomposability in terms of the rank of a certain system of linear equalities and inequalities determined by the facets of the polytope. A clear picture of the situation and new proofs of these results can be obtained if one uses a certain diagram technique (polyhedral representations); see McMullen (1973) and also McMullen (1979), Section 5. Fourneau (1979) used McMullen's representations to define a certain metric related to decomposability.

Kallay (1982) studied the decomposability of a polytope in connection with the decomposability of its (geometric) edge graph. Smilansky (1986) gave an example of an indecomposable 4-polytope all of whose facets are decomposable.

Although some sufficient criteria for the indecomposability of a polytope are of a combinatorial nature, such as Corollary 3.2.13, decomposability is not a combinatorial property. W. Meyer (1969, 1970b) and Kallay (1982) both gave examples of two combinatorially equivalent 3-polytopes, one decomposable and the other not. Decomposability of polytopes is, however, a projective property: If T is a projective transformation of $\mathbb{E}^n$ that is admissible for the polytope P, then the cones $\mathcal{S}(P)$ and $\mathcal{S}(TP)$ have the same dimension. This was shown by Kallay (1984); see also Smilansky (1987), pp. 40–1.

Smilansky (1987) studied the decomposability of polytopes (and polyhedral sets) by investigating the space of affine dependences of the vertices of the polar polytope. He obtained new proofs for results of W. Meyer (1974) and McMullen (1973) and several new results, in particular on combinatorial decomposability, for instance: If a 3-polytope has more vertices than facets, than it is decomposable. If a 3-polytope has at most three triangular facets, then it is decomposable.

9. *Summands and symmetry*. Shephard (1966b) called a convex body K with centre o *irreducible* if it is not the difference body of some body that is not homothetic to K; otherwise K is called *reducible* (this notion of reducibility is different from the one in Section 3.1, Note 13). He proved that K is reducible if and only if it has a summand without a centre of symmetry, and gave examples of reducible and of irreducible convex bodies. An application of Shephard's investigation to Banach space geometry can be found in Payá & Yost (1988). A thorough study of irreducible polytopes, including various necessary and sufficient criteria, is due to Yost (1991). One of his results states that each n-dimensional convex polytope, for $n \geqq 3$, with less than $4n$ vertices is irreducible; the bound $4n$ is best possible.

For $K \in \mathcal{K}_0^n$, Tennison (1967) defined $f(K) := \sup V(L)/V(K)$, where the supremum is taken over the centrally symmetric summands L ($\{o\}$ admitted) of K. He showed that f has many of the properties usually associated with a measure of symmetry, except possibly continuity. However, it follows from Corollary 3.2.13 and the denseness of simplicial polytopes that f is not continuous.

10. *Asymmetry classes*. Symmetric summands also play a role in the following question. For $K_1, K_2 \in \mathcal{K}^n$ write $K_1 \alpha K_2$ if there exist centrally symmetric convex bodies $S_1, S_2 \in \mathcal{K}^n$ such that $K_1 + S_1 = K_2 + S_2$. Thus $K_1 \alpha K_2$ if and only if the support functions of K_1 and K_2 have identical odd parts. The relation α is an equivalence relation on $\mathcal{K}^n$, and the corresponding classes are called *asymmetry classes*. Ewald & Shephard (1966) posed the question of whether each asymmetry class has a minimal member, where $M \in [K]$ (the α-equivalence class of K) is called *minimal* if each other element $L \in [K]$ can be expressed in the form

$$L = M + S$$

with a centrally symmetric body $S \in \mathcal{K}^n$. Schneider (1974e) proved that for $n = 2$ the answer is in the affirmative, but not for $n \geqq 3$. The latter follows from the result that a polytope P is a minimal member of its asymmetry class if and only if there does not exist a direction u for which $F(P, u)$ and $F(P, -u)$ are parallel edges.

11. *Relative indecomposability*. As explained in Section 3.2, indecomposable convex bodies correspond to extreme points of the compact convex set $\mathcal{K}_s^n$. More generally, if $\mathcal{M}$ is a nonempty, compact, convex subset of $\mathcal{K}^n$, we say

that $K \in \mathcal{M}$ is *extreme in* $\mathcal{M}$ if $K = (1-\lambda)K_1 + \lambda K_2$ with $K_1, K_2 \in \mathcal{M}$ and $\lambda \in (0, 1)$ implies that $K_1 = K_2$. For a closed convex subset $\mathcal{M} \subset \mathcal{K}^n$ we also say that $K \in \mathcal{M}$ is *indecomposable relative to* $\mathcal{M}$ if $K = (1-\lambda)K_1 + \lambda K_2$ with $K_1, K_2 \in \mathcal{M}$ and $\lambda \in (0, 1)$ implies that K_1 and K_2 are homothetic to K. The following cases have been considered. Kallay (1974, 1975) gave a complete characterization, in the plane, of all convex bodies that are extreme in the class of bodies with a given width function. (This corrected an erroneous statement by Silverman (1973b).) In particular, a convex body $K \in \mathcal{K}^2$ of constant width unity is indecomposable relative to the class of bodies of constant width if and only if its radius of curvature function (which exists almost everywhere on S^1) is almost everywhere either 0 or 1. In higher dimensions, it seems unknown whether the relatively indecomposable bodies of constant width have a similar characterization, and whether they are dense in the set of all bodies of constant width.

McMullen (1984) determined (and used for an extremum problem) the convex bodies that are indecomposable relative to the set of bodies of revolution.

Grzaślewicz (1984) determined the convex bodies that are extreme in $\mathcal{K}(C)$ (the set of convex bodies contained in C) for a strictly convex set $C \in \mathcal{K}_0^2$. He showed that K is extreme in $\mathcal{K}(C)$ if and only if either (1) K is a line segment with endpoints on bd C or (2) int $K \neq \emptyset$ and each connected component of bd $K \setminus$bd C contains at most one extreme point of K.

12. Most of the preceding treatment of decompositions $K = L + M$ was restricted to convex bodies K, L, M in $\mathbb{E}^n$. For unbounded convex sets and in infinite-dimensional vector spaces, many new problems arise. A detailed study was begun by Bair (1975, 1976b, 1977a, b) and Bair & Jongmans (1977).
13. *Decomposition of non-convex sets.* In this note only, 'indecomposable' will have a different meaning. The set $Z \subset \mathbb{E}^n$ (not necessarily convex) is called *indecomposable* if it cannot be represented as a vector sum $Z = Z_1 + Z_2$ with $Z_1, Z_2 \subset \mathbb{E}^n$ where each Z_i contains more than one point. Milka (1973) investigated the indecomposability of convex surfaces (i.e., boundaries of convex sets with interior points). Milka showed that every (relative) neighbourhood of the set of extreme points of a closed convex surface in $\mathbb{E}^3$ is indecomposable and that a closed strictly convex hypersurface in $\mathbb{E}^n$ is indecomposable. Among further results, he proved: If $K \subset \mathbb{E}^3$ is closed and convex, with interior points and with recession cone rec K a ray, and if bd K is decomposable in two different ways, then bd K is a paraboloid.
14. *Extremal convex functions* are, in some sense, analogous to indecomposable convex bodies. They were investigated by, e.g., Bronshtein (1978b).

3.3. Approximation and addition

Our objective in this section is the treatment of approximation results for convex bodies, with particular emphasis on the relation to Minkowski addition. First we describe an approximation procedure by smooth bodies that has particularly useful properties with regard to addition. Then we study briefly the possibility of approximating a convex body by Minkowski sums of bodies from a given class.

It is often necessary to approximate a convex body by sufficiently smooth bodies. We can do this by applying to its support function a suitable regularization process (the usual convolution, with a minor modification to retain homogeneity).

Theorem 3.3.1. *Let* $\varepsilon > 0$ and *let* $\varphi: [0, \infty) \to [0, \infty)$ *be a function of class* C^∞ *with support in* $[\varepsilon/2, \varepsilon]$ *and with*

$$\int_{\mathbb{E}^n} \varphi(|z|)\, \mathrm{d}z = 1.$$

If $f: \mathbb{E}^n \to \mathbb{R}$ *is a support function, then the function* $\tilde{f}$ *defined by*

$$\tilde{f}(x) := \int_{\mathbb{E}^n} f(x + |x|z)\varphi(|z|)\, \mathrm{d}z \quad \textit{for } x \in \mathbb{E}^n$$

is a support function of class C^∞ *on* $\mathbb{E}^n \setminus \{o\}$. *The map* $T: \mathcal{K}^n \to \mathcal{K}^n$ *defined by* $h_{TK} = \tilde{h}_K$ *has the following properties:*

(a) $T(K + L) = TK + TL$ *and* $T(\lambda K) = \lambda TK$ *for* $K, L \in \mathcal{K}^n$ *and* $\lambda \geqq 0$;

(b) $T(gK) = gTK$ *for* $K \in \mathcal{K}^n$ *and each rigid motion* g *of* $\mathbb{E}^n$;

(c) $\delta(K, TK) \leqq R\varepsilon$ *for all convex bodies* $K \subset RB^n$ *(where* $R > 0$*)*;

(d) $\delta(TK, TL) \leqq (1 + \varepsilon)\delta(K, L)$ *for* $K, L \in \mathcal{K}^n$.

Proof. Let φ be as in the theorem and let f be a support function. It follows from the definition that $\tilde{f}$ is positively homogeneous. Let $z \in \mathbb{E}^n$ and put

$$g_z(x) := f(x + |x|z) + f(x - |x|z) \quad \text{for } x \in \mathbb{E}^n.$$

For $x, y \in \mathbb{E}^n$ and $\alpha \in [0, 1]$ we have

$$\begin{aligned} g_z(x + y) &= f(x + y + |x + y|z) + f(x + y - |x + y|z) \\ &\leqq f(x + \alpha|x + y|z) + f(y + (1 - \alpha)|x + y|z) \\ &\quad + f(x - \alpha|x + y|z) + f(y - (1 - \alpha)|x + y|z). \end{aligned}$$

Without loss of generality we may assume that x and y are linearly independent. Put

$$\alpha := \frac{|x|}{|x| + |y|}, \quad \beta := \frac{|x + y|}{|x|}, \quad \gamma := \frac{|x + y|}{|y|};$$

then $1 - \alpha\beta > 0$ and $1 - (1 - \alpha)\gamma > 0$. From

$$\begin{aligned} 2(x + \alpha|x + y|z) &= (1 + \alpha\beta)(x + |x|z) + (1 - \alpha\beta)(x - |x|z), \\ 2(x - \alpha|x + y|z) &= (1 - \alpha\beta)(x + |x|z) + (1 + \alpha\beta)(x - |x|z) \end{aligned}$$

we infer that

$$\begin{aligned} f(x + \alpha|x + y|z) + f(x - \alpha|x + y|z) &\leqq f(x + |x|z) + f(x - |x|z) \\ &= g_z(x). \end{aligned}$$

Similarly we obtain

$$f(y + (1 - \alpha)|x + y|z) + f(y - (1 - \alpha)|x + y|z) \leqq g_z(y).$$

Together with the inequality for $g_z(x + y)$ as obtained above this yields

$$g_z(x + y) \leqq g_z(x) + g_z(y).$$

Since

$$\tilde{f}(x) = \tfrac{1}{2} \int_{\mathbb{E}^n} g_z(x)\varphi(|z|)\,\mathrm{d}z$$

and $\varphi \geqq 0$, the convexity of $\tilde{f}$ follows. By Theorem 1.7.1, $\tilde{f}$ is a support function.

For $u \in S^{n-1}$ we have

$$\tilde{f}(u) = \int_{\mathbb{E}^n} f(u + z)\varphi(|z|)\,\mathrm{d}z = \int_{\mathbb{E}^n} f(y)\varphi(|y - u|)\,\mathrm{d}y$$

and hence, for $x \in \mathbb{E}^n \setminus \{o\}$,

$$\tilde{f}(x) = |x| \int_{\mathbb{E}^n} f(y)\varphi\left(\left|y - \frac{x}{|x|}\right|\right)\mathrm{d}y,$$

from which we see that $\tilde{f}$ is of class C^∞ on $\mathbb{E}^n \setminus \{o\}$.

The map $T: \mathcal{K}^n \to \mathcal{K}^n$ defined by $h_{TK} = \tilde{h}_K$ clearly has property (a) of the theorem. For $t \in \mathbb{E}^n$ and $u \in S^{n-1}$ we have

$$h_{T\{t\}}(u) = \int_{\mathbb{E}^n} \langle u + z, t\rangle \varphi(|z|)\,\mathrm{d}z = \langle u, t\rangle = h_{\{t\}}(u),$$

hence $T\{t\} = \{t\}$. Together with (a) this yields $T(K + t) = TK + T\{t\} = TK + t$, thus T commutes with translations. Let ρ be a rotation of $\mathbb{E}^n$, proper or improper. Defining ρf by $(\rho f)(x) := f(\rho^{-1}x)$ for $x \in \mathbb{E}^n$, we obtain

$$\begin{aligned}(\rho\tilde{f})(x) &= \int_{\mathbb{E}^n} f(\rho^{-1}x + |\rho^{-1}x|z)\varphi(|z|)\,\mathrm{d}z \\ &= \int_{\mathbb{E}^n} f(\rho^{-1}(x + |x|\rho z))\varphi(|\rho z|)\,\mathrm{d}z \\ &= \int_{\mathbb{E}^n} (\rho f)(x + |x|z)\varphi(|z|)\,\mathrm{d}z,\end{aligned}$$

hence $(\widetilde{\rho f}) = \rho\tilde{f}$. This shows that $T(\rho K) = \rho TK$ for $K \in \mathcal{K}^n$, which completes the proof of part (b).

For $K \in \mathcal{K}^n$ with $K \subset RB^n$ and for $u \in S^{n-1}$ we obtain

$$\begin{aligned}|h_{TK}(u) - h_K(u)| &= \left|\int_{\mathbb{E}^n} h_K(u + z)\varphi(|z|)\,\mathrm{d}z - h_K(u)\right| \\ &\leqq \int_{\mathbb{E}^n} |h_K(u + z) - h_K(u)|\varphi(|z|)\,\mathrm{d}z \\ &\leqq \int_{\mathbb{E}^n} R|z|\varphi(|z|)\,\mathrm{d}z \leqq R\varepsilon,\end{aligned}$$

where Lemma 1.8.10 together with $\varphi(|z|) = 0$ for $|z| > \varepsilon$ was used. Since $u \in S^{n-1}$ was arbitrary, this proves (c). In a similar way assertion (d) is obtained, again using Lemma 1.8.10. ∎

Although the bodies TK obtained in Theorem 3.3.1 have support functions of class C^∞, they are not necessarily smooth since they can have singular points. This can easily be remedied by defining $T'K := TK + \varepsilon B^n$ for $K \in \mathcal{K}^n$. Then $T'K$ is of class C^∞_+, and the map T' has properties that are easily derived from those of the map T.

If a convex body K is approximated by the use of Theorem 3.3.1, then the approximating bodies inherit many properties from K. For instance, TK has (at least) the same symmetries as K, and if K is of constant width w, then $TK - TK = T(K - K) = T(wB^n)$ is a ball by (b); hence TK is of constant width.

The approximation procedure of Theorem 3.3.1 can be refined further, as we now briefly describe. For this, we use spherical harmonics (see the Appendix for the necessary explanations and definitions). By π_m we denote the orthogonal projection from the space $C(S^{n-1})$ of the real continuous functions on S^{n-1} onto the subspace of spherical harmonics of degree m. Let $K \in \mathcal{K}^n$ be a convex body of class C^∞_+. Since $\bar{h}_K$ (the restriction of the support function of K to S^{n-1}) is of class C^∞, we have

$$\bar{h}_K = \sum_{m=0}^{\infty} \pi_m \bar{h}_K$$

with convergence in the maximum norm. If, for some integer $k \geqq 2$, the partial sum $\sum_{m=0}^{k} \pi_m \bar{h}_K$ happens to be the restriction of a support function, we denote the convex body which it determines by $S_k K$. For given k, the map S_k is defined only on a subset of $\mathcal{K}^n$, but if K and L are in its domain, then it follows from the properties of spherical harmonics that

$$S_k(K + L) = S_k K + S_k L,$$
$$S_k(gK) = g S_k K$$

for each rigid motion of $\mathbb{E}^n$. Now for K of class C^∞_+, the body $S_k K$ does, in fact, exist for all sufficiently large k (see Appendix).

Now let $K \in \mathcal{K}^n$ and $\varepsilon > 0$ be given. We can choose a map T according to Theorem 3.3.1 so that $\delta(K, TK) < \varepsilon$. Then the body $TK + \varepsilon B^n$ is of class C^∞_+, hence k can be chosen so that $T_k K := S_k(TK + \varepsilon B^n)$ exists and satisfies $\delta(TK + \varepsilon B^n, T_k K) < \varepsilon$. Thus the body $T_k K$ satisfies $\delta(K, T_k K) < 3\varepsilon$ and is of class C^∞_+ (for sufficiently large k). Finally, the support function of $T_k K$ is, in fact, algebraic, since $\pi_m h$ (for any $h \in C(S^{n-1})$) is the restriction to S^{n-1} of a homogeneous polynomial on $\mathbb{E}^n$. Hence, $h(T_k K, x) = |x| P(x/|x|)$, where $P(y)$ is a polynomial in the cartesian coordinates of y. The map T_k commutes with rigid motions, and if K and L are in its domain, then $T_k(K + L) = T_k K + T_k L$ for $k > 0$.

Our second topic in this section is the approximation of convex bodies by Minkowski sums of bodies from a given class. This vaguely formulated program gives rise to many interesting questions. Here we treat only two special aspects.

If $K \in \mathcal{K}^n$ is a convex body, we say that K' is a *rotation mean* of K if there are a number $m \in \mathbb{N}$ and rotations $\rho_1, \ldots, \rho_m \in SO(n)$ such that

$$K' = \frac{1}{m} (\rho_1 K + \ldots + \rho_m K).$$

Theorem 3.3.2 (Hadwiger). *For each convex body $K \in \mathcal{K}^n$ with* $\dim K > 0$ *there is a sequence of rotation means of K converging to a ball.*

Proof. For $L \in \mathcal{K}^n$ let $d(L) := \min \{\lambda \geqq 0 | L \subset \lambda B^n\}$. Clearly the function d is continuous. A ball λB^n containing K contains all rotation means of K, hence the family $\mathcal{R}(K)$ of rotation means of K is bounded. The function d attains a minimum $d_0 > 0$ on the compact set $\operatorname{cl} \mathcal{R}(K)$, say at L. Assume that $L \neq d_0 B^n$. Then there is a vector $u_0 \in S^{n-1}$ with $h(L, u_0) < d_0$, hence $h(L, u) < d_0$ for all u in a suitable neighbourhood U of u_0 on S^{n-1}. Since S^{n-1} is compact, we can find finitely many rotations $\rho_1, \ldots, \rho_m \in SO(n)$ such that $\bigcup_{i=1}^m \rho_i U = S^{n-1}$. Put

$$\bar{L} := \frac{1}{m} (\rho_1 L + \ldots + \rho_m L).$$

Let $u \in S^{n-1}$. There is a number $i \in \{1, \ldots, m\}$ with $u \in \rho_i U$, hence $\rho_i^{-1} u \in U$ and thus $h(L, \rho_i^{-1} u) < d_0$, which implies

$$h(\bar{L}, u) = \frac{1}{m} \sum_{j=1}^{m} h(\rho_j L, u) = \frac{1}{m} \sum_{j=1}^{m} h(L, \rho_j^{-1} u) < d_0.$$

By the continuity of $h(\bar{L}, \cdot)$, this yields $d(\bar{L}) < d_0$. There is a sequence $(K_j)_{j \in \mathbb{N}}$ in $\mathcal{R}(K)$ converging to L, and we deduce

$$\bar{K}_j := \frac{1}{m} (\rho_1 K_j + \ldots + \rho_m K_j) \to \bar{L}$$

for $j \to \infty$, hence $d(\bar{K}_j) < d_0$ for large j. Since $\bar{K}_j \in \mathcal{R}(K)$, this contradicts the minimality of d_0. Thus L is a ball, which proves the theorem. ■

To put the foregoing result into a more general context, we say that a convex body K is *approximable by the class* $\mathcal{A} \subset \mathcal{K}^n$, if K is the limit of a sequence of bodies of the form

$$\lambda_1 K_1 + \ldots + \lambda_r K_r \quad \text{with } K_i \in \mathcal{A}, \quad \lambda_i \geqq 0 \ (i = 1, \ldots, r), \quad r \in \mathbb{N}.$$

Theorem 3.3.2 shows that a ball is always approximable by a class $\mathcal{A}$ if this class contains the congruent images of some convex body of positive dimension. On the contrary, the following theorem shows that for many convex bodies, namely the indecomposable ones, such an approximation is only possible in an essentially trivial way.

***Theorem* 3.3.3.** *If the indecomposable body* $K \in \mathcal{K}^n$ *is approximable by the class* $\mathcal{A} \subset \mathcal{K}^n$, *then the closure of* $\mathcal{A}$ *contains a homothetic copy of* K.

Before the proof of this theorem, a few remarks are in order. The theorem was proved for the case of indecomposable polytopes by Shephard (1964a); this proof is also given in Grünbaum [15] (Section 15.2). To formulate a consequence, call a subset $\mathcal{A} \subset \mathcal{K}^n$ a *universal approximating class* if any convex body in $\mathcal{K}^n$ is approximable by $\mathcal{A}$. For $n = 2$, the set of all triangles and all line segments is such a universal approximating class, since each convex polygon is a sum of triangles and line segments. For $n \geqq 3$, however, the set of indecomposable polytopes is dense in $\mathcal{K}^n$, by Theorem 3.2.14. Hence, for $n \geqq 3$, the only closed, homothety invariant, universal approximating class in $\mathcal{K}^n$ is $\mathcal{K}^n$ itself.

Shephard's proof seems to be restricted to polytopes. In its general form, Theorem 3.3.3 is due to Berg (1969a), who observed that it is an immediate consequence of Milman's converse to the Krein–Milman theorem (see, e.g., Phelps 1966, p. 9) applied to the set $\varphi(\mathcal{K}_s^n)$ (see Section 3.2), which is a compact convex set in $C(S^{n-1})$. Since we have not treated this theorem about convexity in topological vector spaces, we present here an elementary version of the proof, adapted to our particular situation.

Proof of Theorem 3.3.3. Let $K \in \mathcal{K}^n$ be an indecomposable body of positive dimension that is approximable by a class $\mathcal{A}$. We can assume that $\mathcal{A}$ is closed.

By assumption, there is a sequence $(K_i)_{i \in \mathbb{N}}$ converging to K where

$$K_i = \sum_{j=1}^{m_i} \lambda_{ij} K_{ij}, \quad K_{ij} \in \mathcal{A}, \quad \lambda_{ij} \geqq 0. \tag{3.3.1}$$

For $L \in \mathcal{K}^n$ with $\dim L > 0$ we denote its normalized homothet by $\bar{L}$, that is, $\bar{L} = b(L)^{-1}[L - s(L)]$. Then $K_i \to K$ implies $\bar{K}_i \to \bar{K}$ for $i \to \infty$, and we have

$$\bar{K}_i = \sum_j \bar{\lambda}_{ij} \bar{K}_{ij} \quad \text{with } \bar{\lambda}_{ij} = \frac{\lambda_{ij} b(K_{ij})}{b(K_i)},$$

so that

$$\sum_j \bar{\lambda}_{ij} = 1.$$

Since indecomposability is invariant under homothety, we may, therefore, assume from the beginning that $\mathcal{A} \subset \mathcal{K}_s^n$, $K \in \mathcal{K}_s^n$ and that (3.3.1) holds with

$$\sum_j \lambda_{ij} = 1.$$

We assert now that $K \in \mathcal{A}$. Suppose this were false. Then for each $A \in \mathcal{A}$ we have $\delta(A, K) > 0$, hence the set

$$\mathcal{U}(A) := \{L \in \mathcal{K}^n \mid \delta(A, L) \leqq \tfrac{1}{2}\delta(A, K)\}$$

has interior points. For $i \in \mathbb{N}$, let $\mu_i(\mathcal{U}(A))$ be the sum of those coefficients λ_{ij} for which $K_{ij} \in \mathcal{U}(A)$. We assert that

$$\limsup_{i \to \infty} \mu_i(\mathcal{U}(A)) > 0$$

for at least one $A \in \mathcal{A}$. Suppose this were false. The compact set $\mathcal{A}$ can be covered by finitely many sets $\mathcal{U}(A_1), \ldots, \mathcal{U}(A_p)$. Then

$$\sum_{r=1}^{p} \mu_i(\mathcal{U}(A_r)) \geqq \sum_{j=1}^{m_i} \lambda_{ij} = 1,$$

but on the other hand, $\mu_i(\mathcal{U}(A_r)) \to 0$ for $i \to \infty$, a contradiction. Hence, there is some $A_o \in \mathcal{A}$ for which $\limsup \mu_i(\mathcal{U}(A_o)) > 0$, and (taking a subsequence and changing the notation) we may assume that

$$\lim_{i \to \infty} \mu_i(\mathcal{U}(A_o)) > 0. \tag{3.3.2}$$

Now for $i \in \mathbb{N}$ we define

$$K_i' := \frac{\sum' \lambda_{ij} K_{ij}}{\sum' \lambda_{ij}}, \quad K_i'' := \frac{\sum'' \lambda_{ij} K_{ij}}{\sum'' \lambda_{ij}},$$

where the sums $\sum'$ extend over those j for which $K_{ij} \in \mathcal{U}(A_0)$ and the sums $\sum''$ extend over the remaining j (with the convention that $K_i' = \{o\}$ if $\sum' \lambda_{ij} = 0$, and similarly for K_i''). Then we have

$$K_i = \tau_i K_i' + (1 - \tau_i) K_i''$$

with

$$\tau_i = \sum' \lambda_{ij} = \mu_i(\mathcal{U}(A_0)). \tag{3.3.3}$$

Since K_i', K_i'' belong to the compact set $\mathcal{K}_s^n \cup \{\{o\}\}$ and τ_i belongs to the compact interval $[0, 1]$, we may assume, after selecting subsequences and changing the notation, that $K_i' \to K'$, $K_i'' \to K''$, with $K', K'' \in \mathcal{K}_s^n \cup \{\{o\}\}$ and $\tau_i \to \tau \in [0, 1]$, and thus

$$K = \tau K' + (1 - \tau)K''.$$

From (3.3.2) and (3.3.3) we have $\tau > 0$, in particular $K_i' \neq \{o\}$ for large i. For these i, K_i' is a convex combination of elements from $\mathcal{U}(A_0)$, hence $K_i' \in \mathcal{U}(A_0)$ and thus $K' \in \mathcal{U}(A_0)$, which implies $K' \neq K$. Since $K', K'' \in \mathcal{K}_s^n$, K' and K'' are not homothetic to K, thus K is decomposable. This contradiction shows that $K \in \mathcal{A}$, which proves the theorem. ∎

In the preceding proofs, some closed convex subsets of the space of convex bodies have appeared. In studies related to Minkowski addition and approximation, it is convenient to have a special notion for these sets. By an *M-class* of convex bodies in $\mathbb{E}^n$ we understand a subset $\mathcal{M} \subset \mathcal{K}^n$ that is closed in the Hausdorff metric and closed under Minkowskian convex combinations, that is, $K_1, \ldots, K_m \in \mathcal{M}$ implies

$$\lambda_1 K_1 + \ldots + \lambda_m K_m \in \mathcal{M} \quad \text{for } \lambda_i \geqq 0, \ \sum \lambda_i = 1.$$

Such an M-class will be called *homothety invariant* if together with a convex body K it contains all homothetic images of K; similarly for other types of transformations of $\mathbb{E}^n$. Examples of similarity invariant M-classes are the set of all one-pointed convex bodies, which we call the trivial M-class, the set of all centrally symmetric convex bodies, the set of all bodies of constant width, the set of all closed balls, etc. The set $\mathcal{K}_s^n$ of normalized convex bodies is a rotation invariant M-class. The set of convex bodies K for which $K - K$ is equal to a fixed convex body S is an example of a translation invariant M-class. The set of all summands of a given convex body is an M-class that, in general, has no invariance properties. In this terminology, Theorem 3.3.2 tells us that each non-trivial rotation invariant M-class contains a ball.

If $\mathcal{A} \subset \mathcal{K}^n$ is any set of convex bodies and G is a set of transformations of $\mathbb{E}^n$, let us consider the smallest M-class (smallest G-invariant M-class) containing $\mathcal{A}$; we call this the M-class (G-invariant M-class) *generated by* $\mathcal{A}$. Theorem 3.3.3 implies that the M-class generated by a homothety invariant subset $\mathcal{A} \subset \mathcal{K}^n$ is equal to $\mathcal{K}^n$ only if $\mathcal{A}$ is dense in $\mathcal{K}^n$ $(n \geqq 3)$.

In Section 3.5 we shall return to this topic; in particular we shall consider the similarity invariant M-class generated by a segment.

Notes for Section 3.3

1. *Regularization and approximation*. A regularization process very similar to that of Theorem 3.3.1 was first proposed by Radon (1936), but it appears that

his paper remained unnoticed for a long time. Radon considered (in a more general form) the transformation defined by

$$\tilde{f}(x) := \int_{\mathbb{E}^n} [f(x - |x|z) + |x| f(z)] \varphi(|z|) \, dz$$

and showed that it yields a support function if applied to such a function. He used this to show that a convex body can be approximated by convex bodies with analytic support functions.

The map T of Theorem 3.3.1 appeared in essence in Schneider (1974a), where it was considered for a different reason, and where the convexity of f was proved on the basis of a communication from W. Weil. As a special approximation procedure for convex bodies, the method of Theorem 3.3.1 was used by Weil (1975b). He investigated for which pairs of polytopes $P, Q \in \mathcal{P}_0^n$ with $P \subset Q$ it is possible to find a convex body of class C_+^∞ with $P \subset K \subset Q$. He proved that this is possible if and only if each point of $\operatorname{bd} P \cap \operatorname{bd} Q$ is a vertex of P and an internal point of some facet of Q.

Schneider (1984) extended the method, in the way sketched above, by using spherical harmonics. A regularization process close to that of Theorem 3.3.1 was independently proposed by Saint Pierre (1985).

As mentioned, the approximation process of Theorem 3.3.1 automatically yields bodies of constant width if applied to such a body. This does not hold for other procedures, described in the literature, that involve approximation by bodies with very smooth boundaries or support functions; see Minkowski (1903), §2, Bonnesen & Fenchel [8], p. 36, and Hammer (1963) and Firey (1974b). Special approximations for plane ovals of constant width were constructed by Tanno (1976) and Wegner (1977).

A thorough study of the approximation of centrally symmetric bodies (mainly in the plane) by polynomial bodies was made by Faro Rivas (1986).

2. *Rotation means*. Apparently the first proof of Theorem 3.3.2 appeared (for $n = 3$) in Hadwiger [17], p. 27; see also Hadwiger [18], Section 4.5.3. The basic idea of the proof seems to be older; for instance, it is very similar to an idea in the usual construction of an invariant mean on a compact topological group; compare Pontrjagin (1957).

Interesting problems arise if one asks for quantitative improvements in Theorem 3.3.2: How many rotations are required to reach a prescribed degree of accuracy of the approximation? The following result was proved by Bourgain, Lindenstrauss & Milman (1989), Theorem 6.3: Let $K \in \mathcal{K}^n$ and let $0 < \varepsilon < 1$. There are a constant $r = r(K)$ and rotations $\rho_1, \ldots, \rho_N$ of $\mathbb{E}^n$ with $N \leqq cn\varepsilon^{-2} \log \varepsilon^{-1}$ (c an absolute constant) so that

$$(1 - \varepsilon) rB^n \subset \frac{1}{N} (\rho_1 K + \ldots + \rho_N K) \subset (1 + \varepsilon) rB^n.$$

Bourgain, Lindenstrauss & Milman (1988) have a number of similarly interesting and deep results. We mention only one of these on random Minkowski symmetrizations. For $u \in S^{n-1}$, let π_u denote the orthogonal reflection at the hyperplane through zero with normal vector u. For $K \in \mathcal{K}^n$, the convex body $\frac{1}{2}(K + \pi_u K)$ is said to be obtained from K by a *Minkowski symmetrization* (or Blaschke symmetrization; see Blaschke [5], §22, VII). This Minkowski symmetrization is said to be random if u is chosen randomly on S^{n-1} with rotation invariant distribution. Now Theorem 14 of Bourgain, Lindenstrauss & Milman (1988) says: Let K be a centrally symmetric convex body in $\mathbb{E}^n$ and let $\varepsilon > 0$. If $n > n_0(\varepsilon)$ and if we perform $N = cn \log n + c(\varepsilon)n$ independent random Minkowski symmetrizations on K we obtain with probability $1 - \exp[-\tilde{c}(\varepsilon)n]$ a body $\tilde{K}$ that satisfies

$$(1 - \varepsilon)rB^n \subset \widetilde{K} \subset (1 + \varepsilon)rB^n$$

for suitable $r = r(K)$.

3. Theorem 3.3.2 can be used, in a similar way to other symmetrization procedures, to prove extremal properties of the ball; see, for example, Hadwiger (1955b) and Schneider (1967b).
4. *M-classes generated by lower-dimensional bodies*. Let $k \in \{1, \ldots, n\}$ and let $\mathcal{S}_k$ be the set of all simplices in $\mathbb{E}^n$ of dimension at most k. The M-class generated by $\mathcal{S}_k$ was investigated by Ricker (1981). He found an integral representation for its elements in terms of $\mathbb{E}^n$-valued measures, which generalizes an earlier result for the case $k = 1$ (for this, see Section 3.5).

   Let $k \in \{2, \ldots, n-1\}$ and let $\mathcal{K}_k$ be the set of all convex bodies in $\mathbb{E}^n$ of dimension at most k. Hildenbrand & Neyman (1982) proved a result that implies that the M-class generated by $\mathcal{K}_k$ is nowhere dense in $\mathcal{K}^n$. This follows also from Theorem 3.3.3 and the denseness of the indecomposable bodies for $n \geqq 3$.

3.4. Additive maps

Since Minkowski addition is a basic structure on the set $\mathcal{K}^n$ of convex bodies, the maps defined on $\mathcal{K}^n$ and compatible with this addition deserve special interest. Let φ be a map from $\mathcal{K}^n$ into some abelian group. (If the natural range is an abelian semigroup with cancellation law, we embed it into an abelian group.) According to Section 1.7, φ is called *Minkowski additive* if

$$\varphi(K + L) = \varphi(K) + \varphi(L) \quad \text{for } K, L \in \mathcal{K}^n.$$

If φ takes its values in a real vector space or in $\mathcal{K}^n$, we say that φ is *Minkowski linear* if it is Minkowski additive and satisfies

$$\varphi(\lambda K) = \lambda\varphi(K) \quad \text{for } K \in \mathcal{K}^n \text{ and } \lambda \geqq 0.$$

Examples of Minkowski linear maps are the mean width, the Steiner point, the map $K \mapsto F(K, u)$ for fixed $u \in S^{n-1}$ and the map $K \mapsto h_K$ that associates with each convex body its support function.

In the first part of this section, we study Minkowski additive maps with particular properties and prove characterization theorems. In the second part, we relate Minkowski additivity to a more general notion of additivity that will play a role in later chapters.

From a geometric point of view, the Minkowski additive maps enjoying an invariance property are of particular interest. Under additional continuity assumptions, one can obtain uniqueness results.

Theorem 3.4.1. *Let $n \geqq 2$. If the map $\varphi: \mathcal{K}^n \to \mathbb{R}$ is Minkowski additive, invariant under proper rotations and continuous at the unit ball B^n then φ is a constant multiple of the mean width.*

A map φ from $\mathcal{K}^n$ into $\mathbb{E}^n$ or into $\mathcal{K}^n$ is called *equivariant* under a group G of transformations of $\mathbb{E}^n$ if $\varphi(gK) = g\varphi(K)$ for all $g \in G$.

Theorem 3.4.2. *Let $n \geqq 2$. If the map $\varphi: \mathcal{K}^n \to \mathbb{E}^n$ is Minkowski linear, equivariant under proper rigid motions and continuous at the unit ball B^n, then φ is the Steiner point map.*

Remark 3.4.3. Let φ be a Minkowski additive map from $\mathcal{K}^n$ into $\mathbb{R}$ or $\mathbb{E}^n$. For $K \in \mathcal{K}^n$ we have $2K = K + K$, hence $\varphi(2K) = 2\varphi(K)$, and induction gives $\varphi(kK) = k\varphi(K)$ for $k \in \mathbb{N}$. For $k, m \in \mathbb{N}$ one then obtains $k\varphi(K) = \varphi(kK) = \varphi(m(k/m)K) = m\varphi((k/m)K)$, thus $\varphi(qK) = q\varphi(K)$ for rational $q > 0$. If now φ is assumed to be continuous on $\mathcal{K}^n$, then we deduce $\varphi(\lambda K) = \lambda\varphi(K)$ for real $\lambda \geqq 0$. Thus the assumption 'Minkowski linear and continuous at B^n' in Theorem 3.4.2 is weaker than the assumption 'Minkowski additive and continuous'.

Theorem 3.4.1 and its proof are essentially due to Hadwiger (e.g., [18], p. 213). We give a slightly modified version.

Proof of Theorem 3.4.1. For $x \in \mathbb{E}^n$ we can choose $\rho \in SO(n)$ with $\rho x = -x$, hence $\varphi(\{o\}) = \varphi(\{x\} + \rho\{x\}) = 2\varphi(\{x\})$. This yields $\varphi(\{o\}) = 0$ and then $\varphi(\{x\}) = 0$ in general. Let $K \in \mathcal{K}^n$ and $\dim K > 0$. If

$$K' = \lambda_1\rho_1 K + \ldots + \lambda_m\rho_m K$$

with positive rational numbers $\lambda_1, \ldots, \lambda_m$ and rotations $\rho_1, \ldots, \rho_m \in SO(n)$, then $\varphi(K') = (\lambda_1 + \ldots + \lambda_m)\varphi(K)$ by the properties of φ and Remark 3.4.3. Since the mean width has the same properties as φ, we deduce that $\varphi(K)/b(K) = \varphi(K')/b(K')$. Now, it follows from Theorem 3.3.2 that the integer m, the rotations ρ_i and the rational numbers λ_i can be chosen so that $\delta(K', B^n)$ is smaller than a given $\varepsilon > 0$. Since φ is continuous at B^n, we deduce that $\varphi(K)/b(K) = \varphi(B^n)/2$. ■

The following elegant proof of Theorem 3.4.2 is due to Positsel'skii (1973), except that he assumed equivariance under improper rigid motions also.

Proof of Theorem 3.4.2. Since the rotation group is compact we can, for every $\varepsilon > 0$, decompose $SO(n)$ into finitely many nonempty Borel sets $\Delta_{1,\varepsilon}, \ldots, \Delta_{m(\varepsilon),\varepsilon}$ of diameter less than ε. We choose $\rho_{k,\varepsilon} \in \Delta_{k,\varepsilon}$ and write $\nu_{k,\varepsilon} := \nu(\Delta_{k,\varepsilon})$. Then the usual estimate shows that

$$\lim_{\varepsilon\to 0} \sum_{k=1}^{m(\varepsilon)} f(\rho_{k,\varepsilon})\nu_{k,\varepsilon} = \int_{SO(n)} f \, d\nu$$

for any continuous real function f on $SO(n)$.

Let $K \in \mathcal{K}^n$. If $K = \{x\}$, then $\varphi(K) = x = s(K)$ by the rigid motion equivariance of φ. Assume, therefore, that $\dim K > 0$. We choose vectors $v \in \mathbb{E}^n$, $x \in S^{n-1}$ and a number $c > |v|$ and define for $\varepsilon > 0$ a convex body K_ε by

$$K_\varepsilon := \sum_{k=1}^{m(\varepsilon)} [c + \langle v, \rho_{k,\varepsilon} x \rangle] \nu_{k,\varepsilon} \rho_{k,\varepsilon}^{-1} K.$$

Applying the map φ to this body and using the properties of φ we get, for an arbitrary vector $y \in S^{n-1}$,

$$\langle \varphi(K_\varepsilon), y \rangle = \sum_{k=1}^{m(\varepsilon)} [c + \langle v, \rho_{k,\varepsilon} x \rangle] \langle \varphi(K), \rho_{k,\varepsilon} y \rangle \nu_{k,\varepsilon}$$

and hence

$$\lim_{\varepsilon \to 0} \langle \varphi(K_\varepsilon), y \rangle = \int_{SO(n)} [c + \langle v, \rho x \rangle] \langle \varphi(K), \rho y \rangle \, d\nu(\rho)$$

$$= \int_{SO(n)} \langle v, \rho x \rangle \langle \varphi(K), \rho y \rangle \, d\nu(\rho),$$

since $\int_{SO(n)} \langle \varphi(K), \rho y \rangle \, d\nu(\rho)$, as a function of y, is odd and rotation invariant, and hence zero.

On the other hand, the support function of K_ε is given by

$$h(K_\varepsilon, y) = \sum_{k=1}^{m(\varepsilon)} [c + \langle v, \rho_{k,\varepsilon} x \rangle] h(K, \rho_{k,\varepsilon} y) \nu_{k,\varepsilon}$$

and hence satisfies

$$\lim_{\varepsilon \to 0} h(K_\varepsilon, y) = \int_{SO(n)} [c + \langle v, \rho x \rangle] h(K, \rho y) \, d\nu(\rho).$$

Now

$$\int_{SO(n)} c h(K, \rho y) \, d\nu(\rho) = r$$

is a positive number (the integral is invariant under translations of K and clearly positive if $o \in \operatorname{relint} K$) that depends only on K and v. The integral

$$I(x, y) := \int_{SO(n)} \langle v, \rho x \rangle h(K, \rho y) \, d\nu(\rho)$$

satisfies $I(x, y) = I(y, x)$ if $n \geqq 3$, since $I(\tau x, \tau y) = I(x, y)$ for each rotation $\tau \in SO(n)$, and for $n \geqq 3$ we can choose τ such that $\tau x = y$ and $\tau y = x$. Writing

$$\int_{SO(n)} \rho^{-1} v h(K, \rho x) \, d\nu(\rho) := z$$

for $n \geqq 3$, we thus have $I(x, y) = \langle z, y \rangle$. For $n = 2$, $I(x, y)$ is not symmetric in x and y, but in this case we can write

$$I(x, y) = \frac{1}{2\pi}\int_0^{2\pi} \langle v, u(\xi + \alpha)\rangle h(K, u(\eta + \alpha))\,\mathrm{d}\alpha$$

with $u(\alpha) := (\cos\alpha)e_1 + (\sin\alpha)e_2$, where (e_1, e_2) is an orthonormal basis of $\mathbb{E}^2$, and $x = u(\xi)$, $y = u(\eta)$. An elementary computation now yields

$$I(x, y) = (A_1\cos\xi + A_2\sin\xi)\cos\eta + (A_1\sin\xi - A_2\cos\xi)\sin\eta$$

where the A_i depend only on K and v; hence again we have $I(x, y) = \langle z, y\rangle$ with some vector z depending only on K, v and x. Thus in both cases we have arrived at

$$\lim_{\varepsilon\to 0} h(K_\varepsilon, y) = r + \langle z, y\rangle = h(B(z, r), y),$$

where $B(z, r)$ is the ball with centre z and radius r. This holds for each $y \in S^{n-1}$, hence $\lim_{\varepsilon\to 0} K_\varepsilon = B(z, r)$ in the Hausdorff metric (observe that pointwise convergence of support functions implies uniform convergence on S^{n-1}, by Theorem 1.8.12). From this we get $r^{-1}(K_\varepsilon - z) \to B^n$ for $\varepsilon \to 0$ and hence $\varphi(r^{-1}(K_\varepsilon - z)) \to \varphi(B^n)$ by the assumed continuity of φ at B^n. Since $\varphi(B^n) = o$ by the rotation equivariance of φ, we arrive at $\varphi(K_\varepsilon) \to z$ and thus

$$\lim_{\varepsilon\to 0}\langle \varphi(K_\varepsilon), y\rangle = \langle z, y\rangle$$

for $y \in S^{n-1}$. We have proved that

$$\int_{SO(n)} \langle v, \rho x\rangle \langle \varphi(K), \rho y\rangle \,\mathrm{d}\nu(\rho) = \langle z, y\rangle,$$

which depends only on K, v, x, y and not on φ. Since the Steiner point map has all the properties of φ, this yields

$$\int_{SO(n)} \langle v, \rho x\rangle \langle \varphi(K) - s(K), \rho y\rangle \,\mathrm{d}\nu(\rho) = 0.$$

The choices $v := \varphi(K) - s(K)$ and $x = y$ now yield $\varphi(K) = s(K)$. ■

Remark 3.4.4. Inspection of the proofs of Theorems 3.4.1 and 3.4.2 shows that in neither case is it necessary to assume that φ is defined on all of $\mathcal{K}^n$. In fact, the methods work even if the domain of φ is a similarity invariant M-class.

It is, however, not known whether similar characterization theorems hold for maps that are only defined on polytopes. Here, of course, the continuity at the ball must be replaced by some other assumption. To formulate a definite problem, suppose that the map $\varphi: \mathcal{P}^n \to \mathbb{R}$ is Minkowski additive, invariant under rigid motions and continuous. Is it then true that $\varphi(P) = cb(P)$ for all $P \in \mathcal{P}^n$, with some constant c? A similar question can be posed for the Steiner point.

Although the continuity assumptions in Theorems 3.4.1 and 3.4.2 are rather weak, they cannot be omitted. Examples showing this can be constructed in different ways. Here we refer to the fact that in Chapter 4 (see especially Section 4.3) we shall associate with each convex body $K \in \mathcal{K}^n$ a certain finite positive Borel measure $S_1(K, \cdot)$ on the unit sphere S^{n-1}, its area measure of order one. The dependence on K is Minkowski linear, that is,

$$S_1(K + L, \cdot) = S_1(K, \cdot) + S_1(L, \cdot),$$
$$S_1(\lambda K, \cdot) = \lambda S_1(K, \cdot)$$

for $K, L \in \mathcal{K}^n$ and $\lambda \geqq 0$. If $S_1^s(K, \cdot)$ denotes the singular part (with respect to spherical Lebesgue measure) of $S_1(K, \cdot)$, then it follows from the uniqueness of the Lebesgue decomposition of a measure into its singular and its absolutely continuous part that the map $K \mapsto S_1^s(K, \cdot)$ is also Minkowski linear. Furthermore, the map $K \mapsto S_1(K, \cdot)$ is translation invariant and rotation equivariant, that is

$$S_1(K + t, \cdot) = S_1(K, \cdot),$$
$$S_1(\rho K, \rho\omega) = S_1(K, \omega)$$

for any translation vector $t \in \mathbb{E}^n$, any rotation $\rho \in SO(n)$ and any Borel set $\omega \subset S^{n-1}$. Again, these properties carry over to S_1^s. Now we can define

$$b'(K) := S_1^s(K, S^{n-1}),$$
$$s'(K) := \int_{S^{n-1}} u \, dS_1^s(K, u) + s(K)$$

for $K \in \mathcal{K}^n$. Then b' and s' are Minkowski linear functions, respectively invariant and equivarient under rigid motions, but b' is not a multiple of b and s' is not equal to s. This can be shown by constructing suitable convex bodies (as will become clear in Chapter 4), and it implies that b' and s' are not continuous.

Having found that real-valued and vector-valued Minkowski additive maps satisfying certain natural invariance and continuity properties are essentially unique, we may go a step further and consider body-valued Minkowski additive maps. We call a map $\varphi: \mathcal{K}^n \to \mathcal{K}^n$ an *endomorphism* of $\mathcal{K}^n$ if it is Minkowski additive, rigid motion equivariant and continuous. Thus such a map is compatible with the most basic geometric structures on $\mathcal{K}^n$. Examples of endomorphisms are the maps T occurring in Theorem 3.3.1. They show that there is a great variety of endomorphisms, and, hence, that simple ones can only be singled out by imposing additional assumptions. We shall present, without proofs, some of the known results in this direction.

The two-dimensional case differs considerably from the higher-dimensional one, as a result of the commutativity of the rotation group in $\mathbb{E}^2$. A particular example of an endomorphism $\varphi: \mathcal{K}^2 \to \mathcal{K}^2$ is given by

$$\varphi(K) := \left\{\sum_{i=1}^{m} \lambda_i \rho_i [K - s(K)]\right\} + s(K)$$

for $K \in \mathcal{K}^2$, where $\lambda_1, \ldots, \lambda_m \geqq 0$ and $\rho_1, \ldots, \rho_m \in SO(2)$. One can show that the general endomorphism is a limit of such rotation averages. For a convenient formulation, let (e_1, e_2) be an orthonormal basis of $\mathbb{E}^2$ and write $u(\alpha) := (\cos\alpha)e_1 + (\sin\alpha)e_2$ for $\alpha \in [0, 2\pi)$. For $K \in \mathcal{K}^2$ we then write $h(K, \alpha)$ instead of $h(K, u(\alpha))$. The following was proved by Schneider (1974c).

***Theorem* 3.4.5.** *If φ is an endomorphism of $\mathcal{K}^2$, then there exists a positive measure μ on the Borel subsets of $[0, 2\pi)$ such that*

$$h(\varphi(K), \alpha) = \int_0^{2\pi} h(K - s(K), \alpha + \beta)\, \mathrm{d}\mu(\beta) + \langle s(K), u(\alpha)\rangle \tag{3.4.1}$$

for $\alpha \in [0, 2\pi)$ and all $K \in \mathcal{K}^2$.

Of course, any positive Borel measure μ on $[0, 2\pi)$ defines an endomorphism φ by means of (3.4.1). If φ is given, the measure μ in (3.4.1) is unique up to the indefinite integral of $\langle c, u(\alpha)\rangle$ with a constant vector c. By subtracting a suitable measure of this kind from μ we can obtain a signed measure $\tilde{\mu}$ with

$$\int_0^{2\pi} \cos\alpha\, \mathrm{d}\tilde{\mu}(\alpha) = 1, \quad \int_0^{2\pi} \sin\alpha\, \mathrm{d}\tilde{\mu}(\alpha) = 0,$$

and then (3.4.1) can be written in the more concise form

$$h(\varphi(K), \alpha) = \int_0^{2\pi} h(K, \alpha + \beta)\, \mathrm{d}\tilde{\mu}(\beta).$$

This normalized measure $\tilde{\mu}$ is unique but not, in general, positive.

From Theorem 3.4.5 one can deduce (see also Schneider 1974a):

***Corollary* 3.4.6.** *The only surjective endomorphisms φ of $\mathcal{K}^2$ are of the form*

$$\varphi(K) = \lambda\rho[K - s(K)] + s(K) \textit{ for } K \in \mathcal{K}^2 \tag{3.4.2}$$

with fixed $\lambda > 0$ and $\rho \in SO(2)$.

The endomorphisms described in this corollary can also be characterized in a different way. For endomorphisms φ, ψ we define

$$(\varphi + \psi)(K) := \varphi(K) + \psi(K) - s(K),$$
$$(\lambda\varphi)(K) := \lambda\varphi(K) + (1 - \lambda)s(K)$$

for $K \in \mathcal{K}^2$ and $\lambda \geqq 0$; then $\varphi + \psi$ and $\lambda\varphi$ are endomorphisms. Now an endomorphism φ may be called *indecomposable* if $\varphi = \varphi_1 + \varphi_2$ with endomorphisms φ_1, φ_2 implies that $\varphi_i = \lambda_i\varphi$ with $\lambda_i \geqq 0$ $(i = 1, 2)$. Then Theorem 3.4.5 can be used to show that the indecomposable endomorphisms of $\mathcal{K}^2$ are precisely those of the form (3.4.2).

In dimensions greater than two, no general description of endomorphisms like the one given by Theorem 3.4.5 is known. It is, however, possible to characterize special endomorphisms by means of simple additional assumptions. The following results were obtained by Schneider (1974a). Their proof uses a combination of elementary facts from harmonic analysis (for the rotation group $SO(n)$ acting on the sphere S^{n-1}) with convexity arguments. The first theorem restricts the endomorphisms by imposing assumptions on the images.

Theorem 3.4.7.

(a) *The only surjective endomorphisms φ of $\mathcal{K}^n$ $(n \geqq 3)$ are given by*

$$\varphi(K) = \lambda[K - s(K)] + s(K) \text{ for } K \in \mathcal{K}^n$$

with $\lambda \neq 0$.

(b) *Every endomorphism of $\mathcal{K}^n$ is uniquely determined by the image of one suitably chosen convex body, for example, a triangle with at least one irrational angle.*

(c) *Let φ be an endomorphism of $\mathcal{K}^n$. If the image under φ of some convex body of positive dimension is zero-dimensional, then $\varphi(K) = \{s(K)\}$ for $K \in \mathcal{K}^n$. If the image under φ of some convex body is a segment, then*

$$\varphi(K) = \lambda[K - s(K)] + \mu[-K + s(K)] + s(K) \quad \text{for } K \in \mathcal{K}^n$$

with real numbers λ, $\mu \geqq 0$, $\lambda + \mu > 0$.

Another possible way of reducing the variety of endomorphisms drastically is to replace the rigid motion equivariance by a stronger assumption:

Theorem 3.4.8. *Let $\varphi: \mathcal{K}_0^n \to \mathcal{K}^n$ be a Minkowski additive and continuous map such that $\varphi(\alpha K) = \alpha\varphi(K)$ for every non-singular affine transformation α of $\mathbb{E}^n$. Then*

$$\varphi(K) = K + \lambda[K - K] \quad \text{for } K \in \mathcal{K}_0^n,$$

where $\lambda \geqq 0$ is a real constant.

Observe that in this theorem the map φ need only be defined on the set $\mathcal{K}_0^n$ of convex bodies with interior points (otherwise the theorem would be of little interest, cf. Schneider 1974a, pp. 76–7).

We turn now to a notion of additivity which is much less restrictive than Minkowski additivity and which plays a role in later chapters, and also in other contexts that are outside the scope of the present book.

A function φ defined on a family $\mathcal{S}$ of sets and with values in an abelian group is called *additive* or a *valuation* if

$$\varphi(K \cup L) + \varphi(K \cap L) = \varphi(K) + \varphi(L)$$

whenever K, L, $K \cup L$, $K \cup L \in \mathcal{S}$. If $\emptyset \in \mathcal{S}$, we moreover assume that $\varphi(\emptyset) = 0$.

For functions on $\mathcal{K}^n$ this is, in fact, a weaker property than Minkowski additivity:

Lemma 3.4.9. *Every Minkowski additive map on $\mathcal{K}^n$ is additive.*

Proof. Let K, $L \in \mathcal{K}^n$ be convex bodies such that $K \cup L$ is convex. Then (3.1.4) states that

$$(K \cup L) + (K \cap L) = K + L.$$

If now φ is a Minkowski additive map on $\mathcal{K}^n$, then

$$\varphi(K \cup L) + \varphi(K \cap L) = \varphi(K) + \varphi(L),$$

which shows that φ is additive. ■

A stronger result (Theorem 3.4.13) will be proved below. Another useful remark relating Minkowski addition and the valuation property is the following lemma.

Lemma 3.4.10. *Let φ be a valuation on $\mathcal{K}^n$. If $C \in \mathcal{K}^n$ is a fixed convex body and if*

$$\varphi_C(K) := \varphi(K + C) \quad \textit{for } K \in \mathcal{K}^n,$$

then φ_C is a valuation on $\mathcal{K}^n$.

Proof. Applying φ to (3.1.1) and to (3.1.3) and adding, we obtain the assertion. ■

Additivity of a function on the space $\mathcal{K}^n$ of convex bodies is an important property in view of possible extensions of this function to certain non-convex sets. Let φ be a valuation on a family $\mathcal{S}$ of sets. This family is called *intersectional* if $K, L \in \mathcal{S}$ implies $K \cap L \in \mathcal{S}$. If $\mathcal{S}$ is intersectional, we denote by $U(\mathcal{S})$ the lattice consisting of all finite unions of elements of $\mathcal{S}$. It is a question of general interest whether a valuation φ on an intersectional family $\mathcal{S}$ can be extended, as a valuation, to the lattice $U(\mathcal{S})$. If this is the case (and if the extension is denoted by the same symbol φ), then the *inclusion–exclusion principle*

$$\varphi(K_1 \cup \ldots \cup K_m) = \sum_{r=1}^{m} (-1)^{r-1} \left[\sum_{i_1 < \ldots < i_r} \varphi(K_{i_1} \cap \ldots \cap K_{i_r}) \right]$$

holds for $K_1, \ldots, K_m \in U(\mathcal{S})$. This follows by an obvious induction argument, using the additivity property. To write this equality in a more concise form, we let $S(m)$ denote the set of nonempty subsets of $\{1, \ldots, m\}$, and for $v \in S(m)$ we write $|v| = \text{card } v$. If $K_1, \ldots, K_m$ are given, then we use the abbreviation

$$K_v := K_{i_1} \cap \ldots \cap K_{i_r} \quad \text{for } v = \{i_1, \ldots, i_r\} \in S(m).$$

The inclusion–exclusion principle now takes the form

$$\varphi(K_1 \cup \ldots \cup K_m) = \sum_{v \in S(m)} (-1)^{|v|-1} \varphi(K_v). \tag{3.4.3}$$

In general, of course, this formula cannot be used to define an additive extension of φ from $\mathcal{S}$ to $U(\mathcal{S})$, since the representation of an element of $U(\mathcal{S})$ in the form $K_1 \cup \ldots \cup K_m$ with $K_i \in \mathcal{S}$ will not be unique. But if such an extension exists, then (3.4.3) gives its values, so that the extension is unique.

Let us say that a function φ from an intersectional family $\mathcal{S}$ of sets into an abelian group is *fully additive* if it satisfies the inclusion–exclusion principle on $\mathcal{S}$, that is, if (3.4.3) holds whenever $K_1, \ldots, K_m \in \mathcal{S}$ and $K_1 \cup \ldots \cup K_m \in \mathcal{S}$. If φ is fully additive and if $K \in U(\mathcal{S})$ has the representations $K = K_1 \cup \ldots \cup K_m$ and $K = L_1 \cup \ldots \cup L_k$ with $K_i, L_j \in \mathcal{S}$, then

$$\begin{aligned}
\sum_{v \in S(m)} (-1)^{|v|-1} \varphi(K_v) &= \sum_{v \in S(m)} (-1)^{|v|-1} \varphi\left(\bigcup_{j=1}^{k} (K_v \cap L_j) \right) \\
&= \sum_{v \in S(m)} (-1)^{|v|-1} \sum_{w \in S(k)} (-1)^{|w|-1} \varphi(K_v \cap L_w) \\
&= \sum_{w \in S(k)} (-1)^{|w|-1} \varphi\left(\bigcup_{i=1}^{m} (K_i \cap L_w) \right) \\
&= \sum_{w \in S(k)} (-1)^{|w|-1} \varphi(L_w).
\end{aligned}$$

Hence, in this case φ can be unambiguously defined on the lattice $U(\mathcal{S})$ by means of (3.4.3). This extension is additive on $U(\mathcal{S})$, as can be shown by some combinatorial argument (see Perles & Sallee 1970). We do not prove this here, since no use will be made of it, but we formulate it as a theorem.

Theorem 3.4.11. *Let φ be a valuation on an intersectional family $\mathcal{S}$ of sets. Then φ has an additive extension to the lattice $U(\mathcal{S})$ if and only if φ is fully additive. The extension is unique.*

Of particular interest for us is the lattice $U(\mathcal{K}^n)$ generated by the intersectional family $\mathcal{K}^n \cup \{\emptyset\}$. Thus $U(\mathcal{K}^n)$ consists of all subsets of $\mathbb{E}^n$ that can be represented as finite unions of convex bodies. $U(\mathcal{K}^n)$ was much studied by Hadwiger, who called it the 'Konvexring'. We shall use the disputable translation *convex ring*. The following result is fundamental in a study of extensions of valuations to the convex ring.

Theorem 3.4.12. *There is a unique valuation χ on the convex ring $U(\mathcal{K}^n)$, which satisfies*

$$\chi(K) = 1 \quad \text{for } K \in \mathcal{K}^n. \tag{3.4.4}$$

That (3.4.4) defines a valuation on $\mathcal{K}^n$ is clear, since $K_1 \cap K_2 \neq \emptyset$ if $K_1, K_2 \in \mathcal{K}^n$ and $K_1 \cup K_2 \in \mathcal{K}^n$. The uniqueness of an additive extension to $U(\mathcal{K}^n)$ follows from (3.4.3). The existence is well known since the Euler characteristic as defined, for example, in singular homology theory, yields a valuation on $U(\mathcal{K}^n)$. For this reason, the function χ on $U(\mathcal{K}^n)$ given by Theorem 3.4.12 is, of course, called the *Euler characteristic*. We shall present Hadwiger's elementary existence proof (Hadwiger 1955a; see also [18], p. 239).

Proof of Theorem 3.4.12. The proof proceeds by induction with respect to the dimension. The existence for $n = 0$ being trivial, suppose that the existence of χ has been proved in dimension $n - 1$. Choose $u \in S^{n-1}$ and define

$$\chi(K) := \sum_{\lambda \in \mathbb{R}} [\chi(K \cap H_{u,\lambda}) - \lim_{\mu \downarrow \lambda} \chi(K \cup H_{u,\mu})] \tag{3.4.5}$$

for $K \in U(\mathcal{K}^n)$, where on the right-hand side χ denotes the (unique) Euler characteristic, which by the inductive assumption, exists in each hyperplane $H_{u,\lambda}$ $(\lambda \in \mathbb{R})$. Obviously, $\chi(K) = 1$ for $K \in \mathcal{K}^n$. For $K = K_1 \cup \ldots \cup K_m$ with $K_i \in \mathcal{K}^n$ we have from (3.4.3)

$$\chi(K \cap H_{u,\lambda}) = \sum_{v \in S(m)} (-1)^{|v|-1} \chi(K_v \cap H_{u,\lambda}).$$

Since $\lambda \mapsto \chi(K_v \cap H_{u,\lambda})$ is the indicator function of a compact interval, it is clear that in (3.4.5) the limits exist and the sum is finite. It is also clear (using the inductive assumption) that χ, thus defined on $U(\mathcal{K}^n)$, is additive. ∎

In a similar way, for $K \in U(\mathcal{K}^n)$ and for $x \in \mathbb{E}^n \setminus \{o\}$ we define, following Mani (1971),

$$h(K, x) := \sum_{\lambda \in \mathbb{R}} \lambda[\chi(K \cap H_{x,\lambda}) - \lim_{\mu \downarrow \lambda} \chi(K \cap H_{x,\mu})]. \quad (3.4.6)$$

For $K \in \mathcal{K}^n$, the right-hand side evidently gives the value of the support function of K at x, thus the definition is consistent. Since the additivity of χ carries over to $h(\cdot, x)$, we see that (3.4.6) extends the map $K \mapsto h(K, \cdot)$ as a valuation from $\mathcal{K}^n$ to $U(\mathcal{K}^n)$. Also for $K \in U(\mathcal{K}^n)$, we call $h(K, \cdot)$, as defined by (3.4.6), the *support function* of K. It should, however, be observed that it no longer determines K uniquely; for instance, $h(\{o\}, \cdot) = 0 = h(\emptyset, \cdot)$. Observe also that

$$h(K + t, x) = h(K, x) + \chi(K)\langle t, x \rangle$$

for $t \in \mathbb{E}^n$.

We can now use formulae (1.7.2) and (1.7.3) to extend the mean width and the Steiner point additively to $U(\mathcal{K}^n)$, that is, we define

$$b(K) := \frac{2}{\omega_n} \int_{S^{n-1}} h(K, u) \, d\mathcal{H}^{n-1}(u),$$

$$s(K) := \frac{1}{\kappa_n} \int_{S^{n-1}} h(K, u) u \, d\mathcal{H}^{n-1}(u)$$

for $K \in U(\mathcal{K}^n)$. We still have $b(K + t) = b(K)$ for $t \in \mathbb{E}^n$, but

$$s(K + t) = s(K) + \chi(K)t.$$

The additive extension of the functions b and s from $\mathcal{K}^n$ to $U(\mathcal{K}^n)$ was achieved here by means of explicit formulae. The mere existence of such extensions would also follow from Theorem 3.4.11 and the following result concerning arbitrary Minkowski additive functions.

Theorem 3.4.13. *Every Minkowski additive function on $\mathcal{K}^n$ is fully additive.*

Proof. By formula (3.4.6), the support function was extended as a valuation to $U(\mathcal{K}^n)$, hence it satisfies the inclusion–exclusion principle. In particular, for $K \in \mathcal{K}^n$ of the form $K = K_1 \cup \ldots \cup K_m$ with $K_i \in \mathcal{K}^n$ $(i = 1, \ldots, m)$ this implies that

$$h(K, \cdot) = \sum_{v \in S(m)} (-1)^{|v|-1} h(K_v, \cdot),$$

hence

$$h(K, \cdot) + \sum_{\substack{v \in S(m) \\ |v| \text{even}}} h(K_v, \cdot) = \sum_{\substack{v \in S(m) \\ |v| \text{odd}}} h(K_v, \cdot).$$

Since $h(\emptyset, \cdot) = 0$ by (3.4.6), it follows that

$$K + \sum_{\substack{v \in S(m) \\ |v| \text{even}}}' K_v = \sum_{\substack{v \in S(m) \\ |v| \text{odd}}}' K_v,$$

where the sums $\sum'$ extend only over those $v \in S(m)$ for which $K_v \neq \emptyset$. If now φ is a Minkowski additive function on $\mathcal{K}^n$, we apply it to both sides of this equation. Defining $\varphi(\emptyset) := 0$, we obtain

$$\varphi(K) = \sum_{v \in S(m)} (-1)^{|v|-1} \varphi(K_v),$$

which shows that φ is fully additive. ∎

Finally we remark that there is no point in studying Minkowski additive maps on $\mathcal{C}^n$ instead of $\mathcal{K}^n$. In fact, let φ be a Minkowski additive function on $\mathcal{C}^n$. From Theorem 3.1.2 we know that

$$A + n \operatorname{conv} A = \operatorname{conv} A + n \operatorname{conv} A$$

for $A \subset \mathbb{E}^n$; hence for $A \in \mathcal{C}^n$ and any Minkowski additive map φ on $\mathcal{C}^n$ we obtain

$$\varphi(A) + \varphi(n \operatorname{conv} A) = \varphi(\operatorname{conv} A) + \varphi(n \operatorname{conv} A)$$

and thus $\varphi(A) = \varphi(\operatorname{conv} A)$. Thus φ is completely known if we know its restriction to $\mathcal{K}^n$.

Notes for Section 3.4

1. *Characterizations of the Steiner point.* For the early history of the Steiner point we refer to the notes for Section 5.4. The problem of characterizing the Steiner point by some of its properties was first posed by Grünbaum [37], p. 239, who asked whether Minkowski additivity and similarity equivariance are sufficient to characterize the Steiner point. Sallee (1971) constructed an example showing that one needs a continuity assumption. The example given above after the proof of Theorem 3.4.2 is taken from Schneider (1974a). By the properties of Minkowski additivity, rigid motion equivariance and continuity, the Steiner point was characterized by Shephard (1968b) for $n = 2$ and by Schneider (1971a) for $n \geqq 3$. Before that, W. Meyer (1970a) had proved a slightly weaker version of the characterization, by assuming uniform continuity. Two earlier attempts (Schmitt 1968, Hadwiger 1969a) contained errors. For dimensions two and three, interesting elementary proofs were given by Hadwiger (1971) and Berg (1971).

   Positsel'skii (1973), whose proof we presented above in essence, assumed equivariance under improper motions also; that proper motions suffice was pointed out (for $n \geqq 3$) by Saint Pierre (1985).
2. *Abstract Steiner points of polytopes.* If $P \in \mathcal{P}^n$ is a convex polytope, then its Steiner point can be represented by

$$s(P) = \sum_{i=1}^{f_0(P)} \omega_n^{-1} \mathcal{H}^{n-1}(N(P, v_i) \cap S^{n-1}) v_i,$$

where $v_1, \ldots, v_{f_0(P)}$ are the vertices of P. This is a special case of (5.4.11). Motivated by this, Berg (1971) defined

$$s_\psi(P) := \sum_{i=1}^{f_0(P)} \psi(N(P, v_i) \cap S^{n-1})v_i,$$

where ψ is a real-valued simple valuation on spherical polytopes (that is, $\psi(Q_1 \cup Q_2) = \psi(Q_1) + \psi(Q_2)$ if Q_1, Q_2 are intersections of S^{n-1} with convex polyhedral cones having convex union and without common interior points). It is not difficult to show that the map s_ψ is Minkowski linear, and that it is translation equivariant if $\psi(S^{n-1}) = 1$. Berg (1971) calls a map $\varphi: \mathcal{P}^n \to \mathbb{E}^n$ an *abstract Steiner point* if it is Minkowski additive and equivariant under (proper and improper) similarities. The definition of s_ψ above yields an abstract Steiner point if ψ is invariant under rotations and reflections and satisfies $\psi(S^{n-1}) = 1$. It is not known whether every abstract Steiner point is of this form, but Berg (1971) proved this for $n = 2$ and $n = 3$. As a consequence, every abstract Steiner point on $\mathcal{P}^n$, for $n = 2$ or 3, that is bounded on bounded sets of polytopes, is the usual Steiner point.

3. *Additive maps preserving some functional*. Further results in the style of Theorems 3.4.7 and 3.4.8 can be obtained if one considers only endomorphisms preserving some functional such as volume or surface area. For example, the following theorem was proved in Schneider (1974a) (the quermassintegrals W_k are defined in Section 4.2 below).

Theorem. If φ is an endomorphism of $\mathcal{K}^n$ satisfying $W_k(\varphi(K)) = W_k(K)$ for some $k \in \{0, 1, \ldots, n-2\}$ and all $K \in \mathcal{K}^n$, then

$$\varphi(K) = \lambda[K - s(K)] + s(K) \quad \text{for } K \in \mathcal{K}^n,$$

where $\lambda \in \{1, -1\}$.

The following result of Schneider (1974d) requires no continuity assumption. Here V_n denotes the volume.

Theorem. Let $n \geqq 2$. If $\varphi: \mathcal{K}^n \to \mathcal{K}^n$ is a Minkowski additive map satisfying $V_n(\varphi(K)) = V_n(K)$ for $K \in \mathcal{K}^n$, then there exists a volume-preserving affine map $\alpha: \mathbb{E}^n \to \mathbb{E}^n$ such that $\varphi(K)$ is a translate of αK for each $K \in \mathcal{K}^n$.

4. *Characterization of Hammer body maps*. According to Theorem 3.4.8, every Minkowski additive and continuous map $\varphi: \mathcal{K}^n \to \mathcal{K}^n$ that commutes with affine transformations is of the form $\varphi(K) = K + \lambda DK$ with a fixed $\lambda \geqq 0$. Thus φ associates with each K its outer Hammer body $K(\lambda + 1)$ (see Section 3.1, Note 13). The question arises of whether it is possible to characterize the map that associates with each convex body K its inner Hammer body $K(1 - \lambda)$ (where $\lambda \in [0, 1/n]$) in a similar way. Such a characterization was obtained by Valette (1974). He proved (in a more general form): The family of all maps $\varphi: \mathcal{K}^n \to \mathcal{K}^n$ that are continuous, commute with affine transformations and satisfy

$$\varphi(K + L) \supset \varphi(K) + \varphi(L) \quad \text{for } K, L \in \mathcal{K}^n,$$

$$\varphi(P) \subset (1 - 2\lambda)(P - p) + p,$$

where P is some parallelotope with centre p and where $\lambda \in [0, 1/n]$, has a largest element (with respect to inclusion of images); this is precisely the map given by $\varphi(K) = K(1 - \lambda) = K \sim \lambda DK$.

5. *The Euler characteristic*. The elementary proof given by Hadwiger (1955a) for the existence of the Euler characteristic on $\mathcal{K}^n$ (Theorem 3.4.11) appears in the literature in various forms: see Hadwiger (1959, 1968a, 1969c) and Hadwiger & Mani (1972); see also Hadwiger (1974) for an elementary

treatment of the Euler characteristic for polygons in the plane. Applications of the Euler characteristic on the convex ring to some questions of combinatorial geometry were treated by Hadwiger (1947, 1955a, 1968a) and Klee (1963b). Klee's paper put the Euler characteristic in a lattice-theoretic setting, and the general treatment of the Euler characteristic in combinatorial and algebraic terms was continued by Rota (1971); see also Rota (1964). Eckhoff (1980) solved a problem posed by Hadwiger & Mani (1974), namely to determine sharp lower and upper bounds for the Euler characteristic of a union of k convex bodies in $\mathcal{K}^n$.

The recursive definition (3.4.5) of the Euler characteristic can also be used for non-closed convex sets. In particular, let $\mathcal{P}^n_{\mathrm{ro}}$ denote the set of all relative interiors of convex polytopes in $\mathbb{E}^n$. Using (3.4.5) for $K \in U(\mathcal{P}^n_{\mathrm{ro}})$, one obtains a valuation χ on the lattice $U(\mathcal{P}^n_{\mathrm{ro}})$ that evidently satisfies

$$\chi(P) = (-1)^{\dim P} \quad \text{for } P \in \mathcal{P}^n_{\mathrm{ro}}.$$

This extended Euler characteristic was considered by Lenz (1970) and Groemer (1972) and in special cases (and with a different sign for odd-dimensional P) also by Hadwiger (1969c, 1973). If one decomposes a compact convex polytope $P \in \mathcal{P}^n$ into the relative interiors of its faces (Theorem 2.1.2) and applies the valuation χ to this representation of P, one immediately obtains the well-known *Euler relation*

$$\sum_{r=0}^{n} (-1)^r f_r(P) = 1,$$

where $f_r(P)$ is the number of r-faces of P.

6. *Extensions of valuations.* The first extension theorem of the type of Theorem 3.4.11, which is due to Perles & Sallee (1970), was obtained by Volland (1957). He proved that every valuation on the class $\mathcal{P}^n$ of polytopes in $\mathbb{E}^n$ admits a unique additive extension to the class $U(\mathcal{P}^n)$. (Related but simpler extension results can be found in Hadwiger [18], p. 81, and in Böhm & Hertel (1980), p. 47.) Volland's method can be extended to establish further results. Let $\mathcal{P}^n_{\mathrm{ro}}$ be the set of all relatively open convex polytopes in $\mathbb{E}^n$. A function φ from $\mathcal{S}$ into some abelian group is called a *weak valuation* if either $\mathcal{S} = \mathcal{P}^n$ and

$$\varphi(P) = \varphi(P \cap H^+) + \varphi(P \cap H^-) - \varphi(P \cap H) \quad \text{for } P \in \mathcal{P}^n$$

or $\mathcal{S} = \mathcal{P}^n_{\mathrm{ro}}$ and

$$\varphi(P) = \varphi(P \cap H_0^+) + \varphi(P \cap H_0^-) + \varphi(P \cap H) \quad \text{for } P \in \mathcal{P}^n_{\mathrm{ro}},$$

for any hyperplane $H \subset \mathbb{E}^n$, where H^+, H^- and H_0^+, H_0^- are respectively the closed and open halfspaces bounded by H. Schneider (1987c) proved: Every weak valuation on either $\mathcal{P}^n$ or $\mathcal{P}^n_{\mathrm{ro}}$ can be uniquely extended to a valuation on $U(\mathcal{P}^n_{\mathrm{ro}})$.

If φ is extended in this way, its value at a relatively open polytope is given by

$$\varphi(\operatorname{relint} P) = \sum_F (-1)^{\dim P - \dim F} \varphi(F)$$

for $P \in \mathcal{P}^n$, where the sum extends over all faces of P (including P). This can be proved by induction with respect to $\dim P$ if one uses the Euler relation for faces $F \subset G$ of a polytope (see, e.g., Brøndsted [11], p. 103). For a given valuation φ on $\mathcal{P}^n$, Sallee (1968) defined

$$\varphi^*(P) := \sum_F (-1)^{\dim F} \varphi(F) \quad \text{for } P \in \mathcal{P}^n.$$

Hence, $\varphi^*(P) = (-1)^{\dim P} \varphi(\operatorname{relint} P)$. Using the additivity of φ on $U(\mathcal{P}^n_{\mathrm{ro}})$ it is easy to see that φ^* is a valuation on $\mathcal{P}^n$. Sallee proved this in a different way.

The function φ on $\mathcal{P}^n$ is said to satisfy an *Euler-type relation* if $\varphi^*(P) = \varepsilon\varphi(\eta P)$ for all $P \in \mathcal{P}^n$, where $\varepsilon, \eta \in \{1, -1\}$ are fixed. Shephard (1966a, 1968a) showed that the Steiner point and the mean width satisfy Euler-type relations; see also Shephard (1968c). A general result of this kind is due to McMullen (1975c, 1977), who proved that valuations satisfying certain invariance and homogeneity properties automatically fulfil an Euler-type relation. For further information, see Sallee (1982) and §12 of McMullen & Schneider [40].

7. *Valuations and linear functionals*. In studying extensions of valuations, the following viewpoint can be useful. Let $\mathcal{S}$ be a nonempty class of subsets of some nonempty set S. For $K \in \mathcal{S}$, let K^* be the characteristic function of K^*, that is,

$$K^*(x) := \begin{cases} 1 \text{ for } x \in K, \\ 0 \text{ for } x \in S \backslash K. \end{cases}$$

Let $V(\mathcal{S})$ denote the real vector space generated by the functions K^*, $K \in \mathcal{S}$. If $\mathcal{S}$ is intersectional, then $X^* \in V(\mathcal{S})$ if and only if X belongs to the ring $R(\mathcal{S})$ generated by $\mathcal{S}$ (this is the smallest subclass of S containing $\mathcal{S}$ that, together with A and B, contains $A \cup B$ and $A \backslash B$). This result is due to Groemer (1987b).

If $K, L, K \cup L, K \cap L \in \mathcal{S}$, then $(K \cup L)^* + (K \cap L)^* = K^* + L^*$, thus the map $K \mapsto K^*$ is a valuation. In particular, if $\mathcal{S}$ is intersectional, then it follows from this or from Groemer's result that $K^* \in V(\mathcal{S})$ for all K in the lattice $U(\mathcal{S})$.

Now suppose that $\bar{\varphi}$ is a homomorphism from the additive group of $V(\mathcal{S})$ into some abelian group. Defining $\varphi(K) := \bar{\varphi}(K^*)$ for those $K \subset S$ for which $K^* \in V(\mathcal{S})$, we obtain

$$\begin{aligned} \varphi(K \cup L) + \varphi(K \cap L) &= \bar{\varphi}((K \cup L)^*) + \bar{\varphi}((K \cap L)^*) \\ &= \bar{\varphi}((K \cup L)^* + (K \cap L)^*) = \bar{\varphi}(K^* + L^*) \\ &= \bar{\varphi}(K^*) + \bar{\varphi}(L^*) = \varphi(K) + \varphi(L), \end{aligned}$$

provided that $\bar{\varphi}$ is defined in each case. Thus φ is a valuation. If $\mathcal{S}$ is intersectional, then this procedure yields a valuation on $U(\mathcal{S})$.

These general remarks can be used to extend the notion of an Euler characteristic. First, Groemer (1974) has proved an abstract theorem from which results on Euler characteristics in special concrete situations can be deduced. The class $\mathcal{S}$ of subsets of S is called *separable* if to any two disjoint sets $A, B \in \mathcal{S}$ there exists a pair $X, Y \subset S$ such that $X \cap C \in \mathcal{S}$ and $Y \cap C \in \mathcal{S}$ for every $C \in \mathcal{S}$, $A \subset X$, $A \cap Y = \emptyset$, $B \subset Y$, $B \cap X = \emptyset$, $X \cup Y = S$, and $Z \cap X \neq \emptyset$, $Z \cap Y \neq \emptyset$ for $Z \in \mathcal{S}$ only if $Z \cap X \cap Y \neq \emptyset$. Groemer proved the following:

Theorem. Let $\mathcal{S}$ be a separable intersectional class of subsets of $\mathcal{S}$. There exists exactly one linear functional χ on the vector space $V(\mathcal{S})$ such that $\chi(K^*) = 1$ for every nonempty set K of $\mathcal{S}$.

Second, Hadwiger's recursive procedure for the introduction of the Euler characteristic can also be modified to extend the Euler characteristic to a linear functional on $V(\mathcal{S})$, for suitable $\mathcal{S}$. For instance, consider the class $\mathcal{S} = \mathcal{P}^n \cup \{\emptyset\}$ of (possibly empty) polytopes in $\mathbb{E}^n$. We want to show the existence of a linear functional $\bar{\chi}$ on $V(\mathcal{P}^n)$ such that $\bar{\chi}(P^*) = 1$ for $P \in \mathcal{P}^n$. For $n = 1$, such a linear functional $\bar{\chi}_1$ is evidently given by

$$\bar{\chi}_1(f) := \sum_{y \in \mathbb{R}} [f(\lambda) - \lim_{\mu \downarrow \lambda} f(\mu)] \quad \text{for } f \in V(\mathcal{P}^1).$$

Let $n \geqq 2$ and suppose that the existence of the required linear functional $\bar{\chi}_{n-1}$ in dimension $n-1$ has already been proved. Consider $\mathbb{E}^{n-1}$ as a linear subspace of $\mathbb{E}^n$ and let $u \in S^{n-1} \backslash \mathbb{E}^{n-1}$. For given $f \in V(\mathcal{P}^n)$, define

$$\tilde{f}(x, \lambda) := f(x + \lambda u) \quad \text{for } x \in \mathbb{E}^{n-1} \text{ and } \lambda \in \mathbb{R}.$$

One can proceed in either of two ways: (a) define the projection $\pi_1 f$ of f onto $\mathbb{E}^{n-1}$ by

$$(\pi_1 f)(x) := \bar{\chi}_1(\tilde{f}(x, \cdot)) \quad \text{for } x \in \mathbb{E}^{n-1}$$

and then put $\bar{\chi}_n(f) := \bar{\chi}_{n-1}(\pi_1 f)$; or (b) define the projection $\pi_2 f$ of f on to $\mathbb{R}$ by

$$(\pi_2 f)(\lambda) := \bar{\chi}_{n-1}(\tilde{f}(\cdot, \lambda)) \quad \text{for } \lambda \in \mathbb{R}$$

and put $\bar{\chi}_n(f) := \bar{\chi}_1(\pi_2 f)$. In each case it is easy to see that $\bar{\chi}_n$ has the desired properties.

A procedure equivalent to method (a) was employed by Hadwiger (1960) and later by Groemer (1972), and a procedure equivalent to method (b) by Lenz (1970). Lenz further generalized this procedure and extended the Euler characteristic, as a valuation, from $\mathcal{K}^n$ to the system of all subsets of $\mathbb{E}^n$.

Further investigations of an extended notion of the Euler characteristic in certain situations are due to Groemer (1973, 1975).

We turn now to more general valuations on $\mathcal{S}$ and their extension to linear functionals. Here we assume that φ is a function from $\mathcal{S}$ into some real vector space $\mathcal{X}$. By a *linear extension* of φ to $V(\mathcal{S})$ one understands a linear map $\bar{\varphi}$ from $V(\mathcal{S})$ into $\mathcal{X}$ for which $\bar{\varphi}(K^*) = \varphi(K)$ for $K \in \mathcal{S}$. If φ admits such an extension, it is clearly a valuation. The following was proved by Groemer (1978).

Theorem. Let φ be a function from the intersectional class $\mathcal{S}$ into a real vector space so that $\varphi(\emptyset) = 0$. Then the following statements are equivalent.

(a) φ has an additive extension to $U(\mathcal{S})$;
(b) φ has a linear extension to $V(\mathcal{S})$;
(c) $\alpha_1 K_1^* + \ldots + \alpha_m K_m^* = 0$ with $m \in \mathbb{N}$, $K_i \in \mathcal{S}$, $\alpha_i \in \mathbb{R}$ for $i = 1, \ldots, m$ implies $\alpha_1 \varphi(K_1) + \ldots + \alpha_m \varphi(K_m) = 0$.

Using this theorem, Groemer also proved an extension theorem for valuations on $\mathcal{K}^n$. Let us say that the function φ from $\mathcal{K}^n$ into a topological (Hausdorff) vector space is σ-continuous if for every decreasing sequence $(K_i)_{i \in \mathbb{N}}$ in $\mathcal{K}^n$ one has

$$\lim_{i \to \infty} \varphi(K_i) = \varphi\left(\bigcap_{i \in \mathbb{N}} K_i\right).$$

Then Groemer (1978) proved that every σ-continuous valuation on $\mathcal{K}^n$ (with values in some topological vector space) has an additive extension to $U(\mathcal{K}^n)$. It seems to be unknown whether every valuation on $\mathcal{K}^n$ has an additive extension to $U(\mathcal{K}^n)$.

A thorough study of the extension of the Euler characteristic, and of many operations over convex sets to multilinear mappings on linear spaces by replacing convex sets by their characteristic functions was made by Przesławski (1988).

8. *Minkowski additivity and full additivity*. Theorem 3.4.13 was first formulated in McMullen & Schneider [40], Theorem (5.20). It generalizes earlier results of Sallee (1966), Mani (1971) and Spiegel (1976b) referring to special cases.

For certain additive functions on $\mathcal{P}^n$ satisfying additional assumptions one can deduce Minkowski additivity; see Spiegel (1976a) and §10 of McMullen & Schneider [40].

9. *An identity for polytopes*. An Euler-type relation (see Note 6) is satisfied, in particular, by the support function. If $P \in \mathcal{P}^n$ and $u \in \mathbb{E}^n$, then

$$\sum_{j=0}^{n}(-1)^j \sum_{F \in \mathcal{F}_j(P)} h(F, u) = -h(P, -u),$$

as proved by Shephard (1968c). This can be written as an identity for Minkowski sums, namely

$$-P + \sum_{2|\dim F} F = \sum_{2 \nmid \dim F} F,$$

where the sums extend over all faces of P of respectively even and odd dimensions (including 0 and n).

3.5. Zonoids and other classes of convex bodies

At the end of Section 3.3 we introduced the notion of an M-class of convex bodies. In some sense the simplest non-trivial M-class is that generated by the segments; it is also the smallest non-trivial affine-invariant M-class. Its elements are called *zonoids*. Thus a zonoid in $\mathbb{E}^n$ is a convex body that can be approximated by finite sums of line segments. From the viewpoint of Minkowski addition, the zonoids are generated in a particularly simple way and deserve, therefore, special attention. It turns out that zonoids appear in several different contexts. Here we collect some basic remarks on zonoids, and we also consider a more general concept.

A Minkowski sum of (finitely many) segments is called a *zonotope*. Thus a zonotope in $\mathbb{E}^n$ is a polytope of the form $Z = S_1 + \ldots + S_k$ with $k \in \mathbb{N}$ and segments $S_1, \ldots, S_k$. After applying suitable translations to the S_i (which results in one translation applied to Z) we may assume that each S_i has its centre at the origin; hence Z is centrally symmetric. Let $u \in S^{n-1}$. By Theorem 1.7.5,

$$F(Z, u) = F(S_1, u) + \ldots + F(S_k, u),$$

hence each face of a zonotope is itself a zonotope and, in particular, is centrally symmetric. This fact characterizes zonotopes:

Theorem 3.5.1. *A convex polytope is a zonotope if and only if all its two-dimensional faces are centrally symmetric.*

Proof. Only the sufficiency has to be proved. Let $P \in \mathcal{P}^n$ be a polytope of dimension at least two all of whose 2-faces are centrally symmetric. Let S be an edge of P, and let $u \in S^{n-1}$ be orthogonal to S. By Π we denote the orthogonal projection from P on to a hyperplane orthogonal to S. Let p be a vertex of $\Pi F(P, u)$. In the polytope ΠP we can find an edge path $(E_1, \ldots, E_m)$ connecting the vertex ΠS with p. The set $\Pi^{-1}(E_1)$ is a 2-face of P and hence centrally symmetric and has, therefore, an edge $S' \neq S$ that is a translate of S. The projections of S and S' under Π are the endpoints of E_1. Applying the same argument to E_2, and so on, we deduce that $\Pi^{-1}(p)$ is a translate of S. Thus the face $F(P, u)$ contains a translate of S. Since $u \perp S$ was arbitrary, it follows from Theorem 3.2.8 that S is a summand of P, hence $P = S + P'$ with a polytope P'. Since $F(P, u) = F(S, u) + F(P', u)$ for $u \in S^{n-1}$, it is clear that each 2-face of P' is centrally symmetric and that P' has no edge parallel to S. Repeating the argument, we conclude after finitely many steps that P is a sum of segments. ∎

If $P = S + P'$ with a segment S, then each support set $F(P, u)$ with $u \in S^{n-1}$ orthogonal to S contains a translate of S. The union of all these sets $F(P, u)$ makes up a 'zone' in the boundary of P; this explains the name 'zonotope'.

If $Z = S_1 + \ldots + S_k$ with $S_i = \operatorname{conv}\{\alpha_i v_i, -\alpha_i v_i\}$ where $v_i \in S^{n-1}$ and $\alpha_i > 0$ for $i = 1, \ldots, k$, then the support function of the zonotope Z is given by

$$h(Z, \cdot) = \sum_{i=1}^{k} \alpha_i |\langle \cdot, v_i \rangle|. \tag{3.5.1}$$

Conversely, a convex body Z with such a support function is a zonotope with centre at the origin.

By definition, a *zonoid* in $\mathbb{E}^n$ is a convex body that can be approximated, in the Hausdorff metric, by a sequence of zonotopes. Thus each zonoid has a centre of symmetry. The representation (3.5.1) is generalized as follows. By a signed measure (a measure) on S^{n-1} we understand a real-valued σ-additive (and non-negative) function on the σ-algebra $\mathcal{B}(S^{n-1})$ of Borel subsets of S^{n-1}. A signed measure on S^{n-1} and a function on S^{n-1} are called *even* if they are invariant under reflection in the origin.

Theorem 3.5.2. *A convex body $K \in \mathcal{K}^n$ is a zonoid with centre at o if and only if its support function can be represented in the form*

$$h(K, x) = \int_{S^{n-1}} |\langle x, v \rangle| \, \mathrm{d}\rho(v) \quad \text{for } x \in \mathbb{E}^n \tag{3.5.2}$$

with some even measure ρ on S^{n-1}.

Proof. Suppose that (3.5.2) holds. For $k \in \mathbb{N}$ we decompose S^{n-1} into finitely many nonempty Borel sets $\Delta_1^k, \ldots, \Delta_{m(k)}^k$ of diameter less than $1/k$ and choose $v_i^k \in \Delta_i^k$. Then

$$\lim_{k \to \infty} \sum_{i=1}^{m(k)} |\langle x, v_i^k \rangle| \rho(\Delta_i^k) = \int_{S^{n-1}} |\langle x, v \rangle| \, \mathrm{d}\rho(v),$$

uniformly for $x \in S^{n-1}$. Hence, the zonotopes Z_k defined by

$$Z_k := \sum_{i=1}^{m(k)} \rho(\Delta_i^k) \operatorname{conv}\{v_i^k, -v_i^k\}$$

satisfy $h(Z_k, \cdot) \to h(K, \cdot)$ and thus $Z_k \to K$ for $k \to \infty$. By definition, K is a zonoid.

For the converse, observe that (3.5.1) can also be written in the form (3.5.2), with an even measure ρ concentrated in finitely many points. Hence, we may assume that

$$h(Z_k, \cdot) = \int_{S^{n-1}} |\langle \cdot, v \rangle| \, \mathrm{d}\rho_k(v) \tag{3.5.3}$$

with an even measure ρ_k on S^{n-1} ($k \in \mathbb{N}$) and that $Z_k \to K$ for $k \to \infty$. Integrating (3.5.3) over S^{n-1} and using Fubini's theorem, we obtain

$$\int_{S^{n-1}} h(Z_k, u) \, \mathrm{d}\mathcal{H}^{n-1}(u) = c(n) \rho_k(S^{n-1}).$$

Since the left-hand side converges for $k \to \infty$, the sequence $(\rho_k(S^{n-1}))_{k \in \mathbb{N}}$ is bounded. In the space of signed measures on S^{n-1} with the total variation norm, which can be identified with the dual space of $C(S^{n-1})$, bounded sets are relatively weak*-compact. Hence, some subsequence $(\rho_{k_i})_{i \in \mathbb{N}}$ is weak*-convergent to a signed measure ρ, necessarily an even measure, and we get

$$h(K, x) = \lim_{i \to \infty} h(Z_{k_i}, x) = \lim_{i \to \infty} \int_{S^{n-1}} |\langle x, v \rangle| \, \mathrm{d}\rho_{k_i}(v)$$

$$= \int_{S^{n-1}} |\langle x, v \rangle| \, \mathrm{d}\rho(v)$$

for each $x \in \mathbb{E}^n$. ■

If the support function of a zonoid K with centre at the origin is represented in the form (3.5.2) with an even measure ρ, we say that ρ is the *generating measure* of K. The next theorem shows that in fact this generating measure is unique. It also shows that a representation of the form (3.5.2) with a signed measure ρ is always possible if K is centrally symmetric with respect to o and has a support function that is sufficiently often differentiable.

***Theorem* 3.5.3.** *If ρ is an even signed measure on S^{n-1} with*

$$\int_{S^{n-1}} |\langle u, v \rangle| \, \mathrm{d}\rho(v) = 0 \quad \textit{for } u \in S^{n-1} \tag{3.5.4}$$

then $\rho = 0$. If G is an even real function on S^{n-1} of differentiability class C^k, where $k \geqq n+2$ is even, there exists an even continuous function g on S^{n-1} such that

$$G(u) = \int_{S^{n-1}} |\langle u, v\rangle| g(v)\, d\mathcal{H}^{n-1}(v) \tag{3.5.5}$$

for $u \in S^{n-1}$.

Proof. We use spherical harmonics and refer to the Appendix for some of the details. In the following, all integrations without specified domain are over S^{n-1}. Let Y_m by a spherical harmonic of order m on S^{n-1}. The Funk–Hecke theorem yields

$$\int |\langle u, v\rangle| Y_m(v)\, d\mathcal{H}^{n-1}(v) = \lambda_m Y_m(u) \tag{3.5.6}$$

with

$$\lambda_m = \frac{(-1)^m \pi^{(n-1)/2}}{2^{m-1}\Gamma(m + (n-1)/2)} \int_{-1}^{1} |t| \left(\frac{d}{dt}\right)^m (1 - t^2)^{m+(n-3)/2}\, dt$$

$$= \frac{(-1)^{(m-2)/2} \pi^{(n-1)/2} \Gamma(m-1)}{2^{m-2}\Gamma(m/2)\Gamma((m+n+1)/2)} \quad \text{for even } m \tag{3.5.7}$$

and $\lambda_m = 0$ for odd m. If now (3.5.4) holds, we have

$$0 = \iint |\langle u, v\rangle|\, d\rho(v)\, Y_m(u)\, d\mathcal{H}^{n-1}(u)$$

$$= \lambda_m \int Y_m(v)\, d\rho(v)$$

and hence

$$\int Y_m(v)\, d\rho(v) = 0.$$

For even m this holds since $\lambda_m \neq 0$, and for odd m it holds since Y_m is odd and ρ is even. From the completeness of the system of spherical harmonics it now follows that $\rho = 0$.

Now let G satisfy the assumptions of the theorem. To solve the integral equation (3.5.5), we develop G into a series of spherical harmonics:

$$G \sim \sum_{m=0}^{\infty} Y_m \quad \text{with } Y_m := \pi_m G.$$

If the series

$$g(v) := \sum_{\substack{m=0 \\ m \text{ even}}}^{\infty} \lambda_m^{-1} Y_m(v) \tag{3.5.8}$$

converges uniformly on S^{n-1}, the continuous function g solves (3.5.5), by (3.5.6). For the uniform convergence of (3.5.8), it is sufficient to prove that

$$||\lambda_m^{-1} Y_m|| = \mathrm{O}(m^{-2}). \tag{3.5.9}$$

Now from (3.5.7) we get, using Stirling's formula, that

$$|\lambda_m^{-1}| = \mathrm{O}(m^{(n+2)/2}) \tag{3.5.10}$$

for even $m \to \infty$. By Appendix equation (A.12), to which we refer for the notation,

$$||Y_m|| \leqq \left[\frac{N(n, m)}{\omega_n}\right]^{1/2} ||Y_m||_2. \tag{3.5.11}$$

From Appendix equation (A.1),

$$\left[\frac{N(n, m)}{\omega_n}\right]^{1/2} = \mathrm{O}(m^{(n-2)/2}). \tag{3.5.12}$$

It remains to estimate $||Y_m||_2$. Let $(Y_{mj})_{j=1}^{N(n,m)}$ be an orthonormal basis of $\mathcal{S}^m$ ($m \in \mathbb{N}_0$), and put $a_{mj} := (G, Y_{mj})$, so that

$$Y_m = \pi_m G = \sum_{j=1}^{N(n,m)} a_{mj} Y_{mj},$$

$$||Y_m||_2^2 = (G, Y_m) = \sum_{j=1}^{N(n,m)} a_{mj}^2 =: a_m^2.$$

Using $\Delta_S Y_m = -m(m+n-2) Y_m$ and applying Green's formula $(f, \Delta_S g) = (\Delta_S f, g)$ on the sphere $k/2$ times, we get

$$a_m^2 = (G, Y_m) = \left[\frac{-1}{m(m+n-2)}\right]^{k/2} (\Delta_S^{k/2} G, Y_m)$$

for $m > 0$. Writing

$$b_{mj} := (\Delta_S^{k/2} G, Y_{mj}),$$

$$b_m^2 := \sum_{j=1}^{N(n,m)} b_{mj}^2 = ||\pi_m \Delta_S^{k/2} G||_2^2,$$

we deduce, using the Cauchy–Schwarz inequality, that

$$|a_m| \leqq m^{-k} |b_m|.$$

By the Parseval relation,

$$\sum_{m=0}^{\infty} b_m^2 = ||\Delta_S^{k/2} G||_2^2 < \infty,$$

hence $|a_m| = \mathrm{o}(m^{-k})$. This, together with (3.5.10), (3.5.11) and (3.5.12) yields (3.5.9), which completes the proof. ∎

***Remark* 3.5.4.** For $n = 2$, the solution of the integral equation (3.5.5) can easily be given explicitly. It is convenient to write the argument of G as an angle. If $G: [0, 2\pi] \to \mathbb{R}$ is a periodic function of class C^2, then

$$G(\varphi) = \frac{1}{4}\int_0^{2\pi} |\cos(\varphi - \psi)| \left[G\left(\psi - \frac{\pi}{2}\right) + G''\left(\psi - \frac{\pi}{2}\right)\right] \mathrm{d}\psi,$$

as one may check by partial integration.

The existence part of Theorem 3.5.3 is a motivation for considering also convex bodies $K \in \mathcal{K}^n$ whose support function can be represented in the form

$$h(K, u) = \int_{S^{n-1}} |\langle u, v\rangle| \,\mathrm{d}\rho(v)$$

with an even signed measure ρ. Such a body K and each of its translates is called a *generalized zonoid*. First, information on these bodies is obtained from the following lemma, showing that each support set of a generalized zonoid is itself a generalized zonoid.

For $e \in S^{n-1}$ we write

$$\Omega_e := \{v \in S^{n-1} | \langle e, v\rangle > 0\},$$
$$\omega_e := \{v \in S^{n-1} | \langle e, v\rangle = 0\},$$

and for $K \in \mathcal{K}^n$ we abbreviate the support set $F(K, e)$ by K_e.

***Lemma* 3.5.5.** *If the support function of the convex body $K \in \mathcal{K}^n$ is given by*

$$h(K, u) = \int_{S^{n-1}} |\langle u, v\rangle| \,\mathrm{d}\rho(v) \quad \text{for } u \in \mathbb{E}^n$$

with an even signed measure ρ, then, for $e \in S^{n-1}$,

$$h(K_e, u) = \int_{\omega_e} |\langle u, v\rangle| \,\mathrm{d}\rho(v) + \langle v_e, u\rangle \quad \text{for } u \in \mathbb{E}^n,$$

where

$$v_e = 2\int_{\Omega_e} v \,\mathrm{d}\rho(v).$$

Proof. Theorem 1.7.2 tells us that

$$h(K_e, u) = h'_K(e; u)$$
$$= \lim_{\lambda \downarrow 0} \lambda^{-1}[h(K, e + \lambda u) - h(K, e)].$$

The assertion of the lemma is true for $u = \pm e$, because $h(K_e, \pm e) = \pm h(K, e)$; hence we may assume that u and e are linearly independent. Put

$$A := \{v \in S^{n-1} | \langle e, v\rangle > 0, \quad \langle e + \lambda u, v\rangle > 0\},$$
$$B := \{v \in S^{n-1} | \langle e, v\rangle \leqq 0, \quad \langle e + \lambda u, v\rangle > 0\},$$
$$C := \{v \in S^{n-1} | \langle e, v\rangle > 0, \quad \langle e + \lambda u, v\rangle \leqq 0\}.$$

Then $\Omega_{e+\lambda u} = A \cup B$ and $\Omega_e = A \cup C$, and we obtain

$$h(K_e, u) = \lim \lambda^{-1} \left[\int_{S^{n-1}} |\langle e + \lambda u, v\rangle| \, \mathrm{d}\rho(v) - \int_{S^{n-1}} |\langle e, v\rangle| \, \mathrm{d}\rho(v) \right]$$

$$= 2 \lim \lambda^{-1} \left[\int_{A \cup B} \langle e + \lambda u, v\rangle \, \mathrm{d}\rho(v) - \int_{A \cup C} \langle e, v\rangle \, \mathrm{d}\rho(v) \right]$$

$$= 2 \lim \lambda^{-1} \left[\int_B \langle e, v\rangle \, \mathrm{d}\rho(v) - \int_C \langle e, v\rangle \, \mathrm{d}\rho(v) \right]$$

$$+ 2 \lim \int_{A \cup B} \langle u, v\rangle \, \mathrm{d}\rho(v),$$

where all limits refer to $\lambda \downarrow 0$. For $v \in B$ we have $|\langle e, v\rangle| \leqq c\lambda$ with a constant c independent of λ. Writing

$$B' := \{v \in S^{n-1} | \langle e, v\rangle < 0, \langle e + \lambda u, v\rangle > 0\},$$

we obtain

$$\left| \lambda^{-1} \int_B \langle e, v\rangle \, \mathrm{d}\rho(v) \right| = \left| \lambda^{-1} \int_{B'} \langle e, v\rangle \, \mathrm{d}\rho(v) \right| \leqq c|\rho|(B'),$$

where $|\rho|$ denotes the total variation measure of ρ. Since (in the set-theoretic sense) $\lim B' = \emptyset$, we have

$$\lim |\rho|(B') = 0,$$

hence

$$\lim \lambda^{-1} \int_B \langle e, v\rangle \, \mathrm{d}\rho(v) = 0.$$

From $\lim C = \emptyset$ we similarly find

$$\lim \lambda^{-1} \int_C \langle e, v\rangle \, \mathrm{d}\rho(v) = 0.$$

Further, $\lim A = \Omega_e$ and

$$\lim B = D := \{v \in S^{n-1} | \langle e, v\rangle = 0, \ \langle u, v\rangle > 0\}.$$

This yields

$$h(K_e, u) = 2 \int_{\Omega_e} \langle u, v\rangle \, \mathrm{d}\rho(v) + 2 \int_D \langle u, v\rangle \, \mathrm{d}\rho(v)$$

$$= 2 \left\langle u, \int_{\Omega_e} v \, \mathrm{d}\rho(v) \right\rangle + \int_{\omega_e} |\langle u, v\rangle| \mathrm{d}\rho(v),$$

which completes the proof of the lemma. ■

Corollary 3.5.6. *Each support set of a generalized zonoid is itself a generalized zonoid and is, in particular, centrally symmetric. Each* *support set of a zonoid is a summand of the zonoid. A polytope is a generalized zonoid only if it is a zonotope. In the set of centrally symmetric convex bodies in* $\mathbb{E}^n$, *the class of generalized zonoids is dense but not closed, and the class of zonoids is closed and nowhere dense.*

Proof. If

$$h(K, u) = \int_{S^{n-1}} |\langle u, v\rangle| \, \mathrm{d}\rho(v)$$

with an even signed measure ρ, then Lemma 3.5.5 shows that

$$h(K_e - v_e, u) = \int_{S^{n-1}} |\langle u, v\rangle| \, \mathrm{d}\rho_e(v),$$

where ρ_e is the restriction of ρ to ω_e; hence K_e is a generalized zonoid. If ρ is a measure, then

$$h(K_e - v_e, u) + \int_{S^{n-1}} |\langle u, v\rangle| \, \mathrm{d}(\rho - \rho_e)(v) = h(K, u).$$

Since $\rho - \rho_e \geqq 0$, the integral defines a support function, hence K_e is a summand of K. If the polytope P is a generalized zonoid, then each of its faces is centrally symmetric, hence P is a zonotope by Theorem 3.5.1. That the class of generalized zonoids is dense in the class of centrally symmetric convex bodies follows from Theorem 3.5.3, since the set of convex bodies with support function of class C^∞ is dense by Theorem 3.3.1. A convex body with a non-symmetric support set cannot be a generalized zonoid; hence the set of generalized zonoids is not closed, and the set of zonoids, which is closed by definition, is nowhere dense in the set of centrally symmetric convex bodies. ∎

Generalized zonoids were obtained by extending the integral representation (3.5.2) from measures to signed measures. The fact that the generalized zonoids are dense in the class of centrally symmetric convex bodies but do not exhaust this class is a reason for a further extension, namely to distributions. It was shown by Weil (1976b) that for every convex body $K \in \mathcal{K}^n$ with centre at the origin there exists an even distribution T_K on S^{n-1}, whose domain can be extended to include the functions $|\langle u, \cdot\rangle|$, $u \in S^{n-1}$, such that $T_K(|\langle u, \cdot\rangle|) = h(K, u)$. The properties of the correspondence $K \mapsto T_K$ and some applications were studied further by Weil (1976b, 1979a). We shall not pursue this matter here.

We conclude this section with a brief look at a different type of M-class. An M-class that has been thoroughly studied is the class of bodies of constant width. These bodies are, in a certain sense, a counterpart to the centrally symmetric convex bodies: A convex body is centrally symmetric if and only if its support function is essentially (that is, up to a linear function) an even function, and a body is of constant width if and only if its support function is essentially (meaning, here, up to the support function of a ball) an odd function. For the various aspects from which the bodies of constant width have been studied, we refer the reader to the comprehensive survey article of Chakerian & Groemer [31].

There is a special way of looking at bodies of constant width that readily suggests a generalization. A body of constant width can be freely turned inside a suitable cube while always touching the facets of the cube. More generally, let $P \subset \mathbb{E}^n$ be a polyhedral set, that is, an n-dimensional intersection of finitely many closed halfspaces. A convex body $K \in \mathcal{K}_0^n$ is said to be a *rotor* of P if for each rotation ρ of $\mathbb{E}^n$ there exists a translation vector $t \in \mathbb{E}^n$ such that $\rho K + t$ is contained in P and touches each facet of P. Clearly the set of all rotors of a given polyhedral set is a motion invariant M-class. If it is not empty, then it contains a ball, by Theorem 3.3.2. Thus P can have a rotor only if it has an inscribed ball. If the exterior normal vectors of the facets of P are linearly independent, then obviously every convex body is a rotor of P. Leaving this trivial case aside, we ask for the polyhedral sets admitting non-spherical rotors. A complete classification is possible. Of course, if K is a rotor of the polyhedral set

$$P = \bigcap_{i \in I} H^-(P, u_i),$$

where the u_i for $i \in I$ (a finite index set) are the exterior normal vectors of the facets of P, then K is also a rotor of

$$P' := \bigcap_{i \in I'} H^-(P, u_i)$$

whenever I' is a subset of I. Such a polyhedral set P' is said to be *derived from* P. The following theorem, which we quote without proof, lists all the non-trivial pairs of polyhedral sets and corresponding rotors. In the formulation of this theorem, support functions are restricted to S^{n-1}.

Theorem 3.5.7. *Let $P \subset \mathbb{E}^n$ be a polyhedral set whose facets have linearly dependent normal vectors. Suppose that K is a non-spherical rotor of P. Then one of the following assertions holds.*

(a) *P is derived from a parallelepiped with equal distances between parallel facets; K is a body of constant width.*

(b) $n = 2$; *P is derived from a regular k-gon with* $k \in \{3, 4, \ldots\}$; *the Fourier coefficients* a_i, b_i *of the support function of K are zero except for* $i = 0$ *and* $i \equiv \pm 1 \pmod k$.

(c) $n = 3$; *P is a regular tetrahedron; the support function of K is a sum of spherical harmonics of orders* 0, 1, 2, 5.

(d) $n = 3$; *P is derived from a regular octahedron, but is not a tetrahedron; the support function of K is a sum of spherical harmonics of orders* 0, 1, 5.

(e) $n = 3$; *P is congruent to the cone*

$$\{x \in \mathbb{E}^3 | \langle x, e_i \rangle \leqq 0 \quad \text{for } i = 1, 2, 3, 4\},$$

where

$$e_1 = \frac{1}{\sqrt{7}}(\sqrt{6}, 0, 1), \quad e_2 = \frac{1}{\sqrt{7}}(-\sqrt{6}, 0, 1),$$

$$e_3 = \frac{1}{\sqrt{7}}(0, \sqrt{6}, 1), \quad e_4 = \frac{1}{\sqrt{7}}(0, -\sqrt{6}, 1)$$

with respect to some orthonormal basis; the support function of K is a sum of spherical harmonics of orders 0, 1, 4.

(f) $n \geqq 4$; *P is a regular simplex; the support function of K is a sum of spherical harmonics of orders* 0, 1, 2.

Vice versa, each of the cases listed here does really occur, so that no further reduction is possible. We wish to point out that cases (c) to (f), unlike (a) and (b), describe M-classes that depend only on finitely many real parameters.

References are given in Note 8 below.

Notes for Section 3.5

1. *Zonotopes and generalizations.* The characterization of zonotopes contained in Theorem 3.5.1 must have been folklore for a long time, at least for $n = 3$; compare Blaschke (1923), p. 250, and Coxeter (1963), Section 2.8. Proofs for n-dimensional space were given by Bolker (1969) and Schneider (1970a).

Some results have been proved for polytopes with centrally symmetric faces of dimension greater than two. Aleksandrov (1933) (see also Burckhardt 1940) proved that a polytope of dimension at least three all of whose facets have a centre of symmetry is itself centrally symmetric. As a generalization of the zonotopes, one may consider the class $\mathcal{P}^n_k$ of n-polytopes all of whose k-faces are centrally symmetric, where $k \in \{2, \ldots, n\}$. Aleksandrov's theorem implies that $\mathcal{P}^n_k \subset \mathcal{P}^n_{k+1} \subset \ldots \subset \mathcal{P}^n_n$ for $k \geqq 2$; see also Shephard (1967) for further results and McMullen (1976b). McMullen (1970) showed that $\mathcal{P}^n_{n-2} \neq \mathcal{P}^n_{n-1}$ for $n \geqq 4$ and proved that $\mathcal{P}^n_k$ for $2 \leqq k \leqq n-2$ is equal to the class of n-dimensional zonotopes.

Various combinatorial and geometric results on zonotopes have been obtained by McMullen (1971) and Shephard (1974a); for a survey and further references see McMullen (1979), §7.

Zonotopes are also related to space-filling polytopes, that is, polytopes that tile space by translations. Every space-filling polytope in $\mathbb{E}^3$ is a zonotope, with the additional property that every zone contains four or six facets. Coxeter (1962) used this fact for a simple classification of the space-filling polytopes in $\mathbb{E}^3$. In higher dimensions, the space-filling zonotopes were investigated by Shephard (1974b) and McMullen (1975b).

Hadwiger (1952) proved for $n = 3$ and Mürner (1975) proved for general n that every space-filling n-polytope is T-equidecomposable (in the sense of scissors congruence) to a cube, where T denotes the group of translations. In three-space, this property leads to a characterization of zonotopes, and, more generally: An n-polytope is T-equidecomposable to a cube if and only if it belongs to the class $\mathcal{P}^n_{n-1}$ defined above. This was conjectured by Hadwiger and proved by Mürner (1977).

For lattice zonotopes in $\mathbb{E}^n$ (i.e., zonotopes whose vertices belong to the integer lattice $\mathbb{Z}^n$), Betke & Gritzmann (1986) have proved two interesting inequalities of discrete geometry, which for general polytopes are respectively not true and unknown.

Several aspects of the geometry of zonotopes are discussed in Martini (1987). Some extremum problems for zonotopes were treated by Linhart (1986, 1987) and Filliman (1988).

2. *The sum of a zonotope and an ellipsoid*. For a convex body $K \subset \mathbb{E}^n$, consider the following property: (A) For each $u \in S^{n-1}$, there exists $\varepsilon(u) > 0$ such that the section $[H(K, u) - \alpha u] \cap K$ is centrally symmetric for $0 < \alpha < \varepsilon(u)$. Generalizing a classical characterization of the ellipsoid, Burton (1976) proved the following remarkable result: An n-dimensional convex body K ($n \geqq 3$) has property (A) if and only if K is the Minkowski sum of a (not necessarily n-dimensional) zonotope and an (n-dimensional) ellipsoid.
3. *The integral equation for zonoids*. Proofs of Theorem 3.5.2 may be found in Bolker (1969), p. 336, Schneider (1970a), Lindquist (1975a) and Matheron (1975), p. 94.

   Theorem 3.5.3 goes back (at least) to Blaschke [5]. For the uniqueness part, two proofs for $n = 3$ and special measures appear on pp. 152 and 154–5 of that book. The general uniqueness theorem was first proved by Aleksandrov (1937b), §8, and rediscovered by Petty (1961a), Rickert (1967a, b) and Matheron (1974a) (reproduced in Matheron 1975, Section 4.5). Aleksandrov, Petty and Rickert use spherical harmonics for the uniqueness proof, while Matheron's proof is motivated by probability theory. Clearly the uniqueness assertion of Theorem 3.5.3 is equivalent to the fact that the real vector space spanned by the functions $u \mapsto |\langle u, v\rangle|$, $v \in S^{n-1}$, is dense in the space $C_e(S^{n-1})$ of even continuous real functions on S^{n-1} with the maximum norm. A simple elegant proof of this fact was given by Choquet (1969b, p. 53; 1969c, p. 171).

   Matheron (1975), p. 73, had conjectured a generalization of the uniqueness assertion of Theorem 3.5.3, with the sphere replaced by Grassmannians. This was disproved by Goodey & Howard (1990).

   The fact that the integral equation

$$G(u) = \int_{S^{n-1}} |\langle u, v\rangle| g(v) \, \mathrm{d}\mathcal{H}^{n-1}(v), \tag{3.5.13}$$

   where G is a given even function on S^{n-1}, can be solved by means of spherical harmonics was mentioned by Blaschke [5], p. 154 (for $n = 3$ and without mentioning the necessary smoothness assumptions). The general argument given above was carried out in Schneider (1967c). Geometric applications of a solution of (3.5.13) appear in Chakerian (1967b), Firey (1967a), Petty (1967) and Schneider (1967c). Schneider (1975c) proved a (weak) stability result for the solutions of (3.5.13), namely $\|g\| \leqq \|G\|_{2k}$, where $\|\cdot\|$ denotes the maximum norm and $\|\cdot\|_{2k}$ is a certain norm involving derivatives up to order $2k$. This was used to prove the existence of zonoids, other than ellipsoids, whose polar bodies (with respect to the centre) are also zonoids.

For special functions G, an explicit solution of the integral equation (3.5.13) may be found. In this way (and using the uniqueness result) one can obtain explicit examples of centrally symmetric convex bodies of class C_+^∞ that are not zonoids, for instance the body $K \in \mathcal{K}^3$ with support function given by

$$h(K, u) = 1 + \alpha P_2(\langle e, u \rangle) \quad \text{for } u \in S^2,$$

where $e \in S^2$ is fixed, $P_2(t) = (3t^2 - 1)/2$ denotes the second Legendre polynomial and $\alpha \in [-2/5, -1/4]$ is a real constant (see Schneider 1970a, p. 69).

4. *Approximation problems for zonoids*. By definition, a zonoid can be approximated arbitrarily closely by finite sums of segments. Interesting problems arise if one asks for the minimal number of segments necessary to reach a given degree of approximation. For a zonoid Z in $\mathbb{E}^n$ (with centre at the origin) and for $\varepsilon > 0$ (sufficiently small, in dependence on the cases considered below) let $N(Z, \varepsilon)$ be the smallest number N such that there is a sum P_N of N segments satisfying

$$Z \subset P_N \subset (1 + \varepsilon)Z.$$

Improving a slightly weaker result of Figiel, Lindenstrauss & Milman (1977), Gordon (1985) showed for the ball B^n that

$$N(B^n, \varepsilon) \leqq c\varepsilon^{-2}n,$$

where c is an absolute constant. The remarkable fact here is the linear dependence of the upper bound on n. For general zonoids Z, Bourgain, Lindenstrauss & Milman (1989, announced 1986) obtained results that are only slightly weaker. For $\tau > 0$ they showed that

$$N(Z, \varepsilon) \leqq c(\tau)\varepsilon^{-(2+\tau)}(\log n)^3 n$$

with a constant $c(\tau)$ depending only on τ, and

$$N(Z, \varepsilon) \leqq c(\tau, \delta)\varepsilon^{-(2+\tau)} n$$

if the zonoid Z is the unit ball of a uniformly convex norm, δ being the degree of uniform convexity. For given n, the above bounds are not optimal in ε, but here for the ball sharper estimates are known. There are constants $c_i(n)$, depending only on n, such that

$$c_1(n)\varepsilon^{-2(n-1)/(n+2)} \leqq N(B^n, \varepsilon) \leqq c_2(n)(\varepsilon^{-2}|\log \varepsilon|)^{(n-1)/(n+2)}.$$

The left-hand inequality is due to Bourgain, Lindenstrauss & Milman (1989), Theorem 6.5, and the right-hand one to Bourgain & Lindenstrauss (1988c). The latter authors also have similar, but in general weaker, results for arbitrary zonoids. Weaker inequalities of the above type were obtained earlier by Betke & McMullen (1983) and Linhart (1989).

5. *Characterizations of zonoids*. Theorem 3.5.1 characterizes zonotopes in a very simple and intuitive way. An equally simple characterization of zonoids apparently does not exist. In particular, it is not possible to characterize zonoids by a strictly local criterion. The question of such a characterization was posed repeatedly; see Blaschke (1923), p. 250, Blaschke & Reidemeister

(1922), pp. 81–2 and Bolker (1971). However, the following was shown by Weil (1977). There exists a convex body $K \in \mathcal{K}^n$ ($n \geqq 3$), arbitrarily smooth, that is not a zonoid but has the following property: For each $u \in S^{n-1}$ there exists a zonoid Z with centre at the origin and a neighbourhood U of u in S^{n-1} such that the boundaries of K and Z coincide at all points where the exterior unit normal vectors belong to U. Thus, no characterization of zonoids is possible that involves only arbitrarily small neighbourhoods of boundary points. Instead, Weil (1977) proposed the following conjecture: Let $K \in \mathcal{K}^n$ be a convex body such that to any great sphere $\sigma \subset S^{n-1}$ there exists a zonoid Z and a neighbourhood U of σ in S^{n-1} such that the boundaries of K and Z coincide at all points where the exterior unit vector belongs to U; then K is a zonoid. Affirmative answers for even dimensions were given independently by Panina (1988, 1989a, b) and Goodey & Weil (1990+).

It is also not possible to characterize zonoids by means of their projections: Weil (1982c) constructed a convex body $K \in \mathcal{K}^n$ ($n \geqq 3$) that is not a zonoid but has the property that all its projections onto hyperplanes are zonoids.

An approach to characterizations of zonoids by means of a different integral representation was proposed by Ambartzumian (1987).

Non-trivial characterizations of zonoids are possible if systems of inequalities satisfied by support functions are taken into account. Some definitions are needed. A function $f: \mathbb{E}^n \to \mathbb{R}$ is called *positive definite* if

$$\sum_{i,j=1}^{k} f(x_i - x_j)\alpha_i\alpha_j \geqq 0$$

holds for all $k \in \mathbb{N}$, $x_1, \ldots, x_k \in \mathbb{E}^n$, $\alpha_1, \ldots, \alpha_k \in \mathbb{R}$, and it is called *conditionally positive definite* if these inequalities are only assumed for $\sum \alpha_i = 0$. The function f is *of negative type* if for every $t > 0$ the function e^{-tf} is positive definite. A real normed vector space $(V, \|\cdot\|)$ is called *hypermetric* if

$$\sum_{i,j=1}^{k} \|x_i - x_j\|\alpha_i\alpha_j \leqq 0$$

for all $k \in \mathbb{N}$, $x_1, \ldots, x_k \in V$ and all integers $\alpha_1, \ldots, \alpha_k$ satisfying $\sum \alpha_i = 1$. With these definitions, the following properties of an n-dimensional convex body $Z \in \mathcal{K}^n$ with centre at the origin are equivalent:

(a) Z is a zonoid;
(b) $h(Z, \cdot)$ is of negative type;
(c) $e^{-h(Z,\cdot)}$ is positive definite;
(d) $\mathbb{E}^n$ with norm $\|\cdot\|_Z = h(Z, \cdot)$ is hypermetric;
(e) $-h(Z, \cdot)$ is conditionally positive definite.

Here the equivalence of (a) and (b) is a classical theorem due to Lévy (1937), §63; see also Choquet (1969c), p. 173. The equivalence of (b) and (c) is trivial; for the equivalence of (c), (d) and (e) see Witsenhausen (1973). Witsenhausen (1978) and Assouad (1980) also obtained characterizations of zonotopes among the polytopes by means of norm inequalities.

6. *Relations of zonoids to other fields*. Zonoids appear, and are useful, in various contexts. We mention briefly some of their occurrences; for more information we refer to the survey articles of Bolker (1969) and Schneider & Weil [44]; we have already quoted from the latter above.

A well-known theorem of Liapounoff (1940) says that the range of a non-atomic $\mathbb{E}^n$-valued measure is compact and convex; shorter proofs are due to Halmos (1948), Blackwell (1951), Lindenstrauss (1966) and others. It is not difficult to show that the convex bodies occurring as such ranges are precisely the zonoids; see Rickert (1967b) and Bolker (1969).

In Banach space theory, zonoids occur as follows. Let $Z \subset \mathbb{E}^n$ be a convex body with o as centre and interior point. Then $\|\cdot\|_Z = h(Z, \cdot)$ defines a norm on $\mathbb{E}^n$, for which the polar body Z^* is the unit ball. The body Z is a zonoid if and only if the space $\mathbb{E}^n$ with norm $\|\cdot\|_Z$ is isometric to a subspace of $L_1 = L_1([0, 1])$ (a proof and further discussion can be found in Bolker 1969). Thus, questions about the isometric embedding of finite-dimensional Banach spaces in L_1 reduce to problems on zonoids. For instance, Bolker (1969) pointed out that the space l_p^n (which is $\mathbb{R}^n$ endowed with the norm defined by $\|(\xi_1, \ldots, \xi_n)\| = (\sum_{i=1}^n |\xi_i|^p)^{1/p}$) embeds in L_1 isometrically if $1 \leqq p \leqq 2$, and he conjectured that this is not the case if $n \geqq 3$ and $p > 2$. This was proved for $p > 2.7$ by Witsenhausen (1973) and for all $p > 2$ by Dor (1976). Bolker (1969, 1971) also formulated the following question: If E is an n-dimensional Banach space ($n \geqq 3$) such that E and its dual space both embed isometrically in L_1, is E isometric to l_2^n? The result of Schneider (1975c) mentioned in Note 3 provides counterexamples.

In stochastic geometry, zonoids have proved useful as auxiliary bodies. If ρ is an even measure on S^{n-1}, which can be, e.g., the orientation distribution of a random hyperplane or a measure associated in some way with a stochastic process of geometric objects, one defines an associated zonoid K by means of equation (3.5.2). Some parameters describing the stochastic situation can then be expressed in terms of functionals of K, so that, for instance, inequalities for convex bodies can be used to solve extremal problems of stochastic geometry. This method was initiated by Matheron (1974b, 1975), who called the associated zonoid a 'Steiner compact' (for reasons that are not entirely convincing). Later applications were given by Schneider (1982a, b, 1987a), Thomas (1984), Wieacker (1984, 1986, 1989) and Weil (1988).

Alexander (1988) made an interesting contribution to Hilbert's fourth problem, using zonoids and generalized zonoids. In particular, the equivalence of (a) and (d) in Note 5 plays an essential role in his argument.

An application of zonoids to mathematical economics may be found in Hildenbrand (1981).

7. *Generalized zonoids*. Lemma 3.5.5 is taken from Schneider (1970a).

The existence of generalized zonoids that are not zonoids motivates two questions: to characterize the even signed measure ρ on S^{n-1} for which

$$h(K, \cdot) := \int_{S^{n-1}} |\langle \cdot, v \rangle| \, \mathrm{d}\rho(v) \tag{3.5.14}$$

yields a support function, and to characterize the convex bodies K obtainable in this way. Aiming to answer these questions, Weil (1976a) developed formulae for mixed area measures of generalized zonoids (Section 5.3). With their aid and using results from Weil (1974a) he obtained the following result. For $u_1, \ldots, u_j \in S^{n-1}$ let $D_j(u_1, \ldots, u_j)$ denote the j-dimensional volume of the parallelepiped spanned by $u_1, \ldots, u_j$. Further, define a partial map $T\colon (S^{n-1})^{n-1} \to S^{n-1}$ by the normalized vector product. Then (3.5.14) defines a support function if and only if

$$T\left[\int_{(\cdot)} D_{n-1} \, \mathrm{d}(\rho^j \times \omega^{n-j-1})\right] \geqq 0$$

for $j = 1, \ldots, n-1$ (where ω denotes the normalized rotation invariant measure on S^{n-1}). A different characterization of the signed measures ρ that generate convex bodies by means of (3.5.14) can be obtained by considering projections. For a signed measure ρ on S^{n-1} and for a j-dimensional great subsphere σ_j of S^{n-1}, Weil (1982c) defined a projection of ρ onto σ_j such that, for any generalized zonoid K with generating measure ρ and any $(j+1)$-dimensional linear subspace E_{j+1} of $\mathbb{E}^n$, the orthogonal projection of K onto E_{j+1} has the projection of ρ on $E_{j+1} \cap S^{n-1}$ as its generating measure.

He deduced, for instance, that ρ generates a convex body if and only if all its projections onto great circles are non-negative. Another application generalizes a convexity criterion of Lindquist (1975b).

Several characterizations of zonoids, generalized zonoids and centrally symmetric convex bodies, expressed in terms of inequalities involving mixed volumes, were investigated by Weil (1976b, 1979a) and Goodey (1977, 1984a, b) (some of the assertions in Goodey (1984a, b) have to be modified; see the corrections in Goodey & Weil (1991)). See also Section 5.1, Note 9.

8. *Rotors*. Special rotors in regular polygons were considered, often in connection with kinematic considerations, by Reuleaux (1875), Meissner (1909), Wunderlich (1939), Goldberg (1948, 1957), Schaal (1963) and others. All the rotors of the regular polygons were determined by Meissner (1909). Case (b) of Theorem 3.5.7 is due to Fujiwara (1915); simpler proofs were given by Kameneckii (1947) and Schaal (1962).

Meissner (1918) determined the rotors of the three-dimensional regular polyhedra. Models of such rotors are shown in the survey article by Goldberg (1960), who also discussed the problem of the existence of rotors in higher dimensions and for non-regular polytopes. The complete classification given in Theorem 3.5.7 is due to Schneider (1971b).

Since the class of rotors of a regular k-gon depends on infinitely many parameters, it is sufficiently interesting to be the subject of extremal problems. In particular, one may ask which rotors of a given regular k-gon have the least area. The case $k = 4$, where the Reuleaux triangle is the solution, was solved independently by Lebesgue (1914) and Blaschke (1915). The survey [31] (p. 67) lists several later proofs, of which the one by Chakerian (1966) is especially elegant. The case $k = 3$ was settled by Fujiwara & Kakeya (1917); alternative proofs were given by Weissbach (1972, 1977a), see also Jaglom & Boltjanski [20]. For arbitrary $k \geqq 3$, the problem was treated by Focke (1969a) under special symmetry assumptions and solved completely by Klötzler (1975).

Some other extremal problems for rotors of regular polygons were investigated by Fujiwara & Kakeya (1917), Fujiwara (1919), Schaal (1963), Focke (1969b), Focke & Gensel (1971) and Weissbach (1977b).

Other M-classes in the plane, more general than rotors, are made up by the so-called U_k-curves, for which all circumscribed equiangular k-gons have the same perimeter, or by the P_k-curves, for which all circumscribed equiangular k-gons are regular. These classes have been studied by Meissner (1909), Jaglom & Boltjanski [20], Goldberg (1966) and Weissbach (1977a).

4

Curvature measures and quermassintegrals

4.1. Local parallel sets

For a convex body $K \in \mathcal{K}^n$ and a number $\rho > 0$, the Minkowski sum $K + \rho B^n$ is called the *outer parallel body* of K at distance ρ. According to the well-known *Steiner formula* (to be proved in Section 4.2), its volume can be expressed as a polynomial of degree at most n in the parameter ρ:

$$V_n(K + \rho B^n) = \sum_{i=0}^{n} \rho^i \binom{n}{i} W_i(K). \tag{4.1.1}$$

The coefficients defined by this expansion turn out to be important quantities, carrying basic geometric information on the body K. Formula (4.1.1) is only a very special case of a similar expansion valid for a general Minkowski sum $\lambda_1 K_1 + \ldots + \lambda_m K_m$; this leads to the notion of mixed volumes and will be extensively studied in later chapters. The particular case of adding a ball deserves a separate study, as its strong dependence on the Euclidean metric results in a theory of different character and leads to a number of notions and results related to the Euclidean geometry of convex bodies.

The Minkowski sum $K + \rho B^n$ is at the same time the set of all points of $\mathbb{E}^n$ having distance $d(K, \cdot)$ from K at most ρ. This point of view leads naturally to the more general notion of a local parallel set: Given a subset $\beta \subset K$, we may consider the set of all points $x \in \mathbb{E}^n$ for which $d(K, x) \leqq \rho$ and for which the nearest point $p(K, x)$ belongs to β. Alternatively, we may prescribe a set $\beta \subset S^{n-1}$ of unit vectors and then consider the set of all $x \in \mathbb{E}^n$ for which $d(K, x) \leqq \rho$ and for which the unit vector $u(K, x)$ pointing from $p(K, x)$ to x (see Section 1.2) belongs to β. Restricting β to Borel sets, we find that the Lebesgue measure of the local parallel set thus defined is again a polynomial in the parameter

ρ. The coefficients now depend on K and β, and as functions of the latter argument they are measures. These are the curvature or area measures studied in the present chapter. Their basic properties will be derived from corresponding properties of the local parallel sets as established in this section. Instead of treating the two types of local parallel sets separately, we consider a common generalization.

We need some more notation. The product space $\mathbb{E}^n \times S^{n-1}$ is denoted by Σ. If $K \in \mathcal{K}^n$, a pair $(x, u) \in \Sigma$ is called a *support element* of K if $x \in \operatorname{bd} K$ and u is an outer unit normal vector of K at x. In particular, if $x \in \mathbb{E}^n \backslash K$, then the pair $(p(K, x),\ u(K, x))$ is a support element of K. The set of all support elements of K, also called the *generalized normal bundle* of K, will be denoted by $\operatorname{Nor} K$. For a topological space X, the σ-algebra of Borel subsets of X is denoted by $\mathcal{B}(K)$. 'Measurable' without further comment means Borel measurable.

Now let a convex body $K \in \mathcal{K}^n$ and a number $\rho > 0$ be given. The outer parallel body $K + \rho B^n$ is abbreviated by K_ρ. The map

$$\begin{aligned} f_\rho\colon\ & K_\rho \backslash K \to \Sigma \\ & x \mapsto (p(K, x),\ u(K, x)) \end{aligned}$$

is continuous and hence measurable. We may, therefore, consider the image measure $\mu_\rho(K, \cdot)$ under f_ρ of the Lebesgue measure. Thus $\mu_\rho(K, \cdot)$ is a finite measure on $\mathcal{B}(\Sigma)$, and for a Borel set $\eta \in \mathcal{B}(\Sigma)$, $\mu_\rho(K, \eta)$ is the Lebesgue measure of the *local parallel set*

$$\begin{aligned} M_\rho(K, \eta) &:= f_\rho^{-1}(\eta) \\ &= \{x \in \mathbb{E}^n \mid 0 < d(K, x) \leqq \rho \ \text{ and } \ (p(K, x),\ u(K, x)) \in \eta\}. \end{aligned}$$

We establish some basic properties of the map $\mu_\rho\colon \mathcal{K}^n \times \mathcal{B}(\Sigma) \to \mathbb{R}$. For the measure-theoretic notions and results to be used we refer, e.g., to Ash (1972) and Bauer (1974). By $\xrightarrow{\text{w}}$ we denote weak convergence of Borel measures.

Theorem 4.1.1. *Let $(K_j)_{j \in \mathbb{N}}$ be a sequence in $\mathcal{K}^n$ with $K_j \to K$ for $j \to \infty$. Then $\mu_\rho(K_j, \cdot) \xrightarrow{\text{w}} \mu_\rho(K, \cdot)$ for $j \to \infty$.*

Proof. Let $\eta \subset \Sigma$ be open, and let $x \in M_\rho(K, \eta)$ be a point with $d(K, x) < \rho$. From Lemma 1.8.9 it follows that $d(K_j, x) \to d(K, x)$ and $(p(K_j, x),\ u(K_j, x)) \to (p(K, x),\ u(K, x))$ for $j \to \infty$. Hence, for almost all j we have $d(K_j, x) < \rho$ and $(p(K_j, x), u(K_j, x)) \in \eta$, thus $x \in M_\rho(K_j, \eta)$. This shows that

$$M_\rho(K, \eta) \backslash \operatorname{bd} K_\rho \subset \liminf_{j \to \infty} M_\rho(K_j, \eta)$$

and hence that

$$\begin{aligned}\mu_\rho(K, \eta) &= \mathcal{H}^n(M_\rho(K, \eta)\backslash \operatorname{bd} K_\rho)\\ &\leqq \mathcal{H}^n(\liminf_{j\to\infty} M_\rho(K_j, \eta))\\ &\leqq \liminf_{j\to\infty} \mathcal{H}^n(M_\rho(K_j, \eta))\\ &= \liminf_{j\to\infty} \mu_\rho(K_j, \eta).\end{aligned}$$

This holds for all open sets $\eta \subset \Sigma$. Further, we have $\mu_\rho(K, \Sigma) = \mathcal{H}^n(K_\rho \backslash K) = \mathcal{H}^n(K_\rho) - \mathcal{H}^n(K)$. The continuity of the volume on $\mathcal{K}^n$ (Theorem 1.8.16) together with the continuity of Minkowski addition shows that

$$\mu_\rho(K_j, \Sigma) \to \mu_\rho(K, \Sigma) \quad \text{for} \quad j \to \infty.$$

The assertion follows (see, e.g., Ash 1972, p. 196). ∎

Theorem 4.1.2. *For each $\eta \in \mathcal{B}(\Sigma)$, the function $\mu_\rho(\cdot, \eta): \mathcal{K}^n \to \mathbb{R}$ is measurable.*

Proof. In the preceding proof it was shown that $K_j \to K$ implies

$$\liminf_{j\to\infty} \mu_\rho(K_j, \eta) \geqq \mu_\rho(K, \eta)$$

for open η, hence for such η the function $\mu_\rho(\cdot, \eta)$ is lower semicontinuous and thus measurable. Let $\mathcal{A}$ be the family of all sets $\eta \in \mathcal{B}(\Sigma)$ for which $\mu_\rho(\cdot, \eta)$ is measurable. We show that $\mathcal{A}$ is a Dynkin system. For $\eta_1, \eta_2 \in \mathcal{A}$ with $\eta_2 \subset \eta_1$ we have $M_\rho(K, \eta_2) \subset M_\rho(K, \eta_1)$ and

$$M_\rho(K, \eta_1 \backslash \eta_2) = M_\rho(K, \eta_1) \backslash M_\rho(K, \eta_2),$$

hence

$$\mu_\rho(K, \eta_1 \backslash \eta_2) = \mu_\rho(K, \eta_1) - \mu_\rho(K, \eta_2)$$

for $K \in \mathcal{K}^n$, thus $\eta_1 \backslash \eta_2 \in \mathcal{A}$. For a sequence $(\eta_j)_{j\in\mathbb{N}}$ of pairwise disjoint elements of $\mathcal{A}$ we have

$$\mu_\rho\left(K, \bigcup_{j=1}^{\infty} \eta_j\right) = \sum_{j=1}^{\infty} \mu_\rho(K, \eta_j)$$

for $K \in \mathcal{K}^n$, since $\mu_\rho(K, \cdot)$ is a measure. It follows that $\bigcup_{j=1}^{\infty} \eta_j \in \mathcal{A}$. Thus $\mathcal{A}$ is a Dynkin system that contains the open sets and, therefore, the σ-algebra generated by them (Bauer 1974, p. 20). Hence $\mathcal{B}(\Sigma) \subset \mathcal{A}$, which was the assertion. ∎

Theorem 4.1.3. *For each $\eta \in \mathcal{B}(\Sigma)$, the function $\mu_\rho(\cdot, \eta)$ is additive, that is,*

$$\mu_\rho(K_1 \cup K_2, \eta) + \mu_\rho(K_1 \cap K_2, \eta) = \mu_\rho(K_1, \eta) + \mu_\rho(K_2, \eta)$$

if $K_1, K_2, K_1 \cup K_2 \in \mathcal{K}^n$.

Proof. Let $K_1, K_2 \in \mathcal{K}^n$ be convex bodies such that $K_1 \cup K_2 \in \mathcal{K}^n$. By $I_\rho(K, \eta, \cdot)$ we denote the characteristic function of the set $M_\rho(K, \eta)$. Let $x \in \mathbb{E}^n$ be given and suppose that $y := p(K_1 \cup K_2, x) \in K_1$, without loss of generality. Then clearly

$$p(K_1 \cup K_2, x) = p(K_1, x).$$

Let $p(K_2, x) =: z$. Since $K_1 \cup K_2$ is convex, there is a point $a \in [z, y]$ for which $a \in K_1 \cap K_2$. Since $y = p(K_1 \cup K_2, x)$, we have $|y - x| \leqq |z - x|$ and hence $|a - x| \leqq |z - x|$. Here strict inequality is impossible; hence $a = z$ and thus $z \in K_1 \cap K_2$. It follows that

$$p(K_1 \cap K_2, x) = p(K_2, x).$$

We deduce that

$$d(K_1 \cup K_2, x) = d(K_1, x), \quad d(K_1 \cap K_2, x) = d(K_2, x)$$
$$u(K_1 \cup K_2, x) = u(K_1, x), \quad u(K_1 \cap K_2, x) = u(K_2, x),$$

from which we get

$$I_\rho(K_1 \cup K_2, \eta, x) = I_\rho(K_1, \eta, x),$$
$$I_\rho(K_1 \cap K_2, \eta, x) = I_\rho(K_2, \eta, x).$$

As x was arbitrary, we conclude that

$$I_\rho(K_1 \cup K_2, \eta, \cdot) + I_\rho(K_1 \cap K_2, \eta, \cdot) = I_\rho(K_1, \eta, \cdot) + I_\rho(K_2, \eta, \cdot).$$

Integration of this equality with respect to $\mathcal{H}^n$ yields the assertion. ■

4.2. Curvature measures and area measures

We shall now show that the measure $\mu_\rho(K, \eta)$ of the local parallel set $M_\rho(K, \eta)$ can be expressed as a polynomial in the parameter ρ. First let $P \in \mathcal{K}^n$ be a polytope. For $x \in \mathbb{E}^n \backslash P$, the nearest point $p(P, x)$ belongs to the relative interior of a unique face of P. For a given face F of P, the Lebesgue measure of the set

$$M_\rho(P, \eta) \cap p(P, \cdot)^{-1}(\text{relint}\, F),$$

where $\eta \in \mathcal{B}(\Sigma)$ and $\rho > 0$, can be computed by means of Fubini's theorem. For this, we observe that

$$(K_\rho \backslash K) \cap p(P, \cdot)^{-1}(\text{relint}\, F)$$

is, up to a set of measure zero, equal to the direct sum

$$F \oplus (N(P, F) \cap \rho B^n),$$

where $N(P, F)$ is the normal cone of P at F. If F is of dimension $m \in \{0, \ldots, n-1\}$, we obtain

$$\mathcal{H}^n(M_\rho(P, \eta) \cap p(P, \cdot)^{-1}(\operatorname{relint} F))$$
$$= \int_F \mathcal{H}^{n-m}\left\{z \in N(P, F) | 0 < |z| \leqq \rho, \left(y, \frac{z}{|z|}\right) \in \eta\right\} \mathrm{d}\mathcal{H}^m(y)$$
$$= \int_F \mathcal{H}^{n-m}\{\lambda u | 0 < \lambda \leqq \rho, u \in N(P, F) \cap S^{n-1}, (y, u) \in \eta\} \mathrm{d}\mathcal{H}^m(y)$$
$$= \int_F \frac{1}{n-m} \rho^{n-m} \mathcal{H}^{n-m-1}(N(P, F) \cap \eta_y) \mathrm{d}\mathcal{H}^m(y),$$

where

$$\eta_y := \{u \in S^{n-1} | (y, u) \in \eta\}.$$

Thus we arrive at

$$\mu_\rho(P, \eta)$$
$$= \sum_{m=0}^{n-1} \rho^{n-m} \frac{1}{n-m} \sum_{F \in \mathcal{F}_m(P)} \int_F \mathcal{H}^{n-1-m}(N(P, F) \cap \eta_y) \mathrm{d}\mathcal{H}^m(y). \tag{4.2.1}$$

We define

$$\binom{n-1}{m} \Theta_m(P, \eta) := \sum_{F \in \mathcal{F}_m(P)} \int_F \mathcal{H}^{n-1-m}(N(P, F) \cap \eta_y) \mathrm{d}\mathcal{H}^m(y) \tag{4.2.2}$$

for $m = 0, \ldots, n-1$, so that

$$\mu_\rho(P, \eta) = \frac{1}{n} \sum_{m=0}^{n-1} \rho^{n-m} \binom{n}{m} \Theta_m(P, \eta). \tag{4.2.3}$$

This representation can be extended to arbitrary convex bodies:

Theorem 4.2.1. *For every convex body $K \in \mathcal{K}^n$ there exist finite positive measures $\Theta_0(K, \cdot), \ldots, \Theta_{n-1}(K, \cdot)$ on $\mathcal{B}(\Sigma)$ such that, for every $\eta \in \mathcal{B}(\Sigma)$ and every $\rho > 0$, the measure $\mu_\rho(K, \eta)$ of the local parallel set $M_\rho(K, \eta)$ is given by*

$$\mu_\rho(K, \eta) = \frac{1}{n} \sum_{m=0}^{n-1} \rho^{n-m} \binom{n}{m} \Theta_m(K, \eta). \tag{4.2.4}$$

The mapping $K \mapsto \Theta_m(K, \cdot)$ (from $\mathcal{K}^n$ into the space of Borel measures on Σ) is weakly continuous and additive, that is,

$$K_j \to K \quad \text{implies} \quad \Theta_m(K_j, \cdot) \xrightarrow{\mathrm{w}} \Theta_m(K, \cdot),$$

and $K_1, K_2, K_1 \cup K_2 \in \mathcal{K}^n$ implies

$$\Theta_m(K_1 \cup K_2, \cdot) + \Theta_m(K_1 \cap K_2, \cdot) = \Theta_m(K_1, \cdot) + \Theta_m(K_2, \cdot).$$

For each $\eta \in \mathcal{B}(\Sigma)$, the function $\Theta_m(\cdot, \eta)$ (from $\mathcal{K}^n$ to $\mathbb{R}$) *is measurable.*

Proof. First let $P \in \mathcal{K}^n$ be a polytope. Let $\eta \in \mathcal{B}(\Sigma)$ and $\rho > 0$ be given. Writing down equation (4.2.3) successively for $\rho = 1, \ldots, n$, we obtain a system of linear equations that can be solved for $\Theta_0(P, \eta), \ldots, \Theta_{n-1}(P, \eta)$. We get representations of the form

$$\Theta_m(P, \eta) = \sum_{k=1}^{n} a_{mk}\mu_k(P, \eta)$$

with certain constants a_{mk}. With these constants we define

$$\Theta_m(K, \cdot) := \sum_{k=1}^{n} a_{mk}\mu_k(K, \cdot) \tag{4.2.5}$$

for arbitrary convex bodies $K \in \mathcal{K}^n$. Then $\Theta_m(K, \cdot)$ is a finite measure on $\mathcal{B}(\Sigma)$, for the moment possibly a signed one. From Theorem 4.1.1 it follows that $K_j \to K$ (for $K_j, K \in \mathcal{K}^n$) implies $\Theta_m(K_j, \cdot) \xrightarrow{w} \Theta_m(K, \cdot)$. Applying this to a sequence of polytopes converging to K, we see that $\Theta_m(K, \cdot)$ is a positive measure and that (4.2.3) can be extended to give (4.2.4). The final assertions of the theorem follow from (4.2.5) and Theorems 4.1.3, 4.1.2. ∎

The measure $\Theta_m(K, \cdot)$ is called the mth *generalized curvature measure* of K, for a reason to be explained later. Although we have defined the measure $\Theta_m(K, \cdot)$ on the whole space Σ, it is clearly concentrated on the set $\operatorname{Nor} K$ of support elements of K: If $x \in M_\rho(K, \eta)$, then $(p(K, x), u(K, x)) \in \eta \cap \operatorname{Nor} K$, hence $\mu_\rho(K, \eta) = \mu_\rho(K, \eta \cap \operatorname{Nor} K)$ and, therefore, by (4.2.5)

$$\Theta_m(K, \eta) = \Theta_m(K, \eta \cap \operatorname{Nor} K) \tag{4.2.6}$$

for $\eta \in \mathcal{B}(\Sigma)$ and $m = 0, \ldots, n-1$.

The existence of a polynomial expansion of the type (4.2.4) carries over to the generalized curvature measures. For $\rho > 0$ we define a map $t_\rho: \Sigma \to \Sigma$ by $t_\rho(x, u) := (x + \rho u, u)$. Then we have:

Theorem 4.2.2. *Let $K \in \mathcal{K}^n$, $\eta \in \mathcal{B}(\Sigma)$, $\rho > 0$ and $m \in \{0, \ldots, n-1\}$. Then*

$$\Theta_m(K_\rho, t_\rho\eta) = \sum_{j=0}^{m} \rho^j \binom{m}{j} \Theta_{m-j}(K, \eta). \tag{4.2.7}$$

Proof. Let $x \in \mathbb{E}^n \setminus K_\rho$. There is a unique point $y \in [x, p(K, x)] \cap \operatorname{bd} K_\rho$. Assume that $p(K_\rho, x) \neq y$. Then $|x - p(K_\rho, x)| < |x - y|$. Let $p(K, p(K_\rho, x)) =: z$. From $p(K_\rho, x) \in \operatorname{bd} K_\rho$ and $y \in \operatorname{bd} K_\rho$ it follows that

$$|p(K_\rho, x) - z| = \rho \leqq |y - p(K, x)|$$

and hence

$$\begin{aligned}|x - z| &\leqq |x - p(K_\rho, x)| + |p(K_\rho, x) - z| \\ &< |x - y| + |y - p(K, x)| \\ &= |x - p(K, x)|,\end{aligned}$$

a contradiction. Hence $p(K_\rho, x) = y$. This implies $u(K_\rho, x) = u(K, x)$, $d(K_\rho, x) = d(K, x) - \rho$ and $p(K_\rho, x) = p(K, x) + \rho u(K, x)$. Now we obtain the disjoint decomposition

$$M_{\rho+\lambda}(K, \eta) = M_\rho(K, \eta) \cup M_\lambda(K_\rho, t_\rho\eta)$$

for $\lambda > 0$ and hence the equality

$$\mu_{\rho+\lambda}(K, \eta) = \mu_\rho(K, \eta) + \mu_\lambda(K_\rho, t_\rho\eta).$$

Here we insert (4.2.4) and compare the coefficients of equal powers of λ to obtain the assertion (4.2.7). ■

From the measures $\Theta_m(K, \cdot)$, which are defined on sets of support elements, we derive two series of measures defined either on sets of boundary points or on sets of normal vectors. For $m = 0, \ldots, n-1$ we put

$$\begin{aligned}C_m(K, \beta) &:= \Theta_m(K, \beta \times S^{n-1}) \quad \text{for } \beta \in \mathcal{B}(\mathbb{E}^n), \\ S_m(K, \omega) &:= \Theta_m(K, \mathbb{E}^n \times \omega) \quad \text{for } \omega \in \mathcal{B}(S^{n-1}).\end{aligned}$$

The measures $C_0(K, \cdot), \ldots, C_{n-1}(K, \cdot)$ are called the *curvature measures* of K, and $S_0(K, \cdot), \ldots, S_{n-1}(K, \cdot)$ are called the *area measures* of K. Observe that the curvature measures are Borel measures on $\mathbb{E}^n$, while the area measures are Borel measures on the unit sphere S^{n-1}. For the reader's convenience, we repeat the definition of these measures by specializing formula (4.2.4) to these cases. Writing

$$A_\rho(K, \beta) := \{x \in \mathbb{E}^n \,|\, 0 < d(K, x) \leqq \rho \text{ and } p(K, x) \in \beta\}$$

for $\beta \in \mathcal{B}(\mathbb{E}^n)$ and $\rho > 0$, we have

$$\mathcal{H}^n(A_\rho(K, \beta)) = \frac{1}{n}\sum_{m=0}^{n-1} \rho^{n-m}\binom{n}{m} C_m(K, \beta), \tag{4.2.8}$$

and with

$$B_\rho(K, \omega) := \{x \in \mathbb{E}^n \,|\, 0 < d(K, x) \leqq \rho \text{ and } u(K, x) \in \omega\}$$

for $\omega \in \mathcal{B}(S^{n-1})$ we obtain

$$\mathcal{H}^n(B_\rho(K, \omega)) = \frac{1}{n}\sum_{m=0}^{n-1} \rho^{n-m}\binom{n}{m} S_m(K, \omega). \tag{4.2.9}$$

It is evident that the need to introduce two distinct series of measures is a result of the fact that a convex body can have singular boundary points and singular normal vectors. The following lemma expresses one aspect under which both types of measures are related to each other.

Lemma 4.2.3. Let $m \in \{0, \dots, n-1\}$, *let* $\omega \subset S^{n-1}$ *and* $\beta \subset \mathbb{E}^n$ *be closed. Then*

$$C_m(K, \tau(K, \omega) \cap \operatorname{reg} K) \leqq S_m(K, \omega) \leqq C_m(K, \tau(K, \omega)),$$
$$S_m(K, \sigma(K, \beta) \cap \operatorname{regn} K) \leqq C_m(K, \beta) \leqq S_m(K, \sigma(K, \beta)).$$

Proof. Let $(x, u) \in \operatorname{Nor} K$ be a support element of K for which $x \in \tau(K, \omega) \cap \operatorname{reg} K$; then $u \in N(K, x)$. Since $x \in \tau(K, \omega)$, there is a normal vector of K at x belonging to ω, and this is equal to u since $x \in \operatorname{reg} K$. Thus $u \in \omega$. This shows that

$$([\tau(K, \omega) \cap \operatorname{reg} K] \times S^{n-1}) \cap \operatorname{Nor} K \subset (\mathbb{E}^n \times \omega) \cap \operatorname{Nor} K.$$

The sets occurring here are Borel sets, since $\tau(K, \omega)$ and $\operatorname{Nor} K$ are closed and $\operatorname{bd} K \backslash \operatorname{reg} K$ is an F_σ-set. We deduce that

$$\Theta_m(K, ([\tau(K, \omega) \cap \operatorname{reg} K] \times S^{n-1}) \cap \operatorname{Nor} K) \leqq \Theta_m(K, (\mathbb{E}^n \times \omega) \cap \operatorname{Nor} K)$$

and hence, using (4.2.6), that

$$\begin{aligned} C_m(K, \tau(K, \omega) \cap \operatorname{reg} K) &= \Theta_m(K, ([\tau(K, \omega) \cap \operatorname{reg} K] \times S^{n-1}) \cap \operatorname{Nor} K) \\ &\leqq \Theta_m(K, (\mathbb{E}^n \times \omega) \cap \operatorname{Nor} K) \\ &= S_m(K, \omega). \end{aligned}$$

Let $(x, u) \in \operatorname{Nor} K$ be a support element of K for which $u \in \omega$; then $x \in \tau(K, \omega)$. Thus

$$(\mathbb{E}^n \times \omega) \cap \operatorname{Nor} K \subset [\tau(K, \omega) \times S^{n-1}] \cap \operatorname{Nor} K$$

and hence

$$\begin{aligned} S_m(K, \omega) &= \Theta_m(K, (\mathbb{E}^n \times \omega) \cap \operatorname{Nor} K) \\ &\leqq \Theta_m(K, [\tau(K, \omega) \times S^{n-1}] \cap \operatorname{Nor} K) \\ &= C_m(K, \tau(K, \omega)). \end{aligned}$$

This proves the first two inequalities of the lemma. The proof of the last two is completely analogous. ■

If $K \in \mathcal{K}^n$ is smooth, then the spherical image map σ_K is defined on all of $\operatorname{bd} K$, and $\tau(K, \omega) = \sigma_K^{-1}(\omega)$ for $\omega \subset S^{n-1}$. If ω is a Borel set, this shows that $\tau(K, \omega)$ is a Borel set, since σ_K is continuous. From Lemma 4.2.3 we obtain, first for closed sets and then for general $\omega \in \mathcal{B}(S^{n-1})$,

$$C_m(K, \sigma_K^{-1}(\omega)) = C_m(K, \tau(K, \omega)) = S_m(K, \omega).$$

An analogous result holds for strictly convex bodies; hence we can state:

Theorem 4.2.4. *If $K \in \mathcal{K}^n$ is smooth, then $S_m(K, \cdot)$ is the image measure of $C_m(K, \cdot)$ under the spherical image map σ_K. If K is strictly convex, then $C_m(K, \cdot)$ is the image measure of $S_m(K, \cdot)$ under the reverse spherical image map τ_K $(m = 0, \ldots, n-1)$.*

Thus for convex bodies K that are both smooth and strictly convex there is in fact no essential difference between both types of measures, since

$$C_m(K, \beta) = S_m(K, \sigma_K(\beta \cap \operatorname{bd} K)) \quad \text{for } \beta \in \mathcal{B}(\mathbb{E}^n).$$

In the literature, curvature measures and area measures also appear with a different normalization. One defines $\Phi_m(K, \cdot)$ and $\Psi_m(K, \cdot)$ by

$$n\kappa_{n-m}\Phi_m(K, \cdot) = \binom{n}{m} C_m(K, \cdot), \tag{4.2.10}$$

$$n\kappa_{n-m}\Psi_m(K, \cdot) = \binom{n}{m} S_m(K, \cdot) \tag{4.2.11}$$

for $m = 0, \ldots, n-1$ (recall that κ_j is the j-dimensional volume of the j-dimensional unit ball). It is reasonable to supplement the first definition by

$$\Phi_n(K, \beta) := \mathcal{H}^n(K \cap \beta) \quad \text{for } \beta \in \mathcal{B}(\mathbb{E}^n).$$

Neither of the two nomenclatures deserves unique preference over the other. In studies related to differential geometry the C_m, S_m appear more natural, while in integral geometry the Φ_m, Ψ_m yield slightly less clumsy formulae. One advantage of the second normalization is that $\Phi_m(K, \beta)$ depends only on K and β and not on the dimension of the surrounding space in which it is computed. This follows from (4.2.17) below.

It is clear that the properties established for the generalized curvature measures carry over to curvature measures and area measures. The functions $K \mapsto C_m(K, \cdot)$ and $K \mapsto S_m(K, \cdot)$ are weakly continuous and additive; the function $C_m(\cdot, \beta)$ is measurable for each $\beta \in \mathcal{B}(\mathbb{E}^n)$, and $S_m(\cdot, \omega)$ is measurable for each $\omega \in \mathcal{B}(S^{n-1})$. The following invariance and homogeneity properties are easily obtained from (4.2.8), (4.2.9): If g is a rigid motion of $\mathbb{E}^n$ and g_0 denotes the corresponding rotation, then

$$C_m(gK, g\beta) = C_m(K, \beta), \tag{4.2.12}$$

$$S_m(gK, g_0\omega) = S_m(K, \omega). \tag{4.2.13}$$

If $\lambda > 0$, then

$$C_m(\lambda K, \lambda\beta) = \lambda^m C_m(K, \beta), \tag{4.2.14}$$

$$S_m(\lambda K, \omega) = \lambda^m S_m(K, \omega). \tag{4.2.15}$$

The measure $C_m(K, \cdot)$ is concentrated on the boundary of K, since $A_\rho(K, \beta) = \emptyset$ if $\beta \cap \operatorname{bd} K = \emptyset$. It is *defined locally*, in the following sense: If $\beta \subset \mathbb{E}^n$ is open and if $K_1, K_2 \in \mathcal{K}^n$ are such that $K_1 \cap \beta = K_2 \cap \beta$, then $C_m(K_1, \beta') = C_m(K_2, \beta')$ for every Borel set $\beta' \subset \beta$. This follows from the observation that $p(K_i, x) \in \beta$ ($i = 1$ or 2) for an open set β, together with $K_1 \cap \beta = K_2 \cap \beta$, implies $p(K_1, x) = p(K_2, x)$, hence $A_\rho(K_1, \beta') = A_\rho(K_2, \beta')$ for $\beta' \subset \beta$. The assertion then follows from (4.2.8). Also, the measure $S_m(K, \cdot)$ is *defined locally*, which is now meant in the following sense. If $\omega \in \mathcal{B}(S^{n-1})$ (not necessarily open in this case) and if $K_1, K_2 \in \mathcal{K}^n$ are such that $\tau(K_1, \omega) = \tau(K_2, \omega)$, then $S_m(K_1, \omega) = S_m(K_2, \omega)$. In fact, the local parallel set $B_\rho(K_i, \omega)$ depends only on the reverse spherical image $\tau(K_i, \omega)$. Finally we remark that, for area measures, formula (4.2.7) takes the simple form

$$S_m(K_\rho, \omega) = \sum_{j=0}^{m} \rho^j \binom{m}{j} S_{m-j}(K, \omega) \tag{4.2.16}$$

for $\omega \in \mathcal{B}(S^{n-1})$ and $\rho > 0$.

We turn to the consideration of special cases where the curvature and area measures can be given a more direct and intuitive intepretation. First let $P \in \mathcal{K}^n$ be a polytope. From (4.2.2), (4.2.10) and the definition of the external angle $\gamma(F, P)$ (see Section 2.4) we obtain

$$\Phi_m(P, \beta) = \sum_{F \in \mathcal{F}_m(P)} \gamma(F, P)\, \mathcal{H}^m(F \cap \beta) \tag{4.2.17}$$

for $\beta \in \mathcal{B}(\mathbb{E}^n)$ and $m = 0, \ldots, n$. Since $\gamma(F, P)$ depends only on P and F, as remarked in Section 2.4, we see that $\Phi_m(P, \beta)$ does not depend on the dimension of the surrounding space, that is, it can be computed in any affine subspace containing P. Similarly we get

$$\Psi_m(P, \omega) = \sum_{F \in \mathcal{F}_m(P)} \frac{\mathcal{H}^{n-1-m}(N(P, F) \cap \omega)}{\omega_{n-m}}\, \mathcal{H}^m(F) \tag{4.2.18}$$

for $\omega \in \mathcal{B}(S^{n-1})$ and $m = 0, \ldots, n-1$.

If K is of class C^2_+, then a comparison of (2.5.31) with (4.2.8) and (4.2.9) shows that

$$C_m(K, \beta) = \int_{\beta \cap \operatorname{bd} K} H_{n-1-m}\, d\mathcal{H}^{n-1}, \tag{4.2.19}$$

$$S_m(K, \omega) = \int_\omega s_m\, d\mathcal{H}^{n-1} \tag{4.2.20}$$

for $\beta \in \mathcal{B}(\mathbb{E}^n)$, $\omega \in \mathcal{B}(S^{n-1})$, $m = 0, \ldots, n-1$ (first for open sets β, ω, but then for Borel sets, since both sides of (4.2.19) and (4.2.20) are measures in respectively β and ω).

Formula (4.2.19) is, of course, the reason for the name 'curvature measure'. These measures replace, for general convex bodies, the elementary symmetric functions of the principal curvatures that can be defined in the C^2 case. Similarly, the measures $S_m(K, \cdot)$ replace the elementary symmetric functions of the principal radii of curvature (as functions on the spherical image). The name 'area measures' for this series of measures comes from the fact that $S_{n-1}(K, \omega)$ is the area of the reverse spherical image of K at ω (see (4.2.24) below for the general case) and that $S_m(K, \omega)$ can be obtained from this by means of the Steiner-type formula

$$S_{n-1}(K_\rho, \cdot) = \sum_{m=0}^{n-1} \rho^{n-1-m} \binom{n-1}{m} S_m(K, \cdot),$$

which is a special case of (4.2.16). $S_m(K, \cdot)$ is also called the *area measure of order* m.

We return to general convex bodies and establish the intuitive meaning of the measures C_m, S_m in the extreme cases $m = 0$ and $n - 1$, whereby we shall show that they are, in fact, suitable Hausdorff measures.

Theorem 4.2.5. *Let* $K \in \mathcal{K}^n$, $\beta \in \mathcal{B}(\mathbb{E}^n)$ *and* $\omega \in \mathcal{B}(S^{n-1})$. *Then*

$$C_0(K, \beta) = \mathcal{H}^{n-1}(\sigma(K, \beta)), \tag{4.2.21}$$

$$S_0(K, \omega) = \mathcal{H}^{n-1}(\omega). \tag{4.2.22}$$

If K is n-dimensional, then

$$C_{n-1}(K, \beta) = \mathcal{H}^{n-1}(\beta \cap \operatorname{bd} K), \tag{4.2.23}$$

$$S_{n-1}(K, \omega) = \mathcal{H}^{n-1}(\tau(K, \omega)). \tag{4.2.24}$$

Proof. First we remark that the formulae are valid if K is a polytope, as follows from (4.2.17) and (4.2.18).

For $K \in \mathcal{K}^n$ and $\beta \in \mathcal{B}(\mathbb{E}^{n-1})$ we define

$$\kappa(K, \beta) := \mathcal{H}^{n-1}(\sigma(K, \beta)).$$

By Lemma 2.2.10, $\sigma(K, \beta)$ is a Lebesgue measurable subset of S^{n-1}. In the proof of that lemma it was shown that $\beta_1 \cap \beta_2 = \emptyset$ implies that

$$\mathcal{H}^{n-1}(\sigma(K, \beta_1) \cap \sigma(K, \beta_2)) = 0.$$

Since $\mathcal{H}^{n-1}$, restricted to the Lebesgue measurable subsets of S^{n-1}, is σ-additive, it follows that $\kappa(K, \cdot)$ is a measure on $\mathcal{B}(\mathbb{E}^n)$.

Let $(K_j)_{j\in\mathbb{N}}$ be a sequence in $\mathcal{K}^n$ converging to a convex body K. We take an open set $\beta \subset \mathbb{E}^n$ and assume that $u \in \sigma(K, \beta) \cap \operatorname{regn} K$. There is a unique point $x \in F(K, u)$; it belongs to β. For $j \in \mathbb{N}$ choose $x_j \in F(K_j, u)$. Any accumulation point y of the sequence $(x_j)_{j\in\mathbb{N}}$ belongs to K and lies in $H(K, u)$, hence $y = x$. Thus $x_j \to x$ for $j \to \infty$ and, therefore, $x_j \in \beta$ and $u \in \sigma(K_j, \beta)$ for almost all j. This shows that

$$\sigma(K, \beta) \cap \operatorname{regn} K \subset \liminf_{j\to\infty} \sigma(K_j, \beta),$$

and in view of Theorem 2.2.9 we deduce

$$\begin{aligned}\kappa(K, \beta) &= \mathcal{H}^{n-1}(\sigma(K, \beta) \cap \operatorname{regn} K)\\ &\leqq \mathcal{H}^{n-1}(\liminf_{j\to\infty} \sigma(K_j, \beta))\\ &\leqq \liminf_{j\to\infty} \mathcal{H}^{n-1}(\sigma(K_j, \beta))\\ &= \liminf_{j\to\infty} \kappa(K_j, \beta).\end{aligned}$$

Since this holds for all open sets $\beta \subset \mathbb{E}^n$ and since $\kappa(K_j, \mathbb{E}^n) = \mathcal{H}^{n-1}(S^{n-1}) = \kappa(K, \mathbb{E}^n)$, we have proved the weak convergence $\kappa(K_j, \cdot) \xrightarrow{w} \kappa(K, \cdot)$. Now the equality $\kappa(K, \cdot) = C_0(K, \cdot)$ follows immediately if K is approximated by a sequence of polytopes, since it is true for polytopes and both sides are weakly continuous. This proves (4.2.21).

Equality (4.2.22) holds since it is true for polytopes and $S_0(K, \cdot)$ depends weakly continuously on K.

For the proof of (4.2.23) we put

$$\eta(K, \beta) := \mathcal{H}^{n-1}(\beta \cap \operatorname{bd} K)$$

for $K \in \mathcal{K}_0^n$ and $\beta \in \mathcal{B}(\mathbb{E}^n)$. Then $\eta(K, \cdot)$ is a measure, and as above it suffices to prove that $K_j \to K$ (for $K_j, K \in \mathcal{K}_0^n$) implies $\eta(K_j, \cdot) \xrightarrow{w} \eta(K, \cdot)$.

We may suppose that $o \in \operatorname{int} K$. Define

$$\varphi: \mathbb{E}^n \setminus \{o\} \to S^{n-1} \quad \text{by} \quad \varphi(x) := \frac{x}{|x|}.$$

Let $\rho(K, \cdot)$ be the radial function of K (see Section 1.7) so that, for $y \in S^{n-1}$, the point $\rho(K, y)y$ is in the boundary of K. By $n(K, y)$ we denote an arbitrary outer unit normal vector of K at this point. Theorem 2.2.4 tells us that $n(K, y)$ is unique for $\mathcal{H}^{n-1}$-almost all $y \in S^{n-1}$. We can now write $\eta(K, \beta)$ as the integral

$$\eta(K, \beta) = \int_{\varphi(\beta \cap \operatorname{bd} K)} \frac{\rho(K, y)^{n-1}}{\langle y, n(K, y)\rangle} \, d\mathcal{H}^{n-1}(y). \tag{4.2.25}$$

This can be proved with the aid of Theorem 3.2.3 in Federer (1969). If now the sequence $(K_j)_{j\in\mathbb{N}}$ of convex bodies converges to K, we may assume that $o \in \operatorname{int} K_j$ for all j. We have $\rho(K_j, \cdot) \to \rho(K, \cdot)$, $n(K_j, \cdot) \to n(K, \cdot)$ almost everywhere on S^{n-1}, and the functions $\langle \cdot, n(K_j, \cdot)\rangle$ are bounded from below by a positive constant. Using Fatou's lemma and the bounded convergence theorem, it is now easy to see that $\eta(K_j, \cdot) \xrightarrow{w} \eta(K, \cdot)$ for $j \to \infty$. This completes the proof of (4.2.23).

For the proof of (4.2.24) we put

$$F(K, \omega) := \mathcal{H}^{n-1}(\tau(K, \omega))$$

for $K \in \mathcal{K}_0^n$ and $\omega \in \mathcal{B}(S^{n-1})$. Again we may assume that $o \in \operatorname{int} K$. From Lemma 2.2.11 we know that $\varphi(\tau(K, \omega))$ is Lebesgue measurable on S^{n-1}. If $\omega_1 \cap \omega_2 = \emptyset$, then

$$\mathcal{H}^{n-1}(\tau(K, \omega_1) \cap \tau(K, \omega_2)) = 0,$$

as a consequence of Theorem 2.2.4. We deduce that $F(K, \cdot)$ is a measure on $\mathcal{B}(S^{n-1})$, and from (4.2.25) we have

$$F(K, \omega) = \int_{\varphi(\tau(K,\omega))} \frac{\rho(K, y)^{n-1}}{\langle y, n(K, y)\rangle} \, \mathrm{d}\mathcal{H}^{n-1}(y).$$

As above, this can be used to show that $K_j \to K$ (with $K \in \mathcal{K}_0^n$) implies $F(K_j, \cdot) \xrightarrow{w} F(K, \cdot)$ for $j \to \infty$. Using this, (4.2.24) is extended from polytopes to general convex bodies. ■

If the body $K \in \mathcal{K}^n$ is of dimension less than n, it is easy to see (e.g., from (4.2.8), (4.2.9)) how the equations (4.2.23), (4.2.24) have to be modified: If $\dim K = n - 1$, then

$$C_{n-1}(K, \beta) = 2\mathcal{H}^{n-1}(\beta \cap K),$$

$$S_{n-1}(K, \omega) = 2\mathcal{H}^{n-1}(K), \quad \mathcal{H}^{n-1}(K), \text{ or } 0,$$

according to whether both, one, or none of the unit normal vectors of the affine hull of K belongs to ω. If $\dim K < n - 1$, then $C_{n-1}(K, \beta) = 0 = S_{n-1}(K, \omega)$ (so that (4.2.23), (4.2.24) are true again).

Next we consider the total curvature measures

$$\Theta_m(K, \Sigma) = C_m(K, \mathbb{E}^n) = S_m(K, S^{n-1}),$$

for which special notation is common. These functionals were in use long before the curvature measures. The traditional notation is

$$W_i(K) := \frac{1}{n} \Theta_{n-i}(K, \Sigma)$$

for $i = 1, \ldots, n$, and $W_0(K) := V_n(K)$, the volume of K. The functions $W_0, \ldots, W_n$ are called the *Minkowski functionals* or *quermassintegrals*. The latter term comes from the German 'Quermaß', which can be the measure of either a cross-section or a projection. The reason for this terminology will be clear when certain integral-geometric interpretations of the functions W_i have been obtained in Section 4.5.

We denote the total values of the renormalized curvature measures by

$$V_m(K) := \Phi_m(K, \mathbb{E}^n) = \Psi_m(K, S^{n-1}),$$

so that

$$\kappa_{n-m} V_m(K) = \binom{n}{m} W_{n-m}(K) \tag{4.2.26}$$

for $m = 0, \ldots, n$. McMullen (1975a) proposed that $V_m(K)$ should be called the *intrinsic m-volume* of K since, if K is m-dimensional, then $V_m(K)$ is the ordinary m-dimensional volume of K.

In order to illustrate the meaning of quermassintegrals, we recall some of the earlier formulae for this special case. The representation

$$V_n(K + \rho B^n) = \sum_{i=1}^{n} \rho^i \binom{n}{i} W_i(K) = \sum_{i=1}^{n} \rho^{n-i} \kappa_{n-i} V_i(K), \tag{4.2.27}$$

a very special case of (4.2.4), is the classical *Steiner formula*, by which quermassintegrals and intrinsic volumes can be defined. If K is of class C^2_+, then

$$W_i(K) = \frac{1}{n} \int_{\mathrm{bd}\, K} H_{i-1} \,\mathrm{d}\mathcal{H}^{n-1} \tag{4.2.28}$$

$$W_i(K) = \frac{1}{n} \int_{S^{n-1}} s_{i-1} \,\mathrm{d}\mathcal{H}^{n-1} \tag{4.2.29}$$

for $i = 1, \ldots, n$. If P is a polytope, then

$$V_m(P) = \sum_{F \in \mathcal{F}_m(P)} \gamma(F, P) \mathcal{H}^m(F) \tag{4.2.30}$$

for $m = 0, \ldots, n$. For general convex bodies, we list the following special cases:

$W_0 = V_n,$	the volume;
$nW_1 = 2V_{n-1},$	the surface area;
$\dfrac{2}{\kappa_n} W_{n-1} = \dfrac{2\kappa_{n-1}}{\omega_n} V_1,$	the mean width (formula (5.3.12));
$\dfrac{1}{\kappa_n} W_n = V_0 = 1,$	the Euler characteristic.

Because of (4.2.28), $nW_2(K)$ is sometimes called the *integral of the mean curvature* (even if K is not of class C^2_+).

As a function on $\mathcal{K}^n$, the quermassintegral W_i is rigid motion invariant, additive and continuous. This follows from the corresponding properties of the curvature measures; the continuity is clear from (4.2.27) and from the continuity of both the volume functional and Minkowski addition. The following celebrated result shows that these properties are sufficient to characterize the linear combinations of quermassintegrals.

Theorem 4.2.6 (Hadwiger). *If* $\varphi: \mathcal{K}^n \to \mathbb{R}$ *is a function that is invariant under rigid motions, additive and continuous, then*

$$\varphi(K) = \sum_{i=0}^{n} c_i W_i(K)$$

for $K \in \mathcal{K}^n$, *where* $c_0, \ldots, c_n$ *are real constants.*

There is a counterpart to Theorem 4.2.6 with continuity replaced by monotoneity. The quermassintegrals are *monotone* (or *increasing*), that is, $K \subset L$ implies $W_i(K) \leqq W_i(L)$. This will become clear later, either by means of the integral-geometric interpretation (Section 4.5) of the W_i or from their interpretation as mixed volumes (Chapter 5).

Theorem 4.2.7 (Hadwiger). *If* $\varphi: \mathcal{K}^n \to \mathbb{R}$ *is a function that is invariant under rigid motions, additive and monotone, then*

$$\varphi(K) = \sum_{i=0}^{n} c_i W_i(K)$$

for $K \in \mathcal{K}^n$, *where* $c_0, \ldots, c_n$ *are non-negative real constants.*

Both theorems characterize the series of quermassintegrals by simple and natural properties and thus illustrate the importance and basic character of these functionals.

The only known proof of these two theorems is the original one given by Hadwiger. It requires some results from the dissection theory of polytopes and would not fit well into our present context. We refer the reader to the book of Hadwiger [18].

A result of McMullen (1977; see also [40], Theorem (11.5)) permits us to deduce Theorem 4.2.7 from Theorem 4.2.6.

Notes for Section 4.2

1. *The Steiner formula.* The Steiner formula goes back to Steiner (1840b), who derived it in $\mathbb{E}^2$ and $\mathbb{E}^3$, for polytopes and surfaces of class C^2_+. For general convex bodies in $\mathbb{E}^n$, this formula is just a special case of more general expansion formulae for mixed volumes (see Chapter 5). For the volume of the 'tube' around a smooth submanifold of Euclidean or spherical space, Weyl (1939) derived a similar expansion, now called 'Weyl's tube formula'. The essential point here is that the coefficients can be expressed in terms of the Riemann curvature tensor of the submanifold and thus depend only on its intrinsic metric. Counterparts to the original Steiner formula in spaces of constant curvature were treated by Herglotz (1943), Vidal Abascal (1947), Allendoerfer (1948) and Santaló (1950). The extension of Weyl's tube formula to general Riemannian spaces was studied by Gray & Vanhecke (1981); see also Gray (1990). Hadwiger (1945, 1946a, b, c) used integral-geometric methods to obtain Steiner-type formulae for certain non-convex domains also. Other analogues, partly in the form of inequalities, are due to Hadwiger (1946d), Ohmann (1955b), Sz.-Nagy (1959), Makai (1959) and P. Meyer (1977).

2. *Steiner formulae for Minkowski subtraction*. A special case of (4.2.16) is the system of Steiner formulae for the quermassintegrals, that is,
$$W_p(K + \rho B^n) = \sum_{i=0}^{n-p} \rho^i \binom{n-p}{i} W_{p+i}(K)$$
for $K \in \mathcal{K}^n$, $\rho \geqq 0$ and $p = 0, \ldots, n$. If B^n is a summand of K, then it is not difficult to see that
$$W_p(K \sim \rho B^n) = \sum_{i=0}^{n-p} (-\rho)^i \binom{n-p}{i} W_{p+i}(K) \tag{4.2.31}$$
for $0 \leqq \rho \leqq 1$ and for $p = 0, \ldots, n$. This was pointed out (in a more general form, with B^n replaced by an arbitrary convex body and with the quermassintegrals replaced by corresponding mixed volumes) by Matheron (1978b). He showed, conversely, that the validity of (4.2.31) for $\rho \in (0, 1)$ and $p = 0, \ldots, n$ implies that B^n is a summand of K. He conjectured that this is already true if (4.2.31) holds for $\rho \in (0, 1)$ and $p = 0$ (the case of the volume); he proved this for $n = 2$.
3. *Sets of positive reach*. A natural common generalization of convex sets and of smooth submanifolds, for which local parallel sets can be defined in such a way that their measure has a polynomial expansion, yielding a Steiner formula, are the sets of positive reach. A compact set $A \subset \mathbb{E}^n$ is said to be *of positive reach* if there exists a number $\rho > 0$ such that, for each point $x \in \mathbb{E}^n$ whose distance from A is less than ρ, there is a unique point in A that is nearest to x. For sets A of positive reach, Federer (1959) introduced the curvature measures $\Phi_m(A, \cdot)$. He made an extensive study and proved many of the relevant results which, in this book, are considered only for convex bodies.

   For extensions of the curvature measures to other classes of sets, see Section 4.4 and the notes therein.
4. The simpler introduction of the curvature measures for the case of convex bodies and the derivation of their properties, as presented here, is taken from Schneider (1978a). For the area measures (which are related to the theory of mixed volumes, see Chapter 5), one should see the beautiful paper by Fenchel & Jessen (1938). The generalized measures on sets of support elements appear in Schneider (1979a).
5. *Differentiation of curvature measures*. Let $K \in \mathcal{K}^n$ be a convex body of class C^2_+. If $x \in \operatorname{bd} K$ and $(\beta_i)_{i\in\mathbb{N}}$ is a sequence of Borel subsets of $\operatorname{bd} K$ with positive measure and shrinking to x, then it follows from (4.2.19) that
$$\lim_{i\to\infty} \frac{C_m(K, \beta_i)}{\mathcal{H}^{n-1}(\beta_i)} = H_{n-1-m}(x).$$
   One may ask whether this relation holds true for a general convex body K if x is a normal point (see Section 2.5), so that $H_{n-1-m}(x)$ is defined. Under certain (not too restrictive) assumptions on the sequence $(\beta_i)_{i\in\mathbb{N}}$ this can, in fact, be proved; see Aleksandrov (1939c) for $m = 0$ and Schneider (1979a) for $m = 0, \ldots, n - 1$.
6. *Integral and current representation of Federer's curvature measures*. In a paper under this title, Zähle (1986a) defined (for sets of positive reach) principal curvatures as functions on the unit normal bundle and expressed the curvature measures as $\mathcal{H}^{n-1}$-integrals of elementary symmetric functions of such curvatures over the unit normal bundle. She also interpreted these integrals as values of associated rectifiable currents on specially chosen differential forms. This work was continued in Zähle (1987a). (See also the notes for Section 4.4.)
7. Arguments leading to special cases of Theorem 4.2.5 can be found at several

places in the work of Aleksandrov (1937a, §1, 1939b, 1939c, §6, [1], Chapter V, §2).

8. *Valuations on* $\mathcal{K}^n$. The important theorems 4.2.6 and 4.2.7 of Hadwiger characterize the linear combinations of quermassintegrals as the only valuations on $\mathcal{K}^n$ that are rigid motion invariant and either continuous or monotone. These theorems were one of the starting points for a series of investigations on general valuations on the space of convex bodies or subspaces thereof, aiming at establishing general properties as well as characterization and classification results. We refer the reader to the survey given by McMullen & Schneider [40], in particular to Chapter IV. More recent contributions are due to Spiegel (1982), Goodey & Weil (1984), Betke & Goodey (1984) (this paper contains a mistake, see Goodey & Weil (1991)), Betke & Kneser (1985) and McMullen (1990).
9. *Axiomatic characterizations of curvature measures and area measures*. Linear combinations of curvature measures or of area measures can be characterized by some of their properties in a manner analogous to Hadwiger's theorem 4.2.6.

Theorem. Let φ be a map from $\mathcal{K}^n$ into the set of finite Borel measures on $\mathbb{E}^n$ satisfying the following conditions (where $\varphi(K)(\beta)$ is written as $\varphi(K, \beta)$).

(a) φ is rigid motion invariant, that is, $\varphi(gK, g\beta) = \varphi(K, \beta)$ for $K \in \mathcal{K}^n$, $\beta \in \mathcal{B}(\mathbb{E}^n)$ and every rigid motion g of $\mathbb{E}^n$.
(b) φ is additive.
(c) φ is weakly continuous, that is, $K_j \to K$ in $\mathcal{K}^n$ implies $\varphi(K_j, \cdot) \xrightarrow{w} \varphi(K, \cdot)$ for $j \to \infty$.
(d) φ is defined locally, which means: If $\beta \subset \mathbb{E}^n$ is open and $K \cap \beta = L \cap \beta$, then $\varphi(K, \beta') = \varphi(L, \beta')$ for every Borel set $\beta' \subset \beta$.

Under these assumptions, there are non-negative real constants $c_0, \ldots, c_n$ such that

$$\varphi(K, \beta) = \sum_{i=0}^{n} c_i \Phi_i(K, \beta)$$

for $K \in \mathcal{K}^n$ and $\beta \in \mathcal{B}(\mathbb{E}^n)$.

This theorem was proved by Schneider (1978a). It gives a partial answer to a question of Federer (1959, Remark 5.17), who had asked for a characterization, similar to Hadwiger's characterization theorem for quermassintegrals, of his curvature measures on sets of positive reach. A similar result for area measures was obtained by Schneider (1975e):

Theorem. Let ψ be a map from $\mathcal{K}^n$ into the set of finite signed Borel measures on S^{n-1} satisfying the following conditions (where $\psi(K)(\omega)$ is written as $\psi(K, \omega)$).

(a) ψ is rigid motion invariant, that is, $\psi(gK, g_0\omega) = \psi(K, \omega)$ for $K \in \mathcal{K}^n$, $\omega \in \mathcal{B}(S^{n-1})$ and every rigid motion g of $\mathbb{E}^n$, where g_0 denotes the rotation corresponding to g.
(b) ψ is additive.
(c) ψ is weakly continuous.
(d) ψ is defined locally, which means: If $\omega \in \mathcal{B}(S^{n-1})$ and $\tau(K, \omega) = \tau(L, \omega)$, then $\psi(K, \omega) = \psi(L, \omega)$.

Under these assumptions, there are real constants $c_0, \ldots, c_{n-1}$ such that

$$\psi(K, \omega) = \sum_{i=0}^{n-1} c_i \psi_i(K, \omega)$$

for $K \in \mathcal{K}^n$ and $\omega \in \mathcal{B}(S^{n-1})$.

A far-reaching generalization of these theorems was proved by Zähle (1990).

4.3. The area measure of order one

The area measure $S_m(K, \cdot)$ of order m, for a convex body K, is a special case of the mixed area measure to be treated in Section 5.1 and is, therefore, closely related to the theory of mixed volumes. The area measure of order one, however, exhibits some special features; this suggests a separate treatment that is independent of mixed volumes. The essential special property of S_1 is its Minkowski linearity, that is, the equality

$$S_1(\lambda K + \mu L, \cdot) = \lambda S_1(K, \cdot) + \mu S_1(L, \cdot) \tag{4.3.1}$$

for $K, L \in \mathcal{K}^n$ and $\lambda, \mu \geqq 0$. For convex bodies of class C_+^2 this follows from (4.2.20) and the linearity of s_1, which is evident from (2.5.19); the general case is then obtained by approximation, using the weak continuity of the area measures. In view of (4.3.1), the area measure S_1 is closer to Chapter 3 than to Chapters 5 and 6 and will, therefore, be considered briefly in the present section.

First we show that $S_1(K, \cdot)$ determines K uniquely up to a translation. It appears that the shortest proof of this is the one using spherical harmonics.

Theorem 4.3.1. *If $K, L \in \mathcal{K}^n$ are convex bodies with $S_1(K, \cdot) = S_1(L, \cdot)$, then K and L are translates of each other.*

Proof. Let Y_m be a spherical harmonic of degree m. First let K be of class C_+^2. Then it follows from (4.2.20) and (2.5.23) that

$$\int_{S^{n-1}} Y_m(u) \mathrm{d}S_1(K, u) = \int_{S^{n-1}} Y_m \left(h_K + \frac{1}{n-1} \Delta_S h_K \right) \mathrm{d}\sigma,$$

where σ denotes spherical Lebesgue measure on S^{n-1}. Using Green's formula $(Y_m, \Delta_S h_K) = (\Delta_S Y_m, h_K)$ and $\Delta_S Y_m = -m(m+n-2)Y_m$ (Appendix equation (A.2)), we obtain

$$\int_{S^{n-1}} Y_m(u)\, \mathrm{d}S_1(K, u) = \lambda_{n,m} \int_{S^{n-1}} h_K Y_m \, \mathrm{d}\sigma \tag{4.3.2}$$

with a real number $\lambda_{n,m}$ satisfying $\lambda_{n,m} \neq 0$ for $m \neq 1$. By approximation, (4.3.2) extends to arbitrary convex bodies. The assumption of the theorem now yields

$$\lambda_{n,m} \int_{S^{n-1}} (h_K - h_L) Y_m \, \mathrm{d}\sigma = 0.$$

Since $m \in \{0, 1, \ldots\}$ was arbitrary, it follows from the completeness of the system of spherical harmonics that $\bar{h}_K - \bar{h}_L$ is a spherical harmonic of degree 1, hence $h_K - h_L = \langle \cdot, t \rangle$ with $t \in \mathbb{E}^n$. This is equivalent to $K = L + t$. ■

In the plane, the uniqueness result holds even locally. This is not surprising, in view of equation (2.5.22) which is valid in the C^2_+ case. To formulate the general case, we understand by an *open convex arc* a nonempty, relatively open and connected proper subset C of the boundary of a convex body $K \in \mathcal{K}^2_0$; we also assume that C is not contained in a line. The spherical image $\sigma(K, C)$ of K at C depends only on C; it is called the spherical image of C and denoted by σ_C. The definition of the measure S_1 can be extended to open convex arcs by putting

$$S_1(C, \omega) := \mathcal{H}^1(\tau(K, \omega) \cap C)$$

for Borel sets $\omega \subset \sigma_C$, in agreement with (4.2.24). We call $S_1(C, \cdot)$ the *length measure* of C.

Theorem 4.3.2. *Assume that $C, C' \subset \mathbb{E}^2$ are open convex arcs such that $\sigma_C = \sigma_{C'}$, this set is contained in a closed semicircle of S^1, and $S_1(C, \cdot) = S_1(C', \cdot)$. Then C and C' are translates of each other.*

Proof. The open convex arc C has two well-defined endpoints (not belonging to it), say x and y. Let $\bar{C}$ be the image of C under reflection in the point $(x + y)/2$. Since σ_C is contained in a semicircle of S^1, it is easy to see that $C \cup \bar{C} \cup \{x, y\}$ is the boundary of a convex body, say K. Let K' be constructed from C' in the same way. Then $S_1(K, \omega) = S_1(K', \omega)$ if ω is a Borel subset of σ_C or of its reflection $\bar{\sigma}_C$ in the origin, and $S_1(K, \omega) = 0 = S_1(K', \omega)$ if ω is a Borel subset of $S^1 \backslash (\sigma_C \cup \bar{\sigma}_C)$. Thus $S_1(K, \cdot) = S_1(K', \cdot)$, from which we deduce that K' is a translate of K. It is then clear that C' must be a translate of C. Here the following has to be observed. If, say, x is the endpoint of a segment S contained in C, with outer unit normal vector u, then $u \in \sigma_C$, u is an endpoint of σ_C and $S_1(C, \{u\})$ is the length of S. It follows that C' contains a parallel segment of the same length. ∎

Clearly, Theorem 4.3.2 can also be applied to open convex arcs having spherical images not contained in a semicircle, by covering a given arc with smaller arcs and applying the theorem to each of these.

In the plane, one also has an existence theorem for convex bodies with prescribed measure S_1:

Theorem 4.3.3. *Let φ be a measure on S^1 satisfying*

$$\int_{S^1} u \, d\varphi(u) = o.$$

Then there exists a convex body $K \in \mathcal{K}^2$ for which $S_1(K, \cdot) = \varphi$.

This is just the two-dimensional case of Minkowski's existence theorem (see Section 7.1). Therefore, we do not give a proof here but simply mention that the polygonal case (the two-dimensional case of Theorem 7.1.1) is almost trivial (compare the construction after (3.2.3)).

The n-dimensional existence problem for S_1 consists of finding necessary and sufficient conditions for a Borel measure φ on S^{n-1} in order that there exist a convex body $K \in \mathcal{K}^n$ for which $S_1(K, \cdot) = \varphi$. In the smooth version, one asks which conditions a real function f on S^{n-1} must satisfy in order that there exist a (sufficiently smooth) convex body K such that f is equal to $s_1(K, \cdot)$, the mean radius of curvature of K as a function of the outer unit normal vector. The three-dimensional case was treated by Christoffel (1865); the general case is now called *Christoffel's problem*. In the smooth case, this problem comes down to solving the elliptic linear differential equation

$$\frac{1}{n-1} \Delta_S h + h = f \tag{4.3.3}$$

on S^{n-1} and expressing the condition that the solution h has to be the restriction of a support function in terms of conditions on the given function f (the latter was not done by Christoffel). The first complete solution was given by Firey (1967b). Instead of using (4.3.3), he based his treatment on the differential equation

$$\Delta \xi(u) = \operatorname{grad} f(u), \tag{4.3.4}$$

where $\xi(u)$ is defined as in Section 2.5 and f is continuously differentiable and is extended to $\mathbb{E}^n \setminus \{o\}$ as a positively homogeneous function of degree -1. Solutions for the general Christoffel problem were given, independently, by Firey (1968) and Berg (1969b). We formulate Firey's result without proof. He introduces the fundamental singularity g_1, defined for $u, u' \in \mathbb{E}^n$, $u \neq u'$, by

$$g_1(u', u) := f_n(s(u', u)),$$

where

$$f_3(s) := \frac{1}{2\pi} \log s,$$

$$f_n(s) := -\frac{1}{(n-3)\omega_{n-1}} s^{3-n}, \quad \text{if } n > 3,$$

$$s(u', u) := \arccos \frac{\langle u', u \rangle}{|u'||u|},$$

and the Green's function G associated with the Poisson equation (4.3.4) by

$$G(u', u) := \gamma(s(u', u)),$$

where

$$\gamma(s) := -\frac{1}{\omega_n}\int_{\pi/2}^{s}\operatorname{cosec}^{n-2} t\left(\int_{\pi}^{t}\sin^{n-2}\tau\,\mathrm{d}\tau\right)\mathrm{d}t.$$

Theorem 4.3.4 (Firey). *A Borel measure φ on S^{n-1} is the first area measure $S_1(K,\cdot)$ of a convex body $K \in \mathcal{K}^n$ if and only if it satisfies*

$$\int_{S^{n-1}} u\,\mathrm{d}\varphi(u) = o,$$

$$\left|\int_{S^{n-1}} g_1(u', u)\,\mathrm{d}\varphi(u)\right| < \infty$$

for all $u' \in S^{n-1}$, and

$$\int_{S^{n-1}} \Lambda(u', v', u)\,\mathrm{d}\varphi(u) \geqq 0$$

for all $u', v' \in \mathbb{E}^n \setminus \{o\}$, where

$$\Lambda(u', v', u) := \Gamma(u', u) + \Gamma(v', u) - \Gamma(u' + v', u),$$

$$\Gamma(u', u) := (n-2)\langle u', u\rangle G(u', u) - \langle u', \operatorname{grad} G(u', u)\rangle.$$

Berg (1969b) developed a theory of potentials and subharmonic functions on the sphere S^{n-1} and within this theory finally formulated a necessary and sufficient condition for a measure φ to be the first area measure of a convex body. In the course of his investigation, he constructed real functions g_n, for $n = 2, 3, \ldots$, on $[-1, 1]$ such that

$$h(K, u) = \int_{S^{n-1}} g_n(\langle u, v\rangle)\mathrm{d}S_1(K, v) + \langle s(K), u\rangle \tag{4.3.5}$$

for $K \in \mathcal{K}^n$ and $u \in S^{n-1}$. From this result and the fact that

$$\int_{S^{n-1}} |g_n(\langle u, v\rangle)|\mathrm{d}\mathcal{H}^{n-1}(v) < \infty, \tag{4.3.6}$$

also shown by Berg, a stability version of the uniqueness theorem 4.3.1 can be derived:

Theorem 4.3.5. *If $K, K' \in \mathcal{K}^n$ are convex bodies with coinciding Steiner points and satisfying*

$$|S_1(K, \omega) - S_1(K', \omega)| \leqq \varepsilon\mathcal{H}^{n-1}(\omega) \tag{4.3.7}$$

for all $\omega \in \mathcal{B}(S^{n-1})$ and a given number $\varepsilon > 0$, then

$$\delta(K, K') \leqq c_n\varepsilon$$

with some constant c_n depending only on the dimension n.

Proof. From (4.3.5) and the assumption $s(K) = s(K')$ we obtain

$$|h(K, u) - h(K', u)| \leqq \int_{S^{n-1}} |g_n(\langle u, v \rangle)| \mathrm{d}\nu(v),$$

where ν is the variation of the signed measure $S_1(K, \cdot) - S_1(K', \cdot)$. The assumption (4.3.7) yields $\nu(\omega) \leqq \varepsilon \mathcal{H}^{n-1}(\omega)$. The existence of the constant c_n now follows from (4.3.6). ■

Notes for Section 4.3

1. *Christoffel's problem.* It is clear from the Introduction in Christoffel (1865) that the surfaces that he studied are assumed to be convex ('allenthalben gewölbt'), but in the existence part of his investigation no such property is discussed (an erroneous statement to the contrary was made in Bonnesen & Fenchel [8], p. 123). Favard (1933b) and Süss (1932b, 1933) claimed to prove that the condition $\int u \mathrm{d}\varphi(u) = o$ is sufficient for the existence of a convex body K with $S_1(K, \cdot) = \varphi$ (under smoothness assumptions). However, Aleksandrov (1937c, 1938a) gave examples of positive measures on $\mathcal{B}(S^{n-1})$, some even with analytic densities, which satisfy that necessary condition and are not the first area measure $S_1(K, \cdot)$ of any convex body K. A sufficient, but not necessary, condition was found by Pogorelov (1953).

   The proof of the uniqueness result, Theorem 4.3.1, by means of spherical harmonics goes back essentially to Hurwitz (1902). His proof for the three-dimensional case is reproduced in Blaschke (1924), §95, and Pogorelov [27], and was extended to higher dimensions by Kubota (1925b).

   A brief review of the various treatments of Christoffel's problem was given by Firey (1981). An application of Firey's (1967b) existence result to the study of surfaces of constant width appears in Fillmore (1969).

   In the centrally symmetric case, Goodey & Weil (1990+) found a connection between Christoffel's problem and Radon transforms.
2. *Christoffel's problem for polytopes.* The necessary and sufficient conditions for first-order area measures found by Firey and Berg are not easy to handle in concrete cases. For instance, it seems difficult to derive from these criteria a description of the first-order area measures of polytopes. However, Christoffel's problem for polytopes can be given an independent treatment by direct elementary methods. This has been done by Schneider (1977a).
3. *The length measure.* In the plane, the map $K \mapsto S_1(K, \cdot)$ establishes an isomorphism between the convex cone of convex bodies with Steiner point o and the cone of measures φ on $\mathcal{B}(S^{n-1})$ with $\int u \mathrm{d}\varphi(u) = o$. This fact is sometimes useful for the treatment of decompositions of plane convex bodies. Schneider (1974e) used the length measure in the investigation of the asymmetry classes of convex bodies. Kallay (1975) characterized the extreme convex sets K in the set of bodies in $\mathcal{K}^2$ with a given width function in terms of a property of the Radon–Nikodym derivative of $S_1(K, \cdot)$.

   A detailed study of the length measure was also made by Letac (1983).
4. *Stability results.* Theorem 4.3.5, for the special case of three-dimensional convex bodies with twice continuously differentiable support functions, is due to Pogorelov [27], p. 502. The general case is in Schneider [43], Theorem (9.8).

Sen'kin (1966a, b) proved the estimate

$$|w(K, u) - w(K', u)| \leqq \frac{1}{\pi} \sup_{\omega \subset S^2} |S_1(K, \omega) - S_1(K', \omega)|$$

for the width $w(\cdot, u)$ in any direction $u \in S^2$, for convex bodies $K, K' \in \mathcal{K}^3$. He used an elementary argument for strongly isomorphic polytopes and then Aleksandrov's approximation theorem 2.4.14.

4.4. Additive extension

The generalized curvature measure $\Theta_m(K, \cdot)$ is an additive function of K, by Theorem 4.2.1. In Section 3.4 we introduced the convex ring $U(\mathcal{K}^n)$, consisting of all finite unions of convex bodies in $\mathbb{E}^n$, and collected some results about the additive extension of valuations from $\mathcal{K}^n$ to $U(\mathcal{K}^n)$. In the present section we shall treat the additive extension of the generalized curvature measures Θ_m to the convex ring. While the existence of such an extension could be deduced from general results on valuations (see Section 3.4, Note 7), we prefer to give an explicit construction that provides additional insight and at the same time extends the Steiner formula to local 'parallel sets with multiplicity'. The availability of curvature measures for sets more general than convex bodies is useful for applications, and the additivity of the extension permits the almost automatic generalization of some integral-geometric formulae to be proved in the next section.

First let $K \in \mathcal{K}^n$ be a convex body, let $\rho > 0$ and $\eta \in \mathcal{B}(\Sigma)$. In Section 4.1 we defined the local parallel set $M_\rho(K, \eta)$, and equation (4.2.4) states that

$$\mathcal{H}^n(M_\rho(K, \eta)) = \frac{1}{n} \sum_{m=0}^{n-1} \rho^{n-m} \binom{n}{m} \Theta_m(K, \eta).$$

In order to obtain a Steiner formula such as the above for non-convex sets $K \in U(\mathcal{K}^n)$ also, we consider, instead of the parallel set $M_\rho(K, \eta)$, its characteristic function $c_\rho(K, \eta, \cdot)$ and extend this additively to $U(\mathcal{K}^n)$. The value $c_\rho(K, \eta, \cdot)$ is then interpreted as counting the multiplicity with which the point x belongs to the parallel set of K with respect to η.

Recall that the Euler characteristic χ is the (unique) valuation on the convex ring $U(\mathcal{K}^n)$ satisfying $\chi(K) = 1$ for $K \in \mathcal{K}^n$; see Theorem 3.4.12. Now, for $K \in U(\mathcal{K}^n)$ and points $q, x \in \mathbb{E}^n$ we define the *index of K at q with respect to x* by

$$j(K, q, x) := \begin{cases} 1 - \lim_{\delta \downarrow 0} \lim_{\varepsilon \downarrow 0} \chi(K \cap B(x, |x - q| - \varepsilon) \cap B(q, \delta)) & \text{if } q \in K, \\ 0 & \text{if } q \notin K. \end{cases}$$

If K is convex, then clearly

$$j(K, q, x) = \begin{cases} 1 & \text{if } q = p(K, x), \\ 0 & \text{otherwise.} \end{cases} \tag{4.4.1}$$

To prove the existence of the limits, choose a representation $K = \bigcup_{i=1}^{r} K_i$ with $K_i \in \mathcal{K}^n$. Without loss of generality, we may assume that $q \in K_i$ for $i = 1, \ldots, m$ and $q \notin K_i$ for $i > m$, where $1 \leqq m \leqq r$. We can choose $\delta_0 > 0$ such that $K_i \cap B(q, \delta) = \emptyset$ for $i \in \{m+1, \ldots, r\}$ and $0 < \delta < \delta_0$. If $\delta < \delta_0$ is fixed, then for all sufficiently small $\varepsilon > 0$ the inequality

$$K_v \cap B(x, |x - q| - \varepsilon) \cap B(q, \delta) \neq \emptyset$$

holds for all $v \in S(m)$ with $j(K_v, q, x) = 0$ (where we use the notation introduced in Section 3.4). In fact, for $v \in S(m)$ we have $q \in K_v$, and if $j(K_v, q, x) = 0$, then (4.4.1) implies that K_v and hence $K_v \cap B(q, \delta)$ contains a point whose distance from x is smaller than that of q. For sufficiently small $\varepsilon > 0$ we thus have

$$j(K_v, q, x) = 1 - \chi(K_v \cap B(x, |x - q| - \varepsilon) \cap B(q, \delta)).$$

Using the additivity of χ, we deduce that

$$\begin{aligned} 1 - \chi(K \cap B(x, |x - q| - \varepsilon) \cap B(q, \delta)) \\ &= \sum_{v \in S(m)} (-1)^{|v|-1} [1 - \chi(K_v \cap B(x, |x - q| - \varepsilon) \cap B(q, \delta))] \\ &= \sum_{v \in S(m)} (-1)^{|v|-1} j(K_v, q, x) \\ &= \sum_{v \in S(r)} (-1)^{|v|-1} j(K_v, q, x). \end{aligned}$$

The right-hand side does not depend on ε or δ, which shows the existence of the limits.

The index function j thus defined is additive in its first argument, that is, for fixed $q, x \in \mathbb{E}^n$ and for $K, L \in U(\mathcal{K}^n)$ we have

$$j(K \cup L, q, x) + j(K \cap L, q, x) = j(K, q, x) + j(L, q, x).$$

This follows from the definition; in fact, for $q \notin K \cup L$ both sides are zero and for $q \in K \cap L$ it follows from the additivity of χ. If $q \in K \backslash L$ (and similarly for $q \in L \backslash K$) one has to observe that $(K \cup L) \cap B(q, \delta) = K \cap B(q, \delta)$ for all sufficiently small $\delta > 0$.

Now for $K \in U(\mathcal{K}^n)$, $\eta \in \mathcal{B}(\Sigma)$, $\rho > 0$ and $x \in \mathbb{E}^n$ we define

$$c_\rho(K, \eta, x) := \sum_{\substack{q \in \mathbb{E}^n \backslash \{x\} \\ (q, \overline{x - q}) \in \eta}} j(K \cap B(x, \rho), q, x)$$

where $\overline{x-q} := (x-q)/|x-q|$. Actually, the sum is finite, since $j(K, q, x) \neq 0$ for $K = \bigcup_{i=1}^{r} K_i$ with $K_i \in \mathcal{K}^n$ implies, by the additivity of $j(\cdot, q, x)$, that there is some $v \in S(r)$ for which $j(K_v, q, x) \neq 0$ and hence $q = p(K_v, x)$.

If K is convex, then (4.4.1) obviously implies that

$$c_\rho(K, \eta, x) = \begin{cases} 1 & \text{if } 0 < d(K, x) \leqq \rho \text{ and } (p(K, x), u(K, x)) \in \eta, \\ 0 & \text{otherwise.} \end{cases}$$

Thus $c_\rho(K, \eta, \cdot)$ is the characteristic function of the local parallel set $M_\rho(K, \eta)$. From the additivity of $j(\cdot, q, x)$ it follows that $c_\rho(\cdot, \eta, x)$ is additive on $U(\mathcal{K}^n)$. Hence, for $K = \bigcup_{i=1}^{r} K_i$ with $K_i \in \mathcal{K}^n$ we have

$$c_\rho(K, \eta, \cdot) = \sum_{v \in S(r)} (-1)^{|v|-1} c_\rho(K_v, \eta, \cdot).$$

Since the right-hand side is a finite sum of integrable functions, we can define

$$\mu_\rho(K, \eta) := \int_{\mathbb{E}^n} c_\rho(K, \eta, x) \, d\mathcal{H}^n(x)$$

for $K \in U(\mathcal{K}^n)$ and $\eta \in \mathcal{B}(\Sigma)$. The notation is consistent with that of Section 4.1, and $\mu_\rho(\cdot, \eta)$ is an additive function on $U(\mathcal{K}^n)$. Hence, for $K = \bigcup_{i=1}^{r} K_i$ with $K_i \in \mathcal{K}^n$ we have

$$\begin{aligned} \mu_\rho(K, \eta) &= \sum_{v \in S(r)} (-1)^{|v|-1} \mu_\rho(K_v, \eta) \\ &= \sum_{v \in S(r)} (-1)^{|v|-1} \frac{1}{n} \sum_{m=1}^{n-1} \rho^{n-m} \binom{n}{m} \Theta_m(K_v, \eta) \\ &= \frac{1}{n} \sum_{m=0}^{n-1} \rho^{n-m} \binom{n}{m} \sum_{v \in S(r)} (-1)^{|v|-1} \Theta_m(K_v, \eta). \end{aligned}$$

Since the left-hand side depends only on the point set K and not on its chosen representation as a finite union of convex bodies, the same is true for the coefficients of the polynomial in ρ on the right-hand side. Hence, we are now in a position to define

$$\Theta_m(K, \eta) := \sum_{v \in S(r)} (-1)^{|v|-1} \Theta_m(K_v, \eta)$$

for $m = 0, \ldots, n-1$. Thus $\Theta_m(K, \cdot)$ is a finite signed measure on $\mathcal{B}(\Sigma)$. From this representation (or from the additivity of $\mu_\rho(\cdot, \eta)$) we deduce that $\Theta_m(\cdot, \eta)$ is additive on $U(\mathcal{K}^n)$. In this way, we obtain the (unique) additive extension of the generalized curvature measures to the convex ring, and for the generalized parallel volume $\mu_\rho(K, \eta)$ we arrive at the Steiner formula

$$\mu_\rho(K, \eta) = \frac{1}{n} \sum_{m=0}^{n-1} \rho^{n-m} \binom{n}{m} \Theta_m(K, \eta).$$

As in Section 4.2, we specialize the generalized curvature measures by putting

$$C_m(K, \beta) = \frac{n\kappa_{n-m}}{\binom{n}{m}} \Phi_m(K, \beta) = \Theta_m(K, \beta \times S^{n-1}),$$

$$S_m(K, \omega) = \frac{n\kappa_{n-m}}{\binom{n}{m}} \Psi_m(K, \omega) = \Theta_m(K, \mathbb{E}^n \times \omega)$$

for $K \in U(\mathcal{K}^n)$, Borel sets $\beta \in \mathcal{B}(\mathbb{E}^n)$, $\omega \in \mathcal{B}(S^{n-1})$ and for $m = 0, \ldots, n-1$; further

$$\Phi_n(K, \beta) := \mathcal{H}^n(K \cap \beta).$$

Of these additive extensions of the curvature measures and area measures to the convex ring, it will mainly be the signed measures C_m (or Φ_m) that will be used in the sequel. We shall first extend the interpretation of C_{n-1} given by (4.2.23).

Theorem 4.4.1. *If $K \in U(\mathcal{K}^n)$ is the closure of its interior, then*

$$C_{n-1}(K, \beta) = \mathcal{H}^{n-1}(\beta \cap \operatorname{bd} K)$$

for $\beta \in \mathcal{B}(\mathbb{E}^n)$.

Proof. Let $K = \bigcup_{i=1}^r K_i$ with $K_i \in \mathcal{K}^n$. If among the convex bodies $K_1, \ldots, K_r$ there were one of dimension less than n not covered by the union of the others, then K would not be the closure of its interior. Hence, lower-dimensional bodies in the representation may be omitted, and we can assume from the beginning that $\dim K_i = n$ for $i = 1, \ldots, r$.

For $L \in \mathcal{K}^n$, let $\varphi(L, \cdot)$ denote the characteristic function of the boundary of L. Let $x \in \operatorname{bd} K$. If, say, $x \in K_i$ precisely for $i = 1, \ldots, m$, then $x \in \operatorname{bd} K_v$ for $v \in S(m)$ and hence

$$\sum_{v \in S(r)} (-1)^{|v|-1} \varphi(K_v, x) = \sum_{v \in S(m)} (-1)^{|v|-1} \times 1 = 1.$$

Let $\beta \in \mathcal{B}(\mathbb{E}^n)$ and $v \in S(r)$. Theorem 4.2.5 together with the remark after its proof tells us that $C_{n-1}(K_v, \beta) = \mathcal{H}^{n-1}(\beta \cap \operatorname{bd} K_v)$, provided that $\dim K_v \neq n-1$. If $\dim K_v = n-1$, then $\operatorname{relint} K_v \subset \operatorname{int} K$, hence $C_{n-1}(K_v, \beta \cap \operatorname{bd} K) = 0 = \mathcal{H}^{n-1}(\beta \cap \operatorname{bd} K \cap \operatorname{bd} K_v)$. Further, $C_{n-1}(K, \cdot)$ is concentrated on the boundary of K, since

$c_\rho(K, \beta \times S^{n-1}, \cdot) = c_\rho(K, (\beta \cap \text{bd } K) \times S^{n-1}, \cdot)$ by the definition of c_ρ. For $\beta \in \mathcal{B}(\mathbb{E}^n)$ we conclude that

$$\begin{aligned} C_{n-1}(K, \beta) &= C_{n-1}(K, \beta \cap \text{bd } K) \\ &= \sum_{v \in S(r)} (-1)^{|v|-1} C_{n-1}(K_v, \beta \cap \text{bd } K) \\ &= \sum_{v \in S(r)} (-1)^{|v|-1} \mathcal{H}^{n-1}(\beta \cap \text{bd } K \cap \text{bd } K_v) \\ &= \sum_{v \in S(r)} (-1)^{|v|-1} \int_{\beta \cap \text{bd} K} \varphi(K_v, x) \text{d}\mathcal{H}^{n-1}(x) \\ &= \mathcal{H}^{n-1}(\beta \cap \text{bd } K). \end{aligned}$$

■

For convex bodies $K \in \mathcal{K}^n$, formula (4.2.21) interprets the curvature measure C_0 as the content of the spherical image, namely

$$C_0(K, \beta) = \mathcal{H}^{n-1}(\sigma(K, \beta)).$$

If we want to extend this representation to the elements of the convex ring, we have to count the points of the spherical image with a suitable multiplicity. For this, we use a similar approach to the above and introduce another index function. For $K \in U(\mathcal{K}^n)$, a point $q \in \mathbb{E}^n$ and a vector $u \in S^{n-1}$ we define

$$i(K, q, u) := \begin{cases} 1 - \lim\limits_{\delta \downarrow 0} \lim\limits_{\varepsilon \downarrow 0} \chi(K \cap B(q + (\delta + \varepsilon)u, \delta)) & \text{if } q \in K, \\ 0 & \text{if } q \notin K. \end{cases}$$

If K is convex, then clearly

$$i(K, q, u) = \begin{cases} 1 & \text{if } (q, u) \in \text{Nor } K, \\ 0 & \text{otherwise.} \end{cases} \tag{4.4.2}$$

As for the index function $j(K, q, \cdot)$ one sees that $i(K, q, u)$ is well defined and that $i(\cdot, q, u)$ is an additive function on the convex ring $U(\mathcal{K}^n)$. For $K \in U(\mathcal{K}^n)$, $\beta \in \mathcal{B}(\mathbb{E}^n)$ and $u \in S^{n-1}$ we define

$$c(K, \beta, u) := \sum_{q \in \beta} i(K, q, u).$$

Here we have to allow infinite values, but these can be neglected: Let $K = \bigcup_{i=1}^r K_i$ with $K_i \in \mathcal{K}^n$. If $u \in \bigcap_{i=1}^r \text{regn } K_i$ and $v \in S(r)$, there is at most one point q for which $(q, u) \in \text{Nor } K_v$ and thus $i(K_v, q, u) \neq 0$. It follows from Theorem 2.2.9 that $c(K, \beta, \cdot)$ is finite $\mathcal{H}^{n-1}$-almost everywhere on S^{n-1}. If K is convex and $u \in \text{regn } K$, then (4.4.2) implies that $c(K, \beta, \cdot)$ is the characteristic function of the spherical image

$\sigma(K, \beta)$, hence (4.2.21) can be written as

$$C_0(K, \beta) = \int_{S^{n-1}} c(K, \beta, u)\, d\mathscr{H}^{n-1}(u). \tag{4.4.3}$$

Since both $C_0(\cdot, \beta)$ and $c(\cdot, \beta, u)$ are additive on $U(\mathcal{K}^n)$, the equality extends immediately to sets K of the convex ring. This is the desired interpretation of the curvature measure $C_0(K, \cdot)$ as the $\mathscr{H}^{n-1}$-integral of a multiplicity function for the spherical image.

Finally we remark that

$$\frac{1}{\omega_n} C_0(K, \mathbb{E}^n) = V_0(K) = \chi(K) \tag{4.4.4}$$

for $K \in U(\mathcal{K}^n)$. This is true for convex bodies, and the general case follows by additivity. In the same way, a corresponding result for the index functions is obtained. This is in analogy to well-known index sum formulae for smooth submanifolds. For $K \in U(\mathcal{K}^n)$ and $x \in \mathbb{E}^n \backslash K$ we have

$$\sum_{q \in \mathbb{E}^n} j(K, q, x) = \chi(K), \tag{4.4.5}$$

and if $u \in S^{n-1}$ is regular for K, which means that $u \in \bigcap_{i=1}^{r} \operatorname{regn} K_i$ for some representation $K = \bigcup_{i=1}^{r} K_i$ with $K_i \in \mathcal{K}^n$, we have

$$\sum_{q \in \mathbb{E}^n} i(K, q, u) = \chi(K). \tag{4.4.6}$$

Notes for Section 4.4

1. *Index functions*. The additive extension of the generalized curvature measures to the convex ring by means of an index function was carried out by Schneider (1980a). This had previously been done for the curvature measure C_0 in Schneider (1977d), extending some work of Hadwiger (1969c) and Banchoff (1967) for polyhedra and cell complexes. The index $i(K, q, u)$ is also used in Zähle (1987b).

   As mentioned at the end of Section 4.4, the index functions that were introduced are analogues of index functions known in differential geometry. Calling the point q a critical point of $K \in U(\mathcal{K}^n)$ with respect to the height function $\langle \cdot, u \rangle$ if $i(K, q, u) \neq 0$, one may consider equality (4.4.6) as an analogue of the critical point theorem for smooth submanifolds. Similarly, (4.4.4) is an analogue of the Gauss–Bonnet theorem.

   Suppose that $K \in U(\mathcal{K}^n)$ is the point set of a polyhedral cell complex of which Δ^k is the set of k-dimensional cells. Then the index $i(K, q, u)$ satisfies

$$i(K, q, u) = \sum_{k=0}^{n} (-1)^k \sum_{Z \in \Delta^k} i(Z, q, -u). \tag{4.4.7}$$

   This was proved by Shephard (1968c) for the special case of the boundary complex of a convex polytope, and it can be extended to the general case by means of an argument due to Perles & Sallee (1970). A definition equivalent to (4.4.7) was used by Banchoff (1967, 1970) in his investigation of critical point theory, curvature and the Gauss–Bonnet theorem for polyhedra.

2. *Absolute curvature measures on the convex ring.* If the additivity assumption is dropped, other extensions of the curvature measures to the convex ring are possible. Of particular interest are non-negative, or absolute, curvature measures. A non-negative extension of Federer's curvature measures to the convex ring was proposed by Matheron (1975), pp. 119ff. His construction was extended in Schneider (1980a), as follows. For given $K \in U(\mathcal{K}^n)$ and $x \in \mathbb{E}^n$, a point $q \in \mathbb{E}^n$ is called a *projection* of x in K if $q \in K$ and there exists a neighbourhood N of q such that $|x - y| > |x - q|$ for all $y \in K \cap N$, $y \neq q$. The set $\Pi(K, x)$ of all projections of x in K is finite. For $\eta \in \mathcal{B}(\Sigma)$ and $\rho > 0$ let

$$\bar{c}_\rho(K, \eta, x) := \operatorname{card}\{q \in \Pi(K, x) | 0 < |x - q| \leqq \rho \text{ and } (q, \overline{x - q}) \in \eta\}$$

and

$$\bar{\mu}_\rho(K, \eta) := \int_{\mathbb{E}^n} \bar{c}_\rho(K, \eta, x) \, d\mathcal{H}^n(x).$$

One can show that $\bar{\mu}_\rho(K, \eta)$ satisfies a Steiner formula, that is, a polynomial representation

$$\bar{\mu}_\rho(K, \eta) = \frac{1}{n} \sum_{m=0}^{n-1} \rho^{n-m} \binom{n}{m} \bar{\Theta}_m(K, \eta),$$

and that this defines (positive) measures $\bar{\Theta}_0(K, \cdot), \ldots, \bar{\Theta}_{n-1}(K, \cdot)$ on $\mathcal{B}(\Sigma)$, which for $K \in \mathcal{K}^n$ coincide with respectively $\Theta_0(K, \cdot), \ldots, \Theta_{n-1}(K, \cdot)$. The specialization

$$\bar{C}_m(K, \beta) := \bar{\Theta}_m(K, \beta \times S^{n-1}), \quad \beta \in \mathcal{B}(\mathbb{E}^n),$$

yields the measures introduced by Matheron.

The measure $\bar{C}_0(K, \cdot)$ can also be interpreted as follows. For $K \in U(\mathcal{K}^n)$ and $x \in \operatorname{bd} K$, a unit vector $u \in S^{n-1}$ is called a *normal vector* of K at x if there exists a neighbourhood N of x such that $\langle x, u \rangle \geqq \langle y, u \rangle$ for all $y \in K \cap N$. For $\beta \in \mathcal{B}(\mathbb{E}^n)$, let $\bar{c}(K, \beta, u)$ be the number (possibly infinite) of points $x \in K \cap \beta$ for which u is a normal vector. Then

$$\bar{C}_0(K, \beta) = \int_{S^{n-1}} \bar{c}(K, \beta, u) \, d\mathcal{H}^{n-1}(u).$$

Thus $\bar{C}_0(K, \cdot)$ can be interpreted as the measure of the 'spherical image with multiplicity'. The normalized total measure $\bar{C}_0(K, \mathbb{E}^n)/\omega_n$ was called the *convexity number* of K by Matheron (1975). It is equal to one for convex bodies, but not only for these.

3. *Other generalizations of curvature measures.* The treatment of curvature measures given in this section was restricted to convex bodies and the sets of the convex ring. Its extension to more general classes of sets is possible, but requires in general deeper methods from geometric measure theory. We give only some brief hints to the literature. Federer's (1959) curvature measures for sets of positive reach were further extended and investigated by Zähle (1984a, b, 1986a, 1987a, b). The extension is to finite unions of sets of positive reach and later to so-called second-order rectifiable sets. Applications to cell complexes and mosaics appear in Zähle (1987c, 1988), Weiss (1986) and Weiss & Zähle (1988). Fu (1989) relates curvature measures to generalized Morse theory.

The approach to curvature measures via so-called normal cycles, as begun by Wintgen (1982) and Zähle (1986a, 1987a, b), is further extended in work by Fu (1990+a, b). General versions of absolute curvature measures were introduced by Baddeley (1980) and Zähle (1989).

Further different approaches have been followed by Stachó (1979), who constructed generalizations of Federer's curvature measures to arbitrary sets (but with weaker properties), and by Kuiper (1971), who used singular homology to define a sequence of curvature measures, which includes a generalization of C_0.

Curvature measures for (non-convex) polyhedra were considered by Flaherty (1973). A thorough study of curvatures for piecewise linear spaces was made by Cheeger, Müller & Schrader (1984, 1986). The latter paper is also related to some (more general) work of Zähle quoted above. Part of the work of Cheeger, Müller & Schrader can be simplified; see Budach (1989) and also the remark at the end of Section 2 in Schneider (1990c).

Approximation results, of different kinds, for curvature measures appear in Brehm & Kühnel (1982), Cheeger, Müller & Schrader (1984), Lafontaine (1987) and Zähle (1990).

4. *Curvature measures in spaces of constant curvature*. Kohlmann (1988) introduced curvature measures for sets of positive reach in Riemannian space forms. Among other results, he obtained integral representations, Minkowski-type integral formulae (see Section 5.3, Note 3) and characterizations of balls (see Section 7.2, Note 4).
5. *Groemer's extension of the quermassintegrals*. The following extension of the quermassintegrals, which apparently does not generalize to curvature measures, was proposed by Groemer (1972). He introduced a vector space A^n of real functions on $\mathbb{E}^n$ with a pseudonorm such that A^n contains $V(\mathcal{P}^n)$ (see Section 3.4, Note 7) as a proper dense subspace. The elements of A^n are called 'approximable' functions. The system $\mathcal{S}_A$ of subsets of $\mathbb{E}^n$ whose characteristic functions are approximable contains the convex ring $U(\mathcal{K}^n)$ and, for example, the relative interiors of convex bodies. Groemer showed that the quermassintegrals can be extended from $\mathcal{K}^n$ to continuous linear functionals on A^n. In particular, this yields an additive extension of the quermassintegral W_i to the class $\mathcal{S}_A$.

4.5. Integral-geometric formulae

Some integral-geometric formulae can be proved for curvature measures and area measures. Generally speaking, integral geometry is concerned with the computation and application of mean values for geometrically defined functions with respect to invariant measures. In this section we shall treat local versions of some intersection and projection formulae, which in their global versions, namely for the Minkowski functionals, are central results of classical integral geometry. We shall also consider mean value formulae for Minkowski sums, and counterparts to the intersection formulae in the form of kinematic formulae involving the distances between non-intersecting bodies, or between bodies and flats. This yields, among other results, an integral-geometric interpretation of generalized curvature measures.

We assume that the reader is familiar with the topological groups $SO(n)$ of proper rotations of $\mathbb{E}^n$ and G_n of rigid motions of $\mathbb{E}^n$, and with their Haar measures. We normalize the Haar measure ν on $SO(n)$

by imposing that $\nu(SO(n)) = 1$. Every rigid motion $g \in G_n$ determines uniquely a rotation $\rho \in SO(n)$ and a translation vector $t \in \mathbb{E}^n$ so that $gx = \rho x + t$ for $x \in \mathbb{E}^n$, and we write $g = g_{t,\rho}$. The map

$$\gamma: \mathbb{E}^n \times SO(n) \to G_n$$
$$(t, \rho) \mapsto g_{t,\rho}$$

is a homeomorphism (products are always equipped with the product topology). We can define a Haar measure μ on G_n by the image measure

$$\mu := \gamma(\lambda_n \otimes \nu),$$

where λ_n denotes the restriction of $\mathcal{H}^n$ to the Borel sets of $\mathbb{E}^n$; this implies a definite normalization of μ. The measure μ, defined on the σ-algebra $\mathcal{B}(G_n)$ of Borel subsets of G_n, is left invariant and right invariant, thus the motion group is unimodular.

On $A(n, k)$, the set of k-dimensional affine subspaces of $\mathbb{E}^n$, where $k \in \{0, \ldots, n\}$, the motion group G_n operates in the natural way, making $A(n, k)$ into a homogenous space. To facilitate the handling of its invariant measure, we choose a k-dimensional linear subspace E_k of $\mathbb{E}^n$ and denote its orthogonal complement by $E_k^\perp$. The map

$$\gamma_k: E_k^\perp \times SO(n) \to A(n, k)$$
$$(t, \rho) \mapsto \rho(E_k + t)$$

is surjective. The usual topology of $A(n, k)$ is the finest topology for which γ_k is continuous. With this topology, $A(n, k)$ is a locally compact, second countable Hausdorff space, the natural operation of G_n on $A(n, k)$ is continuous and $A(n, k)$, with this operation, is a homogeneous G_n-space. The image measure

$$\mu_k := \gamma_k(\lambda_{n-k} \otimes \nu)$$

of the product measure $\lambda_{n-k} \otimes \nu$, where λ_{n-k} is the restriction of $\mathcal{H}^{n-k}$ to the Borel sets of $E_k^\perp$, is a G_n-invariant measure on $A(n, k)$ with a suitable normalization. It does not depend on the special choice of E_k.

In a similar, even simpler, way we could obtain the $SO(n)$-invariant measure ν_k on the compact homogeneous space $G(n, k)$ of k-dimensional linear subspaces of $\mathbb{E}^n$, but we shall not need it in what follows.

Below, we shall sometimes need to know that certain sets of rotations, rigid motions or flats are of measure zero. Some of these facts follow from the results of Section 2.3; a more elementary one is contained in the following lemma.

Two linear subspaces E, F of $\mathbb{E}^n$ are said to be *in special position* if

$$\operatorname{lin}(E \cup F) \neq \mathbb{E}^n \quad \text{and} \quad E \cap F \neq \{o\}.$$

Lemma 4.5.1. *Let $E \in G(n, p)$, $F \in G(n, q)$, $0 \leqq p$, $q \leqq n-1$. The set of all rotations $\rho \in SO(n)$ for which E and ρF are in special position is of ν-measure zero.*

To avoid computations requiring an explicit description of the Haar measure ν, we give a proof that uses only its invariance and finiteness.

Proof of Lemma 4.5.1. For given E, F let X be the set of all $\rho \in SO(n)$ for which E and ρF are in special position. We prove the assertion $\nu(X) = 0$ by induction on p. The case $p = 0$ is trivial; we assume that $p \geqq 1$ and the assertion has been proved for $\dim E < p$. Let $\dim E = p$. We choose a linear subspace $U \in G(n, n-p+1)$ and denote its orthogonal complement by $U^\perp$. Further, we choose a number $k \in \mathbb{N}$ and k vectors $u_1, \ldots, u_k \in U$ of which any $n-p+1$ are linearly independent. Put $E_i := \mathrm{lin}\,(U^\perp \cup \{u_i\})$; then $\dim E_i = p$. Let X_i be the set of all $\rho \in SO(n)$ for which E_i and ρF are in special position. As $E_i = \rho_i E$ for suitable $\rho_i \in SO(n)$, we have $X_i = \rho_i X$ and hence $\nu(X_i) = \nu(X)$. Let Y be the set of all $\rho \in SO(n)$ for which $U^\perp$ and ρF are in special position. Since $\dim U^\perp = p-1$, we have $\nu(Y) = 0$ by the inductive hypothesis. We assert that each $\rho \in SO(n)\backslash Y$ lies in at most $n-p$ of the sets $X_1, \ldots, X_k$. Let $\rho \in SO(n)\backslash Y$ and, say, $\rho \in X_i$ for $i = 1, \ldots, r$. Thus E_i and ρF are in special position for $i = 1, \ldots, r$. In particular, $\mathrm{lin}\,(E_i \cup \rho F) \neq \mathbb{E}^n$ and hence $\mathrm{proj}_U\, \rho F \neq U$, since $U^\perp \subset E_i$. $U^\perp$ and ρF are not in special position since $\rho \notin Y$; however, E_i and ρF are in special position. This is only possible if $u_i \in \mathrm{proj}_U\, \rho F$. Since the vectors $u_1, \ldots, u_{n-p+1}$ are linearly independent and $\dim \mathrm{proj}_U\, \rho F \leqq n-p$, we deduce that $r \leqq n-p$. Let ζ_i be the characteristic function of X_i on $SO(n)$. We have proved that $\sum_{i=1}^k \zeta_i(\rho) \leqq n-p$ for $\rho \in SO(n)\backslash Y$. Since $\nu(Y) = 0$, integration yields

$$k\nu(X) = \sum_{i=1}^{k} \nu(X_i) = \int_{SO(n)} \sum_{i=1}^{k} \zeta_i(\rho)\, \mathrm{d}\nu(\rho) \leqq n-p.$$

Letting $k \to \infty$, we conclude that $\nu(X) = 0$. ∎

Our first major goal in this section is the proof of the following theorem.

Theorem 4.5.2. *Let $K, K' \in U(\mathcal{K}^n)$ be sets of the convex ring, let β, $\beta' \in \mathcal{B}(\mathbb{E}^n)$ and $j \in \{0, \ldots, n\}$. Then*

$$\int_{G_n} \Phi_j(K \cap gK', \beta \cap g\beta')\, \mathrm{d}\mu(g) = \sum_{k=j}^{n} \alpha_{njk} \Phi_k(K, \beta) \Phi_{n+j-k}(K', \beta') \tag{4.5.1}$$

with

$$\alpha_{njk} = \frac{\binom{k}{j}\kappa_k\kappa_{n+j-k}}{\binom{n}{k-j}\kappa_j\kappa_n} = \frac{\Gamma\left(\frac{k+1}{2}\right)\Gamma\left(\frac{n+j-k+1}{2}\right)}{\Gamma\left(\frac{j+1}{2}\right)\Gamma\left(\frac{n+1}{2}\right)}.$$

The special case $\beta = \beta' = \mathbb{E}^n$ reduces to what Hadwiger [18] (with different notation) called the 'complete system of kinematic formulae', namely

$$\int_{G_i} V_j(K \cap gK')\,\mathrm{d}\mu(g) = \sum_{k=j}^{n} \alpha_{njk} V_k(K) V_{n+j-k}(K'). \qquad (4.5.2)$$

The further specialization $j = 0$ can be written in the form

$$\int_{G_n} \chi(K \cap gK')\,\mathrm{d}\mu(g) = \frac{1}{\kappa_n} \sum_{k=0}^{n} \frac{\kappa_k \kappa_{n-k}}{\binom{n}{k}} V_k(K) V_{n-k}(K'). \qquad (4.5.3)$$

This result is known as the 'principal kinematic formula' (for convex bodies). It expresses the total measure of the set of all rigid motions g for which the moving body gK' meets K in terms of the intrinsic volumes or in terms of the Minkowski functionals of K and K'.

The proof of Theorem 4.5.2 is divided into several steps. As a preliminary, we remark that it suffices to prove the theorem for convex bodies K, K'. In fact, if (4.5.1) is true in this case, then one can utilize the fact that both sides of (4.5.1), as functions of K, are additive on $U(\mathcal{K}^n)$. By additivity, (4.5.1) remains true if K is replaced by a finite union of convex bodies. In a second step, K' can similarly be replaced by a finite union of convex bodies. Hence, from now on we assume that $K, K' \in \mathcal{K}^n$.

First we have to show that the integrand in (4.5.1) is, in fact, a measurable function of g. For this, let $K, K' \in \mathcal{K}^n$ and $\beta, \beta' \in \mathcal{B}(\mathbb{E}^n)$ be given. Let $X(K, K')$ be the set of all rigid motions $g \in G_n$ for which K and gK' cannot be separated by a hyperplane. For given $\varepsilon > 0$, we define a function $h: G_n \to \mathbb{R}$ by

$$h(g) := U_\varepsilon(K \cap gK', \beta \cap g\beta'),$$

where $U_\varepsilon(L, \alpha) := \mu_\varepsilon(L, \alpha \times S^{n-1}) = \mathcal{H}^n(A_\varepsilon(L, \alpha))$ for $L \in \mathcal{K}^n$, $\alpha \in \mathcal{B}(\mathbb{E}^n)$ (μ_ε was defined in Section 4.1), and we define a function $f: X(K, K') \times \mathbb{E}^n \to \mathbb{R}$ by

$$f(g, x) := \zeta_\beta(x)\zeta_{\beta'}(g^{-1}x)U_\varepsilon(K \cap gK', \mathbb{E}^n),$$

where ζ_α is the characteristic function of α on $\mathbb{E}^n$. The function $(g, x) \mapsto g^{-1}x$ is continuous on $G_n \times \mathbb{E}^n$ and the function $g \mapsto U_\varepsilon(K \cap gK', \mathbb{E}^n)$ is continuous on $X(K, K')$ by Theorem 1.8.8 and by the continuity of $U_\varepsilon(\cdot, \mathbb{E}^n)$ (which follows from Theorem

1.8.16). Since ζ_β, $\zeta_{\beta'}$ are measurable on $\mathbb{E}^n$, the function f is measurable. Let

$$P(g, \alpha) := U_\varepsilon(K \cap gK', \alpha)/U_\varepsilon(K \cap gK', \mathbb{E}^n)$$

for $g \in X(K, K')$ and $\alpha \in \mathcal{B}(\mathbb{E}^n)$. Since the function $g \mapsto K \cap gK'$ is continuous on $X(K, K')$ and $U_\varepsilon(\cdot, \alpha)$ is measurable on $\mathcal{K}^n$ by Theorem 4.1.2, it follows that $P(\cdot, \alpha)$ is measurable on $X(K, K')$. For fixed g, $P(g, \cdot)$ is a probability measure on $\mathcal{B}(\mathbb{E}^n)$. Since, for $g \in X(K, K')$,

$$\begin{aligned} h(g) &= \int_{\mathbb{E}^n} \zeta_{\beta \cap g\beta'}(x) U_\varepsilon(K \cap gK', \mathrm{d}x) \\ &= \int_{\mathbb{E}^n} f(g, x) P(g, \mathrm{d}x), \end{aligned}$$

it follows (e.g., Neveu 1969, p. 95) that h is measurable on $X(K, K')$. Now $h(g) = 0$ if $K \cap gK' = \emptyset$, and the (closed) set M of all $g \in G_n \backslash X(K, K')$ for which $K \cap gK' \neq \emptyset$ satisfies

$$\mu(M) = \int_{SO(n)} \int_{T(\rho)} \mathrm{d}\mathcal{H}^n \, \mathrm{d}\nu(\rho), \tag{4.5.4}$$

where $T(\rho)$ is the set of all $t \in \mathbb{E}^n$ for which $\rho K'$ and K meet but can be separated by a hyperplane. It is easy to see that $T(\rho) = \mathrm{bd}\,(K - \rho K')$ and hence $\mu(M) = 0$. Thus h is measurable. Since $\varepsilon > 0$ was arbitrary, we deduce from (4.2.5) that the function $g \mapsto \Phi_j(K \cap gK', \beta \cap g\beta')$ is measurable if $j \in \{0, \ldots, n-1\}$; for $j = n$ this is easily seen directly.

Now we enter into the essential part of the proof of Theorem 4.5.2, by writing the integral of (4.5.1) in the form

$$\begin{aligned} &\int_{G_n} \Phi_j(K \cap gK', \beta \cap g\beta') \, \mathrm{d}\mu(g) \\ &= \int_{SO(n)} \int_{\mathbb{E}^n} \Phi_j(K \cap (\rho K' + t), \beta \cap (\rho\beta' + t)) \, \mathrm{d}t \, \mathrm{d}\nu(\rho). \end{aligned} \tag{4.5.5}$$

This is possible by the description of the invariant measure μ given initially and by Fubini's theorem (this argument was already used in (4.5.4)). We first consider the inner integral for the case of polytopes, where an explicit computation is possible.

Some preparatory definitions are needed. First, for $A \subset \mathbb{E}^n$ and $x \in \mathbb{E}^n$ we often abbreviate $A + x$ by A_x. For linear subspaces L, L' of $\mathbb{E}^n$ we define a number $[L, L']$ as follows. We choose an orthonormal basis of $L \cap L'$ and extend it to an orthonormal basis of L as well as to an orthonormal basis of L'. The resulting vectors span a parallelepiped of dimension $\dim \operatorname{lin}(L \cup L')$; let $[L, L']$ be its n-dimensional volume. This definition does not depend on the choice of the bases. Next, let $P, P' \in \mathcal{P}^n$ be convex polytopes, let F be a face of P and F' a face of P'. By $L(F) := \operatorname{aff} F - \operatorname{aff} F$ we denote the linear subspace parallel to aff F. Then we define $[F, F'] := [L(F), L(F')]$. We also define an external angle

$$\gamma(F, F', P, P') := \gamma(F \cap F'_x, P \cap P'_x),$$

where $x \in \mathbb{E}^n$ is chosen such that $\operatorname{relint} F \cap \operatorname{relint} F'_x \neq \emptyset$; this definition does not depend on the choice of x. Finally, we write

$$\lambda_F(\beta) := \mathcal{H}^{\dim F}(F \cap \beta) \quad \text{for } \beta \in \mathcal{B}(\mathbb{E}^n),$$

so that λ_F is a measure with support F.

The faces F, F' are said to be *in special position* if the linear subspaces $L(F)$, $L(F')$ are in special position.

As an intermediate result, we state the following 'principal translative formula' for polytopes.

Theorem 4.5.3. *Let $P, P' \in \mathcal{P}^n$ be polytopes; let $\beta, \beta' \in \mathcal{B}(\mathbb{E}^n)$ and $j \in \{0, \ldots, n\}$. Then*

$$\int_{\mathbb{E}^n} \Phi_j(P \cap P'_x, \beta \cap \beta'_x)\,\mathrm{d}x = \Phi_j(P, \beta)\Phi_n(P', \beta') + \Phi_n(P, \beta)\Phi_j(P', \beta')$$

$$+ \sum_{k=j+1}^{n-1} \sum_{F \in \mathcal{F}_k(P)} \sum_{F' \in \mathcal{F}_{n+j-k}(P')} \gamma(F, F', P, P')[F, F']\lambda_F(\beta)\lambda_{F'}(\beta').$$

Proof. By (4.2.17) we have

$$I := \int_{\mathbb{E}^n} \Phi_j(P \cap P'_x, \beta \cap \beta'_x)\,\mathrm{d}x$$

$$= \int_{\mathbb{E}^n} \sum_{F \in \mathcal{F}_j(P \cap P'_x)} \gamma(F, P \cap P'_x)\lambda_F(\beta \cap \beta'_x)\,\mathrm{d}x.$$

In computing this integral, we may neglect those translation vectors x for which a k-face F of P meets an m-face F'_x of P'_x with $k + m < n$ or with F, F' in special position, since

$$\{x \in \mathbb{E}^n \mid F \cap F'_x \neq \emptyset\} = F - F'$$

is of measure zero in each case. As the vectors x with $F \cap F'_x \neq \emptyset$ but $\operatorname{relint} F \cap \operatorname{relint} F'_x = \emptyset$ also form a set of measure zero, we may assume that each j-face of $P \cap P'_x$ is the intersection of a k-face of P and an $(n + j - k)$-face of P'_x, for some $k \in \{j, \ldots, n\}$. It follows that

$$I = \sum_{k=j}^{n} \sum_{F \in \mathcal{F}_k(P)} \sum_{F' \in \mathcal{F}_{n+j-k}(P')} \int_{\mathbb{E}^n} \gamma(F \cap F'_x, P \cap P'_x) \lambda_{F \cap F'_x}(\beta \cap \beta'_x) \, dx$$

$$= \sum_{k=j}^{n} \sum_{F \in \mathcal{F}_k(P)} \sum_{F' \in \mathcal{F}_{n+j-k}(P')} \gamma(F, F', P, P') J(F, F', \beta, \beta')$$

with

$$J(F, F', \beta, \beta') := \int_{\mathbb{E}^n} \lambda_{F \cap F'_x}(\beta \cap \beta'_x) \, dx.$$

We fix $k \in \{j, \ldots, n\}$, $F \in \mathcal{F}_k(P)$, $F' \in \mathcal{F}_{n+j-k}(P')$ and write $J(F, F', \beta, \beta') = J$. For the computation of J we may assume that F and F' are not in special position (otherwise $J = 0$) and that $o \in \operatorname{aff} F \cap \operatorname{aff} F'$. We put $L(F) \cap L(F') := L_1$, $L_1^\perp \cap L(F) := L_2$, $L_1^\perp \cap L(F') := L_3$ and write $x \in \mathbb{E}^n$ uniquely in the form $x = x_1 + x_2 + x_3$ with $x_i \in L_i$ $(i = 1, 2, 3)$. Writing $F \cap \beta = \alpha$, $F' \cap \beta' = \alpha'$ we obtain, using the definition of $[F, F']$,

$$J = [F, F'] \int_{L_3} \int_{L_2} \int_{L_1} \mathcal{H}^j(\alpha \cap \alpha'_x) \, d\mathcal{H}^j(x_1) \, d\mathcal{H}^{k-j}(x_2) \, d\mathcal{H}^{n-k}(x_3).$$

Since $(\alpha \cap \alpha'_x) - x_2 \subset L_1$, we obtain

$$\int_{L_1} \mathcal{H}^j(\alpha \cap \alpha'_x) \, d\mathcal{H}^j(x_1)$$

$$= \int_{L_1} \mathcal{H}^j((\alpha - x_2) \cap (\alpha' + x_3 + x_1) \cap L_1) \, d\mathcal{H}^j(x_1)$$

$$= \mathcal{H}^j((\alpha - x_2) \cap L_1) \mathcal{H}^j((\alpha' + x_3) \cap L_1),$$

by Fubini's theorem. Also by Fubini's theorem,

$$\int_{L_2} \mathcal{H}^j(\alpha \cap (L_1 + x_2)) \, d\mathcal{H}^{k-j}(x_2) = \mathcal{H}^k(\alpha),$$

$$\int_{L_3} \mathcal{H}^j(\alpha' \cap (L_1 - x_3)) \, d\mathcal{H}^{n-k}(x_3) = \mathcal{H}^{n+j-k}(\alpha'),$$

thus

$$J = [F, F'] \lambda_F(\beta) \lambda_{F'}(\beta').$$

If $F' \in \mathcal{F}_n(P')$, we have $\gamma(F, F', P, P') = \gamma(F, P)$, $[F, F'] = 1$ and $\lambda_{F'} = \Phi_n(P', \cdot)$, thus

$$\sum_{F \in \mathcal{F}_j(P)} \sum_{F' \in \mathcal{F}_n(P')} \gamma(F, F', P, P') J(F, F', \beta, \beta') = \Phi_j(P, \beta) \Phi_n(P', \beta').$$

This holds trivially if $\dim P' < n$, since both sides are then zero. A similar remark concerns the case $F \in \mathcal{F}_n(P)$. This completes the proof of Theorem 4.5.3. ■

To perform the outer integration in (4.5.5) for the case of polytopes, we need the following mean value formula.

Lemma 4.5.4. Let $P, P' \in \mathcal{P}^n$ *be polytopes,* $j \in \{0, \ldots, n-2\}$, $k \in \{j+1, \ldots, n-1\}$, $F \in \mathcal{F}_k(P)$, $F' \in \mathcal{F}_{n+j-k}(P')$, $\beta, \beta' \in \mathcal{B}(\mathbb{E}^n)$. *Then*

$$\int_{SO(n)} \gamma(F, \rho F', P, \rho P')[F, \rho F'] \, \mathrm{d}\nu(\rho) = \alpha_{njk} \gamma(F, P) \gamma(F', P')$$

with a real constant α_{njk} *depending only on n, j, k (its explicit value is given by* (4.5.13)).

Proof. Since a direct computation of the integral appears difficult, we proceed in an indirect way.

Denoting by σ^m the normalized spherical Lebesgue measure on m-dimensional great subspheres of S^{n-1}, we have $\gamma(F, P) = \sigma^{n-k-1}(N(P, F) \cap S^{n-1})$ by the definition of outer angles. By Lemma 4.5.1 we may neglect those rotations ρ for which F and $\rho F'$ are in special position. Hence, we may assume that

$$\gamma(F, \rho F', P, \rho P') = \sigma^{n-j-1}([N(P, F) + \rho N(P', F')] \cap S^{n-1}),$$

where Theorem 2.2.1(b) was used. Now we generalize the integral to be determined. For $\omega \subset S^{n-1}$, let $C(\omega) := \{\lambda x \mid x \in \omega, \lambda \geqq 0\}$ be the cone spanned by ω. For arbitrary Borel sets $\omega \subset L(F)^\perp \cap S^{n-1}$ and $\omega' \subset L(F')^\perp \cap S^{n-1}$ we write

$$J(\omega, \omega') := \int_{SO(n)} \sigma^{n-j-1}\{[C(\omega) + \rho C(\omega')] \cap S^{n-1}\}[F, \rho F'] \, \mathrm{d}\nu(\rho)$$

(the measurability of the integrand is proved in a standard way). First we fix ω' and consider the function $J(\cdot, \omega')$. For ν-almost all ρ, $C(\omega)$ and $\rho C(\omega')$ lie in complementary subspaces. For such ρ,

$$\left[C\left(\bigcup_m \omega_m\right) + \rho C(\omega')\right] \cap S^{n-1} = \bigcup_m [C(\omega_m) + \rho C(\omega')] \cap S^{n-1}$$

for any sequence $(\omega_m)_{m \in \mathbb{N}}$ of Borel sets in $L(F)^\perp \cap S^{n-1}$. If the sets of this sequence are pairwise disjoint, the union on the right is disjoint up to a set of σ^{n-j-1}-measure zero. From the σ-additivity of σ^{n-j-1} and the monotone convergence theorem it follows that $J(\cdot, \omega')$ is a finite Borel measure on the sphere $L(F)^\perp \cap S^{n-1}$. A rotation τ of S^{n-1} that carries $L(F)^\perp$ into itself and keeps $L(F)$ pointwise fixed satisfies $C(\tau\omega) + \rho C(\omega') = \tau[C(\omega) + \tau^{-1}\rho C(\omega')]$ and $[F, \rho F'] = [F, \tau^{-1}\rho F']$; hence the invariance of σ^{n-j-1} and ν implies that $J(\tau\omega, \omega') = J(\omega, \omega')$. Thus $J(\cdot, \omega')$ must be a constant multiple of the invariant measure on the sphere $L(F)^\perp \cap S^{n-1}$. Interchanging the roles of ω and ω', we arrive at

$$J(\omega, \omega') = \alpha_{njk} \sigma^{n-k-1}(\omega) \sigma^{k-j-1}(\omega')$$

with a constant α_{njk} that evidently depends only on n, j, k. The particular choice $\omega = N(P, F) \cap S^{n-1}$, $\omega' = N(P', F') \cap S^{n-1}$ now yields the assertion of the lemma. ■

If we now put equality (4.5.5), Theorem 4.5.3 and Lemma 4.5.4 together, we arrive at formula (4.5.1) for the case where K, K' are polytopes. The coefficients are those of Lemma 4.5.4.

For general convex bodies, Theorem 4.5.2 will now be proved by approximation. For given $g \in G_n$, we define a map $T_g: \mathbb{E}^n \to \mathbb{E}^n \times \mathbb{E}^n$ by $T_g(x) := (x, g^{-1}x)$. For convex bodies $K, K' \in \mathcal{K}^n$, we denote by $\varphi_j(g, K, K', \cdot)$ the image measure of $\Phi_j(K \cap gK', \cdot)$ under the (continuous) map T_g; then

$$\varphi_j(g, K, K', \beta \times \beta') = \Phi_j(K \cap gK', \beta \cap g\beta')$$

for β, $\beta' \in \mathcal{B}(\mathbb{E}^n)$. The set $\mathcal{A}$ of all $\alpha \in \mathcal{B}(\mathbb{E}^n \times \mathbb{E}^n)$ for which $\varphi_j(\cdot, K, K', \alpha)$ is measurable is a Dynkin system. Since $\mathcal{A}$ contains the product sets $\beta \times \beta'$ $(\beta, \beta' \in \mathcal{B}(\mathbb{E}^n))$, it contains the Dynkin system generated by these products, which is equal to the generated σ-algebra $\mathcal{B}(\mathbb{E}^n \times \mathbb{E}^n)$. Thus $\varphi_j(\cdot, K, K', \alpha)$ is measurable for each $\alpha \in \mathcal{B}(\mathbb{E}^n \times \mathbb{E}^n)$, and we can define a measure on $\mathcal{B}(\mathbb{E}^n \times \mathbb{E}^n)$ by

$$\varphi_j(K, K', \cdot) := \int_{G_n} \varphi_j(g, K, K', \cdot)\, d\mu(g).$$

This measure is finite since

$$\varphi_j(g, K, K', \cdot) \leqq V_j(K \cap gK'). \tag{4.5.6}$$

In $\mathcal{K}^n$ we consider convergent sequences $K_i \to K$, $K_i' \to K'$. For μ-almost all g, either $K \cap gK' = \emptyset$ or K and gK' cannot be separated by a hyperplane, which by Theorem 1.8.8 implies that $K_i \cap gK_i' \to K \cap gK$ for $i \to \infty$. For these g, it follows from the weak continuity of the curvature measures that

$$\liminf_{i\to\infty} \Phi_j(K_i \cap gK_i', T_g^{-1}(\alpha)) \geqq \Phi_j(K \cap gK', T_g^{-1}(\alpha))$$

if $\alpha \subset \mathbb{E}^n \times \mathbb{E}^n$ is open and hence also $T_g^{-1}(\alpha)$ is open. Integration with respect to g and application of Fatou's lemma yields

$$\liminf_{i\to\infty} \varphi_j(K_i, K_i', \alpha) \geqq \varphi_j(K, K', \alpha).$$

Similarly, we obtain

$$\lim_{i\to\infty} \varphi_j(K_i, K_i', \mathbb{E}^n \times \mathbb{E}^n) = \varphi_j(K, K', \mathbb{E}^n \times \mathbb{E}^n)$$

if we use the bounded convergence theorem, which can be applied because of (4.5.6), and the fact that $V_j(K_i \cap gK_i') \to V_j(K \cap gK')$ for

μ-almost all g. Thus we have proved that

$$\varphi_j(K_i, K_i', \cdot) \xrightarrow{w} \varphi_j(K, K', \cdot) \quad \text{for } i \to \infty.$$

This can be used to show that

$$\varphi_j(K, K', \cdot) = \sum_{k=j}^{n} \alpha_{njk} \Phi_k(K, \cdot) \otimes \Phi_{n+j-k}(K', \cdot) \tag{4.5.7}$$

if $K, K' \in \mathcal{K}^n$. In fact, both sides are finite measures on $\mathcal{B}(\mathbb{E}^n \times \mathbb{E}^n)$. If K, K' are polytopes, both sides are equal on product sets $\beta \times \beta'$ ($\beta, \beta' \in \mathcal{B}(\mathbb{E}^n)$), since (4.5.1) holds for polytopes, and thus are equal on an intersection stable generating system of $\mathcal{B}(\mathbb{E}^n \times \mathbb{E}^n)$; hence they are equal on $\mathcal{B}(\mathbb{E}^n \times \mathbb{E}^n)$. Since both sides are weakly continuous functions of (K, K'), approximation by polytopes shows that (4.5.7) holds for arbitrary convex bodies K, K'. Formula (4.5.1) is the special case obtained by applying both sides to $\beta \times \beta'$. This completes the proof of Theorem 4.5.2, except that we still have to determine the explicit values of the constants α_{njk}.

Before we compute the numbers α_{njk}, it is convenient to deduce from Theorem 4.5.2 another integral-geometric formula for the case where the moving convex body gK' is replaced by a moving flat.

Theorem 4.5.5. *Let $K \in U(\mathcal{K}^n)$ be a set of the convex ring, let $\beta \in \mathcal{B}(\mathbb{E}^n)$, $k \in \{0, \ldots, n\}$ and $j \in \{0, \ldots, k\}$. Then*

$$\int_{A(n,k)} \Phi_j(K \cap E, \beta \cap E)\, \mathrm{d}\mu_k(E) = \alpha_{njk} \Phi_{n+j-k}(K, \beta)$$

(where α_{njk} is the same number as in Theorem 4.5.2).

The special case $\beta = \mathbb{E}^n$ gives

$$\int_{A(n,k)} V_j(K \cap E)\, \mathrm{d}\mu_k(E) = \alpha_{njk} V_{n+j-k}(K). \tag{4.5.8}$$

These equalities are called *Crofton's intersection formulae*. In particular, one has

$$\int_{A(n,k)} \chi(K \cap E)\, \mathrm{d}\mu_k(E) = \alpha_{n0k} V_{n-k}(K), \tag{4.5.9}$$

and if K is convex, this gives

$$\alpha_{n0k} V_{n-k}(K) = \mu_k(\{E \in A(n, k) | K \cap E \neq \emptyset\}). \tag{4.5.10}$$

We note also the special case

$$\alpha_{n0(n-j)} \Phi_j(K, \beta) = \int_{A(n,n-j)} \Phi_0(K \cap E, \beta \cap E)\, \mathrm{d}\mu_{n-j}(E) \tag{4.5.11}$$

giving a mean value interpretation of the curvature measure Φ_j in terms of Φ_0, which has a simple intuitive meaning.

Proof of Theorem 4.5.5. We choose a k-dimensional linear subspace E_k of $\mathbb{E}^n$ and recall that the invariant measure μ_k can be defined as the image measure of $\lambda_{n-k} \otimes \nu$ under the map

$$\gamma_k \colon E_k^\perp \times SO(n) \to A(n, k)$$
$$(t, \rho) \mapsto \rho(E_k + t),$$

where λ_{n-k} is the restriction of $\mathcal{H}^{n-k}$ to $\mathcal{B}(E_k^\perp)$. Similarly, we denote by λ_k the restriction of $\mathcal{H}^k$ to $\mathcal{B}(E_k)$.

We choose a bounded Borel set $\alpha \subset E_k$ with $\lambda_k(\alpha) = 1$ and consider the integral

$$J := \int_{G_n} \Phi_j(E_k \cap gK, \alpha \cap g\beta)\, \mathrm{d}\mu(g).$$

We can choose a convex body $K' \subset E_k$ such that $\alpha \subset \operatorname{relint} K'$ and that every $g \in G_n$ with $gK \cap \alpha \neq \varnothing$ satisfies $E_k \cap gK = K' \cap gK$. Therefore, the integral J can be computed by means of Theorem 4.5.2, and we obtain

$$J = \sum_{i=j}^{n} \alpha_{nji} \Phi_i(K', \alpha) \Phi_{n+j-i}(K, \beta).$$

Because $\alpha \subset \operatorname{relint} K'$ we have $\Phi_i(K', \alpha) = \lambda_k(\alpha) = 1$ if $i = k$ and 0 otherwise; hence

$$J = \alpha_{njk} \Phi_{n+j-k}(K, \beta). \tag{4.5.12}$$

In the following, we write $t \in \mathbb{E}^n$ in the form $t = t_1 + t_2$ with $t_1 \in E_k^\perp$ and $t_2 \in E_k$. We find

$$J = \int_{SO(n)} \int_{\mathbb{E}^n} \Phi_j(E_k \cap (\rho K + t), \alpha \cap (\rho\beta + t))\, \mathrm{d}t\, \mathrm{d}\nu(\rho)$$

$$= \int_{SO(n)} \int_{E_k^\perp} \int_{E_k} \Phi_j(E_k \cap (\rho K + t_1 + t_2), \alpha \cap (\rho\beta + t_1 + t_2))$$
$$\times\, \mathrm{d}\lambda_k(t_2)\, \mathrm{d}\lambda_{n-k}(t_1)\, \mathrm{d}\nu(\rho)$$

$$= \int_{E_k^\perp \times SO(n)} \int_{E_k} \Phi_j(E_k \cap (\rho K + t_1 + t_2), \alpha \cap (\rho\beta + t_1 + t_2))$$
$$\times \mathrm{d}\lambda_k(t_2)\, \mathrm{d}(\lambda_{n-k} \otimes \nu)(t_1, \rho).$$

For the computation of the inner integral, put

$$\Phi_j(E_k \cap (\rho K + t_1), \cdot\,) =: \varphi,$$
$$\rho\beta + t_1 =: \beta';$$

then

$$\Phi_j(E_k \cap (\rho K + t_1 + t_2), \alpha \cap (\rho\beta + t_1 + t_2))$$
$$= \Phi_j(E_k \cap (\rho K + t_1), (\alpha - t_2) \cap \beta')$$
$$= \varphi((\alpha - t_2) \cap \beta').$$

Denoting the characteristic function of $A \subset \mathbb{E}^n$ by ζ_A, we obtain

$$\int_{E_k} \Phi_j(E_k \cap (\rho K + t_1 + t_2), \alpha \cap (\rho\beta + t_1 + t_2))\, \mathrm{d}\lambda_k(t_2)$$

$$= \int_{E_k} \varphi((\alpha - t_2) \cap \beta')\, \mathrm{d}\lambda_k(t_2)$$

$$= \int_{E_k}\int_{\mathbb{E}^n} \zeta_\alpha(x + t_2)\zeta_{\beta'}(x)\, \mathrm{d}\varphi(x)\, \mathrm{d}\lambda_k(t_2)$$

$$= \int_{\mathbb{E}^n}\int_{E_k} \zeta_{\alpha - x}(t_2)\zeta_{\beta'}(x)\, \mathrm{d}\lambda_k(t_2)\, \mathrm{d}\varphi(x)$$

$$= \int_{\mathbb{E}^n} \zeta_{\beta'}(x)\, \mathrm{d}\varphi(x) = \varphi(\beta')$$

$$= \Phi_j(E_k \cap (\rho K + t_1), \rho\beta + t_1)$$

$$= \Phi_j(\rho^{-1}(E_k - t_1) \cap K, \beta)$$

$$= \Phi_j(K \cap \gamma_k(-t_1, \rho^{-1}), \beta).$$

This yields

$$J = \int_{E_k^\perp \times SO(n)} \Phi_j(K \cap \gamma_k(-t_1, \rho^{-1}), \beta)\, \mathrm{d}(\lambda_{n-k} \otimes \nu)(t_1, \rho)$$

$$= \int_{E_k^\perp \times SO(n)} \Phi_j(K \cap \gamma_k(t, \rho), \beta)\, \mathrm{d}(\lambda_{n-k} \otimes \nu)(t, \rho)$$

$$= \int_{A(n,k)} \Phi_j(K \cap E, \beta)\, \mathrm{d}\mu_k(E).$$

Together with (4.5.12) this proves the assertion of Theorem 4.5.5. ∎

We are now in a position to compute the constants α_{njk} appearing in Theorems 4.5.2 and 4.5.5. For the ball rB^n of radius $r > 0$ we have by (4.2.27), for $\varepsilon > 0$,

$$\sum_{j=0}^{n} \varepsilon^{n-j}\kappa_{n-j}V_j(rB^n) = V_n((r + \varepsilon)B^n) = \sum_{j=0}^{n}\binom{n}{j} r^j \varepsilon^{n-j}\kappa_n,$$

hence

$$V_j(rB^n) = \binom{n}{j}\frac{\kappa_n}{\kappa_{n-j}}\, r^j.$$

By Theorem 4.5.5 with $K = B^n$ and $\beta = \mathbb{E}^n$ we have

$$\alpha_{njk}\binom{n}{k-j}\frac{\kappa_n}{\kappa_{k-j}} = \alpha_{njk}V_{n+j-k}(K)$$

$$= \int_{A(n,k)} V_j(K \cap E)\, \mathrm{d}\mu_k(E)$$

$$= \int_{SO(n)}\int_{E_k^\perp} V_j(K \cap \rho(E_k + t))\, \mathrm{d}\lambda_{n-k}(t)\, \mathrm{d}\nu(\rho).$$

Since $K \cap \rho(E_k + t)$ is a k-dimensional ball of radius $(1 - |t|^2)^{1/2}$, we obtain

$$\int_{E_k^\perp} V_j(K \cap \rho(E_k + t))\, \mathrm{d}\lambda_{n-k}(t) = \binom{k}{j} \frac{\kappa_k}{\kappa_{k-j}} \int_{E_k^\perp} (1 - |t|^2)^{j/2}\, \mathrm{d}\lambda_{n-k}(t)$$

$$= \binom{k}{j} \frac{\kappa_k}{\kappa_{k-j}} \frac{\kappa_{n+j-k}}{\kappa_j}.$$

We deduce that

$$\alpha_{njk} = \frac{\binom{k}{j}\kappa_k \kappa_{n+j-k}}{\binom{n}{k-j}\kappa_j \kappa_n}. \tag{4.5.13}$$

The identity

$$\frac{\Gamma(n+1)}{\Gamma\left(\frac{n}{2}+1\right)} = \frac{2^n \Gamma\left(\frac{n+1}{2}\right)}{\Gamma\left(\frac{1}{2}\right)}, \quad \text{or} \quad n!\kappa_n = 2^n \pi^{(n-1)/2} \Gamma\left(\frac{n+1}{2}\right),$$

can be used to bring (4.5.13) into the form given in Theorem 4.5.2. The proof of Theorem 4.5.2 is now complete.

We come now to integral-geometric formulae of a different type. An essential feature of the kinematic formulae of Theorem 4.5.2 and the Crofton formulae of Theorem 4.5.5 is that they refer to mean values of functions defined on the intersection of a fixed set and a moving set. We will now replace intersection by other geometric operations. These are the 'linear' operations of Minkowski addition and projection, which in the context of the present book are of particular interest. For the rest of this section, the investigations are restricted to convex bodies. First we consider rotational mean values for area measures of Minkowski sums.

Theorem 4.5.6. *Let $K, K' \in \mathcal{K}^n$ be convex bodies, $\omega, \omega' \in \mathcal{B}(S^{n-1})$ and $j \in \{0, \ldots, n-1\}$. Then*

$$\int_{SO(n)} S_j(K + \rho K', \omega \cap \rho\omega')\, \mathrm{d}\nu(\rho) = \frac{1}{\omega_n} \sum_{k=0}^{j} \binom{j}{k} S_k(K, \omega) S_{j-k}(K', \omega'). \tag{4.5.14}$$

Here we prefer to work with the area measures S_i and not with their renormalized versions Ψ_i, since in this way we can avoid complicated coefficients.

Proof of Theorem 4.5.6. The proof of the measurability of the integrand is similar to that for the integrand in (4.5.1). For given $\varepsilon > 0$, we define

$$V_\varepsilon(L, \alpha) := \mu_\varepsilon(L, \mathbb{E}^n \times \alpha) = \mathcal{H}^n(B_\varepsilon(L, \alpha))$$

for $L \in \mathcal{K}^n$ and $\alpha \in \mathcal{B}(S^{n-1})$,

$$h(\rho) := V_\varepsilon(K + \rho K', \omega \cap \rho\omega')$$

for $\rho \in SO(n)$ and

$$f(\rho, u) := \zeta_\omega(u)\zeta_{\omega'}(\rho^{-1}u)V_\varepsilon(K + \rho K', S^{n-1})$$

for $\rho \in SO(n)$ and $u \in S^{n-1}$, where ζ_A denotes the characteristic function of A on S^{n-1}. Then f is measurable. If we define

$$P(\rho, \alpha) := V_\varepsilon(K + \rho K', \alpha)/V_\varepsilon(K + \rho K', S^{n-1})$$

for $\rho \in SO(n)$ and $\alpha \in \mathcal{B}(S^{n-1})$, then $P(\cdot, \alpha)$ is measurable and $P(\rho, \cdot)$ is a probability measure, and from

$$h(\rho) = \int_{S^{n-1}} \zeta_{\omega\cap\rho\omega'}(u)V_\varepsilon(K + \rho K', \mathrm{d}u)$$

$$= \int_{S^{n-1}} f(\rho, u)P(\rho, \mathrm{d}u)$$

we deduce that h is measurable. From (4.2.5) we now infer that the function $\rho \mapsto S_j(K + \rho K', \omega \cap \rho\omega')$ is measurable.

For the proof of (4.5.14), we first consider the special case where $P, P' \in \mathcal{P}^n$ are n-dimensional polytopes and $j = n - 1$. Polytopes P, P' are said to be in *general relative position* if no pair of faces F of P and F' of P' is in special position.

Let $\rho \in SO(n)$ be a rotation for which P and $\rho P'$ are in general relative position. By Lemma 4.5.1, we thus exclude only a set of rotations ρ of ν-measure zero. Let G be a facet of the polytope $P + \rho P'$; let u be the outward unit normal vector of G. Then $G = F + \rho F'$, where $F = F(P, u)$ and $\rho F' = F(\rho P', u)$. Let $\dim F = i$ and $\dim F' = k$; then $i + k = n - 1$ since P and $\rho P'$ are in general relative position. By Theorem 2.2.1(a) we have

$$u \in N(P, F) \cap N(\rho P', \rho F').$$

Vice versa, if F is an i-face of P and F' is an $(n - 1 - i)$-face of P' with $N(P, F) \cap N(\rho P', \rho F') \neq \emptyset$, then $F + \rho F'$ is a j-face of $P + \rho P'$ with $j \leqq n - 1$. Again excluding a set of rotations of measure zero, we may assume that $j = n - 1$. Hence, for almost all ρ,

$$S_{n-1}(P + \rho P', \omega \cap \rho\omega') = \sum_{\substack{G \in \mathcal{F}_{n-1}(P+\rho P') \\ N(P+\rho P', G)\cap\omega\cap\rho\omega' \neq \emptyset}} \mathcal{H}^{n-1}(G)$$

$$= \underbrace{\sum_{i=0}^{n-1} \sum_{F \in \mathcal{F}_i(P)} \sum_{F' \in \mathcal{F}_{n-1-i}(P')}}_{A(\rho, F, F') \neq \emptyset} \mathcal{H}^{n-1}(F + \rho F')$$

with

$$A(\rho, F, F') := [N(P, F) \cap \omega] \cap \rho[N(P', F') \cap \omega'].$$

Observing that, for the rotations and faces considered, the set $A(\rho, F, F')$ is either one-pointed or empty, we arrive at

$$\int_{SO(n)} S_{n-1}(P + \rho P', \omega \cap \rho\omega')\, \mathrm{d}\nu(\rho)$$

$$= \sum_{i=0}^{n-1} \sum_{F \in \mathcal{F}_i(P)} \sum_{F' \in \mathcal{F}_{n-1-i}(P')} \int_{SO(n)} \mathcal{H}^{n-1}(F + \rho F')\, \mathrm{card}\, A(\rho, F, F')\, \mathrm{d}\nu(\rho).$$

We fix $i \in \{0, \ldots, n-1\}$, $F \in \mathcal{F}_i(P)$ and $F' \in \mathcal{F}_{n-1-i}(P')$. For almost all ρ we have $\dim(F + \rho F') = n - 1$ and hence

$$\mathcal{H}^{n-1}(F + \rho F') = [F, \rho F']\mathcal{H}^i(F)\mathcal{H}^{n-1-i}(F').$$

It remains, therefore, to compute

$$\int_{SO(n)} \mathrm{card}\, A(\rho, F, F')[F, \rho F']\, \mathrm{d}\nu(\rho).$$

We argue in the same way as in the proof of Lemma 4.5.4, defining a more general integral by

$$J(\beta, \beta') := \int_{SO(n)} \mathrm{card}\,(\beta \cap \rho\beta')[F, \rho F']\, \mathrm{d}\nu(\rho)$$

for Borel sets $\beta \subset L(F)^\perp \cap S^{n-1}$, $\beta' \subset L(F')^\perp \cap S^{n-1}$. If $\tau \in SO(n)$ is a rotation that maps $L(F)^\perp$ into itself and keeps $L(F)$ pointwise fixed, then $\mathrm{card}\,(\tau\beta \cap \rho\beta') = \mathrm{card}\,(\beta \cap \tau^{-1}\rho\beta')$ and $[F, \rho F'] = [F, \tau^{-1}\rho F']$ and hence $J(\tau\beta, \beta') = J(\beta, \beta')$ by the invariance of ν. Since $J(\cdot, \beta')$ is obviously non-negative, finite and σ-additive, it follows from the uniqueness of spherical Lebesgue measure that $J(\beta, \beta') = c\sigma^{n-1-i}(\beta)$ where the constant c depends on β'. In combination with the same argument with the roles of β and β' interchanged, this shows that

$$J(\beta, \beta') = \alpha_{ni}\sigma^{n-1-i}(\beta)\sigma^i(\beta'),$$

where the constant α_{ni} evidently depends only on n and i. The special case $\beta = N(P, F) \cap \omega$, $\beta' = N(P', F') \cap \omega'$ now yields

$$\int_{SO(n)} S_{n-1}(P + \rho P', \omega \cap \rho\omega')\, \mathrm{d}\nu(\rho) = \sum_{k=0}^{n-1} \gamma_{nk} S_k(P, \omega) S_{n-1-i}(P', \omega')$$

with constants γ_{nk}, where (4.2.18) has been used. By an approximation argument strictly analogous to that used in the proof of Theorem 4.5.2 we extend this formula to general convex bodies K, K' instead of polytopes P, P'. The explicit values of the constants γ_{nk} are found by choosing balls of radii 1 and r for K and K' respectively, and $\omega = \omega' = S^{n-1}$, and then comparing equal powers of r. In the resulting formula we now replace K by $K + \varepsilon B^n$ with $\varepsilon > 0$, develop both sides according to formula (4.2.16) and compare equal powers of ε. This yields formula (4.5.14) and thus completes the proof of Theorem 4.5.6. ■

Formula (4.5.14) can serve as the source for several other integral-geometric results. These include extensions to generalized curvature measures, projection formulae and kinematic formulae for non-intersecting convex bodies. We consider results of the latter type first.

For convex bodies $K, K' \in \mathcal{K}^n$ we denote by

$$d(K, K') := \min\{|x - x'| \mid x \in K, x' \in K'\}$$

their Euclidean distance. The principal kinematic formula provides an explicit expression, in terms of Minkowski functionals of K and K', for the invariant measure of the set $\{g \in G_n | d(K, gK') = 0\}$. From this one can deduce a similar expression for the measure of the set $\{g \in G_n | 0 < d(K, gK') \leqq \varepsilon\}$ for given $\varepsilon > 0$, by applying the principal kinematic formula to $K + \varepsilon B_n$ and to K, and subtracting. More generally, we are interested in the measure of the set of all rigid motions g for which $0 < d(K, gK') \leqq \varepsilon$ and for which the points $x \in K$, $x' \in gK'$ defining the distance belong to certain preassigned sets. We can impose a further condition on the set of rigid motions to be measured. If $x \in K$, $x' \in K'$ are such that

$$|x - x'| = d(K, K') > 0,$$

we define a unit vector by

$$u(K, K') := \frac{x' - x}{|x' - x|}.$$

The pair (x, x') is not necessarily unique but one readily verifies that this vector, pointing from K to K' 'in the most direct way', is unique. Clearly we have $(x, u(K, K')) \in \operatorname{Nor} K$, the set of support elements of K, and $(x', -u(K, K')) \in \operatorname{Nor} K'$. In addition, the vectors $u(K, gK')$ may be restricted to lie in preassigned sets. More generally, the following theorem treats the case where shortest distances between a fixed convex body and a disjoint moving convex body are realized within given sets of support elements.

Theorem 4.5.7. *Let $K, K' \in \mathcal{K}^n$ be convex bodies and let $\eta, \eta' \in \mathcal{B}(\Sigma)$. For $\varepsilon > 0$, let $N_\varepsilon(K, K', \eta, \eta')$ be the set of all rigid motions $g \in G_n$ for which there exist points $x \in K$ and $x' \in gK'$ such that*

$$0 < |x - x'| = d(K, gK') \leqq \varepsilon,$$

$$(x, u(K, gK')) \in \eta,$$

$$(x', -u(K, gK')) \in g\eta'.$$

Then

$$\mu(N_\varepsilon(K, K', \eta, \eta'))$$
$$= \frac{1}{n\omega_n} \sum_{i=1}^{n} \varepsilon^i \binom{n}{i} \sum_{k=0}^{n-i} \binom{n-i}{k} \Theta_k(K, \eta)\Theta_{n-i-k}(K', \eta'). \quad (4.5.15)$$

Before the proof, some preparations are necessary. In the following, measurability for functions on $SO(n)$ and G_n and for subsets of these spaces refers to the completions of the measure spaces $(SO(n), \mathcal{B}(SO(n)), \nu)$ and $(G_n, \mathcal{B}(G_n), \mu)$, respectively, The completed measures are denoted by the same symbols.

If $(x, u) \in \Sigma = \mathbb{E}^n \times S^{n-1}$ and $g = g_{t,\rho}$ is the rigid motion defined by $gx = \rho x + t$ with $\rho \in SO(n)$ and $t \in \mathbb{E}^n$, we define $g(x, u) := (gx, \rho u)$; also $-(x, u) := (-x, -u)$. Let $K, K' \in \mathcal{K}^n$ be convex bodies. If $(x, u) \in \operatorname{Nor} K$ and $(x', u) \in \operatorname{Nor} K'$, then $(x + x', u) \in \operatorname{Nor}(K + K')$. Vice versa, let $(y, u) \in \operatorname{Nor}(K + K')$. The point $y \in K + K'$ is of the form $y = x + x'$ with $x \in K$, $x' \in K'$, and necessarily $(x, u) \in \operatorname{Nor} K$ and $(x', u) \in \operatorname{Nor} K'$. In general, x and x' are not uniquely determined. If $y = z + z'$ with $(z, u) \in \operatorname{Nor} K$ and $(z', u) \in \operatorname{Nor} K'$ is a different representation, then $z - x = x' - z'$, hence K and K' contain parallel segments lying in supporting hyperplanes with the same outer normal vector u. In this case we say that K and K' are in *singular relative position*.

We define an operation for sets of support elements which is adapted to Minkowski addition. For $\eta, \eta' \subset \Sigma$ let

$$\eta * \eta' := \{(x + x', u) \in \Sigma | (x, u) \in \eta, (x', u) \in \eta'\}.$$

In particular, for $\beta, \beta' \subset \mathbb{E}^n$ and $\omega, \omega' \subset S^{n-1}$ we have

$$(\beta \times \omega) * (\beta' \times \omega') = (\beta + \beta') \times (\omega \cap \omega').$$

If $\eta \subset \operatorname{Nor} K$ and $\eta' \subset \operatorname{Nor} K'$, then $\eta * \eta' \subset \operatorname{Nor}(K + K')$.

Since the sum of two Borel sets is in general not a Borel set, it is clear that $\eta * \eta'$ need not be a Borel set if $\eta, \eta' \in \mathcal{B}(\Sigma)$. The following lemma ensures that this will cause no problems.

Lemma 4.5.8. *For convex bodies $K, K' \in \mathcal{K}^n$ and Borel sets $\eta \subset \operatorname{Nor} K$, $\eta' \subset \operatorname{Nor} K'$, the set $\eta * \rho\eta'$ is a Borel set for ν-almost all $\rho \in SO(n)$.*

Proof. Let R denote the set of all rotations $\rho \in SO(n)$ for which K and $\rho K'$ are not in singular relative position. Let $\rho \in R$. For $(y, u) \in \operatorname{Nor}(K + \rho K')$ we have $y = x + \rho x'$ with suitable $(x, u) \in \operatorname{Nor} K$ and $(\rho x', u) \in \operatorname{Nor} \rho K'$. Since $\rho \in R$, the points x and x' are uniquely determined. Hence, we can define maps

$$f: \operatorname{Nor}(K + \rho K') \to \operatorname{Nor} K \quad \text{by } f(y, u) := (x, u),$$
$$\bar{f}: \operatorname{Nor}(K + \rho K') \to \operatorname{Nor} K' \quad \text{by } \bar{f}(y, u) := (x', \rho^{-1}u).$$

It is easy to see that these maps are continuous, hence

$$\eta * \rho\eta' = f^{-1}(\eta) \cap \bar{f}^{-1}(\eta')$$

is a Borel set. By Corollary 2.3.11 we have $\nu(SO(n)\backslash R) = 0$. This proves the assertion. ■

Proof of Theorem 4.5.7. Let $K, K' \in \mathcal{K}^n$ and $\varepsilon > 0$ be given. We are going to show that the set $N_\varepsilon(K, K', \eta, \eta')$ defined in Theorem 4.5.7 is measurable for arbitrary Borel sets $\eta, \eta' \in \mathcal{B}(\Sigma)$. The following proposition will be useful.

Proposition 1. Let φ be a map from $\mathcal{B}(\Sigma)$ into the set of all subsets of G_n that has the following properties:

(a) If $\eta_1, \eta_2 \in \mathcal{B}(\Sigma)$ and $\eta_1 \cap \eta_2 = \emptyset$, then $\varphi(\eta_1) \cap \varphi(\eta_2)$ is of measure zero;

(b) If $\eta_i \in \mathcal{B}(\Sigma)$ for $i \in \mathbb{N}$, then $\varphi(\bigcup_{i\in\mathbb{N}} \eta_i) = \bigcup_{i\in\mathbb{N}} \varphi(\eta_i)$;

(c) If $\eta \in \mathcal{B}(\Sigma)$ is closed, then $\varphi(\eta)$ is measurable.

Under these assumptions, the set $\varphi(\eta)$ is measurable for each $\eta \in \mathcal{B}(\Sigma)$.

In fact, it follows from (a), (b) and (c) that the system of all sets $\eta \in \mathcal{B}(\Sigma)$ for which $\varphi(\eta)$ is measurable is a σ-algebra in Σ that contains the closed sets; hence this system coincides with $\mathcal{B}(\Sigma)$. This proves Proposition 1.

Now let $\eta, \eta' \subset \Sigma$ be closed and write

$$\bar{N} := N_\varepsilon(K, K', \eta, \eta') \cup N_0(K, K'),$$

where $N_0(K, K')$ is the set of all $g \in G_n$ for which gK' touches K (which means that K and gK' intersect, but can be separated by a hyperplane). We want to show that $\bar{N}$ is closed. Let $(g_i)_{i\in\mathbb{N}}$ be a sequence in $\bar{N}$ that converges to some $g \in G_n$. Then the sequence $(g_iK')_{i\in\mathbb{N}}$ converges to gK'. Since each body g_iK' can be separated from K by a hyperplane, the same is true of gK'. Further, for each $i \in \mathbb{N}$ there exist points $x_i \in \operatorname{bd} K$, $x_i' \in \operatorname{bd} K'$ such that

$$|x_i - g_i x_i'| = d(K, g_iK') \leqq \varepsilon,$$

$$\left.\begin{array}{l}(x_i, u(K, g_iK')) \in \eta, \\ (g_ix_i', -u(K, g_iK')) \in g_i\eta'\end{array}\right\} \quad \text{if } d(K, g_iK') > 0.$$

Since $\operatorname{bd} K$, $\operatorname{bd} K'$ are compact, there exists a sequence $(i_k)_{k\in\mathbb{N}}$ such that $(x_{i_k})_{k\in\mathbb{N}}$ converges to a point $x \in \operatorname{bd} K$ and $(x'_{i_k})_{k\in\mathbb{N}}$ converges to a point

$x' \in \operatorname{bd} K'$. It follows that

$$|x - gx'| = d(K, gK') \leqq \varepsilon,$$

$$\left.\begin{array}{l}(x, u(K, gK')) \in \eta, \\ (gx', -u(K, gK')) \in g\eta'\end{array}\right\} \text{ if } d(K, gK') > 0.$$

Thus $g \in \bar{N}$, which shows that $\bar{N}$ is closed.

Now define $\varphi(\eta) := N_\varepsilon(K, K', \eta, \eta')$ for $\eta \in \mathcal{B}(\Sigma)$, where $\eta' \subset \Sigma$ is a fixed closed set. Let $\eta_1, \eta_2 \in \mathcal{B}(\Sigma)$ be disjoint and suppose that $g \in \varphi(\eta_1) \cap \varphi(\eta_2)$. For $i = 1, 2$ there is a point $x_i \in K$ at distance $d(K, gK')$ from gK' such that $(x_i, u(K, gK')) \in \eta_i$. Since $\eta_1 \cap \eta_2 = \emptyset$, necessarily $x_1 \neq x_2$; hence K and gK' are in singular relative position. From Corollary 2.3.11 it follows that the set $\varphi(\eta_1) \cap \varphi(\eta_2)$ is of measure zero. If η is closed, then $N_\varepsilon(K, K', \eta, \eta') \cup N_0(K, K')$ is closed, as shown above, and since $N_0(K, K')$ is of measure zero (by the argument used after (4.5.4)), it follows that $N_\varepsilon(K, K', \eta, \eta')$ is measurable. Thus the map φ satisfies the assumptions of Proposition 1 (where (b) is satisfied trivially). Hence, for each $\eta \in \mathcal{B}(\Sigma)$ the set $N_\varepsilon(K, K', \eta, \eta')$ is measurable. Now we fix $\eta \in \mathcal{B}(\Sigma)$ and define $\psi(\eta') := N_\varepsilon(K, K', \eta, \eta')$. For η' closed, we have shown that $\psi(\eta')$ is measurable. Proposition 1 yields that $N_\varepsilon(K, K', \eta, \eta')$ is a measurable set, where now η, $\eta' \in \mathcal{B}(\Sigma)$ may be arbitrary Borel sets.

We can now apply Fubini's theorem and obtain

$$\mu(N_\varepsilon(K, K', \eta, \eta')) = \int_{SO(n)} \int_{T(\rho)} \mathrm{d}\mathcal{H}^n \, \mathrm{d}\nu(\rho), \tag{4.5.16}$$

where

$$T(\rho) := \{t \in \mathbb{E}^n \mid g_{t,\rho} \in N_\varepsilon(K, K', \eta, \eta')\}.$$

This set can be described in a different way. Suppose that $t \in T(\rho)$. Then there exist points $x \in K$, $x' \in \rho' K' + t$ such that

$$0 < |x - x'| = d(K, \rho K' + t) \leqq \varepsilon,$$

$$(x, u(K, \rho K' + t)) \in \eta,$$

$$(x', -u(K, \rho K' + t) \in \rho\eta' + t.$$

The vector $u = u(K, \rho K' + t)$ satisfies $(x, u) \in \operatorname{Nor} K$ and $(x', -u) \in \operatorname{Nor}(\rho K' + t)$. Writing $x' = \rho y' + t$, we have

$$(x, u) \in \eta \cap \operatorname{Nor} K, \quad (\rho y', u) \in \rho[\eta' \cap \operatorname{Nor} K'].$$

With $d = d(K, \rho K' + t)$ we get $x' = x + du$ and hence $t = x - \rho y' + du$. It is obvious that $p(K - \rho K', t) = x - \rho y'$ and $u(K - \rho K', t) = u$, thus

$$(p(K - \rho K', t), u(K - \rho K', t)) = (x - \rho y', u)$$
$$\in [\eta \cap \operatorname{Nor} K] * -\rho[\eta' \cap \operatorname{Nor} K'],$$

which shows that

$$t \in M_\varepsilon(K - \rho K', [\eta \cap \operatorname{Nor} K] * -\rho[\eta' \cap \operatorname{Nor} K']) \quad (4.5.17)$$

(with M_ε as in Section 4.1).

If we now assume that K and $-\rho K'$ are not in singular relative position, then we can reverse the argument and conclude that (4.5.17) implies $t \in T(\rho)$. Thus we have

$$T(\rho) = M_\varepsilon(K - \rho K', [\eta \cap \operatorname{Nor} K] * -\rho[\eta' \cap \operatorname{Nor} K'])$$

for a set of rotations $\rho \in SO(n)$ which by Corollary 2.3.11 has full measure. Lemma 4.5.8 tells us that for almost all ρ the set $[\eta \cap \operatorname{Nor} K] * -\rho[\eta' \cap \operatorname{Nor} K']$ is a Borel set, so that (4.2.4) can be applied to compute the measure $\mathcal{H}^n(T(\rho))$. Now (4.5.16) yields

$$\mu(N_\varepsilon(K, K', \eta, \eta')) = \frac{1}{n}\sum_{i=0}^{n-1} \varepsilon^{n-i}\binom{n}{i}$$

$$\times \int_{SO(n)} \Theta_i(K - \rho K', [\eta \cap \operatorname{Nor} K] * -\rho[\eta' \cap \operatorname{Nor} K'])\, \mathrm{d}\nu(\rho). \quad (4.5.18)$$

To compute the integrals, we first consider strictly convex bodies. Let $\pi_2 \colon \mathbb{E}^n \times S^{n-1} \to S^{n-1}$ denote the projection onto the second factor. If K is strictly convex and $\eta \subset \Sigma$, then

$$\eta \cap \operatorname{Nor} K = [\mathbb{E}^n \times \pi_2(\eta \cap \operatorname{Nor} K)] \cap \operatorname{Nor} K,$$

since for $u \in S^{n-1}$ there is exactly one $x \in \mathbb{E}^n$ for which $(x, u) \in \operatorname{Nor} K$. Using (4.2.6) and the definition of S_m, we get, for $\eta \in \mathcal{B}(\Sigma)$,

$$\begin{aligned}\Theta_m(K, \eta) &= \Theta_m(K, \eta \cap \operatorname{Nor} K)\\ &= \Theta_m(K, [\mathbb{E}^n \times \pi_2(\eta \cap \operatorname{Nor} K)] \cap \operatorname{Nor} K)\\ &= \Theta_m(K, \mathbb{E}^n \times \pi_2(\eta \cap \operatorname{Nor} K))\\ &= S_m(K, \pi_2(\eta \cap \operatorname{Nor} K)).\end{aligned}$$

Now let $K, K' \in \mathcal{K}^n$ be strictly convex; then $K - \rho K'$ is strictly convex for all $\rho \in SO(n)$. Let $\eta, \eta' \in \mathcal{B}(\Sigma)$. Observing that

$$[\eta \cap \operatorname{Nor} K] * -\rho[\eta' \cap \operatorname{Nor} K'] \subset \operatorname{Nor}(K - \rho K'),$$

$$\pi_2([\eta \cap \operatorname{Nor} K] * -\rho[\eta' \cap \operatorname{Nor} K']) = \pi_2(\eta \cap \operatorname{Nor} K) \cap -\rho\pi_2(\eta' \cap \operatorname{Nor} K')$$

and $\Theta_m(-K', -\eta') = \Theta_m(K', \eta')$, we deduce from (4.5.18) and (4.5.14) that (4.5.15) is true.

Equality (4.5.15) in the general case is now proved by approximation. Again let $K, K' \in \mathcal{K}^n$ be general convex bodies.

Proposition 2. For fixed $\eta' \in \mathcal{B}(\Sigma)$, the function $\mu(N_\varepsilon(K, K', \cdot, \eta'))$ is a finite measure on $\mathcal{B}(\Sigma)$.

For the proof, only the σ-additivity has to be shown. For η_1, $\eta_2 \in \mathcal{B}(\Sigma)$ with $\eta_1 \cap \eta_2 = \emptyset$ we have seen above that

$$\mu(N_\varepsilon(K, K', \eta_1, \eta') \cap N_\varepsilon(K, K', \eta_2, \eta')) = 0,$$

hence the additivity of μ implies that $\mu(N_\varepsilon(K, K', \cdot, \eta'))$ is additive. Let $(\eta_i)_{i\in\mathbb{N}}$ be a sequence in $\mathcal{B}(\Sigma)$ such that $\eta_i \downarrow \emptyset$. Then

$$N_\varepsilon(K, K', \eta_i, \eta') \downarrow N := \bigcap_{j\in\mathbb{N}} N_\varepsilon(K, K', \eta_j, \eta')$$

for $i \to \infty$, hence

$$\lim_{i\to\infty} \mu(N_\varepsilon(K, K', \eta_i, \eta')) = \mu(N)$$

by the σ-additivity of μ. Let $g \in N$. For each $j \in \mathbb{N}$ there is a point $x_j \in \operatorname{bd} K$ at distance $d(K, gK')$ from gK' such that $(x_j, u(K, gK')) \in \eta_j$. Since $\bigcap_{j\in\mathbb{N}} \eta_j = \emptyset$, not all the points x_j coincide, thus K and gK' are in singular relative position. By Corollary 2.3.11, $\mu(N) = 0$. Thus $\mu(N_\varepsilon(K, K', \cdot, \eta'))$ is $\emptyset$-continuous and hence σ-additive. This proves Proposition 2.

Proposition 3. If the sequence $(K_i)_{i\in\mathbb{N}}$ in $\mathcal{K}^n$ converges to K, then

$$\liminf_{i\to\infty} \mu(N_\varepsilon(K_i, K', \eta, \eta')) \geqq \mu(N_\varepsilon(K, K', \eta, \eta'))$$

for all open sets $\eta, \eta' \subset \Sigma$, and

$$\lim_{i\to\infty} \mu(N_\varepsilon(K_i, K', \Sigma, \Sigma)) = \mu(N_\varepsilon(K, K', \Sigma, \Sigma)).$$

For the proof, let $\eta, \eta' \subset \Sigma$ be open and denote by $N_\varepsilon^0(K, K', \eta, \eta')$ the set of all $g \in N_\varepsilon(K, K', \eta, \eta')$ for which $d(K, gK') \neq \varepsilon$ and K and gK' are not in singular relative position. Let $g \in N_\varepsilon^0(K, K', \eta, \eta')$. For each $i \in \mathbb{N}$ choose points $x_i \in \operatorname{bd} K_i$ and $x_i' \in \operatorname{bd} gK'$ such that $|x_i - x_i'| = d(K_i, gK')$. Let $x \in \operatorname{bd} K$, $x' \in \operatorname{bd} gK'$ be the unique points for which $|x - x'| = d(K, gK')$. It is easy to see that $x_i \to x$ and $x_i' \to x'$ for $i \to \infty$. It follows that $0 < d(K_i, gK) < \varepsilon$ for all i sufficiently large. Since $(x, u(K, gK')) \in \eta$, $(x', -u(K, gK')) \in g\eta'$ and η, η' are open, we also have $(x_i, u(K_i, gK')) \in \eta$ and $(x_i', -u(K_i, gK')) \in g\eta'$ for almost all i; thus $g \in N_\varepsilon(K_i, K', \eta, \eta')$ for almost all i. We have proved that

$$N_\varepsilon^0(K, K', \eta, \eta') \subset \liminf_{i\to\infty} \mu(N_\varepsilon(K_i, K', \eta, \eta'))$$

and hence that

$$\begin{aligned}\mu(N_\varepsilon(K, K', \eta, \eta')) &= \mu(N_\varepsilon^0(K, K', \eta, \eta')) \\ &\leqq \mu(\liminf_{i\to\infty} N_\varepsilon(K_i, K', \eta, \eta')) \\ &\leqq \liminf_{i\to\infty} \mu(N_\varepsilon(K_i, K', \eta, \eta')),\end{aligned}$$

which is the first assertion of Proposition 3. Since $N_\varepsilon(K_i, K', \Sigma, \Sigma)$ is the set of all $g \in G_n$ for which gK' meets $K_i + \varepsilon B^n$ but not K_i, its measure can, by means of the principal kinematic formula, be expressed in terms of Minkowski functionals, from which its continuous dependence on K_i is clear. This proves Proposition 3.

Of course, results similar to Propositions 2 and 3 are valid with the roles of the pairs (K, η) and (K', η') interchanged.

We are now in a position to prove formula (4.5.15) for general convex bodies. Let us write the assertion (4.5.15) in the form

$$\mu(N_\varepsilon(K, K', \eta, \eta')) = \sum_{r,s=0}^{n-1} a_{rs}\Theta_r(K, \eta)\Theta_s(K', \eta'), \quad (4.5.19)$$

where ε is fixed and the coefficients a_{rs} are those resulting from rearranging (4.5.15). Then equality (4.5.19) is already established if K and K' are strictly convex bodies and $\eta, \eta' \in \mathcal{B}(\Sigma)$.

Let $K \in \mathcal{K}^n$ be strictly convex and $K' \in \mathcal{K}^n$ arbitrary. Let $(K_i')_{i\in\mathbb{N}}$ be a sequence of strictly convex bodies converging to K'. By Proposition 3,

$$\liminf_{i\to\infty} \mu(N_\varepsilon(K, K_i', \eta, \Sigma)) \geqq \mu(N_\varepsilon(K, K', \eta, \Sigma))$$

for open $\eta \subset \Sigma$ and

$$\lim_{i\to\infty} \mu(N_\varepsilon(K, K_i', \Sigma, \Sigma)) = \mu(N_\varepsilon(K, K', \Sigma, \Sigma)).$$

Thus $\mu(N_\varepsilon(K, K_i', \cdot, \Sigma)) \xrightarrow{w} \mu(N_\varepsilon(K, K', \cdot, \Sigma))$ for $i \to \infty$. Since (4.5.19) can be applied to K and K_i', it follows that

$$\mu(N_\varepsilon(K, K', \cdot, \Sigma)) = \sum_{r,s} a_{rs}\Theta_r(K, \cdot)\Theta_s(K', \Sigma). \quad (4.5.20)$$

Let K and K_i' be as before and let $\eta \subset \Sigma$ be open. By Proposition 3 we have

$$\liminf_{i\to\infty} \mu(N_\varepsilon(K, K_i', \eta, \eta')) \geqq \mu(N_\varepsilon(K, K_i, \eta, \eta'))$$

for open $\eta' \subset \Sigma$, and from (4.5.20) we obtain

$$\lim_{i\to\infty} \mu(N_\varepsilon(K, K_i', \eta, \Sigma)) = \mu(N_\varepsilon(K, K', \eta, \Sigma)).$$

Thus $\mu(N_\varepsilon(K, K_i', \eta, \cdot)) \xrightarrow{w} \mu(N_\varepsilon(K, K', \eta, \cdot))$ for $i \to \infty$. From (4.5.19) and the weak continuity of the generalized curvature measures we infer that

$$\mu(N_\varepsilon(K, K', \eta, \eta')) = \sum_{r,s} a_{rs}\Theta_s(K, \eta)\Theta_s(K', \eta') \quad (4.5.21)$$

for all $\eta' \in \mathcal{B}(\Sigma)$. Since for fixed K, K', η' both sides, as functions of η, are measures the equality holds for arbitrary Borel sets η.

Now let $K, K' \in \mathcal{K}^n$ both be arbitrary. Choosing a sequence $(K_i)_{i\in\mathbb{N}}$

of strictly convex bodies converging to K, we deduce from Proposition 3 that

$$\mu(N_\varepsilon(K_i, K', \cdot, \Sigma)) \xrightarrow{w} \mu(N_\varepsilon(K, K', \cdot, \Sigma))$$

and then from (4.5.20) that

$$\mu(N_\varepsilon(K, K', \cdot, \Sigma)) = \sum_{r,s} a_{rs} \Theta_r(K, \cdot)\Theta_s(K', \Sigma).$$

Here the roles of K and K' may be interchanged, hence

$$\mu(N_\varepsilon(K, K', \Sigma, \cdot)) = \sum_{r,s} a_{rs} \Theta_r(K, \Sigma)\Theta_s(K', \cdot). \tag{4.5.22}$$

Let K_i be as before, and let $\eta' \subset \Sigma$ be open. From Proposition 3 and from (4.5.22) we see that

$$\mu(N_\varepsilon(K_i, K', \cdot, \eta')) \xrightarrow{w} \mu(N_\varepsilon(K, K', \cdot, \eta')).$$

From (4.5.21) and the weak continuity of the generalized curvature measures we get

$$\mu(N_\varepsilon(K, K', \eta, \eta')) = \sum_{r,s} a_{rs} \Theta_r(K, \eta)\Theta_s(K', \eta')$$

for $\eta \in \mathcal{B}(\Sigma)$. Since both sides are measures as functions of η', we deduce the equality for arbitrary $\eta' \in \mathcal{B}(\Sigma)$. This completes the proof of Theorem 4.5.7. ■

Comparing the coefficients of ε^{n-j} in formulae (4.5.15) and (4.5.18), we obtain the following theorem as a corollary.

Theorem 4.5.9. *Let $K, K' \in \mathcal{K}^n$ be convex bodies, let $\eta \subset \operatorname{Nor} K$, $\eta' \subset \operatorname{Nor} K'$ be Borel sets and let $j \in \{0, \ldots, n-1\}$. Then*

$$\int_{SO(n)} \Theta_j(K + \rho K', \eta * \rho\eta')\,\mathrm{d}\nu(\rho) = \frac{1}{\omega_n} \sum_{k=0}^{j} \binom{j}{k} \Theta_k(K, \eta)\Theta_{j-k}(K', \eta'). \tag{4.5.23}$$

This formula generalizes (4.5.14). Another special case of (4.5.23) is the equality

$$\int_{SO(n)} C_j(K + \rho K', \beta + \rho\beta')\,\mathrm{d}\nu(\rho) = \frac{1}{\omega_n} \sum_{k=0}^{j} \binom{j}{k} C_k(K, \beta)C_{j-k}(K', \beta'), \tag{4.5.24}$$

which is valid for Borel sets $\beta \subset \operatorname{bd} K$, $\beta' \subset \operatorname{bd} K'$. This follows from (4.5.23) since $C_m(K, \beta) = \Theta_m(K, \beta \times S^{n-1})$ and

$$(\beta + \rho\beta') \times S^{n-1} = (\beta \times S^{n-1}) * \rho(\beta' \times S^{n-1}).$$

From (4.5.23) we shall now derive integral-geometric formulae for projections.

If $E \subset \mathbb{E}^n$ is a linear subspace, we denote by $x|E$ the image of $x \in \mathbb{E}^n$ under orthogonal projection onto E, and we use a similar notation for subsets of $\mathbb{E}^n$. For sets $\eta \subset \Sigma$ we define

$$\eta|E := \{(x|E, u)|(x, u) \in \eta \text{ and } u \in E\}.$$

Theorem 4.5.10. *Let $K \in \mathcal{K}^n$ be a convex body, let $\eta \subset \operatorname{Nor} K$ be a Borel set and let $E \subset \mathbb{E}^n$ be a k-dimensional linear subspace, where $k \in \{1, \ldots, n-1\}$. Then*

$$\int_{SO(n)} \Theta'_j(K|\rho E, \eta|\rho E)\, \mathrm{d}\nu(\rho) = \frac{\omega_k}{\omega_n} \Theta_j(K, \eta) \tag{4.5.25}$$

for $j \in \{0, \ldots, k-1\}$, where Θ'_j is the generalized curvature measure taken with respect to the subspace ρE.

Special cases of (4.5.25) are the formulae

$$\int_{SO(n)} S'_j(K|\rho E, \omega \cap \rho E)\, \mathrm{d}\nu(\rho) = \frac{\omega_k}{\omega_n} S_j(K, \omega) \tag{4.5.26}$$

for $\omega \in \mathcal{B}(S^{n-1})$ and

$$\int_{SO(n)} C'_j(K|\rho E, \beta|\rho E)\, \mathrm{d}\nu(\rho) = \frac{\omega_k}{\omega_n} C_j(K, \beta) \tag{4.5.27}$$

for Borel sets $\beta \subset \operatorname{bd} K$.

Proof of Theorem 4.5.10. We choose an $(n-k)$-dimensional convex body $K' \subset E^\perp$ with $\mathcal{H}^{n-k}(K') = 1$ and let η' be the set of all support elements (x, u) of K' with $x \in \operatorname{relint} K'$. From (4.2.4) it is then clear that

$$\Theta_j(K', \eta') = \begin{cases} n\kappa_k \binom{n}{k}^{-1} & \text{for } j = n-k \\ 0 & \text{for } j \neq n-k. \end{cases} \tag{4.5.28}$$

For given $\varepsilon > 0$, let $x \in M_\varepsilon(K + K', \eta * \eta')$. Then

$$0 < d(K + K', x) \leqq \varepsilon,$$
$$(p(K + K', x), u(K + K', x)) \in \eta * \eta'.$$

Writing $u := u(K + K', x)$, we have $p(K + K', x) = y + y'$ with $(y, u) \in \eta \subset \operatorname{Nor} K$ and $(y', u) \in \eta' \subset \operatorname{Nor} K'$. This implies $u \in E$ by the choice of η' and $(y|E, u) \in \eta|E$. Obviously we have $|x|E - y|E| = d(K + K', x)$, $p(K|E, x|E) = y|E$ and $u(K|E, x|E) = u$; thus

$x|E \in M'_\varepsilon(K|E, \eta|E)$, where M'_ε is defined with respect to the subspace E. Vice versa, one readily verifies that each point in $M'_\varepsilon(K|E, \eta|E)$ is the projection of a point in $M_\varepsilon(K + K', \eta * \eta')$; thus

$$M_\varepsilon(K + K', \eta * \eta')|E = M'_\varepsilon(K|E, \eta|E).$$

If K and K' are not in singular relative position, then the set of points in $M_\varepsilon(K + K', \eta * \eta')$ that project into the same point of E is a translate of relint K'; furthermore, $\eta|E$ is a Borel set. Hence, Fubini's theorem gives

$$\mathcal{H}^n(M_\varepsilon(K + K', \eta * \eta')) = \mathcal{H}^k(M'_\varepsilon(K|E, \eta|E)),$$

and (4.2.4) yields

$$\frac{1}{n}\sum_{i=0}^{n-1} \varepsilon^{n-i}\binom{n}{i}\Theta_i(K + K', \eta * \eta') = \frac{1}{k}\sum_{j=0}^{k-1} \varepsilon^{k-j}\binom{k}{j}\Theta'_j(K|E, \eta|E).$$

For ν-almost all ρ, this can be applied to ρE, $\rho K'$, $\rho\eta'$ instead of E, K', η'. If we then integrate over all rotations ρ, use (4.5.23) and (4.5.28) and compare the coefficients of equal powers of ε, we complete the proof of formula (4.5.25). ∎

Theorem 4.5.7 has a counterpart where the moving convex body K' is replaced by a moving k-flat. Let $K \in \mathcal{K}^n$ be a convex body and $E \in A(n, k)$ be a k-dimensional affine subspace of $\mathbb{E}^n$, where $k = \{0, \ldots, n-1\}$. If $x \in K$ and $x' \in E$ are points between which the distance is a minimum, we write $d(K, E) := |x - x'|$ and $u(K, E) := (x' - x)/|x' - x|$.

Theorem 4.5.11. *Let $K \in \mathcal{K}^n$ be a convex body and let $\eta \in \mathcal{B}(\Sigma)$ be a Borel set; let $k \in \{0, \ldots, n-1\}$. For $\varepsilon > 0$, Let $N^k_\varepsilon(K, \eta)$ be the set of all k-flats $E \in A(n, k)$ for which there exist points $x \in K$ and $x' \in E$ such that*

$$0 < |x - x'| = d(K, E) \leqq \varepsilon,$$
$$(x, u(K, E)) \in \eta.$$

Then

$$\mu_k(N^k_\varepsilon(K, \eta)) = \frac{\kappa_{n-k}}{\omega_n}\sum_{i=1}^{n-k} \varepsilon^i\binom{n-k}{i}\Theta_{n-k-i}(K, \eta). \quad (4.5.29)$$

Proof. Some of the subsequent arguments, not surprisingly, are similar to those in the proof of Theorem 4.5.7. The convex body K and the k-flat E are said to be *in singular relative position* if some translate of E supports K (that is, meets K and lies in a supporting hyperplane of K) and contains more than one point of K. We fix a k-dimensional linear subspace L_k of $\mathbb{E}^n$ and use the fact that the invariant measure ν_k can be written as the image measure of $\lambda_{n-k} \otimes \nu$ under the map $(t, \rho) \mapsto \rho(L_k + t)$ from $L_k^\perp \times SO(n)$ into $A(n, k)$, where λ_{n-k} is the restriction of $\mathcal{H}^{n-k}$ to the Borel sets in $L_k^\perp$. For $N_0^k(K)$, the set of all k-flats supporting K, we therefore obtain

$$\mu_k(N_0^k(K)) = \int_{SO(n)} \int_{T(\rho)} \mathrm{d}\mathcal{H}^{n-k}\, \mathrm{d}\nu(\rho)$$

with

$$T(\rho) := \{t \in L_k^\perp | \rho(L_k + t) \in N_0^k(K)\}.$$

Since $T(\rho)$ is the relative boundary of the image of $\rho^{-1}K$ under orthogonal projection onto $L_k^\perp$, we have $\mathcal{H}^{n-k}(T(\rho)) = 0$ and thus $\mu_k(N_0^k(K)) = 0$.

We complete the measure space $(A(n, k), \mathcal{B}(A(n, k)), \mu_k)$, and measurability refers to this completion. First we show that $N_\varepsilon^k(K, \eta)$ is measurable. Let $\eta \subset \Sigma$ be closed. Then $\bar{N} := N_\varepsilon^k(K, \mu) \cup N_0^k(K)$ is closed. In fact, let $(E_i)_{i \in \mathbb{N}}$ be a sequence in $\bar{N}$ converging to some k-flat E. Then $d(K, E) \leqq \varepsilon$, and if $d(K, E) = 0$ then E lies in a supporting hyperplane of K. For each $i \in \mathbb{N}$ there exist points $x_i \in \mathrm{bd}\, K$, $x_i' \in E_i$ such that

$$|x_i - x_i'| = d(K, E_i) \leqq \varepsilon,$$
$$(x_i, u(K, E_i)) \in \eta \quad \text{if } d(K, E_i) > 0.$$

If $d(K, E) > 0$, then by compactness some subsequence of $((x_i, u(K, E_i))_{i \in \mathbb{N}}$ converges to an element $(x, u) \in \eta$, and clearly $x \in K$ and $u = u(K, E)$. Thus $E \in \bar{N}$, which shows that $\bar{N}$ is closed. Since $\mu_k(N_0^k(K)) = 0$, the set $N_\varepsilon(K, \eta)$ is measurable.

Let $\eta_1, \eta_2 \in \mathcal{B}(\Sigma)$ be sets with $\eta_1 \cap \eta_2 = \emptyset$. Suppose that $E \in N_\varepsilon^k(K, \eta_1) \cap N_\varepsilon^k(K, \eta_2)$. Then there are points $x_1, x_2 \in K$ at distance $d(K, E)$ from E and such that $(x_i, u(K, E)) \in \eta_i$ for $i = 1, 2$. Since $\eta_1 \cap \eta_2 = \emptyset$, we must have $x_1 \neq x_2$. Thus the flat $E - d(K, E)u(K, E)$ supports K and contains more than one point of K; in other words, K and E are in singular relative position. From Corollary 2.3.11 it follows that $N_\varepsilon^k(K, \eta_1) \cap N_\varepsilon^k(K, \eta_2)$ has μ_k-measure zero.

By an argument analogous to that in the proof of Theorem 4.5.7 we now conclude that $N_\varepsilon^k(K, \eta)$ is measurable for each $\eta \in \mathcal{B}(\Sigma)$.

By Fubini's theorem,

$$\mu_k(N_\varepsilon^k(K, \eta)) = \int_{SO(n)} \int_{T(\rho)} \mathrm{d}\mathcal{H}^{n-k}\, \mathrm{d}\nu(\rho),$$

where now

$$T(\rho) := \{t \in L_k^\perp | \rho(L_k + t) \in N_\varepsilon^k(K, \eta)\}.$$

Let $t \in T(\rho)$. There are points $x \in K$, $x' \in \rho(L_k + t)$ such that $0 < |x - x'| = d(K, \rho(L_k + t)) \leqq \varepsilon$ and $(x, u(K, \rho(L_k + t))) \in \eta$. Write $d(K, \rho(L_k + t)) = d$ and $u(K, \rho(L_k + t)) = u$. From $x' = \rho(y' + t)$ with $y' \in L_k$ and $x' = x + du$ we obtain

$$t = \rho^{-1}x + d\rho^{-1}u - y' = z + d\rho^{-1}u$$

with $z := \rho^{-1}x - y'$. Because u is orthogonal to $\rho(L_k + t)$, we have $\rho^{-1}u \in L_k^\perp$ and thus $z \in L_k^\perp$. Since $y' \in L_k$, the point z is obtained from $\rho^{-1}x$ by orthogonal projection onto $L_k^\perp$, thus $z \in \rho^{-1}K|L_k^\perp$. From $(x, u) \in \eta \cap \operatorname{Nor} K$ we get $(\rho^{-1}x, \rho^{-1}u) \in \rho^{-1}\eta \cap \operatorname{Nor} \rho^{-1}K$ and hence

$$(z, \rho^{-1}u) \in (\rho^{-1}\eta|L_k^\perp) \cap \operatorname{Nor}(\rho^{-1}K|L_k^\perp).$$

Thus $t = z + d\rho^{-1}u$ satisfies

$$t \in M'_\varepsilon(\rho^{-1}K|L_k^\perp, \rho^{-1}\eta|L_k^\perp),$$

where M'_ε is constructed in $L_k^\perp$. If we assume that K and $\rho(L_k + t)$ are not in singular relative position, then the argument can be reversed and we arrive at

$$T(\rho) = M'_\varepsilon(\rho^{-1}K|L_k^\perp, \rho^{-1}\eta|L_k^\perp).$$

We will show that the projection $\rho^{-1}\eta|L_k^\perp$ is a Borel set for almost all ρ. Let R be the set of all $\rho \in SO(n)$ for which K and ρL_k are not in singular relative position. By Corollary 2.3.11, $\nu(SO(n)\setminus R) = 0$. Let $\rho \in R$. For $(y, u) \in \operatorname{Nor}(K|\rho L_k^\perp)$ we have $y = x|\rho L_k^\perp$ with suitable $(x, u) \in \operatorname{Nor} K$. Since $\rho \in R$, the point x is uniquely determined. Hence we can define a map

$$f\colon \operatorname{Nor}(K|\rho L_k^\perp) \to \operatorname{Nor} K \quad \text{by } f(y, u) := (x, u).$$

This map is continuous, hence the set $\eta|\rho L_k^\perp = f^{-1}(\eta)$ is a Borel set.

For almost all ρ, we can apply (4.2.4) to compute $\mathscr{H}^{n-k}(T(\rho))$. Integrating over $SO(n)$ and applying Theorem 4.5.10, we obtain

$$\begin{aligned}
&\mu(N_\varepsilon^k(K, \eta)) \\
&= \frac{1}{n-k} \sum_{j=0}^{n-k-1} \varepsilon^{n-k-j} \binom{n-k}{j} \int_{SO(n)} \Theta'_j(K|\rho L_k^\perp, \eta|\rho L_k^\perp)\, \mathrm{d}\nu(\rho) \\
&= \frac{\kappa_{n-k}}{\omega_n} \sum_{i=1}^{n-k} \varepsilon^i \binom{n-k}{i} \Theta_{n-k-i}(K, \eta).
\end{aligned}$$

This completes the proof of Theorem 4.5.11. ∎

As a consequence of formula (4.5.29) we note the limit relation

$$\lim_{\varepsilon \downarrow 0} \frac{1}{\varepsilon} \mu_k(N_\varepsilon^k(K, \eta)) = \frac{\omega_{n-k}}{\omega_n} \Theta_{n-k-1}(K, \eta). \tag{4.5.30}$$

It can be viewed as an integral-geometric interpretation of the generalized curvature measures.

Notes for Section 4.5

1. *Integral geometry.* General reference works for integral geometry are the books by Blaschke (1937a) (see also Volume 2 of his collected works, Blaschke (1985a)), Santaló (1953), Hadwiger [18] (Chapter 6), Stoka (1968) and Santaló (1976). In style and attitude these books differ considerably from each other. In contrast with other treatments of integral geometry, which consider smooth submanifolds and use differential-geometric methods, this section of the present volume is restricted to convex bodies (and finite unions of them) and follows a measure-theoretic approach. It thus tries to achieve a synthesis of the viewpoints of Hadwiger [18] and Federer (1959).
2. *Intersection formulae for curvature measures.* Federer (1959) developed a theory of curvature measures for sets of positive reach and in this general case proved the formulae of Theorems 4.5.2 and 4.5.5. A short proof was given by Rother & Zähle (1990). For convex bodies, simpler approaches are possible. Schneider (1978a) proved (4.5.1) as a consequence of the characterization theorem for curvature measures quoted in Section 4.2, Note 9. A slightly simpler proof was given in Schneider (1980b). The proof presented above is taken from Schneider & Weil (1986) and Lemma 4.5.1 is taken from Goodey & Schneider (1980).

   As an example of recent developments leading to very general versions of the kinematic formula (and requiring considerably deeper techniques) we mention the work of Fu (1990).
3. *The classical principal kinematic formula.* The specialization to total measures, that is, formula (4.5.2) for intrinsic volumes or its equivalent formulation in terms of quermassintegrals, includes the convex case of the classical principal kinematic formula of Blaschke, Santaló & Chern (see Santaló 1976). For convex bodies and sets of the convex ring, Hadwiger (1950a, 1951, 1956, [18]) developed an elegant method of proof. It goes back to ideas of Blaschke and uses Hadwiger's characterization theorem for quermassintegrals, quoted as Theorem 4.2.6. This method allows us to prove a general version of the principal kinematic formula (Hadwiger [18], 6.3.5, with different notation):

   Theorem. Let φ be an additive and continuous real functional on $\mathcal{K}^n$, and define associated functionals by
$$\varphi_{n-k}(K) := \frac{1}{\alpha_{n0k}} \int_{A(n,k)} \varphi(K \cap E)\,\mathrm{d}\mu_k(E)$$
   for $K \in \mathcal{K}^n$. Then
$$\int_{G_n} \varphi(K \cap gK')\,\mathrm{d}\mu(g) = \frac{1}{\kappa_n} \sum_{k=0}^{n} \frac{\kappa_k \kappa_{n-k}}{\binom{n}{k}} \varphi_{n-k}(K) V_k(K')$$
   for $K, K' \in \mathcal{K}^n$.

   A short proof of formula (4.5.3) (which extends to a proof of (4.5.2), but apparently not of (4.5.1)) was given by Mani-Levitska (1988).
4. *Convex cylinders.* The method by which Theorem 4.5.5 was deduced from Theorem 4.5.2 is taken from Federer (1959). By the same method, intersection formulae for a convex body and a moving convex cylinder can be obtained. Let $q \in \{0, \ldots, n\}$, let $E_q \subset \mathbb{E}^n$ be a q-dimensional linear subspace and let $E_q^\perp$ be its orthogonal complement. Then for $K \in \mathcal{K}^n$, $\beta \in \mathcal{B}(\mathbb{E}^n)$, a convex body $C \subset E_q^\perp$, a Borel set $\eta \subset E_q^\perp$ and $j \in \{0, \ldots, n\}$, the formula

$$\int_{SO(n)} \int_{E_q^\perp} \Phi_j(K \cap \rho(E_q + C + t), \beta \cap \rho(E_q + \eta + t)) \, d\mathcal{H}^{n-q}(t) \, d\nu(\rho)$$

$$= \sum_{k=j}^{n} \alpha_{njk} \Phi_{k-q}(C, \eta) \Phi_{n+j-k}(K, \beta),$$

with α_{njk} as in Theorem 4.5.2, is valid (Schneider 1980b).

5. *Translative integral geometry*. The simple form of the kinematic formulae (4.5.1), in particular the fact that on the right-hand side the convex bodies K and K' appear separately, is a result of the integration over the full group of rigid motions. If one integrates only over the translations, one still has a formula of the type

$$\int_{\mathbb{E}^n} \Phi_j(K \cap (K' + t), \beta \cap (\beta' + t)) \, dt = \sum_{k=j}^{n} \Phi_k^{(j)}(K, K', \beta \times \beta') \tag{4.5.31}$$

with finite measures $\Phi_k^{(j)}(K, K', \cdot)$ on $\mathcal{B}(\mathbb{E}^n \times \mathbb{E}^n)$, where

$$\Phi_j^{(j)}(K, K', \beta \times \beta') = \Phi_j(K, \beta)\Phi_n(K', \beta'),$$
$$\Phi_n^{(j)}(K, K', \beta \times \beta') = \Phi_n(K, \beta)\Phi_j(K', \beta'),$$

but the other mixed functionals $\Phi_k^{(j)}$ do not simplify in a similar way. Theorem 4.5.3 contains an explicit representation for them in the case of polytopes. From that theorem, one can derive the existence of a representation (4.5.31) for general convex bodies and can prove some properties of the functions $\Phi_k^{(j)}$ such as weak continuity, additivity and homogeneity in each of the first two arguments; see Schneider & Weil (1986), where one may also find analogous results for cylinders and flats. The cases $j = n$ and $j = n - 1$ of (4.5.31), where no proper mixed measures $\Phi_k^{(j)}$ occur, can be deduced from more general results in measure theory; see Groemer (1977b, 1980a) and Schneider (1981b).

With the notation

$$\Phi_k^{(j)}(K, K', \mathbb{E}^n \times \mathbb{E}^n) = V_{k,n+j-k}^{(j)}(K, K'),$$

the case $\beta = \beta' = \mathbb{E}^n$ of (4.5.31) can be written in the form

$$\int_{\mathbb{E}^n} V_j(K \cap (K' + t)) \, dt$$
$$= V_j(K)V_n(K') + \sum_{k=j+1}^{n-1} V_{k,n+j-k}^{(j)}(K, K') + V_n(K)V_j(K'). \tag{4.5.32}$$

Special cases are

$$\int_{\mathbb{E}^n} V_n(K \cap (K' + t)) \, dt = V_n(K)V_n(K'),$$

$$\int_{\mathbb{E}^n} V_{n-1}(K \cap (K' + t)) \, dt = V_{n-1}(K)V_n(K') + V_n(K)V_{n-1}(K').$$

Further results can be formulated in terms of mixed volumes. The latter will be treated in Chapter 5; we use here the results and notation of that chapter without explanation. For $j = 0$ the left-hand side of (4.5.32) is equal to

$$\mathcal{H}^n(\{t \in \mathbb{E}^n | K \cap (K' + t) \neq \emptyset\}) = V_n(K - K'),$$

hence

$$\int_{\mathbb{E}^n} V_0(K \cap (K' + t)) \, dt = \sum_{k=0}^{n} \binom{n}{k} V(K[k], -K'[n-k]). \tag{4.5.33}$$

For the remaining cases, $1 \leqq j \leqq n - 2$, the expressions appearing in (4.5.32) can be expressed as Crofton-type integrals involving mixed volumes, namely

$$V^{(j)}_{k,n+j-k}(K, K') = \frac{\binom{n}{k-j}}{\alpha_{n0j}} \int_{A(n,n-j)} V(K \cap E[k-j], -K'[n+j-k]) \, d\mu_{n-j}(E). \quad (4.5.34)$$

For dimensions two and three, translative integral geometry was treated by Blaschke (1937b), Berwald & Varga (1937) and Miles (1974) (see also Firey 1977). For (4.5.33) and extensions see Groemer (1977b); for (4.5.34) and other representations of $V^{(j)}_{k,n+j-k}$ in the case of centrally symmetric convex bodies, see Goodey & Weil (1987). These authors also study the counterpart to (4.5.32) for the case where K' is a convex cylinder. The special case of a flat reads

$$\int_{E_k^\perp} V_j(K \cap (E_k + t)) \, d\mathcal{H}^{n-k}(t) = \frac{1}{\kappa_{k-j}} \binom{n}{k-j} V(K[n+j-k], B_k[k-j]) \quad (4.5.35)$$

for $K \in \mathcal{K}^n$ and $j \in \{0, \ldots, k\}$. Here $k \in \{1, \ldots, n-1\}$, E_k is a k-dimensional linear subspace of $\mathbb{E}^n$ and B_k is a k-dimensional unit ball in E_k. A proof of (4.5.35) appears in Schneider (1981a); see also Section 5.3.

6. *Iterations*. The kinematic formulae (4.5.1) can clearly be extended to the intersections of a fixed convex body and a finite number of independently moving convex bodies. In particular, from (4.5.2) one deduces by induction that

$$\int_{G_n} \cdots \int_{G_n} V_j(K_0 \cap g_1 K_1 \cap \ldots \cap g_p K_p) \, d\mu(g_1) \cdots d\mu(g_p) = \sum_{\substack{k_0, \ldots, k_p = j \\ k_0 + \ldots + k_p = pn + j}}^{n} c^{(j)}_{k_0, \ldots, k_p} V_{k_0}(K_0) \cdots V_{k_p}(K_p)$$

with

$$c^{(j)}_{k_0, \ldots, k_p} = \frac{\prod_{i=0}^{p} k_i! \kappa_{k_i}}{j! \kappa_j (n! \kappa_n)^p}$$

for $K_0, K_1, \ldots, K_p \in \mathcal{K}^n$, $p \in \mathbb{N}$, and $j \in \{0, \ldots, n\}$. The possibility of iterating stems from the fact that the same series of functionals appear on the left-hand side and on the right-hand side of (4.5.2). See also Streit (1970).

A similar iteration of the translative formula (4.5.32) can be obtained if either $n = 2$ (Blaschke 1937b, Miles 1974) or $j = n$ or $n - 1$ (Schneider 1970b, Streit 1973, 1975). In the other cases, new classes of mixed functionals appear; see Blaschke (1937b) and Berwald & Varga (1937) for $n = 3$ and Weil (1990a) for a thorough study of the general case (short surveys are given in Weil (1989b, c)). However, explicit descriptions of the mixed functionals that appear are still known only in special cases.

7. *Applications of intersection formulae in stochastic geometry*. The integral-geometric formulae for the intersection of a fixed and a moving set have remarkable applications in stochastic geometry, where they yield expectation formulae for the functional densities that can be associated with certain random sets or particle processes. We first explain the relevant concepts and formulae and then give some hints about the literature.

A *random closed set* X in $\mathbb{E}^n$ (see Matheron 1975) is a random variable (a measurable map from some probability space) with values in the space of closed subsets of $\mathbb{E}^n$ that has the topology of closed convergence and the induced Borel structure. It is called *stationary (isotropic)* if $t \circ X$ and X have the same distribution, where t is a translation (rotation) of $\mathbb{E}^n$. The *extended convex ring* $LU(\mathcal{K}^n)$ consists of all sets M in $\mathbb{E}^n$ that are locally finite unions of convex bodies, in the sense that $M \cap K \in U(\mathcal{K}^n)$ for each $K \in \mathcal{K}^n$. For $K \in U(\mathcal{K}^n)$, there exist $m \in \mathbb{N}$ and convex bodies $K_1, \ldots, K_m$ such that $K = K_1 \cup \ldots \cup K_m$, and we denote by $N(K)$ the smallest number m for which such a representation is possible. For the purposes of this note, we now say that X is a *standard random closed set* if it is a stationary random closed set with values in $LU(\mathcal{K}^n)$ and satisfies

$$\mathrm{E}2^{N(X\cap K)} < \infty \quad \text{for all } K \in \mathcal{K}^n. \tag{4.5.36}$$

Here E denotes mathematical expectation. (4.5.36) is a type of integrability condition, ensuring the existence of the expectations appearing below. To show this, a theorem of Eckhoff (1980) can be used.

For a standard random closed set X, densities corresponding to the intrinsic volumes (or the quermassintegrals) can be introduced in several equivalent ways. First, one can show that

$$\lim_{r\to\infty} \frac{\mathrm{E}V_j(X \cap rK)}{V_n(rK)} := D_j(X) \tag{4.5.37}$$

exists for each convex body $K \in \mathcal{K}^n$ with $o \in \operatorname{int} K$ and is independent of K. Second, let $\beta \subset \mathbb{E}^n$ be a bounded Borel set. Choose a convex body K with $\beta \subset \operatorname{int} K$. Then $\Phi_j(X, \beta) := \Phi_j(X \cap K, \beta)$ is independent of K and defines a random variable. One can show that

$$\mathrm{E}\Phi_j(X, \beta) = D_j(X)\mathcal{H}^n(\beta). \tag{4.5.38}$$

Finally, let

$$C := \{(\xi_1, \ldots, \xi_n) \in \mathbb{E}^n | 0 \leqq \xi_i \leqq 1, \quad i = 1, \ldots, n\},$$
$$\partial^+ C := \{(\xi_1, \ldots, \xi_n) \in \mathbb{E}^n | \max\{\xi_1, \ldots, \xi_n\} = 1\}$$

(observe that $\partial^+ C$ is an element of the convex ring $U(\mathcal{K}^n)$). Then

$$\mathrm{E}[V_j(X \cap C) - V_j(X \cap \partial^+ C)] = D_j(X). \tag{4.5.39}$$

The densities $D_j(X)$ appear in the following counterpart to the kinematic formula (4.5.2).

Theorem. Let X be an isotropic standard random closed set in $\mathbb{E}^n$, let $K \in \mathcal{K}^n$ and let $j \in \{0, \ldots, n\}$. Then

$$\mathrm{E}V_j(K \cap X) = \sum_{k=j}^{n} \alpha_{njk} V_k(K) D_{n+j-k}(X). \tag{4.5.40}$$

The Crofton intersection formula has the following analogue.

Theorem. Let X be an isotropic standard random closed set in $\mathbb{E}^n$, let $L \in A(n, k)$ be a k-flat and let $j \in \{0, \ldots, k\}$. Then

$$D_j(X \cap L) = \alpha_{njk} D_{n+j-k}(X). \tag{4.5.41}$$

Of particular importance are the standard random closed sets arising as union sets of stationary Poisson processes in the space $\mathcal{K}^n$ (the so-called *Boolean models*). A stationary Poisson process Y in $\mathcal{K}^n$ (to be interpreted as a random field of convex particles) is determined by a positive number γ, its *intensity*, and by P_0, the *distribution of the typical particle* or *grain distribution* for the process Y; P_0 is a probability measure on the space $\mathcal{K}^n_c$ of convex bodies with circumcentre (centre of the smallest circumscribed ball) at the origin. For a bounded Borel set $A \in \mathcal{B}(\mathcal{K}^n)$, the probability that exactly k particles of Y belong to A is then given by

$$\operatorname{prob}\{\operatorname{card} Y \cap A = k\} = e^{-\Lambda(A)} \frac{\Lambda(A)^k}{k!},$$

where Λ (the intensity measure of the Poisson process Y) is the image measure of $\gamma P_0 \otimes \mathcal{H}^n$ under the map $(K, t) \mapsto K + t$ from $\mathcal{K}_c^n \times \mathbb{E}^n$ onto $\mathcal{K}^n$.

For a stationary Poisson process Y on $\mathcal{K}^n$, the V_j-density can be defined by

$$\Delta_j(Y) := \gamma \int_{\mathcal{K}_c^n} V_j(K) \, dP_0(K).$$

It satisfies

$$\lim_{r \to \infty} \frac{\mathrm{E} \sum_{L \in Y} V_j(L \cap rK)}{V_n(rK)} = \Delta_j(Y),$$

for each convex body $K \in \mathcal{K}^n$ with $o \in \operatorname{int} K$, and also analogues of (4.5.38) and (4.5.39).

If the process Y is isotropic, it satisfies versions of the kinematic and Crofton formulae that are analogous to (4.5.40) and (4.5.41).

The union set of the process Y is defined by $X_Y := \bigcup_{L \in Y} L$. Assuming a suitable integrability condition, X_Y is a standard random closed set, and it is of interest to know how the densities $D_j(X_Y)$ and $\Delta_j(Y)$ are related. Due to the Poisson assumption, the following general result can be obtained. If f is a translation invariant valuation on $U(\mathcal{K}^n)$ and $K \in \mathcal{K}^n$ is a convex body, then

$$\begin{aligned} &\mathrm{E} f(K \cap X_Y) \\ &= \sum_{k=1}^{\infty} (-1)^{k-1} \frac{\gamma^k}{k!} \int_{\mathcal{K}_c^n} \cdots \int_{\mathcal{K}_c^n} F(K, K_1, \ldots, K_k) \, dP_0(K_1) \cdots dP_0(K_k) \end{aligned}$$

where

$$\begin{aligned} &F(K, K_1, \ldots, K_k) \\ &:= \int_{\mathbb{E}^n} \cdots \int_{\mathbb{E}^n} f(K \cap (K_1 + x_1) \cap \ldots \cap (K_k + x_k)) \, dx_1 \cdots dx_k. \end{aligned}$$

Evidently, further evaluation requires iterated intersection formulae as discussed in Note 6. In the isotropic case, one can thus obtain explicit expressions for $D_j(X_Y)$ in terms of $\Delta_j(Y), \ldots, \Delta_n(Y)$. For $n = 2$, for example, these relations are

$$\begin{aligned} D_2(X_Y) &= 1 - e^{-\Delta_2(Y)}, \\ D_1(X_Y) &= e^{-\Delta_2(Y)} \Delta_1(Y), \\ D_0(X_Y) &= e^{-\Delta_2(Y)} [\Delta_0(Y) - \frac{1}{16\pi} \Delta_1(Y)^2]. \end{aligned}$$

Now we give some hints to the literature. Much of the motivation for the considerations described above came from the need to establish mathematical models and solid foundations for some methods and formulae of stereology. For an introduction to stereology, see Stoyan, Kendall & Mecke (1987), Chapter 11, and for a survey emphasizing the connections with integral geometry, Weil (1983a); see also Weil (1983b). All the results mentioned in the present note can be found with proofs in Weil (1984) and Weil & Wieacker (1984). These papers unify and extend previous results of varying degrees of generality obtained, using different methods and partly under stronger assumptions, by Matheron (1975), Miles (1976), Davy (1976,

1978), Wieacker (1982), A.M. Kellerer (1983, see also 1985) and H.G. Kellerer (1984). Weil & Wieacker (1984, 1988) show (in a refined version) that each random set with values in $LU(\mathcal{K}^n)$ can be written as the union set of a point process Y on $\mathcal{K}^n$. Extensions of the considerations in this note, for instance to cylinders, non-isotropic random sets and particle processes, are treated by Weil (1987). The approach to stochastic geometry as described here is further developed (but mostly without proof) in Mecke, Schneider, Stoyan & Weil (1990). Similar investigations for more general classes of random sets and point processes are due to Zähle (1982, 1986b). More recent objects of study are $LU(\mathcal{K}^n)$-valued random sets, which are stationary, but not necessarily isotropic; see Weil (1988, 1989b, c, 1990a, b) and Betke & Weil (1990).

8. *Rotational mean values*. Theorem 4.5.6 is due to Schneider (1975f). The proof given here is essentially that of Schneider (1986b). In that paper, the more general version of Theorem 4.5.9 is obtained. The special case (4.5.24) was proved by Weil (1979c). Also, the projection formula of Theorem 4.5.10 appears in Schneider (1986b). Its special case (4.5.26) had previously been proved, in Schneider (1975f), and case (4.5.27) in Weil (1979c). A certain extension of the projection formula (4.5.25) to sets of the convex ring, for which multiplicities have to be taken into account, is treated in Schneider (1990c). A more general integral-geometric formula derived from Theorem 4.5.6 is applied in Papaderou-Vogiatzaki & Schneider (1988) to a question on geometric collision probabilities.

   The proof of Theorem 4.5.7 is adapted from Schneider (1986b) and Schneider (1978d). The special case $\eta = \beta \times S^{n-1}$ of Theorem 4.5.11 was first proved in Schneider (1978a); the proof given above is an immediate extension. The integral-geometric interpretation (4.5.30) of generalized curvature measures was mentioned in Schneider (1980a); its specialization to area measures is due to Firey (1972) and its specialization to Federer's curvature measures appears in Schneider (1978a).

9. *Distance formulae*. Integral-geometric intersection formulae such as the principal kinematic formula and its generalizations involve integrations over the set of rigid motions g for which a moving set gK' intersects a fixed set K. If the distance between K and gK' is taken into account, one can also obtain formulae for non-intersecting sets. Let $K, K' \in \mathcal{K}^n$ be convex bodies and let $g \in G_n$ be a rigid motion for which $K \cap gK' = \emptyset$. Let $x \in K$, $x' \in gK'$ be points between which the distance is minimal. Before Theorem 4.5.7 we defined $d(K, gK') := |x' - x|$ and $u(K, gK') := (x' - x)/|x' - x|$; we also put $x(K, gK') := x$ if the pair (x, x') is unique, which is the case for μ-almost all g.

   Theorem. Let $f\colon (0, \infty) \times \operatorname{bd} K \times \operatorname{bd} K' \to \mathbb{R}$ be a measurable function for which the integrals in (4.5.42) are finite. Then

$$\int_{K \cap gK' = \emptyset} f(d(K, gK'), x(K, gK'), g^{-1}x(gK', K))\, d\mu(g)$$
$$= \omega_n \sum_{j=0}^{n-1} \sum_{k=0}^{j} \binom{n-1}{j} \binom{j}{k}$$
$$\times \int_0^\infty \int_{\operatorname{bd} K} \int_{\operatorname{bd} K'} f(r, x, x')\, dC_k(K, x)\, dC_{j-k}(K', x') r^{n-j-1}\, dr. \qquad (4.5.42)$$

   Let $f\colon (0, \infty) \times S^{n-1} \times S^{n-1} \to \mathbb{R}$ be a measurable function for which the integrals in (4.5.43) are finite; then

$$\int_{K\cap gK'=\emptyset} f(d(K, gK'), u(K, gK'), g_0^{-1}u(gK', K))\,\mathrm{d}\mu(g)$$
$$= \omega_n \sum_{j=0}^{n-1}\sum_{k=0}^{j}\binom{n-1}{j}\binom{j}{k}$$
$$\times \int_0^\infty \int_{S^{n-1}}\int_{S^{n-1}} f(r, u, u')\,\mathrm{d}S_k(K, u)\,\mathrm{d}S_{j-k}(K', u')r^{n-j-1}\,\mathrm{d}r. \tag{4.5.43}$$

Here we avoid some complicated coefficients by using C_k, S_k instead of Φ_k, Ψ_k. A common generalization of (4.5.42) and (4.5.43) can be formulated, if support elements and the generalized curvature measures Θ_k are used.

The following result is obtained as a special case. Let $f: [0, \infty) \to [0, \infty)$ be a measurable function for which $f(0) = 0$ and

$$M_k(f) := k\int_0^\infty f(r)r^{k-1}\,\mathrm{d}r < \infty \quad \text{for } k = 1, \ldots, n.$$

Then

$$\int_{G_n} f(d(K, gK'))\,\mathrm{d}\mu(g) = \frac{1}{\kappa_n}\sum_{k=1}^{n}\sum_{j=n+1-k}^{n}\binom{n}{k}\binom{k}{n-j}M_{k+j-n}(f)W_k(K)W_j(K').$$

This formula was first proved by Hadwiger (1975a). Analogues for moving flats are due to Bokowski, Hadwiger & Wills (1976), and various generalizations to Groemer (1980b). Special formulae of the type (4.5.43) were first obtained by Hadwiger (1975b); see Schneider (1977b) for a short proof. A thorough study leading to the results above, extensions, translative versions and analogues for flats was made by Weil (1979b, c, 1981a).

10. *Contact measures and touching probabilities*. The consequences of Theorems 4.5.7 and 4.5.11 can be interpreted in terms of touching or collision probabilities for convex bodies. To explain this, let us first formulate an intuitive and unprecise question. Let convex bodies $K, K' \in \mathcal{K}^n$, and subsets $\beta \subset \mathrm{bd}\,K$, $\beta' \subset \mathrm{bd}\,K'$ of their respective boundaries, be given. A random motion g is applied to K' in such a way that gK' touches K. What is the probability that K and gK' touch at a point of $\beta \cap g\beta'$? Of course, this question does not make sense as long as no underlying probability space is specified. For this, one would need a 'natural' probability measure on the set $G_0(K, K')$ of rigid motions $g \in G_n$ for which gK' touches K. Such a natural measure can, in fact, be deduced from the Haar measure, in the following way.

Let $K, L \in \mathcal{K}^n$ be disjoint convex bodies. There is a unique translation, denoted by $\tau(K, L)$, by a vector of length $d(K, L)$, such that $K \cap \tau(K, L)L \neq \emptyset$. For a Borel set $\alpha \in \mathcal{B}(G_n)$ and for $\varepsilon > 0$, define

$$A_\varepsilon(K, L, \alpha) := \{g \in G_n \mid 0 < d(K, gL) \leqq \varepsilon,\ \tau(K, gL)\circ g \in \alpha\}.$$

Then $A_\varepsilon(K, L, \alpha)$ is a Borel set, and

$$\mu(A_\varepsilon(K, L, \alpha)) = \frac{1}{n}\sum_{j=1}^{n-1}\varepsilon^{n-j}\binom{n}{j}\int_{SO(n)} C_j(K - \rho L, T(\alpha, \rho))\,\mathrm{d}\nu(\rho)$$

with

$$T(\alpha, \rho) := \{t \in \mathbb{E}^n \mid \gamma(t, \rho) \in \alpha \cap G_0(K, L)\}.$$

It follows that

$$\varphi(K, L, \alpha) := \lim_{\varepsilon\to 0}\frac{1}{\varepsilon}\mu(A_\varepsilon(K, L, \alpha)) = \int_{SO(n)} C_{n-1}(K - \rho L, T(\alpha, \rho))\,\mathrm{d}\nu(\rho).$$

This defines $\varphi(K, L, \cdot)$, a finite measure concentrated on the set $G_0(K, L)$

of rigid motions bringing L into a contact position with K; it is called the *contact measure* of K and L.

Assuming now that convex bodies $K, K' \in \mathcal{K}^n$ and Borel sets $\beta, \beta' \in \mathcal{B}(\mathbb{E}^n)$ are given, we can choose the completion of the probability space

$$(G_0(K, K'), \mathcal{B}(G_0(K, K')), \varphi(K, K', \cdot)/\varphi(K, K', G_0(K, K')))$$

as a model for the question posed at the beginning. Then it follows from Theorem 4.5.7 that the probability for a collision at a point of $\beta \cap g\beta'$ is given by

$$\frac{\sum_{j=0}^{n-1} \binom{n-1}{j} C_j(K, \beta) C_{n-1-j}(K', \beta')}{\sum_{j=0}^{n-1} \binom{n-1}{j} C_j(K, K) C_{n-1-j}(K', K')}.$$

The contact measure as defined above was introduced by Weil (1979b, c). Touching probabilities have been found in several cases, including the case where bodies touch at prescribed sets of normal vectors, the case where flats or cylinders touch and the case where there is inner contact (Firey 1974a, 1979, McMullen 1974a, Molter 1986, Schneider 1975e, f, 1976, 1978d, 1980b, Schneider & Wieacker 1984, Weil 1979b, c, 1981a, 1982a, 1989a). See also Burton (1980b) for a result on associated measurability problems.

11. *Absolute curvature measures*. Integral-geometric formulae for absolute curvature measures of sets of positive reach are treated by Rother & Zähle (1992).

12. *Mean measure of shadow boundaries*. An interesting integral-geometric result on convex bodies, of an entirely different type, was proved by Steenaerts (1985). In spirit, it is related to the boundary behaviour considered, from various aspects, in Chapter 2. For $K \in \mathcal{K}_0^n$ and a vector $u \in S^{n-1}$, let $\Sigma(K, u)$ be the shadow boundary of K in direction u, that is,

$$\Sigma(K, u) := \{x \in \operatorname{bd} K \mid x + \lambda u \notin \operatorname{int} K \quad \text{for all } \lambda \in \mathbb{R}\}.$$

Define

$$\alpha(K) := \frac{1}{\omega_n} \int_{S^{n-1}} \mathcal{H}^{n-2}(\Sigma(K, u)) \, d\mathcal{H}^{n-1}(u)$$

and

$$\beta(K) := \frac{1}{\omega_n} \int_{S^{n-1}} \mathcal{H}^{n-2}(\operatorname{relbd} K \mid u^\perp) \, d\mathcal{H}^{n-1}(u).$$

Then Steenaerts proved, confirming a conjecture of McMullen, that

$$1 \leqq \frac{\alpha(K)}{\beta(K)} \leqq \frac{\omega_n}{\pi \kappa_{n-1}}$$

for smooth K. If K is a ball, the left equality sign holds, and for K a polytope the right equality sign is valid.

13. *Mean intersection bodies*. In Crofton's intersection formula (4.5.8) one integrates an intrinsic volume of the intersection $K \cap E$ over all k-flats $E \in A(n, k)$, where the integration is with respect to the invariant measure μ_k. One may also integrate the intersections themselves (in the sense of Minkowski addition): For $K \in \mathcal{K}^n$ and $k \in \{1, \ldots, n-1\}$, there is a unique convex body $M_k(K)$ for which

$$h(M_k(K), \cdot) = \int_{A(n,k)} h(K \cap E, \cdot) \, d\mu_k(E).$$

This 'kth mean intersection body' $M_k(K)$ was investigated by Goodey & Weil (1991+). In particular, they showed that $M_1(K)$ is always a ball, whereas $M_2(K)$ determines K uniquely, up to a translation.

4.6. Local behaviour of curvature measures

The notion of curvature, in its various forms existing in geometry, is designed to describe and measure, in one sense or other, local geometric shapes. In the present section we ask for the information of this kind that can be gained from knowledge of the curvature measures or area measures of a convex body.

A natural first question to ask is the following. What does it mean geometrically if the mth curvature measure $C_m(K, \beta)$ vanishes on a relatively open subset β of the boundary of K? The answer is given below by the description of the support of $C_m(K, \cdot)$. The *support* of a Borel measure μ, denoted by $\operatorname{supp} \mu$, is the complement of the largest open set on which the measure vanishes.

Theorem 4.6.1. *Let $K \in \mathcal{K}_0^n$ and $m \in \{0, \ldots, n-1\}$. The support of the mth curvature measure $C_m(K, \cdot)$ is the closure of the m-skeleton of K:*

$$\operatorname{supp} C_m(K, \cdot) = \operatorname{cl} \operatorname{ext}_m K.$$

Proof. Let $K \in \mathcal{K}_0^n$ and $m \in \{0, \ldots, n-1\}$ be given. First we show that

$$C_m(K, \operatorname{bd} K \setminus \operatorname{ext}_m K) = 0. \tag{4.6.1}$$

Put $\beta_m := \operatorname{bd} K \setminus \operatorname{ext}_m K$ (which is a Borel set by Section 2.1). From (4.2.21) we have

$$C_0(K, \beta_0) = \mathcal{H}^{n-1}(\sigma(K, \beta_0)) = 0,$$

where the latter follows from Theorem 2.2.9, since $u \in \sigma(K, \beta_0)$ implies that $F(K, u)$ contains a segment; hence u is a singular normal vector of K. Now we use a special case of Theorem 4.5.5 that can be written as

$$C_m(K, \beta) = a_{nm} \int_{A(n,n-m)} C_0(K \cap E, \beta) \, d\mu_{n-m}(E) \tag{4.6.2}$$

with a positive constant a_{nm}. For each $E \in A(n, n-m)$ we have

$$\beta_m \cap E \subset \operatorname{relbd}(K \cap E) \setminus \operatorname{ext}_0(K \cap E),$$

since a point $x \in \beta_m \cap E$ is the centre of an $(m+1)$-dimensional ball contained in K and hence is the centre of a segment contained in $K \cap E$. It follows that

$$C_0(K \cap E, \beta_m) = C_0(K \cap E, \beta_m \cap E) = 0,$$

by the result proved above, applied to $K \cap E$. From (4.6.2) we deduce that

$$C_m(K, \text{bd}\, K \backslash \text{ext}_m\, K) = 0 \tag{4.6.3}$$

and hence that

$$\text{supp}\, C_m(K, \cdot) \subset \text{cl}\, \text{ext}_m\, K.$$

For the proof of the oppposite inclusion we need a quantitative improvement of the case $m = 0$.

Proposition. Let $\beta \subset \mathbb{E}^n$ be an open ball with centre $x \in K$ and radius ρ. If $C_0(K, \beta) = 0$, then x is the centre of a segment of length $2\rho/n$ contained in K.

For the proof we first show that

$$x \in \text{conv}\,(K \cap \text{bd}\, \beta). \tag{4.6.4}$$

Suppose this were false. Clearly $K \cap \text{bd}\, \beta \neq \emptyset$, since otherwise $K \subset \beta$ and hence $C_0(K, \beta) > 0$. Thus $\text{conv}\,(K \cap \text{bd}\, \beta)$ is a convex body not containing x. By Theorem 1.3.4, there exists a hyperplane H strongly separating x and $K \cap \text{bd}\, \beta$. Let H^+ be the open halfspace bounded by H that contains x. Since $x \in K \cap H^+$, it is clear that the outer unit normal vectors of K at points of $K \cap H^+$ fill a neighbourhood of the inner normal vector of H^+ on S^{n-1}. From (4.2.21) it follows that

$$C_0(K, \beta) \geqq C_0(K, \text{bd}\, K \cap H^+) > 0,$$

a contradiction. Thus (4.6.4) holds.

By Carathéodory's theorem, there are $k \leqq n + 1$ points $y_1, \ldots, y_k \in K \cap \text{bd}\, \beta$ such that

$$x = \sum_{j=1}^{k} \alpha_j y_j \quad \text{with} \quad \alpha_j > 0 \quad (j = 1, \ldots, k), \quad \sum_{j=1}^{k} \alpha_j = 1.$$

From

$$\sum_{j=1}^{k} \alpha_j = \frac{1}{k-1} \sum_{j=1}^{k} (1 - \alpha_j)$$

we deduce that there exists an index i for which

$$\frac{\alpha_i}{1 - \alpha_i} \geqq \frac{1}{k-1}.$$

For

$$y := \frac{x - \alpha_i y_i}{1 - \alpha_i},$$

then

$$y = \frac{\alpha_1 y_1 + \ldots + \alpha_{i-1} y_{i-1} + \alpha_{i+1} y_{i+1} + \ldots + \alpha_k y_k}{1 - \alpha_i}$$
$$\in \text{conv}\,\{y_1, \ldots, y_k\} \subset K.$$

Since $x = \alpha_i y_i + (1 - \alpha_i)y$,

$$|x - y_i| = \rho,\ |x - y| = \frac{\alpha_i}{1 - \alpha_i}|x - y_i| \geqq \frac{\rho}{n},$$

the proposition is proved.

Now let $\beta \subset \mathbb{E}^n$ be an open set for which $C_m(K, \beta) = 0$. From (4.6.2) we have

$$\int_{A(n,n-m)} C_0(K \cap E, \beta)\, \mathrm{d}\mu_{n-m}(E) = 0,$$

hence $C_0(K \cap E, \beta) = 0$ for μ_{n-m}-almost all planes $E \in A(n, n-m)$.

Let $x \in K \cap \beta$ and let $E \in A(n, n-m)$ be an $(n-m)$-plane through x that meets $\operatorname{int} K$. There exists a sequence $(E_j)_{j\in\mathbb{N}}$ of $(n-m)$-planes converging to E such that $C_0(K \cap E_j, \beta) = 0$ for all j. For j sufficiently large, we may choose $x_j \in K \cap E_j \cap \beta$ such that $x_j \to x$ for $j \to \infty$, and we may further assume that, for some fixed $\rho > 0$, the open ball with centre x_j and radius ρ is contained in β. By the proposition, x_j is the centre of a segment $[a_j, b_j]$ of length $2\rho/(n-m)$ contained in $K \cap E_j$. By selecting suitable subsequences, we infer that x is the centre of a segment $[a, b] \subset K \cap E$. We have proved that every $(n-m)$-dimensional plane through x that meets $\operatorname{int} K$, meets K in a segment of length $2\rho/(n-m)$ with centre x. This is only possible if x is the centre of an $(m+1)$-dimensional ball contained in K. Thus $x \notin \operatorname{ext}_m K$, and we have proved that

$$\mathbb{E}^n \backslash \operatorname{supp} C_m(K, \cdot) \subset \mathbb{E}^n \backslash \operatorname{ext}_m K,$$

hence

$$\operatorname{cl} \operatorname{ext}_m K \subset \operatorname{supp} C_m(K, \cdot).$$

This proves Theorem 4.6.1. ∎

In a similar way, we can obtain a counterpart to Theorem 4.6.1 for area measures. Since the proof is slightly more complicated, we formulate part of it as a lemma. For $u \in S^{n-1}$ and $0 < t < 1$ we define

$$D(u, t) := \{v \in S^{n-1} | \langle u, v \rangle > t\};$$

this is the open *spherical cap* with centre u and spherical radius $\arccos t$.

Lemma 4.6.2. *Let $K \in \mathcal{K}_0^n$, $n \geqq 2$, $u \in S^{n-1}$, $t \in (0, 1)$ and $x \in F(K, u)$ and suppose that*

$$S_{n-1}(K, D(u, t)) = 0.$$

Then there are normal vectors $u_1, u_2 \in N(K, x) \cap S^{n-1}$ such that

$$u \in \operatorname{relint} \operatorname{pos} \{u_1, u_2\}$$

and $\langle u, u_i \rangle \leqq c(n, t)$ *for* $i = 1, 2$, *where* $c(n, t) < 1$ *is a constant depending only on n and t.*

Proof. First we prove that

$$u \in \operatorname{pos}(N(K, x) \cap \operatorname{bd} D), \tag{4.6.5}$$

where $D = D(u, t)$, and $\operatorname{bd} D$ denotes the boundary of D relative to S^{n-1}. Suppose this were false. Then there exists a hyperplane through o that strongly separates u and $\operatorname{conv}(N(K, x) \cap \operatorname{bd} D)$. Hence there are a vector $v \in S^{n-1}$ with $\langle u, v \rangle > 0$ and a number $\eta > 0$ such that $\langle w, v \rangle < -\eta$ for all $w \in N(K, x) \cap \operatorname{bd} D$. All elements w of the compact set $\{w \in \operatorname{bd} D \mid \langle w, v \rangle \geqq -\eta\}$ satisfy $w \notin N(K, x)$ and hence $\langle x, w \rangle < h(K, w)$. Therefore, we can choose a number $\varepsilon > 0$ with $\langle x, w \rangle < h(K, w) - \varepsilon$ for all $w \in \operatorname{bd} D$ satisfying $\langle w, v \rangle \geqq -\eta$. Put

$$y := x + \alpha u + \beta v \quad \text{with } \alpha := \frac{\varepsilon \eta}{\eta + t}, \quad \beta := \frac{\varepsilon t}{\eta + t},$$

then $\langle y, u \rangle > \langle x, u \rangle$ and hence $y \notin K$. Let $w \in \operatorname{bd} D$. If $\langle w, v \rangle < -\eta$, then

$$\begin{aligned} \langle y, w \rangle &= \langle x, w \rangle + \alpha \langle u, w \rangle + \beta \langle v, w \rangle \\ &< h(K, w) + \alpha t - \beta \eta \\ &= h(K, w). \end{aligned}$$

If $\langle w, v \rangle \geqq -\eta$, then

$$\begin{aligned} \langle y, w \rangle &= \langle x, w \rangle + \alpha \langle u, w \rangle + \beta \langle v, w \rangle \\ &< h(K, w) - \varepsilon + \alpha + \beta \\ &= h(K, w). \end{aligned}$$

Let $w \in S^{n-1} \setminus \operatorname{cl} D$ and $w \neq -u$. We can choose $w_0 \in \operatorname{bd} D$ and $\lambda, \mu > 0$ such that $w_0 = \lambda u + \mu w$. If $\langle y, w \rangle \geqq h(K, w)$, we obtain

$$\begin{aligned} h(K, w_0) > \langle y, w_0 \rangle &= \lambda \langle y, u \rangle + \mu \langle y, w \rangle \\ &> \lambda h(K, u) + \mu h(K, w) \\ &\geqq h(K, \lambda u + \mu w) \\ &= h(K, w_0), \end{aligned}$$

a contradiction; hence $\langle y, w \rangle < h(K, w)$. For $w = -u$ this holds trivially.

Let $B \subset \operatorname{int} K$ be a ball and put

$$A := \operatorname{conv}(B \cup \{y\}) \cap \operatorname{bd} K.$$

For each $w \in S^{n-1} \setminus D$ we have proved that $\langle y, w \rangle < h(K, w)$, hence $F(K, w) \cap A = \emptyset$. This shows that $A \subset \tau(K, D)$ and hence, by (4.2.24), that

$$S_{n-1}(K, D) = \mathcal{H}^{n-1}(\tau(K, D)) \geqq \mathcal{H}^{n-1}(A) > 0,$$

contradicting the assumption of the Lemma. Thus (4.6.5) is true.

From (4.6.5) and Carathéodory's theorem we deduce the existence of $k \leqq n$ vectors $v_1, \ldots, v_k \in N(K, x) \cap \text{bd}\, D$ such that

$$u = \sum_{j=1}^{k} \alpha_j v_j \quad \text{with} \quad \alpha_j > 0 \quad (j = 1, \ldots, k).$$

From $\langle v_j, u \rangle = t$ for $j = 1, \ldots, k$ we get $(\alpha_1 + \ldots + \alpha_k)t = 1$ and thus $kt^2(\alpha_1^2 + \ldots + \alpha_k^2) \geqq 1$, hence there exists a positive constant $c_1 = c_1(k, t) < 1$ for which

$$\sum_{j=1}^{k} (1 - \alpha_j t)^2 \leqq c_1^2 \sum_{j=1}^{k} (1 - 2\alpha_j t + \alpha_j^2).$$

We deduce the existence of a number $i \in \{1, \ldots, k\}$ for which

$$\frac{\langle u, u - \alpha_i v_i \rangle}{|u - \alpha_i v_i|} = \frac{1 - \alpha_i t}{\sqrt{1 - 2\alpha_i t + \alpha_i^2}} \leqq c_1.$$

Put

$$u_1 := v_i, \quad u_2 := \frac{u - \alpha_i v_i}{|u - \alpha_i v_i|}$$

and

$$c(n, t) := \max \{c_1(2, t), \ldots, c_1(n, t), t\}.$$

Then $u_1 \in N(K, x) \cap S^{n-1}$,

$$\begin{aligned}|u - \alpha_i v_i| u_2 &= \alpha_1 v_1 + \ldots + \alpha_{i-1} v_{i-1} + \alpha_{i+1} v_{i+1} + \ldots + \alpha_k v_k \\ &\in N(K, x)\end{aligned}$$

and thus $u_2 \in N(K, x) \cap S^{n-1}$; finally

$$u = \alpha_i u_1 + |u - \alpha_i v_i| u_2 \in \text{relint pos}\, \{u_1, u_2\}$$

and $\langle u, u_1 \rangle = t \leqq c(n, t)$, $\langle u, u_2 \rangle \leqq c_1(k, t) \leqq c(n, t)$. This completes the proof of Lemma 4.6.2. ■

We can now describe the support of the mth area measure.

Theorem 4.6.3. *Let $K \in \mathcal{K}_0^n$ and $m \in \{0, \ldots, n-1\}$. The support of the mth area measure $S_m(K, \cdot)$ is the closure of the set of all $(n-1-m)$-extreme unit normal vectors of K.*

Proof. Let $E \in G(n, m+1)$ be an $(m+1)$-dimensional linear subspace of $\mathbb{E}^n$. As a special case of (4.5.26) we have

$$S_m(K, \omega) = b_{nm} \int_{SO(n)} S'_m(K|\rho E, \omega \cap \rho E)\, \mathrm{d}\nu(\rho) \qquad (4.6.6)$$

for $\omega \in \mathcal{B}(S^{n-1})$ with a positive constant b_{nm}; here S'_m is computed in

ρE. Let $\omega \subset S^{n-1}$ be an open set for which $S_m(K, \omega) = 0$. Then (4.6.6) shows that

$$S'_m(K|\rho E, \omega \cap \rho E) = 0$$

for ν-almost all rotations $\rho \in SO(n)$. Let $u \in \omega$ and let $\rho \in SO(n)$ be such that $u \in \rho E$. We can choose a sequence $(\rho_j)_{j \in \mathbb{N}}$ of rotations converging to ρ such that

$$S'_m(K|\rho_j E, \omega \cap \rho_j E) = 0 \quad \text{for } j \in \mathbb{N}.$$

Writing $u_j = \rho_j \rho^{-1} u$ for $j \in \mathbb{N}$, we have $u_j \to u$ for $j \to \infty$ and $u_j \in \rho_j E$. Since ω is open, we can assume that $D(u_j, t) \subset \omega$ for all $j \in \mathbb{N}$ with some fixed $t < 1$. Choosing $x_j \in F(K, u_j)$ for $j \in \mathbb{N}$, we have

$$x_j|\rho_j E \in F(K|\rho_j E, u_j).$$

By Lemma 4.6.2 applied to $K|\rho_j E$, there are vectors

$$\begin{aligned} v_j, w_j &\in N(K|\rho_j E, x_j|\rho_j E) \cap (S^{n-1} \cap \rho_j E) \\ &= N(K, x_j) \cap S^{n-1} \cap \rho_j E \end{aligned}$$

such that

$$u_j \in \operatorname{relint} \operatorname{pos}\{v_j, w_j\} \tag{4.6.7}$$

and

$$\langle u_j, v_j \rangle \leqq c(m+1, t), \ \langle u_j, w_j \rangle \leqq c(m+1, t). \tag{4.6.8}$$

After suitable selection of subsequences and a change of notation, we may assume that $x_j \to x$, $v_j \to v$, $w_j \to w$ for $j \to \infty$, with a certain point $x \in \operatorname{bd} K$ and vectors $u, v \in S^{n-1}$. Since $u_j \in N(K, x_j)$, we have $h(K, u_j) = \langle x_j, u_j \rangle$ and hence $h(K, u) = \langle x, u \rangle$; thus $u \in N(K, x)$. Similarly we see that $v, w \in N(K, x)$. From $v_j, w_j \in \rho_j E$ we get v, $w \in \rho E$. Relation (4.6.7) implies $u \in \operatorname{pos}\{v, w\}$, and from (4.6.8) we obtain $\langle u, v \rangle \leqq c(m+1, t)$, $\langle u, w \rangle \leqq c(m+1, t) < 1$; hence $u \in \operatorname{relint} \operatorname{pos}\{v, w\}$. Thus we have proved: Any $(m+1)$-dimensional linear subspace ρE through the given vector $u \in \omega$ contains a two-dimensional subcone of $N(K, x)$, where $x \in F(K, u)$, which contains u in its relative interior. In other words, each $(m+1)$-dimensional linear subspace through u intersects the normal cone $N(K, F(K, u))$ in a cone of dimension at least two containing u in its relative interior. This is only possible if the face $T(K, u)$ of $N(K, F(K, u))$ (see Section 2.2) containing u in its relative interior has dimension at least $n + 1 - m$. By definition, this means that u is not an $(n - 1 - m)$-extreme normal vector of K. Denoting the set of k-extreme normal vectors of K by $\operatorname{extn}_k K$, we have proved that any open set $\omega \subset S^{n-1} \backslash \operatorname{supp} S_m(K, \cdot)$ satisfies $\omega \subset S^{n-1} \backslash \operatorname{extn}_{n-1-m} K$, hence

$$\operatorname{cl} \operatorname{extn}_{n-1-m} K \subset \operatorname{supp} S_m(K, \cdot).$$

Vice versa, let ω be an open subset of $S^{n-1}\backslash \operatorname{extn}_{n-1-m} K$. Then each vector $u \in \omega$ satisfies $\dim T(K, u) \geqq n+1-m$. If $m = n-1$, we deduce from (4.2.24) and Theorem 2.2.4 that

$$S_{n-1}(K, \omega) = \mathcal{H}^{n-1}(\tau(K, \omega)) = 0,$$

since $\tau(K, \omega)$ contains only singular points of K. Using this result and formula (4.6.6), we obtain $S_m(K, \omega) = 0$ for $m = 1, \ldots, n-2$. This completes the proof of Theorem 4.6.3. ∎

In special cases, the support of the area measure $S_m(K, \cdot)$ tells us even more about the structure of K. If $P \in \mathcal{P}^n$ is a polytope, then formula (4.2.18) shows that the area measure $S_m(P, \cdot)$ is concentrated on the union of the spherical images of the relative interiors of the m-faces of P. The following theorem implies that only polytopes can have area measures with supports of this kind.

Theorem 4.6.4. *Let $K \in \mathcal{K}_0^n$ and $m \in \{1, \ldots, n-1\}$, and suppose that the support of the area measure $S_m(K, \cdot)$ can be covered by finitely many $(n-1-m)$-dimensional great spheres. Then K is a polytope.*

Proof. First let $m = n-1$. Then $S_{n-1}(K, \cdot)$ is concentrated on finitely many points $u_1, \ldots, u_k \in S^{n-1}$. Let

$$P := \bigcap_{i=1}^{k} H^{-}(K, u_i).$$

Then P is a polytope containing K. Suppose P has a vertex x that is not a point of K. The set ω of unit normal vectors (pointing towards x) of the hyperplanes strongly separating x and K is disjoint from $\{u_1, \ldots, u_k\}$ and satisfies

$$S_{n-1}(K, \omega) = \mathcal{H}^{n-1}(\tau(K, \omega)) > 0,$$

a contradiction. Hence each vertex of P is a point of K and we conclude that $K = P$.

We may, therefore, assume that $m \leqq n-2$. Let $A_1, \ldots, A_k$ be $(n-1-m)$-dimensional great spheres covering $\operatorname{supp} S_m(K, \cdot)$. We put

$$A := \bigcup_{i=1}^{k} A_i \quad \text{and} \quad \omega := S^{n-1}\backslash A;$$

then $S_m(K, \omega) = 0$. From formula (4.6.6) we deduce

$$S'_m(K|\rho E, \omega \cap \rho E) = 0 \tag{4.6.9}$$

for ν-almost all rotations $\rho \in SO(n)$, where $E \in G(n, m+1)$ is arbitrary but fixed.

Let $i \in \{1, \ldots, k\}$ and suppose that $\rho \in SO(n)$ is a rotation for which ρE meets A_i in more than two points. Then ρE meets $\operatorname{lin} A_i$ in a subspace of dimension at least two. Since $\dim \operatorname{lin} A_i = n - m$, this means that ρE and $\operatorname{lin} A_i$ are in special position. From Lemma 4.5.1 we know that this happens only for the rotations ρ in a set of measure zero. Hence we deduce that for ν-almost all $\rho \in SO(n)$ the subspace ρE meets A in at most $2k$ points. Together with (4.6.9) this shows that, for ν-almost all ρ, the support of $S'_m(K|\rho E, \cdot)$ contains only $2k$ points and hence (by the result proved above, applied to $K|\rho E$) $K|\rho E$ is an $(m+1)$-polytope with at most $2k$ facets.

If $\rho \in SO(n)$ is arbitrary, we can choose a sequence $(\rho_j)_{j\in\mathbb{N}}$ in $SO(n)$ converging to ρ such that each projection $K|\rho_j E$ is an $(m+1)$-polytope with at most $2k$ facets. Since $K|\rho_j E \to K|\rho E$ (in the Hausdorff metric) for $j \to \infty$, it follows that $K|\rho E$ is an $(m+1)$-polytope with at most $2k$ facets. By a well-known theorem, which we do not prove here (see Klee (1959a), Corollary 4.4), P is itself a polytope. ■

Another aspect of the local behaviour of curvature measures is expressed by the fact that they cannot be positive on sets where the Hausdorff dimension is too small. For instance, for $m > 0$ the measure $C_m(K, \cdot)$ cannot have point masses. This is a consequence of the more general estimates expressed in the following theorem. Its proof requires some facts from geometric measure theory; for these (as well as for the notion of rectifiability) we refer to Federer (1969).

Theorem 4.6.5. *Let $K \in \mathcal{K}^n$ and $m \in \{1, \dots, n-1\}$. Then*

$$C_m(K, \beta) \leqq a_1 \mathcal{H}^m(\beta) \tag{4.6.10}$$

for every Borel set $\beta \subset \mathbb{E}^n$, with some constant a_1 depending only on n and m, and

$$S_m(K, \omega) \leqq a_2 \mathcal{H}^{n-m-1}(\omega) \tag{4.6.11}$$

for each $(\mathcal{H}^{n-m-1}, n-m-1)$ rectifiable set $\omega \subset S^{n-1}$, with some constant a_2 depending only on n, m and K.

Proof. From (4.6.2) we have

$$\begin{aligned} C_m(K, \beta) &= a_{nm} \int_{A(n,n-m)} C_0(K \cap E, \beta)\, \mathrm{d}\mu_{n-m}(E) \\ &\leqq a_{nm} C_0(K \cap E, \mathbb{E}^n) \int_{A(n,n-m)} \operatorname{card}(E \cap \beta)\, \mathrm{d}\mu_{n-m}(E) \\ &\leqq a_1 \mathcal{H}^m(\beta) \end{aligned}$$

by Federer (1969, 2.10.16 and 3.3.13). Similarly from (4.6.6) we obtain (with $E \in G(n, m+1)$)

$$S_m(K, \omega) = b_{nm}\int_{SO(n)} S'_m(K|\rho E, \omega \cap \rho E)\, \mathrm{d}\nu(\rho)$$

$$\leqq b_{nm} \max_E S'_m(K|\rho E, S^{n-1} \cap \rho E)\int_{SO(n)} \mathrm{card}\,(\omega \cap \rho E)\, \mathrm{d}\nu(\rho)$$

$$\leqq a_2 \mathcal{H}^{n-m-1}(\omega)$$

by Federer (1969, Theorem 3.2.48). ∎

Notes for Section 4.6

1. Theorem 4.6.1 was proved by Schneider (1978a). Theorem 4.6.3 was conjectured by Weil (1973), p. 356, who showed it for $m = 1$, and proved in general by Schneider (1975f). The latter theorem implies, in particular, that

$$\mathrm{supp}\, S_q(K, \cdot) \subset \mathrm{supp}\, S_p(K, \cdot) \quad \text{for } q > p,$$

which had been conjectured by Weil (1973); see also Firey (1975). Theorem 4.6.4 is due to Goodey & Schneider (1980); for $m = 1$, Weil (1973), Satz (Theorem) 4.4, proved a stronger result. A special case of the estimate (4.6.10) appears in Federer (1959), p. 489. The general case and the estimate (4.6.11) were noted in Schneider [43], §8. Firey (1970a) proved the inequality

$$S_m(K, D(u, \cos\alpha)) \leqq AD(K)^m \sin^{n-m-1}\alpha \sec\alpha$$

for $K \in \mathcal{K}^n$, $m \in \{1, \dots, n-1\}$ and $\alpha \in (0, \pi/2)$, where $D(K)$ denotes the diameter of K and A is a constant depending only on n and m. From this one can also deduce an inequality of the type (4.6.11), namely

$$S_m(K, \omega) \leqq A_{nm} D(K)^m \mathcal{H}^{n-m-1}(\omega)$$

for $\omega \in \mathcal{B}(S^{n-1})$.

2. In $\mathbb{R}^3$, Aleksandrov (1942b) considered the *specific curvature* $C_0(K, \cdot)/C_2(K, \cdot)$, defined on the Borel sets β with $C_2(K, \beta) > 0$, and its influence on the local shape of bd K. He proved: If the specific curvature of a convex body $K \in \mathcal{K}_0^3$ is bounded on a neighbourhood of $x \in \mathrm{bd}\, K$, then either bd K is differentiable at x, or x is an internal point of a perfect 1-face of K. Consequences and further results in this spirit can be found in Busemann [12], Section 5.
3. Weil (1973, Satz 4.7) showed: If the convex body $K \in \mathcal{K}^n$ satisfies $S_1(K, \cdot) \leqq c\sigma$ with some constant c, where σ is the spherical Lebesgue measure on S^{n-1}, then K is a summand of some ball (the converse assertion is trivial); in particular, K is strictly convex.

5

Mixed volumes and related concepts

5.1. Mixed volumes and mixed area measures

The concept of mixed volumes, which forms a central part of the Brunn–Minkowski theory of convex bodies, arises naturally if one combines the two fundamental concepts of Minkowski addition and volume. In this section, we introduce mixed volumes and the closely related mixed area measures, and we establish their fundamental properties. This is most easily done by first considering strongly isomorphic polytopes, and then using the approximation theorem, 2.4.14.

As before, we use the symbol V_n for the volume of n-dimensional convex bodies, and we recall that $V_i(K)$, for an i-dimensional convex body K, is its i-dimensional volume.

***Lemma* 5.1.1.** *Let $P \in \mathcal{P}_0^n$ be an n-dimensional polytope, let $F_1, \ldots, F_N$ be its facets and let u_i be the outer unit normal vector of P at F_i ($i = 1, \ldots, N$). Then*

$$\sum_{i=1}^{N} V_{n-1}(F_i) u_i = o \tag{5.1.1}$$

and

$$V_n(P) = \frac{1}{n} \sum_{i=1}^{N} h(P, u_i) V_{n-1}(F_i). \tag{5.1.2}$$

Proof. Let $z \in \mathbb{E}^n \setminus \{o\}$. If π_z denotes the orthogonal projection onto $H_{z,0}$, then obviously

$$V_{n-1}(\pi_z P) = \sum_{\langle z, u_i \rangle \geqq 0} \langle z, u_i \rangle V_{n-1}(F_i) = - \sum_{\langle z, u_i \rangle < 0} \langle z, u_i \rangle V_{n-1}(F_i),$$

hence $\sum_{i=1}^{N} \langle z, u_i \rangle V_{n-1}(F_i) = 0$. Since z was arbitrary, (5.1.1) follows.

From (5.1.1) we see that the right-hand side of (5.1.2) does not change under translations of P. Since this is also true for the left-hand side, we may assume that $o \in \operatorname{int} P$. Then P is the union of the pyramids $\operatorname{conv}(F_i \cup \{o\})$, $i = 1, \ldots, N$, which have pairwise disjoint interiors. Formula (5.1.2) is an immediate consequence. ■

In the following, we assume that $\mathcal{A}$ is a given a-type of n-dimensional simple polytope, as defined in Section 2.4. By $u_1, \ldots, u_N$ we denote the unit normal vectors (of the facets) of the a-type $\mathcal{A}$. For $P \in \mathcal{A}$, we use the abbreviations

$$F_i := F(P, u_i),$$
$$F_{ij} := F_i \cap F_j,$$
$$h_i := h(P, u_i).$$

The numbers $h_1, \ldots, h_N$, which determine P uniquely, are called the *support numbers* of P. Further, we define

$$J := \{(i, j) | i, j \in \{1, \ldots, N\}, \dim F_{ij} = n - 2\}.$$

Observe that J depends only on the a-type $\mathcal{A}$. For $(i, j) \in J$, let Θ_{ij} be the angle between u_i and u_j, and let $v_{ij} \perp u_i$ be the unit normal vector of the $(n-1)$-polytope F_i at its $(n-2)$-face F_{ij}. We write

$$h_{ij} := h(F_i, v_{ij}).$$

Using the obvious relation

$$u_j = u_i \cos \Theta_{ij} + v_{ij} \sin \Theta_{ij},$$

we take the inner product of u_j with some vector in F_{ij} and obtain

$$h_{ij} = h_j \operatorname{cosec} \Theta_{ij} - h_i \cot \Theta_{ij}. \tag{5.1.3}$$

Lemma 5.1.2. *The volume of $P \in \mathcal{A}$ can be represented in the form*

$$V_n(P) = \sum a_{j_1 \ldots j_n} h_{j_1} \cdots h_{j_n}, \tag{5.1.4}$$

where the sum extends over $j_1, \ldots, j_n \in \{1, \ldots, N\}$ and where the coefficients $a_{j_1 \ldots j_n}$ are symmetric and depend only on the a-type $\mathcal{A}$.

Proof. We use induction on n. For $n = 1$, the assertion is trivial. We assume that $n > 1$ and that the assertion has been proved in dimension $n - 1$. Let $i \in \{1, \ldots, N\}$. By the induction hypothesis, there is a representation

$$V_{n-1}(F_i) = \sum_{(i,k_r) \in J} a^{(i)}_{k_1 \ldots k_{n-1}} h_{ik_1} \cdots h_{ik_{n-1}}. \tag{5.1.5}$$

Here the coefficients $a^{(i)}_{k_1 \ldots k_{n-1}}$ depend only on the a-type $\mathcal{A}_i$ of F_i and thus, by Lemma 2.4.10, only on the a-type $\mathcal{A}$. By (5.1.3),

$$h_{ik_r} = h_{k_r} \operatorname{cosec} \Theta_{ik_r} - h_i \cot \Theta_{ik_r}. \tag{5.1.6}$$

Inserting (5.1.6) into (5.1.5) and the latter into (5.1.2), we obtain a representation of the form

$$V_n(P) = \sum a_{j_1 \dots j_n} h_{j_1} \cdots h_{j_n}.$$

Introducing additional zero coefficients, we may assume that the summation extends formally over all $j_1, \dots, j_n \in \{1, \dots, N\}$. Clearly the coefficients can be assumed to be symmetric in their indices. Since the numbers $a^{(i)}_{k_1 \dots k_{n-1}}$, as well as the angles Θ_{ij}, depend only on $\mathcal{A}$ the same is true for the coefficients $a_{j_1 \dots j_n}$. ■

Now we assume that strongly isomorphic polytopes $P_1, \dots, P_n \in \mathcal{A}$ are given. We define $F^{(r)}_i$, $F^{(r)}_{ij}$, $h^{(r)}_i$, $h^{(r)}_{ij}$ for P_r in the same way as F_i, F_{ij}, h_i, h_{ij} were defined for P. Then we introduce the *mixed volume* of $P_1, \dots, P_n$ by

$$V(P_1, \dots, P_n) := \sum a_{j_1 \dots j_n} h^{(1)}_{j_1} \cdots h^{(n)}_{j_n}, \tag{5.1.7}$$

where the coefficients $a_{j_1 \dots j_n}$ are those of (5.1.4). Thus $V(P_1, \dots, P_n)$ is symmetric in its arguments, and

$$V(P, \dots, P) = V_n(P). \tag{5.1.8}$$

Let polytopes $P_1, \dots, P_m \in \mathcal{A}$ and numbers $\lambda_1, \dots, \lambda_m \geqq 0$ with $\sum \lambda_i > 0$ be given. Then $\lambda_1 P_1 + \dots + \lambda_m P_m \in \mathcal{A}$ by Lemma 2.4.10 and the definition of strongly isomorphic polytopes, and

$$h(\lambda_1 P_1 + \dots + \lambda_m P_m, u_i) = \lambda_1 h^{(1)}_i + \dots + \lambda_m h^{(m)}_i.$$

From (5.1.4) and (5.1.7) we immediately obtain

$$V_n(\lambda_1 P_1 + \dots + \lambda_m P_m) = \sum_{i_1, \dots, i_n = 1}^{m} \lambda_{i_1} \cdots \lambda_{i_n} V(P_{i_1}, \dots, P_{i_n}). \tag{5.1.9}$$

The following lemma expresses the mixed volume explicitly in terms of Minkowski sums.

Lemma 5.1.3. *For $P_1, \dots, P_n \in \mathcal{A}$,*

$$V(P_1, \dots, P_n) = \frac{1}{n!} \sum_{k=1}^{n} (-1)^{n+k} \sum_{i_1 < \dots < i_k} V_n(P_{i_1} + \dots + P_{i_k}). \tag{5.1.10}$$

Proof. Denote the right-hand side of (5.1.10) by $f(P_1, \dots, P_n)$, including the case where one of the P_i is equal to $\{o\}$. For $\lambda_1, \dots, \lambda_n > 0$ it follows from (5.1.9) that $f(\lambda_1 P_1, \dots, \lambda_n P_n)$ is a homogeneous polynomial of degree n in $\lambda_1, \ldots, \lambda_n$. Now

$$(-1)^{n+1} n!\, f(\{o\}, P_2, \ldots, P_n)$$

$$= \sum_{2 \leqq i \leqq n} V_n(P_i) - \left[\sum_{2 \leqq j \leqq n} V_n(\{o\} + P_j) + \sum_{2 \leqq i < j \leqq n} V_n(P_i + P_j) \right]$$

$$+ \left[\sum_{2 \leqq j < k \leqq n} V_n(\{o\} + P_j + P_k) + \sum_{2 \leqq i < j < k \leqq n} V_n(P_i + P_j + P_k) \right]$$

$$- \ldots$$

$$= 0;$$

thus $f(0P_1, \lambda_2 P_2, \ldots, \lambda_n P_n)$ is identically zero for all $\lambda_2, \ldots, \lambda_n$. Hence, in the polynomial $f(\lambda_1 P_1, \ldots, \lambda_n P_n)$ all monomials $\lambda_{i_1} \cdots \lambda_{i_n}$ with $1 \notin \{i_1, \ldots, i_n\}$ have zero coefficients. Since 1 can be replaced by each of the numbers $2, \ldots, n$, we conclude that only the monomial $\lambda_1 \cdots \lambda_n$ has a non-zero coefficient, and this is obviously equal to $V(P_1, \ldots, P_n)$. ■

We note that (5.1.10) implies, in particular, that $V(P_1, \ldots, P_n)$ does not change under arbitrary translations of any of the P_i.

Formula (5.1.2) extends to mixed volumes. To see this, let $P_1', \ldots, P_{n-1}'$ be strongly isomorphic $(n-1)$-polytopes. They lie in $(n-1)$-dimensional affine subspaces that are all parallel, say to H. We can define $v(P_1', \ldots, P_{n-1}')$ as the mixed volume, relative to H, of arbitrary translates of $P_1', \ldots, P_{n-1}'$ contained in H. By the translation invariance property just noted, this mixed volume is well defined. Similarly, we can define the $(n-2)$-dimensional mixed volume $v^{(n-2)}$ for strongly isomorphic $(n-2)$-polytopes.

Lemma 5.1.4. *For* $P_1, \ldots, P_n \in \mathcal{A}$,

$$V(P_1, \ldots, P_n) = \frac{1}{n} \sum_{i=1}^{N} h_i^{(1)} v(F_i^{(2)}, \ldots, F_i^{(n)}). \qquad (5.1.11)$$

Proof. We use induction over n. The case $n = 1$ being trivial, assume that $n > 1$ and the assertion is true in dimension $n - 1$.

Denote the right-hand side of (5.1.11) by $W(P_1, \ldots, P_n)$. Applying (5.1.2) to $\lambda_1 P_1 + \ldots + \lambda_n P_n$ and using (5.1.9) in dimension $n - 1$, we obtain

$$V_n(\lambda_1 P_1 + \ldots + \lambda_n P_n) = \sum_{i_1, \ldots, i_n = 1}^{n} \lambda_{i_1} \cdots \lambda_{i_n} W(P_{i_1}, \ldots, P_{i_n})$$

for arbitrary $\lambda_1, \ldots, \lambda_n > 0$. Hence, it suffices to show that $W(P_1, \ldots, P_n)$ is symmetric in its arguments. By the inductive hypothesis,

$$W(P_1, \dots, P_n)$$
$$= \frac{1}{n} \sum_{i=1}^{n} h_i^{(1)} \frac{1}{n-1} \sum_{(i,j)\in J} h_{ij}^{(2)} v^{(n-2)}(F_{ij}^{(3)}, \dots, F_{ij}^{(n)})$$
$$= \frac{1}{n(n-1)} \sum_{\substack{(r,s)\in J \\ r<s}} [h_r^{(1)} h_{rs}^{(2)} + h_s^{(1)} h_{sr}^{(2)}] v^{(n-2)}(F_{rs}^{(3)}, \dots, F_{rs}^{(n)})$$

(for $n = 2$, the expressions $v^{(n-2)}$ have to be replaced by 1). By (5.1.3),

$$h_r^{(1)} h_{rs}^{(2)} + h_s^{(1)} h_{sr}^{(2)} = h_r^{(1)}[h_s^{(2)} \operatorname{cosec} \Theta_{rs} - h_r^{(2)} \cot \Theta_{rs}] + h_s^{(1)}[h_r^{(2)} \operatorname{cosec} \Theta_{rs} - h_s^{(2)} \cot \Theta_{rs}],$$

and this expression is symmetric in the upper indices 1 and 2. Thus $W(P_1, P_2, P_3, \dots, P_n) = W(P_2, P_1, P_3, \dots, P_n)$, and since $W(P_1, P_2, P_3, \dots, P_n)$ is symmetric in its last $n-1$ arguments, as follows from its definition and the symmetry of mixed volumes in dimension $n-1$, the proof of the lemma is complete. ■

Mixed volumes are closely related to mixed area measures, which we now define for strongly isomorphic polytopes. Recall that the area measure $S_{n-1}(K, \cdot)$ of a convex body $K \in \mathcal{K}_0^n$ is, by Theorem 4.2.5, the Borel measure on the unit sphere S^{n-1} for which

$$S_{n-1}(K, \omega) = \mathcal{H}^{n-1}(\tau(K, \omega)) \quad \text{for } \omega \in \mathcal{B}(S^{n-1}).$$

In particular, for a polytope P (not necessarily with interior points) with normal vectors $u_1, \dots, u_N$ we have

$$S_{n-1}(P, \omega) = \sum_{u_i \in \omega} V_{n-1}(F(P, u_i)). \tag{5.1.12}$$

For strongly isomorphic polytopes $P_1, \dots, P_{n-1} \in \mathcal{A}$ we define their *mixed area measure* by

$$S(P_1, \dots, P_{n-1}, \omega) := \sum_{u_i \in \omega} v(F_i^{(1)}, \dots, F_i^{(n-1)}). \tag{5.1.13}$$

Thus $S(P_1, \dots, P_{n-1}, \cdot)$ is a finite measure on $\mathcal{B}(S^{n-1})$. For $P_1, \dots, P_m \in \mathcal{A}$ and for $\lambda_1, \dots, \lambda_m \geqq 0$ we obviously have

$$S_{n-1}(\lambda_1 P_1 + \dots + \lambda_m P_m, \cdot)$$
$$= \sum_{i_1, \dots, i_{n-1}=1}^{m} \lambda_{i_1} \cdots \lambda_{i_{n-1}} S(P_{i_1}, \dots, P_{i_{n-1}}, \cdot), \tag{5.1.14}$$

in analogy to equation (5.1.9) and as a consequence of it, applied in dimension $n-1$.

In the same way as Lemma 5.1.3 was proved (or from that lemma and from (5.1.13)) one obtains the following result, which expresses the mixed area measure as a linear combination of area measures of Minkowski sums.

Lemma 5.1.5. *For $P_1, \ldots, P_{n-1} \in \mathcal{A}$,*

$$S(P_1, \ldots, P_{n-1}, \cdot) = \frac{1}{(n-1)!} \sum_{k=1}^{n-1} (-1)^{n+k-1} \sum_{i_1 < \ldots < i_k} S_{n-1}(P_{i_1} + \ldots + P_{i_k}, \cdot).$$

We note that formula (5.1.11), valid for $P_1, \ldots, P_n \in \mathcal{A}$, can be written in the form

$$V(P_1, \ldots, P_n) = \frac{1}{n} \int_{S^{n-1}} h(P_1, u) \, \mathrm{d}S(P_2, \ldots, P_n, u). \quad (5.1.15)$$

It is now easy to extend all this to general convex bodies.

Theorem 5.1.6 (and Definition). *There is a non-negative symmetric function $V : (\mathcal{K}^n)^n \to \mathbb{R}$, the mixed volume, such that*

$$V_n(\lambda_1 K_1 + \ldots + \lambda_m K_m) = \sum_{i_1, \ldots, i_n = 1}^{m} \lambda_{i_1} \cdots \lambda_{i_n} V(K_{i_1}, \ldots, K_{i_n}) \quad (5.1.16)$$

for arbitrary convex bodies $K_1, \ldots, K_m \in \mathcal{K}^n$ and numbers $\lambda_1, \ldots, \lambda_m \geqq 0$. Further, there is a symmetric map S from $(\mathcal{K}^n)^{n-1}$ into the space of finite Borel measures on S^{n-1}, the mixed area measure, such that

$$S_{n-1}(\lambda_1 K_1 + \ldots + \lambda_m K_m, \cdot) = \sum_{i_1, \ldots, i_{n-1} = 1}^{m} \lambda_{i_1} \cdots \lambda_{i_{n-1}} S(K_{i_1}, \ldots, K_{i_{n-1}}, \cdot) \quad (5.1.17)$$

for $K_1, \ldots, K_m \in \mathcal{K}^n$ and $\lambda_1, \ldots, \lambda_m \geqq 0$, where $S(K_1, \ldots, K_{n-1}, \cdot) := S(K_1, \ldots, K_{n-1})(\cdot)$. The equality

$$V(K_1, \ldots, K_n) = \frac{1}{n} \int_{S^{n-1}} h(K_1, u) \, \mathrm{d}S(K_2, \ldots, K_n, u) \quad (5.1.18)$$

holds for $K_1, \ldots, K_n \in \mathcal{K}^n$.

Proof. We define

$$V(K_1, \ldots, K_n) := \frac{1}{n!} \sum_{k=1}^{n} (-1)^{n+k} \sum_{i_1 < \ldots < i_k} V_n(K_{i_1} + \ldots + K_{i_k}) \quad (5.1.19)$$

for arbitrary $K_1, \ldots, K_n \in \mathcal{K}^n$, which by Lemma 5.1.3 is consistent with the definition already given for the mixed volume of strongly isomorphic polytopes.

From the continuity of Minkowski addition and the volume functional we deduce that V is a continuous function on $(\mathcal{K}^n)^n$. Let $K_1, \ldots, K_n \in \mathcal{K}^n$ be given. By Theorem 2.4.14 we can find sequences $(P_{1i})_{i \in \mathbb{N}}, \ldots, (P_{ni})_{i \in \mathbb{N}}$ of polytopes such that $P_{ri} \to K_r$ for $i \to \infty$ $(r = 1, \ldots, n)$ and such that, for each $i \in \mathbb{N}$, the polytopes $P_{1i}, \ldots, P_{ni}$ are strongly isomorphic. From (5.1.9) and the continuity of volume and mixed volume we deduce that (5.1.16) holds.

Similarly, we define

$$S(K_1, \ldots, K_{n-1}, \cdot)$$
$$:= \frac{1}{(n-1)!} \sum_{k=1}^{n-1} (-1)^{n+k-1} \sum_{i_1<\ldots<i_k} S_{n-1}(K_{i_1} + \ldots + K_{i_k}, \cdot) \quad (5.1.20)$$

for $K_1, \ldots, K_{n-1} \in \mathcal{K}^n$, which again is consistent with the former definition, because of Lemma 5.1.5. Thus $S(K_1, \ldots, K_{n-1}, \cdot)$ is a signed measure on $\mathcal{B}(S^{n-1})$. From the weak continuity of the area measure, a special case of Theorem 4.2.1, we see that the map S thus defined is weakly continuous; that is, $K_{1i} \to K_1, \ldots, K_{n-1,i} \to K_{n-1}$ for $i \to \infty$ implies

$$S(K_{1i}, \ldots, K_{n-1,i}, \cdot) \xrightarrow{w} S(K_1, \ldots, K_{n-1}, \cdot).$$

An approximation argument analogous to that above now shows that (5.1.14) leads to (5.1.17).

Formula (5.1.18) is obtained from (5.1.15) by approximating $K_1, \ldots, K_n$ by strongly isomorphic polytopes and using the continuity of the mixed volume, the weak continuity of the mixed area measure and the fact that support functions are continuous.

It remains to show that V and S are non-negative. That V is non-negative on strongly isomorphic polytopes follows by an obvious induction argument, using (5.1.11) (where $o \in P_1$ and hence $h_i^{(1)} \geqq 0$ can be assumed). The non-negativity of the mixed area measure on strongly isomorphic polytopes is then a consequence of their definition (5.1.13). Finally, approximation by strongly isomorphic polytopes yields the non-negativity of V and S in general. ∎

We remark that equality (5.1.13) and Lemma 5.1.4, valid so far for strongly isomorphic polytopes, can now be shown to hold for arbitrary polytopes. Let $P_1, \ldots, P_n \in \mathcal{P}^n$. Applying (5.1.12) to $P = \lambda_1 P_1 + \ldots + \lambda_{n-1} P_{n-1}$ with $\lambda_1, \ldots, \lambda_{n-1} \geqq 0$ and using (5.1.14) on one hand and (5.1.12), (5.1.9) (in dimension $n-1$) on the other, we find, comparing the coefficients of $\lambda_1 \cdots \lambda_{n-1}$, that

$$S(P_1, \ldots, P_{n-1}, \omega) = \sum_{u\in\omega} v(F(P_1, u), \ldots, F(P_{n-1}, u)) \quad (5.1.21)$$

for $\omega \in \mathcal{B}(S^{n-1})$. Here the sum extends formally over all $u \in \omega$, but in fact only over the finitely many normal vectors of the facets of $P_1 + \ldots + P_{n-1}$ contained in ω. In view of (5.1.21), formula (5.1.18) reads

$$V(P_1, \ldots, P_n) = \frac{1}{n} \sum_{u\in S^{n-1}} h(P_1, u) v(F(P_2, u), \ldots, F(P_n, u)), \quad (5.1.22)$$

where the sum extends over the normal vectors of the facets of $P_2 + \ldots + P_n$.

We have already shown and used in the proof of Theorem 5.1.6 that the mixed volume is continuous and the mixed area measure is weakly continuous. We shall now establish some further properties of these maps.

From (5.1.19) and (5.1.20) it follows that the mixed volume and the mixed area measure do not change under an arbitrary translation of any of their arguments. Further, it follows that

$$V(K, \ldots, K) = V_n(K),$$
$$V(\alpha K_1, \ldots, \alpha K_n) = V(K_1, \ldots, K_n)$$

for any volume-preserving affine map $\alpha: \mathbb{E}^n \to \mathbb{E}^n$, and that

$$S(K, \ldots, K, \cdot) = S_{n-1}(K, \cdot),$$
$$S(\rho K_1, \ldots, \rho K_{n-1}, \rho\omega) = S(K_1, \ldots, K_{n-1}, \omega)$$

for any rotation $\rho \in SO(n)$.

Let $K, L \in \mathcal{K}^n$ be convex bodies such that $K \subset L$, and let $K_2, \ldots, K_n \in \mathcal{K}^n$ be arbitrary. From (5.1.18) we obtain

$$\begin{aligned} V(K, K_2, \ldots, K_n) &- V(L, K_2, \ldots, K_n) \\ &= \frac{1}{n} \int_{S^{n-1}} [h(K, u) - h(L, u)] \, \mathrm{d}S(K_2, \ldots, K_n, u) \\ &\leqq 0, \end{aligned}$$

with equality if and only if

$$h(K, u) = h(L, u) \quad \text{for each } u \in \operatorname{supp} S(K_2, \ldots, K_n, \cdot).$$

Thus

$$K \subset L \Rightarrow V(K, K_2, \ldots, K_n) \leqq V(L, K_2, \ldots, K_n). \tag{5.1.23}$$

By symmetry, the mixed volume is monotone in each of its arguments. Unfortunately, no geometric description of the support of the mixed area measure $S(K_2, \ldots, K_n, \cdot)$ is known (see, however, Conjecture 6.6.13). Therefore, the complete characterization of the equality cases in (5.1.23) is an open problem. A special case is treated in Theorem 6.6.16.

Easier to decide is the question of when the mixed volume is strictly positive.

Theorem 5.1.7. *For* $K_1, \ldots, K_n \in \mathcal{K}^n$, *the following assertions are equivalent*:

(a) $V(K_1, \ldots, K_n) > 0$;

(b) *there are segments* $S_i \subset K_i$ $(i = 1, \ldots, n)$ *with linearly independent directions*;

(c) $\dim(K_{i_1} + \ldots + K_{i_k}) \geqq k$ *for each choice of indices* $1 \leqq i_1 < \ldots < i_k \leqq n$ *and for all* $k \in \{1, \ldots, n\}$.

Proof. Suppose (a) holds. We prove (b) by induction. The case $n = 1$ being trivial, suppose that $n > 1$ and that the assertion is true in smaller dimensions. Since a convex body can be approximated by polytopes contained in it and since the mixed volume is continuous, there are polytopes $P_i \subset K_i$ $(i = 1, \ldots, n)$ such that $V(P_1, \ldots, P_n) > 0$. By (5.1.22),

$$0 < V(P_1, \ldots, P_n) = \frac{1}{n} \sum h(P_1, u_i) v(F(P_2, u_i), \ldots, F(P_n, u_i)),$$

where the sum extends over finitely many unit vectors u_i. Assuming that $o \in \operatorname{relint} P_1$, all the summands are non-negative; hence there is some j for which

$$h(P_1, u_j) > 0,$$
$$v(F(P_2, u_j), \ldots, F(P_n, u_j)) > 0.$$

By the induction hypothesis, the latter implies the existence of segments $S_i \subset F(P_i, u_j)$, $i = 2, \ldots, n$, with independent directions. Since $h(P_1, u_j) > 0$, there is a segment $S_1 \subset P_1$ that is not orthogonal to u_j. Thus $S_i \subset K_i$ for $i = 1, \ldots, n$, and the directions of $S_1, \ldots, S_n$ are linearly independent. Thus (b) holds.

Suppose that (b) holds and $S_i \subset K_i$ $(i = 1, \ldots, n)$ are segments with independent directions. The monotoneity of the mixed volume in each argument implies $V(K_1, \ldots, K_n) \geqq V(S_1, \ldots, S_n)$. By (5.1.19),

$$n! V(S_1, \ldots, S_n) = V_n(S_1 + \ldots + S_n) > 0,$$

which shows that (a) is true.

It is trivial that (b) implies (c). Vice versa, if (c) holds, we may assume that $o \in \operatorname{relint} K_i$ for $i = 1, \ldots, n$, and then the truth of (b) is an immediate consequence of the subsequent lemma. ■

Lemma 5.1.8. *Let* $L_1, \ldots, L_n$ *be linear subspaces of* $\mathbb{E}^n$. *If*

$$\dim(L_{i_1} + \ldots + L_{i_k}) \geqq k$$

for each choice of indices $1 \leqq i_1 < \ldots < i_k \leqq n$ *and for all* $k \in \{1, \ldots, n\}$, *then there are lines* $G_i \subset L_i$ $(i = 1, \ldots, n)$ *such that* $\dim(G_1 + \ldots + G_n) = n$.

Proof. For each m-tuple $(i_1, \ldots, i_m)$ from $\{2, \ldots, n\}$ we have either (a) $L_1 \subset L_{i_1} + \ldots + L_{i_m}$ or (b) $\dim(L_1 \cap (L_{i_1} + \ldots + L_{i_m})) < \dim L_1$. Since L_1 is not covered by finitely many lower-dimensional subspaces, we can choose a line $G_1 \subset L_1$ such that in case (b) we always have

$$G_1 \not\subset L_{i_1} + \ldots + L_{i_m}.$$

Let $(i_2, \ldots, i_k)$ be a $(k-1)$-tuple from $\{2, \ldots, n\}$. Then either $L_1 \subset L_{i_2} + \ldots + L_{i_k}$ and hence

$$\dim(G_1 + L_{i_2} + \ldots + L_{i_k}) = \dim(L_1 + L_{i_2} + \ldots + L_{i_k}) \geqq k,$$

or $L_1 \not\subset L_{i_2} + \ldots + L_{i_k}$ and thus

$$\begin{aligned}\dim(G_1 + L_{i_2} + \ldots + L_{i_k}) &= 1 + \dim(L_{i_2} + \ldots + L_{i_k})\\ &\geqq 1 + (k-1) = k.\end{aligned}$$

Repeating the procedure with L_2, and so on, we complete the proof of the lemma. ∎

An important property of the mixed volume is its Minkowski linearity in each argument: for $K, L, K_2, \ldots, K_n \in \mathcal{K}^n$ and $\lambda, \mu \geqq 0$ we have

$$\begin{aligned}&V(\lambda K + \mu L, K_2, \ldots, K_m)\\ &\quad = \lambda V(K, K_2 \ldots, K_m) + \mu V(L, K_2, \ldots, K_m).\end{aligned} \tag{5.1.24}$$

This follows from (5.1.18). By symmetry, V is Minkowski linear in each of its arguments. The same holds for the mixed area measure:

$$\begin{aligned}&S(\lambda K + \mu L, K_2, \ldots, K_{n-1}, \cdot)\\ &\quad = \lambda S(K, K_2, \ldots, K_{n-1}, \cdot) + \mu S(L, K_2, \ldots, K_{n-1}, \cdot).\end{aligned} \tag{5.1.25}$$

For polytopes, this is clear from (5.1.21) and the linearity of the mixed volume (in dimension $n-1$); the general case follows by approximation.

The polynomial expansion (5.1.16) can be written in a more concise form. Introducing the abbreviation

$$V(\underbrace{K_1, \ldots, K_1}_{r_1}, \ldots, \underbrace{K_k, \ldots, K_k}_{r_k}) =: V(K_1[r_1], \ldots, K_k[r_k])$$

and the multinomial coefficient

$$\binom{n}{r_1 \ldots r_m} := \begin{cases} \dfrac{n!}{r_1! \cdots r_m!} & \text{if } \sum_{j=1}^{m} r_j = n \text{ and } r_j \in \{0, 1, \ldots, n\},\\ 0 & \text{otherwise,}\end{cases}$$

we find, using a standard combinatorial argument, for $K_1, \ldots, K_m \in \mathcal{K}^n$ and $\lambda_1, \ldots, \lambda_m \geqq 0$,

$$\begin{aligned}&V_n(\lambda_1 K_1 + \ldots + \lambda_m K_m)\\ &\quad = \sum_{r_1, \ldots, r_m = 0}^{n} \binom{n}{r_1 \ldots r_m} \lambda_1^{r_1} \cdots \lambda_m^{r_m} V(K_1[r_1], \ldots, K_m[r_m]).\end{aligned} \tag{5.1.26}$$

Similar polynomial expansions are obtained for the mixed volume if some of its arguments are held fixed. Let an integer $p \in \{1, \ldots, n\}$ and convex bodies $C_{p+1}, \ldots, C_n \in \mathcal{K}^n$ be given. Then

$$V(\lambda_1 K_1 + \ldots + \lambda_m K_m[p], C_{p+1}, \ldots, C_n)$$
$$= \sum_{r_1, \ldots r_m = 0}^{p} \binom{p}{r_1 \ldots r_m} \lambda_1^{r_1} \cdots \lambda_m^{r_m} V(K_1[r_1], \ldots, K_m[r_m], C_{p+1}, \ldots, C_n). \tag{5.1.27}$$

For the proof, we develop both sides of the identity

$$V_n(\mu(\lambda_1 K_1 + \ldots + \lambda_m K_m) + \mu_{p+1} C_{p+1} + \ldots + \mu_n C_n)$$
$$= V_n(\mu\lambda_1 K_1 + \ldots + \mu\lambda_m K_m + \mu_{p+1} C_{p+1} + \ldots + \mu_n C_n)$$

for $\mu, \mu_{p+1}, \ldots, \mu_n \geqq 0$, using (5.1.26), and compare the coefficients of $\mu^p \mu_{p+1} \cdots \mu_n$ in the two expressions thus obtained.

It is clear that the polynomial expansion (5.1.17) leads to exactly analogous results for the mixed area measure; for these, we use the analogous notation. The case $p = 1$ of (5.1.27) and the corresponding formula for the mixed area measure expresses the known fact that both maps are Minkowski linear in each argument.

A further additivity property is to be noted. For given $p \in \{1, \ldots, n\}$ and convex bodies $C_{p+1}, \ldots, C_n \in \mathcal{K}^n$, the function f defined by

$$f(K) := V(K[p], C_{p+1}, \ldots, C_n), \quad K \in \mathcal{K}^n$$

is additive on $\mathcal{K}^n$; thus

$$V(K \cup L[p], C_{p+1}, \ldots, C_n) + V(K \cap L[p], C_{p+1}, \ldots, C_n)$$
$$= V(K[p], C_{p+1}, \ldots, C_n) + V(L[p], C_{p+1}, \ldots, C_n)$$

for $K, L \in \mathcal{K}^n$ such that $K \cup L \in \mathcal{K}^n$. In fact, for such K, L and for $\mu, \mu_{p+1}, \ldots, \mu_n \geqq 0$, we have

$$V_n(\mu(K \cup L) + \mu_{p+1} C_{p+1} + \ldots + \mu_n C_n)$$
$$+ V_n(\mu(K \cap L) + \mu_{p+1} C_{p+1} + \ldots + \mu_n C_n)$$
$$= V_n(\mu K + \mu_{p+1} C_{p+1} + \ldots + \mu_n C_n)$$
$$+ V_n(\mu L + \mu_{p+1} C_{p+1} + \ldots + \mu_n C_n)$$

by Lemma 3.4.10. Developing both sides according to (5.1.26) and comparing the coefficients of $\mu^p \mu_{p+1} \cdots \mu_n$, we arrive at the assertion.

Similarly, from the additivity of the area measure (a special case of Theorem 4.2.1) we infer that the map

$$K \mapsto S(K[p], C_{p+1}, \ldots, C_{n-1}, \cdot)$$

is additive on $\mathcal{K}^n$.

Finally, we note that the mixed area measure, when considered as defining a mass distribution on the sphere, always has its centroid at the origin:

$$\int_{S^{n-1}} u \,\mathrm{d}S(K_1, \ldots, K_{n-1}, u) = o. \tag{5.1.28}$$

In fact, from (5.1.18) and the translation invariance of the mixed volume we have

$$\begin{aligned} 0 &= V(K_1, \ldots, K_{n-1}, K_n + t) - V(K_1, \ldots, K_{n-1}, K_n) \\ &= \frac{1}{n} \int_{S^{n-1}} \langle u, t \rangle \,\mathrm{d}S(K_1, \ldots, K_{n-1}, u) \end{aligned}$$

for arbitrary $t \in \mathbb{E}^n$, from which (5.1.28) follows.

Notes for Section 5.1

1. The theory of mixed volumes was created by Minkowski (1903, 1911). A presentation of its early development and history can be found in Bonnesen & Fenchel [8]. The theory attained a more general and elegant form when the mixed area measure was introduced, independently by Aleksandrov (1937a) and by Fenchel & Jessen (1938).

   If the mixed volume has previously been defined, then formula (5.1.18) can serve as a starting point for the definition of the mixed area measure: Let $K_1, \ldots, K_{n-1} \in \mathcal{K}^n$ be given. For $K \in \mathcal{K}^n$, let $h := \bar{h}_K$ and $f(h) := V(K, K_1, \ldots, K_{n-1})$. This defines an additive functional f on a subset of $C(S^{n-1})$, the real vector space of continuous real functions on S^{n-1} with the maximum norm. It can be shown (see Section 5.2) that f has a unique extension to a continuous linear functional on $C(S^{n-1})$. By the Riesz representation theorem, there exists a unique measure $S(K_1, \ldots, K_{n-1}, \cdot)$ on $\mathcal{B}(S^{n-1})$ for which, in particular,

$$V(K, K_1, \ldots, K_{n-1}) = \frac{1}{n} \int_{S^{n-1}} h(K, u) \,\mathrm{d}S(K_1, \ldots, K_{n-1}, u)$$

   for all $K \in \mathcal{K}^n$. The properties of S can be deduced from those of V. By this method, the mixed area measure was introduced by Aleksandrov (1937a), and in a similar way by Fenchel & Jessen (1938). For a geometric interpretation, Aleksandrov first defined $S_{n-1}(K, \cdot)$, essentially by formula (4.2.24), and then used his definition of the mixed area measure to prove the expansion (5.1.17). Later, Aleksandrov (1939b) simplified part of his reasoning by working with weak convergence. Fenchel & Jessen (1938) used the formula

$$V(L, K, \ldots, K) = \frac{1}{2} \int_{S^{n-1}} h(L, u) \,\mathrm{d}S_{n-1}(K, u)$$

   to define $S_{n-1}(K, \cdot)$ and later obtained an intuitive interpretation by proving formula (4.2.9).
2. Mixed volumes can be introduced in different ways, but at some point, a symmetry property such as that of $W(P_1, \ldots, P_n)$ in the proof of Lemma 5.1.4 has to be shown. For the case of polytopes, one may compare Minkowski (1911), §21, and Aleksandrov [2], p. 411; for analytic proofs, see Bonnesen & Fenchel [8], p. 57 and the references on p. 61.
3. Lemma 5.1.8 is taken from Leichtweiß [23], pp. 176–7. A refinement is proved and used in Schneider (1988a).

4. *Polynomial expansions for valuations*. The existence of the crucial polynomial expansion (5.1.16) can be shown for functions that are considerably more general than the volume:

   Theorem. Let φ be a real-valued translation invariant valuation on $\mathcal{P}^n$. Then for $P_1, \ldots, P_k \in \mathcal{P}^n$ and rational $\lambda_1, \ldots, \lambda_k \geqq 0$, $\varphi(\lambda_1 P_1 + \ldots + \lambda_k P_k)$ is a polynomial in $\lambda_1, \ldots, \lambda_k$ of degree at most n. The coefficient of $\lambda_1^{r_1} \cdots \lambda_k^{r_k}$ is a translation invariant valuation in P_j that is homogeneous of degree r_j ($j = 1, \ldots, k$).

   If, moreover, φ is defined and continuous on $\mathcal{K}^n$, then the expansion extends to $\lambda_1 K_1 + \ldots + \lambda_k K_k$ with $K_1, \ldots, K_k \in \mathcal{K}^n$ and real $\lambda_1, \ldots, \lambda_k \geqq 0$. The question whether such a result might hold was posed by McMullen in 1974. He developed a theory yielding the above theorem and more general results (McMullen 1975c, 1977); see also McMullen & Schneider [40], §10. Using different approaches, McMullen's question was answered by Meier (1977) and Spiegel (1978).

   Meier's approach to mixed valuations was via 'mixed polytopes'; for these, see also McMullen & Schneider [40], pp. 200–1, and, as part of a quite general theory unifying many previous results, McMullen (1989).
5. *The mixed volume as a distribution*. The representation (5.1.18) exhibits the linearity of the mixed volume in its first argument, but not the symmetry of the mixed volume. To obtain a more symmetric representation, Weil (1981b) expressed the mixed volume as a distribution. More generally, such a representation is possible for multilinear and continuous functions on $\mathcal{K}^n$. The following theorem is due to Goodey & Weil (1984):

   Theorem. Let $\varphi: (\mathcal{K}^n)^n \to \mathbb{R}$ be multilinear and continuous. Then there is a unique distribution T on $(S^{n-1})^n$ that can be extended to a Banach space of functions on $(S^{n-1})^n$ containing the tensor products $h_{K_1} \otimes \ldots \otimes h_{K_n}$, for all $K_1, \ldots, K_n \in \mathcal{K}^n$, in such a way that

$$\varphi(K_1, \ldots, K_n) = T(h_{K_1} \otimes \ldots \otimes h_{K_n}).$$

   In fact, there are real functions f_j of class C^∞ on $(S^{n-1})^n$ for $j = 1, 2, \ldots$ such that

$$\int_{S^{n-1}} \cdots \int_{S^{n-1}} h(K_1, u_1) \cdots h(K_n, u_n) f_j(u_1, \ldots, u_n) \,\mathrm{d}\sigma(u_1) \cdots \mathrm{d}\sigma(u_n) \to \varphi(K_1, \ldots, K_n)$$

   as $j \to \infty$, uniformly for all K_i in any fixed ball (σ is the spherical Lebesgue measure).
6. *Characterization theorems*. As mentioned in Section 4.2, Hadwiger's characterization theorems 4.2.6 and 4.2.7 for linear combinations of quermassintegrals were the starting point for several similar investigations. Characterizations of functionals related to mixed volumes by means of their additivity properties require strong additional assumptions. Fáry (1961) characterized, for a given convex body $U \in \mathcal{K}^n$, the functionals

$$\varphi: K \mapsto \sum_{i=0}^{n} c_i V(K[i], U[n-i]),$$

   where $c_0, \ldots, c_n \in \mathbb{R}$, as the translation invariant, continuous valuations φ satisfying $\varphi(K) = \varphi(L)$ whenever $V(K[i], U[n-i]) = V(L[i], U[n-i])$ for $i = 0, \ldots, n$. Firey (1976) showed that an increasing Minkowski linear function $\varphi: \mathcal{K}^n \to \mathbb{R}$ that is zero on one-pointed sets must be of the form

$$\varphi(K) = V(K, L[p-1], S_{p+1}, \ldots, S_n)$$

with an essentially unique convex body L and pairwise orthogonal unit segments $S_{p+1}, \ldots, S_n$ that span the orthogonal complement of the affine hull of L.

The following theorem was proved by McMullen (1980): Let φ be a continuous, translation invariant valuation on $\mathcal{K}^n$ that is homogeneous of degree $n-1$. Then there is a continuous function $g: S^{n-1} \to \mathbb{R}$ such that

$$\varphi(K) = \int_{S^{n-1}} g(u) \, \mathrm{d}S_{n-1}(K, u) \quad \text{for } K \in \mathcal{K}^n.$$

From this result, McMullen deduced that for any such valuation φ there exist sequences $(L_j)_{j \in \mathbb{N}}, (M_j)_{j \in \mathbb{N}}$ in $\mathcal{K}^n$ such that

$$\varphi(K) = \lim_{j \to \infty} [V(K[n-1], L_j) - V(K[n-1], M_j)]$$

for $K \in \mathcal{K}^n$.

The following counterpart is due to Goodey & Weil (1984): A functional $\varphi: \mathcal{K}^n \to \mathbb{R}$ is a continuous, translation invariant valuation that is homogeneous of degree 1 if and only if there are convex bodies L_j, M_j for $j = 1, 2, \ldots$ such that

$$\varphi(K) = \lim_{j \to \infty} [V(K, L_j[n-1]) - V(K, M_j[n-1])]$$

uniformly for all $K \in \{K \in \mathcal{K}^n | K \subset mB^n\}$ and for any $m > 0$.

McMullen (1990) showed: If $\varphi: \mathcal{K}^n \to R$ is a monotone, translation invariant valuation that is homogeneous of degree $n-1$, then there is a convex body $L \in \mathcal{K}^n$ such that

$$\varphi(K) = V(K[n-1], L)$$

for all $K \in \mathcal{K}^n$.

7. *An identity for mixed volumes*. If $K, L, K_3, \ldots, K_n \in \mathcal{K}^n$ are convex bodies such that $K \cup L$ is convex, then

$$V(K, L, K_3, \ldots, K_n) = V(K \cup L, K \cap L, K_3, \ldots, K_n).$$

This follows from the additivity property of the mixed volume. It was noted by Groemer (1977a), p. 160, and proved in a shorter way in McMullen & Schneider [40], p. 179. An abstract version was given by McMullen (1989), Theorem 15.

8. *Mixed area under separate rotations*. The mixed volume $V(K_1, \ldots, K_n)$ is invariant under simultaneous rigid motions of the arguments, but not, of course, under separate motions applied to the arguments. Görtler (1937a) investigated the pairs K, L in $\mathcal{K}^2$ for which $V(K, L)$ is invariant under arbitrary rotations of one of the arguments.

9. *Characterizations of convex bodies in terms of mixed volumes*. A convex body is determined, up to a translation, by the values of its mixed volumes with sufficiently many other convex bodies. As an example, the following result was proved in Schneider (1974d): Let $i \in \{1, \ldots, n-1\}$ and let $K, K' \in \mathcal{K}^n$ be convex bodies of dimension $\geqq i+1$. If

$$V(K[i], B^n[n-i-1], L) = V(K'[i], B^n[n-i-1], L)$$

for all two-dimensional convex bodies L, then K is a translate of K'. In fact, it is sufficient to assume the equality only for those convex bodies L that are congruent to a fixed triangle having at least one angle that is an irrational multiple of π.

In view of such uniqueness results, it is not suprising that it is possible to characterize certain classes of convex bodies in terms of relations satisfied by mixed volumes. For example, Weil (1974b) showed that $L \in \mathcal{K}^n$ is contained in some translate of $K \in \mathcal{K}^n$ if and only if

$$V(L, K_2, \ldots, K_n) \leqq V(K, K_2, \ldots, K_n)$$

for all $K_2, \ldots, K_n \in \mathcal{K}^n$. Furthermore, each of the following conditions (a), (b), (c) is equivalent to the fact that L is a summand of K:

(a) (Weil, 1974b) For each $j \in \{1, \ldots, n-1\}$, the real functional

$$T_{K,L,j} = \sum_{k=0}^{j} \binom{j}{k} (-1)^{j-k} V(K[k], L[j-k], B^n[n-1-j], \cdot)$$

is monotone on $\mathcal{K}^n$.

(b) (Matheron, 1978b) For each $j \in \{0, \ldots, n\}$ and $\lambda \in (0, 1)$,

$$V(K \sim \lambda L[n-j], L[j]) = \sum_{k=0}^{n-j} \binom{n-j}{k} (-1)^k \lambda^k V(K[n-k-j], L[k+j]).$$

(c) (McMullen, 1990) For all polytopes $P, Q \in \mathcal{K}^n$ with $P \subset Q$,

$$V(Q[n-1], K) - V(Q[n-1], L) - V(P[n-1], K) + V(P[n-1], L) \geqq 0.$$

Other results in a similar spirit concern zonoids, generalized zonoids and centrally symmetric convex bodies. For $K \in \mathcal{K}^n$ and $u \in S^{n-1}$, let $V_{n-1}(K^u)$ denote the $(n-1)$-dimensional volume of the orthogonal projection of K onto the hyperplane $H_{u,0}$; here $K^u = K|H_{u,0}$. Weil (1976b) proved that a centrally symmetric convex body K is a zonoid if and only if

$$V(K, L[n-1]) \leqq V(K, M[n-1])$$

for all centrally symmetric $L, M \in \mathcal{K}^n$ satisfying $V_{n-1}(L^u) \leqq V_{n-1}(M^u)$ for all $u \in S^{n-1}$. Goodey (1977) showed that a centrally symmetric convex body $K \in \mathcal{K}^n$ is a generalized zonoid if and only if there exists a constant C such that

$$|V(K, L[n-1]) - V(K, M[n-1])| \leqq C \sup_{u \in S^{n-1}} |V_{n-1}(L^u) - V_{n-1}(M^u)|$$

for all centrally symmetric $L, M \in \mathcal{K}^n$. He also showed that $K \in \mathcal{K}^n$ is centrally symmetic if and only if

$$V(K, L[n-1]) = V(K, M[n-1])$$

for all $L, M \in \mathcal{K}^n$ satisfying $V_{n-1}(L^u) = V_{n-1}(M^u)$ for $u \in S^{n-1}$. Similar and more general results were proved by Weil (1979a) and by Goodey (1984a, b); for some necessary modifications of the latter results, see §6 of Goodey & Weil (1991).

5.2. Extensions of mixed volumes

The mixed volume, as defined in the preceding section, is a function on n-tuples of convex bodies that is Minkowski-linear in each variable. In the present section we shall collect some information on the possibilities for extending the mixed volume to other kinds of argument, while preserving or generalizing its linearity properties.

Our first observation is a negative one: under any extension to non-convex compact sets, the mixed volume would lose its essential properties. More precisely, the following result holds (Weil 1975a).

Theorem 5.2.1. *Let $\mathcal{K}' \subset \mathcal{C}^n$ be a class of compact sets containing $\mathcal{K}^n$ and closed under Minkowski addition. If there exists a function*

$V: (\mathcal{K}')^n \to \mathbb{R}$ that is Minkowski-additive in each variable and for which $V(K, \dots, K) = \mathcal{H}^n(K)$ for $K \in \mathcal{K}'$, then $\mathcal{K}' = \mathcal{K}^n$.

Proof. Suppose that $\mathcal{K}'$ and V satisfy the assumptions. Since $V(\cdot, K_2, \dots, K_n)$ (where $K_2, \dots, K_n \in \mathcal{K}'$ are fixed) is Minkowski-additive, the observation at the end of Section 3.4 shows that

$$V(K_1, K_2, \dots, K_n) = V(\operatorname{conv} K_1, K_2, \dots, K_n)$$

for $K_1 \in \mathcal{K}'$. Similarly,

$$V(\operatorname{conv} K_1, K_2, K_3, \dots, K_n) = V(\operatorname{conv} K_1, \operatorname{conv} K_2, K_3, \dots, K_n),$$

and so on. In particular, for $M \in \mathcal{K}'$ we have $\mathcal{H}^n(M) = \mathcal{H}^n(\operatorname{conv} M)$. If $\dim \operatorname{conv} M = n$, this implies $M = \operatorname{conv} M$. If $\dim \operatorname{conv} M < n$, we choose a convex body K in an affine subspace complementary to $\operatorname{aff} M$ such that $\dim(K + \operatorname{conv} M) = n$. For $M' := K + M$ we have $\mathcal{H}^n(M') = \mathcal{H}^n(\operatorname{conv} M')$ and thus $K + M = M' = \operatorname{conv} M' = K + \operatorname{conv} M$. Since K and $\operatorname{conv} M$ lie in complementary subspaces, this implies $M = \operatorname{conv} M$. Thus each element of $\mathcal{K}'$ is convex. ■

If, however, we consider the mixed volume not as a Minkowski multi-additive function on convex bodies, but as a multi-additive functional defined on support functions, then a natural and sometimes useful extension is possible. In the following, we identify convex bodies with their support functions, restricted to S^{n-1}, and hence also write

$$V(K_1, \dots, K_n) = V(\bar{h}_{K_1}, \dots, \bar{h}_{K_n}) \tag{5.2.1}$$

for $K_1, \dots, K_n \in \mathcal{K}^n$.

The real vector space $C(S^{n-1})$ of real continuous functions on the sphere S^{n-1} contains as a subspace the vector space $D(S^{n-1})$ spanned by restrictions of support functions. If $g \in D(S^{n-1})$ has the representations $g = \bar{h}_K - \bar{h}_L = \bar{h}_{K'} - \bar{h}_{L'}$ with $K, L, K', L' \in \mathcal{K}^n$, then $K + L' = K' + L$; hence it follows from the additivity of the mixed volume that

$$V(g, K_2, \dots, K_n) := V(K, K_2, \dots, K_n) - V(L, K_2, \dots, K_n)$$

does not depend on the special representation. Clearly $V(\cdot, K_2, \dots, K_n)$ thus defined is a linear functional on $D(S^{n-1})$. Similar extensions are possible in the other arguments. Explicitly, for $g_i \in D(S^{n-1})$ with the representations $g_i = h_i^0 - h_i^1$, where h_i^0, h_i^1 are restrictions of support functions ($i = 1, \dots, n$), we can define the mixed volume of $g_1, \dots, g_n$ by

$$V(g_1, \dots, g_n) = \sum_{\nu_1, \dots, \nu_n \in \{0,1\}} (-1)^{\nu_1 + \dots + \nu_n} V(h_1^{\nu_1}, \dots, h_n^{\nu_n}).$$

Then V is symmetric and n-linear on $D(S^{n-1})$, and satisfies (5.2.1). Similarly, the mixed area measure has an $(n-1)$-linear extension to $D(S^{n-1})$ given by

$$S(g_1, \ldots, g_{n-1}, \cdot) = \sum_{v_1,\ldots,v_{n-1}\in\{0,1\}} (-1)^{v_1+\ldots+v_{n-1}} S(h_1^{v_1}, \ldots, h_{n-1}^{v_{n-1}}, \cdot).$$

Formula (5.1.18) extends trivially to give

$$V(g_1, \ldots, g_n) = \frac{1}{n}\int_{S^{n-1}} g_1(u)\,\mathrm{d}S(g_2, \ldots, g_n, u)$$

for $g_1, \ldots, g_n \in D(S^{n-1})$.

For given convex bodies $K_2, \ldots, K_n \in \mathcal{K}^n$, we can immediately obtain an extension of the functional $V(\cdot, K_2, \ldots, K_n)$ to the space $C(S^{n-1})$ (with the maximum norm) by defining

$$V(f, K_2, \ldots, K_n) := \frac{1}{n}\int_{S^{n-1}} f(u)\,\mathrm{d}S(K_2, \ldots, K_n, u) \quad (5.2.2)$$

for $f \in C(S^{n-1})$. Then $V(\cdot, K_2, \ldots, K_n)$ is a continuous linear functional on $C(S^{n-1})$, with norm $V(B^n, K_2, \ldots, K_n)$. The mixed area measure $S(K_2, \ldots, K_n, \cdot)$ turns out to be the unique measure representing this linear functional according to the Riesz representation theorem. (This remark has played a role in the literature; cf. Section 5.1, Note 1).

It should, however, be pointed out that a further extension of the mixed volume to a multilinear continuous functional on $C(S^{n-1})$ is not possible. It suffices to show this for $n = 2$.

Theorem 5.2.2. *There is no bilinear function* $V: C(S^1)^2 \to \mathbb{R}$ *that is continuous in each variable and satisfies* $V(\bar{h}_K, \bar{h}_L) = V(K, L)$ *for* $K, L \in \mathcal{K}^2$.

Proof. Suppose V exists. We identify S^1 with $[0, 2\pi]$ via $(\cos\alpha, \sin\alpha) \mapsto \alpha$. For a continuous function f on $[0, 2\pi]$ and a convex body $K \in \mathcal{K}^2$ formula (5.1.18) extends to give

$$V(f, \bar{h}_K) = \frac{1}{2}\int_0^{2\pi} f(\alpha)\,\mathrm{d}S(K, \alpha).$$

For the proof, we approximate f uniformly by differences of support functions (which is possible by Lemma 1.7.9) and use the linearity and continuity of V in its first argument.

Now choose $f(\alpha) = 1 - |\sin\alpha|^{1/2}$ and let $0 < \varepsilon < 1$. Let K be the segment with endpoints $\pm(0, a)$, where $a = 1/\varepsilon$. The measure $S(K, \cdot)$ is concentrated at the points $\alpha = 0$ and $\alpha = \pi$ and associates a weight $2a$ with each of these points. Let L be obtained from K by a rotation around its centre by an angle φ, where $\sin\varphi = \varepsilon^2$. Then K and L are Hausdorff distance ε apart. On the other hand,

$$V(f, \bar{h}_K) - V(f, \bar{h}_L) = 2a - 2a(1 - \sqrt{\sin\varphi}) = 2.$$

Thus the function $g_\varepsilon := \bar{h}_K - \bar{h}_L$ satisfies $\|g_\varepsilon\| = \varepsilon$ and $V(f, g_\varepsilon) = 2$, which for $\varepsilon \to 0$ contradicts the continuity of V in its second variable. ∎

If convex bodies are identified not with their support functions but with their characteristic functions, then a different extension problem arises for the mixed volume. We shall describe briefly the theory developed by Groemer (1977a) for this case. For $K \in \mathcal{K}^n$, let K^* be the characteristic function of K, and let $V(\mathcal{K}^n)$ be the real vector space spanned by these functions for $K \in \mathcal{K}^n$. (Compare Section 3.4, Note 7. Since Groemer's notation K^* is used only in that note and in the present section, no confusion with the polar body should arise.) The first step is to show that there exists a unique bilinear map $\psi: V(\mathcal{K}^n)^2 \to V(\mathcal{K}^n)$ such that

$$\psi(K^*, L^*) = (K + L)^* \quad \text{for } K, L \in \mathcal{K}^n.$$

The construction of this map uses the Euler characteristic on $V(\mathcal{K}^n)$ as established by Groemer (1972, 1974). The vector space $V(\mathcal{K}^n)$ together with the multiplication $\times$ defined by $f \times g = \psi(f, g)$ is a commutative algebra over $\mathbb{R}$ with unit element. The Euler characteristic is an algebra homomorphism. To every affine map $\alpha: \mathbb{E}^n \to \mathbb{E}^n$ there exists a unique linear map $\bar{\alpha}: V(\mathcal{K}^n) \to V(\mathcal{K}^n)$ such that $\bar{\alpha}(K^*) = (\alpha K)^*$ for $K \in \mathcal{K}^n$. In the special case $\alpha x = \lambda x$ with $\lambda \geqq 0$ one writes $\bar{\alpha}(f) := \lambda \circ f$ for $f \in V(\mathcal{K}^n)$. Obviously there is a unique linear functional $\bar{V}_n$ on $V(\mathcal{K}^n)$ such that $\bar{V}_n(K^*) = V_n(K)$ for $K \in \mathcal{K}^n$. Groemer showed that (5.1.26) can be generalized as follows. For non-negative integers $r_1, \ldots, r_n$ with $r_1 + \ldots + r_n = n$ there exists exactly one n-linear map

$$v_{(r_1,\ldots,r_n)}: V(\mathcal{K}^n)^n \to \mathbb{R}$$

such that

$$v_{(r_1,\ldots,r_n)}(K_1^*, \ldots, K_n^*) = V(K_1[r_1], \ldots, K_n[r_n])$$

for $K_1, \ldots, K_n \in \mathcal{K}^n$. For $f_1, \ldots, f_n \in V(\mathcal{K}^n)$ and $\lambda_1, \ldots, \lambda_n \geqq 0$ we have

$$\bar{V}_n((\lambda_1 \circ f_1) \times \ldots \times (\lambda_n \circ f_n))$$
$$= \sum \binom{n}{r_1 \ldots r_n} \lambda_1^{r_1} \cdots \lambda_n^{r_n} v_{(r_1,\ldots,r_n)}(f_1, \ldots, f_n).$$

For the further investigation of these generalized mixed volumes we refer to the original article of Groemer (1977a).

Finally we consider a very special situation where the mixed volume can be extended in a natural way. This extension will be particularly useful for the proof of the Aleksandrov–Fenchel inequality in Section 6.3.

Let $\mathcal{A}$ be a given a-type of strongly isomorphic simple n-dimensional polytopes. As in Section 5.1, we let $u_1, \ldots, u_N$ be the normal vectors of the facets of any $P \in \mathcal{A}$, and we use the notation of the first part of Section 5.1. The N-tuple

$$\bar{P} := (h_1, \ldots, h_N) \in \mathbb{R}^N$$

is called the *support vector* of P. We recall that the support numbers of the facet F_i are given by

$$h_{ij} = h_j \operatorname{cosec} \Theta_{ij} - h_i \cot \Theta_{ij} \quad \text{for } (i, j) \in J,$$

and furthermore that

$$V(P_1, \ldots, P_n) = \sum a_{j_1 \ldots j_n} h_{j_1}^{(1)} \cdots h_{j_n}^{(n)}, \tag{5.2.3}$$

$$v(F_i^{(1)}, \ldots, F_i^{(n-1)}) = \sum a_{k_1 \ldots k_{n-1}}^{(i)} h_{ik_1}^{(1)} \cdots h_{ik_{n-1}}^{(n-1)} \tag{5.2.4}$$

for $P_1, \ldots, P_n \in \mathcal{A}$, where the coefficients are symmetric and depend only on the a-type $\mathcal{A}$. The summations in (5.2.3) and (5.2.4) can be considered to extend respectively over all $j_1, \ldots, j_n$ and all $k_1, \ldots, k_{n-1}$, from 1 to N, if zero coefficients are introduced and, say, $h_{ij}^{(r)} := 0$ if $(i, j) \notin J$.

We can consider V, as given by (5.2.3), to be defined for n-tuples of support vectors, and we now extend this to arbitrary N-tuples

$$X_r = (x_1^{(r)}, \ldots, x_N^{(r)}) \in \mathbb{R}^N, \quad r = 1, \ldots, n,$$

by putting

$$V(X_1, \ldots, X_n) := \sum a_{j_1 \ldots j_n} x_{j_1}^{(1)} \cdots x_{j_n}^{(n)}. \tag{5.2.5}$$

Then V is an N-linear function on $\mathbb{R}^N$. Furthermore, for $X = (x_1, \ldots, x_N) \in \mathbb{R}^N$ we define

$$x_{ij} := \begin{cases} x_j \operatorname{cosec} \Theta_{ij} - x_i \cot \Theta_{ij} & \text{if } (i, j) \in J, \\ 0 \text{ if } (i, j) \notin J \end{cases}$$

and

$$\Lambda_i X := (x_{i1}, \ldots, x_{iN}), \tag{5.2.6}$$

so that $\Lambda_i : \mathbb{R}^N \to \mathbb{R}^N$ is a linear map. Then we put

$$v(\Lambda_i X_1, \ldots, \Lambda_i X_{n-1}) := \sum a_{k_1 \ldots k_{n-1}}^{(i)} x_{ik_1}^{(1)} \cdots x_{ik_{n-1}}^{(n-1)}. \tag{5.2.7}$$

Lemma 5.1.4 now extends to give

$$V(X_1, \ldots, X_n) = \frac{1}{n} \sum_{i=1}^{N} x_i^{(1)} v(\Lambda_i X_2, \ldots, \Lambda_i X_n). \tag{5.2.8}$$

This follows from the facts that the equality is true for support vectors, $\bar{P} + \varepsilon X_r$ is a support vector if $P \in \mathcal{A}$ and ε is in a suitable neighbourhood of 0 (as follows from Lemma 2.4.12), and both sides are n-linear functions.

Note for Section 5.2

The multilinear extension of V and S to differences of support functions appears in Aleksandrov (1937a). It was further investigated (and used for characterizations of convex functions within the vector space of differences of support functions) by Weil (1974a); see also Weil (1974b).

The assertion of Theorem 5.2.2 was first proved, in a different way, by Meier in 1982 (unpublished). The construction used above is taken from Schneider (1974c), in a modified form.

5.3. Special formulae for mixed volumes and quermassintegrals

The purpose of this section is to collect a number of special formulae for mixed volumes and mixed area measures and, in particular, for quermassintegrals. These formulae relate mixed volumes to the notions of Section 2.5, that is, to local curvature functions in the case of convex bodies of class C^2_+, and also to the integral-geometric considerations of Chapter 4. Moreover, some special representations are treated that are valid in restricted cases or for particular convex bodies, such as zonoids.

In the first part of this section, it is supposed that the convex bodies $K, L, K_i \in \mathcal{K}^n_0$ are all of class C^2_+. First we recall that in this case, by (4.2.19) and (4.2.20), the curvature measures and area measures have the explicit representations

$$C_m(K, \beta) = \int_{B \cap \mathrm{bd} K} H_{n-1-m} \, \mathrm{d}\mathcal{H}^{n-1}, \tag{5.3.1}$$

$$S_m(K, \omega) = \int_\omega s_m \, \mathrm{d}\mathcal{H}^{n-1} \tag{5.3.2}$$

for $\beta \in \mathcal{B}(\mathbb{E}^n)$, $\omega \in \mathcal{B}(S^{n-1})$ and $m = 0, 1, \ldots, n-1$. Applying (5.3.2) for $m = n-1$ to the body $K = \lambda_1 K_1 + \ldots + \lambda_{n-1} K_{n-1}$ with $\lambda_1, \ldots, \lambda_{n-1} \geqq 0$ and using (5.1.17) and (2.5.33), we find

$$S(K_1, \ldots, K_{n-1}, \omega) = \int_\omega s(K_1, \ldots, K_{n-1}, u) \, \mathrm{d}\mathcal{H}^{n-1}(u). \tag{5.3.3}$$

Formula (5.1.18) for the mixed volume now reads

$$V(K_1, \ldots, K_n) = \frac{1}{n} \int_{S^{n-1}} h(K_1, u) s(K_2, \ldots, K_n, u) \, \mathrm{d}\mathcal{H}^{n-1}(u). \tag{5.3.4}$$

Since the mixed volume is symmetric in its arguments, this yields integral formulae such as

$$\int_{S^{n-1}} h(K_1, u) s(K_2, K_3, \ldots, K_n, u) \, \mathrm{d}\mathcal{H}^{n-1}(u)$$
$$= \int_{S^{n-1}} h(K_2, u) s(K_1, K_3, \ldots, K_n, u) \, \mathrm{d}\mathcal{H}^{n-1}(u). \quad (5.3.5)$$

The area measures of lower order are obtained from the mixed area measure by taking a special case: in analogy to formula (2.5.35), namely

$$s_j(K, u) = s(\underbrace{K, \ldots, K}_{j}, \underbrace{B^n, \ldots, B^n}_{n-1-j}, u), \quad (5.3.6)$$

we have

$$S_j(K, \omega) = S(\underbrace{K, \ldots, K}_{j}, \underbrace{B^n, \ldots, B^n}_{n-1-j}, \omega). \quad (5.3.7)$$

This follows immediately from (5.3.3), or (for general convex bodies) from a comparison of the local Steiner formula (4.2.16) for $m = n - 1$, that is,

$$S_{n-1}(K + \rho B^n, \omega) = \sum_{j=0}^{n-1} \rho^{n-1-j} \binom{n-1}{j} S_j(K, \omega),$$

with the polynomial expansion (5.1.17). Consequently, the quermassintegrals or Minkowski functionals $W_1, \ldots, W_n$, which in Section 4.2 were introduced by

$$W_i(K) := \frac{1}{n} S_{n-i}(K, S^{n-1}), \quad i = 1, \ldots, n,$$

turn out to be special mixed volumes: by (5.3.7) and (5.1.18),

$$W_i(K) = V(\underbrace{K, \ldots, K}_{n-i}, \underbrace{B^n, \ldots, B^n}_{i}) \quad (5.3.8)$$

for $i = 1, \ldots, n$, but also for $i = 0$, since $W_0(K) = V_n(K)$, by definition. Of course, (5.3.8) is valid for arbitrary convex bodies K.

From (5.3.2) and (5.3.1) we see that

$$W_i(K) = \frac{1}{n} \int_{S^{n-1}} s_{n-i} \, \mathrm{d}\mathcal{H}^{n-1}, \quad (5.3.9)$$

$$W_i(K) = \frac{1}{n} \int_{\mathrm{bd}\, K} H_{i-1}\, \mathrm{d}\mathscr{H}^{n-1} \tag{5.3.10}$$

for $i = 1, \dots, n$. Using (5.3.5), we obtain another representation, namely

$$W_i(K) = \frac{1}{n} \int_{S^{n-1}} h_K s_{n-i-1}\, \mathrm{d}\mathscr{H}^{n-1} \tag{5.3.11}$$

for $i = 0, \dots, n-1$. From (5.3.10) we see again that $nW_1(K)$ is the surface area of K, and the case $i = n-1$ of (5.3.11), together with $h_K(u) + h_K(-u) = w(K, u)$, shows that

$$W_{n-1}(K) = \frac{\kappa_n}{2}\, b(K), \tag{5.3.12}$$

where $b(K)$ is the mean width of K (by approximation, this extends to general convex bodies).

Using (2.5.30) and (2.5.28), we can transform formula (5.3.11) into

$$W_i(K) = \frac{1}{n} \int_{\mathrm{bd}\, K} q_K H_i\, \mathrm{d}\mathscr{H}^{n-1}, \tag{5.3.13}$$

where $q_K(x) := h_K(\nu(x)) = \langle x, \nu(x) \rangle$ for $x \in \mathrm{bd}\, K$.

In differential geometry, the equalities resulting from (5.3.9), (5.3.11) and from (5.3.10), (5.3.13) respectively,

$$\int_{S^{n-1}} s_j\, \mathrm{d}\mathscr{H}^{n-1} = \int_{S^{n-1}} h_K s_{j-1}\, \mathrm{d}\mathscr{H}^{n-1}, \tag{5.3.14}$$

and

$$\int_{\mathrm{bd}\, K} H_{j-1}\, \mathrm{d}\mathscr{H}^{n-1} = \int_{\mathrm{bd}\, K} q_K H_j\, \mathrm{d}\mathscr{H}^{n-1}, \tag{5.3.15}$$

for $j = 1, \dots, n-1$, are known as *Minkowskian integral formulae*.

The case $i = n-2$ of (5.3.11) can be given a form that is quite useful. By (2.5.23) we know that

$$s_1 = h_K + \frac{1}{n-1} \Delta_S h_K,$$

where Δ_S is the spherical Laplace operator on S^{n-1}. This yields

$$W_{n-2}(K) = \frac{1}{n} \int_{S^{n-1}} h_K \left(h_K + \frac{1}{n-1} \Delta_S h_K \right) \mathrm{d}\mathscr{H}^{n-1}. \tag{5.3.16}$$

More generally, it follows from (5.3.4) and (5.3.6) that

$$V(K, L, B^n, \dots, B^n) = \frac{1}{n} \int_{S^{n-1}} h_K \left(h_L + \frac{1}{n-1} \Delta_S h_L \right) \mathrm{d}\mathscr{H}^{n-1}. \tag{5.3.17}$$

Using Green's formula on S^{n-1}, we can rewrite (5.3.16) as

$$W_{n-2}(K) = \frac{1}{n}\int_{S^{n-1}} \left(h_K^2 - \frac{1}{n-1}\nabla_S h_K\right) \mathrm{d}\mathcal{H}^{n-1}, \qquad (5.3.18)$$

where ∇_S denotes the second Beltrami operator (square of the gradient) on S^{n-1}. Taking a local parametrization $N: M \to \mathbb{E}^n$ of the sphere S^{n-1} and using the notation of Section 2.5, in particular $X = \bar{\xi} \circ N$ and $e_{ij} = \langle N_i, N_j \rangle$, we have, for $h = h_K \circ N$,

$$(\nabla_S h_K) \circ N = \sum_{i,j=1}^{n-1} e_{ij} h^i h^j$$

with

$$h^i := \sum_{j=1}^{n-1} e^{ij} h_j.$$

One easily finds

$$X = hN + \sum_{j=1}^{n-1} h^j N_j$$

and thus

$$(\nabla_S h_K) \circ N = \left|\sum h^j N_j\right|^2 = |X - hN|^2 = |X|^2 - h^2.$$

Thus $(\nabla_S h_K)(u)$ is equal to the squared distance between the point where the support plane $H(K, u)$ touches K and the foot of the perpendicular to $H(K, u)$ through o.

Heil (1987) pointed out that formula (5.3.18) holds for arbitrary convex bodies K. Observe that in $\mathbb{E}^3$ this is a formula for the surface area.

In the second part of this section we now treat the relationship of mixed volumes and of mixed area measures to the integral-geometric considerations of Section 4.5. In particular, we collect some observations on translative integral geometry and on rotational mean values. The convex bodies K, K', K_i occurring in the following are arbitrary.

First we recall the principal kinematic formula (the case $j = 0$ and $\beta = \beta' = \mathbb{E}^n$ of Theorem 4.5.2), written in terms of quermassintegrals:

$$\int_{G_n} \chi(K \cap gK')\,\mathrm{d}\mu(g) = \frac{1}{\kappa_n}\sum_{k=0}^{n}\binom{n}{k} W_k(K) W_{n-k}(K'). \qquad (5.3.19)$$

If here the motion group G_n is replaced by the group of translations, which may be identified with $\mathbb{E}^n$, we obtain

$$\begin{aligned}\int_{\mathbb{E}^n} \chi(K \cap (K' + t))\,\mathrm{d}t &= \mathcal{H}^n\{t \in \mathbb{E}^n \mid K \cap (K' + t) \neq \emptyset\} \\ &= V_n(K - K') \\ &= \sum_{k=0}^{n}\binom{n}{k} V(K[k], -K'[n-k]). \qquad (5.3.20)\end{aligned}$$

Thus mixed volumes appear inevitably in translative integral geometry. (See also Section 4.5, Note 5.)

Since the proof given in Section 4.5 for the principal kinematic formula first treated the translative case, it yields, incidentally, a new representation for the mixed volume in the special case where the arguments are two polytopes. Suppose that $P, P' \in \mathcal{P}^n$ are polytopes. Theorem 4.5.3, for $j = 0$, $\beta = \beta' = \mathbb{E}^n$, gives

$$\int_{\mathbb{E}^n} \chi(P \cap (P' + t))\,\mathrm{d}t = \chi(P)V_n(P') + V_n(P)\chi(P') + \sum_{k=1}^{n-1} \sum_{F \in \mathcal{F}_k(P)} \sum_{F' \in \mathcal{F}_{n-k}(P')} \gamma(F, F', P, P')[F, F']V_k(F)V_{n-k}(F').$$

Using (5.3.20) for $K = P$, $K' = P'$ and then replacing P by λP in both equations and comparing the coefficients of equal powers of λ, we obtain

$$\binom{n}{k} V(P[k], -P'[n-k]) = \sum_{F \in \mathcal{F}_k(P)} \sum_{F' \in \mathcal{F}_{n-k}(P')} \gamma(F, F', P, P')[F, F']V_k(F)V_{n-k}(F'). \quad (5.3.21)$$

At present it is not clear whether this is a special case of some formula valid for general convex bodies.

Another translative integral formula, but now referring to subspaces, is obtained as follows. Let $E_k \subset \mathbb{E}^n$ be a k-dimensional linear subspace, $k \in \{1, \ldots, n-1\}$, and let U_k be a convex body contained in E_k. By $v^{(k)}$, $W_j^{(k)}$, $s^{(k)}$, $S_j^{(k)}$ we denote respectively the mixed volume, jth quermassintegral, mixed area measure and area measure of order j, each computed in a k-dimensional affine subspace. For $\lambda \geqq 0$ we have

$$\begin{aligned}\sum_{j=0}^{n} \binom{n}{j} \lambda^j V(K[n-j], U_k[j]) &= V_n(K + \lambda U_k)\\ &= \int_{E_k^\perp} V_k((K + \lambda U_k) \cap (E_k + t))\,\mathrm{d}\mathcal{H}^{n-k}(t)\\ &= \int_{E_k^\perp} V_k([K \cap (E_k + t)] + \lambda U_k)\,\mathrm{d}\mathcal{H}^{n-k}(t)\\ &= \int_{K|E_k^\perp} \sum_{j=0}^{k} \binom{k}{j} \lambda^j v^{(k)}(K \cap (E_k + t)[k-j], U_k[j])\,\mathrm{d}\mathcal{H}^{n-k}(t).\end{aligned}$$

By comparison, the following two consequences are obtained. Choosing for U_k the k-dimensional unit ball $B_k = B^n \cap E_k$, we find

$$\int_{E_k^\perp} W_j^{(k)}(K \cap (E_k + t))\,\mathrm{d}\mathcal{H}^{n-k}(t) = \binom{n}{j}\binom{k}{j}^{-1} V(K[n-j], B_k[j]) \quad (5.3.22)$$

for $j = 0, \ldots, k-1$. This is another example where a formula in translative integral geometry requires mixed volumes. The case $j = k$ yields

$$\binom{n}{k} V(K[n-k], U_k[k]) = V_k(U_k) V_{n-k}(K | E_k^\perp)$$

Here we choose $V_k(U_k) = 1$, replace K by $\sum \lambda_i K_i$ and compare the coefficients. Interchanging the roles of E_k and $E_k^\perp$, we arrive at

$$v^{(k)}(K_1 | E_k, \ldots, K_k | E_k) = \binom{n}{k} V(K_1, \ldots, K_k, U_{n-k}[n-k]), \tag{5.3.23}$$

which is valid if $U_{n-k} \subset E_k^\perp$ and $V_{n-k}(U_{n-k}) = 1$. Thus k-dimensional mixed volumes of orthogonal projections onto E_k can be expressed by mixed volumes in $\mathbb{E}^n$.

We turn to rotational mean values. If in (5.3.20) we replace K' by $-\rho\lambda K'$ with $\rho \in SO(n)$ and $\lambda \geqq 0$, integrate over $SO(n)$ with respect to the invariant measure ν, use (5.3.19) and finally compare the coefficients of equal powers of λ, we end up with the equality

$$\int_{SO(n)} V(K[m], \rho K'[n-m]) \, d\nu(\rho)$$
$$= \frac{1}{\kappa_n} V(K[m], B^n[n-m]) V(B^n[m], K'[n-m]) \tag{5.3.24}$$

for $m = 0, \ldots, n$. Here we replace K, K' by $\sum \lambda_i K_i$ and $\sum \mu_j K'_j$, respectively, expand both sides and compare the coefficients to obtain (after changing the notation)

$$\int_{SO(n)} V(K_1, \ldots, K_m, \rho K_{m+1}, \ldots, \rho K_n) \, d\nu(\rho)$$
$$= \frac{1}{\kappa_n} V(K_1, \ldots, K_m, B^n[n-m]) V(B^n[m], K_{m+1}, \ldots, K_n). \tag{5.3.25}$$

A mean value formula for projections is obtained as follows. Let E_k be a k-dimensional linear subspace of $\mathbb{E}^n$ ($k \in \{1, \ldots, n-1\}$) and choose a convex body $U_{n-k} \subset E_k^\perp$ with $V_{n-k}(U_{n-k}) = 1$. Then

$$\int_{SO(n)} v^{(k)}(K_1 | \rho E_k, \ldots, K_k | \rho E_k) \, d\nu(\rho)$$
$$= \int_{SO(n)} \binom{n}{k} V(K_1, \ldots, K_k, \rho U_{n-k}[n-k]) \, d\nu(\rho)$$
$$= \frac{\binom{n}{k}}{\kappa_n} V(K_1, \ldots, K_k, B^n[n-k]) V(B^n[k], U_{n-k}[n-k]).$$

Here the formulae (5.3.23) and (5.3.25) were used. Since

$$V(B^n[k], U_{n-k}[n-k]) = W_k(U_{n-k}) = \frac{\kappa_k}{\binom{n}{k}} V_{n-k}(U_{n-k}) = \frac{\kappa_k}{\binom{n}{k}}$$

by (4.2.26), we arrive at the formula

$$\int_{SO(n)} v^{(k)}(K_1|\rho E_k, \ldots, K_k|\rho E_k)\, \mathrm{d}\nu(\rho)$$
$$= \frac{\kappa_k}{\kappa_n} V(K_1, \ldots, K_k, B^n[n-k]). \quad (5.3.26)$$

As a special case, we note that

$$\int_{SO(n)} W^{(k)}_{k-j}(K|\rho E_k)\, \mathrm{d}\nu(\rho) = \frac{\kappa_k}{\kappa_n} W_{n-j}(K) \quad (5.3.27)$$

for $j = 0, \ldots, k$, which is known as *Kubota's integral recursion.* The further specialization $k = j = n-1$ is *Cauchy's surface area formula*.

Similar formulae for mixed area measures are obtained if we use the equality (a special case of Theorem 4.5.6)

$$\int_{SO(n)} S_{n-1}(K + \rho K', \omega \cap \rho\omega')\, \mathrm{d}\nu(\rho)$$
$$= \frac{1}{\omega_n} \sum_{k=0}^{n-1} \binom{n-1}{k} S_k(K, \omega) S_{n-1-k}(K', \omega'),$$

valid for Borel sets $\omega, \omega' \in \mathcal{B}(S^{n-1})$. In the usual way, by 'mixing', we deduce

$$\int_{SO(n)} S(K_1, \ldots, K_m, \rho K_{m+1}, \ldots, \rho K_{n-1}, \omega \cap \rho\omega')\, \mathrm{d}\nu(\rho)$$
$$= \frac{1}{\omega_n} S(K_1, \ldots, K_m, B^n[n-1-m], \omega)$$
$$\times S(B^n[m], K_{m+1}, \ldots, K_{n-1}, \omega') \quad (5.3.28)$$

for $m = 0, \ldots, n-1$.

For projections onto a k-dimensional linear subspace E_k, formula (4.5.26) states that

$$\int_{SO(n)} S_j^{(k)}(K|\rho E_k, \omega \cap \rho E_k)\, \mathrm{d}\nu(\rho) = \frac{\omega_k}{\omega_n} S_j(K, \omega) \quad (5.3.29)$$

for $\omega \in \mathcal{B}(S^{n-1})$ and $j = 0, \ldots, k-1$. From the case $j = k-1$ we deduce, again in the usual way, the rotational mean value formula

$$\int_{SO(n)} s^{(k)}(K_1|\rho E_k, \ldots, K_{k-1}|\rho E_k, \omega \cap \rho E_k)\, \mathrm{d}\nu(\rho)$$
$$= \frac{\omega_k}{\omega_n} S(K_1, \ldots, K_{k-1}, B^n[n-k], \omega). \quad (5.3.30)$$

Of particular importance are projections onto hyperplanes. For these, we introduce some simplified notation. For $K \in \mathcal{K}^n$ and a unit vector $u \in S^{n-1}$ we denote by K^u the image of K under orthogonal projection onto the linear subspace orthogonal to u; thus $K^u = K|H_{u,0}$. For the $(n-1)$-dimensional mixed volume in a hyperplane we write v instead of $v^{(n-1)}$. If U is a segment of unit length parallel to u, then

$$v(K_1^u, \ldots, K_{n-1}^u) = nV(K_1, \ldots, K_{n-1}, U) \tag{5.3.31}$$

by (5.3.23). We may assume that U has centre o; then $h(U, v) = \frac{1}{2}|\langle u, v\rangle|$ and hence, by (5.1.18),

$$v(K_1^u, \ldots, K_{n-1}^u) = \frac{1}{2}\int_{S^{n-1}} |\langle u, v\rangle| \,\mathrm{d}S(K_1, \ldots, K_{n-1}, v). \tag{5.3.32}$$

By integration we immediately obtain

$$\int_{S^{n-1}} v(K_1^u, \ldots, K_{n-1}^u)\,\mathrm{d}\mathcal{H}^{n-1}(u) = n\kappa_{n-1}V(K_1, \ldots, K_{n-1}, B^n), \tag{5.3.33}$$

which is a special case of (5.3.26) in a slightly different formulation.

Formula (5.3.32) shows that the function $u \mapsto v(K_1^u, \ldots, K_{n-1}^u)$ is always the restriction of a support function to S^{n-1}. In particular, for $K \in \mathcal{K}^n$ there is a convex body $\Pi K \in \mathcal{K}^n$ for which

$$h(\Pi K, u) = \frac{1}{2}\int_{S^{n-1}} |\langle u, v\rangle| \,\mathrm{d}S_{n-1}(K, v) \tag{5.3.34}$$

for $u \in \mathbb{E}^n$. Thus ΠK is a zonoid (see Section 3.5) with centre at the origin; it is characterized by

$$h(\Pi K, u) = V_{n-1}(K^u) \quad \text{for } u \in S^{n-1}.$$

We call ΠK the *projection body* of K.

In the last part of this section we shall now derive formulae for mixed volumes, mixed area measures and quermassintegrals of zonoids. It is sometimes useful to have explicit expressions for these functions in terms of the generating measures.

For $u_1, \ldots, u_k \in S^{n-1}$ we denote by $D_k(u_1, \ldots, u_k)$ the k-dimensional volume of the parallelepiped spanned by $u_1, \ldots, u_k$; thus

$$D_k(u_1, \ldots, u_k) = V_k(\operatorname{conv}\{o, u_1\} + \ldots + \operatorname{conv}\{o, u_k\}).$$

For a linearly independent $(n-1)$-tuple $(u_1, \ldots, u_{n-1}) \in (S^{n-1})^{n-1}$ let $T(u_1, \ldots, u_{n-1}) \in S^{n-1}$ be the unit vector orthogonal to $u_1, \ldots, u_{n-1}$ and such that the n-tuple $(u_1, \ldots, u_{n-1}, T(u_1, \ldots, u_{n-1}))$ has the same orientation as some fixed basis of $\mathbb{E}^n$. Then T is continuous and hence can be extended (trivially) to a Borel measurable map from $(S^{n-1})^{n-1}$ into S^{n-1}. In particular, for a Borel measure φ on $(S^{n-1})^{n-1}$ the image measure $T(\int D_{n-1}\,\mathrm{d}\varphi)$ of the indefinite integral $\int D_{n-1}\,\mathrm{d}\varphi$ is well defined by

$$T\left(\int D_{n-1}\,\mathrm{d}\varphi\right)(\omega) = \int_{T^{-1}(\omega)} D_{n-1}\,\mathrm{d}\varphi$$

for $\omega \in \mathcal{B}(S^{n-1})$.

Theorem 5.3.1. *For $i = 1, \ldots, n$, let $Z_i \in \mathcal{K}^n$ be a zonoid with generating measure ρ_i (and centre o); thus*

$$h(Z_i, u) = \int_{S^{n-1}} |\langle u, v\rangle|\,\mathrm{d}\rho_i(v) \quad \text{for } u \in \mathbb{E}^n.$$

Then

$$V(Z_1, \ldots, Z_n) = \frac{2^n}{n!}\int_{S^{n-1}} \cdots \int_{S^{n-1}} D_n(u_1, \ldots, u_n)\,\mathrm{d}\rho_1(u_1)\cdots \mathrm{d}\rho_n(u_n) \quad (5.3.35)$$

and

$$S(Z_1, \ldots, Z_{n-1}, \cdot) = \frac{2^n}{(n-1)!}\, T\left(\int D_{n-1}\,\mathrm{d}(\rho_1 \otimes \ldots \otimes \rho_{n-1})\right). \quad (5.3.36)$$

Proof. We first assume that $Z_1 = \ldots = Z_n = Z$ and Z is a zonotope, say $Z = S_1 + \ldots + S_k$, where

$$S_i = \operatorname{conv}\{\alpha_i v_i, -\alpha_i v_i\} \quad \text{with } v_i \in S^{n-1},\ \alpha_i \geqq 0.$$

Then

$$h(Z, \cdot) = \sum_{i=1}^{k} \alpha_i |\langle \cdot, v_i\rangle| = \int_{S^{n-1}} |\langle \cdot, v\rangle|\,\mathrm{d}\rho(v),$$

where ρ is concentrated at $\pm v_i$ and assigns mass $\alpha_i/2$ to each of these points. For an even function f on S^{n-1} one has $\int f\,\mathrm{d}\rho = \sum \alpha_i f(v_i)$ and hence

$$\begin{aligned}\int_{S^{n-1}} \cdots \int_{S^{n-1}} & D_n(u_1, \ldots, u_n)\,\mathrm{d}\rho(u_1)\cdots \mathrm{d}\rho(u_n) \\ &= \sum_{i_1,\ldots,i_n=1}^{k} D_n(v_{i_1}, \ldots, v_{i_n})\alpha_{i_1}\cdots\alpha_{i_n} \\ &= \frac{n!}{2^n}\sum_{1 \leqq i_1 < \ldots < i_n \leqq k} V_n(S_{i_1} + \ldots + S_{i_n}).\end{aligned}$$

On the other hand

$$V_n(S_1 + \ldots + S_k) = \sum_{1 \leqq i_1 < \ldots < i_n \leqq k} V_n(S_{i_1} + \ldots + S_{i_n}). \quad (5.3.37)$$

This is easily proved by induction if one uses the relation

$$V_n(S_1 + \ldots + S_{k+1}) = V_n(Z + S_{k+1}) = V_n(Z) + |S_{k+1}|V_{n-1}(\pi Z),$$

where $|S_{k+1}|$ denotes the length of the segment S_{k+1} and π is the orthogonal projection onto a hyperplane orthogonal to S_{k+1}. Thus the formula

$$V_n(Z) = \frac{2^n}{n!}\int_{S^{n-1}} \cdots \int_{S^{n-1}} D_n(u_1, \ldots, u_n)\, d\rho(u_1) \cdots d\rho(u_n) \quad (5.3.38)$$

is proved for a zonotope Z with generating measure ρ.

The area measure of Z is, according to (5.1.12), given by

$$S_{n-1}(Z, \omega) = \sum_{u\in\omega} V_{n-1}(F(Z, u))$$

for $\omega \in \mathcal{B}(S^{n-1})$. The face $F(Z, u)$ is a translate of the sum of the faces of the segments $S_1, \ldots, S_k$ that are orthogonal to u, hence from (5.3.37) in dimension $n-1$ we obtain

$$\begin{aligned} S_{n-1}(Z, \omega) &= \sum_{u\in\omega} \sum_{\substack{1\leq i_1<\ldots<i_{n-1}\leq k \\ v_{i_1},\ldots,v_{i_{n-1}}\perp u}} V_{n-1}(S_{i_1} + \ldots + S_{i_{n-1}}) \\ &= 2^{n-1} \sum_{u\in\omega} \sum_{\substack{1\leq i_1<\ldots<i_{n-1}\leq k \\ v_{i_1},\ldots,v_{i_{n-1}}\perp u}} D_{n-1}(v_{i_1}, \ldots, v_{i_{n-1}})\alpha_{i_1} \cdots \alpha_{i_{n-1}} \\ &= \frac{2^n}{(n-1)!} \sum_{\substack{i_1,\ldots,i_{n-1} \\ T(v_{i_1},\ldots,v_{i_{n-1}})\in\omega}}^{k} D_{n-1}(v_{i_1}, \ldots, v_{i_{n-1}})\alpha_{i_1} \cdots \alpha_{i_{n-1}}, \end{aligned}$$

hence

$$S_{n-1}(Z, \omega) = \frac{2^n}{(n-1)!}\, T\left(\int D_{n-1}\, d\rho^{n-1}\right)(\omega). \quad (5.3.39)$$

Now let Z be a general zonoid. We can approximate Z by a sequence $(Z_i)_{i\in\mathbb{N}}$ of zonotopes whose generating measures ρ_i converge weakly to the generating measure ρ of Z (compare the proof of Theorem 3.5.2). Then the product measures ρ_i^n converge weakly to ρ^n. Hence, equality (5.3.38) extends to arbitrary zonoids.

Formula (5.3.39) can also be extended in this way. For any continuous real function f on S^{n-1},

$$\int_{S^{n-1}} f(x)\, dT\left(\int D_{n-1}\, d\rho_i^{n-1}\right)(x) =$$

$$\int_{S^{n-1}} \cdots \int_{S^{n-1}} f(T(u_1, \ldots, u_{n-1}))D_{n-1}(u_1, \ldots, u_{n-1})\, d\rho_i(u_1) \cdots d\rho_i(u_{n-1}),$$

and the function $f(T(\cdot, \ldots, \cdot))D_{n-1}(\cdot, \ldots, \cdot)$ is continuous on $(S^{n-1})^{n-1}$, since D_{n-1} is continuous and vanishes where T is discontinuous. Hence, the sequence $(T(\int D_{n-1}\, d\rho_i^{n-1}))_{i\in\mathbb{N}}$ converges weakly to $T(\int D_{n-1}\, d\rho^{n-1})$. Since $(S_{n-1}(Z_i, \cdot))_{i\in\mathbb{N}}$ converges weakly to $S_{n-1}(Z, \cdot)$, equality (5.3.39) follows.

Finally, if Z_i has generating measure ρ_i, then $\lambda_1 Z_1 + \ldots + \lambda_n Z_n$ ($\lambda_i \geqq 0$) has generating measure $\lambda_1\rho_1 + \ldots + \lambda_n\rho_n$. Inserting this into (5.3.38) and (5.3.39) and comparing the coefficients, we complete the proof. ■

If some of the Z_i in (5.3.35) are balls, one can reduce the number of integrations, using the following formula. Here we denote spherical Lebesgue measure on a k-dimensional unit sphere by σ_k.

Lemma 5.3.2. *If $L \subset \mathbb{E}^n$ is a k-dimensional linear subspace ($0 < k < n$) and $\varphi(u)$ is the angle between $u \in \mathbb{E}^n\setminus\{o\}$ and L, then*

$$\int_{S^{n-1}} \sin\varphi(u)\,\mathrm{d}\sigma_{n-1}(u) = \frac{\omega_{n-k}\omega_{n+1}}{\omega_{n-k+1}}. \qquad (5.3.40)$$

Proof. For $x \in B^n$ let $y := p(L, x)$ and $s(x) := |x - y|$. Then

$$\int_{B^n} s(x)\,\mathrm{d}\mathcal{H}^n(x) = \int_{L\cap B^n}\int_{B(y)} s(x)\,\mathrm{d}\mathcal{H}^{n-k}(x)\,\mathrm{d}\mathcal{H}^k(y)$$

with $B(y) := (L^\perp + y) \cap B^n$, which is an $(n-k)$-dimensional ball of radius $R = (1-\rho^2)^{1/2}$, $\rho = |y|$. With an obvious transformation,

$$\begin{aligned}\int_{B(y)} s(x)\,\mathrm{d}\mathcal{H}^{n-k}(x) &= \int_{S^{n-k-1}}\int_0^R r^{n-k}\,\mathrm{d}r\,\mathrm{d}\sigma_{n-k-1}\\ &= \frac{\omega_{n-k}}{n-k+1}(1-\rho^2)^{(n-k+1)/2},\end{aligned}$$

hence, with a similar transformation,

$$\begin{aligned}\int_{B^n} s(x)\,\mathrm{d}\mathcal{H}^n(x) &= \frac{\omega_{n-k}}{n-k+1}\int_{S^{k-1}}\int_0^1 (1-\rho^2)^{(n-k+1)/2}\rho^{k-1}\,\mathrm{d}\rho\,\mathrm{d}\sigma_{k-1}\\ &= \frac{\omega_{n-k}}{\omega_{n-k+1}}\kappa_{n+1}.\end{aligned}$$

On the other hand,

$$\begin{aligned}\int_{B^n} s(x)\,\mathrm{d}\mathcal{H}^n(x) &= \int_{B^n} |x| \sin\varphi(x)\,\mathrm{d}\mathcal{H}^n(x)\\ &= \int_{S^{n-1}}\int_0^1 r\sin\varphi(u) r^{n-1}\,\mathrm{d}r\,\mathrm{d}\sigma_{n-1}(u)\\ &= \frac{1}{n+1}\int_{S^{n-1}} \sin\varphi(u)\,\mathrm{d}\sigma_{n-1}(u),\end{aligned}$$

from which the assertion follows. ■

This allows us, in particular, to simplify the formula for the quermassintegrals of a zonoid.

Theorem 5.3.3. *If $Z \subset \mathbb{E}^n$ is a zonoid with generating measure ρ, then*

$$W_j(Z) = \frac{2^{n-j} j! \kappa_j}{n!} \int_{S^{n-1}} \cdots \int_{S^{n-1}} D_{n-j}(u_1, \ldots, u_{n-j}) \, \mathrm{d}\rho(u_1) \cdots \mathrm{d}\rho(u_{n-j}) \tag{5.3.41}$$

for $j = 0, \ldots, n$.

Proof. By (5.3.8) and (5.3.35),

$$W_j(Z) = \frac{2^{n-j}}{n! \kappa_{n-1}^j} \int_{S^{n-1}} \cdots \int_{S^{n-1}} D_n(u_1, \ldots, u_n) \times \mathrm{d}\rho(u_1) \cdots \mathrm{d}\rho(u_{n-j}) \, \mathrm{d}\sigma(u_{n-j+1}) \cdots \mathrm{d}\sigma(u_n),$$

since B^n is a zonoid with generating measure $\sigma/2\kappa_{n-1}$, where $\sigma = \sigma_{n-1}$ is spherical Lebesgue measure on S^{n-1}. Now

$$D_{k+1}(u_1, \ldots, u_{k+1}) = D_k(u_1, \ldots, u_k) \sin \varphi$$

if φ denotes the angle between u_{k+1} and $\operatorname{lin}\{u_1, \ldots, u_k\}$. Repeated application of (5.3.40) now yields the assertion (the Legendre relation $(m - \frac{1}{2})! m! = 2^{-2m} \pi^{1/2} (2m)!$ can be used to simplify the numerical factor). ■

Notes for Section 5.3

1. For dimensions two and three, many of the special formulae of this section can be found in Blaschke [5]; see also Bonnesen & Fenchel [8]. For the integral-geometric formulae, references are found in the notes for Section 4.5. Theorems 5.3.1 and 5.3.3 are taken from Weil (1976a), where they are proved for generalized zonoids; special cases appear in Blaschke [5], Weil (1971) and Matheron (1975). Weil (1976a) also gave a formula for the density of $S(Z_1, \ldots, Z_{n-1}, \cdot)$ in the case where the generating measures $\rho_1, \ldots, \rho_{n-1}$ have densities (with respect to σ).
2. *Differentiation of mixed area measures.* Let $K_1, \ldots, K_{n-1} \in \mathcal{K}^n$ be convex bodies of class C_+^2. If $u \in S^{n-1}$ and $(\omega_i)_{i \in \mathbb{N}}$ is a sequence of Borel subsets of S^{n-1} with $\mathcal{H}^{n-1}(\omega_i) > 0$ shrinking to u, then (5.3.3) shows that
$$\lim_{i \to \infty} \frac{S(K_1, \ldots, K_{n-1}, \omega_i)}{\mathcal{H}^{n-1}(\omega_i)} = s(K_1, \ldots, K_{n-1}, u).$$
Aleksandrov (1939c) extended this to general convex bodies $K_1, \ldots, K_{n-1}$: Almost everywhere on S^{n-1} their support functions are simultaneously twice differentiable. At each point u where this happens, $s(K_1, \ldots, K_{n-1}, u)$ can be defined by (2.5.36), and then the limit relation above is valid for suitably chosen sequences $(\omega_i)_{i \in \mathbb{N}}$.
3. *Minkowskian integral formulae.* The formulae
$$\int_{S^{n-1}} s_j \, \mathrm{d}\mathcal{H}^{n-1} = \int_{S^{n-1}} h_K s_{j-1} \, \mathrm{d}\mathcal{H}^{n-1} \tag{5.3.14}$$
and
$$\int_{\mathrm{bd}\,K} H_{j-1} \, \mathrm{d}\mathcal{H}^{n-1} = \int_{\mathrm{bd}\,K} q_K H_j \, \mathrm{d}\mathcal{H}^{n-1} \tag{5.3.15}$$

are the prototypes for a number of integral formulae that have proved quite useful in global differential geometry. We give a simple example of a typical application. Suppose that K is a convex body of class C^2_+ whose kth elementary symmetric function of the principal radii of curvature is constant, say

$$s_k = 1,$$

for some $k \in \{1, \ldots, n-1\}$. Newton's inequalities for elementary symmetric functions yield

$$s_{k-1}^{1/(k-1)} \geqq s_k^{1/k} = 1$$

if $k \geqq 2$, hence $s_{k-1} \geqq s_0$, and

$$s_1 \geqq s_k^{1/k} = 1 = s_k.$$

We may assume that $o \in \operatorname{int} K$, hence $h_K > 0$. Applying (5.3.14) twice, we obtain (integrating over S^{n-1} with respect to $\mathscr{H}^{n-1}$)

$$\int s_k = \int h_K s_{k-1} \geqq \int h_K s_0 = \int s_1 \geqq \int s_k,$$

hence $s_1 \equiv s_k^{1/k}$. If $k = 1$, we have

$$s_1 = s_1^2 \geqq s_2$$

and

$$\int s_1 \geqq \int s_2 = \int h_K s_1 = \int h_K = \int s_1,$$

hence $s_1 \equiv s_2^{1/2}$. In both cases, we deduce that every boundary point of K is an umbilic, and this is known to imply that K is a ball. Similarly, (5.3.15) can be used to show that K (of class C^2_+) must be a ball if H_k is constant for some $k \in \{1, \ldots, n-1\}$.

Apparently this simple argument was first used (in the more general setting of relative differential geometry) by Süss (1929).

Formulae (5.3.14) and (5.3.15) express the symmetry of the mixed volume in special cases. As mentioned in Section 5.1, Note 2, this symmetry can be proved analytically. In differential geometry, the usual way to prove such integral formulae is to apply Stokes' theorem to a suitably chosen differential form. Various such choices yield useful generalizations and analogues of the Minkowskian integral formulae. A systematic exposition of the derivation and application of integral formulae in global differential geometry can be found in Chapter 3 of Huck *et al.* (1973). To illustrate the wide applicability of generalized Minkowskian integral formulae we mention a few papers only: Chern (1959) (pairs of convex hypersurfaces), Simon (1967) (a systematic study), Schneider (1967d) (affine differential geometry), Katsurada (1962) (Riemannian spaces).

The extension of the Minkowskian integral formula (5.3.15) to convex bodies without differentiability assumptions, also in spaces of constant curvature, was studied and applied by Kohlmann (1988).

Other extensions of the Minkowskian integral formulae yield representations of the quermassintegrals of the types (5.3.9) and (5.3.11), but involving higher powers of the support function (and other expressions). General versions of such formulae were derived by Bokowski & Heil (1986), who extended and unified earlier results of Simon (1967), Shahin (1968), Yano & Tani (1969) and Firey (1974d). The formulae are complicated, but Bokowski & Heil (1986) made an elegant application to inequalities for quermassintegrals.

4. *Quermassintegrals of bodies of constant width*. The quermassintegrals $W_0, W_1, \ldots, W_n$ of a convex body of constant width b satisfy $[(n+1)/2]$ independent linear relations, namely
$$2W_{n-k} = \sum_{i=0}^{k-1} (-1)^i \binom{k}{i} b^{k-i} W_{n-i} \quad \text{for } 0 \leqq k \leqq n,\ k \text{ odd}.$$
For $n = 3$, this includes Blaschke's relation
$$V_3 = \tfrac{1}{2}bF - \tfrac{1}{3}\pi b^3,$$
where F is the surface area. Proofs for the general case were given by Dinghas (1940b), Santaló (1946) and Debrunner (1955).
5. *Wills' functional*. The special linear combination of quermassintegrals given by
$$W(K) = \sum_{i=0}^{n} \binom{n}{i} \frac{1}{\kappa_i} W_i(K) = \sum_{j=0}^{n} V_j(K)$$
has some remarkable properties; for instance,
$$W(K) = \int_{\mathbb{E}^n} \mathrm{e}^{-\pi d(K,x)^2}\, \mathrm{d}x$$
and, hence, $W(K \oplus L) = W(K)W(L)$ for an orthogonal direct sum $K \oplus L$. For these and further properties, see Hadwiger (1975c). The functional W attracted considerable interest after Wills (1973) had conjectured that $G(K) \leqq W(K)$ for $K \in \mathcal{K}^n$, where $G(K)$ denotes the number of points with integer coordinates (with respect to some orthonormal basis of $\mathbb{E}^n$) in K. The conjecture is easily established for $n = 2$ and was proved for $n = 3$ by Overhagen (1975), but Hadwiger (1979) found counterexamples for $n \geqq 441$. Related results and many connections between discrete and convex geometry can be found in the survey article of Wills (1990a). Much of the work described there is concerned with relations between lattice point inequalities and quermassintegrals, topics both of which were created by Minkowski. Wills (1990b) establishes inequalities between the zeros of the polynomial $\sum_{j=0}^{n} V_j(K)(-x)^j$ and the successive minima of K, as defined by Minkowski in the geometry of numbers.
6. *Further integral-geometric formulae*. Goodey & Weil (1987) and Weil (1990b) obtained some Crofton-type formulae for mixed volumes, of which we mention the following. Let $j \in \{1, \ldots, n-1\}$ and $K, L \in \mathcal{K}^n$. Then
$$\int_{A(n,j)} V(K \cap E[j], L[n-j])\, \mathrm{d}\mu_j(E) = \frac{\kappa_j \kappa_{n-j}}{\binom{n}{j}^2 \kappa_n} V_n(K) V_{n-j}(L).$$
If K and L are centrally symmetric, then
$$\int_{A(n,n-j+1)} \int_{A(n,j+1)} V(K \cap E[j], L \cap F[n-j])\, \mathrm{d}\mu_{j+1}(E)\, \mathrm{d}\mu_{n-j+1}(F)$$
$$= \frac{n(n-1)\alpha_{n0(j+1)}}{4\binom{n}{j}\kappa_{n-2}} V(\Pi K, \Pi L, B^n, \ldots, B^n),$$
where ΠK denotes the projection body of K.
7. *A formula for mixed volumes of two polytopes*. For polytopes $P, Q \in \mathcal{P}^n$ and for $k \in \{1, \ldots, n-1\}$, the mixed volume $V(P[k], Q[n-k])$ can be represented in a way similar to (5.3.21), but depending on a vectorial parameter. Suppose that $x \in \mathbb{E}^n \setminus \{o\}$ is a vector in general position with respect to P and Q, which means that there do not exist faces F of P and G of Q such that $\dim F + \dim G > n$ and
$$x \in \operatorname{lin}[N(P, F) + N(Q, G)].$$

Then, as proved by Betke (1992),

$$\binom{n}{k} V(P[k], Q[n-k]) = \underbrace{\sum_{F \in \mathcal{F}_k(P)} \sum_{G \in \mathcal{F}_{n-k}(Q)}}_{x \in N(P,F) - N(Q,G)} [F, G] V_k(F) V_{n-k}(G).$$

Here, as indicated, the summation extends only over those pairs F, G for which x is in the difference of the corresponding normal cones. By integration over all unit vectors x (almost all of which are in general position with respect to P and Q), formula (5.3.21) is again obtained.

5.4. Moment vectors and curvature centroids

In combining the notion of volume with the Minkowski addition of convex bodies, one is led naturally to mixed volumes and, as a special case, to quermassintegrals. In a similar way, a theory of mixed moment vectors can be developed, which by specialization leads to curvature centroids. We give a brief outline of such a theory, but without proofs. The latter can essentially be modelled on those for mixed volumes, and can also be found in Schneider (1972a).

For $K \in \mathcal{K}^n$ and for $r \in \{\dim K, \ldots, n\}$ we define

$$z_{r+1}(K) := \int_K x \, \mathrm{d}\mathcal{H}^r(x)$$

and call $z_{r+1}(K)$, for $r = \dim K$, the *moment vector* of K. This vector depends on the position of K relative to the origin; in fact

$$z_{r+1}(K + t) = z_{r+1}(K) + V_r(K)t$$

for $t \in \mathbb{E}^n$. If $P \in \mathcal{P}_0^n$ is an n-dimensional polytope with facets $F_1, \ldots, F_N$ and corresponding outer unit normal vectors $u_1, \ldots, u_N$, then

$$z_{n+1}(P) = \frac{1}{n+1} \sum_{i=1}^{N} h(P, u_i) z_n(F_i). \tag{5.4.1}$$

There is a symmetric function $z: (\mathcal{K}^n)^{n+1} \to \mathbb{E}^n$ such that

$$z_{n+1}(\lambda_1 K_1 + \ldots + \lambda_m K_m) = \sum_{i_1, \ldots, i_{n+1}}^{m} \lambda_{i_1} \cdots \lambda_{i_{n+1}} z(K_{i_1}, \ldots, K_{i_{n+1}}) \tag{5.4.2}$$

for $K_1, \ldots, K_m \in \mathcal{K}^n$ and $\lambda_1, \ldots, \lambda_m \geqq 0$. The map z is called the *mixed moment vector*. We have

$$z(K_1, \ldots, K_{n+1}) = \frac{1}{(n+1)!} \sum_{k=1}^{n+1} (-1)^{n+k+1} \sum_{i_1 < \ldots < i_k} z_{n+1}(K_{i_1} + \ldots + K_{i_k}) \tag{5.4.3}$$

and, with notation similar to that in Section (5.1),

$$z(\lambda_1 K_1 + \ldots + \lambda_m K_m[p], \bar{K}_{p+1}, \ldots, \bar{K}_{n+1}) =$$
$$\sum_{r_1,\ldots,r_m=0}^{p} \binom{p}{r_1 \ldots r_m} \lambda_1^{r_1} \cdots \lambda_m^{r_m} z(K_1[r_1], \ldots, K_m[r_m], \bar{K}_{p+1}, \ldots, \bar{K}_{n+1}) \tag{5.4.4}$$

for $1 \leqq p \leqq n+1$.

We shall list some properties of the map z. For $K, K_1, \ldots, K_{n+1} \in \mathcal{K}^n$ we have

$$z(K, \ldots, K) = z_{n+1}(K),$$
$$z(\alpha K_1, \ldots, \alpha K_{n+1}) = |\det \alpha| \alpha z(K_1, \ldots, K_{n+1})$$

for any linear map $\alpha: \mathbb{E}^n \to \mathbb{E}^n$. Under translations, z behaves according to

$$z(K_1 + t_1, \ldots, K_{n+1} + t_{n+1}) = z(K_1, \ldots, K_{n+1})$$
$$+ \frac{1}{n+1}[V(K_2, \ldots, K_{n+1})t_1 + \ldots + V(K_1, \ldots, K_n)t_{n+1}] \tag{5.4.5}$$

for $t_1, \ldots, t_{n+1} \in \mathbb{E}^n$. Furthermore, z is continuous, and for fixed $p \in \{1, \ldots, n+1\}$ and $K_{p+1}, \ldots, K_{n+1}$ the map

$$K \mapsto z(K[p], K_{p+1}, \ldots, K_{n+1})$$

is additive on $\mathcal{K}^n$. If $V(K_1, \ldots, \check{K}_i, \ldots, K_{n+1}) = 0$ (K_i deleted) for $i = 1, \ldots, n+1$, then $z(K_1, \ldots, K_{n+1}) = o$.

In analogy with the quermassintegral $W_r(K) = V(K[n-r], B^n[r])$ we define the rth *quermassvector* of K by

$$q_r(K) := \frac{n+1}{n+1-r} z(K[n+1-r], B^n[r]) \tag{5.4.6}$$

for $r = 0, \ldots, n$; in particular, $q_0(K) = z_{n+1}(K)$. From (5.4.4), we have the Steiner-type formula

$$z_{n+1}(K + \lambda B^n) = \sum_{r=0}^{n} \binom{n}{r} \lambda^r q_r(K), \tag{5.4.7}$$

since $z_{n+1}(B^n) = o$. It is consistent with the earlier definition of $z_{r+1}(K)$ to extend z_{r+1} to convex bodies K with $\dim K > r$ as follows:

$$z_{r+1}(K) := \frac{\binom{n}{r}}{\kappa_{n-r}} q_{n-r}(K) \tag{5.4.8}$$

for $r = 0, \ldots, n$. The vector $z_{r+1}(K)$ is called the *intrinsic* $(r+1)$-*moment* of K. If $P \in \mathcal{P}^n$ is a polytope, then

$$z_{r+1}(P) = \sum_{F \in \mathcal{F}_r(P)} \gamma(F, P) z_{r+1}(F), \tag{5.4.9}$$

in analogy with (4.2.30). From (5.4.8) and (5.4.9) we obtain a representation for $q_r(P)$, which in view of (4.2.17) can be written in the form

$$q_r(P) = \frac{1}{n} \int_{\mathbb{E}^n} x \, \mathrm{d}C_{n-r}(P, x).$$

Using approximation and the weak continuity of the curvature measures, we deduce that

$$q_r(K) = \frac{1}{n} \int_{\mathbb{E}^n} x \, \mathrm{d}C_{n-r}(K, x) \tag{5.4.10}$$

for arbitrary $K \in \mathcal{K}^n$.

Let $\mathcal{K}^n_r := \{K \in \mathcal{K}^n | \dim K \geqq n - r\}$. For $r = 0, \ldots, n$ and $K \in \mathcal{K}^n_r$ we define

$$p_r(K) := \frac{q_r(K)}{W_r(K)} = \frac{(n+1)z(K[n+1-r], B^n[r])}{(n+1-r)V(K[n-r], B^n[r])}.$$

Thus $p_0(K)$ is the ordinary centroid (centre of gravity) of K, and for $r = 1, \ldots, n$,

$$p_r(K) = \frac{\int x \, \mathrm{d}C_{n-r}(K, x)}{\int \mathrm{d}C_{n-r}(K, x)} \tag{5.4.11}$$

is the centroid of the mass distribution given by the curvature measure $C_{n-r}(K, \cdot)$. The points $p_1(K), \ldots, p_n(K)$ are, therefore, called the *curvature centroids* of K. In particular, $p_1(K)$ is the *surface-area centroid* of K. The centroid of the Gaussian curvature measure $C_0(K, \cdot)$ is the Steiner point, that is,

$$p_n(K) = s(K) \tag{5.4.12}$$

for $K \in \mathcal{K}^n$. This can be seen as follows. First let K be of class C^2_+. We use the notation of Section 2.5. By (2.5.8), $\xi(u) = \operatorname{grad} h_K(u)$. We integrate this over the unit ball B^n. (More precisely, we apply the divergence theorem to each of the vector fields $h_K e_i$, where $(e_1, \ldots, e_n)$ is an orthonormal basis of $\mathbb{E}^n$, and to the domain $\{x \in \mathbb{E}^n | r \leqq |x| \leqq 1\}$, and then let $r > 0$ tend to 0.) The result is

$$\int_{B^n} \xi(u) \, \mathrm{d}u = \int_{S^{n-1}} h_K(u) u \, \mathrm{d}\mathcal{H}^{n-1}(u) = \kappa_n s(K),$$

according to the definition (1.7.3). On the other hand, using $\xi(\lambda u) = \xi(u)$ for $\lambda > 0$,

$$\begin{aligned} \int_{B^n} \xi(u) \, \mathrm{d}u &= \frac{1}{n} \int_{S^{n-1}} \xi(u) \, \mathrm{d}\mathcal{H}^{n-1}(u) \\ &= \frac{1}{n} \int_{\mathrm{bd}\, K} x H_{n-1}(x) \, \mathrm{d}\mathcal{H}^{n-1}(x) \end{aligned}$$

by (2.5.30). This yields

$$\kappa_n s(K) = \frac{1}{n} \int x \, \mathrm{d}C_0(K, x) = q_n(K),$$

which by approximation extends to arbitrary $K \in \mathcal{K}^n$ and thus proves (5.4.12).

We mention some properties of the curvature centroids. From (5.4.11) it is clear that

$$p_r(K) \in \operatorname{relint} K. \tag{5.4.13}$$

For each similarity $\alpha: \mathbb{E}^n \to \mathbb{E}^n$, we have $p_r(\alpha K) = \alpha p_r(K)$. The map p_r is continuous on $\mathcal{K}_r^n$, and $W_r p_r = q_r$ is additive. For $\lambda \geqq 0$,

$$p_r(K + \lambda B^n) = \frac{\sum_{i=r}^{n} \binom{n-r}{i-r} \lambda^{i-r} W_i(K) p_i(K)}{\sum_{i=r}^{n} \binom{n-r}{i-r} \lambda^{i-r} W_i(K)}, \tag{5.4.14}$$

so that $p_r(K + \lambda B^n)$ is a convex combination of $p_r(K), \ldots, p_n(K)$.

In analogy with Hadwiger's axiomatic characterization of the quermassintegrals, quoted as Theorem 4.2.6, one can characterize the quermassvectors of the curvature centroids by some of their properties. Proofs of the following theorems appear in Schneider (1972b).

Theorem 5.4.1. *Let $n \geqq 2$. If the map $\varphi: \mathcal{K}^n \to \mathbb{E}^n$ is additive, equivariant under rotations, continuous and such that $\varphi(K + t) - \varphi(K)$ is parallel to t for all $t \in \mathbb{E}^n$ then there are real constants $c_0, \ldots, c_n$ such that*

$$\varphi(K) = \sum_{r=0}^{n} c_r q_r(K) \quad \text{for } K \in \mathcal{K}^n.$$

Theorem 5.4.2. *Let $n \geqq 2$ and $r \in \{0, \ldots, n\}$. If the map $\varphi: \mathcal{K}_r^n \to \mathbb{E}^n$ is equivariant under rigid motions, continuous and such that $W_r \varphi$ is additive on $\mathcal{K}_r^n$ then*

$$\varphi(K) = p_r(K) \quad \text{for } K \in \mathcal{K}_r^n.$$

Notes for Section 5.4

1. *Steiner point and curvature centroids*. The point later called the Steiner point was first introduced, as a curvature centroid, by Steiner (1840a), for planar convex bodies that are either a polygon or sufficiently smooth. The characterization given by equation (A.7) in the Appendix appears, for $n = 2$, in Kubota (1918) and implicitly in Meissner (1909), p. 3.11. (A.7) is essentially the same as (1.7.3). Equality (5.4.12), that is, the equivalence of definition (5.4.11) for $r = n$ with (1.7.3), was proved for some special cases by Bose & Roy (1935c), Gericke (1940a) and Shephard (1966a). Bose & Roy (1935c) and S. N. Roy (1936) have similar integral representations for the other curvature centroids in $\mathbb{E}^3$. An extension of the Steiner point to not necessarily convex hypersurfaces was treated by Flanders (1966). Some related integral identities are derived in Hadwiger & Meier (1973).

2. *Extremal properties of the Steiner point*. Originally, Steiner was led to the Steiner point by considering certain extremal problems. Other proofs of the extremal properties established by him as well as further such properties are found in Ferrers (1861), Meissner (1909), Nakajima (1921), Hayashi (1924), Su (1927a, b) and Arnold (1989).
3. *Curvature centroids of parallel bodies*. Equation (5.4.14) appears in special cases (mostly for $n = 2$) in Kubota (1918); Blaschke (1933, 1937a) and Bose & Roy (1935a, b, c). Related observations on the relative positions of the various curvature centroids of a convex body and its parallel bodies were made by Duporcq (1896, 1897), Kubota (1918), Nicliborc (1932), Blaschke (1933) and Bose & Roy (1935a, b, c).
4. *Curvature centroids of bodies of constant width*. In analogy with Note 4 of Section 5.3, the curvature centroids $p_0, \ldots, p_n$ of a convex body of constant width b satisfy
$$2W_{n-k}p_{n-k} - \sum_{i=0}^{k-1}(-1)^i \binom{k}{i} b^{k-i}W_{n-i}p_{n-i} = o$$
for $1 \leqq k \leqq n$, k odd. For $n = 2$ this states that the Steiner point and perimeter centroid coincide; this was noted by Meissner (1909) and proved in a different way by Bose & Roy (1935a). For $n = 3$ one finds that the Steiner point coincides with the mean curvature centroid (see Bose & Roy 1935c) and that the centroid, surface-area centroid and Steiner point are collinear. A proof of the general relations appears in Schneider (1972a).
5. *Characterization theorems*. As a counterpart to Theorem 3.4.2, the following characterization of the Steiner point, with Minkowski additivity replaced by additivity, was proved in Schneider (1972b):

Theorem. Let $n \geqq 2$. If the map $\varphi: \mathcal{K}^n \to \mathbb{E}^n$ is additive, equivariant under rigid motions and continuous then φ is the Steiner point map.

This result was used to prove Theorem 5.4.2 in Schneider (1972b) and to deduce Theorem 5.4.1 in Hadwiger & Schneider (1971) and in Schneider (1972b).

The special case of Theorem 5.4.2 referring to the ordinary centroid states: If $\varphi: \mathcal{K}_0^n \to \mathbb{E}^n$ (where $n \geqq 2$) is a continuous map commuting with rigid motions and such that $V_n\varphi$ is additive, then $\varphi(K)$ is the centroid of K, for each $K \in \mathcal{K}_0^n$. Another characterization of the centroid, in this case for polyhedra, was proved in Schneider (1973): Let $\varphi: U(\mathcal{P}_0^n) \to \mathbb{E}^n$ be a translation equivariant map such that $\varphi(P) \in \operatorname{conv} P$ for $P \in U(\mathcal{P}_0^n)$ and $V_n\varphi$ is simply additive. Then $\varphi(P)$ is the centroid of P, for each $P \in U(\mathcal{P}_0^n)$.
6. *Integral-geometric formulae for quermassvectors*. The special case $\beta' = \mathbb{E}^n$ of the integral-geometric formula (4.5.1) states that
$$\int_{G_n} \Phi_j(K \cap gK', \beta)\, \mathrm{d}\mu(g) = \sum_{k=j}^{n} \alpha_{njk} V_{n+j-k}(K')\Phi_k(K, \beta)$$
for $\beta \in \mathcal{B}(\mathbb{E}^n)$. As functions of β, both sides of this equality define finite measures on $\mathcal{B}(\mathbb{E}^n)$. Integrating the identity function on $\mathbb{E}^n$ with respect to these measures and observing (4.2.10) and (5.4.10), one obtains integral-geometric formulae for the quermassvectors, which are analogous to (4.5.2). In a similar way, Theorem 4.5.5 and formula (4.5.24) (with $\beta' = \mathbb{E}^n$) lead to formulae for quermassvectors or, equivalently, for curvature centroids. In a different way, such formulae were derived by Hadwiger & Schneider (1971) and by Schneider (1972b); special cases for low dimensions appeared earlier in Blaschke (1937a) and H. R. Müller (1953).

7. *Position of the curvature centroids*. Let $K \in \mathcal{K}_0^n$ be a convex body with centroid o. Then a result going back to Minkowski (1897) (cf. Bonnesen & Fenchel [8], p. 53) states that

$$\frac{1}{n+1} w(K, u) \leqq h(K, u) \leqq \frac{n}{n+1} w(K, u)$$

for $u \in S^{n-1}$. For a convex body with centroid c this implies, in particular, that

$$B\left(c, \frac{1}{n+1} \Delta(K)\right) \subset K \subset B\left(c, \frac{n}{n+1} D(K)\right).$$

Similar estimates for the position of the curvature centroids seem to be unknown, except for the surface area centroid; see G. v. Sz.-Nagy (1949).

Rather precise information on the position of the curvature centroids is available for convex bodies with curvature restrictions. Let $K \in \mathcal{K}^n$ be of class C_+^2 and assume that there are constants R_1, R_2 such that $0 < R_1 \leqq \rho \leqq R_2$ whenever ρ is a radius of curvature of bd K. Then

$$B(c, R_1) \subset K \subset B(c, R_2)$$

whenever c is one of the curvature centroids $p_1(K), \ldots, p_n(K)$. This was proved by Schneider (1988b), generalizing a result of Goodman (1985) for the perimeter centroid in the plane.

8. *Centroids in stereology*. Possible stereologic applications of curvature centroids in dimensions two and three were discussed by Davy (1980a, b, 1981).

6

Inequalities for mixed volumes

6.1. The Brunn–Minkowski theorem

Studying the behaviour of volume under Minkowski addition, one is led to the notion of mixed volumes on one hand and to the Brunn–Minkowski theorem on the other. In its simplest form, this theorem says that for two convex bodies $K, L \in \mathcal{K}^n$ each of volume 1 the Minkowski sum $\frac{1}{2}(K + L)$ has volume at least 1, and its volume is equal to 1 only if K and L are translates. This theorem is the starting point for a rich theory of geometric inequalities, with several applications to extremal, uniqueness and other problems. The present section is devoted to proofs, extensions and analogues of this important theorem.

Theorem 6.1.1 (Brunn–Minkowski). *For convex bodies $K_0, K_1 \in \mathcal{K}^n$ and for $0 \leqq \lambda \leqq 1$,*

$$V_n((1 - \lambda)K_0 + \lambda K_1)^{1/n} \geqq (1 - \lambda)V_n(K_0)^{1/n} + \lambda V_n(K_1)^{1/n}. \quad (6.1.1)$$

Equality for some $\lambda \in (0, 1)$ holds if and only if K_0 and K_1 either lie in parallel hyperplanes or are homothetic.

If $K_0, K_1 \in \mathcal{K}^n$ are given, we often write

$$K_\lambda := (1 - \lambda)K_0 + \lambda K_1 \quad \text{for } 0 \leqq \lambda \leqq 1. \quad (6.1.2)$$

For $\sigma, \tau \in [0, 1]$ and $0 \leqq \lambda \leqq 1$ we have

$$(1 - \lambda)K_\sigma + \lambda K_\tau = (1 - \alpha)K_0 + \alpha K_1$$

with $\alpha = (1 - \lambda)\sigma + \lambda\tau$. Applying Theorem 6.1.1 to K_σ and K_τ, where $\sigma, \tau \in [0, 1]$ are arbitrary, we deduce that the function

$$\lambda \mapsto V_n((1 - \lambda)K_0 + \lambda K_1)^{1/n}$$

is concave on $[0, 1]$, and linear only in the cases described in Theorem 6.1.1. (As is usual in this context, we say 'linear' instead of 'affine'.)

The first proof we shall give for the Brunn–Minkowski theorem is a classical version due to Kneser & Süss (1932), which is also reproduced in Bonnesen & Fenchel [8]. It still has some advantages.

Proof of Theorem 6.1.1. If K_0 and K_1 lie in parallel hyperplanes, then K_λ lies in a hyperplane and hence $V_n(K_\lambda) = 0$ for $0 \leqq \lambda \leqq 1$. Thus equality holds in (6.1.1). This is also true, trivially, if K_0 and K_1 are homothetic.

Suppose that $\dim K_0 < n$ and $\dim K_1 < n$. Then (6.1.1) is true, and if equality holds, then K_λ lies in a hyperplane, hence K_0 and K_1 lie in parallel hyperplanes.

If, say, $\dim K_0 < n$ and $\dim K_1 = n$, then the inclusion $K_\lambda \supset (1-\lambda)x + \lambda K_1$, for arbitrary $x \in K_0$, implies

$$V_n(K_\lambda) \geqq V_n((1-\lambda)x + \lambda K_1) = \lambda^n V_n(K_1),$$

with equality if and only if $K_0 = \{x\}$. In that case, K_0 and K_1 are homothetic.

Hence, from now on we can assume that $\dim K_0 = \dim K_1 = n$. We can also assume that $V_n(K_0) = V_n(K_1) = 1$; if (6.1.1) is proved under this assumption and if $K_0, K_1 \in \mathcal{K}_0^n$ are arbitrary, we put

$$\bar{K}_i := V_n(K_i)^{-1/n} K_i \quad \text{for } i = 0, 1,$$

$$\bar{\lambda} := \frac{\lambda V_n(K_1)^{1/n}}{(1-\lambda)V_n(K_0)^{1/n} + \lambda V_n(K_1)^{1/n}}.$$

From $V_n((1-\bar{\lambda})\bar{K}_0 + \bar{\lambda}\bar{K}_1)^{1/n} \geqq 1$ we then obtain (6.1.1); the results for the case of equality can also be generalized.

Theorem 6.1.1 is proved by induction with respect to the dimension. The case $n = 1$ being trivial, we assume that $n \geqq 2$ and that the theorem is true in dimension $n-1$. We choose $u \in S^{n-1}$ and write $H_{u,\alpha} =: H(\alpha)$, $H^-_{u,\alpha} =: H^-(\alpha)$ and, furthermore, $\alpha_\lambda := -h(K_\lambda, -u)$ and $\beta_\lambda := h(K_\lambda, u)$. For $\zeta \in \mathbb{R}$ and $i = 0, 1$ define

$$v_i(\zeta) := V_{n-1}(K_i \cap H(\zeta)),$$

$$w_i(\zeta) := V_n(K_i \cap H^-(\zeta)),$$

so that

$$w_i(\zeta) = \int_{\alpha_i}^{\zeta} v_i(t)\,\mathrm{d}t.$$

On (α_i, β_i) the function v_i is continuous, hence w_i is differentiable and

$$w_i'(\zeta) = v_i(\zeta) > 0 \quad \text{for } \alpha_i < \zeta < \beta_i.$$

Let z_i be the inverse function of w_i; then

$$z_i'(\tau) = \frac{1}{v_i(z_i(\tau))} \quad \text{for } 0 < \tau < 1.$$

Writing

$$k_i(\tau) := K_i \cap H(z_i(\tau)),$$
$$z_\lambda(\tau) := (1-\lambda)z_0(\tau) + \lambda z_1(\tau),$$

we have

$$K_\lambda \cap H(z_\lambda(\tau)) \supset (1-\lambda)k_0(\tau) + \lambda k_1(\tau).$$

We deduce that

$$\begin{aligned} V_n(K_\lambda) &= \int_{\alpha_\lambda}^{\beta_\lambda} V_{n-1}(K_\lambda \cap H(\zeta))\,\mathrm{d}\zeta = \int_0^1 V_{n-1}(K_\lambda \cap H(z_\lambda(\tau)))z_\lambda'(\tau)\,\mathrm{d}\tau \\ &\geqq \int_0^1 V_{n-1}((1-\lambda)k_0(\tau) + \lambda k_1(\tau))\left[\frac{1-\lambda}{v_0(z_0(\tau))} + \frac{\lambda}{v_1(z_1(\tau))}\right]\mathrm{d}\tau \\ &\geqq \int_0^1 [(1-\lambda)v_0^{1/(n-1)} + \lambda v_1^{1/(n-1)}]^{n-1}\left[\frac{1-\lambda}{v_0} + \frac{\lambda}{v_1}\right]\mathrm{d}\tau, \end{aligned}$$

where in the final line we have used the induction hypothesis and abbreviated $v_i(z_i(\tau))$ by v_i. The inequality

$$[(1-\lambda)v_0^p + \lambda v_1^p]^{1/p}\left[\frac{1-\lambda}{v_0} + \frac{\lambda}{v_1}\right] \geqq 1$$

holds for $v_0, v_1, p > 0$ and $0 < \lambda < 1$. For the proof, one takes logarithms and notes that log is concave and increasing. Equality holds only for $v_0 = v_1$.

It follows that $V_n(K_\lambda) \geqq 1$, and this completes the inductive proof of the Brunn–Minkowski inequality.

Suppose that the equality $V_n(K_\lambda) = 1$ holds for some $\lambda \in (0,1)$. Then $v_0(z_0(\tau)) = v_1(z_1(\tau))$ and hence $z_0'(\tau) = z_1'(\tau)$ for $0 \leqq \tau \leqq 1$; thus $z_1(\tau) - z_0(\tau)$ is constant. We may assume that K_0 and K_1 have their centroids at the origin. Then

$$0 = \int_{K_i} \langle x, u\rangle\,\mathrm{d}x = \int_{\alpha_i}^{\beta_i} V_{n-1}(K_i \cap H(\zeta))\zeta\,\mathrm{d}\zeta = \int_0^1 z_i(\tau)\,\mathrm{d}\tau$$

for $i = 0, 1$ and hence $z_0(\tau) = z_1(\tau)$ for $0 \leqq \tau \leqq 1$. This yields $\beta_0 = \beta_1$, thus $h(K_0, u) = h(K_1, u)$. Since $u \in S^{n-1}$ was arbitrary, we conclude that $K_0 = K_1$. ■

One of the advantages of the proof given above may be seen in the fact that it can be modified to yield improved versions of the Brunn–Minkowski theorem. In particular, it leads to an associated stability result. The uniqueness assertion of Theorem 6.1.1 states that two n-dimensional convex bodies K_0, K_1 that satisfy (6.1.1) with equality, for some $\lambda \in (0,1)$, must be homothetic. If equality holds only approximately, can one assert that K_0 and K_1 are nearly homothetic? Explicit estimates giving positive answers are usually called stability results. To formulate a more perspicuous special case, assume that $V_n(K_0) = V_n(K_1) = 1$ and $0 < \lambda < 1$. Then Theorem 6.1.1 reduces to the assertion that

$$V_n((1-\lambda)K_0 + \lambda K_1) \geqq 1, \tag{6.1.3}$$

with equality only if K_0 and K_1 are translates, hence only if $\delta(\widetilde{K}_0, \widetilde{K}_1) = 0$, where $\widetilde{K}_i$ denotes the translate of K_i having its centroid at the origin. A stability result for the inequality (6.1.3) would ensure that $V_n((1-\lambda)K_0 + \lambda K_1) \leqq 1 + \varepsilon$ implies $\delta(\widetilde{K}_0, \widetilde{K}_1) \leqq f(\varepsilon)$, with some explicitly given function f (possibly involving limitations on the size or degeneracy of K_0, K_1) that satisfies $f(\varepsilon) \to 0$ for $\varepsilon \to 0$. The following result of this type is due to Groemer (1988a) (it is a special case of his Theorem 3).

Theorem 6.1.2. *Let* $K_0, K_1 \in \mathcal{K}_0^n$ *be convex bodies with* $V_n(K_0) = V_n(K_1) = 1$, *and define* $D := \max\{D(K_0), D(K_1)\}$. *If*

$$V_n((1-\lambda)K_0 + \lambda K_1)^{1/n} \leqq 1 + \varepsilon \tag{6.1.4}$$

for some $\varepsilon > 0$ *and* $\lambda \in (0, 1)$, *then*

$$\delta(\widetilde{K}_0, \widetilde{K}_1) \leqq \eta_n\left(\frac{1}{\sqrt{\lambda(1-\lambda)}} + 2\right) D\varepsilon^{1/(n+1)} \tag{6.1.5}$$

with

$$\eta_n = 6.00025n. \tag{6.1.6}$$

We refer to Groemer (1988a) for the proof, which follows the approach of Kneser & Süss but introduces explicit estimates. There one also finds some consequences of the theorem, for instance, a formulation for convex bodies of arbitrary volumes. We remark also that Theorem 6.1.2 can be formulated as a strengthened version of (6.1.3), namely as the inequality

$$V((1-\lambda)K_0 + \lambda K_1)^{1/n} \geqq 1 + \left\{\frac{\sqrt{\lambda(1-\lambda)}}{\eta_n[1 + 2\sqrt{\lambda(1-\lambda)}]D}\right\}^{n+1} \delta(\widetilde{K}_0, \widetilde{K}_1)^{n+1}, \tag{6.1.7}$$

which is valid for $V_n(K_0) = V_n(K_1) = 1$ and $0 \leqq \lambda \leqq 1$. This inequality is obtained from (6.1.5) by setting $\varepsilon = V_n((1-\lambda)K_0 + \lambda K_1)^{1/n} - 1$.

We shall now sketch a second proof of the Brunn–Minkowski theorem, essentially following Knothe (1957a). As above, it is sufficient to consider two convex bodies $K, L \in \mathcal{K}^n$ satisfying $V_n(K) = V_n(L) = 1$. We write $x \in \mathbb{E}^n$ in the form $x = (x_1, \ldots, x_n)$, where the x_i are the coordinates of x which respect to a given orthonormal basis, and define

$$K_{x_1,\ldots,x_j} := \{y \in K \mid y_i = x_i \quad \text{for } 1 \leqq i \leqq j\}$$

for $j = 1, \ldots, n$, similarly for L. On $\operatorname{int} K$ we can define a function f_1 by

$$\int_{-\infty}^{x_1} V_{n-1}(K_t)\,\mathrm{d}t = \int_{-\infty}^{f_1(x)} V_{n-1}(L_t)\,\mathrm{d}t.$$

If $j \in \{2, \ldots, n\}$ and $f_1, \ldots, f_{j-1}$ have already been defined, we define f_j by

$$\int_{-\infty}^{x_j} V_{n-j}(K_{x_1,\ldots,x_{j-1},t})\,\mathrm{d}t$$
$$= \frac{V_{n-j+1}(K_{x_1,\ldots,x_{j-1}})}{V_{n-j+1}(L_{f_1(x),\ldots,f_{j-1}(x)})} \int_{-\infty}^{f_j(x)} V_{n-j}(L_{f_1(x),\ldots,f_{j-1}(x),t})\,\mathrm{d}t.$$

Then $f_j(x)$ depends only on $x_1, \ldots, x_j$, and for $x \in \operatorname{int} K$ we obviously have

$$\frac{\partial f_j}{\partial x_j}(x) = \frac{V_{n-j}(K_{x_1,\ldots,x_j})}{V_{n-j}(L_{f_1(x),\ldots,f_j(x)})} \frac{V_{n-j+1}(L_{f_1(x),\ldots,f_{j-1}(x)})}{V_{n-j+1}(K_{x_1,\ldots,x_{j-1}})} > 0$$

and thus

$$\prod_{j=1}^{n} \frac{\partial f_j}{\partial x_j} = 1. \tag{6.1.8}$$

The map $F : \operatorname{int} K \to \operatorname{int} L$ defined by $F(x) = (f_1(x), \ldots, f_n(x))$ is bijective.

Now let $\lambda \in [0, 1]$. Then $[(1-\lambda)\mathrm{id} + \lambda F]K \subset K_\lambda$ and hence

$$\begin{aligned} V_n(K_\lambda) &\geqq V_n([(1-\lambda)\mathrm{id} + \lambda F]K) \\ &= \underbrace{\int \cdots \int}_{\operatorname{int} K} |\mathrm{Jac}\,[(1-\lambda)\mathrm{id} + \lambda F]|\,\mathrm{d}x_1 \cdots \mathrm{d}x_n \\ &= \int \cdots \int \prod_{j=1}^{n} \left[(1-\lambda) + \lambda \frac{\partial f_j}{\partial x_j}(x)\right] \mathrm{d}x_1 \cdots \mathrm{d}x_n. \end{aligned}$$

(In writing down the Jacobian, we do not need to know whether $\partial f_j/\partial x_i$ exists for $i < j$. In fact, since $f_j(x)$ depends only on $x_1, \ldots, x_j$, the transition from the first to the third line does not require the general transformation rule for multiple integrals, but is obtained by using successively the substitution rule for one-variable integrals.)

Now

$$\log\left[(1-\lambda) + \lambda \frac{\partial f_j}{\partial x_j}\right] \geqq \lambda \log \frac{\partial f_j}{\partial x_j}$$

since log is a concave function, and together with (6.1.8) this gives

$$V_n(K_\lambda) \geqq 1.$$

If equality holds, we must have

$$\frac{\partial f_j}{\partial x_j}(x) = 1 \quad \text{for } x \in \operatorname{int} K \text{ and } j = 1, \ldots, n$$

and thus $f_1(x) = x_1 + a_1$ with a constant a_1. Since any coordinate axis may play the role of the first one, we deduce that L is a translate of K.

The preceding proof of the Brunn–Minkowski theorem can be used to obtain extensions of that theorem, with the volume replaced by the integral of a function with suitable properties (see Note 1 below).

Notes for Section 6.1

1. *The Brunn–Minkowski theorem for convex bodies*. The theorem now named after Brunn and Minkowski was discovered (for dimensions $\leqq 3$) by Brunn (1887, 1889). Its importance was recognized by Minkowski, who gave an analytic proof for the n-dimensional case (Minkowski 1910) and characterized the equality case; for the latter, see also Brunn (1894). Alternative proofs were given by Blaschke [5], §22, who used symmetrization, and, for $n = 3$, Hilbert (1910) who applied his theory of integral equations on the sphere. An improved version is due to Bonnesen [7], Chapter VI. The first proof presented above was given by Kneser & Süss (1932). All this can be found in Bonnesen & Fenchel [8]. The proof by Kneser and Süss was extended by Hadwiger [18], Section 6.4.5, to yield a more general result. More recently, different proofs have been given by Gromov & Milman (1987), Appendix, and Pisier (1989).

   Knothe (1957a) extended his proof, sketched above, to yield the following more general result. For each $K \in \mathcal{K}_0^n$, let $\rho(K, \cdot)$ be a real function on K and suppose that ρ is non-negative and continuous in both variables, that
   $$\rho(\lambda K + a, \lambda x + a) = \lambda^m \rho(K, x)$$
   for $\lambda > 0$ and $a \in \mathbb{E}^n$, where $m > 0$ is a given number, and furthermore that
   $$\log \rho(K_\lambda, (1-\lambda)x_0 + \lambda x_1) \geqq (1-\lambda)\log \rho(K_0, x_0) + \lambda \log \rho(K_1, x_1)$$
   for $x_0 \in K_0$, $x_1 \in K_1$, $\lambda \in [0, 1]$. (An example, with $m = 1$, is given by $\rho(K, x) := d(x, \operatorname{bd} K)$ for $x \in K$, where d denotes the distance.) Under these assumptions, Knothe showed that
   $$\left(\int_{K_\lambda} \rho(K_\lambda, x)\, \mathrm{d}x\right)^{1/(n+m)} \geqq (1-\lambda)\left(\int_{K_0} \rho(K_0, x)\, \mathrm{d}x\right)^{1/(n+m)} + \lambda\left(\int_{K_1} \rho(K_1, x)\, \mathrm{d}x\right)^{1/(n+m)}$$
   for $K_0, K_1 \in \mathcal{K}_0^n$ and $\lambda \in [0, 1]$. Related results, also for non-convex sets, are treated by Dinghas (1957b).

   Knothe's method is also used in Bourgain & Lindenstrauss (1988b) (see also Bourgain 1987). These authors call the map F appearing in the proof the Knothe map. Dinghas (1957b) called a closely related map the Brunn–Minkowski–Schmidt map. He had previously employed it in Dinghas (1944), where he referred to it as an oral communication by E. Schmidt. For more history, see Dinghas (1957b), footnote 18.
2. *Stability estimates*. Alongside Groemer's stability estimate for the Brunn–Minkowski theorem, which was quoted as Theorem 6.1.2, a similar result is known due to Diskant (1973a): Suppose that $K, L \in \mathcal{K}^n$ satisfy $V_n(K) =$

$V_n(L) = 1$. There are constants $\varepsilon_0 > 0$ and $c > 0$, depending on n and the inradius and circumradius of K, such that the inequality

$$V_n((1-\lambda)K + \lambda L) \leqq 1 + \varepsilon \quad \text{for } 0 \leqq \lambda \leqq 1$$

with some $\varepsilon \in [0, \varepsilon_0]$ implies $\delta(K', L) \leqq c\varepsilon^{1/n}$ for a suitable translate K' of K.

A survey over the stability properties of geometric inequalities, including the present one and several others, is given by Groemer (1990a).

3. *Non-convex sets*. The Brunn–Minkowski inequality is not restricted to convex sets. There are, in fact, very general versions of it, even for non-measurable sets. Its development is connected with the names of Lusternik, Henstock and Macbeath, Dinghas, Hadwiger, Ohmann. A description of their methods and results would lead us too far from our topic, the geometry of convex bodies; hence we refer only to Hadwiger [18], Dinghas (1961) and Burago & Zalgaller (1988) for details and references and to Uhrin (1984, 1987) for some more recent achievements.
4. *Logarithmic concave measures*. Taking logarithms on both sides of the Brunn–Minkowski inequality

$$V_n((1-\lambda)K + \lambda L)^{1/n} \geqq (1-\lambda)V_n(K)^{1/n} + \lambda V_n(L)^{1/n}$$

and using the concavity and monotoneity of the log function, one obtains the inequality

$$V_n((1-\lambda)K + \lambda L) \geqq V_n(K)^{1-\lambda} V_n(L)^{\lambda}.$$

Generally, a measure P defined on the Lebesgue-measurable subsets of $\mathbb{E}^n$ is called logarithmic concave if

$$P((1-\lambda)K + \lambda L) \geqq P(K)^{1-\lambda} P(L)^{\lambda}$$

for every pair K, L of convex subsets of $\mathbb{E}^n$ and for all $\lambda \in (0, 1)$. Prékopa (1971, 1973) and Leindler (1972) established theorems, using the Brunn–Minkowski theorem, which assert that certain integrals yield logarithmic concave measures.

The following formulation of the Prékopa–Leindler theorem appears in Brascamp & Lieb (1976). If f, g are non-negative measurable functions on $\mathbb{E}^n$ and if k is defined by

$$k(x) := \sup_{y \in \mathbb{E}^n} f\left(\frac{x-y}{1-\lambda}\right)^{1-\lambda} g\left(\frac{y}{\lambda}\right)^{\lambda},$$

then

$$\|k\|_1 \geqq \|f\|_1^{1-\lambda} \|g\|_1^{\lambda} \tag{6.1.9}$$

for $\lambda \in (0, 1)$. If $A, B \subset \mathbb{E}^n$ are nonempty measurable sets and if f and g are the characteristic functions of A and B respectively, then k is the characteristic function of $(1-\lambda)A + \lambda B$. Hence, this set has (Lebesgue) measure at least unity if A and B have measure unity. Thus (6.1.9) implies the Brunn–Minkowski inequality.

Proofs, extensions, analogues and hints about the applications of such results are found in Brascamp & Lieb (1976), Das Gupta (1980) and Uhrin (1987).

5. A counterpart to the Brunn–Minkowski inequality

$$V_n(K + L)^{1/n} \geqq V_n(K)^{1/n} + V_n(L)^{1/n}$$

concerns the Minkowski difference $K \sim L$. If it is not empty, then

$$V_n(K \sim L)^{1/n} \leqq V_n(K)^{1/n} - V_n(L)^{1/n}.$$

This result (which is not restricted to convex bodies) was found by E.

Schmidt. It can be deduced from the Brunn–Minkowski theorem; cf. Hadwiger (1950b) and [18], p. 159. Related inequalities were studied by Ohmann (1953, 1956a).

6. *Applications*. The Brunn–Minkowski theorem is fundamental for much of the development in later sections. It has many other applications, either direct or indirect, in different fields. We mention only a few examples: it has proved useful in the investigation of (i) other geometric volume estimates (Mityagin 1969), (ii) the characterization of parallelotopes (Guggenheimer & Lutwak 1976), (iii) the equilibrium shape of crystals (Dinghas 1944) and (iv) the shapes of worn stones (Firey 1974d). Applications to probability and multivariate statistics are found in the surveys by Das Gupta (1980) and Buldygin & Kharazishvili (1985) and in the references given there, and still other applications are discussed in Zalgaller (1967) and Vitale (1990, 1991).
7. *Busemann's theorem*. A well-known consequence (or reformulation) of the Brunn–Minkowski inequality is the following assertion: Let $K \in \mathcal{K}_0^n$ and let P be a two-dimensional half-plane bounded by the line L. Let the hyperplane H normal to L intersect K in the nonempty set $H \cap K$. If $V_{n-1}(H \cap K)^{1/(n-1)}$ is laid off from the point $H \cap L$ on the ray $H \cap P$, then the resulting curve in P is convex.

   Busemann (1949a) proved an analogous theorem on the intersection of K by pencils of non-parallel hyperplanes. In his original formulation, this theorem reads:

   In $\mathbb{E}^n$ let K be a convex body with interior points, L an $(n-2)$-dimensional space that intersects K, and P a two-dimensional space normal to L at a point 0. Any half-hyperplane H bounded by L intersects K in a nonempty set $H \cap K$. If the $(n-1)$-dimensional volume $V_{n-1}(H \cap K)$ of $H \cap K$ is laid off from 0 on the ray $H \cap P$ then the resulting curve C is convex.

   Busemann also pointed out the following corollary. Let K be a convex body in $\mathbb{E}^n$ with o as interior point and centre. If for any hyperplane H through o the volume $V_{n-1}(H \cap K)$ is laid off from o on the normal to H at o (in both directions), then the resulting surface is convex.

   In other words: $|x|/V_{n-1}(K \cap x^{\perp})$ defines a norm on $\mathbb{E}^n \backslash \{o\}$. A generalization (to sections of higher codimension) is proved in Milman & Pajor (1989), Theorem 3.9.

   Busemann's proof of his theorem (which is important for Busemann's theory of area in Minkowski spaces) used the Brunn–Minkowski theorem as well as ideas employed in its proofs. Barthel (1959a) extended Busemann's theorem by weakening the convexity assumptions, reformulated it as an affine invariant inequality and clarified the equality case (see also Leichtweiß [23], pp. 257–67). Barthel & Franz (1961) gave a geometric proof without any convexity assumptions and simplified the equality discussion.
8. *Milman's inverse Brunn–Minkowski inequality*. If $K, L \in \mathcal{K}^n$ are convex bodies of volume 1, the Brunn–Minkowski inequality says that their sum $K + L$ has volume at least 2^n. The question for an upper bound does not make sense: the volume of $K + L$ can be arbitrarily large. However, the following is true: If $K, L \in K_0^n$ are centrally symmetric bodies with interior points, there exists a volume-preserving linear transformation α of $\mathbb{E}^n$ such that

$$V_n(K + \alpha L)^{1/n} \leqq C[V_n(K)^{1/n} + V_n(\alpha L)^{1/n}],$$

   where C is a constant independent of n. This theorem is due to Milman (1986). Different approaches to its proof and relations to other theorems with applications in the local theory of Banach spaces are described in the book by Pisier (1989).

9. Lutwak (1979b) interpreted the Brunn–Minkowski inequality for convex bodies as a complementary Minkowski inequality for special functions.

6.2. The Minkowski and isoperimetric inequalities

We now have two fundamental pieces of information about the volume of the Minkowski convex combination $(1-\lambda)K_0 + \lambda K_1$ of two convex bodies K_0, K_1: it is a polynomial in λ, and its nth root is a concave function. These facts taken together have consequences for mixed volumes.

Let $K_0, K_1 \in \mathcal{K}_0^n$ be n-dimensional convex bodies and let $K_\lambda := (1-\lambda)K_0 + \lambda K_1$ for $0 \leqq \lambda \leqq 1$. By (5.1.26),

$$V(K_\lambda) = \sum_{i=0}^{n} \binom{n}{i} (1-\lambda)^{n-i} \lambda^i V_i(K_0, K_1)$$

with

$$V_i(K_0, K_1) := V(K_0[n-i], K_1[i]) =: V_{(i)}. \tag{6.2.1}$$

When we use the (classical) notation $V_{(i)}$, it will be clear from the context that K_0, K_1 are given and held fixed. Since $V_i(K_0, K_1)$ has two arguments, it cannot be confused with an intrinsic volume.

The function f defined by

$$f(\lambda) := V_n(K_\lambda)^{1/n} - (1-\lambda)V_n(K_0)^{1/n} - \lambda V_n(K_1)^{1/n}$$

for $0 \leqq \lambda \leqq 1$ is concave, by Theorem 6.1.1, and satisfies $f(0) = f(1) = 0$. Hence, its derivative at 0,

$$f'(0) = V_{(0)}^{-(n-1)/n} [V_{(1)} - V_{(0)}^{(n-1)/n} V_{(n)}^{1/n}],$$

fulfils $f'(0) \leqq 0$, and $f'(0) = 0$ only if f is identically 0. The latter implies equality in (6.1.1). Furthermore,

$$f''(0) = -(n-1)V_{(0)}^{-(2n-1)/n} [V_{(1)}^2 - V_{(0)}V_{(2)}] \leqq 0.$$

This yields:

Theorem 6.2.1 (Minkowski's inequalities). *For n-dimensional convex bodies K, L* $\in \mathcal{K}_0^n$,

$$V(K, \ldots, K, L)^n \geqq V_n(K)^{n-1} V_n(L). \tag{6.2.2}$$

Equality holds if and only if K and L are homothetic. Further,

$$V(K, \ldots, K, L)^2 \geqq V_n(K) V(K, \ldots, K, L, L). \tag{6.2.3}$$

One should note that so far no information on the validity of the equality sign in (6.2.3) has been obtained (equality holds, of course, if

K and L are homothetic). More general quadratic inequalities for mixed volumes are proved in the next section. These again imply (6.2.2). However, (6.2.3) is required for their proof.

By considering the special case where K or L is equal to a ball, Minkowski's inequality (6.2.2) reduces to well-known and fundamental geometric inequalities for convex bodies. Recall that

$$S(K) := nW_1(K) = nV(K[n-1], B^n)$$

is the surface area of K and

$$b(K) = \frac{2}{\kappa_n} W_{n-1}(K) = \frac{2}{\kappa_n} V(K, B^n[n-1])$$

is the mean width of K. Hence, (6.2.2) implies the *isoperimetric inequality*, which states that the volume V_n and surface area S of an n-dimensional convex body satisfy

$$\left(\frac{S}{\omega_n}\right)^n \geqq \left(\frac{V_n}{\kappa_n}\right)^{n-1}, \tag{6.2.4}$$

with equality if and only if K is a ball. It also implies *Urysohn's inequality* for volume V_n and mean width,

$$\left(\frac{b}{2}\right)^n \geqq \frac{V_n}{\kappa_n}, \tag{6.2.5}$$

with the same equality cases. A weaker form is the *isodiametric* (or *Bieberbach's*) *inequality*

$$\left(\frac{D}{2}\right)^n \geqq \frac{V_n}{\kappa_n} \tag{6.2.6}$$

for the diameter D of K, which follows trivially since $b(K) \leqq D(K)$.

The improved version of the Brunn–Minkowski theorem given by Theorem 6.1.2 also leads to a stability result for Minkowski's inequality (6.2.2). For a convex body $K \in \mathcal{K}_0^n$, we use the normalization defined by

$$\tilde{K} := V_n(K)^{-1/n}[K - c(K)], \tag{6.2.7}$$

where $c(K)$ denotes the centroid of K. Thus $\tilde{K}$ has centroid o and volume 1.

Theorem 6.2.2. *Let $K, L \in \mathcal{K}_0^n$ and set $D := \max\{D(K), D(L)\}$. Then*

$$\frac{V(K, \ldots, K, L)^n}{V_n(K)^{n-1}V_n(L)} - 1 \geqq \frac{\gamma_n}{D^{n+1}} \delta(\tilde{K}, \tilde{L})^{n+1}, \tag{6.2.8}$$

where $\gamma_n = 2n/(4\eta_n)^{n+1}$ and η_n is given by (6.1.6).

Although we did not reproduce Groemer's proof of Theorem 6.1.2, we give the simple argument of Groemer (1990b) to deduce (6.2.8).

Proof. First suppose that $V_n(K) = V_n(L) = 1$ and $c(K) = c(L) = o$, and put $f(\lambda) := V_n((1-\lambda)K + \lambda L)^{1/n}$ and

$$u(\lambda) := \left\{\frac{\sqrt{\lambda(1-\lambda)}}{\eta_n[1+2\sqrt{\lambda(1-\lambda)}]D}\right\}^{n+1} \delta(K, L)^{n+1}.$$

Then $f(\lambda) \geqq 1 + u(\lambda)$ by (6.1.7), and $f(0) = f(1) = 1$. By the Brunn–Minkowski theorem, f is concave, hence $f'(0) \geqq [f(h) - f(0)]/h$ for $0 < h \leqq 1$ and in particular $f'(0) \geqq 2[f(\frac{1}{2}) - 1] \geqq 2u(\frac{1}{2})$. Since $f'(0) = V(K, \dots, K, L) - 1$, we obtain

$$V(K, \dots, K, L) \geqq 1 + \frac{\gamma_n}{nD^{n+1}} \delta(K, L)^{n+1}$$

and thus

$$V(K, \dots, K, L)^n \geqq 1 + \frac{\gamma_n}{D^{n+1}} \delta(K, L)^{n+1}.$$

Inequality (6.2.8) for convex bodies of arbitrary volume follows immediately. ■

Next we formulate a theorem that strengthens Minkowski's inequality (6.2.2) in several respects. It may look rather complicated at first sight, but it has interesting consequences and admits specializations of intuitive appeal.

Theorem 6.2.3. *Let $K, L \in \mathcal{K}_0^n$ be convex bodies, and let $\Omega \subset S^{n-1}$ be a closed set such that*

$$K = \bigcap_{u \in \Omega} H^-(K, u). \tag{6.2.9}$$

If $\bar{L}$ is defined by

$$\bar{L} := \bigcap_{u \in \Omega} H^-(L, u), \tag{6.2.10}$$

then

$$V_1(K, L)^{n/(n-1)} - V_n(K)V_n(\bar{L})^{1/(n-1)} \geqq [V_1(K, L)^{1/(n-1)} - r(K, L)V_n(\bar{L})^{1/(n-1)}]^n, \tag{6.2.11}$$

where $r(K, L)$ denotes the inradius of K relative to L.

This theorem is due to Diskant (1973b); the special case $\Omega = S^{n-1}$ (in which $\bar{L} = L$) had been treated previously by Diskant (1972). We shall present Diskant's proof in Section 6.5.

Solving inequality (6.2.11) for $r(K, L)$ and choosing $\Omega = S^{n-1}$, we obtain

$$r(K, L) \geqq \left(\frac{V_1(K, L)}{V_n(L)}\right)^{1/(n-1)} - \frac{[V_1(K, L)^{n/(n-1)} - V_n(K)V_n(L)^{1/(n-1)}]^{1/n}}{V_n(L)^{1/(n-1)}}, \tag{6.2.12}$$

which will be used in the proof of Theorem 7.2.2.

To interpret inequality (6.2.11), we suppose that the assumptions of Theorem 6.2.3 are satisfied. We may assume that $r(K, L)L \subset K$. If $x \in \mathbb{E}^n \setminus K$, there exists $u \in \Omega$ with $x \notin H^-(K, u)$ and hence

$$x \notin H^-(r(K, L)L, u) = H^-(r(K, L)\bar{L}, u).$$

It follows that

$$r(K, L)\bar{L} \subset K.$$

Further, we remark that

$$V_1(K, L) = V_1(K, \bar{L}).$$

This is easy to prove, but we refer to Section 6.5, where the result is obtained in the course of the proof of Theorem 6.2.3. Using Theorem 6.2.1, we obtain

$$V_1(K, L)^n = V_1(K, \bar{L})^n \geqq V_n(K)^{n-1}V_n(\bar{L}) \geqq r(K, L)^{n(n-1)}V_n(\bar{L})^n.$$

Hence, for the right-hand side of (6.2.11) we have

$$V_1(K, L)^{1/(n-1)} - r(K, L)V_n(\bar{L})^{1/(n-1)} \geqq 0,$$

with equality if and only if K is homothetic to $\bar{L}$. Since $V_n(\bar{L}) \geqq V_n(L)$, it is now clear that inequality (6.2.11) is an improvement of Minkowski's inequality (6.2.2).

Inequality (6.2.11) can be used to obtain some stability results; see Note 1 and Theorem 7.2.2. Here we discuss only the special case obtained by taking for L the unit ball $B = B^n$.

Let $K \in \mathcal{K}_0^n$ be given. Denoting its volume by V_n, its surface area by S and its inradius by r, we obtain from (6.2.11) the inequality

$$\left(\frac{S}{n}\right)^{n/(n-1)} - V_n(\bar{B})^{1/(n-1)}V_n \geqq \left[\left(\frac{S}{n}\right)^{1/(n-1)} - rV_n(\bar{B})^{1/(n-1)}\right]^n. \tag{6.2.13}$$

Two further special cases are of interest. We take $\Omega = S^{n-1}$ and use the inequality $x^{n-1} - y^{n-1} \geqq (x-y)^{n-1}$ (for $x \geqq y \geqq 0$) to obtain

$$S^n - n^n\kappa_n V_n^{n-1} \geqq (S^{1/(n-1)} - \omega_n^{1/(n-1)}r)^{n(n-1)}. \tag{6.2.14}$$

Since $\omega_n r^{n-1}$ is the surface area of the inball of K, it is clear that the right-hand side of (6.2.14) is non-negative and is zero only if K is a ball. Thus (6.2.14) estimates the 'isoperimetric deficit' $S^n - n^n\kappa_n V_n^{n-1}$ in terms of a quantity that has a simple geometric meaning and is obviously positive if K is not a ball.

Next, if Ω is a closed set satisfying (6.2.9), we note that (6.2.13) implies

$$S^n/V_n^{n-1} \geqq n^n V_n(\bar{B}), \tag{6.2.15}$$

with equality if and only if K is homothetic to $\bar{B}$. One says (see Section 6.5) that the convex body K is *determined by* Ω if (6.2.9) holds. Thus, among all convex bodies determined by a given closed set $\Omega \subset S^{n-1}$, precisely the bodies circumscribed about a ball, that is, homothetic to

$$\bigcap_{u\in\Omega} H^-(B^n, u),$$

have the smallest 'isoperimetric ratio' S^n/V_n^{n-1}. For finite sets Ω, this is a theorem first proved (for $n = 3$) by Lindelöf (1869) and in a different way by Minkowski (1897); for general sets Ω it was proved by Aleksandrov (1938a).

The smallest closed set Ω that one can take for a given convex body K so that (6.2.9) holds is the closure of the set of outer unit normal vectors at regular boundary points of K. If Ω is this set, then

$$K_* := \bigcap_{u\in\Omega} H^-(B^n, u)$$

is called the *form body* of K. Inequality (6.2.15) now reads

$$S^n/V_n^{n-1} \geqq n^n V_n(K_*), \tag{6.2.16}$$

with equality if and only if K is homothetic to its form body. Using Theorem 2.2.8, one sees that this is the case if and only if K is a tangential body of a ball. Inequality (6.2.16) strengthens the isoperimetric inequality for convex bodies with singularities.

Notes for Section 6.2

1. *Minkowski's inequalities.* In the plane $\mathbb{E}^2$, there is only one Minkowskian inequality, namely

$$V(K_1, K_2)^2 \geqq V_2(K_1)V_2(K_2)$$

for the *mixed area* $V(K_1, K_2)$, with equality if and only if K_1 and K_2 are homothetic. There are several proofs and improvements in the literature; see Bonnesen & Fenchel [8], Section 51, and Note 4 below.

Upper bounds for the mixed area in terms of perimeters were obtained by Betke & Weil (1990). For convex bodies $K_1, K_2 \in \mathcal{K}^2$ with perimeters $L(K_1)$, $L(K_2)$ they showed that

$$V(K_1, K_2) \leqq \tfrac{1}{8}L(K_1)L(K_2),$$

with equality if and only if K_1 and K_2 are orthogonal segments (possibly degenerate), and

$$V(K, -K) \leqq \frac{\sqrt{3}}{18} L(K)^2.$$

Here the equality sign holds if K is an equilateral triangle, and probably only in this case, but this has only been proved if K is a polygon.

For $n = 3$, the cubic inequality (6.2.2) is a consequence of the quadratic inequalities of type (6.2.3). These inequalities were first proved by Minkowski (1903) (announced in (1901a, b); see Bonnesen & Fenchel [8], Section 52, for more information). For general dimension n, (6.2.2) was proved by Süss (1931b), using the Brunn–Minkowski theorem, and Bonnesen (1932) noticed that it can be obtained from the Brunn–Minkowski theorem in the simple way described in the proof of Theorem 6.2.1.

The *stability problem* for the Minkowski inequality (6.2.2) requires us to find explicit estimates for the deviation of suitable homothets of convex bodies $K, L \in \mathcal{K}_0^n$ in terms of the 'deficit'

$$\frac{V_1(K, L)^n}{V_n(K)^{n-1}V_n(L)} - 1.$$

Theorem 6.2.2 gives a result of this type, which is due to Groemer (1990b). The first result of this kind had already been obtained by Minkowski (1903), §6, who proved his inequality (6.2.2), for $n = 3$, in the following sharper form. Let $K, L \in \mathcal{K}_0^3$ have coinciding centroids, and let D be the maximum of

$$\frac{\sqrt[3]{V_3(K)}h(L, \cdot)}{\sqrt[3]{V_3(L)}h(K, \cdot)}$$

over S^2; then (inequality (76) of Minkowski 1903)

$$\frac{V(K, K, L)}{\sqrt[3]{V_3(K)^2V_3(L)}} - 1 \geqq \frac{1}{2^{10}3^4 7^{4/3}} \frac{(D - 1)^6}{D^5}.$$

Stability estimates for the Minkowski inequality (6.2.2) in dimension n were obtained by Volkov (1963) and Diskant (1972); see also Diskant (1973a) and Bourgain & Lindenstrauss (1988b).

Minkowski's inequality (6.2.2) is a useful tool for the solution of many different geometric extremum problems. Some examples are found in Ohmann (1958).

2. *The Blaschke diagram*. For a three-dimensional convex body $K \in \mathcal{K}^3$ one often uses the classical notation $V(K, K, K) = V$ (volume), $3V(K, K, B^3) = S$ (surface area), $3V(K, B^3, B^3) = M$ ('integral of mean curvature'). The quadratic Minkowskian inequalities, special cases of (6.2.3), now state that

$$M^2 \geqq 4\pi S,$$

where equality holds if and only if K is a ball, and

$$S^2 \geqq 3VM,$$

where equality holds if and only if K is a cap body of a ball (equality conditions are discussed in Section 6.6). These quadratic inequalities imply the two cubic ones,

$$S^3 \geqq 36\pi V^2, \quad M^3 \geqq 48\pi^2 V.$$

On the other hand, the two quadratic inequalities are not a complete system of inequalities satisfied by the three functionals V, S, M of convex bodies. Following Blaschke, one considers in an (xy)-plane all points with coordinates

$$x = \frac{4\pi S}{M^2}, \quad y = \frac{48\pi^2 V}{M^3},$$

for all possible three-dimensional convex bodies and finds that part of the boundary of the resulting 'Blaschke diagram' must correspond to a third sharp inequality, of the form $V \geqq f(S, M)$, which is still unknown. This problem is neatly discussed in Hadwiger [17], §§28–9. Later contributions resulted in the non-sharp inequality

$$V \geqq \frac{S}{12\pi^2 M}\left(\frac{\pi^3}{2} S - M^2\right)$$

given by Groemer (1965) (extended to higher dimensions by Firey 1964) and in the interesting conjecture, due to Sangwine-Yager (1989), that

$$\frac{S}{M} \leqq \frac{8}{\pi^2}\,\frac{M}{4\pi} + \left(1 - \frac{8}{\pi^2}\right)\frac{3V}{S}.$$

3. *The isoperimetric inequality*. The isoperimetric inequality has many facets, and with all its versions and ramifications it can easily fill a book by itself. In fact, isoperimetric inequalities are one of the central themes of the book by Burago & Zalgaller (1988), which a reader interested in a broad synopsis of isoperimetric problems should consult first.

Minkowski's discovery that the isoperimetric inequality for convex bodies can be deduced from the Brunn–Minkowski theorem has strongly influenced the later development. Minkowski's inequality (6.2.2) can be interpreted as an isoperimetric inequality for a suitably defined notion of relative surface area, and versions of the Brunn–Minkowski theorem for non-convex sets lead to corresponding isoperimetric inequalities (for a brief version of the proof, including the Brunn–Minkowski inequality, see Federer (1969), 3.2.41–3.2.43). Here the uniqueness question often poses major problems. We refer the reader to the treatments in Hadwiger [18] and Dinghas (1961). Further references are Busemann (1949b), Ohmann (1952a), Baebler (1957) and Barthel & Bettinger (1961, 1963).

Isoperimetric problems in Minkowski spaces are the subject of papers by Busemann (1947, 1949b), Barthel (1959b) and Holmes & Thompson (1979).

An impressive survey of the various ramifications of the isoperimetric inequality, mainly from the viewpoint of analysis and differential geometry, and its applications to these fields is given by Osserman (1978).

For isoperimetric inequalities in mathematical physics and in analysis that are analogous to the geometric one, we refer to the classical book of Pólya & Szegö (1951) and to the more recent one of Bandle (1980).

4. *Diskant's inequality*. Diskant's (1973b) improvement (6.2.11) of Minkowski's inequality extends and unifies many earlier (and even some later) results.

The two-dimensional case, for $\Omega = S^1$, can be written in the form

$$A_{01}^2 - A_0 A_1 \geqq (A_{01} - rA_1)^2 \tag{6.2.17}$$

with $A_{01} := V(K, L)$, $A_0 := V_2(K)$, $A_1 := V_2(L)$, $r := r(K, L)$; an equivalent formulation is

$$A_0 - 2rA_{01} + r^2 A_1 \leqq 0. \tag{6.2.18}$$

If $K \in \mathcal{K}_0^2$ has area A, perimeter L and inradius r, a special case of (6.2.17) gives

$$L^2 - 4\pi A \geqq (L - 2\pi r)^2. \tag{6.2.19}$$

For various proofs, see the references in Note 3, but also Bonnesen & Fenchel [8], p. 113, and Hadwiger (1944, 1948).

Inequality (6.2.14) was first proved by Hadwiger (1949c) for $n = 3$ and by Dinghas (1948) for general n; a short proof is given in Hadwiger [18], Section 6.5.2. Dinghas (1948) had already obtained a slightly weaker inequality of the form (6.2.13).

Inequality (6.2.16) was proved in Hadwiger [18], Section 6.5.5. Several special cases had been treated before by Bol (1940), Bol & Knothe (1949) and Dinghas (1940c, 1942, 1943a, b, 1949b, c, d), but the easier deductibility from Minkowski's inequalities was not always noticed.

The improvement of the isoperimetric inequality by (6.2.16) is strong if K

has sharp singularities. Improvements that are strong if K has large flat pieces in the boundary were treated by Hadwiger (1949b); see also Ohmann (1958).

5. *Bonnesen-style inequalities*. Let $n = 2$, $K_0, K_1 \in \mathcal{K}_0^2$, $V(K_0, K_1) =: A_{01}$, $V_2(K_0) =: A_0$, $V_2(K_1) := A_1$, $r(K_0, K_1) := r$ and $r(K_1, K_0) := 1/R$, so that R is the circumradius of K_0 relative to K_1; in other words, $K_0 \subset RK_1'$ for a suitable translate K_1' of K_1, and R is the smallest number with this property. With these notations, inequality (6.2.18) can be generalized to
$$A_0 - 2\rho A_{01} + \rho^2 A_1 \leqq 0, \tag{6.2.20}$$
or equivalently
$$A_{01}^2 - A_0 A_1 \geqq (A_{01} - \rho A_1)^2, \tag{6.2.21}$$
for all numbers ρ in the interval $[r, R]$. The two cases $\rho = r$ and $\rho = R$ together with $(a^2 + b^2)/2 \geqq [(a + b)/2]^2$ yield
$$A_{01}^2 - A_0 A_1 \geqq \frac{A_1^2}{4}(R - r)^2. \tag{6.2.22}$$
If $K_1 = B^2$, these inequalities read
$$A - \rho L + \pi\rho^2 \leqq 0, \tag{6.2.23}$$
$$L^2 - 4\pi A \geqq (L - 2\pi\rho)^2 \tag{6.2.24}$$
for $r \leqq \rho \leqq R$ and
$$L^2 - 4\pi A \geqq \pi^2(R - r)^2; \tag{6.2.25}$$
where r, R are respectively the usual inradius and circumradius. The last inequality has a sharper version, namely
$$L^2 - 4\pi A \geqq 4\pi(R_0 - r_0)^2, \tag{6.2.26}$$
where r_0 and R_0, $r_0 \leqq R_0$, are the radii of two concentric circles enclosing bd K_0 such that $R_0 - r_0$ is minimal.

All these results are known as Bonnesen's inequalities. Inequalities (6.2.24) and thus (6.2.25) were proved by Bonnesen (1921) and are reproduced in Bonnesen [7], pp. 60–3 (see also Eggleston [13], pp. 108–10); for (6.2.26), see Bonnesen (1924) and [7], pp. 70–4. An elegant integral-geometric proof of (6.2.25) due to Santaló can be found in Blaschke (1937a), §11. Osserman (1979) gives a comprehensive survey of inequalities equivalent to (6.2.24) and their extensions, including non-convex curves and curves on surfaces. New treatments of (6.2.26) are given by Gage (1990) and by Fuglede (1991), who considers non-convex curves.

An integral-geometric proof of (6.2.22) appears in Blaschke (1937a), §15, where the inequality itself is ascribed to Bonnesen. Also related to integral geometry are the proofs given by Hadwiger (1941), Fejes Tóth (1950) and Flanders (1968). Fourier series are used by Bol (1939) and Wallen (1987). Bol added the equality condition: Equality in (6.2.22) holds if and only if K_0 and K_1 are either homothetic or are parallelograms with parallel sides.

In higher dimensions, the inequality (6.2.14), that is,
$$S^n - n^n \kappa_n V_n^{n-1} \geqq (S^{1/(n-1)} - \omega_n^{1/(n-1)} r)^{n(n-1)}$$
can be considered as a generalization of (6.2.24) for the value $\rho = r$, the inradius. An extension of (6.2.23) for $\rho = r$ is the inequality
$$V_n - rS + (n - 1)\kappa_n r^n \leqq 0, \tag{6.2.27}$$
where equality holds only for the ball. (6.2.27) was conjectured by Wills (1970) and proved independently by Bokowski (1973) and Diskant (1973c) and in a slightly sharper form by Osserman (1979). In each case, the inequality is deduced from (6.2.13).

For $n = 2$, inequality (6.2.27) remains true if the inradius r is replaced by the circumradius R, giving

$$A - RL + \pi R^2 \leqq 0; \tag{6.2.28}$$

however, this does not hold in dimensions $n \geqq 3$. Instead, one has

$$(n-1)V_n - 2RS + (n+1)\kappa_n R^n \geqq 0 \tag{6.2.29}$$

for $n \geqq 2$ (note the reversed inequality sign). This was proved by Bokowski & Heil (1986), who obtained the following more general result.

Theorem. For an arbitrary convex body $K \in \mathcal{K}^n$, $n \geqq 2$, with circumradius R and quermassintegrals $W_0, \ldots, W_n$, the inequalities

$$c_{ijk}R^iW_i + c_{jki}R^jW_j + c_{kij}R^kW_k \geqq 0 \tag{6.2.30}$$

hold for $0 \leqq i < j < k \leqq n$ and $c_{pqr} := (r-q)(p+1)$.

(6.2.29) is a special case of (6.2.30), as is the inequality

$$iW_{i-1} - 2(i+1)RW_i + (i+2)R^2W_{i+1} \geqq 0$$

for $i = 1, \ldots, n-1$.

For $n = 2$, (6.2.30) reduces to

$$A - 2RL + 3\pi R^2 \geqq 0, \tag{6.2.31}$$

with equality only for circles; this is due to Favard (1929). A survey over inequalities for three functionals of a planar convex body was given by Santaló (1961).

An extension of (6.2.28) to higher dimensions that does hold is the inequality

$$W_{n-2} - \beta RW_{n-1} + (\beta - 1)R^2W_n \leqq 0 \tag{6.2.32}$$

with $\beta = n/(n-1)$. More generally, Sangwine-Yager (1988b) established the inequalities

$$V_n(K) - nrV_1(K, L) + (n-1)r^2V_2(K, L) \leqq 0 \tag{6.2.33}$$

and

$$(n-1)V_{n-2}(K, L) - nRV_{n-1}(K, L) + R^2V_n(L) \leqq 0 \tag{6.2.34}$$

for $K, L \in \mathcal{K}^n$, with $r = r(K, L)$ and $R = 1/r(L, K)$. Equality in (6.2.33) holds if K is an $(n-2)$-tangential body of L, and in (6.2.34) if L is an $(n-2)$-tangential body of K. The special case $n = 3$, $L = B^3$ of (6.2.33) has already been treated by Sangwine-Yager (1988a), where the author was able to show that equality holds only for cap bodies of balls.

Heil (1987) investigated whether (6.2.32) can be improved for $n \geqq 3$ by increasing β and, by an interesting application of the calculus of variations, he found that this is the case, at least if the body considered is centrally symmetric or if R is replaced by half the diameter.

Teissier (1982) discussed Bonnesen-type inequalities in relation to algebraic geometry; see also Oda (1988), p. 188.

6. *Stability estimates for the isoperimetric inequality.* If the isoperimetric inequality

$$S^n - n^n\kappa_nV_n^{n-1} \geqq 0$$

for the surface area S and the volume V_n of an n-dimensional convex body $K \in \mathcal{K}_0^n$ holds with equality, then K is a ball. A stability estimate for this inequality would be any result of the kind

$$S^n - n^n\kappa_nV_n^{n-1} \leqq \varepsilon \Rightarrow \Delta(K, B_K) \leqq f(\varepsilon),$$

where Δ is some metric on the space of convex bodies (or some other measure for the deviation of two convex bodies), B_K is a suitable ball and f

is an explicitly known function, non-decreasing and satisfying $\lim_{\varepsilon \downarrow 0} f(\varepsilon) = f(0) = 0$. This function may involve constants that depend on some given bounds for ε and for the inradius and circumradius of K as well as on the dimension.

In the plane, the Bonnesen inequalities (6.2.25) and (6.2.26) immediately yield such stability estimates. The difference $R - r$ of circumradius and inradius is, of course, a satisfactory measure for the deviation of a convex body from a suitable ball. Inequality (6.2.25) has no immediate extension to higher dimensions, since the example of a convex body close to a segment shows that for $n \geqq 3$ the isoperimetric deficit $S^n - n^n \kappa_n V_n^{n-1}$ can be arbitrarily small while $R - r \geqq 1$, say. However, one can show that

$$S^n - n^n \kappa_n V_n^{n-1} \geqq c(n) r^{n(n-1)} \left(\frac{R - r}{r} \right)^{(n+3)/2} \tag{6.2.35}$$

for $K \in \mathcal{K}^n$, where $c(n)$ is an explicitly known constant depending only on n. On the class of convex bodies with $r \geqq c_0 > 0$, c_0 given, (6.2.35) provides a stability estimate for the isoperimetric inequality. It was obtained by Groemer & Schneider (1991), as a consequence of more general results (see also Section 6.6, in particular (6.6.11)).

Some results in the literature can be expressed in the form

$$S^n - n^n \kappa_n V_n^{n-1} \geqq c \delta(K, B_K)^\alpha, \tag{6.2.36}$$

where B_K denotes a suitable ball, c may depend on n and bounds for inradius and circumradius of K, and α depends only on n. An inequality of type (6.2.36) becomes sharper as its exponent α becomes smaller. The specializations of the stability results for the Minkowski inequality due to Volkov (1963) and Diskant (1972, 1973a) mentioned in Note 1, as well as an argument of Osserman (1987) using (6.2.14), yield estimates of this kind, all with $\alpha \geqq n$. In Groemer & Schneider (1991), an estimate with $\alpha = (n + 3)/2$ was achieved, and it was shown that one cannot have $\alpha < (n + 1)/2$. By analytic methods, Fuglede (1986b, 1989) obtained stability estimates for the isoperimetric inequality and under additional assumptions also for non-convex sets, which in a certain sense are of optimal orders.

For the sharp inequality bounding the area of a plane domain of given constant width from below, stability estimates were obtained by Groemer (1988b).

7. *Improved versions of the isoperimetric inequality for restricted classes of convex bodies*. It is clear that on any class of convex bodies that does not contain bodies arbitrarily close to balls, a strengthening of the isoperimetric inequality must be possible (in principle). Simple examples can be found in Hadwiger [18], 6.5.3, 6.5.4, 6.5.6. A beautiful inequality of this type was proved by Fejes Tóth (1948):

Theorem. The surface area S and volume V of a three-dimensional convex polytope with k facets satisfy

$$\frac{S^3}{V^2} \geqq 54(k - 2) \tan \alpha_k (4 \sin^2 \alpha_k - 1), \quad \alpha_k = \frac{\pi}{6} \frac{k}{k - 2}.$$

Equality holds precisely for the simple regular polytopes (tetrahedron, cube, dodecahedron).

Further related results (and some interesting problems) are found in the survey article by Florian [46].

Another example is Hadwiger's improvement of the isoperimetric inequality for 'half-bodies'. A convex body $K \in \mathcal{K}_0^n$ is called a half-body if it has a supporting hyperplane H such that the union of K and its image under reflection in H is convex. Hadwiger (1968b) proved that the volume and surface area of such a body satisfy

$$2S^n \geqq n^n(\kappa_n + 2\kappa_{n-1})V_n^{n-1}.$$

Equality holds if K is the union of a half-ball of radius ρ and a right circular cylinder of height ρ attached to the equator $(n-1)$-ball of the half-ball. A different proof will be given in Section 6.7, Note 3.

8. *The ratio of volume and surface area*. While we are on the topic of volume and surface area, we mention that the ratio V_n/S has the property of quasi-monotoneity: If $K, L \in \mathcal{K}^n$ and $K \subset L$, then

$$\frac{V_n(K)}{S(K)} \leqq n \frac{V_n(L)}{S(L)},$$

and here the factor n cannot be replaced by a smaller number. This was proved by Wills (1970).

9. *Quermassintegrals and other functionals*. Bonnesen-type inequalities are, as described in Note 5, a class of inequalities involving volume, surface area and possibly other quermassintegrals, together with some other functional of convex bodies, such as inradius or circumradius. There are several inequalities of a similar character, that is, connecting some of the quermassintegrals with additional geometric quantities, for example minimal width, diameter, or the sizes of suitably chosen projections or plane sections. Some of these results are strengthenings of isoperimetric inequalities; others stress various different viewpoints. We refer to the survey of Santaló (1961) concerning the case of the plane and mention results by Benson (1970), Boček (1983), Bonnesen [7], Chapter 6, Chakerian (1971), Favard (1933a), Firey (1965b, 1966, 1969), Hadwiger [18], p. 292, Knothe (1957a), inequalities (41), (43), Nádenîk (1965, 1967) and Petermann (1967).

6.3. The Aleksandrov–Fenchel inequality

We have seen in Section 6.2 that the Brunn–Minkowski theorem implies the quadratic inequality

$$V(K_0, K_1, \ldots, K_1)^2 \geqq V(K_0, K_0, K_1, \ldots, K_1)V_n(K_1).$$

This is merely a special case of a system of quadratic inequalities satisfied by general mixed volumes.

Theorem 6.3.1 (Aleksandrov–Fenchel inequality). *For* $K_1, K_2, K_3, \ldots, K_n \in \mathcal{K}^n$,

$$V(K_1, K_2, K_3, \ldots, K_n)^2 \geqq V(K_1, K_1, K_3, \ldots, K_n)V(K_2, K_2, K_3, \ldots, K_n). \quad (6.3.1)$$

Clearly, equality holds in (6.3.1) if K_1 and K_2 are homothetic. However, examples show that this is by no means the only possibility for equality. The complete classification of the equality cases is an unsolved problem. We shall return to this question and some partial results in Section 6.6.

The proof we shall give for Theorem 6.3.1 is due to Aleksandrov (1937b). It uses strongly isomorphic polytopes and approximation.

As in the last part of Section 5.2 we assume that an a-type, $\mathcal{A}$, of strongly isomorphic simple n-dimensional polytopes is given. We use the notation of Sections 5.1 and 5.2; in particular, $u_1, \ldots, u_N$ are the unit normal vectors corresponding to the facets of $\mathcal{A}$, $\bar{P} = (h_1, \ldots, h_N)$ with $h_i := h(P, u_i)$ is the support vector of $P \in \mathcal{A}$, and the mixed volume of N-tuples $X_1, \ldots, X_n \in \mathbb{R}^N$ is defined by (5.2.5). We shall often identify $P \in \mathcal{A}$ with its support vector $\bar{P}$; that is, if in $V(X_1, \ldots, X_n)$ or $v(\Lambda_i X_1, \ldots, \Lambda_i X_{n-1})$ (defined by (5.2.6) and (5.2.7)) one of the arguments X_r is a support vector $\bar{P}_r$, we replace X_r by P_r and $\Lambda_i X_r$ by $F_i^{(r)} = F(P_r, u_i)$. Further, we say that $Z = (\zeta_1, \ldots, \zeta_N) \in \mathbb{R}^N$ is the support vector of a point $z \in \mathbb{E}^n$ if $\zeta_i = h(\{z\}, u_i)$ for $i = 1, \ldots, N$, thus if

$$Z = (\langle z, u_1 \rangle, \ldots, \langle z, u_N \rangle).$$

The following theorem is a sharper version of a special case of Theorem 6.3.1, and the latter can be deduced from it.

Theorem 6.3.2. *If $P, P_3, \ldots, P_n$ are strongly isomorphic polytopes of the simple a-type $\mathcal{A}$ and if $Z \in \mathbb{R}^N$ then*

$$V(Z, P, P_3, \ldots, P_n)^2 \geqq V(Z, Z, P_3, \ldots, P_n)V(P, P, P_3, \ldots, P_n).$$

The equality sign holds if and only if $Z = \lambda\bar{P} + A$, where $\lambda \in \mathbb{R}$ and A is the support vector of a point.

Taking for Z the support vector of another polytope in $\mathcal{A}$, we obtain inequality (6.3.1) for the special case of simple strongly isomorphic polytopes. The general case then follows from the approximation theorem 2.4.14 and the continuity of the mixed volume. The limit procedure is responsible for the fact that the cases of equality cause problems.

Proof of Theorem 6.3.2. We introduce a symmetric bilinear form Φ on $\mathbb{R}^N$ by

$$\Phi(X, Y) := V(X, Y, P_3, \ldots, P_n) \quad \text{for } X, Y \in \mathbb{R}^N$$

if $n \geqq 2$ (with $P_3, \ldots, P_n$ omitted if $n = 2$). It suffices to prove the following proposition.

Proposition 1. If $\Phi(Z, P) = 0$, then $\Phi(Z, Z) \leqq 0$, and equality holds if and only if Z is the support vector of a point.

In fact, suppose that Proposition 1 is true. If $Z \in \mathbb{R}^N$ is given, define

$$\lambda := \frac{\Phi(Z, P)}{\Phi(P, P)} \quad \text{and} \quad Z' := Z - \lambda\bar{P}$$

(observe that $\Phi(P, P) = V(P, P, P_3, \ldots, P_n) > 0$). Then $\Phi(Z', P) = 0$ and hence $\Phi(Z', Z') \leqq 0$, with equality if and only if Z' is the support vector of a point. From

$$\Phi(Z', Z') = \Phi(Z, Z) - \frac{\Phi(Z, P)^2}{\Phi(P, P)}$$

the assertion of the theorem follows.

To prove Proposition 1, we first consider the special case $P_3 = \ldots = P_n = P$ of the theorem.

Proposition 2. The inequality

$$V(Z, P, \ldots, P)^2 \geqq V(Z, Z, P, \ldots, P)V_n(P) \qquad (6.3.2)$$

holds. If $n = 2$, equality holds if and only if $Z = \lambda\bar{P} + A$, where $\lambda \in \mathbb{R}$ and A is the support vector of a point.

For the proof, we note that, by Lemma 2.4.12, $\bar{P} + \varepsilon Z$ is the support vector of a polytope $Q \in \mathcal{A}$ if $|\varepsilon| > 0$ is sufficiently small. From Minkowski's inequality (6.2.3) we infer that

$$\begin{aligned} 0 &\leqq V(Q, P, \ldots, P)^2 - V(Q, Q, P, \ldots, P)V_n(P) \\ &= \varepsilon^2[V(Z, P, \ldots, P)^2 - V(Z, Z, P, \ldots, P)V_n(P)]. \end{aligned}$$

This proves (6.3.2). If $n = 2$, then by Theorem 6.2.1 (first part) equality in (6.3.2) holds if and only if Q and P are homothetic. This completes the proof of Proposition 2.

We prove Proposition 1 by induction with respect to the dimension. For $n = 2$, the assertion is true by Proposition 2. We assume that $n \geqq 3$ and that the assertion of Proposition 1 is valid in smaller dimensions.

For each $i \in \{1, \ldots, N\}$ we define a symmetric bilinear form φ_i on $\mathbb{R}^N$ by

$$\varphi_i(X, Y) := v(\Lambda_i X, \Lambda_i Y, F_i^{(4)}, \ldots, F_i^{(n)}) \quad \text{for } X, Y \in \mathbb{R}^N$$

(with $F_i^{(4)}, \ldots, F_i^{(n)}$ omitted if $n = 3$).

Proposition 3. $Z \in \mathbb{R}^N$ is an eigenvector of the bilinear form Φ with eigenvalue 0 if and only if Z is the support vector of a point.

For the proof, we first note that

$$\Phi(X, Y) = \frac{1}{n} \sum_{i=1}^{N} x_i \varphi_i(Y, P_3)$$

by (5.2.8). Since $\varphi_i(\cdot, P_3)$ is linear, it is of the form

$$\varphi_i(Y, P_3) = \sum_{j=1}^{N} b_{ij} y_j,$$

thus

$$\Phi(X, Y) = \frac{1}{n} \sum_{i,j=1}^{N} b_{ij} x_i y_j. \tag{6.3.3}$$

Here $b_{ij} = b_{ji}$, because Φ is symmetric. Stating that $Z = (\zeta_1, \ldots, \zeta_N) \neq 0$ is an eigenvector of Φ corresponding to the eigenvalue 0 means that

$$\sum_{j=1}^{N} b_{ij} \zeta_j = 0 \quad \text{for } i = 1, \ldots, N,$$

or equivalently that

$$\varphi_i(Z, P_3) = 0 \quad \text{for } i = 1, \ldots, N. \tag{6.3.4}$$

If Z is the support vector of the point z, then

$$\varphi_i(Z, P_3) = v(\{z\}, F_i^{(3)}, \ldots, F_i^{(n)}) = 0.$$

Suppose, conversely, that (6.3.4) holds. By the induction hypothesis, this implies $\varphi_i(Z, Z) \leqq 0$. Without loss of generality, we may assume that $h(P_3, u_i) > 0$; then

$$\begin{aligned} 0 &= \frac{1}{n} \sum_{i=1}^{N} \zeta_i \varphi_i(Z, P_3) = \Phi(Z, Z) \\ &= V(Z, Z, P_3, \ldots, P_n) = V(P_3, Z, Z, P_4, \ldots, P_n) \\ &= \frac{1}{n} \sum_{i=1}^{N} h(P_3, u_i) \varphi_i(Z, Z) \leqq 0 \end{aligned}$$

and hence $\varphi_i(Z, Z) = 0$. By the induction hypothesis this implies that $\Lambda_i Z$ is the support vector, relative to the a-type of F_i, of a point z_i. Explicitly, this means that

$$\Lambda_i Z = (\langle z_i, v_{i1} \rangle, \ldots, \langle z_i, v_{iN} \rangle),$$

with $v_{ij} := o$ for $(i, j) \notin J$ (where we use the notation of Section 5.1). We choose $\varepsilon \neq 0$ so that $\bar{P}_3 + \varepsilon Z$ is a support vector of a polytope $Q \in \mathcal{A}$. The equality $\Lambda_i(\bar{P}_3 + \varepsilon Z) = \Lambda_i \bar{P}_3 + \varepsilon \Lambda_i Z$ yields

$$\begin{aligned} h(F(Q, u_i), v_{ij}) &= h(F_i^{(3)}, v_{ij}) + \varepsilon \langle z_i, v_{ij} \rangle \\ &= h(F_i^{(3)} + \varepsilon z_i, v_{ij}) \end{aligned}$$

and thus $F(Q, u_i) = F_i^{(3)} + t_i$ with a vector t_i ($= \varepsilon z_i + \alpha_i u_i$ for some $\alpha_i \in \mathbb{R}$). For $(i, j) \in J$ we conclude that the $(n-2)$-face $G := F(Q, u_i) \cap F(Q, u_j)$ satisfies $G = F_{ij}^{(3)} + t_i$ and analogously $G = F_{ij}^{(3)} + t_j$, hence $t_i = t_j$. Since any two facets of P_3 can be joined by a chain of facets such that any two consecutive facets in the chain have an $(n-2)$-dimensional intersection, we conclude that $t_i = t_j$ for all i, j. Thus Q is a translate of P_3 and, hence, Z is the support vector of a point. This completes the proof of Proposition 3.

Besides Φ, we now introduce a second symmetric bilinear form Ψ on $\mathbb{R}^N$ by means of

$$\Psi(X, Y) := \frac{1}{n} \sum_{i=1}^{N} \frac{\varphi_i(P, P_3)}{h(P, u_i)} x_i y_i \quad \text{for } X, Y \in \mathbb{R}^N.$$

Here we assume, without loss of generality, that $h(P, u_i) > 0$ for $i = 1, \ldots, N$. Since $\varphi_i(P, P_3) > 0$, the form Ψ is positive definite.

We consider the eigenvalues $\lambda_1 > \lambda_2 > \ldots$ of Φ relative to Ψ and make use of the fact that

$$\begin{aligned} \lambda_1 &= \max\{\Phi(X, X) | \Psi(X, X) = 1\}, \\ \lambda_2 &= \max\{\Phi(X, X) | \Psi(X, X) = 1 \\ &\qquad \text{and } \Psi(X, Y) = 0 \text{ for all } Y \text{ in the } \lambda_1\text{-eigenspace}\}. \end{aligned} \tag{6.3.5}$$

In analogy with (6.3.3) we write

$$\Psi(X, Y) = \frac{1}{n} \sum_{i,j=1}^{N} c_{ij} x_i y_j$$

where

$$c_{ij} := \begin{cases} \dfrac{\varphi_i(P, P_3)}{h(P, u_i)} & \text{if } i = j, \\ 0 & \text{if } i \neq j. \end{cases}$$

Then $Z = (\zeta_1, \ldots, \zeta_N) \in \mathbb{R}^N$ is an eigenvector of Φ relative to Ψ with eigenvalue λ if and only if

$$\sum_{j=1}^{N} (b_{ij} - \lambda c_{ij}) \zeta_j = 0 \quad \text{for } i = 1, \ldots, N,$$

or equivalently if

$$\varphi_i(Z, P_3) = \lambda \frac{\varphi_i(P, P_3)}{h(P, u_i)} \zeta_i \quad \text{for } i = 1, \ldots, N.$$

In particular, $\lambda = 1$ is an eigenvalue with corresponding eigenvector $Z = \bar{P}$.

Proposition 4. The only positive eigenvalue of Φ relative to Ψ is 1, and it is simple.

For the proof we first assume that $P = P_3 = \ldots = P_n$. Suppose Proposition 4 were false in this case. If there is a positive eigenvalue $\mu \neq 1$, then there exists $Z \in \mathbb{R}^N$ with $\Psi(Z, P) = 0$ and $\Phi(Z, Z) = \mu \Psi(Z, Z) > 0$. If 1 is a multiple eigenvalue, the corresponding eigenspace is at least two-dimensional and hence contains a vector Z with $\Psi(Z, P) = 0$ and $\Phi(Z, Z) = \Psi(Z, Z) > 0$. Thus in either case we conclude from

$$\Psi(Z, P) = \frac{1}{n} \sum_{i=1}^{N} \frac{\varphi_i(P, P)}{h(P, u_i)} \zeta_i h(P, u_i) = V(Z, P, \ldots, P)$$

that $V(Z, P, \ldots, P) = 0$. Since $V(Z, Z, P, \ldots, P) = \Phi(Z, Z) > 0$, this contradicts Proposition 2.

Now let $P_3, \ldots, P_n \in \mathcal{A}$ be arbitrary. For $\vartheta \in [0, 1]$ let $P_r(\vartheta) := (1 - \vartheta)P + \vartheta P_r$, $r = 3, \ldots, n$. The coefficients of the corresponding forms Φ, Ψ depend continuously on ϑ, hence the same is true for the relative eigenvalues. By Proposition 3, the number 0 is always an eigenvalue with multiplicity n. It follows that the sum of the multiplicities of the positive eigenvalues is independent of ϑ. Since it is equal to 1 for $\vartheta = 0$, it must be equal to 1 for $\vartheta = 1$. This proves Proposition 4.

Proposition 4 implies that the eigenspace corresponding to the eigenvalue 1 coincides with $\text{lin}\{\bar{P}\}$ and that the second eigenvalue is not positive, and hence that $\Phi(Z, Z) \leqq 0$ for all Z satisfying $\Psi(Z, P) = 0$; the latter is equivalent to $\Phi(Z, P) = 0$. Thus $\Phi(Z, P) = 0$ implies $\Phi(Z, Z) \leqq 0$. Suppose that we have equality for some $Z \neq 0$. Since at Z the maximum in (6.3.5) is attained, Z is an eigenvector with eigenvalue 0. By Proposition 3, Z is the support vector of a point. This completes the proof of Theorem 6.3.2. ■

Notes for Section 6.3

1. *Proofs of the Aleksandrov–Fenchel inequality*. The proof of Theorem 6.3.1 presented above is due to Aleksandrov (1937b). Aleksandrov (1938b) gave a second proof, which uses bodies of class C^2_+ and Aleksandrov's inequalities for mixed discriminants (see Theorem 6.8.1). Although the underlying classes of convex bodies are quite different, there are certain analogies between the two proofs. The principal ideas can be traced back to the proof given by Hilbert (1910) for Minkowski's quadratic inequalities in three-space. The second of Aleksandrov's proofs is reproduced in Busemann [12], and the first in Leichtweiß [23].

It has become customary to talk of the 'Aleksandrov–Fenchel' inequality, because Fenchel (1936a) also stated the inequality and sketched a proof. We quote from Busemann [12], p. 51: 'Fenchel's proof is very sketchy, a detailed version has never appeared and it is not quite clear what it would involve'. Hadwiger [18], p. 290, characterizes Fenchel's note as 'schwer verständlich' [hard to understand]. In honour of Fenchel, it seems justified to talk of the Aleksandrov–Fenchel inequalities, although it is doubtful whether he had a complete proof.

For coaxial bodies of revolution, Hadwiger [18], Sections 6.5.8, 6.5.9, obtained improved versions of the Aleksandrov–Fenchel inequality in an elementary way.

In recent years, surprising connections have been discovered between the Aleksandrov–Fenchel inequality and algebraic geometry. Independently, Khovanskii, and Teissier (1979, 1981), found algebraic approaches to the Aleksandrov–Fenchel inequality, yielding a new proof via the Hodge index theorem. A detailed version, written by Khovanskii, can be found in Burago & Zalgaller (1988), §27. Here mixed volumes of Newton polyhedra associated with Laurent polynomials (Bernstein 1975) play an essential role. This line of research was carried further by Gromov (1988).

2. *Full sets of inequalities*. If $p \geqq 2$ convex bodies $K_1, \ldots, K_p$ in $\mathbb{E}^n$ are given, there are $N = \binom{n+p-1}{n}$ mixed volumes $V(K_{i_1}, \ldots, K_{i_n})$ $(1 \leqq i_1 \leqq \ldots \leqq i_n \leqq p)$ that can be formed. The Aleksandrov–Fenchel inequality, applied to these values in all possible ways permitted by the symmetry properties, yields a certain system of quadratic inequalities. This set of inequalities is said to be a *full set* if any given set of N quantities satisfying these inequalities can arise as the set of mixed volumes of some system of p convex bodies. Heine (1937) for $n = 2$ and Shephard (1960) for $n \geqq 2$ investigated whether the known inequalities are a full set. The answer depends on p, but is not yet known for all values of p.
3. *Applications*. Several applications of the Aleksandrov–Fenchel inequality to the geometry of convex bodies appear in later sections. Here we mention briefly a few uses of this inequality in other fields. It has been applied to differential-geometric uniqueness theorems (Schneider 1966a), to extremal problems for geometric probabilities (Schneider 1985b) and to combinatorial questions, in particular to showing that certain sequences of combinatorial interest are log concave (Stanley 1981, 1986, Kahn & Saks 1984).

6.4. Consequences and improvements

Since the mixed volume $V(K_1, K_2, \ldots, K_n)$ is symmetric in its arguments, it is clear that from the Aleksandrov–Fenchel inequality

$$V(K_1, K_2, K_3, \ldots, K_n)^2 \geqq V(K_1, K_1, K_3, \ldots, K_n)V(K_2, K_2, K_3, \ldots, K_n)$$

many other inequalities can be deduced, by repeated application. We first mention some of the more frequently occurring of such derived inequalities.

Some preliminary remarks are in order. A finite sequence $(a_0, a_1, \ldots, a_m)$ of real numbers is called *concave* if

$$a_{i-1} - 2a_i + a_{i+1} \leqq 0 \quad \text{for } i = 1, \ldots, m-1,$$

or, equivalently if

$$a_0 - a_1 \leqq a_1 - a_2 \leqq \ldots \leqq a_{m-1} - a_m.$$

For $0 \leqq i < j < k \leqq m$ we deduce, taking arithmetic means, that a concave sequence $(a_0, a_1, \ldots, a_m)$ satisfies

$$\frac{(a_i - a_{i+1}) + \ldots + (a_{j-1} - a_j)}{j - i} \leqq \frac{(a_j - a_{j+1}) + \ldots + (a_{k-1} - a_k)}{k - j},$$

hence

$$\frac{a_i - a_j}{j - i} \leqq \frac{a_j - a_k}{k - j}$$

and thus

$$(k - j)a_i + (i - k)a_j + (j - i)a_k \leqq 0. \tag{6.4.1}$$

Equality holds if and only if

$$a_{r-1} - 2a_r + a_{r+1} = 0 \quad \text{for } r = i+1, \ldots, k-1.$$

The sequence $(a_0, a_1, \ldots, a_m)$ of positive numbers is called *log-concave* if the sequence $(\log a_0, \ldots, \log a_n)$ is concave, or equivalently if

$$a_i^2 \geqq a_{i-1}a_{i+1} \quad \text{for } i = 1, \ldots, m-1.$$

In this case, for $0 \leqq i < j < k \leqq m$,

$$a_i^{k-j} a_j^{i-k} a_k^{j-i} \leqq 1 \tag{6.4.2}$$

or, written with positive exponents,

$$a_j^{k-i} \geqq a_i^{k-j} a_k^{j-i}. \tag{6.4.3}$$

Equality holds if and only if

$$a_r^2 = a_{r-1}a_{r+1} \quad \text{for } r = i+1, \ldots, k-1.$$

If a number $m \in \{1, \ldots, n\}$ and convex bodies $K_0, K_1, K_{m+1}, \ldots, K_n \in \mathcal{K}^n$ are given, we often use the abbreviations

$$\mathcal{C} := (K_{m+1}, \ldots, K_n)$$

and

$$\begin{aligned} V_{(i)} &:= V_i(K_0, K_1, \mathcal{C}) \\ &:= V(K_0[m-i], K_1[i], K_{m+1}, \ldots, K_n) \end{aligned} \tag{6.4.4}$$

for $i = 0, \ldots, m$. This notation extends that of (6.2.1) (see the remarks made there).

If the numbers

$$a_i := V_i(K_0, K_1, \mathcal{C}), \quad i = 0, \ldots, m,$$

are positive, the sequence $(a_0, \ldots, a_m)$ is log-concave by the Aleksandrov–Fenchel inequality, hence

$$V_{(j)}^{k-i} \geqq V_{(i)}^{k-j} V_{(k)}^{j-i} \tag{6.4.5}$$

if $0 \leqq i < j < k \leqq m$. By approximation, this inequality holds also if one of the numbers involved is zero. The case $m = n$, $i = 0$, $j = n-1$, $k = n$ is Minkowski's inequality (6.2.2) again (but without information on the equality case, if derived in this way).

Again taking $m = n$ in (6.4.5), but now taking $K_1 = B^n$, we find that the quermassintegrals $W_0, \ldots, W_n$ of the convex body $K_0 \in \mathcal{K}^n$ satisfy

$$W_j^{k-i} \geqq W_i^{k-j} W_k^{j-i} \tag{6.4.6}$$

for $0 \leqq i < j < k \leqq n$. The case $k = n$ deserves special mention; since $W_n = \kappa_n$, it reduces to an inequality between only two quermassintegrals:

$$\kappa_n^{i-j} W_j^{n-i} \geqq W_i^{n-j} \tag{6.4.7}$$

for $0 \leqq i < j < n$. Assuming $\dim K_0 \geqq n - i$, we have $W_i > 0$. In this case, equality in (6.4.7) holds if and only if $W_r^2 = W_{r-1}W_{r+1}$ for $r = i+1, \ldots, n-1$. We shall see later (Section 6.6) that equality in the inequality $W_{n-1}^2 \geqq W_{n-2}W_n$ holds only if K_0 is a ball. Hence, equality in (6.4.7) (for $\dim K_0 \geqq n-i$) characterizes balls. In particular, among all convex bodies of given (positive) volume, precisely the balls have smallest jth quermassintegral for $j = 1, \ldots, n-1$ (case $i = 0$), and among all convex bodies of given mean width (a fortiori, of given diameter) precisely the balls have greatest ith quermassintegral for $i = 0, \ldots, n-2$ (case $j = n-1$).

Next, we derive from the Aleksandrov–Fenchel inequality an improved version, from which some useful consequences can be drawn. We assume that $n+1$ convex bodies $K_0, K_1, \ldots, K_n \in \mathcal{K}^n$ are given, and we now abbreviate the $(n-2)$-tuple $(K_3, \ldots, K_n)$ by $\mathcal{C}$. For $i, j = 0, 1, 2$ we write

$$U_{ij} := V(K_i, K_j, \mathcal{C}) = V(K_i, K_j, K_3, \ldots, K_n). \tag{6.4.8}$$

For $\lambda_1, \lambda_2 \geqq 0$, the Aleksandrov–Fenchel inequality gives

$$\begin{aligned}
0 &\leqq V(K_1 + \lambda_1 K_0, K_2 + \lambda_2 K_0, \mathcal{C})^2 \\
&\quad - V(K_1 + \lambda_1 K_0, K_1 + \lambda_1 K_0, \mathcal{C})V(K_2 + \lambda_2 K_0, K_2 + \lambda_2 K_0, \mathcal{C}) \\
&= (U_{12} + \lambda_1 U_{02} + \lambda_2 U_{01} + \lambda_1\lambda_2 U_{00})^2 \\
&\quad - (U_{11} + 2\lambda_1 U_{01} + \lambda_1^2 U_{00})(U_{22} + 2\lambda_2 U_{02} + \lambda_2^2 U_{00}) \\
&= \text{absolute} + \text{linear terms} \\
&\quad + \lambda_1^2(U_{02}^2 - U_{00}U_{22}) + \lambda_2^2(U_{01}^2 - U_{00}U_{11}) \\
&\quad + 2\lambda_1\lambda_2(U_{12}U_{00} - U_{01}U_{02})
\end{aligned}$$

(the higher-degree terms cancel). Letting $\lambda_1 \to \infty$, $\lambda_2 \to \infty$ we deduce that the quadratic term is non-negative. Hence, its discriminant is non-positive, which gives

$$(U_{00}U_{12} - U_{01}U_{02})^2 \leqq (U_{01}^2 - U_{00}U_{11})(U_{02}^2 - U_{00}U_{22}). \tag{6.4.9}$$

For a first important consequence, we assume that the inequality

$$U_{01}^2 \geqq U_{00}U_{11} \tag{6.4.10}$$

holds with equality. Then (6.4.9) gives

$$U_{00}U_{12} - U_{01}U_{02} = 0$$

for all convex bodies K_2. By (5.1.18), this implies that the signed measure

$$\mu := U_{00}S(K_1, \mathcal{C}, \cdot) - U_{01}S(K_0, \mathcal{C}, \cdot)$$

satisfies

$$\int_{S^{n-1}} h_K d\mu = 0$$

for all $K \in \mathcal{K}^n$. Using Lemma 1.7.9 we conclude that $\int f \,\mathrm{d}\mu = 0$ for all $f \in C(S^{n-1})$ and hence that $\mu = 0$. If we now assume that

$$U_{01} > 0, \tag{6.4.11}$$

this implies that the measures $S(K_0, \mathcal{C}, \cdot)$ and $S(K_1, \mathcal{C}, \cdot)$ are proportional (by (6.4.11), none of them is zero). Vice versa, suppose that $S(K_0, \mathcal{C}, \cdot) = \alpha S(K_1, \mathcal{C}, \cdot)$ with $\alpha > 0$. Then integration of $h(K_0, \cdot)$ and $h(K_1, \cdot)$ respectively, gives $U_{00} = \alpha U_{01}$ and $U_{01} = \alpha U_{11}$ and hence equality in (6.4.10). Thus we have proved:

Theorem 6.4.1. *Let $K, L, K_3, \ldots, K_n \in \mathcal{K}^n$ be convex bodies and $\mathcal{C} = (K_3, \ldots, K_n)$, and assume that*

$$V(K, L, \mathcal{C}) > 0. \tag{6.4.12}$$

Then equality holds in the inequality

$$V(K, L, \mathcal{C})^2 \geqq V(K, K, \mathcal{C}) V(L, L, \mathcal{C}) \tag{6.4.13}$$

if and only if

$$S(K, \mathcal{C}, \cdot) = \alpha S(L, \mathcal{C}, \cdot) \tag{6.4.14}$$

with some $\alpha > 0$.

Equality (6.4.14) may be considered as a kind of generalized Euler–Lagrange equation for the extremum problem connected with the equality case in (6.4.13).

To derive further consequences of (6.4.9), we now assume that

$$U_{00} > 0,\ U_{01} > 0,\ U_{02} > 0 \tag{6.4.15}$$

and write (6.4.9) in the form

$$(U_{00}U_{12} - U_{01}U_{02})^2 \leqq U_{01}^2 U_{02}^2 \left(1 - \frac{U_{00}U_{11}}{U_{01}^2}\right)\left(1 - \frac{U_{00}U_{22}}{U_{02}^2}\right).$$

Taking square roots (the negative one on the left-hand side) and applying the inequality $4ab \leqq (a+b)^2$, where a, b are the brackets on the right-hand side, we obtain

$$\frac{U_{11}}{U_{01}^2} - \frac{2U_{12}}{U_{01}U_{02}} + \frac{U_{22}}{U_{02}^2} \leqq 0. \tag{6.4.16}$$

This improves the Aleksandrov–Fenchel inequality

$$U_{12}^2 \geqq U_{11}U_{22}. \tag{6.4.17}$$

If equality holds in (6.4.16), then equality must hold in the arithmetic–geometric mean inequality applied in the proof, hence

$$\frac{U_{11}}{U_{01}^2} = \frac{U_{22}}{U_{02}^2}. \tag{6.4.18}$$

Together with (6.4.16) this yields equality in (6.4.17). Vice versa, if (6.4.17) holds with equality, the quadratic equation

$$U_{11}x^2 - 2U_{12}x + U_{22} = 0$$

has only one real root, hence (6.4.16) holds with equality.

We collect the results obtained so far, but change notation ($K_1 = K$, $K_2 = L$, $K_0 = M$).

Theorem 6.4.2. *Let* $K, L, M, K_3, \ldots, K_n \in \mathcal{K}^n$ *be convex bodies and* $\mathcal{C} = (K_3, \ldots, K_n)$, *and suppose that*

$$V(K, M, \mathcal{C}) > 0, \quad V(L, M, \mathcal{C}) > 0, \tag{6.4.19}$$

$$V(M, M, \mathcal{C}) > 0. \tag{6.4.20}$$

Then

$$V(K, L, \mathcal{C})^2 \geqq V(K, K, \mathcal{C})V(L, L, \mathcal{C}) \tag{6.4.21}$$

and

$$\frac{V(K, K, \mathcal{C})}{V(K, M, \mathcal{C})^2} - \frac{2V(K, L, \mathcal{C})}{V(K, M, \mathcal{C})V(L, M, \mathcal{C})} + \frac{V(L, L, \mathcal{C})}{V(L, M, \mathcal{C})^2} \leqq 0. \tag{6.4.22}$$

The following assertions are equivalent:

(a) *equality in* (6.4.21);

(b) *equality in* (6.4.22).

The particular advantage of inequality (6.4.22) lies in the fact that the convex body M is at one's disposal. Observe that (b) for one convex body M implies (a), which in turn implies (b) for all convex bodies M.

For the derivation of inequality (6.4.22) and the equivalence of (a) and (b) we had to assume non-degeneracy in the form (6.4.19), (6.4.20). If (6.4.19) holds, but not necessarily (6.4.20), the inequality (6.4.22) is still true, by approximation. The implication (b) $\Rightarrow$ (a), however, is no longer valid. We shall now investigate equality in (6.4.22) in the case $V(M, M, \mathcal{C}) = 0$. This will play a role in Section 6.7, where M is a segment.

Let K, L, M, $\mathcal{C}$ be as in Theorem 6.4.2, but without assuming (6.4.20). We write $K = C_1$ and $L = C_2$, choose $C_3 \in \mathcal{K}_0^n$ to be arbitrary and define

$$W_{ij} := V(C_i, C_j, \mathcal{C})$$
$$q_i := V(C_i, M, \mathcal{C})$$

for $i, j = 1, 2, 3$,

$$K_\alpha := \sum_{i=1}^{3} x_{\alpha i} C_i \quad \text{with } x_{\alpha i} > 0$$

for $\alpha = 1, 2$, and

$$U_{\alpha\beta} := V(K_\alpha, K_\beta, \mathscr{C}) = \sum_{i,j=1}^{3} x_{\alpha i} x_{\beta j} W_{ij},$$
$$p_\alpha := V(K_\alpha, M, \mathscr{C}) = \sum_{i=1}^{3} x_{\alpha i} q_i. \tag{6.4.23}$$

By (6.4.22), applied to K_1, K_2 instead of K, L, we have

$$\frac{U_{11}}{p_1^2} - \frac{2U_{12}}{p_1 p_2} + \frac{U_{22}}{p_2^2} \leqq 0. \tag{6.4.24}$$

We choose real numbers μ_i and positive numbers ν_i and put $x_{1i} := \nu_i$, $x_{2i} := t\nu_i + \mu_i$, where t is so large that $x_{2i} > 0$ $(i = 1, 2, 3)$. Inserting (6.4.23) into (6.4.24), we obtain

$$\left(\sum W_{ij}\nu_i\nu_j\right)\left(\sum q_i\mu_i\right)^2 + \left(\sum W_{ij}\mu_i\mu_j\right)\left(\sum q_i\nu_i\right)^2$$
$$- 2\left(\sum W_{ij}\nu_i\mu_j\right)\left(\sum q_i\nu_i\right)\left(\sum q_i\mu_i\right) \leqq 0$$

(the terms containing t cancel). Choosing $\nu_1 = 1$, $\mu_1 = 0$ and letting $\nu_2, \nu_3 \to 0$, an inequality results that is quadratic in μ_2 and μ_3. If now equality holds in (6.4.22), the coefficient of μ_2^2 is zero. Since the inequality holds for arbitrary real numbers μ_2, μ_3, the coefficient of $\mu_2\mu_3$ must vanish, too, which gives

$$\frac{W_{11}q_2 - W_{12}q_1}{q_1} = \frac{W_{13}q_2 - W_{23}q_1}{q_3}.$$

Denoting this number by k, we see that k is independent of C_3 and that

$$W_{13}q_2 - W_{23}q_1 - kq_3 = 0.$$

By (5.1.18), this can be written in the form

$$\int_{S^{n-1}} h(C_3, \cdot)\,\mathrm{d}[q_2 S(K, \mathscr{C}, \cdot) - q_1 S(L, \mathscr{C}, \cdot) - kS(M, \mathscr{C}, \cdot)] = 0.$$

Since this holds for all $C_3 \in \mathscr{K}_0^n$, we deduce that

$$q_2 S(K, \mathscr{C}, \cdot) - q_1 S(L, \mathscr{C}, \cdot) = kS(M, \mathscr{C}, \cdot). \tag{6.4.25}$$

Vice versa, suppose that (6.4.25) is satisfied. Integrating the support functions of K and L with this measure, we obtain

$$W_{11}q_2 - W_{12}q_1 = kq_1, \quad W_{12}q_2 - W_{22}q_1 = kq_2,$$

hence

$$W_{11}q_2^2 - 2W_{12}q_1q_2 + W_{22}q_1^2 = 0,$$

which is (6.4.22) with equality.

Condition (6.4.25), which thus has been shown to be equivalent to equality in (6.4.22), can be written in the form

$$\begin{aligned} S(q_2K, \mathscr{C}, \cdot) &= S(q_1L + kM, \mathscr{C}, \cdot) \quad \text{if } k \geqq 0 \\ S(q_1L, \mathscr{C}, \cdot) &= S(q_2K + |k|M, \mathscr{C}, \cdot) \quad \text{if } k \leqq 0. \end{aligned} \tag{6.4.26}$$

We turn to a different class of consequences of the Aleksandrov–Fenchel inequality, which concern a generalization of the Brunn–Minkowski theorem.

Theorem 6.4.3 (the general Brunn–Minkowski theorem). *Let a number $m \in \{1, \ldots, n\}$ and convex bodies $K_0, K_1, K_{m+1}, \ldots, K_n \in \mathcal{K}^n$ be given; define $K_\lambda := (1-\lambda)K_0 + \lambda K_1$ and*

$$f(\lambda) := V(K_\lambda[m], K_{m+1}, \ldots, K_n)^{1/m} \tag{6.4.27}$$

for $0 \leqq \lambda \leqq 1$. Then f is a concave function on $[0, 1]$.

Proof. We have to show that $f''(\lambda) \leqq 0$. It suffices to prove this for $\lambda = 0$: If $0 < \lambda < 1$, we put $\bar{K}_\tau := (1-\tau)K_\lambda + \tau K_1$ and

$$h(\tau) := V(\bar{K}_\tau[m], K_{m+1}, \ldots, K_n)^{1/m}.$$

Then $f(\lambda + \mu) = h(\mu/(1-\lambda))$, hence $f''(\lambda) \leqq 0$ follows from $h''(0) \leqq 0$. Now

$$f''(0) = (m-1)V_{(0)}^{(1/m)-2}[V_{(0)}V_{(2)} - V_{(1)}^2] \leqq 0$$

with $V_{(i)}$ given by (6.4.4). We have assumed that $V_{(0)} > 0$; the general case is then obtained by approximation. ■

For $m = n$, Theorem 6.4.3 reduces to the Brunn–Minkowski theorem, and in that case we know that for n-dimensional convex bodies K_0, K_1 the function f is linear only if K_0 and K_1 are homothetic. An analogous assertion is not true for the general Brunn–Minkowski theorem, but some conditions equivalent to linearity of f can be formulated and are of interest.

In analogy to (6.4.4), that is,

$$V_{(i)} := V(K_0[m-i], K_1[i], \mathscr{C}),$$

we write

$$S_{(i)} := S(K_0[m-1-i], K_1[i], \mathscr{C}, \cdot) \tag{6.4.28}$$

for $i = 0, \ldots, m-1$; here $\mathscr{C} = (K_{m+1}, \ldots, K_n)$.

Theorem 6.4.4. *Under the assumptions of Theorem 6.4.3 and*

$$V_{(0)} > 0, \; V_{(m)} > 0,$$

the following conditions are equivalent:

(a) *the function f is linear;*

(b) $V_{(i)}^2 = V_{(i-1)}V_{(i+1)}$ *for $i = 1, \ldots, m-1$;*

(c) $V_{(1)}^m = V_{(0)}^{m-1} V_{(m)}$;
(d) *the measures* $S_{(0)}$ *and* $S_{(m-1)}$ *are proportional.*

Proof. Suppose that f is linear; hence

$$V(K_\lambda[m], \mathscr{C}) = [(1-\lambda)V_{(0)}^{1/m} + \lambda V_{(m)}^{1/m}]^m$$

for $0 \leqq \lambda \leqq 1$. By (5.1.27), this gives

$$\sum_{i=0}^{m} \binom{m}{i} (1-\lambda)^{m-i} \lambda^i V_{(i)} = \sum_{i=0}^{m} \binom{m}{i} (1-\lambda)^{m-i} \lambda^i V_{(0)}^{(m-i)/m} V_{(m)}^{i/m},$$

hence

$$V_{(i)} = V_{(0)}^{(m-i)/m} V_{(m)}^{i/m} \quad \text{for } i = 0, \ldots, m.$$

From

$$V_{(i)}^{2m} = V_{(0)}^{2(m-i)} V_{(m)}^{2i},$$
$$V_{(i-1)}^m = V_{(0)}^{m-i+1} V_{(m)}^{i-1},$$
$$V_{(i+1)}^m = V_{(0)}^{m-i-1} V_{(m)}^{i+1}$$

we deduce (b).

Since the function f is concave, it satisfies $f'(0) \geqq f(1) - f(0)$, with equality if and only if f is linear. As in the proof of (6.2.2), this yields

$$V_{(1)}^m \geqq V_{(0)}^{m-1} V_{(m)}, \tag{6.4.29}$$

with equality if and only if f is linear. Thus (a) and (c) are equivalent.

If (b) holds, then Theorem 6.4.1 yields that $S_{(i-1)}$ and $S_{(i)}$ are proportional for $i = 1, \ldots, m-1$. This implies (d).

Suppose that (d) holds. After a dilatation applied to K_0 we may assume that $S_{(0)} = S_{(m-1)}$. Integrating $h(K_1, \cdot)$ with respect to this measure, we deduce $V_{(1)} = V_{(m)}$ and hence, by (6.4.29),

$$V_{(m)}^m = V_{(1)}^m \geqq V_{(0)}^{m-1} V_{(m)},$$

thus $V_{(m)} \geqq V_{(0)}$. Interchanging the roles of K_0 and K_1, we obtain $V_{(0)} \geqq V_{(m)}$ and thus $V_{(m)} = V_{(0)}$. This implies equality in (6.4.29), thus (c) holds. ■

By Theorem 6.4.1, equality in the Aleksandrov–Fenchel inequality (6.4.13) implies the equality (6.4.14) between two mixed area measures. In the special case where one of the bodies involved in the $(n-2)$-tuple $\mathscr{C}$ is a ball, the latter equality in turn implies equality in a set of lower-dimensional Aleksandrov–Fenchel inequalities for projections. This opens the way to proofs employing induction with respect to the dimension. We state the crucial step as a lemma.

As in Section 5.3, we use the abbreviation K^u for the projection $K|H_{u,0}$, and we denote the $(n-1)$-dimensional mixed volume by v. If

$\mathcal{C} = (K_1, \ldots, K_m)$, we write $\mathcal{C}^u := (K_1^u, \ldots, K_m^u)$, and the unit ball B^n in $\mathbb{E}^n$ is denoted in brief by B.

Lemma 6.4.5. *Let $n \geqq 3$, let $K, L, K_4, \ldots, K_n \in \mathcal{K}^n$ be convex bodies and $\mathcal{C} = (K_4, \ldots, K_n)$, and suppose that $v(K^u, B^u, \mathcal{C}^u) > 0$, $v(L^u, B^u, \mathcal{C}^u) > 0$. If*

$$S(K, B, \mathcal{C}, \cdot) = S(L, B, \mathcal{C}, \cdot), \tag{6.4.30}$$

then

$$v(K^u, L^u, \mathcal{C}^u)^2 = v(K^u, K^u, \mathcal{C}^u)v(L^u, L^u, \mathcal{C}^u) \tag{6.4.31}$$

for all $u \in S^{n-1}$.

Proof. By inequality (6.4.22), applied in $H_{u,0}$ for any $u \in S^{n-1}$, we have

$$\frac{v(K^u, K^u, \mathcal{C}^u)}{v(K^u, B^u, \mathcal{C}^u)^2} - \frac{2v(K^u, L^u, \mathcal{C}^u)}{v(K^u, B^u, \mathcal{C}^u)v(L^u, B^u, \mathcal{C}^u)} + \frac{v(L^u, L^u, \mathcal{C}^u)}{v(L^u, B^u, \mathcal{C}^u)^2} \leqq 0.$$

By formula (5.3.32), assumption (6.4.30) implies

$$v(K^u, B^u, \mathcal{C}^u) = v(L^u, B^u, \mathcal{C}^u),$$

hence

$$v(K^u, K^u, \mathcal{C}^u) - 2v(K^u, L^u, \mathcal{C}^u) + v(L^u, L^u, \mathcal{C}^u) \leqq 0. \tag{6.4.32}$$

By (5.3.33), this yields

$$V(K, K, B, \mathcal{C}) - 2V(K, L, B, \mathcal{C}) + V(L, L, B, \mathcal{C}) \leqq 0. \tag{6.4.33}$$

Again using (6.4.30), from (5.1.18) we obtain

$$V(K, K, B, \mathcal{C}) = V(K, L, B, \mathcal{C}) = V(L, L, B, \mathcal{C}).$$

Thus (6.4.33) holds with equality, which implies equality in (6.4.32). By Theorem 6.4.2, this implies equality (6.4.31). ■

Notes for Section 6.4

1. *Log-linear inequalities for quermassintegrals.* Inequality (6.4.6) says that the quermassintegrals $W_0, \ldots, W_n$ of a convex body in $\mathcal{K}^n$ satisfy

$$W_i^{j-k} W_j^{k-i} W_k^{i-j} \geqq 1$$

for $0 \leqq i < j < k \leqq n$. Gritzmann (1988) investigated systematically all inequalities of this type. Let $0 \leqq i < j < k \leqq n-1$, $\alpha, \beta, \gamma \in \mathbb{R}$, $\alpha^2 + \beta^2 + \gamma^2 \neq 0$ and $c > 0$. Then Gritzmann showed that there is an inequality of the type

$$W_i^\alpha(K) W_j^\beta(K) W_k^\gamma(K) \geqq c, \tag{6.4.34}$$

valid for all convex bodies $K \in \mathcal{K}_0^n$, if and only if

$$\alpha \leqq 0,\ \alpha + \beta + \gamma \geqq 0,\ \alpha(n-i) + \beta(n-j) + \gamma(n-k) = 0.$$

Furthermore, in each case the constant c can be chosen such that equality holds for the ball, and all the valid inequalities of type (6.4.34) are generated by the $n-1$ special Aleksandrov–Fenchel inequalities

$$W_i(K)^2 \geqq W_{i-1}(K) W_{i+1}(K) \quad \text{for } i = 1, \ldots, n-1.$$

From this result, Gritzmann also derived a characterization of all log-linear inequalities for the diameter D and two quermassintegrals. Examples are Bieberbach's inequality (6.2.6), Kubota's (1925a) inequality

$$\left(\frac{D}{2}\right)^{n-1} \geqq \frac{S}{\omega_n},$$

and the inequality

$$S^{n-1} > \kappa_{n-1} D(nV_n)^{n-2},$$

which was proved in an entirely different way by Gritzmann, Wills & Wrase (1987).

2. Theorem 6.4.1 appears in Aleksandrov (1937b) and Fenchel & Jessen (1938). Aleksandrov's proof is different. As mentioned before, if the question for the equality case in (6.4.13) is considered as a minimum problem with side condition, then equality (6.4.14) can be considered as a generalized Euler–Lagrange equation for this problem. Correspondingly, Aleksandrov employed a variational argument. The method used above goes back to Favard (1933a) and Fenchel (1936b).

   It appears that the algebraic manipulations leading to inequalities (6.4.9), (6.4.16) and finally to Theorem 6.4.2 were, in special cases, first applied by Favard (1930) and slightly extended by Matsumura (1932); see also Bonnesen & Fenchel [8], Section 51. Full use of this method was made by Favard (1933a).

   The proof of the equivalence of the equality in (6.4.22) (without the assumption (6.4.20)) and the equation (6.4.25), as given above, extends an argument due to Favard (1933a).

3. *The general Brunn–Minkowski theorem*. Theorem 6.4.3 is a corollary of the Aleksandrov–Fenchel inequalities and consequently appeared in print right after their proof; see Fenchel (1936b) and Aleksandrov (1937b). Theorem 6.4.4 is essentially taken from the latter paper.

   Stability estimates for some special cases of the general Brunn–Minkowski theorem were obtained by Diskant (1975, 1982). We sketch a method for obtaining more general results.

   With the assumptions and notations of Theorem 6.4.3 and of (6.4.4), put $\mathcal{C} = (K_{m+1}, \ldots, K_n)$ and

$$F_m(K_0, K_1, \mathcal{C}, \lambda) := V(K_\lambda[m], \mathcal{C}) - [(1-\lambda)V_{(0)}^{1/m} + \lambda V_{(m)}^{1/m}]^m.$$

Then $F_m(K_0, K_1, \mathcal{C}, \lambda) \geqq 0$ for $\lambda \in [0, 1]$, by Theorem 6.4.3. Now we assume that

$$F_m(K_0, K_1, \mathcal{C}, \lambda) \leqq \varepsilon$$

for some $\lambda \in (0, 1)$ and some $\varepsilon \in [0, \varepsilon_0]$, where $\varepsilon_0 > 0$ is a given number. In the sum

$$F_m(K_0, K_1, \mathcal{C}, \lambda) = \sum_{i=0}^{m} \binom{m}{i} (1-\lambda)^{m-i} \lambda^i [V_{(i)} - V_{(0)}^{(m-i)/m} V_{(m)}^{i/m}]$$

each summand is non-negative, hence

$$\binom{m}{i} (1-\lambda)^{m-i} \lambda^i [V_{(i)} - V_{(0)}^{(m-i)/m} V_{(m)}^{i/m}] \leqq \varepsilon$$

for $i = 1, \ldots, m-1$. From the cases $i = 1$ and $i = m-1$ we obtain

$$V_{(1)} V_{(m-1)} \leqq V_{(0)} V_{(m)} + c\varepsilon$$

where the constant c depends on n, r, R, λ, ε_0, if the bodies K_0, K_1, K_{m+1}, $\ldots$, K_n have inradius at least $r > 0$ and circumradius at most R. Under different assumptions on the bodies $K_{m+1}, \ldots, K_n$, one can now obtain estimates for the distance between suitable homothetic copies of K_0, K_1 in terms of a function of ε (involving constants that depend on $n, r, R, \lambda, \varepsilon_0$). For example, if $K_{m+1}, \ldots, K_m$ are balls, one can proceed precisely as in the proof of Theorem 7.2.6.

4. The proof of Lemma 6.4.5, which in the form given above appears in Schneider (1985a), extends an argument due to Favard (1933a).
5. *Integral geometric proofs*. For the special case $i = n - 2$ (and hence $j = n - 1$) of inequality (6.4.7), that is,
$$W_{n-1}^2 \geqq \kappa_n W_{n-2},$$
two integral-geometric proofs, due to Santaló and to Blaschke are known for $n = 3$. The n-dimensional extensions can be found in Gericke (1937).
6. *Non-convex sets*. Ohmann (1952b, 1954, 1956b) used an extension of Kubota's integral recursion (formula (5.3.27) for $j = k$) to define quermassintegrals for non-convex sets, and he was able to extend some of the inequalities (6.4.7) to this general situation.

6.5. Generalized parallel bodies

Some of the deeper investigations of inequalities for mixed volumes and of the equality cases make essential use of the method of inner parallel bodies. In this section we study a more general concept and collect some of the necessary tools and auxiliary results. In particular, we give the proof of Theorem 6.2.3 and a result (Lemma 6.5.4) that will be needed in the next section.

Let us assume that a closed subset $\Omega \subset S^{n-1}$ of the unit sphere, not lying in a closed hemisphere, and a positive continuous function $f: S^{n-1} \to \mathbb{R}$ are given. (Only the values of f on Ω will be needed, but without loss of generality we may assume that f is defined on all of S^{n-1}.) The closed convex set

$$K := \bigcap_{u \in \Omega} H^-_{u,f(u)} \tag{6.5.1}$$

is bounded, since Ω positively spans $\mathbb{E}^n$, and it has o as an interior point since the restriction of f to Ω has a positive lower bound. We say that the convex body K is *determined by* (Ω, f) if (6.5.1) holds, and we say that K is *determined by* Ω if K is determined by (Ω, f) for some positive continuous function f.

Let $K, L \in \mathcal{K}_0^n$. In Section 3.1, the system $\{K_\rho\}$ of parallel bodies of K relative to L was defined by

$$K_\rho := \begin{cases} K + \rho L & \text{for } 0 \leqq \rho < \infty, \\ K \sim -\rho L & \text{for } -r(K, L) \leqq \rho > 0, \end{cases}$$

where $r(K, L)$ is the inradius of K relative to L. By (3.1.16) and Theorem 1.7.5(b),

$$K_\rho = \bigcap_{u \in S^{n-1}} H^-_{u,h(K,u)+\rho h(L,u)}$$

for $-r(K, L) \leqq \rho < \infty$. Hence, if we arrange by suitable translations that $o \in \operatorname{int} r(K, L)L \subset K$, then for $\rho > -r(K, L)$ the body K_ρ is determined by $(S^{n-1}, h_K + \rho h_L)$. Thus the operation (6.5.1) is a very general version of the formation of parallel bodies.

In the following, the set Ω is fixed. The functions used to determine a convex body, together with Ω, are always continuous and positive. We shall need some results on the volume of the bodies determined by Ω.

Lemma 6.5.1. *If K is determined by* (Ω, f), *then*

$$S_{n-1}(K, S^{n-1}\backslash\Omega) = 0 \tag{6.5.2}$$

and

$$V_n(K) = \frac{1}{n}\int_\Omega f(u)\,\mathrm{d}S_{n-1}(K, u). \tag{6.5.3}$$

Proof. Let $x \in \operatorname{bd} K$. Then there exists a vector $u \in \Omega$ such that $x \in H_{u,f(u)}$, since otherwise $\langle x, u\rangle < f(u)$ for all $u \in \Omega$ and hence $\langle x, u\rangle + \varepsilon < f(u)$ for all $u \in \Omega$ with some $\varepsilon > 0$, which would imply $x \in \operatorname{int} K$.

Let $v \in S^{n-1}\backslash\Omega$ and $x \in K \cap H(K, v)$. Then x lies in the support plane $H(K, v)$ and, by the preceding remark, also in a different support plane. Thus x is a singular point of K. From (4.2.24) and Theorem 2.2.4 the assertion (6.5.2) follows.

If $w \in \Omega$ is such that $h(K, w) \neq f(w)$, then any point $x \in K \cap H(K, w)$ lies in $H(K, w)$ and in some distinct support plane $H_{u,f(u)}$, hence x is singular. Thus

$$S_{n-1}(K, \{w \in \Omega \mid h(K, w) \neq f(w)\}) = 0. \tag{6.5.4}$$

Now we conclude from

$$V_n(K) = \frac{1}{n}\int_{S^{n-1}} h(K, u)\,\mathrm{d}S_{n-1}(K, u)$$

together with (6.5.2) and (6.5.4) that (6.5.3) holds. ■

Lemma 6.5.2. *If* K_j *is determined by* (Ω, f_j) *for* $j = 0, 1, 2, \ldots$ *and if* $(f_j)_{j\in\mathbb{N}}$ *converges uniformly to* f_0, *then* $(K_j)_{j\in\mathbb{N}}$ *converges to* K_0.

Proof. It is easy to see that each interior point of K_0 is an interior point of K_j for almost all j. The assertion then follows with the aid of Theorem 1.8.7. ■

We write $V(f)$ for the volume of the convex body determined by (Ω, f). If g is a continuous function on S^{n-1} (not necessarily positive), then $f + tg$ is positive, and hence $V(f + tg)$ is defined, for all sufficiently small $|t|$.

Lemma 6.5.3. *Let K be determined by (Ω, f). If $g: S^{n-1} \to \mathbb{R}$ is a continuous function, then*

$$\frac{\mathrm{d}}{\mathrm{d}t} V(f + tg)\bigg|_{t=0} = \int_\Omega g(u)\,\mathrm{d}S_{n-1}(K, u).$$

Proof. For sufficiently small $|t|$, let K_t be the convex body determined by $(\Omega, f + tg)$, and define $V_j(t) := V(K_t[j], K[n-j])$. In particular, $V(f) = V_n(K) = V_0(t) =: V_0$.

From $h(K_t, u) \leqq f(u) + tg(u)$ for $u \in \Omega$ and (6.5.2) we have

$$\int_\Omega [f(u) + tg(u)]\,\mathrm{d}S_{n-1}(K, u) \geqq \int_{S^{n-1}} h(K_t, u)\,\mathrm{d}S_{n-1}(K, u).$$

Together with (6.5.3) this gives

$$\int_\Omega tg(u)\,\mathrm{d}S_{n-1}(K, u) \geqq nV_1(t) - nV_0.$$

Considering first positive t only, we deduce

$$\int_\Omega g(u)\,\mathrm{d}S_{n-1}(K, u) \geqq n \limsup_{t\to 0+} \frac{V_1(t) - V_0}{t}. \tag{6.5.5}$$

Again by (6.5.3),

$$\int_\Omega [f(u) + tg(u)]\,\mathrm{d}S_{n-1}(K_t, u) = nV_n(t).$$

From $h(K, u) \leqq f(u)$ for $u \in \Omega$ and (6.5.2) we have

$$\int_\Omega f(u)\,\mathrm{d}S_{n-1}(K_t, u) \geqq \int_{S^{n-1}} h(K, u)\,\mathrm{d}S_{n-1}(K_t, u).$$

Subtraction gives

$$\int_\Omega tg(u)\,\mathrm{d}S_{n-1}(K_t, u) \leqq nV_n(t) - nV_{n-1}(t).$$

From $K_t \to K$ for $t \to 0$, the weak continuity of S_{n-1}, and (6.5.2) (for K and K_t) we obtain

$$\int_\Omega g(u)\,\mathrm{d}S_{n-1}(K, u) \leqq n \liminf_{t\to 0+} \frac{V_n(t) - V_{n-1}(t)}{t}. \tag{6.5.6}$$

Using Minkowski's inequality (6.2.2), we obtain

$$\begin{aligned}[V_1(t) - V_0] \sum_{k=0}^{n-1} [V_1(t)/V_0]^k &= [V_1(t)^n - V_0^n]/V_0^{n-1} \\ &\geqq [V_0^{n-1}V_n(t) - V_0^n]/V_0^{n-1} = V_n(t) - V_0.\end{aligned}$$

Since $K_t \to K$ for $t \to 0$, we deduce

$$n \limsup_{t\to 0+} \frac{V_1(t) - V_0}{t} \geqq \limsup_{t\to 0+} \frac{V_n(t) - V_0}{t}. \tag{6.5.7}$$

In an analogous way, one finds

$$n \liminf_{t\to 0+} \frac{V_n(t) - V_{n-1}(t)}{t} \leqq \liminf_{t\to 0+} \frac{V_n(t) - V_0}{t}. \tag{6.5.8}$$

Applying successively (6.5.5), (6.5.6), (6.5.8), $\liminf \leqq \limsup$ and (6.5.7) we conclude that in all these inequalities the equality sign is valid, hence

$$\lim_{t\to 0+} \frac{V_n(t) - V_0}{t} = \int_\Omega g(u)\,\mathrm{d}S_{n-1}(K, u).$$

If in this relation we replace t by $-t$ and g by $-g$, we obtain

$$\lim_{t\to 0-} \frac{V_n(t) - V_0}{t} = \int_\Omega g(u)\,\mathrm{d}S_{n-1}(K, u).$$

Both relations together prove the lemma. ■

We are now in a position to present the proof of Theorem 6.2.3, which was postponed.

Proof of Theorem 6.2.3. By assumption, $K, L \in \mathcal{K}_0^n$ are n-dimensional convex bodies, and $\Omega \subset S^{n-1}$ is a closed subset such that

$$K = \bigcap_{u\in\Omega} H^-(K, u).$$

Then Ω is not contained in a closed hemisphere, since K is bounded, and with the terminology introduced above we see that K is determined by Ω. The body $\bar{L}$ is defined by

$$\bar{L} := \bigcap_{u\in\Omega} H^-(L, u).$$

The relative inradius $r(K, L)$ of K relative to L will be abbreviated by r. Without loss of generality, we may assume that $o \in \operatorname{int} rL \subset K$.

If $x \in r\bar{L}$, then $\langle x, u\rangle \leqq h(rL, u) \leqq h(K, u)$ for $u \in \Omega$ and hence $x \in K$. Thus $r\bar{L} \subset K$ and $r(K, \bar{L}) = r$.

For $-r < \lambda \leqq 0$, let K_λ be the convex body determined by $(\Omega, h_K + \lambda h_{\bar{L}})$. By Lemma 6.5.3,

$$\frac{\mathrm{d}V_n(K_\lambda)}{\mathrm{d}\lambda} = \int_\Omega h(\bar{L}, u)\,\mathrm{d}S_{n-1}(K_\lambda, u), \tag{6.5.9}$$

and by (6.5.2),

$$\int_{\Omega} h(\bar{L}, u)\,\mathrm{d}S_{n-1}(K_\lambda, u) = \int_{S^{n-1}} h(\bar{L}, u)\,\mathrm{d}S_{n-1}(K_\lambda, u)$$
$$= nV_1(K_\lambda, \bar{L}). \tag{6.5.10}$$

We assert that

$$\lim_{\lambda \to -r} V_n(K_\lambda) = 0. \tag{6.5.11}$$

For the proof, let $A := \{u \in \Omega | h(K, u) = h(rL, u)\}$. Then $o \in \operatorname{conv} A$, since otherwise one could construct a homothet of L contained in K and larger than rL. There is a finite subset $A' \subset A$ such that $o \in \operatorname{conv} A'$ and hence a finite subset $\Omega' \subset \Omega$ such that

$$P := \bigcap_{u \in \Omega'} H^-(K, u)$$

is a polytope containing no homothet of L larger than rL. If P_λ is determined by $(\Omega', h_K + \lambda h_L)$, then $K_\lambda \subset P_\lambda$ for $-r < \lambda \leqq 0$, and it is easy to see that $V_n(P_\lambda) \to 0$ for $\lambda \to -r$. This proves (6.5.11).

From (6.5.9), (6.5.10), (6.5.11) we now obtain

$$V_n(K) = n \int_{-r}^{0} V_1(K_\lambda, \bar{L})\,\mathrm{d}\lambda. \tag{6.5.12}$$

By the general Brunn–Minkowski theorem, 6.4.3,

$$V_1(K_\lambda + |\lambda| \bar{L}, \bar{L})^{1/(n-1)} \geqq V_1(K_\lambda, \bar{L})^{1/(n-1)} + |\lambda| V_n(\bar{L})^{1/(n-1)}.$$

From $K_\lambda + |\lambda| \bar{L} \subset K$ we have $V_1(K_\lambda + |\lambda| \bar{L}, \bar{L}) \leqq V_1(K, \bar{L})$ and hence

$$V_1(K_\lambda, \bar{L})^{1/(n-1)} \leqq V_1(K, \bar{L})^{1/(n-1)} + \lambda V_n(\bar{L})^{1/(n-1)}.$$

Inserting this into (6.5.12), we obtain

$$V_n(K) \leqq n \int_{-r}^{0} [V_1(K, \bar{L})^{1/(n-1)} + \lambda V_n(\bar{L})^{1/(n-1)}]^{n-1}\,\mathrm{d}\lambda.$$

Since $V_1(K, \bar{L}) = V_1(K, L)$ by (6.5.2), we arrive at

$$V_n(K) V_n(\bar{L})^{1/(n-1)}$$
$$\leqq V_1(K, L)^{n(n-1)} - [V_1(K, L)^{1/(n-1)} - rV_n(\bar{L})^{1/(n-1)}]^n,$$

which is the assertion of Theorem 6.2.3. ■

As we have seen, and shall further see in the proof of Theorem 6.6.18, differentiability assertions with respect to the parameter of a system of inner parallel bodies play an essential role. Another result of that type is needed in the proof of Theorem 6.6.7 and will now be proved.

Let $A, C \in \mathcal{K}_0^n$ be convex bodies with interior points. We say that A is *adapted to* C if to each point $x \in \operatorname{bd} C$ there is a point $y \in \operatorname{bd} A$ such that the normal cones at these points satisfy $N(C, x) \subset N(A, y)$. For example, it follows from Theorem 2.2.1 that A is adapted to C if A is a summand of C, but this is only a very special case.

In Section 3.1, the Minkowski difference of the two convex bodies C and A was defined by

$$C \sim A = \bigcap_{a \in A} (C - a) = \{x \in \mathbb{E}^n | A + x \subset C\}.$$

For $0 \leqq \tau \leqq r(C, A)$ (the inradius of C relative to A) we define

$$C_\tau := (C \sim \tau A) + \tau A$$

and

$$h_u(\tau) := h(C_\tau, u) \quad \text{for } u \in S^{n-1}.$$

Lemma 6.5.4. *If A is adapted to C, then $h'_u(0) = 0$ for each $u \in S^{n-1}$.*

Proof. Let $u \in S^{n-1}$ and $0 < \tau < r(C, A)$ be given. We can choose $z \in \mathbb{E}^n$ such that $\tau A + z \subset C$ and $\langle z, u \rangle$ is maximal. Let

$$U_\tau := \{v \in S^{n-1} | h(C, v) = h(\tau A + z, v)\}.$$

We assert that

$$u \in \text{pos}\, U_\tau. \tag{6.5.13}$$

Assume this is false. Then u and the closed convex cone pos U_τ, which is pointed since $\tau < r(C, A)$, can be separated strongly by a hyperplane. Hence, there is a vector $w \in \mathbb{E}^n$ such that $\langle w, u \rangle > 0$ and $\langle w, v \rangle < 0$ for each $v \in U_\tau \backslash \{o\}$. On the closed hemisphere $S^{n-1} \cap H^+_{w,0}$, the continuous function $h(C, \cdot) - h(\tau A + z, \cdot)$ is positive and hence attains a positive minimum ε. Then

$$h(C, v) - h(\tau A + z, v) \geqq \varepsilon \langle w, v \rangle$$

for all $v \in S^{n-1}$ and hence $\tau A + z + \varepsilon w \subset C$. Since $\langle z + \varepsilon w, u \rangle > \langle z, u \rangle$, this contradicts the choice of z. Hence, the assumption was false, and (6.5.13) holds.

By (6.5.13) and Carathéodory's theorem, we can choose (not necessarily distinct) unit vectors $v_1, \dots, v_n \in U_\tau$ and numbers $\lambda_1, \dots, \lambda_n \geqq 0$ such that

$$u = \sum_{i=1}^{n} \lambda_i v_i.$$

For $x \in C$ we have

$$\langle x, v_i \rangle \leqq h(C, v_i) = h(\tau A + z, v_i),$$

hence

$$\langle x, u \rangle \leqq \sum_{i=1}^{n} \lambda_i h(\tau A + z, v_i).$$

Thus

$$h(C, u) \leqq \sum_{i=1}^{n} \lambda_i h(\tau A + z, v_i).$$

By (3.1.12) we have $h_u(\tau) \leqq h_u(0)$. Further, $\tau A + z \subset C$, hence $z \in C \sim \tau A$ and thus $\tau A + z \subset C_\tau$, from which we infer that

$$\begin{aligned} h_u(0) - h_u(\tau) &\leqq h(C, u) - h(\tau A + z, u) \\ &\leqq \sum_{i=1}^{n} \lambda_i h(\tau A + z, v_i) - h(\tau A + z, u) \\ &= \tau \left[\sum_{i=1}^{n} \lambda_i h(A, v_i) - h(A, u) \right]. \end{aligned}$$

We deduce that

$$0 \geqq \frac{h_u(\tau) - h_u(0)}{\tau} \geqq h(A, u) - \sum_{i=1}^{n} \lambda_i h(A, v_i). \tag{6.5.14}$$

Now we let τ tend to zero. The vectors v_i and numbers λ_i chosen above depend on τ. For each $\tau \in (0, r(C, A))$ we make a definite choice of unit vectors $v_1(\tau), \dots, v_n(\tau)$ and corresponding numbers $\lambda_1(\tau), \dots, \lambda_n(\tau)$. We assert that

$$\lim_{\tau \to 0} \sum_{i=1}^{n} \lambda_i(\tau) h(A, v_i(\tau)) = h(A, u). \tag{6.5.15}$$

Suppose this were false. Then there exists a number $\alpha > 0$ and a sequence $(\tau_j)_{j \in \mathbb{N}}$ such that $\tau_j \to 0$ for $j \to \infty$ and

$$h(A, u) - \sum_{i=1}^{n} \lambda_i(\tau_j) h(A, v_i(\tau_j)) \leqq -\alpha \tag{6.5.16}$$

for $j \in \mathbb{N}$. For each τ_j we choose a vector z_j such that $\tau_j A + z_j \subset C$ and $\langle z_j, u \rangle$ is maximal. After selecting subsequences and changing the notation we may assume that

$$\lim_{j \to \infty} z_j = z, \quad \lim_{j \to \infty} v_i(\tau_j) = v_i \quad \text{for } i = 1, \dots, n,$$

and also that the compact sets U_{τ_j} converge to a set U. Since pos U is pointed, there is a unit vector b and a number $\beta > 0$ such that $\langle b, v \rangle \geqq \beta$ for $v \in U_{\tau_j}$ and all sufficiently large j. From

$$\langle b, u \rangle = \sum_{i=1}^{n} \lambda_i(\tau_j) \langle b, v_i(\tau_j) \rangle$$

it follows that $\lambda_i(\tau_j) \in [0, \langle b, u \rangle / \beta]$ for large j. Hence, we may also assume that

$$\lim_{j \to \infty} \lambda_i(\tau_j) = \lambda_i \quad \text{for } i = 1, \dots, n.$$

Assuming further that $o \in A$, we have $z_j \in \tau_j A + z_j$, and we deduce that

$$z \in F(C, u), \qquad u \in N(C, z),$$
$$v_i \in N(C, z) \qquad \text{for } i = 1, \ldots n,$$
$$u = \sum_{i=1}^{n} \lambda_i v_i.$$

Since A is adapted to C, there is a point $y \in \operatorname{bd} A$ for which $N(C, z) \subset N(A, y)$. Thus $u \in N(A, y)$, $v_i \in N(A, y)$ for $i = 1, \ldots, n$, and

$$h(A, u) - \sum_{i=1}^{n} \lambda_i h(A, v_i) = \langle y, u \rangle - \sum_{i=1}^{n} \lambda_i \langle y, v_i \rangle = 0.$$

But (6.5.16) yields

$$h(A, u) - \sum_{i=1}^{n} \lambda_i h(A, v_i) < 0.$$

This contradiction proves (6.5.15), which together with (6.5.14) completes the proof. ■

Notes for Section 6.5

1. Lemmas 6.5.1–6.5.3 and their proofs are taken from Aleksandrov (1938a). Lemma 6.5.4, with essentially the same proof, appears in Schneider (1990a). The proof of Theorem 6.2.3 given here is due to Diskant (1973b).
2. *Inner parallel bodies*. In the geometry of convex bodies, there are many more applications of inner parallel bodies. Here we give a list of references containing such applications: Bol (1941, 1942, 1943a, b), Chakerian (1972, 1973), Chakerian & Sangwine-Yager (1979), Czipszer (1962), Dinghas (1943a), Diskant (1973b, c), Fáry (1983), Hadwiger [17], [18], Matheron (1978a, b), Oshio (1955, 1958, 1962), Rényi (1946), Sangwine-Yager (1980, 1983, 1988a, b), Schneider (1989a, 1990a) and B. v. Sz.-Nagy (1940). The application by Fáry (1986) contains a serious error (there are counterexamples to the higher-order differentiability properties stated in Theorem 4).
3. *H-convex sets*. The notion of a convex body determined by Ω can be generalized: Let $H \subset S^{n-1}$ be an arbitrary subset. Any halfspace $H^-_{u,\alpha}$ with $u \in H$ and $\alpha \in \mathbb{R}$ is called H-convex, and the intersection of any family of H-convex halfspaces is called an H-convex set. A thorough study, mainly from the combinatorial viewpoint, of the family of H-convex sets, for given H, was made in the book by Boltjanski & Soltan [6].
4. *Generalized outer parallel bodies and Wulff shape*. For $K \in \mathcal{K}^n_0$ and a given continuous non-negative function g on S^{n-1}, let $F_t K$ be the convex body determined by $(S^{n-1}, h_K + tg)$ for $t \geqq 0$, thus

$$F_t K = \bigcap_{u \in S^{n-1}} H^-_{u, h(K,u) + tg(u)}.$$

The map $F: \mathcal{K}^n_0 \times [0, \infty) \to \mathcal{K}^n_0$ defined in this way was studied by Willson (1980). In particular, he proved the semigroup property $F_s F_t K = F_{s+t} K$ and described applications to the Wulff shape in the theory of the growth of physical crystals. For an application of the Brunn–Minkowski theory to the Wulff shape, see also Dinghas (1944).

6.6. Equality cases and stability

The Aleksandrov–Fenchel inequality

$$V(K_1, K_2, K_3, \ldots, K_n)^2 \geqq V(K_1, K_1, K_3, \ldots, K_n)V(K_2, K_2, K_3, \ldots, K_n)$$

holds with equality if K_1 and K_2 are homothetic. For special choices of $K_3, \ldots, K_n$ this is the only case of equality (certainly for $n = 2$, by Theorem 6.2.1). In the following, we shall first study these special cases, and the corresponding stability results and their applications. The second part of the present section is devoted to the (still incomplete) investigation of the cases of equality in general.

The classical result in this connection states that equality implies homothety of K_1 and K_2 if $K_3, \ldots, K_n$ are balls. We give two proofs for this result and start with a lemma.

Lemma 6.6.1. *Let $n \geqq 3$; let $K, L \in \mathcal{K}^n$ be convex bodies such that their projections K^u and L^u are translates, for each $u \in S^{n-1}$. Then K and L are translates.*

Proof. For $u \in S^{n-1}$ we define

$$G(u) := \{z \in \mathbb{E}^n | K^u = (L + z)^u\};$$

then $G(u)$ is a line of direction u. Let $u, v \in S^{n-1}$ be linearly independent. Let $z \in G(u)$, thus $K^u = (L + z)^u$. Since $(L + z)^v$ is a translate of L^v, there is a vector $y \in \mathbb{E}^n$ with $K^v = (L + z + y)^v$. For all $w \in S^{n-1}$ with $w \perp u$ and $w \perp v$ we have $h(K, w) = h(L + z, w)$, because $K^u = (L + z)^u$ and $w \perp u$, and $h(K, w) = h(L + z + y, w)$, because $K^v = (L + z + y)^v$ and $w \perp v$; hence $\langle y, w\rangle = 0$. Since this holds for all w with $w \perp u, v$ we infer that $y = \lambda u + \mu v$ with $\lambda, \mu \in \mathbb{R}$. The vector $x := z + y - \mu v$ satisfies $x \in G(u)$, further $K^v = (L + x)^v$ and hence $x \in G(v)$. Thus we have proved that $G(u) \cap G(v) \neq \emptyset$ for linearly independent u, v.

Now let $u_1, u_2 \in S^{n-1}$ be linearly independent; then $G(u_1) \cap G(u_2) = \{t\}$ for some t. If $v \in S^{n-1}$ is linearly independent of u_1 and u_2, the line $G(v)$, which has to meet $G(u_1)$ and $G(u_2)$, necessarily passes through t. Hence, $K^v = (L + t)^v$ and thus $h(K, w) = h(L + t, w)$ for all $w \perp v$. The set of all w for which this holds is dense in S^{n-1}, hence $K = L + t$. ∎

In the following, the unit ball B^n is simply denoted by B.

Theorem 6.6.2. *If $K, L \in \mathcal{K}_0^n$ are convex bodies for which equality holds in*

$$V(K, L, B, \dots, B)^2 \geqq V(K, K, B, \dots, B)V(L, L, B, \dots, B), \tag{6.6.1}$$

then K and L are homothetic.

Proof. We use induction with respect to the dimension. For $n = 2$, the result is contained in Theorem 6.2.1. Suppose that $n \geqq 3$ and the assertion is true in smaller dimensions. If (6.6.1) holds with equality, then by Theorem 6.4.1 the measures $S(K, \mathcal{B}, \cdot)$ and $S(L, \mathcal{B}, \cdot)$ are proportional, where $\mathcal{B}$ stands for the $(n-2)$-tuple $(B, \dots, B)$. Without loss of generality, we may assume that

$$S(K, \mathcal{B}, \cdot) = S(L, \mathcal{B}, \cdot). \tag{6.6.2}$$

Then Lemma 6.4.5 yields

$$v(K^u, L^u, \mathcal{B}^u[n-3])^2 = v(K^u, K^u, \mathcal{B}^u[n-3])v(L^u, L^u, \mathcal{B}^u[n-3])$$

for $u \in S^{n-1}$. By the induction hypothesis, K^u and L^u are homothetic. From (6.6.2) and (5.3.32) we get $v(K^u, B^u) = v(L^u, B^u)$, hence K^u and L^u are, in fact, translates. By Lemma 6.6.1, K and L are translates. ■

The second proof we shall give for Theorem 6.6.2 yields more, namely a stability version. It depends upon the following analytic inequality, which is closely related to a result known as Wirtinger's lemma (or the Poincaré inequality).

Lemma 6.6.3. *Let f be a real function of class C^2 on S^{n-1} satisfying*

$$\int f \, \mathrm{d}\sigma = 0 \tag{6.6.3}$$

and

$$\int f(u)u \, \mathrm{d}\sigma(u) = o, \tag{6.6.4}$$

where σ denotes spherical Lebesgue measure and the integrations extend over S^{n-1}. Then

$$\int f\left(f + \frac{1}{n-1} \Delta_S f\right) \mathrm{d}\sigma + \frac{n+1}{n-1} \sum f^2 \, \mathrm{d}\sigma \leqq 0. \tag{6.6.5}$$

Proof. We use spherical harmonics (see the Appendix for details). Let $X_m := \pi_m f$ (defined by A.3) for $m = 0, 1, 2, \dots$, then $X_0 = 0$ and $X_1 = 0$ by (6.6.3) and (6.6.4); hence the Fourier series of f with respect to the system of spherical harmonics is

$$f \sim \sum_{m=2}^{\infty} X_m.$$

By Green's formula on the sphere S^{n-1} and by (A.2),

$$(\Delta_S f, X_m) = (f, \Delta_S X_m) = -m(m+n-2)(f, X_m),$$

hence

$$\Delta_S f \sim -\sum_{m=2}^{\infty} m(m+n-2) X_m.$$

The Parseval relation (A.5) gives

$$\int f^2 \, d\sigma = \sum_{m=2}^{\infty} \int X_m^2 \, d\sigma,$$

$$\int f \Delta_S f \, d\sigma = -\sum_{m=2}^{\infty} m(m+n-2) \int X_m^2 \, d\sigma.$$

This yields

$$\int f\left(f + \frac{1}{n-1} \Delta_S f\right) d\sigma + \frac{n+1}{n-1} \int f^2 \, d\sigma$$

$$= \frac{1}{n-1} \sum_{m=2}^{\infty} [2n - m(m+n-2)] \int X_m^2 \, d\sigma \leqq 0. \quad \blacksquare$$

If the preceding result is applied to differences of support functions, one obtains estimates for the deviation of convex bodies; however, these are not immediately in terms of the Hausdorff metric δ but in terms of the L_2 metric δ_2. This is defined by

$$\delta_2(K, L)^2 := \int_{S^{n-1}} |h_K - h_L|^2 \, d\sigma. \tag{6.6.6}$$

Clearly $\delta_2 \leqq \sqrt{\omega_n}\, \delta$. Estimates in the opposite direction are given in the following two lemmas, which we state without proof.

Lemma 6.6.4. *For $K, L \in \mathcal{K}^n$,*

$$\delta_2(K, L)^2 \geqq \alpha_n D(K \cup L)^{1-n} \delta(K, L)^{n+1}$$

with

$$\alpha_n = \frac{\omega_n B(3, n-1)}{B\left(\frac{1}{2}, \frac{n-1}{2}\right)},$$

where $B(\cdot, \cdot)$ is the beta integral.

For a convex body K, we introduce its *Steiner ball* $B(K)$ as the ball that has the same Steiner point and mean width as K, thus

$$B(K) := B(s(K), \tfrac{1}{2} b(K)). \tag{6.6.7}$$

Lemma 6.6.5. *If $K \in \mathcal{K}^n$ and $B(K)$ is the Steiner ball of K, then*

$$\delta_2(K, B(K))^2 \geqq \beta_n \left(\frac{\kappa_n}{W_{n-1}(K)} \right)^{(n-1)/2} \delta(K, B(K))^{(n+3)/2}$$

with $B_n = \gamma_n(\omega_n/2\kappa_{n-1})$, where

$$\gamma_n(c) = \frac{\omega_{n-1}}{(n^2-1)(n+3)} \min \left[\frac{3}{\pi^2 n(n+2)2^n}, \frac{16(c+2)^{(n-3)/2}}{(c+1)^{n-2}} \right].$$

Lemma 6.6.4 was proved (in a more general form) by Vitale (1985a) and Lemma 6.6.5 by Groemer & Schneider (1991). For the special case of the deviation from a suitable ball, Lemma 6.6.5 gives a better estimate, since the exponent $(n+3)/2$ is smaller than the exponent $n+1$ of $\delta(K, L)$ in Lemma 6.6.4. (It is, of course, this exponent that is essential for the strength of the estimate, and not the factor.)

When stability estimates are obtained in terms of the L_2 metric, we shall refrain from translating them into Hausdorff metric estimates, but the reader should keep in mind that this is always possible as a result of Lemmas 6.6.4 and 6.6.5.

To deduce such a stability estimate from Lemma 6.6.3, we associate, with each convex body $K \in \mathcal{K}^n$ of dimension at least one, its normalized homothetic copy $\bar{K}$ defined by

$$\bar{K} := \frac{K - s(K)}{b(K)}.$$

Let $K, L \in \mathcal{K}^n$ be convex bodies of dimension at least one, and put

$$V_{ij} := V(K[i], L[j], B[n-i-j]).$$

Then equality (6.6.1) reads

$$V_{11}^2 - V_{20}V_{02} \geqq 0.$$

The following theorem contains a strengthened version of this inequality.

Theorem 6.6.6. *For convex bodies $K, L \in \mathcal{K}^n$ of dimension $\geqq 1$,*

$$V_{11}^2 - V_{20}V_{02} \geqq \frac{n+1}{n(n-1)} b(K)^2 V_{02} \delta_2(\bar{K}, \bar{L})^2. \tag{6.6.8}$$

Proof. First we assume that K and L are of class C_+^2. The function $f := h(\bar{K}, \cdot) - h(\bar{L}, \cdot)$ satisfies (6.6.3) and (6.6.4), by (A.6) and (A.7). Hence, Lemma 6.6.3 yields

$$\int \left(f^2 + \frac{1}{n-1} f \Delta_S f \right) d\sigma + \frac{n+1}{n-1} \delta_2(\bar{K}, \bar{L})^2 \leqq 0.$$

By formula (5.3.17),

$$\frac{1}{n}\int\left(f^2+\frac{1}{n-1}f\Delta_S f\right)\mathrm{d}\sigma$$
$$= V(\bar{K}, \bar{K}, B[n-2]) - 2V(\bar{K}, \bar{L}, B[n-2]) + V(\bar{L}, \bar{L}, B[n-2])$$
$$= \frac{V_{20}}{b(K)^2} - \frac{2V_{11}}{b(K)b(L)} + \frac{V_{02}}{b(L)^2}.$$

Applying the identity

$$b^2 - ac = -c(a-2b+c) + (b-c)^2$$

with

$$a = \frac{b(L)}{b(K)}V_{20}, \quad b = V_{11}, \quad c = \frac{b(K)}{b(L)}V_{02},$$

we arrive at the inequality

$$V_{11}^2 - V_{20}V_{02} \geqq \frac{n+1}{n(n-1)}b(K)^2V_{02}\delta_2(\bar{K}, \bar{L})^2 + \left(V_{11} - \frac{b(K)}{b(L)}V_{02}\right)^2,$$

of which (6.6.8) is a weaker version. By approximation, this inequality is now extended to general convex bodies K and L. ∎

The special case of (6.6.8) where L is a ball can be written in a more convenient form, using the Steiner ball of K. Since

$$\delta_2(\bar{K}, \bar{B})^2 = b(K)^{-2}\delta_2(K, B(K))^2,$$

as a corollary of Theorem 6.6.6 we obtain, for the quermassintegrals W_{n-1}, W_{n-2} of the convex body $K \in \mathcal{K}^n$, the inequality

$$\frac{1}{\kappa_n}W_{n-1}^2 - W_{n-2} \geqq \frac{n+1}{n(n-1)}\delta_2(K, B(K))^2. \tag{6.6.9}$$

From this, we can deduce a stability result for the more general inequality (6.4.7) between two quermassintegrals of convex bodies. Let $K \in \mathcal{K}_0^n$ and write $W_k = W_k(K)$; let $0 \leqq i < j < n$. Using the special Aleksandrov–Fenchel inequality $W_k^2 \geqq W_{k-1}W_{k+1}$, we obtain

$$\frac{W_{j+1}^{n-j}}{W_j^{n-j-1}W_n} = \left(\frac{W_{j+1}^2}{W_jW_{j+2}}\right)^{n-j-1}\left(\frac{W_{j+2}^2}{W_{j+1}W_{j+3}}\right)^{n-j-2}\cdots\left(\frac{W_{n-1}^2}{W_{n-2}W_n}\right)\cdot$$
$$\geqq \frac{W_{n-1}^2}{W_{n-2}W_n}$$

and

$$\frac{W_j^{j-i+1}}{W_iW_{j+1}^{j-i}} \geqq 1,$$

the latter from (6.4.6). This gives

$$\left(\frac{W_{j+1}^{n-j}}{W_j^{n-j-1}W_n}\right)^{j-i}\left(\frac{W_j^{j-i+1}}{W_iW_{j+1}^{j-i}}\right)^{n-j} \geqq \left(\frac{W_{n-1}^2}{W_{n-2}W_n}\right)^{j-i},$$

hence

$$\frac{W_j^{n-i}}{W_i^{n-j}} \geqq \left(\frac{W_{n-1}^2}{W_{n-2}}\right)^{j-i}$$

and thus

$$\kappa_n^{i-j} W_j^{n-i} - W_i^{n-j} \geqq \frac{W_i^{n-j}}{W_{n-2}^{j-i}} \left[\left(\frac{W_{n-1}^2}{\kappa_n}\right)^{j-i} - W_{n-2}^{j-i}\right].$$

Using the identity

$$\frac{a^k - b^k}{a - b} = \sum_{r=0}^{k-1} a^{k-1-r} b^r =: s_k(a, b)$$

with $k = j - i$, $a = W_{n-1}^2/\kappa_n$, $b = W_{n-2}$ and then (6.6.9), we arrive at

$$\kappa_n^{i-j} W_j^{n-i} - W_i^{n-j} \geqq \gamma_{n,i,j}(K)\delta_2(K, B(K))^2 \qquad (6.6.10)$$

with

$$\gamma_{n,i,j}(K) = \frac{n+1}{n(n-1)} \frac{W_i^{n-j}}{W_{n-2}^{j-i}} s_{j-i}\left(\frac{W_{n-1}^2}{\kappa_n}, W_{n-2}\right).$$

By approximation, inequality (6.6.10) can be extended to convex bodies K of dimension $\geqq 2$.

The case $i = 0$, $j = 1$ of inequality (6.6.10) is a strengthened version of the isoperimetric inequality, namely

$$\left(\frac{S}{\omega_n}\right)^n - \left(\frac{V_n}{\kappa_n}\right)^{n-1} \geqq \frac{n+1}{n(n-1)\kappa_n^{n-1}} \frac{V_n^{n-1}}{W_{n-2}} \delta_2(K, B(K))^2. \qquad (6.6.11)$$

Up to now, the considerations of this section centred around the fact that equality in the Aleksandrov–Fenchel inequality

$$V(K, L, C_1, \ldots, C_{n-2})^2$$
$$\geqq V(K, K, C_1, \ldots, C_{n-2})V(L, L, C_1, \ldots, C_{n-2}), \qquad (6.6.12)$$

under the special assumption $C_1 = \ldots = C_{n-2} = B^n$ (unit ball), holds only if K and L are homothetic. Using this result, we shall now show that the same conclusion can be drawn if we assume only that $C_1, \ldots, C_{n-2}$ are smooth. We first prove a more general result, which states that equality in (6.6.12) is preserved under replacement of the $(n-2)$-tuple $(C_1, \ldots, C_{n-2})$ by any other $(n-2)$-tuple $(A_1, \ldots, A_{n-2})$ of convex bodies such that A_i is adapted to C_i, in the sense defined in Section 6.5.

Theorem 6.6.7. *Let $C_1, \ldots, C_{n-2} \in \mathcal{K}_0^n$ be n-dimensional convex bodies, set $\mathcal{C} = (C_1, \ldots, C_{n-2})$ and let $K, L \in \mathcal{K}^n$ be convex bodies satisfying*

$$V(K, L, \mathcal{C})^2 = V(K, K, \mathcal{C})V(L, L, \mathcal{C}). \qquad (6.6.13)$$

If $\mathcal{A} = (A_1, \ldots, A_{n-2})$ where $A_i \in \mathcal{K}_0^n$ is adapted to C_i for $i = 1, \ldots, n-2$, then also

$$V(K, L, \mathcal{A})^2 = V(K, K, \mathcal{A})V(L, L, \mathcal{A}). \qquad (6.6.14)$$

Proof. Let $n \geqq 3$, without loss of generality. If $\dim K = 0$ or $\dim L = 0$, then K and L are homothetic. If, say, $\dim K = 1$, then $V(K, K, \mathcal{C}) = 0$, hence $V(K, L, \mathcal{C}) = 0$. Since $C_1, \ldots, C_{n-2}$ are n-dimensional, this is only possible if either $\dim L = 0$ or L is a segment parallel to K. Thus K and L are homothetic, and (6.6.14) holds. We may, therefore, assume in the following that $\dim K \geqq 2$ and $\dim L \geqq 2$.

We write

$$C_1 = C,\ A_1 = A \quad \text{and } (C_2, \ldots, C_{n-2}) = \mathcal{C}';$$

thus

$$V(K, L, C, \mathcal{C}')^2 = V(K, K, C, \mathcal{C}')V(L, L, C, \mathcal{C}'). \tag{6.6.15}$$

By Theorem 6.4.1 we may assume, after a dilatation of K, that

$$S(K, C, \mathcal{C}', \cdot) = S(L, C, \mathcal{C}', \cdot). \tag{6.6.16}$$

Let $Q \in \mathcal{K}^n$ be a convex body with $\dim Q \geqq 1$. By (6.4.22), with $\mathcal{C}$ replaced by $(Q, \mathcal{C}')$ and M replaced by C, we have

$$\frac{V(K, K, Q, \mathcal{C}')}{V(K, Q, C, \mathcal{C}')^2} - \frac{2V(K, L, Q, \mathcal{C}')}{V(K, Q, C, \mathcal{C}')V(L, Q, C, \mathcal{C}')} + \frac{V(L, L, Q, \mathcal{C}')}{V(L, Q, C, \mathcal{C}')^2} \leqq 0. \tag{6.6.17}$$

By (6.6.16) and (5.1.18),

$$V(K, Q, C, \mathcal{C}') = V(L, Q, C, \mathcal{C}'), \tag{6.6.18}$$

hence

$$V(K, K, Q, \mathcal{C}') - 2V(K, L, Q, \mathcal{C}') + V(L, L, Q, \mathcal{C}') \leqq 0. \tag{6.6.19}$$

Again by (6.6.16) and (5.1.18),

$$V(K, K, C, \mathcal{C}') - 2V(K, L, C, \mathcal{C}') + V(L, L, C, \mathcal{C}') = 0. \tag{6.6.20}$$

We define

$$\Phi(Q) := V(K, K, Q, \mathcal{C}') - 2V(K, L, Q, \mathcal{C}') + V(L, L, Q, \mathcal{C}')$$

for $Q \in \mathcal{K}^n$ and

$$g(\tau) := \Phi(C_\tau)$$

for $0 < \tau < r(C, A)$, with $C_\tau := (C \sim \tau A) + \tau A$ as in Section 6.5; further

$$f(\tau) := \Phi(C \sim \tau A).$$

Then

$$f(\tau) \leqq 0 \tag{6.6.21}$$

and

$$\Phi(A) \leqq 0 \tag{6.6.22}$$

by (6.6.19). By the linearity of the mixed volume,

$$g(\tau) = f(\tau) + \tau\Phi(A). \tag{6.6.23}$$

We assert that

$$g'(0) = 0. \tag{6.6.24}$$

For the proof, let

$$\varphi(\tau) := V(C_\tau, \bar{\mathscr{C}}),$$

where $\bar{\mathscr{C}}$ is an arbitrary $(n-1)$-tuple of convex bodies. Then, by (5.1.18) and with $h_u(\tau) := h(C_\tau, u)$,

$$\frac{\varphi(\tau) - \varphi(0)}{\tau} = \frac{1}{n}\int_{S^{n-1}} \frac{h_u(\tau) - h_u(0)}{\tau}\, \mathrm{d}S(\bar{\mathscr{C}}, u).$$

We show that the integrand remains bounded for $\tau \to 0$. Choose a positive homothet $\lambda A + z$ contained in C. For given $u \in S^{n-1}$, choose $x \in F(C, u)$. For $0 < \tau < \lambda$, let $\tau A + z_\tau$ be obtained from $\lambda A + z$ by dilatation with centre x. Then $\tau A + z_\tau \subset C$ and hence $h(C_\tau, u) \geqq h(\tau A + z_\tau, u)$. Writing

$$h(C, u) - h(\tau A + z_\tau, u) =: \alpha,$$
$$h(C, u) - h(\lambda A + z, u) =: \beta,$$

we have $\alpha/\beta = \tau/\lambda$ and hence

$$0 \leqq h_u(0) - h_u(\tau) = h(C, u) - h(C_\tau, u) \leqq \alpha = \beta\tau/\lambda \leqq \gamma\tau$$

where the constant γ is independent of u. This proves that the integrand remains bounded, and we can apply the bounded convergence theorem together with Lemma 6.5.4 to deduce that $\varphi'(0) = 0$. Since $\bar{\mathscr{C}}$ was arbitrary, this proves (6.6.24).

From (6.6.23) and (6.6.24) we get $f'(0) + \Phi(A) = 0$, hence $f'(0) \geqq 0$ by (6.6.22). On the other hand, $f(0) = 0$ by (6.6.20) and $f(\tau) \leqq 0$ by (6.6.21). This yields $f'(0) = 0$ and thus $\Phi(A) = 0$, hence

$$V(K, K, A, \mathscr{C}') - 2V(K, L, A, \mathscr{C}') + V(L, L, A, \mathscr{C}') = 0.$$

This, together with (6.6.18) for $Q = A$, shows that there is equality in (6.6.17) for $Q = A$. By Theorem 6.4.2, this implies

$$V(K, L, A, \mathscr{C}')^2 = V(K, K, A, \mathscr{C}')V(L, L, A, \mathscr{C}')$$

or, returning to the former notation,

$$V(K, L, A_1, C_2, \ldots, C_{n-2})^2$$
$$= V(K, K, A_1, C_2, \ldots, C_{n-2})V(L, L, A_1, C_2, \ldots, C_{n-2}).$$

This equality has been deduced under the assumption that (6.6.13) holds and that A_1 is adapted to C_1. In a similar way, each of the bodies C_2,

$\ldots, C_{n-2}$ can be replaced by another one adapted to it. This proves the theorem. ■

Now we can prove

Theorem 6.6.8. *If equality holds in*

$$V(K, L, C_1, \ldots, C_{n-2})^2 \geqq V(K, K, C_1, \ldots, C_{n-2})V(L, L, C_1, \ldots, C_{n-2}), \qquad (6.6.25)$$

where $C_1, \ldots, C_{n-2}$ *are smooth convex bodies, then* K *and* L *are homothetic.*

Proof. If C_i is smooth, then the unit ball B^n is adapted to C_i; hence by Theorem 6.6.7 equality in (6.6.25) implies equality in (6.6.1). By Theorem 6.6.6, K and L are homothetic. ■

We are now in a position to complete the results on the general Brunn–Minkowski theorem that are collected in Theorem 6.4.4, if we introduce smoothness conditions.

Theorem 6.6.9. *Let a number* $m \in \{2, \ldots, n\}$ *and an* $(n-m)$*-tuple* $\mathscr{C} = (K_{m+1}, \ldots, K_n)$ *of smooth convex bodies be given and let* K_0, $K_1 \in \mathcal{K}^n$ *be convex bodies of dimension* $\geqq m$. *Then the conditions* (a)–(d) *of Theorem 6.4.4 hold only if* K_0 *and* K_1 *are homothetic.*

Proof. We use the notation introduced before Theorem 6.4.4. Assume that the equivalent conditions (a)–(d) of that theorem are satisfied. Since (b) holds, that is,

$$V_{(k)}^2 = V_{(k-1)}V_{(k+1)} \quad \text{for } k = 1, \ldots, m-1, \qquad (6.6.26)$$

then by Theorem 6.4.1 the measures $S_{(k)}$ and $S_{(k-1)}$ are proportional. The proportionality factor is obtained by integrating $h(K_0, \cdot)$, and we obtain

$$V_{(k-1)}S_{(k)} = V_{(k)}S_{(k-1)}.$$

Integrating the support function of the unit ball B^n with this measure, we obtain

$$V_{(k-1)}V(K_0[m-1-k], K_1[k], B^n, \mathscr{C}) = V_{(k)}V(K_0[m-k], K_1[k-1], B^n, \mathscr{C}).$$

If we define

$$\bar{V}_{(k)} := V(K_0[m-1-k], K_1[k], B^n, \mathscr{C}),$$

this reads

$$V_{(k-1)}\bar{V}_{(k)} = V_{(k)}\bar{V}_{(k-1)} \quad \text{for } k = 1, \ldots, m-1,$$

hence

$$V_{(k)}\bar{V}_{(k+1)} = V_{(k+1)}\bar{V}_{(k)} \quad \text{for } k = 0, \ldots, m-2.$$

Together with (6.6.26), this yields

$$\bar{V}^2_{(k)} = \bar{V}_{(k-1)}\bar{V}_{(k+1)} \quad \text{for } k = 1, \ldots, m-2.$$

Thus we are again in the situation described by (6.6.26), but with m replaced by $m-1$ and $\mathscr{C}$ replaced by $(B^n, \mathscr{C})$. Repeating the procedure, we end up with the equality

$$V(K_0, K_1, B^n[m-2], \mathscr{C})^2$$
$$= V(K_0, K_0, B^n[m-2], \mathscr{C})V(K_1, K_1, B^n[m-2], \mathscr{C}).$$

By Theorem 6.6.8, this implies that K_0 and K_1 are homothetic. ■

Theorem 6.6.8 generalizes Theorem 6.6.2, in which the bodies $C_1, \ldots, C_{n-2}$ were assumed to be balls. For the latter result, Theorem 6.6.6 gives a strengthened version in the form of a stability result. We shall now investigate the possibilities of improving Theorem 6.6.8 in a similar way. For convex bodies $K, L, C_1, \ldots, C_{n-2} \in \mathscr{K}^n$ we introduce the deficit Δ by

$$\Delta(K, L, C_1, \ldots, C_{n-2})$$
$$:= V(K, L, C_1, \ldots, C_{n-2})^2 - V(K, K, C_1, \ldots, C_{n-2})$$
$$\times V(L, L, C_1, \ldots, C_{n-2}),$$

so that the Aleksandrov–Fenchel inequality can be written as

$$\Delta(K, L, C_1, \ldots, C_{n-2}) \geqq 0. \tag{6.6.27}$$

Inequality (6.6.8) tells us that

$$\Delta(K, L, B^n, \ldots, B^n) \geqq \frac{n+1}{n(n-1)} a(K, L)\, \delta_2(\bar{K}, \bar{L})^2 \tag{6.6.28}$$

(if $\dim K, L \geqq 1$), where

$$a(K, L)$$
$$:= \max\{b(K)^2 V(L, L, B^n[n-2]), b(L)^2 V(K, K, B^n[n-2])\}.$$

For general convex bodies $C_1, \ldots, C_{n-2}$ it is possible (as we shall see later in this section) that equality holds in (6.6.27) without K and L being homothetic. This implies that Theorem 6.6.8 cannot be improved to a stability version of the type (6.6.28) that would hold for all bodies $C_1, \ldots, C_{n-2}$ in a dense subclass of $\mathscr{K}^n$. However, one can show stability for (6.6.27) under the assumption that the convex bodies $C_1, \ldots, C_{n-2}$ are *ρ-smooth*, for some fixed $\rho > 0$. The convex body

$K \in \mathcal{K}_0^n$ is called *ρ-smooth* if ρB^n is a summand of K, equivalently (by Theorem 3.2.2), if to each boundary point x of K there is a ball $\rho B^n + t$ of radius ρ such that $x \in \rho B^n + t \subset K$.

For given numbers $r, R > 0$, we denote by $\mathcal{K}^n(r, R)$ the set of all convex bodies in $\mathcal{K}^n$ that contain some ball of radius r and are contained in some ball of radius R.

Theorem 6.6.10. *Let $\rho, r, R > 0$ and an integer $p \in \{1, \ldots, n-2\}$ be given. If $K, L, C_1, \ldots, C_p \in \mathcal{K}^n(r, R)$ and if $C_1, \ldots, C_p$ are ρ-smooth, then*

$$\Delta(K, L, C_1, \ldots, C_p, B^n, \ldots, B^n) \geqq \gamma \rho^{4(2^p - 1)} \Delta(K, L, B^n, \ldots, B^n)^{2^p}, \quad (6.6.29)$$

where the constant γ depends only on n, p, r, R.

Together with (6.6.28), the inequality (6.6.29), with $p = n-2$, yields a stability estimate for the Aleksandrov–Fenchel inequality (6.6.27), provided that $C_1, \ldots, C_{n-2}$ are ρ-smooth. It is inevitable that the factor $\gamma\Delta$ on the right-hand side of an inequality of the type (6.6.29) tends to zero if ρ tends to zero and hence Theorem 6.6.8 is not a corollary of Theorem 6.6.10.

The gist of the proof of Theorem 6.6.10 is the following lemma.

Lemma 6.6.11. *Let $n \geqq 3$. Let $K, L, M \in \mathcal{K}^n(r, R)$, let $K_4, \ldots, K_n \in \mathcal{K}^n$ be arbitrary convex bodies and set $(K_4, \ldots, K_n) = \mathcal{C}$. If the convex body M' is a summand of M, then*

$$\Delta(K, L, M', \mathcal{C}) \leqq \alpha \sqrt{\Delta(K, L, M, \mathcal{C})} \quad (6.6.30)$$

with

$$\alpha = \left[2\left(\frac{R}{r}\right)^3 + 3\left(\frac{R}{r}\right)\right] R^3 V(B^n[3], \mathcal{C}).$$

The lemma shows, in particular, that $\Delta(K, L, M, \mathcal{C}) = 0$ implies $\Delta(K, L, M', \mathcal{C}) = 0$ if M' is a summand of M. Since a summand M' of M is adapted to M, this is simply a special case of Theorem 6.6.7. This special case, however, admits a stability version in the form of Lemma 6.6.11.

Proof of Lemma 6.6.11. For convex bodies $A, B, C, D \in \mathcal{K}^n$, we use the abbreviations

$$V(A, B, C) := V(A, B, C, K_4, \ldots, K_n),$$
$$\Delta(A, B, C) := \Delta(A, B, C, K_4, \ldots, K_n),$$

$$F(A, B, C; D) := -\frac{V(A, A, C)}{V(A, D, C)^2} + \frac{2V(A, B, C)}{V(A, D, C)V(B, D, C)} - \frac{V(B, B, C)}{V(B, D, C)^2},$$

provided that the denominators are positive.

First we assume that the bodies $K_4, \ldots, K_n$ appearing in the lemma have interior points. Given $K_0, K_1, K_2, K_3 \in \mathcal{K}_0^n$, we use the notation

$$U_{ij} := V(K_i, K_j, K_3) \quad \text{for } i, j \in \{0, 1, 2\}.$$
$$\Delta := \Delta(K_1, K_2, K_3) = U_{12}^2 - U_{11}U_{22},$$
$$F := F(K_1, K_2, K_3; K_0) = -\frac{U_{11}}{U_{01}^2} + \frac{2U_{12}}{U_{01}U_{02}} - \frac{U_{22}}{U_{02}^2}.$$

Then $\Delta \geqq 0$, and by (6.4.16) we have $F \geqq 0$. By (6.4.9),

$$(U_{00}U_{12} - U_{01}U_{02})^2 \leqq (U_{01}^2 - U_{00}U_{11})(U_{02}^2 - U_{00}U_{22}).$$

Rearranging the terms, we deduce that

$$\Delta \leqq \frac{U_{01}^2 U_{02}^2}{U_{00}} F. \tag{6.6.31}$$

From the identity $b^2 - ac = -c(a - 2b + c) + (b - c)^2$ with

$$a = \frac{U_{11}}{U_{01}^2}, \quad b = \frac{U_{12}}{U_{01}U_{02}}, \quad c = \frac{U_{22}}{U_{02}^2},$$

we obtain

$$\frac{\Delta}{U_{01}^2 U_{02}^2} = \frac{U_{22}}{U_{02}^2} F + \left(\frac{U_{12}}{U_{01}U_{02}} - \frac{U_{22}}{U_{02}^2}\right)^2, \tag{6.6.32}$$

and in particular

$$\Delta \geqq U_{01}^2 U_{22} F. \tag{6.6.33}$$

Inequality (6.6.32) together with $F \geqq 0$ yields

$$\left|\frac{U_{12}}{U_{01}U_{02}} - \frac{U_{22}}{U_{02}^2}\right| \leqq \frac{\sqrt{\Delta}}{U_{01}U_{02}}.$$

Multiplying respectively by U_{02}^2/U_{12} and by $U_{01}U_{02}/U_{22}$, we obtain

$$\left|\frac{U_{22}}{U_{12}} - \frac{U_{02}}{U_{01}}\right| \leqq \frac{U_{02}}{U_{01}U_{12}}\sqrt{\Delta} \tag{6.6.34}$$

and

$$\left|\frac{U_{12}}{U_{22}} - \frac{U_{01}}{U_{02}}\right| \leqq \frac{\sqrt{\Delta}}{U_{22}}. \tag{6.6.35}$$

For convex bodies $A, B \in \mathcal{K}_0^n$ we define

$$\alpha(B) := \frac{V(L, M, B)}{V(K, M, B)}$$

and

$$\beta(A, B) := V(K, K, A)\alpha(B) - 2V(K, L, A) + V(L, L, A)\alpha(B)^{-1}.$$

Applying inequality (6.6.34) with $K_1 = K$, $K_2 = L$, $K_3 = M$, we obtain

$$|\alpha(L) - \alpha(K_0)| \leqq \frac{V(K_0, L, M)}{V(K_0, K, M)} \frac{\sqrt{\Delta(K, L, M)}}{V(K, L, M)}, \qquad (6.6.36)$$

and inequality (6.6.35) yields

$$|\alpha(L)^{-1} - \alpha(K_0)^{-1}| \leqq \frac{\sqrt{\Delta(K, L, M)}}{V(L, L, M)}. \qquad (6.6.37)$$

Applying (6.6.33) to $K_0 = L$, $K_1 = K$, $K_2 = L$ and $K_3 = M$, we obtain

$$\beta(M, L) \geqq -\frac{\Delta(K, L, M)}{V(K, L, M)}. \qquad (6.6.38)$$

Now, for $A \in \mathcal{K}_0^n$,

$$|\beta(A, A) - \beta(A, L)|$$
$$\leqq V(K, K, A)|\alpha(A) - \alpha(L)| + V(L, L, A)|\alpha(A)^{-1} - \alpha(L)^{-1}|,$$

hence (6.6.36) and (6.6.37) yield

$$|\beta(A, A) - \beta(A, L)| \leqq \varphi(A)\sqrt{\Delta(K, L, M)} \qquad (6.6.39)$$

with

$$\varphi(A) := \frac{V(K, K, A)V(L, M, A)}{V(K, L, M)V(K, M, A)} + \frac{V(L, L, A)}{V(L, L, M)}. \qquad (6.6.40)$$

By assumption, M' is a summand of M, thus $M = M' + M''$ with a convex body $M'' \in \mathcal{K}^n$. We assume first that M' and M'' have interior points. The inequality $F \geqq 0$, applied to $K_0 = M$, $K_1 = K$, $K_2 = L$ and $K_3 = M''$ yields $\beta(M'', M'') \leqq 0$. Thus $\beta(M'', L) \leqq \beta(M'', L) - \beta(M'', M'')$ and (6.6.39) gives

$$\beta(M'', L) \leqq \varphi(M'')\sqrt{\Delta(K, L, M)}. \qquad (6.6.41)$$

Since $\beta(M', L) = \beta(M, L) - \beta(M'', L)$ by the linearity of the mixed volume, inequalities (6.6.38) and (6.6.41) yield

$$\beta(M', L) \geqq -\frac{\Delta(K, L, M)}{V(K, L, M)} - \varphi(M'')\sqrt{\Delta(K, L, M)}. \qquad (6.6.42)$$

Writing $\beta(M', M') = \beta(M', L) + [\beta(M', M') - \beta(M', L)]$, we deduce from (6.6.42) and (6.6.39) that

$$\beta(M', M') \geqq -\psi \qquad (6.6.43)$$

with

$$\psi := \frac{\Delta(K, L, M)}{V(K, L, M)} + [\varphi(M') + \varphi(M'')]\sqrt{\Delta(K, L, M)}.$$

Dividing (6.6.43) by $V(K, M, M')V(L, M, M')$, we obtain

$$F(K, L, M'; M) \leqq \frac{\psi}{V(K, M, M')V(L, M, M')}.$$

Inequality (6.6.31) with $K_1 = K$, $K_2 = L$, $K_3 = M'$ and $K_0 = M$ yields

$$\Delta(K, L, M') \leqq \frac{V(K, M, M')^2 V(L, M, M')^2}{V(M, M, M')} F(K, L, M'; M).$$

Both inequalities together show that

$$\Delta(K, L, M') \leqq \frac{V(K, M, M')V(L, M, M')}{V(M, M, M')} \times \left\{ \frac{\Delta(K, L, M)}{V(K, L, M)} + [\varphi(M') + \varphi(M'')]\sqrt{\Delta(K, L, M)} \right\}. \tag{6.6.44}$$

This estimate can be simplified if we now use the assumption that $K, L, M \in \mathcal{K}^n(r, R)$. Then K is contained in some translate of $(R/r)M$, hence the monotoneity of the mixed volume gives

$$\frac{V(K, M, M')}{V(M, M, M')} \leqq \frac{R}{r}.$$

Since $M = M' + M''$, we may assume that $M' \subset M$, hence

$$V(L, M, M') \leqq R^3 V(B^n, B^n, B^n).$$

From $\Delta(K, L, M) \leqq V(K, L, M)^2$ we obtain

$$\frac{\Delta(K, L, M)}{V(K, L, M)} \leqq \sqrt{\Delta(K, L, M)}.$$

Since M' is contained in M and, thus, in a translate of $(R/r)K$, we have

$$\varphi(M') \leqq \left(\frac{R}{r}\right)^2 + 1,$$

and the same inequality holds for $\varphi(M'')$. Now (6.6.44) leads to the inequality (6.6.30). By approximation, we can finally extend this inequality to the case where $K_4, \ldots, K_n$ and M', M'' are not necessarily of dimension n. ■

Proof of Theorem 6.6.10. Under the assumptions of Theorem 6.6.10, the ball $M' = \rho B^n$ is a summand of C_1. Hence, Lemma 6.6.11 with $M = C_1$ and

$$\mathcal{C} = (K_4, \ldots, K_n) = (C_2, \ldots, C_p, B^n[n-p-2])$$

yields

$$\Delta(K, L, \rho B^n, \mathcal{C}) \leqq \alpha\sqrt{\Delta(K, L, C_1, \mathcal{C})},$$

hence

$$\Delta(K, L, C_1, \ldots, C_p, B^n[n-p-2]) \geqq \alpha^{-2}\rho^4 \Delta(K, L, C_2, \ldots, C_p, B^n[n-p-1])^2.$$

Repeating this procedure with $C_2, \ldots, C_p$, we arrive at the inequality (6.6.29). ■

In the second part of this section, we shall now investigate the problem of equality in the Aleksandrov–Fenchel inequality

$$V(K, L, \mathcal{C})^2 \geqq V(K, K, \mathcal{C})V(L, L, \mathcal{C}) \tag{6.6.45}$$

in the case where $\mathcal{C} = (C_1, \ldots, C_{n-2})$ and the bodies $C_1, \ldots, C_{n-2}$ are not necessarily smooth. In that case, equality in (6.6.45) may hold even if K and L are not homothetic. An example is given by Corollary 6.6.17 below. Before treating such examples and special cases, we formulate a general conjecture.

Conjecture 6.6.12. *If $K, L \in \mathcal{K}^n$ and $C_1, \ldots, C_{n-2} \in \mathcal{K}_0^n$, then equality in* (6.6.45) *holds if and only if, after applying a suitable homothety to K or L, $h(K, u) = h(L, u)$ for all $(C_1, \ldots, C_{n-2}, B)$-extreme unit vectors u.*

The notion of $(C_1, \ldots, C_{n-1})$-extreme vectors was defined in Section 2.2. In particular, u is $(C_1, \ldots, C_{n-2}, B)$-extreme if and only if there exist $(n-1)$-dimensional linear subspaces $E_1, \ldots, E_{n-2}$ of $\mathbb{E}^n$ such that

$$T(C_i, u) \subset E_i \quad \text{for } i = 1, \ldots, n-2$$

and $\dim E_1 \cap \ldots \cap E_{n-2} = 2$. Here $T(C_i, u)$ is the touching cone of C_i at u.

Without the assumption in Conjecture 6.6.12 that $C_1, \ldots, C_{n-2}$ are n-dimensional, more cases of equality in (6.6.45) are possible. This was pointed out by Ewald (1988). For example, let $n = 3$, let $S_1, S_2 \subset \mathbb{E}^3$ be segments of independent directions and let $M \subset \mathcal{K}^3$ be an arbitrary convex body. Then put $K := M + S_1$, $L := M + S_2$ and $C := S_1 + S_2$. Since $V(M, S_i, S_i) = 0$, one computes $V(K, K, C) = V(L, L, C) = V(K, L, C)$ and thus $V(K, L, C)^2 = V(K, K, C)V(L, L, C)$. The condition of Conjecture 6.6.12, however, is not satisfied.

Of course, if $C_1, \ldots, C_{n-2}$ are smooth, then each vector $u \in S^{n-1}$ is $(C_1, \ldots, C_{n-2}, B)$-extreme, hence Theorem 6.6.8 would follow from the truth of Conjecture 6.6.12. In fact, Theorem 6.6.7 would also follow.

The set of $(\mathcal{C}, B)$-extreme unit vectors appears in Conjecture 6.6.12 since it is probably closely related to the mixed area measure $S(\mathcal{C}, B, \cdot)$, according to another conjecture (of which a special case is verified by Theorem 4.6.3):

Conjecture 6.6.13. *Let $K_1, \ldots, K_{n-1} \in \mathcal{K}^n$ be convex bodies. The support of their mixed area measure,* $\operatorname{supp} S(K_1, \ldots, K_{n-1}, \cdot)$, *is the closure of the set of $(K_1, \ldots, K_{n-1})$-extreme unit vectors.*

Before discussing further the relation between the two conjectures, we state a result on supports of mixed area measures from which further consequences can be drawn.

Lemma 6.6.14. *Let $K, L, C_1, \ldots, C_{n-2} \in \mathcal{K}^n$ and $\mathcal{C} = (C_1, \ldots, C_{n-2})$. If L is smooth and strictly convex, then*

$$\operatorname{supp} S(K, \mathcal{C}, \cdot) \subset \operatorname{supp} S(L, \mathcal{C}, \cdot).$$

Proof. Suppose the assertion were false. Then we can choose a vector $u_0 \in \operatorname{supp} S(K, \mathcal{C}, \cdot) \setminus \operatorname{supp} S(L, \mathcal{C}, \cdot)$ and an open neighbourhood α of u_0 in S^{n-1} such that $S(L, \mathcal{C}, \alpha) = 0$. Let

$$L' := \{x \in L \mid \langle x, u_0 \rangle \leqq h(L, u_0) - \varepsilon\},$$

where $\varepsilon > 0$ is so small that each outer unit normal vector to L at a point of $\operatorname{cl}(\operatorname{bd} L \setminus L')$ belongs to α; this is possible since L is smooth. For $u \in S^{n-1} \setminus \alpha$, the unique boundary point of L at which u is an outer normal vector is also a boundary point of L'. We deduce that $S(L, \mathcal{C}, \beta) = S(L', \mathcal{C}, \beta)$ for every Borel set $\beta \subset S^{n-1} \setminus \alpha$. This yields

$$0 \leqq V(B, L, \mathcal{C}) - V(B, L', \mathcal{C}) = \frac{1}{n} S(L, \mathcal{C}, \alpha) - \frac{1}{n} S(L', \mathcal{C}, \alpha)$$

and hence $S(L', \mathcal{C}, \alpha) = 0$. Thus $S(L, \mathcal{C}, \cdot) = S(L', \mathcal{C}, \cdot)$ and, therefore, $V(K, L, \mathcal{C}) = V(K, L', \mathcal{C})$. On the other hand, we can choose an open neighbourhood β of u_0 such that $h(L, u) > h(L', u)$ for $u \in \beta$. Since $u_0 \in \operatorname{supp} S(K, \mathcal{C}, \cdot)$, we have $S(K, \mathcal{C}, \beta) > 0$, which implies

$$V(K, L, \mathcal{C}) = \frac{1}{n} \int_{S^{n-1}} h(L, u) \, \mathrm{d}S(K, \mathcal{C}, u)$$

$$> \frac{1}{n} \int_{S^{n-1}} h(L', u) \, \mathrm{d}S(K, \mathcal{C}, u) = V(K, L', \mathcal{C}).$$

This contradiction completes the proof. ■

Instead of the condition appearing in Conjecture 6.6.12, let us now assume that the following is satisfied:

$$h(K, u) = h(L, u) \quad \text{for each } u \in \operatorname{supp} S(\mathcal{C}, B, \cdot).$$

Then for any convex body $M \in \mathcal{K}^n$ we have

$$\frac{1}{n}\int_{S^{n-1}} h(M, u)\,\mathrm{d}[S(K, \mathcal{C}, u) - S(L, \mathcal{C}, u)]$$
$$= V(M, K, \mathcal{C}) - V(M, L, \mathcal{C})$$
$$= \frac{1}{n}\int_{S^{n-1}} [h(K, u) - h(L, u)]\,\mathrm{d}S(M, \mathcal{C}, u)$$
$$= 0,$$

because $\operatorname{supp} S(M, \mathcal{C}, \cdot) \subset \operatorname{supp} S(B, \mathcal{C}, \cdot)$ by Lemma 6.6.14. Since M was arbitrary, we deduce that $S(K, \mathcal{C}, \cdot) = S(L, \mathcal{C}, \cdot)$ and thus equality holds in (6.6.45).

One may conjecture that the converse of this observation is also true:

Conjecture 6.6.15. *If $K, L \in \mathcal{K}^n$, $C_1, \dots, C_{n-2} \in \mathcal{K}_0^n$ and equality holds in* (6.6.45), *then, after applying a suitable homothety to K or L,*

$$h(K, u) = h(L, u) \quad \textit{for each } u \in \operatorname{supp} S(\mathcal{C}, B, \cdot).$$

The truth of Conjectures 6.6.13 and 6.6.15 would imply the truth of Conjecture 6.6.12.

The problem of equality in the Aleksandrov–Fenchel inequality is related to another unsolved problem. This is the equality case for (5.1.23), which expresses the monotoneity of mixed volumes. We shall first prove a special result concerning this problem. It is useful in treating a special case of Conjecture 6.6.12, and its corollary gives a first example of non-trivial equality in (6.6.45). The result deals with the special mixed volumes $V_i(K, L) = V(K[n-i], L[i])$; for $L = B$ these are the quermassintegrals $W_i(K)$. If K, L are convex bodies such that $L \subset K$, then the monotoneity (5.1.23) implies

$$V_0(K, L) \geqq V_1(K, L) \geqq \dots \geqq V_n(K, L). \tag{6.6.46}$$

The following theorem characterizes the pairs K, L for which $V_{p-1}(K, L) = V_p(K, L)$ for some p. For $n = 3$, $p = 2$ it goes back to Minkowski (1911); the general case is essentially due to Favard (1933a).

Theorem 6.6.16. *Let $K, L \in \mathcal{K}_0^n$ be convex bodies satisfying $L \subset K$; let $p \in \{1, \dots, n\}$. Then*

$$V_{p-1}(K, L) = V_p(K, L) \tag{6.6.47}$$

holds if and only if K is an $(n-p)$-tangential body of L; in this case, $V_0(K, L) = V_1(K, L) = \dots = V_p(K, L)$.

Proof. Assume that (6.6.47) holds. First let $p = 1$. As remarked before (5.1.23), $V_0(K, L) = V_1(K, L)$ implies

$$h(K, u) = h(L, u) \quad \text{for } u \in \operatorname{supp} S_{n-1}(K, \cdot).$$

By Theorem 4.6.3, $\operatorname{supp} S_{n-1}(K, \cdot)$ is the closure of the set of extreme unit normal vectors of K, hence each extreme supporting hyperplane of K supports L. By definition, K is an $(n-1)$-tangential body of L.

Now let $p \geqq 1$. For $n = 2$, only $p = 1$ is possible, hence the assertion is true. We assume that $n > 2$ and that the assertion (i.e., sufficiency of (6.6.47)) has been proved in dimension $n-1$. We may also assume $p \geqq 2$. From $V_{p-2}(K, L) \geqq V_{p-1}(K, L) = V_p(K, L)$ and the Aleksandrov–Fenchel inequality we have

$$V_{p-1}(K, L)^2 = V_{p-2}(K, L)V_p(K, L).$$

By Theorem 6.4.1 (where (6.6.47) implies $\alpha = 1$) and (5.3.32) this yields

$$v(K^u[n-p+1], L^u[p-2]) = v(K^u[n-p], L^u[p-1]) \tag{6.6.48}$$

for each $u \in S^{n-1}$.

Let $H(K, v)$ be a $(p-1)$-extreme support plane of K. Choose a unit vector u orthogonal to v and, if $\dim T(K, v) \geqq 2$, in the linear hull of the touching cone $T(K, v)$. Then $H' := H(K, v) \cap H_{u,0}$ is a $(p-2)$-extreme support plane of K^u (relative to $H_{u,0}$). Since $L^u \cap K^u$, we deduce from (6.6.48) and the induction hypothesis that K^u is an $[(n-1)-(p-1)]$-tangential body of L^u. Hence, the $(p-2)$-extreme support plane H' of K^u supports L^u. This implies that $H(K, v)$ supports L. Thus K is an $(n-p)$-tangential body of L. This completes the induction.

Vice versa, suppose that K is an $(n-p)$-tangential body of L for some $p \in \{1, \ldots, n\}$. Then

$$h(K, u) = h(L, u) \quad \text{for } u \in \operatorname{extn}_{p-1} K,$$

where $\operatorname{extn}_j K$ denotes the set of j-extreme unit normal vectors of K. For $0 \leqq j \leqq p-1$, we deduce from Lemma 6.6.14 and Theorem 4.6.3 that

$$\begin{aligned} \operatorname{supp} S(K[n-j-1], L[j], \cdot) &\subset \operatorname{supp} S(K[n-j-1], B[j], \cdot) \\ &= \operatorname{cl} \operatorname{extn}_j K \subset \operatorname{cl} \operatorname{extn}_{p-1} K \end{aligned}$$

and hence that $h(K, u) = h(L, u)$ for u in the support of $S(K[n-j-1], L[j], \cdot)$. Thus

$$\int_{S^{n-1}} h(K, u)\,\mathrm{d}S(K[n-j-1], L[j], u) = \int_{S^{n-1}} h(L, u)\,\mathrm{d}S(K[n-j-1], L[j], u),$$

for $j = 0, \ldots, p-1$, which by (5.1.18) gives

$$V_0(K, L) = V_1(K, L) = \ldots = V_p(K, L), \tag{6.6.49}$$

as asserted. ■

Corollary 6.6.17. *Let $K, L \in \mathcal{K}_0^n$. If $i \in \{1, \ldots, n-1\}$ and K is homothetic to an $(n-i-1)$-tangential body of L, then*

$$V_i(K, L)^2 = V_{i-1}(K, L)V_{i+1}(K, L). \tag{6.6.50}$$

Proof. We may assume that rL is a maximal homothetic copy of L contained in K. By assumption, K is an $(n-i-1)$-tangential body of a homothet of L and necessarily of rL. From (6.4.49), with L replaced by rL, we obtain

$$V_0(K, L) = rV_1(K, L) = \ldots = r^{i+1}V_{i+1}(K, L) \tag{6.6.51}$$

and thus (6.6.50). ∎

The next theorem establishes a converse to Corollary 6.6.17, for $i = 1$. It thus settles a special but important case of the general equality problem for the Aleksandrov–Fenchel inequality. It was conjectured (for $n = 3$) by Minkowski (1903) and proved by Bol (1943b). His proof (which we follow) is a nice application of the powerful method of inner parallel bodies.

Theorem 6.6.18. *Let $L \in \mathcal{K}_0^n$ be an n-dimensional convex body and $K \in \mathcal{K}^n$ an arbitrary convex body. In the inequality*

$$V(K, \ldots, K, L)^2 \geqq V_n(K)V(K, \ldots, K, L, L) \tag{6.6.52}$$

equality holds if and only if either $\dim K < n-1$ *or K is homothetic to an $(n-2)$-tangential body of L.*

Proof. If $\dim K < n-1$, both sides of (6.6.52) are zero. If K is homothetic to an $(n-2)$-tangential body of L, then equality in (6.6.52) is a special case of Corollary 6.6.17. We assume now that equality holds in (6.6.52) and that $\dim K \geqq n-1$. Since $\dim K = n-1$ would imply $V_n(K) = 0$ and $V(K, \ldots, K, L) > 0$, this case cannot occur.

We use the system of parallel bodies K_λ of K relative to L, as defined in Section 3.1. In particular, $K_\lambda = K \sim -\lambda L$ for $-r \leqq \lambda \leqq 0$, where $r = r(K, L)$ denotes the inradius of K relative to L. Let $V_i(\lambda) := V_i(K_\lambda, L) = V(K_\lambda[n-i], L[i])$ and $f_i(\lambda) := V_i(\lambda)^{1/(n-i)}$. By the concavity of the system of parallel bodies (Lemma 3.1.9), the monotoneity of mixed volumes and the general Brunn–Minkowski theorem 6.4.3, the function f_i is concave on $[-r, \infty)$. Therefore, $-f_i$ has the properties collected in Theorem 1.5.2, some of which will be used in the following.

We have

$$\frac{\mathrm{d}V_0(\lambda)}{\mathrm{d}\lambda} = nV_1(\lambda), \tag{6.6.53}$$

which is a very special case of Lemma 6.5.3. For the left derivative of V_1 we write

$$\left(\frac{\mathrm{d}V_1(\lambda)}{\mathrm{d}\lambda}\right)_l =: (n-1)V_2^*(\lambda) \tag{6.6.54}$$

for $\lambda > -r$, and we define

$$\Delta(\lambda) := V_1(\lambda)^2 - V_0(\lambda)V_2^*(\lambda).$$

Then

$$V_2^*(\lambda) \geqq V_2(\lambda) \tag{6.6.55}$$

since

$$V_1(K_\lambda + \alpha L, L) = \sum_{i=0}^{n-1}\binom{n-1}{i}\alpha^i V_{i+1}(\lambda)$$

for $\alpha \geqq 0$; furthermore $K_\lambda + \alpha L \subset K_{\lambda+\alpha}$ by (3.1.12) and hence $V_1(K_\lambda + \alpha L, L) \leqq V_1(\lambda + \alpha)$ for $\alpha \geqq 0$ with equality for $\alpha = 0$, and this gives

$$(n-1)V_2^*(\lambda) = \left[\frac{\mathrm{d}}{\mathrm{d}\alpha}V_1(\lambda+\alpha)\right]_l^{\alpha=0} \geqq \left[\frac{\mathrm{d}}{\mathrm{d}\alpha}V_1(\lambda+\alpha)\right]_r^{\alpha=0}$$

$$\geqq \left\{\frac{\mathrm{d}}{\mathrm{d}\alpha}V_1(K_\lambda + \alpha L, L)\right]_r^{\alpha=0} = (n-1)V_2(\lambda).$$

The fact that f_0 is concave and hence satisfies $[f_0'(\lambda)]_l' \leqq 0$ yields

$$\Delta(\lambda) \geqq 0, \tag{6.6.56}$$

by (6.6.53) and (6.6.54).

Proposition. The function Δ is non-decreasing on $(-r, \infty)$.

For the proof, let $-r < \lambda_1 < \lambda_2$ and assume, on the contrary, that $\Delta(\lambda_2) < \Delta(\lambda_1)$. Then $\Delta(\lambda_1) > 0$ by (6.6.56). Put $b := \inf\{\lambda | \lambda > \lambda_1, \Delta(\lambda) < \Delta(\lambda_1)\}$. We write $f_1 := f$, thus

$$V_1(\lambda) = f(\lambda)^{n-1}. \tag{6.6.57}$$

Since f is concave and increasing,

$$f_l'(\lambda) \geqq f_r'(\lambda) \geqq 0 \quad \text{for } \lambda > -r \tag{6.6.58}$$

and

$$f_r'(\alpha_1) \geqq f_l'(\alpha_2) \quad \text{for } -r < \alpha_1 < \alpha_2. \tag{6.6.59}$$

By (6.6.57) and (6.6.54),

$$V_2^*(\lambda) = f(\lambda)^{n-2} f_l'(\lambda). \tag{6.6.60}$$

If $\lambda_1 < \lambda < b$, then (6.6.60), (6.6.58) and (6.6.59) yield

$$\begin{aligned}\Delta(\lambda_1) \leqq \Delta(\lambda) &= V_1(\lambda)^2 - V_0(\lambda)V_2^*(\lambda)\\ &= V_1(\lambda)^2 - V_0(\lambda)f(\lambda)^{n-2}f_1'(\lambda)\\ &\leqq V_1(\lambda)^2 - V_0(\lambda)f(\lambda)^{n-2}f_r'(\lambda)\\ &\leqq V_1(\lambda)^2 - V_0(\lambda)f(\lambda)^{n-2}f_1'(b).\end{aligned}$$

With $\lambda \to b$ we obtain

$$0 < \Delta(\lambda_1) \leqq V_1(b)^2 - V_0(b)f(b)^{n-2}f_1'(b) = \Delta(b). \quad (6.6.61)$$

Now we consider the function Γ defined by

$$\Gamma(\lambda) := V_1(\lambda)^2 - V_0(\lambda)f(\lambda)^{n-2}f_r'(b)$$

for $\lambda > -r$. By (6.6.58) we have

$$\Delta(b) \leqq \Gamma(b), \quad (6.6.62)$$

and by (6.6.59),

$$\Gamma(\lambda) \leqq \Delta(\lambda) \quad \text{for } \lambda > b. \quad (6.6.63)$$

The function Γ has a right derivative at b, for which we obtain, using (6.6.63) and (6.6.59),

$$\begin{aligned}\Gamma_r'(b) &= (2n-2)f(b)^{2n-3}f_r'(b) - nV_1(b)f(b)^{n-2}f_r'(b)\\ &\quad -V_0(b)(n-2)f(b)^{n-3}f_r'(b)^2\\ &= (n-2)V_1(b)^{-1}f(b)^{n-2}f_r'(b)\Gamma(b) > 0\end{aligned}$$

since $f_r'(b) > 0$ (because $f_1'(\lambda) > 0$ for all $\lambda > -r$ by (6.6.60)) and $\Gamma(b) > 0$ by (6.6.62) and (6.6.61). Hence, there is a number $\delta > 0$ such that

$$\Gamma(\lambda) \geqq \Gamma(b) \quad \text{for } b \leqq \lambda \leqq b + \delta. \quad (6.6.64)$$

For $b < \lambda \leqq b + \delta$ we now deduce from (6.6.63), (6.6.64), (6.6.62) and (6.6.61) that

$$\Delta(\lambda) \geqq \Gamma(\lambda) \geqq \Gamma(b) \geqq \Delta(b) \geqq \Delta(\lambda_1),$$

which contradicts the definition of b. This contradiction proves the proposition.

We have assumed equality in (6.6.52); thus $V_1(0)^2 - V_0(0)V_2(0) = 0$. By (6.6.55) and (6.6.56) this implies $\Delta(0) = 0$ and

$$V_2^*(0) = V_2(0). \quad (6.6.65)$$

The proposition yields $\Delta(\lambda) = 0$ for $\lambda \in (-r, 0)$. According to the deduction of (6.6.56), this is only possible if f_0 is a linear function on $(-r, 0)$. For $\lambda \to -r$, $V_0(\lambda)$ converges to 0 (see (6.5.11)), hence

$$V_n(K_\lambda) = \left(1 + \frac{\lambda}{r}\right)^n V_n(K). \quad (6.6.66)$$

K contains a translate of rL, and after a translation we may assume that

$rL \subset K$. Then $(-\lambda)L + [1 + (\lambda/r)]K \subset K$ for $-r \leqq \lambda \leqq 0$ and hence $[1 + (\lambda/r)]K \subset K_\lambda$; thus

$$\left(1 + \frac{\lambda}{r}\right)^n V_n(K) \leqq V_n(K_\lambda).$$

Since equality holds here, each K_λ is homothetic to K. By Lemma 3.1.10, K is homothetic to a tangential body of L.

From (6.6.65) we have

$$V_1(\lambda) = \left(1 + \frac{\lambda}{r}\right)^{n-1} \frac{1}{r} V_n(K),\ V_2^*(\lambda) = \left(1 + \frac{\lambda}{r}\right)^{n-2} \frac{1}{r^2} V_n(K).$$

For $\lambda = 0$, together with $V_i(0) = V_i(K, L)$, (6.6.66) and (6.6.65) this yields

$$V_0(K, rL) = V_1(K, rL) = V_2(K, rL).$$

Since $rL \subset K$, Theorem 6.6.16 now shows that K is an $(n-2)$-tangential body of rL. This completes the proof. ■

A converse to another special case of Corollary 6.6.17 is contained in the following theorem. It concerns the case where L is a ball. Unfortunately, for a technical reason we have to assume central symmetry for K, which is very probably unnecessary.

Theorem 6.6.19. *If* $K \in \mathcal{K}^n$ *is centrally symmetric and* $i \in \{1, \ldots, n-1\}$, *then the equality*

$$W_i(K)^2 = W_{i-1}(K)W_{i+1}(K) \tag{6.6.67}$$

holds if and only if either $\dim K < n - i$ *or* K *is an* $(n - i - 1)$*-tangential body of a ball.*

Proof. Since $W_i(K) > 0$ if and only if $\dim K \geqq n - i$, we may assume that $\dim K \geqq n - i + 1$. That (6.6.67) holds for $(n - i - 1)$-tangential bodies of balls is a special case of Corollary 6.6.17.

We prove by induction with respect to n that (6.6.67) and the assumption $\dim K \geqq n - i + 1$ imply that K is an $(n - i - 1)$-tangential body of a ball $(i = 1, \ldots, n - 1)$. For $n = 2$ the assertion is true by Theorem 6.2.1. Assume that $n \geqq 3$, that the assertion is true in dimensions less than n and that (6.6.67) holds for some $i \in \{1, \ldots, n-1\}$, where $\dim K \geqq n - i + 1$. If $i = 1$, K is an $(n-2)$-tangential body of a ball by Theorem 6.6.18. Let $i \geqq 2$. From (6.6.67) and Theorem 6.4.1 it follows that, after applying a suitable dilatation to K, we may assume that

$$S_{n-i}(K, \cdot) = S_{n-i-1}(K, \cdot). \tag{6.6.68}$$

By Lemma 6.4.5, this implies

$$w_{i-1}(K^u)^2 = w_{i-2}(K^u)w_i(K^u)$$

for $u \in S^{n-1}$, where we have written

$$w_k(K^u) := v(K^u[n-1-k], B^u[k]).$$

By the induction hypothesis, K^u is an $(n-i-1)$-tangential body of a homothet of B^u. Let r_u be the radius of this homothet. Equations (6.6.51) in dimension $n-1$ give

$$r_u^{i-1} w_{i-1}(K^u) = r_u^i w_i(K^u).$$

From (6.6.68) and (5.3.32) we have $w_{i-1}(K^u) = w_i(K^u)$, thus $r_u = 1$.

Let B_r be the maximal ball in K with centre at the symmetry centre of K; here r denotes the radius of B_r. K and B_r have a common supporting hyperplane H, and by symmetry a second one parallel to it, say H'. Let u be a unit vector parallel to H. The intersection $H \cap H_{u,0}$ is a common support plane (in $H_{u,0}$) of K^u and B_r^u and hence an extreme support plane of K^u. The same holds true for $H' \cap H_{u,0}$. Since both $(n-2)$-planes must support some ball in $H_{u,0}$ of radius 1, it follows that $r = 1$. Hence, we may assume that $B_r = B$. Now let $u \in S^{n-1}$ be arbitrary. The projection K^u is an $(n-i-1)$-tangential body of a unit ball and necessarily of B^u.

The argument is completed in a similar way to the proof of Theorem 6.6.16. Let $H(K, v)$ be an i-extreme support plane of K. We choose a unit vector u orthogonal to v and, if $\dim T(K, v) \geqq 2$, in the linear hull of $T(K, v)$. $H' := H(K, v) \cap H_{u,0}$ is then an $(i-1)$-extreme support plane of K^u. Since K^u is an $(n-i-1)$-tangential body of B^u, we deduce that H' supports B^u and hence $H(K, v)$ supports B. Thus K is an $(n-i-1)$-tangential body of B. ■

The general conjecture 6.6.12, with distinct bodies $C_1, \ldots, C_{n-2}$, has so far only been verified under strong restrictions for these bodies. One example is given by the following result, which extends the equality assertion of Theorem 6.3.2.

Theorem 6.6.20. *Conjecture* 6.6.12 *is true if* $C_1, \ldots, C_{n-2}$ *are strongly isomorphic simple polytopes (and* K, L *are arbitrary convex bodies).*

Proof. Let $P_3, \ldots, P_n \in \mathcal{K}_0^n$ be strongly isomorphic simple polytopes. We have to apply some results from Sections 5.1, 5.2 and 6.3 and therefore use the notation introduced there. In particular, $\mathcal{A}$ is the a-type of $P_3, \ldots, P_n$ and $P \in \mathcal{A}$ is an arbitrary member. Further, $u_1, \ldots, u_N$, F_i, F_{ij}, v_{ij}, Θ_{ij}, J, $F_i^{(r)} := F(P_r, u_i)$ etc. are defined as in Section 5.1. We abbreviate $(P_3, \ldots, P_n)$ by $\mathcal{C}$.

Let $K, L \in \mathcal{K}^n$ be convex bodies. After applying a dilatation to K or L we may assume that

$$V(K, K, \mathcal{C}) = V(L, L, \mathcal{C}). \tag{6.6.69}$$

We may also assume that $V(K, L, \mathcal{C}) > 0$, since otherwise K and L are parallel segments (possibly degenerate), and the assertion is trivial. By Theorem 6.4.1, equality in the inequality

$$V(K, L, \mathcal{C})^2 \geqq V(K, K, \mathcal{C})V(L, L, \mathcal{C}) \tag{6.6.70}$$

holds if and only if

$$S(K, \mathcal{C}, \cdot) = S(L, \mathcal{C}, \cdot). \tag{6.6.71}$$

We must, therefore, examine the measure $S(K, P_3, \ldots, P_n, \cdot)$. First,

$$S(K, P_3, \ldots, P_n, \{u_i\}) = v(F(K, u_i), F_i^{(3)}, \ldots, F_i^{(n)}) \tag{6.6.72}$$

for $i = 1, \ldots, N$, by the definition of the mixed area measure. For $(i, j) \in J$ (recall that this means that $F_{ij} := F(P, u_i) \cap F(P, u_j)$ is an $(n-2)$-face of P), we denote by σ_{ij} the spherical image of F_{ij} and by π_{ij} the orthogonal projection onto the two-dimensional linear subspace L_{ij} orthogonal to F_{ij}. Observe that σ_{ij} and π_{ij} depend only on (i, j) and the a-type $\mathcal{A}$.

Let $\omega \in \mathcal{B}(S^{n-1})$ be a Borel set. If $\omega \subset \operatorname{relint} \sigma_{ij}$, then

$$S(K, P_3, \ldots, P_n, \omega) = v^{(n-2)}(F_{ij}^{(3)}, \ldots, F_{ij}^{(n)})s(\pi_{ij}K, \omega \cap L_{ij}), \tag{6.6.73}$$

where s denotes the area measure in L_{ij}. If $\omega \cap \sigma_{ij} = \emptyset$ for all $(i, j) \in J$, then

$$S(K, P_3, \ldots, P_n, \omega) = 0. \tag{6.6.74}$$

The last two equalities are proved by computing $S_{n-1}(\lambda K + \lambda_3 P_3 + \ldots + \lambda_n P_n, \omega)$ for $\lambda, \lambda_3, \ldots, \lambda_n > 0$, using (4.2.24) and Fubini's theorem, and then applying (5.1.17).

Now we assume that equality in (6.6.70) and hence (6.6.71) hold. Suppose that $(i, j) \in J$ and that $\omega \subset \operatorname{relint} \sigma_{ij}$ is a Borel set. Then (6.6.73) yields $s(\pi_{ij}K, \omega) = s(\pi_{ij}L, \omega)$. By the local uniqueness theorem 4.3.2, the convex arcs in $\pi_{ij}K$ and $\pi_{ij}L$ that are the reverse spherical images of $\operatorname{relint} \sigma_{ij}$ must be translates of each other. Hence, there exists a vector $t_{ij} \in \mathbb{E}^n$, orthogonal to F_{ij}, such that

$$h(K, u) - h(L, u) = \langle t_{ij}, u \rangle \quad \text{for } u \in \sigma_{ij} \tag{6.6.75}$$

(first for $u \in \operatorname{relint} \sigma_{ij}$, then by continuity for $u \in \sigma_{ij}$). Define

$$h(K, u_i) - h(L, u_i) = \langle t_{ij}, u_i \rangle =: \alpha_i,$$

$$h(F(K, u_i), v_{ij}) - h(F(L, u_i), v_{ij}) =: \lambda_{ij}$$

for $i \in \{1, \ldots, N\}$ and $(i, j) \in J$. From (6.6.71), (6.6.72) and (5.1.18) we deduce that, for each fixed $i \in \{1, \ldots, N\}$,

$$\sum_{(i,j)\in J} \lambda_{ij} v^{(n-2)}(F_{ij}^{(3)}, \ldots, F_{ij}^{(n)}) = 0. \tag{6.6.76}$$

Taking directional derivatives in (6.6.75), we infer from Theorem 1.7.2 that $\lambda_{ij} = \langle t_{ij}, v_{ij} \rangle$. Since t_{ij} is a linear combination of u_i and v_{ij}, we deduce that $t_{ij} = \alpha_i u_i + \lambda_{ij} v_{ij}$. As $t_{ij} = t_{ji} = \alpha_j u_j + \lambda_{ji} v_{ji}$, multiplication by u_j yields

$$\lambda_{ij} = \alpha_j \operatorname{cosec} \Theta_{ij} - \alpha_i \cot \Theta_{ij} \tag{6.6.77}$$

for $(i, j) \in J$. Writing $Z := (\alpha_1, \ldots, \alpha_N)$, we see from (5.2.6) that

$$\Lambda_i Z = (\lambda_{i1}, \ldots, \lambda_{iN})$$

with $\lambda_{ij} := 0$ if $(i, j) \notin J$. With φ_i as defined in the proof of Theorem 6.3.2, we have

$$\begin{aligned} \varphi_i(Z, P_3) &= v(\Lambda_i Z, F_i^{(3)}, \ldots, F_i^{(n)}) \\ &= \frac{1}{n-1} \sum_{(i,j)\in J} \lambda_{ij} v^{(n-2)}(F_{ij}^{(3)}, \ldots, F_{ij}^{(n)}), \end{aligned}$$

hence $\varphi_i(Z, P_3) = 0$ by (6.6.76). This means that Z is an eigenvector with eigenvalue 0 of the bilinear form Φ used in the proof of Theorem 6.3.2; hence from that proof (Proposition 3) it follows that Z is the support vector of a point. Thus there exists a vector $z \in \mathbb{E}^n$ such that $\alpha_i = \langle z, u_i \rangle$ for $i = 1, \ldots, n$. After applying a suitable translation to K, we may assume that $z = o$. Then $\lambda_{ij} = 0$ by (6.6.77) and hence $t_{ij} = o$; thus (6.6.75) now reads

$$h(K, u) = h(L, u) \quad \text{for } u \in \sigma_{ij}, (i, j) \in J. \tag{6.6.78}$$

The $(P_3, \ldots, P_n, B)$-extreme unit normal vectors are precisely the vectors in $\bigcup_{(i,j)\in J} \sigma_{ij}$.

Vice versa, from (6.6.78) we deduce (6.6.71) and hence equality in (6.6.70). This completes the proof. ■

Another case where the truth of Conjecture 6.6.12 could be established is the following result (Schneider 1988a), which we formulate here without proof.

Theorem 6.6.21. *If* $K, L \in \mathcal{K}_0^n$ *are centrally symmetric and* $C_1, \ldots, C_{n-2} \in \mathcal{K}_0^n$ *are zonoids, then Conjecture* 6.6.12 *holds for these bodies.*

Notes for Section 6.6

1. Lemma 6.6.1 is due to Süss (1932a); proofs also appear in Aleksandrov (1937b) and Leichtweiß [23], pp. 241–3. Corresponding stability results were proved by Groemer (1987a); see also Groemer (1991b).
2. The first proof given above for Theorem 6.6.2 is essentially due to Favard (1933a) (compare also Leichtweiß 1981). The second proof, using Lemma

6.6.3, extends a method of Kubota (1925b). Theorem 6.6.6 was proved by Schneider (1989c) and independently by Goodey & Groemer (1990). The stability estimate (6.6.10) was derived in Groemer & Schneider (1991).

3. Theorems 6.6.7 and 6.6.8 and their proofs are taken from Schneider (1990a). The proof of Theorem 6.6.9 (given Theorem 6.6.8) is modelled on the procedure of Fenchel (1936b). Theorem 6.6.10 appears in Schneider (1990b). It is a quantitative improvement of a result obtained earlier in Schneider (1985a).
4. *Equality cases in the Aleksandrov–Fenchel inequality*. Conjectures 6.6.12, 6.6.13 and 6.6.15 and Lemma 6.6.14 were formulated in Schneider (1985a), where credit for 6.6.15 and a special case of 6.6.14 is given to A. Loritz. Conjecture 6.6.12 states that the equality
$$V(K, L, \mathcal{C})^2 = V(K, K, \mathcal{C})V(L, L, \mathcal{C}),$$
where $K, L \in \mathcal{K}^n$, $\mathcal{C} = (C_1, \ldots, C_{n-2})$ and $C_1, \ldots, C_{n-2} \in \mathcal{K}_0^n$ holds if and only if suitable homothets of K and L have the same $(\mathcal{C}, B)$-extreme supporting hyperplanes. When this conjecture was formulated, the following classical cases in favour of it were known.
   (a) $\mathcal{C}$ consists of balls. This is the case mentioned in Note 2, first proved by Kubota (1925b), with a different proof given by Favard (1933a).
   (b) $\mathcal{C}$ consists of strongly isomorphic simple polytopes, and K, L are polytopes having the same system of normal vectors to the facets as the polytopes of $\mathcal{C}$. This comes out in Aleksandrov's first proof for the Aleksandrov–Fenchel inequality, as presented in Section 6.3. Theorem 6.6.20 extends this result to arbitrary convex bodies K, L.
   (c) $\mathcal{C}$ consists of bodies of class C_+^2, and K, L have support functions of class C^2. This case is a byproduct of Aleksandrov's (1938b) second proof and was also noted by Favard (1938).
   (d) $C_1 = \ldots = C_{n-2} = K$. This is the case treated in Theorem 6.6.18, which is due to Bol (1943b). Corollary 6.6.17 (for $n = 3$) had already been discovered by Minkowski (1903), §7. A special case of Theorem 6.6.18 was treated by Knothe (1949) in a different way.
5. *Tangential bodies*. Theorem 6.6.16, as mentioned, is due to Favard (1933a), but we found it necessary to give a different proof for the necessity of condition (6.6.47).

   If $K \in \mathcal{K}_0^n$ is a convex body with inradius r, then its quermassintegrals satisfy
$$W_{n-2}(K) - rW_{n-1}(K) \geqq 0.$$
   If equality holds here, then Theorem 6.6.16 implies that K is a cap body of a ball. Sangwine-Yager (1991) obtained an explicit estimate showing that K must be close to some cap body of a ball if $W_{n-2}(K) - rW_{n-1}(K)$ is small.

   Theorem 6.6.19 is taken from Schneider (1985a).
6. The proof of Theorem 6.6.20 was given in Schneider (1991+). The second part of the proof, in which it is shown that the vectors t_{ij} associated with the $(n-2)$-faces F_{ij} of P are all equal to a fixed vector, is similar to (but different from) a method of Aleksandrov [2], Chapter XI, §3, which shows the infinitesimal rigidity of convex polytopes in $\mathbb{E}^3$ with the aid of mixed volumes. A connection between rigidity of polytopes and mixed volumes was first exploited by Weyl (1917); his proof is sketched in Efimov (1957), §100.

6.7. Linear inequalities

Under certain circumstances, the quadratic Aleksandrov–Fenchel inequality

$$V(K, L, \mathscr{C})^2 \geqq V(K, K, \mathscr{C})V(L, L, \mathscr{C}), \tag{6.7.1}$$

where $K, L, K_3, \ldots, K_n \in \mathscr{K}^n$ and $\mathscr{C} = (K_3, \ldots, K_n)$, admits the linear improvement

$$2V(K, L, \mathscr{C}) \geqq V(K, K, \mathscr{C}) + V(L, L, \mathscr{C}). \tag{6.7.2}$$

A sufficient condition for the validity of (6.7.2) is that there exist a convex body M fulfilling the conditions

$$V(K, M, \mathscr{C}) = V(L, M, \mathscr{C}) > 0. \tag{6.7.3}$$

This follows immediately from (6.4.22). For example, M could be a segment. If u is a unit vector parallel to it, then, by (5.3.31), condition (6.7.3) is equivalent to

$$v(K^u, \mathscr{C}^u) = v(L^u, \mathscr{C}^u) > 0.$$

A special case of the latter inequality occurs when $K^u = L^u$, that is, K and L have the same circumscribed cylinder of direction u. In this case, the Brunn–Minkowski theorem for K and L also holds, in an improved version: The volume of $(1 - \vartheta)K + \vartheta L$, and not only its nth root, is a concave function of ϑ for $0 \leqq \vartheta \leqq 1$. This is clear from Fubini's theorem and the fact that each line G of direction u satisfies

$$G \cap [(1 - \vartheta)K + \vartheta L] \supset (1 - \vartheta)(G \cap K) + \vartheta(G \cap L).$$

Less immediate is the fact that the general Brunn–Minkowski theorem (Theorem 6.4.3) also admits an improved version of this type. In the following we describe a theory of linear inequalities for mixed volumes of bodies inscribed in a cylinder; this unifies some special results in the literature obtained by other methods.

We assume that a unit vector $u \in S^{n-1}$ and an $(n-1)$-dimensional convex body C in the hyperplane $H_{u,0}$ orthogonal to u are given. Then we define

$$\mathscr{K}_C := \{K \in \mathscr{K}^n | K^u = C\}$$

and call such a set of convex bodies a *canal class*. The $(n-1)$-dimensional mixed volume of convex bodies in $H_{u,0}$ is denoted by v. If U is a segment of length one parallel to u, then

$$v(K_1^u, \ldots, K_{n-1}^u) = nV(K_1, \ldots, K_{n-1}, U) \tag{6.7.4}$$

by (5.3.31). The mixed area measure in $H_{u,0}$ is denoted by s and, without loss of generality, considered as a measure on S^{n-1} that is concentrated on the great subsphere

$$s_u^{n-2} := S^{n-1} \cap H_{u,0}.$$

Then we have

$$s(K_1^u, \ldots, K_{n-2}^u, \cdot) = (n-1)S(K_1, \ldots, K_{n-2}, U, \cdot) \tag{6.7.5}$$

on S^{n-1}. This follows from (5.1.17) and the equality

$$S_{n-1}(K + \lambda U, \cdot) = S_{n-1}(K, \cdot) + \lambda s(K^u, \ldots, K^u, \cdot),$$

which is clear from (4.2.24), if one then replaces K by $\lambda_1 K_1 + \ldots + \lambda_{n-2} K_{n-2}$ and compares the coefficients.

Theorem 6.7.1. *Let $\mathcal{C}$ be an $(n-2)$-tuple of convex bodies in $\mathcal{K}^n$ satisfying*

$$v(C, \mathcal{C}^u) > 0. \tag{6.7.6}$$

For K, L in the canal class $\mathcal{K}_C$, the inequality

$$V(K, K, \mathcal{C}) - 2V(K, L, \mathcal{C}) + V(L, L, \mathcal{C}) \leqq 0 \tag{6.7.7}$$

is true. Here equality holds if and only if

$$S(K, \mathcal{C}, \cdot) = S(L + aU, \mathcal{C}, \cdot) \tag{6.7.8}$$

or

$$S(L, \mathcal{C}, \cdot) = S(K + aU, \mathcal{C}, \cdot) \tag{6.7.9}$$

with some number $a \geqq 0$.

Proof. By (6.7.4) and (6.7.6), each convex body $A \in \mathcal{K}_C$ satisfies

$$nV(A, U, \mathcal{C}) = v(A^u, \mathcal{C}^u) = v(C, \mathcal{C}^u) > 0.$$

In particular, $V(K, U, \mathcal{C}) = V(L, U, \mathcal{C}) > 0$, hence (6.4.22) with $M = U$ yields inequality (6.7.7) (observe that (6.4.20) is not necessary for the validity of (6.4.22), as remarked after the formulation of Theorem 6.4.2). As shown in Section 6.4, equality in (6.4.22) is equivalent to (6.4.26), with a suitable number k. The numbers q_1, q_2 occurring there are, in our present case, given by $q_1 = q_2 = v(C, \mathcal{C}^u)/n$. Hence, the first equation of (6.4.26) is equivalent to (6.7.8) with suitable $a \geqq 0$, and the second equation of (6.4.26) is equivalent to (6.7.9). ■

Next we show that for canal classes the general Brunn–Minkowski theorem 6.4.3 holds in a stronger form, and there is also an analogue of Theorem 6.4.4. Let a number $m \in \{2, \ldots, n\}$, an $(n-m)$-tuple $\mathcal{C} = (K_{m+1}, \ldots, K_n)$ of convex bodies $K_{m+1}, \ldots, K_n \in \mathcal{K}^n$, and two bodies K_0, $K_1 \in \mathcal{K}_C$ be given. We use the notation of Section 6.4, in particular

$$V_{(i)} := V(K_0[m-i], K_1[i], \mathcal{C}) \qquad \text{for } i = 0, \ldots, m,$$

$$S_{(i)} := S(K_0[m-1-i], K_1[i], \mathcal{C}, \cdot) \qquad \text{for } i = 0, \ldots, m-1,$$

and in addition we define a measure γ on S^{n-1} by

$$\gamma := s(C[m-2], \mathcal{C}^u, \cdot).$$

By (6.7.7), the sequence $(V_{(0)}, \dots, V_{(m)})$ is concave, hence (6.4.1) tells us that, for $0 \leqq i < j < k \leqq m$,

$$(k - j)V_{(i)} + (i - k)V_{(j)} + (j - i)V_{(k)} \leqq 0. \tag{6.7.10}$$

This is a stronger version of (6.4.5). The following result is the general Brunn–Minkowski theorem for canal classes.

Theorem 6.7.2. *Let a number $m \in \{2, \dots, n\}$ and convex bodies $K_0, K_1 \in \mathcal{K}_C$, $K_{m+1}, \dots, K_n \in \mathcal{K}^n$ be given and define $\mathscr{C} := (K_{m+1}, \dots, K_n)$, $K_\lambda := (1 - \lambda)K_0 + \lambda K_1$ and*

$$f(\lambda) := V(K_\lambda[m], \mathscr{C})$$

for $0 \leqq \lambda \leqq 1$. Then f is a concave function on $[0, 1]$.

Under the assumption

$$v(C[m - 1], \mathscr{C}^u) > 0$$

the following assertions are equivalent:

(a) *the function f is linear;*

(b) *the sequence $(V_{(0)}, \dots, V_{(m)})$ is linear, that is,*
$$V_{(0)} - V_{(1)} = V_{(1)} - V_{(2)} = \dots = V_{(m-1)} - V_{(m)};$$

(c) $(m - 1)V_{(0)} - mV_{(1)} + V_{(m)} = 0$;

(d) $S_{(k-1)} - S_{(k)} = a\gamma$ *for $k = 1, \dots, m - 1$ with a constant a;*

(e) $S_{(0)} = S_{(m-1)}$ *on $S^{n-1} \backslash s_u^{n-2}$.*

Proof. We have

$$f''(0) = m(m - 1)[V_{(0)} - 2V_{(1)} + V_{(2)}],$$

hence $f''(0) \leqq 0$ by (6.7.7). As in the proof of Theorem 6.4.3, we deduce that f is concave.

The concavity of f yields $f'(0) \geqq f(1) - f(0)$, thus

$$(m - 1)V_{(0)} - mV_{(1)} + V_{(m)} \leqq 0, \tag{6.7.11}$$

with equality if and only if f is linear. Hence (a) and (c) are equivalent.

Inequality (6.7.11) is also a special case of (6.7.10) and hence equality holds here (see the remark after (6.4.1)) if and only if equality holds in $V_{(k-1)} - 2V_{(k)} + V_{(k+1)} \leqq 0$ for $k = 1, \dots, m - 1$. Thus (b) and (c) are equivalent.

Suppose that (b) holds. By Theorem 6.7.1, this implies

$$S_{(k-1)} - S_{(k)} = a_k\gamma \tag{6.7.12}$$

for $k = 1, \dots, m - 1$, with a constant a_k possibly depending on k. Integrating the support function of K_1 with the measure given by (6.7.12) we obtain

$$V_{(k)} - V_{(k+1)} = \frac{n - 1}{n} a_k v(C[m - 1], \mathscr{C}^u).$$

For $k \leqq m-2$ we integrate the support function of K_0 with the measure given by (6.7.12), but with k replaced by $k+1$, and obtain

$$V_{(k)} - V_{(k+1)} = \frac{n-1}{n} a_{k+1} v(C[m-1], \mathscr{C}^u).$$

Thus $a_1 = \ldots = a_{m-1}$, and (d) holds. Trivially, (d) implies (e), since γ is concentrated on s_u^{n-2}.

Suppose that (e) holds. Then

$$\int_{S^{n-1}\setminus s_u^{n-2}} [h(K_0, v) - h(K_1, v)]\, \mathrm{d}(S_{(0)} - S_{(m-1)})(v) = 0.$$

Since $h(K_0, v) = h(K_1, v)$ for $v \in s_u^{n-2}$, we also have

$$\int_{s_u^{n-2}} [h(K_0, v) - h(K_1, v)]\, \mathrm{d}(S_{(0)} - S_{(m-1)})(v) = 0.$$

Thus

$$V_{(0)} - V_{(1)} - V_{(m-1)} + V_{(m)} = 0.$$

Adding (6.7.11) and the inequality

$$(m-1)V_{(m)} - mV_{(m-1)} + V_{(0)} \leqq 0,$$

which also follows from (6.7.10), we see that equality holds in (6.7.11), which is condition (c). This completes the proof of Theorem 6.7.2. ■

There is also a counterpart to Theorem 6.6.9. We say that the convex bodies $K_0, K_1 \in \mathscr{K}_C$ are *equivalent by telescoping* if $K_0 = K_1 + \lambda U$ or $K_1 = K_0 + \lambda U$ with $\lambda \geqq 0$, where U is a unit segment parallel to u.

Theorem 6.7.3. *If the assumptions of Theorem 6.7.2 are satisfied and if $K_{m+1}, \ldots, K_n$ are smooth, then the conditions* (a) *to* (e) *of Theorem 6.7.2 hold if and only if K_0 and K_1 are equivalent by telescoping.*

Proof. It is clear that condition (a) is satisfied if K_0 and K_1 are equivalent by telescoping. To prove the other direction, we use induction over m. Assume that conditions (a)–(e) of Theorem 6.7.2 are fulfilled. First let $m = 2$. Then condition (d) can be written in the form

$$S(K_0, \mathscr{C}, \cdot) - S(K_1, \mathscr{C}, \cdot) = a(n-1)S(U, \mathscr{C}, \cdot).$$

If $a \geqq 0$, this is equivalent to

$$S(K_0, \mathscr{C}, \cdot) = S(K_1 + a(n-1)U, \mathscr{C}, \cdot).$$

By Theorems 6.4.1 and 6.6.8, the bodies K_0 and $K_1 + a(n-1)U$ are homothetic, and since $K_0, K_1 \in \mathscr{K}_C$, this implies that K_0 and K_1 are equivalent by telescoping. If $a < 0$, the argument is similar.

Now assume that $m \geqq 3$ and that the assertion of the theorem is true for $m - 1$. By condition (d),

$$S_{(k-1)} - S_{(k)} = a\gamma \quad \text{for } k \in \{1, \ldots, m-1\}.$$

Integrating the support function of the unit ball B, we obtain

$$V(K_0[m-k], K_1[k-1], B, \mathscr{C}) - V(K_0[m-k-1], K_1[k], B, \mathscr{C})$$
$$= \frac{n-1}{n} av(C[m-2], B^u, \mathscr{C}^u),$$

which is independent of k. Thus, writing

$$\bar{V}_{(k)} := V(K_0[m-1-k], K_1[k], B, \mathscr{C}),$$

we have

$$\bar{V}_{(0)} - \bar{V}_{(1)} = \bar{V}_{(1)} - \bar{V}_{(2)} = \ldots = \bar{V}_{(m-2)} - \bar{V}_{(m-1)}.$$

By the induction hypothesis, this implies that K_0 and K_1 are equivalent by telescoping. ∎

Notes for Section 6.7

1. As remarked above, for $K_0, K_1 \in \mathscr{K}_C$ the improved version of the Brunn–Minkowski theorem is easily obtained; it says that $V_n((1-\vartheta)K_0 + \vartheta K_1)$ is a concave function of ϑ on $[0, 1]$. As in the proof of Theorem 6.2.1, this implies the inequalities

$$(n-1)V_{(0)} - nV_{(1)} + V_{(n)} \leqq 0, \tag{6.7.13}$$
$$V_{(0)} - 2V_{(1)} + V_{(2)} \leqq 0 \tag{6.7.14}$$

for $V_{(i)} := V(K_0[n-i], K_1[i])$; see Bonnesen & Fenchel [8], p. 94. If instead of $K_0^u = K_1^u$ one assumes only that

$$V_{n-1}(K_0^u) = V_{n-1}(K_1^u), \tag{6.7.15}$$

then, as described in Bonnesen & Fenchel [8], p. 95, a volume-preserving symmetrization procedure transforms $K_\vartheta = (1-\vartheta)K_0 + \vartheta K_1$ into K'_ϑ such that $(K'_0)^u = (K'_1)^u$ and

$$V_n(K'_\vartheta) \geqq (1-\vartheta)V_n(K') + \vartheta V_n(K'_1).$$

It follows that (6.7.13) holds also under the assumption (6.7.15). However, $V_n(K'_\vartheta)$ need not be a concave function of ϑ, so that (6.7.14) cannot be deduced if merely (6.7.15) is assumed, contrary to an assertion made in Bonnesen & Fenchel [8], p. 95. This error was pointed out by Diskant (1984), who constructed a counterexample.

2. The results and proofs of this section are taken from a paper (Schneider 1989b) that partially extends ideas of Favard (1933a). The above investigation, together with the remark that (6.7.3) implies (6.7.2), unifies and generalizes several special results on linear improvements of inequalities for mixed volumes scattered in the literature. These are obtained by different methods. Some particular results are not covered by the above. We refer the reader to Bonnesen & Fenchel [8], p. 99, Geppert (1937), Bol (1939), Hadwiger [18], (6.4.4) and p. 279, and Dinghas (1949a).

3. *An isoperimetric problem of Hadwiger*. As mentioned in Section 6.2, Note 7, Hadwiger (1968b) proved that the volume and surface area of a half-body satisfy

$$2S^n \geqq n^n(\kappa_n + 2\kappa_{n-1})V_n^{n-1}, \qquad (6.7.16)$$

and that equality holds if K is the union of a half-ball of radius ρ, say, and a right cylinder of height ρ attached to it. Hadwiger did not prove that equality holds only in this case. The following proof of (6.7.16) proceeds in a different way and settles the equality case. Moreover, it could be generalized to quermassintegrals other than the surface area.

As Hadwiger did, we consider, more generally, the class $\mathcal{K}_S$ of convex bodies K with the property that K has a supporting hyperplane H such that

$$V_{n-1}(K \cap H) \geqq V_{n-1}(K \cap H')$$

for each hyperplane H' parallel to H. Let $K \in \mathcal{K}_S$ be given and let H be as above. By Schwarz symmetrization with respect to a line G orthogonal to H we obtain from K a convex body K_0 with $V_n(K_0) = V_n(K)$ and $S(K_0) \leqq S(K)$; here strict inequality holds unless K is a body of revolution with axis parallel to G. Moreover, $K_0 \in \mathcal{K}_S$; hence K_0 belongs to the canal class $\mathcal{K}_C$, where $C = H \cap K_0$ is an $(n-1)$-dimensional ball, of radius, say, ρ. By K_1 we denote the half-ball of radius ρ with basis C and such that $K_1 \subset H^-$, where H^- is the halfspace bounded by H that contains K_0.

By (6.7.10) and Theorems 6.7.2 and 6.7.3, with $m = n$ and empty $\mathcal{C}$, we have

$$(n-1)V_{(0)} - nV_{(1)} + V_{(n)} \leqq 0,$$

with equality if and only if K_0 and K_1 are equivalent by telescoping. Now

$$V_{(0)} = V_n(K_0),$$
$$V_{(n)} = V_n(K_1) = \tfrac{1}{2}\kappa_n\rho^n,$$
$$V_{(1)} = \frac{1}{n}\int_{S^{n-1}} h(K_1, u)\,\mathrm{d}S_{n-1}(K_0, u)$$
$$= \frac{1}{n}\rho[S(K_0) - \kappa_{n-1}\rho^{n-1}].$$

The latter equality is easily seen if one chooses the origin at the centre of C. With $V_n(K_0) = V_n(K) =: V_n$ and $S(K_0) \leqq S(K) =: S$ we obtain

$$(n-1)V_n - \rho S + \rho^n(\tfrac{1}{2}\kappa_n + \kappa_{n-1}) \leqq 0.$$

Writing the inequality

$$(n-1)a - n\rho b + \rho^n c \leqq 0 \qquad (6.7.17)$$

in the form

$$\rho b \geqq \frac{n-1}{n}a + \frac{1}{n}\rho^n c,$$

taking logarithms and using the concavity of log, we obtain $b^n \geqq a^{n-1}c$, with equality if and only if it holds in (6.7.17) and $a = \rho^n c$. Hence, inequality (6.7.16) follows. If equality holds in (6.7.16), then $S = S(K_0)$, K_0 and K_1 are equivalent by telescoping, and $V_n = \rho^n(\frac{1}{2}\kappa_n + \kappa_{n-1})$. The second condition implies that K_0 is, up to a translation, the union of K_1 and a circular cylinder attached to C, and the third condition implies that the cylinder has height ρ. Finally, $S = S(K_0)$ yields that K must be a translate of K_0.

6.8. Analogous notions and inequalities

In this section we describe, without proofs, a number of inequalities that are in analogy, and often related, to the inequalities for mixed volumes, either in their general form or in their specialized versions for quermassintegrals. The results to be considered are of a rather different type, the unifying viewpoint being simply that they are reminiscent, often strikingly, of the inequalities for mixed volumes.

First we quote Aleksandrov's inequalities for mixed discriminants. (For a definition of the mixed discriminant $D(A_1, \ldots A_n)$ see Section 2.5, in particular (2.5.32).)

Theorem 6.8.1 (Aleksandrov). *Let $A_1, \ldots, A_n$ be real symmetric $n \times n$ matrices, where $A_2, \ldots, A_n$ are positive definite. Then*

$$D(A_1, A_2, A_3, \ldots, A_n)^2 \geqq D(A_1, A_1, A_3, \ldots, A_n)D(A_2, A_2, A_3, \ldots, A_n). \quad (6.8.1)$$

Equality holds if and only if $A_1 = \lambda A_2$ with a real number λ.

Aleksandrov (1938b) proved these inequalities and used them in his second proof of the Aleksandrov–Fenchel inequalities for mixed volumes.

The major part of this section will be concerned with the description of various analogues of notions and inequalities from the classical Brunn–Minkowski theory. There are counterparts to Minkowski addition, mixed volumes, quermassintegrals and to the inequalities connecting these. Often the analogies are surprising, although many of the inequalities go in the opposite direction to that of their analogues, and are easy consequences of familiar analytic inequalities (those of Hölder, Minkowski). But there are also some unsolved problems and interesting connections with the affine geometry of convex bodies (see Section 7.4).

Most of the functionals of convex bodies considered subsequently will not be translation invariant and will require the origin to be an interior point of the body. We introduce, therefore, the class $\mathcal{K}^n_{00}$ of convex bodies that have the origin o as an interior point. A centrally symmetric convex body with centre at the origin will be called *centred*. Two convex bodies $K, L \in \mathcal{K}^n_{00}$ are said to be *equivalent by dilatation* if $K = \lambda L$ with some $\lambda > 0$.

Some of the notions defined below can be applied to lower-dimensional convex bodies and star bodies, but we do not treat these here.

First, we consider extensions or analogues of Minkowski addition. Let $p \geqq 1$ be given. For $K, L \in \mathcal{K}^n_{00}$ and $\lambda, \mu \geqq 0$ $(\lambda + \mu > 0)$ there is a convex body $\lambda \cdot K +_p \mu \cdot L$ for which

$$h(\lambda \cdot K +_p \mu \cdot L, \cdot)^p = \lambda h(K, \cdot)^p + \mu h(L, \cdot)^p. \quad (6.8.2)$$

Such p-sums were introduced by Firey (1962). He proved the extended

Brunn–Minkowski inequality

$$W_i(K +_p L)^{p/(n-i)} \geqq W_i(K)^{p/(n-i)} + W_i(L)^{p/(n-i)} \qquad (6.8.3)$$

for $i \in \{0, 1, \ldots, n-1\}$ and $p \geqq 1$. For $p = 1$ and $i = 0$, this is the ordinary Brunn–Minkowski inequality. For $p > 1$, equality holds in (6.8.3) if and only if K and L are equivalent by dilatation.

Lutwak (1991+a) studied p-sums more systematically. He defined mixed p-quermassintegrals by

$$\frac{n-i}{p} W_{p,i}(K, L) := \lim_{\varepsilon \to 0+} \frac{W_i(K +_p \varepsilon \cdot L) - W_i(K)}{\varepsilon}$$

$(i = 0, \ldots, n-1)$, showed that

$$W_{p,i}(K, L) = \frac{1}{n} \int_{S^{n-1}} h(L, u)^p h(K, u)^{1-p} \, \mathrm{d}S_{n-1-i}(K, u)$$

and deduced (6.8.3) in a new way. He was also able to prove an analogue of Minkowski's existence theorem (Theorem 7.1.2) and a corresponding uniqueness result.

If in definition (6.8.2) one uses gauge functions instead of support functions, then another combination of convex bodies is obtained. It was introduced by Firey (1961a). Let $p \geqq 1$ be given. For $K, L \in \mathcal{K}^n_{00}$ and λ, $\mu \geqq 0$ $(\lambda + \mu > 0)$, there is a convex body $\lambda \bullet K +_p \mu \bullet L)$ for which

$$g(\lambda \bullet K +_p \mu \bullet L, \cdot)^p = \lambda g(K, \cdot)^p + \mu g(L, \cdot)^p. \qquad (6.8.4)$$

Theorem 1.7.6 (together with Theorem 1.6.1) shows that

$$\lambda \bullet K +_p \mu \bullet L = (\lambda \bullet K^* +_p \mu \bullet L^*)^*.$$

Firey proved that

$$V_n(K +_p L)^{-p/n} \geqq V_n(K)^{-p/n} + V_n(L)^{-p/n}, \qquad (6.8.5)$$

with equality if and only if K and L are equivalent by dilatation.

The special case $p = 1$ of (6.8.4) was studied further by Firey (1961b), who showed that

$$W_i(K +_1 L)^{-1/(n-i)} \geqq W_i(K)^{-1/(n-i)} + W_i(L)^{-1/(n-i)} \qquad (6.8.6)$$

for $i = 0, 1, \ldots, n-1$, with the same cases of equality as in (6.8.5).

Next, we consider some analogues of mixed volumes and quermassintegrals. The known inequalities satisfied by these, for which analogues are sought, include the following:

$$V(K_1, \ldots, K_n)^m \geqq \prod_{i=1}^{m} V(K_i, \ldots, K_i, K_{m+1}, \ldots, K_n) \quad (6.8.7)$$

for $m = 2, \ldots, n$, which includes the Aleksandrov–Fenchel inequalities (6.3.1) and is deduced from these by induction (Aleksandrov (1937b), Busemann [12], p. 50); the cyclic inequality (6.4.6), that is,

$$W_j(K)^{k-i} \geqq W_i(K)^{k-j} W_k(K)^{j-i}, \qquad (6.8.8)$$

valid for $0 \leqq i < j < k \leqq n$; in particular

$$\kappa_n^j W_i(K)^{n-j} \leqq \kappa_n^i W_j(K)^{n-i} \tag{6.8.9}$$

for $0 \leqq i < j < n$; and the generalized Brunn–Minkowski inequality

$$W_i(K + L)^{1/(n-i)} \geqq W_i(K)^{1/(n-i)} + W_i(L)^{1/(n-i)} \tag{6.8.10}$$

for $i \in \{0, \ldots, n-1\}$, a special case of Theorem 6.4.3.

Many of the formal properties of the mixed volume (multilinearity, translation invariance, continuity and monotoneity) are shared by the *mixed width integral* introduced by Lutwak (1977). It is defined by

$$A(K_1, \ldots, K_n) := \frac{1}{n2^n} \int_{S^{n-1}} b(K_1, u) \cdots b(K_n, u)\, \mathrm{d}\sigma(u)$$

for $K_1, \ldots, K_n \in \mathcal{K}^n$, where $b(K, u)$ denotes the width of K in direction $u \in S^{n-1}$ and σ is spherical Lebesgue measure. It follows from Hölder's inequality that

$$A(K_1, \ldots, K_n)^m \leqq \prod_{i=1}^m A(K_i, \ldots, K_i, K_{m+1}, \ldots, K_n)$$

for $1 < m \leqq n$. Lutwak also proved that

$$A(K_1, \ldots, K_n)^n \geqq V(K_1) \cdots V(K_n),$$

with equality if and only if $K_1, \ldots, K_n$ are balls.

By specialization, one obtains the width integrals

$$B_i(K) := A(\underbrace{K, \ldots, K}_{n-i}, \underbrace{B^n, \ldots, B^n}_{i})$$

of Lutwak (1975b). They satisfy

$$B_j(K)^{k-i} \leqq B_i(K)^{k-j} B_k(K)^{j-i}$$

for $0 \leqq i < j < k \leqq n$, with equality if and only if K is of constant width; they also satisfy

$$B_i(K + L)^{1/(n-i)} \leqq B_i(K)^{1/(n-i)} + B_i(L)^{1/(n-i)}$$

for $i \in \{0, \ldots, n-1\}$, and other inequalities.

In a similar way as for the mixed width integral, but using the radial function $\rho(K, \cdot)$, one defines the *dual mixed volume*

$$\tilde{V}(K_1, \ldots, K_n) := \frac{1}{n} \int_{S^{n-1}} \rho(K_1, u) \cdots \rho(K_n, u)\, \mathrm{d}\sigma(u) \tag{6.8.11}$$

for $K_1, \ldots, K_n \in \mathcal{K}^n_{00}$. It was introduced by Lutwak (1975a). By specialization, one defines

$$\tilde{V}_i(K, L) := \tilde{V}(\underbrace{K, \ldots, K}_{n-i}, \underbrace{L, \ldots, L}_{i})$$

and the *dual quermassintegrals*

$$\widetilde{W}_i(K) := \widetilde{V}_i(K, B^n) = \frac{1}{n}\int_{S^{n-1}} \rho(K, u)^{n-i}\,\mathrm{d}\sigma(u) \tag{6.8.12}$$

for $i = 0, \ldots, n$. The inequalities

$$\widetilde{V}(K_1, \ldots, K_n)^m \leqq \prod_{i=1}^{m} \widetilde{V}(K_i, \ldots, K_i, K_{m+1}, \ldots, K_n) \tag{6.8.13}$$

$(1 < m \leqq n)$ and

$$\widetilde{W}_j(K)^{k-i} \leqq \widetilde{W}_i(K)^{k-j}\widetilde{W}_k(K)^{j-i} \tag{6.8.14}$$

$(0 \leqq i < j < k \leqq n)$ are valid. Equality holds in (6.8.13) if and only if $K_1, \ldots, K_m$ are equivalent by dilatation, and in (6.8.14) if and only if K is a centred ball.

For the dual quermassintegrals $\widetilde{W}_i$, a representation (or inductive definition) in analogy to Kubota's integral recursion is possible. This was pointed out by Lutwak (1975d). Recall that a special case of Kubota's formula (5.3.27) can be written in the form

$$W_i(K) = \frac{1}{n\kappa_{n-1}} \int_{S^{n-1}} W'_{i-1}(K|H_{u,0})\,\mathrm{d}\sigma(u) \tag{6.8.15}$$

for $i = 1, \ldots, n$, where W'_{i-1} is computed in the $(n-1)$-dimensional subspace $H_{u,0}$. In anology to (6.8.15), the dual quermassintegrals satisfy

$$\widetilde{W}_i(K) = \frac{1}{n\kappa_{n-1}} \int_{S^{n-1}} \widetilde{W}'_{i-1}(K \cap H_{u,0})\,\mathrm{d}\sigma(u), \tag{6.8.16}$$

where $\widetilde{W}'_{i-1}$ is the $(i-1)$th dual quermassintegral in the $(n-1)$-space $H_{u,0}$. Together with $\widetilde{W}_0(K) = V_n(K)$, this can be used for a recursive definition of the dual quermassintegrals. It is also true that

$$\widetilde{W}_{n-k}(K) = \frac{\kappa_n}{\kappa_k} \int_{SO(n)} V_k(K \cap \rho E_k)\,\mathrm{d}\nu(\rho) \tag{6.8.17}$$

(Lutwak (1979a)) for $k = 1, \ldots, n-1$, in analogy to

$$W_{n-k}(K) = \frac{\kappa_n}{\kappa_k} \int_{SO(n)} V_k(K|\rho E_k)\,\mathrm{d}\nu(\rho) \tag{6.8.18}$$

(another special case of (5.3.27)). Here, as in the following, E_k denotes a k-dimensional linear subspace of $\mathbb{E}^n$.

Hadwiger [18], p. 267, has defined *harmonic quermassintegrals* by $\hat{W}_0(K) := V_n(K)$, $\hat{W}_n(K) := \kappa_n$ and

$$\hat{W}_{n-k}(K) = \frac{\kappa_n}{\kappa_k}\left[\int V_k(K|\rho E_k)^{-1}\,\mathrm{d}\nu(\rho)\right]^{-1} \tag{6.8.19}$$

for $K \in \mathcal{K}_0^n$ and $k = 1, \ldots, n-1$, and has shown that

$$\hat{W}_i(K+L)^{1/(n-i)} \geqq \hat{W}_i(K)^{1/(n-i)} + \hat{W}_i(L)^{1/(n-i)}. \tag{6.8.20}$$

Lutwak (1988a) has proved the inequality

$$\kappa_n^j \widehat{W}_i(K)^{n-j} \leqq \kappa_n^i \widehat{W}_j(K)^{n-i}. \tag{6.8.21}$$

for $0 \leqq i < j < n$ and $K \in \mathcal{K}_0^n$; equality holds if and only if K is a ball. A special case is

$$\kappa_n^{i/n} V_n(K)^{(n-i)/n} \leqq \widehat{W}_i(K). \tag{6.8.22}$$

For $i = n-1$, this is the harmonic Urysohn inequality proved by Lutwak (1975a), and for $i = 1$ it is the harmonic isepiphanic inequality of Lutwak (1984).

As special cases of more general averaging processes treated in Lutwak (1975c, 1984), Lutwak (1984) proposed to define *affine quermassintegrals* $\Phi_0(K), \Phi_1(K), \dots, \Phi_n(K)$ for $K \in \mathcal{K}_0^n$ by taking $\Phi_0(K) := V_n(K)$, $\Phi_n(K) := \kappa_n$ and, for $0 < k < n$,

$$\Phi_{n-k}(K) := \frac{\kappa_n}{\kappa_k} \left[\int_{SO(n)} V_k(K|\rho E_k)^{-n} \, \mathrm{d}\nu(\rho) \right]^{-1/n}. \tag{6.8.23}$$

Then

$$\Phi_i(K) \leqq \widehat{W}_i(K) \leqq W_i(K),$$

by Jensen's inequality. Lutwak showed that

$$\Phi_i(K+L)^{1/(n-i)} \geqq \Phi_i(K)^{1/(n-i)} + \Phi_i(L)^{1/(n-i)}. \tag{6.8.24}$$

In analogy to (6.8.9), Lutwak (1988a) conjectured that

$$\kappa_n^j \Phi_i(K)^{n-j} \leqq \kappa_n^i \Phi_j(K)^{n-i} \tag{6.8.25}$$

for $0 \leqq i < j < n$ and $K \in \mathcal{K}_0^n$. A special case would be the inequality

$$\kappa_n^{i/n} V_n(K)^{(n-i)/n} \leqq \Phi_i(K). \tag{6.8.26}$$

The cases $i = n-1$ and $i = 1$ are true; they follow, respectively, from the Blaschke–Santaló inequality and the Petty projection inequality (see Section 7.4), as noted by Lutwak.

The *dual affine quermassintegrals*, also proposed by Lutwak, are defined, for $K \in \mathcal{K}_0^n$, by letting $\widetilde{\Phi}_0(K) := V_n(K)$, $\widetilde{\Phi}_n(K) := \kappa_n$, and, for $0 < k < n$,

$$\widetilde{\Phi}_{n-k}(K) := \frac{\kappa_n}{\kappa_k} \left[\int_{SO(n)} V_k(K \cap \rho E_k)^n \, \mathrm{d}\nu(\rho) \right]^{1/n}. \tag{6.8.27}$$

From Jensen's inequality,

$$\widetilde{W}_i(K) \leqq \widetilde{\Phi}_i(K).$$

Grinberg (1986, 1991) proved the inequality

$$\widetilde{\Phi}_i(K) \leqq \kappa_n^{i/n} V_n(K)^{(n-i)/n} \tag{6.8.28}$$

for $0 < i < n-1$, with equality if and only if K is a centred ellipsoid. For $i = 1$, this is the Busemann intersection inequality (see Section 7.4). Grinberg (1991) also proved that the affine quermassintegrals and the dual affine quermassintegrals are, as the names suggest, invariant under volume-preserving linear transformations (and the affine quermass-integrals, of course, under translations).

Notes for Section 6.8

1. *Aleksandrov's inequalities for mixed discriminants*. Aleksandrov's (1938b) proof of (6.8.1) (which is sketched in Busemann [12]) exhibits some formal analogies to his proof of the inequalities for mixed volumes of strongly isomorphic polytopes. No really simple proof seems to be known (the one given by Schneider (1966b) is erroneous, and the more general result stated there does not hold). New interest in the inequalities for mixed discriminants arose when Egorychev used a special case in his proof of van der Waerden's permanent conjecture; this led to simpler proofs of that special case. We refer the reader to the references given in the survey article of Lagarias (1982).

   Extensions of the inequalities (6.8.1) are treated by Khovanskii (1984) and Panov (1985a, b).
2. *Other types of addition for convex bodies*. Besides the combinations of convex bodies defined by (6.8.2) and (6.8.4), other analogues of Minkowski addition have been introduced and applied. These are radial linear combination and radial Blaschke addition (Lutwak 1988b), harmonic linear combination and harmonic Blaschke addition (Lutwak 1990b, 1991). For ordinary Blaschke addition, see Section 7.1, Note 5. The process defined by (6.8.4) was called harmonic p-combination, and was further studied, by Lutwak (1991+b).
3. *Inequalities for integral means*. Inequalities for power means of the width of a convex body were treated by Lutwak (1975c), and similar inequalities for the brightness by Lutwak (1984). Lutwak (1987a) also obtained various inequalities for the power means of quermassintegrals of projections of convex bodies.
4. *Quermassintegrals of polar bodies*. For $0 \leqq i, j < n$ and $i + j \geqq n - 1$, one has
$$W_i(K)^{n-j} W_j(K^*)^{n-i} \geqq \kappa_n^{2n-i-j}$$
for $K \in \mathcal{K}_{00}^n$. Equality holds precisely for centred balls. For $i + j = n - 1$, this is due to Firey (1973) and for $i = n - 1$ to Lutwak (1975a), and the general result was deduced from Firey's case, independently by Heil (1976) and Lutwak (1976). Firey (1963) proved the sharp inequalities $V(K, K^*) \geqq \kappa_2$ for $n = 2$ and $V(K, K^*, B^3) \geqq \kappa_3$ for $n = 3$.

   For the problem of the infimum of $W_0(K)W_0(K^*)$, see Section 7.4, Note 13.

   Heil (1976) remarks that for $n > 3$ the minimum of $W_1(K)W_1(K^*)$ is not attained by balls.
5. *Mean dual affine quermassintegrals*. In analogy to (6.8.27), one may also define
$$\bar{\Phi}_{n-k}(K) := \frac{\kappa_n}{\kappa_k} \left[\int_{A(n,k)} V_k(K \cap E)^{n+1} \, d\mu_k(E) \right]^{1/(n+1)}$$
for $0 < k < n$ and $K \in \mathcal{K}_0^n$. Then $\bar{\Phi}_{n-k}(K)$ is invariant under unimodular affine transformations of K. Among convex bodies of given positive volume, precisely the ellipsoids attain the maximal value of $\bar{\Phi}_{n-k}$. This was proved by Schneider (1985b), Theorem 1, second part.

7

Selected applications

The Brunn–Minkowski theorem and the inequalities for mixed volumes are useful in many geometric applications. Here we select only a few of them. Minkowski's existence theorem, to be treated in the first section, is, strictly speaking, not necessarily an application of the inequalities for mixed volumes, but it is a classical result of the Brunn–Minkowski theory. The corresponding uniqueness result, on the other hand, is a consequence of the equality condition in Minkowski's inequality and, thus, in the Brunn–Minkowski theorem. The Aleksandrov–Fenchel inequalities are a powerful tool for obtaining extensions of Minkowski's uniqueness result; these are the subject of Section 7.2.

The rest of the chapter will be devoted to some inequalities in the proofs of which the Brunn–Minkowski theorem or the Minkowski inequalities have been used (as well as other techniques such as symmetrization). However, this is only a superficial aspect, and the main reason for presenting these inequalities and the notions involved is that they belong to an interesting and promising part of the geometry of convex bodies; this is the theory of affine-invariant constructions and functionals.

7.1. Minkowski's existence theorem

If φ is the area measure $S_{n-1}(K, \cdot)$ of the n-dimensional convex body $K \in \mathcal{K}_0^n$, then

$$\int_{S^{n-1}} u \, \mathrm{d}\varphi(u) = o, \tag{7.1.1}$$

which is a special case of (5.1.28). From the geometric meaning of the area measure it is clear that it cannot be concentrated on any great subsphere of S^{n-1}. Minkowski's existence theorem says that these trivial necessary conditions are also sufficient in order that φ be the area measure of a convex body $K \in \mathcal{K}_0^n$ (by Theorem 7.2.1, this convex body is unique up to a translation). This beautiful and useful result was proved by Minkowski for special cases, in which, however, the essential ideas were already contained; it was later extended independently by Aleksandrov and by Fenchel and Jessen, when the full notion of area measure was available.

Although Minkowski's existence theorem is not an application of the inequalities for mixed volumes, its proof is motivated by Minkowski's inequality. In fact, if $K \in \mathcal{K}_0^n$ with $S_{n-1}(K, \cdot) = \varphi$ exists, then any convex body $L \in \mathcal{K}_0^n$ satisfies

$$\frac{1}{n}\int_{S^{n-1}} h(L, u)\,\mathrm{d}\varphi(u) = V(L, K, \ldots, K) \geqq V_n(L)^{1/n} V_n(K)^{1-1/n},$$

with equality if and only if L is homothetic to K. Thus, up to homothety the convex body K is characterized by the fact that it minimizes the functional

$$L \mapsto \int_{S^{n-1}} h(L, u)\,\mathrm{d}\varphi(u)$$

under the side condition $V_n(L) = 1$.

Following Minkowski (1897, 1903), we first treat the special case of polytopes.

Theorem 7.1.1 (Minkowski). *Let $u_1, \ldots, u_N \in S^{n-1}$ be pairwise distinct vectors linearly spanning $\mathbb{E}^n$, and let $f_1, \ldots, f_N$ be positive real numbers such that*

$$\sum_{i=1}^{N} f_i u_i = o. \tag{7.1.2}$$

Then there is a polytope $P \in \mathcal{P}_0^n$ having (precisely) $u_1, \ldots, u_N$ as its normal vectors and satisfying

$$V_{n-1}(F(P, u_i)) = f_i \tag{7.1.3}$$

for $i = 1, \ldots, N$.

Proof. For an N-tuple $A = (\alpha_1, \ldots, \alpha_N) \in \mathbb{R}^N$ with non-negative components $\alpha_1, \ldots, \alpha_N$ we define

$$P(A) := \bigcap_{i=1}^{N} H^-_{u_i,\alpha_i}.$$

From $\operatorname{lin}\{u_1, \ldots, u_N\} = \mathbb{E}^n$ and (7.1.2) it follows that $\operatorname{pos}\{u_1, \ldots, u_N\} = \mathbb{E}^n$; hence $P(A)$ is a polytope, with $o \in P(A)$. Its normal vectors are among the vectors $u_1, \ldots, u_N$, and

$$h(P(A), u_i) \leqq \alpha_i \quad \text{for } i = 1, \ldots, N,$$

with equality if $V_{n-1}(F(P(A), u_i)) > 0$. Let

$$M := \{A \in \mathbb{R}^N | \alpha_i \geqq 0, \quad V_n(P(A)) \geqq 1\};$$

then M is closed, since $V_n(P(A))$ depends continuously on A. The linear function Φ defined by

$$\Phi(A) := \frac{1}{n}\sum_{i=1}^{N}\alpha_i f_i, \; A = (\alpha_1, \ldots, \alpha_N),$$

attains a minimum on M, because $f_i > 0$ for all i. Let μ^{n-1} be this minimum and suppose that it is attained at $A^* = (\alpha_1^*, \ldots, \alpha_N^*)$; put $P(A^*) =: P^*$. We assert that μP^* solves the problem.

By (7.1.2), $\Phi(A)$ is not changed under a translation of $P(A)$, hence we may assume that P^* has o as an interior point and, hence, that $\alpha_1^*, \ldots, \alpha_N^* > 0$. Write

$$V_{n-1}(F(P^*, u_i)) := f_i^* \quad \text{for } i = 1, \ldots, N.$$

We have $V_n(P^*) = 1$, since otherwise λP^* with suitable $\lambda < 1$ would satisfy $V_n(\lambda P^*) \geqq 1$ and yield a value less than μ^{n-1} for the function Φ. From

$$V_n(P^*) = \frac{1}{n}\sum_{i=1}^{N} h(P^*, u_i) f_i^*$$

and $h(P^*, u_i) = \alpha_i^*$ if $f_i^* \neq 0$ it follows that

$$\frac{1}{n}\sum_{i=1}^{N}\alpha_i^* f_i^* = 1. \tag{7.1.4}$$

Since $\Phi(A^*) = \mu^{n-1}$, we have

$$\frac{1}{n}\sum_{i=1}^{N}\alpha_i^* f_i = \mu^{n-1}. \tag{7.1.5}$$

Now consider the hyperplanes in $\mathbb{R}^N$ given by

$$H_1 := \left\{A \in \mathbb{R}^N | \; \frac{1}{n}\sum_{i=1}^{N}\alpha_i f_i = \mu^{n-1}\right\},$$

$$H_2 := \left\{A \in \mathbb{R}^N | \; \frac{1}{n}\sum_{i=1}^{N}\alpha_i f_i^* = 1\right\}.$$

From (7.1.4) and (7.1.5) it follows that $A^* \in H_1 \cap H_2$. Suppose that $A = (\alpha_1, \ldots, \alpha_N) \in H_1$, where $\alpha_1, \ldots, \alpha_N \geqq 0$. Then $V_n(P(A)) \leqq 1$, since otherwise μ^{n-1} would not be the minimum of Φ on M. For $0 \leqq \vartheta \leqq 1$ we have $(1 - \vartheta)A^* + \vartheta A \in H_1$. Further, a point $y \in (1 - \vartheta)P^* + \vartheta P(A)$ satisfies

$$\langle y, u_i \rangle \leqq (1 - \vartheta)\alpha_i^* + \vartheta \alpha_i \quad \text{for } i = 1, \ldots, N,$$

hence

$$(1-\vartheta)P^* + \vartheta P(A) \subset P((1-\vartheta)A^* + \vartheta A).$$

It follows that $V_n((1-\vartheta)P^* + \vartheta P(A)) \leqq 1$ for $0 \leqq \vartheta \leqq 1$, which together with $V_n(P^*) = 1$ gives

$$V(P(A), P^*, \ldots, P^*) \leqq 1. \tag{7.1.6}$$

Since $\alpha_1^*, \ldots, \alpha_N^* > 0$, there is a neighbourhood U of A^* in $\mathbb{R}^N$ such that, for $A \in H_1 \cap U$, the polytope $P(A)$ satisfies $\alpha_1, \ldots, \alpha_N > 0$ and

$$V_{n-1}(F(P(A), u_i)) > 0 \quad \text{if } f_i^* > 0.$$

If this holds, then $h(P(A), u_i) = \alpha_i$, hence

$$V(P(A), P^*, \ldots, P^*) = \frac{1}{n}\sum_{i=1}^{N} \alpha_i f_i^*.$$

Inequality (7.1.6) now shows that $H_1 \cap U \subset H_2^-$, where H_2^- is one of the two closed halfspaces bounded by H_2. Since $A^* \in H_1 \cap H_2$, we conclude that $H_1 = H_2$ and hence that

$$f_i = \mu^{n-1} f_i^* = V_{n-1}(F(\mu P^*, u_i))$$

for $i = 1, \ldots, N$. This completes the proof that μP^* is the desired polytope. ■

The general case of Minkowski's existence theorem will now be obtained by approximation.

Theorem 7.1.2. *Let φ be a measure on $\mathcal{B}(S^{n-1})$ with the properties*

$$\int_{S^{n-1}} u \,\mathrm{d}\varphi(u) = o \tag{7.1.7}$$

and $\varphi(s) < \varphi(S^{n-1})$ for each great subsphere s of S^{n-1}. Then there is a convex body $K \in \mathcal{K}_0^n$ for which $S_{n-1}(K, \cdot) = \varphi$.

Proof. For each $k \in \mathbb{N}$, we decompose the sphere S^{n-1} into finitely many pairwise disjoint Borel sets of diameter at most $1/k$ and with spherically convex closure. We fix k and denote by $\Delta_1, \ldots, \Delta_N$ those sets of the decomposition on which φ is positive. For $i = 1, \ldots, N$, let

$$c_i := \frac{1}{\varphi(\Delta_i)} \int_{\Delta_i} u \,\mathrm{d}\varphi(u);$$

then $c_i \neq o$ since Δ_i lies in an open hemisphere, and hence $c_i = f_i u_i$ with $u_i \in S^{n-1}$ and $f_i > 0$. Observing that $c_i \in \operatorname{conv} \Delta_i$, one finds that $1 - (2k^2)^{-1} \leqq f_i \leqq 1$. For $\omega \in \mathcal{B}(S^{n-1})$, let

$$\varphi_k(\omega) := \sum_{u_i \in \omega} \varphi(\Delta_i) f_i.$$

Then φ_k is a measure on $\mathcal{B}(S^{n-1})$ satisfying

$$\int_{S^{n-1}} u \, d\varphi_k(u) = \int_{S^{n-1}} u \, d\varphi(u) = o.$$

For $g \in C(S^{n-1})$ we have

$$\int_{S^{n-1}} g(u) \, d\varphi_k(u) - \int_{S^{n-1}} g(u) \, d\varphi(u) = \sum_{i=1}^{N} \int_{\Delta_i} [f_i g(u_i) - g(u)] \, d\varphi(u)$$

and

$$|f_i g(u_i) - g(u)| \leqq |g(u_i) - g(u)| + \|g\|/2k^2.$$

If $u \in \Delta_i$, then $|u_i - u| \leqq 1/k$, since $u_i \in \operatorname{pos} \Delta_i$. From the uniform continuity of g we deduce that $\int g d\varphi_k \to \int g d\varphi$ for $k \to \infty$, hence

$$\varphi_k \xrightarrow{w} \varphi \quad \text{for } k \to \infty.$$

There is a number $a > 0$ such that

$$\int_{S^{n-1}} \langle u, v \rangle^+ d\varphi(u) \geqq a \quad \text{for } v \in S^{n-1}$$

(where the superscript plus denotes the positive part), since the integral is a continuous function of v that is positive by the properties of φ. The estimate above shows that

$$\lim_{k\to\infty} \int_{S^{n-1}} \langle u, v \rangle^+ d\varphi_k(u) = \int_{S^{n-1}} \langle u, v \rangle^+ d\varphi(u)$$

uniformly in v. Hence, there exists k_0 such that

$$\int_{S^{n-1}} \langle u, v \rangle^+ d\varphi_k(u) > \frac{a}{2}$$

for $k \geqq k_0$ and all $v \in S^{n-1}$. Thus, for $k \geqq k_0$, the measure φ_k is not concentrated on any great subsphere.

By Theorem 7.1.1, for $k \geqq k_0$ there is a polytope $P_k \in \mathcal{P}_0^n$ with $S_{n-1}(P_k, \cdot) = \varphi_k$ and, without loss of generality, $o \in P_k$. Since $\varphi_k(S^{n-1}) \leqq \varphi(S^{n-1})$, the surface areas of the polytopes P_k remain bounded; hence, by the isoperimetric inequality, $V_n(P_k) \leqq b$ with some constant b. Let $x \in P_k$ and write $x = |x| v$ with $v \in S^{n-1}$; then

$$h(P_k, u) \geqq h(\operatorname{conv}\{o, x\}, u) = |x| \langle u, v \rangle^+$$

for $u \in S^{n-1}$ and hence

$$b \geqq V_n(P_k) \geqq \frac{|x|}{n} \int_{S^{n-1}} \langle u, v \rangle^+ d\varphi_k(u) \geqq \frac{|x|}{n} \frac{a}{2},$$

thus $P_k \subset B(o, 2nb/a)$. By the Blaschke selection theorem 1.8.6, there is a subsequence of $(P_k)_{k \geqq k_0}$ converging to some convex body $K \in \mathcal{K}^n$. From the weak convergence of the sequence $(\varphi_k)_{k \geqq k_0}$ to φ we conclude that $S_{n-1}(K, \cdot) = \varphi$. ■

Minkowski's theorem makes it possible to define a new addition for n-dimensional convex bodies: For convex bodies $K_0, K_1 \in \mathcal{K}_0^n$, the sum of their area measures satisfies the conditions of Theorem 7.1.2; hence there exists a convex body M for which

$$S_{n-1}(M, \cdot) = S_{n-1}(K_0, \cdot) + S_{n-1}(K_1, \cdot).$$

By Theorem 7.2.1, the body M is unique up to a translation, and we may assume that it has its area centroid at the origin. This body M is called the *Blaschke sum* of K_0 and K_1 and is denoted by $K_0 \# K_1$. For Blaschke addition, there is a counterpart to the Brunn–Minkowski theorem:

Theorem 7.1.3. *Let $K_0, K_1 \in \mathcal{K}_0^n$ and $\lambda \in (0, 1)$ and let $K_{[\lambda]}$ be a convex body for which*

$$S_{n-1}(K_{[\lambda]}, \cdot) = (1 - \lambda)S_{n-1}(K_0, \cdot) + \lambda S_{n-1}(K_1, \cdot).$$

Then

$$V_n(K_{[\lambda]})^{(n-1)/n} \geqq (1 - \lambda)V_n(K_0)^{(n-1)/n} + \lambda V_n(K_1)^{(n-1)/n}. \quad (7.1.8)$$

Equality in (7.1.8) *holds if and only if K_0 and K_1 are homothetic.*

Proof. Using Minkowski's inequality (6.2.2) in the form

$$V_n(K_{[\lambda]})^{-1/n} V(K_{[\lambda]}, K[n-1]) \geqq V_n(K)^{(n-1)/n},$$

we obtain

$$\begin{aligned} V_n(K_{[\lambda]})^{(n-1)/n} &= V_n(K_{[\lambda]})^{-1/n} \frac{1}{n} \int_{S^{n-1}} h(K_{[\lambda]}, u) \, \mathrm{d}S_{n-1}(K_{[\lambda]}, u) \\ &= V_n(K_{[\lambda]})^{-1/n}((1 - \lambda)V(K_{[\lambda]}, K_0[n-1]) \\ &\quad + \lambda V(K_{[\lambda]}, K_1[n-1])] \\ &\geqq (1 - \lambda)V_n(K_0)^{(n-1)/n} + \lambda V_n(K_1)^{(n-1)/n}, \end{aligned}$$

together with the assertion on the equality sign. ■

Notes for Section 7.1

1. The polytopal case of Minkowski's existence theorem, Theorem 7.1.1, is due (for $n = 3$) to Minkowski (1897); see also Minkowski (1903), §9. We have essentially followed the presentation of the proof given in Bonnesen & Fenchel [8], §60. Related, but slightly different versions of the proof have been given by, e.g., Aleksandrov [2], Chapter VII, §2, and McMullen (1973), §7. Aleksandrov (1939a) found a different proof, using his so-called 'mapping lemma', an application of the 'invariance of domain'. This proof appears also in Aleksandrov [2], Chapter VII, §1, and Pogorelov (1975), §2. Similar existence results for unbounded polyhedra can be found in Aleksandrov [2], Chapter VII, §3.

Minkowski (1903), §10, extended his result to convex bodies more general than polytopes, namely to convex bodies K for which, in present-day terminology, the area measure $S_{n-1}(K, \cdot)$ has a continuous density with respect to spherical Lebesgue measure.

Once the notion of the area measure had been introduced, the general version of Minkowski's result, Theorem 7.1.2, could be deduced from the polytopal case by means of approximation, using the weak continuity of S_{n-1}. This was done by Fenchel & Jessen (1938) (above we followed their approach; see also Fenchel (1938)) and later by Aleksandrov (1939b). Independently of Fenchel and Jessen, Aleksandrov (1938a), §3, gave a proof that generalizes Minkowski's approach via an extremum problem, but in the general case the necessary variation argument (which uses Lemma 6.5.3) is less elementary. A similar attempt, but with an insufficient variation argument, was made by Süss (1931a). Related existence theorems for infinite convex surfaces are stated without proof in Aleksandrov [2], p. 305–6. A similar existence result for convex caps was obtained by Busemann (1959).

2. *Minkowski's problem in differential geometry*. If K is a convex body of class C^2_+, then its area measure $S_{n-1}(K, \cdot)$ has a continuous density (with respect to spherical Lebesgue measure), given by s_{n-1}, the product of the principal radii of curvature. Its reciprocal value is the Gauss–Kronecker curvature, as a function of the outer normal. Thus Minkowski's existence theorem is related to the problem of the existence of a convex hypersurface with Gauss–Kronecker curvature prescribed as a function of the normal vector. Classical contributions to this problem are the papers by Lewy (1938), Miranda (1939) and Nirenberg (1953). If one wants to utilize the general existence result 7.1.2 to solve this problem, one needs regularity results for convex surfaces with sufficiently smooth curvature functions. For such results and for more information on Minkowski's problem and related problems from the viewpoint of differential geometry, we refer to the books of Pogorelov ([27], 1975), the survey article of Gluck (1975), and in particular to Gluck (1972) and Cheng & Yau (1976).
3. *Algorithmic version*. The essential idea of the proof of Theorem 7.1.1 is the minimization of the linear functional Φ over the region $M \subset \mathbb{R}^N$, which is convex by the Brunn–Minkowski theorem. Known iterative methods for such optimization problems can therefore be used to derive algorithms for a constructive solution of Minkowski's problem for polytopes. Such an algorithm in $\mathbb{E}^3$, of interest for computer vision problems, was described by Little (1983). The role of possible applications to robot vision is discussed in Horn (1986), Chapter 16.
4. Applications of Minkowski's existence theorem in stochastic geometry appear in Schneider (1982a, 1987a), Weil (1988) and Wieacker (1989).

   An elegant application of Minkowski's theorem, Blaschke addition and Minkowski's inequalities to a problem in discrete geometry is made in a paper by Böröczky, Bárány, Makai jr. & Pach (1986); another application is in Fáry & Makai jr. (1982).
5. *Blaschke addition*. The addition of convex bodies now called Blaschke addition is mentioned briefly in Blaschke [5], p. 112. For the case of polytopes it had occurred already in the work of Minkowski (1897), p. 117. Theorem 7.1.3 was proved by Kneser & Süss (1932); weaker forms had been proved previously by Herglotz (unpublished) and Süss (1932a).

   Blaschke sums were investigated and used by Firey & Grünbaum (1964), Firey (1965a, 1967a, c), Grünbaum [15], Chapter 15.3, Schneider (1967c, 1987a), Chakerian (1971), Kutateladze & Rubinov (1969), Kutateladze (1973, 1976), Goikhman (1974), Bronshtein (1979), Goodey & Schneider (1980),

McMullen (1980) and Lutwak (1986b, 1987b, 1988b, 1991). An addition of Blaschke type is also used in Bronshtein (1978c).

6. *The existence problem for the area measure of order i.* Necessary and sufficient conditions for a measure φ on $\mathcal{B}(S^{n-1})$ to be the ith area measure $S_i(K, \cdot)$ of some convex body $K \in \mathcal{K}^n$, where $1 < i < n-1$, are not known. The necessary condition

$$\int_{S^{n-1}} u \, d\varphi(u) = o$$

is not sufficient, even if φ has an analytic density, as shown by Aleksandrov (1938a). There is a complete solution of the existence problem for the special case of sufficiently smooth bodies of revolution due to Firey (1970c) and in a special case to Nádenîk (1968). Some necessary conditions follow from results of Firey (1970a) and Weil (1980a) and from Theorem 4.6.3. The latter, together with a result of Larman & Rogers (1970) (applied to the polar body) implies that the support of an ith area measure is arcwise connected if $i < n-1$ (Fedotov 1982).

A special case of the existence problem asks whether the sum of two ith area measures is always an ith area measure, where $i \in \{2, \ldots, n-2\}$. This question was posed, more or less explicitly, by Firey (1967a, 1970b, c, 1975) and Chakerian (1971). Negative answers were given by Fedotov (1979b, 1982, and, in a weaker form, 1978b) and independently by Goodey & Schneider (1980). The latter authors showed that even suitably chosen parallelepipeds with parallel edges provide counterexamples; here Theorem 4.6.4 had to be used. They also proved: Let P, P_0, $P_1 \in \mathcal{P}_0^n$ be convex polytopes, $i \in \{1, \ldots, n-1\}$, and suppose that

$$S_i(P, \cdot) = S_i(P_0, \cdot) + S_i(P_1, \cdot).$$

If π denotes orthogonal projection onto any $(i+1)$-dimensional plane, then πP is the Blaschke sum (relative to the carrying plane) of πP_0 and πP_1.

Since no general criterion for ith area measures is known if $i \in \{2, \ldots, n-2\}$, it is of interest to investigate qualitatively the set $\mathcal{S}_i$ of all ith-order area measures of convex bodies in $\mathcal{K}_0^n$. This was done by Weil (1980a) and Goodey (1981). Weil showed that, for $i < n-1$, $\mathcal{S}_i$ is not dense in $\mathcal{M}$, the set of all finite measures on $\mathcal{B}(S^{n-1})$ with barycentre o, endowed with the weak topology, but that $\mathcal{S}_i - \mathcal{S}_i$ is dense in $\mathcal{M} - \mathcal{M}$. Goodey proved for $i < n-1$ that $\mathcal{S}_i - \mathcal{S}_i$ is of first Baire category in the Banach space $\mathcal{M} - \mathcal{M}$ with the total variation norm. Goodey also investigated weak limits of ith-order area measures.

7. *Mixed bodies.* For convex bodies $K_1, \ldots, K_{n-1} \in \mathcal{K}_0^n$, the mixed area measure $S(K_1, \ldots, K_{n-1}, \cdot)$ satisfies the assumptions of Minkowski's theorem; hence there is a convex body $[K_1, \ldots, K_{n-1}]$, unique up to a translation, for which

$$S_{n-1}([K_1, \ldots, K_{n-1}], \cdot) = S(K_1, \ldots, K_{n-1}, \cdot).$$

$[K_1, \ldots, K_{n-1}]$ is called the *mixed body* of $K_1, \ldots, K_{n-1}$. Mixed bodies were introduced, in special cases, by Firey (1967a), and were extensively studied by Lutwak (1986b), who obtained a number of inequalities for these bodies.

8. *Existence problems for curvature measures.* Existence problems for the curvature measures C_i have apparently only been considered for the case $i = 0$. For a convex body $K \in \mathcal{K}^n$ with $o \in \operatorname{int} K$, let $f(u) := \rho(K, u)u$ for $u \in S^{n-1}$, so that $f(u) \in \operatorname{bd} K$.

Theorem. Let κ be a measure on $\mathcal{B}(S^{n-1})$. There exists a convex body $K \in \mathcal{K}^n$ with $o \in \operatorname{int} K$ for which $C_0(K, f(\omega)) = \kappa(\omega)$ for $\omega \in \mathcal{B}(S^{n-1})$ if and only if the following two conditions are satisfied:

(a) $\kappa(S^{n-1}) = \mathcal{H}^{n-1}(S^{n-1})$,

(b) $\kappa(S^{n-1}\setminus\omega) > \mathcal{H}^{n-1}(\omega^*)$ whenever $\omega \subset S^{n-1}$ is spherically convex and ω^* denotes the set polar to ω.

The body K is unique up to a dilatation.

Existence was proved by Aleksandrov (1939a), first for polytopes and then by approximation in general. The polytopal case is an existence result for polytopes with vertices on given rays through o and preassigned curvatures at these vertices. Uniqueness up to a dilatation was shown in Aleksandrov (1942a).

Analogous results for unbounded convex surfaces and orthogonal projection onto a plane were obtained by Aleksandrov (1942a) and [2], Chapter IX. Oliker (1986) treated, by means of a variational method, an existence problem similar to Aleksandrov's theorem.

The theorems of Minkowski and Aleksandrov are existence results, which a priori give no additional information on the convex bodies whose existence they ensure. It is an interesting problem to find conditions on the given measures from which conclusions on the shape of the solution bodies can be drawn. For Aleksandrov's theorem, such a result was obtained by Treibergs (1990). For $\omega \subset S^{n-1}$ and given α, $0 < \alpha < \pi/2$, let ω_α be the set of all points of S^{n-1} at spherical distance $\leqq \alpha$ from ω. Treibergs showed that if in Aleksandrov's theorem stated above one assumes in addition that

$$\kappa(\omega) \leqq \mathcal{H}^{n-1}(\omega_\alpha) \quad \text{for all } \omega \in \mathcal{B}(S^{n-1}),$$

then the body K satisfies

$$\frac{\sup\{\rho(K, u) | u \in S^{n-1}\}}{\inf\{\rho(K, u) | u \in S^{n-1}\}} \leqq c(\alpha, n)$$

with a real constant $c(\alpha, n)$ depending only on α and n.

7.2. Uniqueness theorems for area measures

The theory of mixed volumes is a powerful tool for obtaining some uniqueness theorems for convex bodies in the style of differential geometry. For convex bodies of class C^2_+ the assumptions of these theorems can be formulated in terms of elementary symmetric functions of principal curvatures or radii of curvature. In the general case they correspondingly involve curvature measures or area measures. In the present section, we prove some of these results and state others without proof.

We start with the uniqueness assertion for the Minkowski problem. Although this theorem will be improved and generalized by later theorems, we give its formulation and proof separately, to show in a basic example the close connection with results on mixed volumes.

Theorem 7.2.1. *If $K, L \in \mathcal{K}^n_0$ are convex bodies with*

$$S_{n-1}(K, \cdot) = S_{n-1}(L, \cdot), \tag{7.2.1}$$

then K and L are translates of each other.

Proof. We integrate the support function $h(K, \cdot)$, using (7.2.1) and (5.1.18), and apply Minkowski's inequality (6.2.2) to obtain

$$V_n(K)^n = V(K, L, \ldots, L)^n \geqq V_n(K)V_n(L)^{n-1}. \qquad (7.2.2)$$

Thus $V_n(K) \geqq V_n(L)$, and analogously we obtain $V_n(L) \geqq V_n(K)$. Hence, equality holds in (7.2.2), which implies that K and L are homothetic. Because of (7.2.1), they must be translates. ■

Suppose that K and L are convex bodies of class C^2_+. Then the assumption (7.2.1) is, by (4.2.20), equivalent to the assumption that the function s_{n-1}, the product of the principal radii of curvature as a function of the outer unit normal vector, is the same for K and L. Equivalently, the Gauss–Kronecker curvatures of the convex hypersurfaces bd K and bd L coincide at points with the same outer normal vector. This is the classical assumption for the uniqueness part of Minkowski's problem in differential geometry.

Along with any uniqueness assertion comes the question for stability: if the assumption enforcing uniqueness is only satisfied approximately, can one ascertain approximate uniqueness? One would want to have explicit estimates, implying the uniqueness assertion, and sharpening it in a quantitative way. The following stability version of Theorem 7.2.1 is due to Diskant (1972). Recall that $\mathcal{K}^n(r, R)$ denotes the set of convex bodies in $\mathbb{E}^n$ with inradius at least r and circumradius at most R.

Theorem 7.2.2. *Let* $0 < r < R$. *There exist numbers* $\varepsilon_0 > 0$ *and* γ, *depending only on* n, r, R, *with the following property. If* $K, L \in \mathcal{K}^n(r, R)$ *are convex bodies satisfying*

$$|S_{n-1}(K, \omega) - S_{n-1}(L, \omega)| \leqq \varepsilon \qquad (7.2.3)$$

for $\omega \in \mathcal{B}(S^{n-1})$ *with some* $\varepsilon \in [0, \varepsilon_0]$, *then*

$$\delta(K, L') \leqq \gamma\varepsilon^{1/n} \qquad (7.2.4)$$

for a suitable translate L' *of* L.

Since the exponent $1/n$ in (7.2.4) is probably not optimal, it is of little interest to look more closely at the size of the constant γ. Similar remarks refer to the stability assertions made below.

Part of the proof will be stated as a lemma, in a more general form since this will be needed later. We assume that a number $m \in \{2, \ldots, n\}$ and convex bodies K, L are given and we use the notation (6.4.4) with an $(n - m)$-tuple of balls, that is

$$V_{(i)} := V(K[m - i], L[i], B^n[n - m]).$$

Lemma 7.2.3. *If $K, L \in \mathcal{K}^n(r, R)$ and*

$$|S_{m-1}(K, \omega) - S_{m-1}(L, \omega)| \leqq \varepsilon \tag{7.2.5}$$

for all $\omega \in \mathcal{B}(S^{n-1})$ with some $\varepsilon \geqq 0$, then

$$|V_{(0)} - V_{(m-1)}| \leqq \frac{2R}{n}\varepsilon, \quad |V_{(m)} - V_{(1)}| \leqq \frac{2R}{n}\varepsilon, \tag{7.2.6}$$

$$0 \leqq V_{(1)} - V_{(0)}^{(m-1)/m} V_{(m)}^{1/m} \leqq \frac{2R}{n}\left(\frac{R}{r} + 1\right)\varepsilon. \tag{7.2.7}$$

Proof. By (5.1.18),

$$V_{(0)} - V_{(m-1)} = \frac{1}{n}\int_{S^{n-1}} h(K, u)\, \mathrm{d}[S_{m-1}(K, u) - S_{m-1}(L, u)].$$

Using the Hahn–Jordan decomposition of the signed measure $S_{m-1}(K, \cdot) - S_{m-1}(L, \cdot)$, we deduce from (7.2.5) the first inequality of (7.2.6), and the second one is obtained similarly. By (6.4.5) with $i = 0$, $j = m - 1$, $k = m$ we have $V_{(m-1)}^m \geqq V_{(m)}^{m-1} V_{(0)}$ (and similarly the left-hand inequality in (7.2.7)). We conclude that

$$\begin{aligned} V_{(1)} - V_{(0)}^{(m-1)/m} V_{(m)}^{1/m} &= V_{(m)} - \left[\frac{V_{(m)}}{V_{(0)}}\right]^{1/m} V_{(m-1)} \\ &\quad + \left[\frac{V_{(m)}}{V_{(0)}}\right]^{1/m} [V_{(m-1)} - V_{(0)}] + [V_{(1)} - V_{(m)}] \\ &\leqq \left(\frac{R}{r} + 1\right)\frac{2R}{n}\varepsilon. \end{aligned}$$

■

Proof of Theorem 7.2.2. We use inequality (6.2.12), which states that

$$r(K, L) \geqq \left[\frac{V_{(1)}}{V_{(n)}}\right]^{1/(n-1)} - \frac{[V_{(1)}^{n/(n-1)} - V_{(0)} V_{(n)}^{1/(n-1)}]^{1/n}}{V_{(n)}^{1/(n-1)}}, \tag{7.2.8}$$

and Lemma 7.2.3 for $m = n$. In the following, $c_1, \ldots, c_6$ denote constants depending only on n, r, R. By (7.2.6), $V_{(n)} - V_{(1)} \leqq c_1\varepsilon$, hence

$$\left[\frac{V_{(1)}}{V_{(n)}}\right]^{1/(n-1)} \geqq (1 - c_2\varepsilon)^{1/(n-1)} \geqq 1 - c_2\varepsilon^{1/n}$$

provided that $\varepsilon \leqq \varepsilon_1 := \min\{1, 1/c_2\}$. By (7.2.7),

$$0 \leqq V_{(1)} - V_{(0)}^{(n-1)/n} V_{(n)}^{1/n} \leqq c_3\varepsilon.$$

Using $x^p - y^p \leqq p x^{p-1}(x - y)$ for $x \geqq y \geqq 0$ and $p > 1$ (Hardy, Littlewood & Pólya 1934, p. 39), we obtain

$$0 \leqq V_{(1)}^{n/(n-1)} - V_{(0)} V_{(n)}^{1/(n-1)} \leqq c_4\varepsilon.$$

Now (7.2.8) yields

$$r(K, L) \geqq 1 - c_5\varepsilon^{1/n} =: \alpha.$$

Since K and L may be interchanged, we also have $r(L, K) \geqq \alpha$. By the definition of $r(L, K)$, there is a translate L' of L such that $\alpha K \subset L'$. Similarly, there is a translate $K + t$ of K such that $L' \subset \alpha^{-1}(K + t)$. Thus $\alpha \leqq 1$, where the trivial case $\alpha = 1$ can be excluded. The bodies αK and $\alpha^{-1}(K + t)$ are homothetic. Choosing the origin at their centre of homothety, we have $t = 0$, thus $\alpha K \subset L' \subset \alpha^{-1}K$, and hence $o \in K$. Now

$$K \subset \alpha^{-1}L' = L' + \frac{1-\alpha}{\alpha} L' \subset L' + \frac{1-\alpha}{\alpha} 2RB^n$$

and similarly $L' \subset K + [(1-\alpha)/\alpha]2RB^n$. This yields

$$\delta(K, L') \leqq 2R \frac{1-\alpha}{\alpha} \leqq c_6 \varepsilon^{1/n},$$

if $\varepsilon \leqq \varepsilon_0 := \min\{\varepsilon, (2c_5)^{-n}\}$, which implies $\alpha \geqq 1/2$. ■

The uniqueness assertion of Theorem 7.2.1 can be generalized considerably, namely to certain mixed area measures.

***Theorem* 7.2.4.** *Let a number $m \in \{2, \ldots, n\}$ and an $(n-m)$-tuple $\mathcal{C} = (K_{m+1}, \ldots, K_n)$ of smooth convex bodies be given. If $K, L \in \mathcal{K}^n$ are convex bodies of dimension $\geqq m$ satisfying*

$$S(K[m-1], \mathcal{C}, \cdot) = S(L[m-1], \mathcal{C}, \cdot), \tag{7.2.9}$$

then K and L are translates of each other.

Proof. The assumption (7.2.9) says that condition (d) in Theorem 6.4.4 is satisfied. By Theorem 6.6.9, this implies that K and L are homothetic. By (7.2.9), they must be translates. ■

The special case $K_{m+1} = \ldots = K_n = B^n$ gives the Aleksandrov–Fenchel–Jessen theorem:

***Corollary* 7.2.5.** *Let $m \in \{2, \ldots, n\}$. If $K, L \in \mathcal{K}^n$ are convex bodies of dimension $\geqq m$ satisfying*

$$S_{m-1}(K, \cdot) = S_{m-1}(L, \cdot), \tag{7.2.10}$$

then K and L are translates of each other.

A special case is the differential-geometric version: If K and L are convex bodies of class C^2_+ such that, at points with the same outer normal, the $(m-1)$th elementary symmetric functions of the principal radii of curvature are the same, then K and L differ only by a translation.

For the cases $m = n$ and $m = 2$ of Corollary 7.2.5 we already know stability versions (Theorems 7.2.2 and 4.3.5). We shall now prove a stability result for the general case.

Theorem 7.2.6. *Let $m \in \{2, \ldots, n\}$ and positive numbers ε_0, r, R be given. If $K, L \in \mathcal{K}^n(r, R)$ are convex bodies satisfying*

$$|S_{m-1}(K, \omega) - S_{m-1}(L, \omega)| \leqq \varepsilon \tag{7.2.11}$$

for all $\omega \in \mathcal{B}(S^{n-1})$ with some $\varepsilon \in [0, \varepsilon_0]$, then

$$\delta(K, L') \leqq \gamma\varepsilon^q \quad \text{with} \quad q = \frac{1}{(n+1)2^{m-2}}$$

for a suitable translate L' of L, where the constant γ depends only on n, ε_0, r, R.

The proof makes use of Lemma 6.6.4, for which we did not reproduce the proof, but the argument involving mixed volumes will be given completely.

Proof. In the following $c_1, \ldots, c_{15}$ denote constants depending only on n, ε_0, r, R. From Lemma 7.2.3 we have

$$V_{(1)} \leqq V_{(0)}^{(m-1)/m} V_{(m)}^{1/m} + c_1\varepsilon.$$

Interchanging K and L we obtain

$$V_{(m-1)} \leqq V_{(m)}^{(m-1)/m} V_{(0)}^{1/m} + c_1\varepsilon,$$

and multiplication leads to

$$V_{(1)} V_{(m-1)} \leqq V_{(0)} V_{(m)} + c_2\varepsilon$$

and thus

$$\frac{V_{(m-1)}}{V_{(m)}} \leqq \frac{V_{(0)}}{V_{(1)}} + c_3\varepsilon. \tag{7.2.12}$$

Combined with the Aleksandrov–Fenchel inequalities, this yields

$$\frac{V_{(0)}}{V_{(1)}} \leqq \frac{V_{(1)}}{V_{(2)}} \leqq \ldots \leqq \frac{V_{(m-1)}}{V_{(m)}} \leqq \frac{V_{(0)}}{V_{(1)}} + c_3\varepsilon.$$

For $j \in \{1, \ldots, m-1\}$ we deduce that

$$\frac{V_{(j)}}{V_{(j+1)}} \leqq \frac{V_{(0)}}{V_{(1)}} + c_3\varepsilon \leqq \frac{V_{(j-1)}}{V_{(j)}} + c_3\varepsilon,$$

and thus

$$V_{(j)}^2 \leqq V_{(j-1)} V_{(j+1)} + c_4\varepsilon. \tag{7.2.13}$$

We use inequality (6.4.9) with the choices $K_0 = K$, $K_1 = L$, $K_2 = B^n$,

$$(K_3, \ldots, K_n) = (\underbrace{K, \ldots, K}_{m-1-j}, \underbrace{L, \ldots, L}_{j-1}, \underbrace{B^n \ldots, B^n}_{n-m})$$

and the notation

$$\bar{V}_{(j)} := V(K[m-1-j], L[j], B^n[n-m+1])$$

to obtain

$$\begin{aligned}
&[V_{(j-1)}\bar{V}_{(j)} - V_{(j)}\bar{V}_{(j-1)}]^2 \\
&\quad \leqq [V_{(j)}^2 - V_{(j-1)}V_{(j+1)}] \\
&\qquad \times [\bar{V}_{(j-1)}^2 - V_{(j-1)}V(K[m-1-j], L[j-1], B^n[n-m+2])] \\
&\quad \leqq c_5\varepsilon,
\end{aligned}$$

where (7.2.13) has been used. This leads to

$$\left| \frac{\bar{V}_{(j-1)}}{\bar{V}_{(j)}} - \frac{V_{(j-1)}}{V_{(j)}} \right| \leqq c_6\varepsilon^{1/2}$$

for $j = 1, \ldots, m-1$, from which we deduce that

$$\begin{aligned}
\frac{\bar{V}_{(m-2)}}{\bar{V}_{(m-1)}} &\leqq \frac{V_{(m-2)}}{V_{(m-1)}} + c_6\varepsilon^{1/2} \\
&\leqq \frac{V_{(0)}}{V_{(1)}} + c_3\varepsilon + c_6\varepsilon^{1/2} \leqq \frac{\bar{V}_{(0)}}{\bar{V}_{(1)}} + c_3\varepsilon + 2c_6\varepsilon^{1/2}.
\end{aligned}$$

Thus we have shown that inequality (7.2.12) implies the inequality

$$\frac{\bar{V}_{(m-2)}}{\bar{V}_{(m-1)}} \leqq \frac{\bar{V}_{(0)}}{\bar{V}_{(1)}} + c_7\varepsilon^{1/2}. \tag{7.2.14}$$

This is the same as (7.2.12), except that on the left-hand side one argument L has been replaced by B^n and on the right-hand side one argument K has been replaced by B^n; furthermore, the error $c_3\varepsilon$ has been replaced by $c_7\varepsilon^{1/2}$. The derivation of (7.2.14) from (7.2.12) applied only general inequalities for convex bodies and not the assumptions of the theorem. Hence, we can repeat the argument and finally arrive at

$$\frac{V_{11}}{V_{02}} \leqq \frac{V_{20}}{V_{11}} + c_8\varepsilon^{(1/2)^{m-2}}$$

with $V_{ij} = V(K[i], L[j], B^n[n-i-j])$, which implies

$$V_{11}^2 - V_{20}V_{02} \leqq c_9\varepsilon^{(1/2)^{m-2}}.$$

From Theorem 6.6.6 and Lemma 6.6.4 we conclude that

$$\delta(\bar{K}, \bar{L}) \leqq c_{10}\varepsilon^q \text{ with } q = \frac{1}{(n+1)2^{m-2}},$$

and where $\bar{K} = [K - s(K)]/b(K)$.

Replacing L by a translate, we may assume that $s(K) = s(L)$. Let

$b(K)/b(L) := \lambda$; then $\delta(K, \lambda L) \leqq \alpha$ with $\alpha = c_{10}b(K)\varepsilon^q$. We have

$$\lambda L \subset K + \alpha B^n \subset (1 + c_{11}\alpha)K,$$

and from (7.2.11) we obtain $|V_{m-1,0} - V_{0,m-1}| \leqq \varepsilon/n$, hence

$$\lambda^{m-1}V_{0,m-1} \leqq (1 + c_{11}\alpha)^{m-1}V_{m-1,0} \leqq (1 + c_{11}\alpha)^{m-1}(V_{0,m-1} + \varepsilon/n).$$

This yields $\lambda \leqq 1 + c_{12}\varepsilon^q$ and therefore

$$K \subset \lambda L + \alpha B^n \subset L + c_{13}\varepsilon^q B^n.$$

Similarly we obtain $L \subset K + c_{14}\varepsilon^q B^n$ and thus $\delta(K, L) \leqq c_{15}\varepsilon^q$. ∎

The theory of mixed volumes is flexible enough to permit also applications to non-closed convex hypersurfaces. As an example, we prove a version of the Aleksandrov–Fenchel–Jessen theorem for convex hypersurfaces with boundary and spherical image in a hemisphere.

Let a unit vector $u \in S^{n-1}$ and an $(n-1)$-dimensional convex body $C \subset H_{u,0}$ be given. By a convex hypersurface of the canal class $\mathcal{K}_C^+$ we mean a point set F in $\mathbb{E}^n$ that by orthogonal projection to $H_{u,0}$ is mapped bijectively onto C and is such that the set

$$K(F) := \{x - \lambda u | x \in F, \lambda \geqq 0\}$$

is convex. Let $\Omega_u^+ := \{x \in S^{n-1} | \langle x, u \rangle > 0\}$. We define the jth-order area measure of F by

$$S_j(F, \omega) := S_j(K(F), \omega) \quad \text{for } \omega \in \mathcal{B}(\Omega_u^+).$$

Thus $S_j(F, \cdot)$ is defined on the half-sphere Ω_u^+. Observe that we do not demand that the hypersurface F be tangential to the right cylinder with cross section C. Thus the relative interiors of two convex hypersurfaces of the canal class $\mathcal{K}_C^+$ may well have different spherical images.

Theorem 7.2.7. *Let F_0, F_1 be convex hypersurfaces of the canal class $\mathcal{K}_C^+$ satisfying*

$$S_{m-1}(F_0, \cdot) = S_{m-1}(F_1, \cdot)$$

for some $m \in \{2, \ldots, n\}$. Then there is a translation parallel to u carrying F_0 into F_1.

Proof. We choose $\lambda \in \mathbb{R}$ with $F_0, F_1 \subset H_{u,\lambda}^+$ and put $K_i := K(F_i) \cap H_{u,\lambda}^+$ for $i = 0, 1$. Then K_0 and K_1 are convex bodies belonging to the canal class $\mathcal{K}_C$ defined in Section 6.7. We have $S_{m-1}(K_0, \omega) = S_{m-1}(K_1, \omega)$ for Borel sets $\omega \subset S^{n-1} \backslash H_{u,0}$. This holds by assumption if $\omega \subset \Omega_u^+$. For $\omega \subset \Omega_u^- := \{x \in S^{n-1} | \langle x, u \rangle < 0\}$ it is true since $F(K_0, u) = F(K_1, u)$ for $u \in \Omega_u^-$. Thus condition (e) in Theorem 6.7.2 (with $K_{m+1} = \ldots = K_n = B^n$) is satisfied. By Theorem 6.7.3, K_0 and K_1 are equivalent by telescoping. This proves the assertion. ∎

We turn now to the extension of some classical differential-geometric characterizations of balls. First, let K be a convex body of class C^2_+ and assume that, for some $j \in \{1, \ldots, n-1\}$, the jth elementary symmetric function of the principal radii of curvature is constant. Then K is a ball. This well-known result can be generalized as follows. The assumption $s_j = \alpha$ with a constant α gives, by integration using (4.2.20), that

$$S_j(K, \omega) = \alpha \mathcal{H}^{n-1}(\omega) \quad \text{for } \omega \in \mathcal{B}(S^{n-1}). \tag{7.2.15}$$

Since $\mathcal{H}^{n-1}(\omega) = S_j(B^n, \omega)$ for $\omega \in \mathcal{B}(S^{n-1})$, condition (7.2.15) is equivalent to $S_j(K, \cdot) = S_j(\alpha^{1/j} B^n, \cdot)$. By the Aleksandrov–Fenchel–Jessen theorem, this assumption implies that K is a ball of radius $\alpha^{1/j}$. This conclusion, namely that a convex body $K \in \mathcal{K}^n_0$ satisfying (7.2.15) with a constant α must be a ball, holds for arbitrary K, and this fact generalizes the classical differential-geometric result.

There is also a stability version of the latter uniqueness result (where assumption and conclusion are stronger than in the corresponding case of Theorem 7.2.6).

Theorem 7.2.8. *There exist numbers ε_0 and γ, depending only on n, with the following property. If $K \in \mathcal{K}^n$ is a convex body satisfying*

$$1 - \varepsilon \leqq \frac{S_j(K, \omega)}{\mathcal{H}^{n-1}(\omega)} \leqq 1 + \varepsilon$$

for each $\omega \in \mathcal{B}(S^{n-1})$ with $\mathcal{H}^{n-1}(\omega) > 0$ and for some $\varepsilon \in [0, \varepsilon_0]$, then K is contained in a ball of radius $1 + \gamma\varepsilon^{1/(n-1)}$ and contains a ball of radius $1 - \gamma\varepsilon^{1/(n-1)}$.

This was proved by Diskant (1971a). The proof uses symmetrization and will not be given here, as it is not closely related to the theory of mixed volumes.

The differential-geometric result mentioned above has a counterpart where radii of curvature are replaced by curvatures. If K is a convex body of class C^2 for which the function H_i, the ith (normalized) elementary symmetric function of the principal curvatures, is constant for some $i \in \{1, \ldots, n-1\}$, then K is a ball. If we want to formulate the condition

$$H_i = \alpha \tag{7.2.16}$$

with a constant α in a way that makes sense for arbitrary convex bodies, we have (at least) two possibilities. By (4.2.19), condition (7.2.16) (for K of class C^2) is equivalent to

$$C_{n-1-i}(K, \beta) = \alpha \mathcal{H}^{n-1}(\beta \cap \operatorname{bd} K)$$

and therefore, in view of (4.2.23), to

$$C_{n-1-i}(K,\ \cdot) = \alpha C_{n-1}(K,\ \cdot). \tag{7.2.17}$$

But if K is of class C^2_+, then by (2.5.28) the condition (7.2.16) is equivalent to

$$s_{n-1-i} = \alpha s_{n-1},$$

and this, by (2.5.27), is equivalent to

$$S_{n-1-i}(K,\ \cdot) = \alpha S_{n-1}(K,\ \cdot). \tag{7.2.18}$$

Thus, aiming at an extension of the differential-geometric result concerning the condition (7.2.16) to general convex bodies, we may take either (7.2.17) or (7.2.18) as an assumption. As it turns out, the results are different. Let us first consider (7.2.18). This condition is not only fulfilled by balls. In fact, suppose that K is an $(n-1-i)$-tangential body of a ball, say of radius one. Then it follows from Theorem 6.6.16 that $W_0(K) = W_1(K) = \ldots = W_{i+1}(K)$, and by Theorem 6.4.1 this implies that

$$S_{n-1}(K,\ \cdot) = S_{n-2}(K,\ \cdot) = \ldots = S_{n-i-1}(K,\ \cdot).$$

The converse is also true:

Theorem 7.2.9. *Let $j \in \{0, \ldots, n-2\}$. If $K \in \mathcal{K}^n_0$ is a convex body satisfying*

$$S_j(K,\ \cdot) = \alpha S_{n-1}(K,\ \cdot) \tag{7.2.19}$$

with some constant α, then K is a j-tangential body of a ball of radius $\alpha^{-1/(n-j-1)}$.

Proof. After replacing K by a suitable homothetic copy, we may assume that $\alpha = 1$. From (7.2.19) and (5.1.18), integrating the support function of either B^n or K, we obtain

$$W_{n-j} = W_1 \quad \text{and} \quad W_{n-j-1} = W_0 \tag{7.2.20}$$

for the quermassintegrals $W_i = W_i(K) = V(K[n-i], B^n[i])$. By the Aleksandrov–Fenchel inequalities,

$$\frac{W_1}{W_0} \geqq \frac{W_2}{W_1} \geqq \ldots \geqq \frac{W_{n-j}}{W_{n-j-1}}. \tag{7.2.21}$$

By (7.2.20), equality holds here throughout; in particular $W_1^2 = W_0 W_2$. By Theorem 6.6.18, K is an $(n-2)$-tangential body of a ball. If this ball has radius r, then Theorem 6.6.16, together with a suitable dilatation, implies $W_0 = rW_1$. Since equality holds in (7.2.21), we have $W_0 = r^{n-j} W_{n-j}$, and together with (7.2.20) this yields $r = 1$. Thus $W_0 = W_1 = \ldots = W_{n-j}$. By Theorem 6.6.16, K is a j-tangential body of B^n. ∎

If, however, condition (7.2.17) is accepted as a substitute for (7.2.16), then a proper generalization of the classical differential-geometric result is obtained:

Theorem 7.2.10. *Let* $j \in \{0, \ldots, n-2\}$. *If* $K \in \mathcal{K}_0^n$ *is a convex body satisfying*

$$C_j(K, \cdot) = \alpha C_{n-1}(K, \cdot) \tag{7.2.22}$$

with some constant α, *then* K *is a ball.*

For the proof, we sketch only the first steps. We may assume that $\alpha = 1$. Let $\omega \subset S^{n-1}$ be a closed set; then

$$\begin{aligned} S_j(K, \omega) &\leqq C_j(K, \tau(K, \omega)) = C_{n-1}(K, \tau(K, \omega)) \\ &= C_{n-1}(K, \tau(K, \omega) \cap \operatorname{reg} K) = C_j(K, \tau(K, \omega) \cap \operatorname{reg} K) \\ &\leqq S_j(K, \omega), \end{aligned}$$

where we have used successively Lemma 4.2.3, the assumption (7.2.22) with $\alpha = 1$, (4.2.23) and Theorem 2.2.4, the assumption (7.2.22), and Lemma 4.2.3. Since the equality sign must hold, we deduce that

$$S_j(K, \omega) = C_{n-1}(K, \tau(K, \omega)) = \mathcal{H}^{n-1}(\tau(K, \omega)) = S_{n-1}(K, \omega)$$

by (4.2.24). The equality $S_j(K, \omega) = S_{n-1}(K, \omega)$ for all closed sets ω permits us to conclude that $S_j(K, \cdot) = S_{n-1}(K, \cdot)$. Now Theorem 7.2.9 shows that K is a j-tangential body of a ball. The rest of the proof consists in showing that among j-tangential bodies of balls only balls satisfy (7.2.22). We omit this part of the proof, since it is not particularly relevant to the theory of mixed volumes. We mention, however, that Theorem 2.5.5 and the result of Section 4.2, Note 5, are essential tools. The proof can be found in Schneider (1979a).

For the case $j = 0$, and only for this case, a stability version of Theorem 7.2.10 is known. This is due to Diskant (1968), and we quote it without proof.

Theorem 7.2.11. *Let* $K \in \mathcal{K}_0^n$, $0 < \varepsilon < 1/2$, *and suppose that*

$$1 - \varepsilon \leqq \frac{C_0(K, \beta)}{C_{n-1}(K, \beta)} \leqq 1 + \varepsilon$$

for each set $\beta \in \mathcal{B}(\mathbb{E}^n)$ *with* $C_{n-1}(K, \beta) > 0$. *Then* K *lies in the* $\gamma\varepsilon$-*neighbourhood of a unit ball, where the constant* γ *depends only on* n.

Finally, the uniqueness part of Aleksandrov's theorem formulated in Section 7.1, Note 8, can be extended to curvature measures of different orders.

Theorem 7.2.12. *Let $j \in \{0, \ldots, n-1\}$. Let $K, L \in \mathcal{K}_0^n$ be convex bodies with o as an interior point, and let the radial projection $f: \operatorname{bd} K \to \operatorname{bd} L$ be defined by $f(x) = \lambda(x)x \in \operatorname{bd} L$ with $\lambda(x) > 0$ for $x \in \operatorname{bd} K$. If*

$$C_j(K, \beta) = C_j(L, f(\beta))$$

for each Borel set $\beta \subset \operatorname{bd} K$, then K and L differ only by a dilatation with centre o if $j = 0$, and they are identical if $j > 0$.

The case $j = 0$ is due to Aleksandrov (1942a), and the case $j > 0$ was proved by Schneider (1978a).

Notes for Section 7.2

1. *Minkowski's uniqueness theorem.* The method of proof for the uniqueness theorem 7.2.1 is the original one found by Minkowski (1897, 1903).

   The first stability result of the type of Theorem 7.2.2 was proved by Volkov (1963). (A sketch of his proof is given in Pogorelov [27], Chapter VII, §10.) His result was slightly weaker than that of Diskant (1972), since it had an exponent of ε equal to $1/(n+2)$ instead of Diskant's $1/n$. Later Diskant (1979) showed that the exponent can be improved to $1/(n-1)$ if one makes the stronger assumption that K and L have continuous radii of curvature and that $|s_{n-1}(K, u) - s_{n-1}(L, u)| < \varepsilon$ for all $u \in S^{n-1}$. The optimal exponent is unknown. A short alternative proof of a weaker form of Theorem 7.2.2 was given by Bourgain & Lindenstrauss (1988b).

   A paper worthy of recommendation that has historical interest is Stoker (1950), which treats Minkowski's and Christoffel's uniqueness theorems and the infinitesimal rigidity for closed convex surfaces in $\mathbb{E}^3$ from a differential-geometric viewpoint, showing clearly the interrelations between these problems and how the methods for their proofs are related to the classical geometry of convex bodies.
2. *The Aleksandrov–Fenchel–Jessen theorem.* Corollary 7.2.5, the classical Aleksandrov–Fenchel–Jessen theorem, was proved independently by Aleksandrov (1937b) and by Fenchel & Jessen (1938) (see also Busemann [12], p. 70, Leichtweiß [23], p. 319, and Leichtweiß (1981)). Theorem 7.2.4 for the special case where the bodies of $\mathcal{C}$ are of class C_+^2 was obtained by Aleksandrov (1938b); the general version for smooth bodies as given here is new. For the special case of the Aleksandrov–Fenchel–Jessen theorem where the bodies K and L are of class C_+^2, a differential-geometric proof was given by Chern (1959).

   A version of the Aleksandrov–Fenchel–Jessen theorem for convex hypersurfaces with boundaries was proved by Busemann (1959). His assumption of identical boundaries is rather strong; it seems more natural to assume that the support functions coincide at the boundary of the spherical image. For differential-geometric uniqueness results of this type, see Aleksandrov (1956/8) and Oliker (1979a, b). The uniqueness theorem 7.2.7 for general convex hypersurfaces with boundaries is due to Schneider (1989b).

   The stability theorem 7.2.6 and its proof are taken from Schneider (1989c). A weaker theorem of this type had previously been proved by Diskant (1985) (for the special case $m-1 = n-2$ also by Diskant 1976). His assumptions are stronger in several respects: He assumed K and L to be smooth and to satisfy an inequality of the form $|S_{n-1}(K, \cdot) - S_{n-1}(L, \cdot)| \leqq \varepsilon \mathcal{H}^{n-1}(\omega)$ for

$\omega \in \mathcal{B}(S^{n-1})$, and he had to impose the assumption that one of the bodies has a projection on some hyperplane that is a ball. Moreover, his exponent for ε is smaller than the one given in Theorem 7.2.6.

Under stronger regularity assumptions, Oliker (1982), Theorem 4, proved a stability theorem involving s_{m-1} for convex hypersurfaces with boundaries and with common spherical image contained in an open hemisphere. Earlier, more specialized results of this type were obtained by Volkov & Oliker (1970) and Oliker (1973).

One may ask whether the Aleksandrov–Fenchel–Jessen theorem can be generalized by imposing more general relations between area measures, for example: If $K, L \in \mathcal{K}_0^n$ are convex bodies satisfying

$$\sum_{j=1}^{n-1} \alpha_j S_j(K, \cdot) = \sum_{j=1}^{n-1} \alpha_j S_j(L, \cdot),$$

where the constants $\alpha_1, \ldots, \alpha_{n-1}$ are non-negative and not all zero, must K be a translate of L? An affirmative answer for $n = 3$ is contained in Schneider (1976).

3. *Stability results via symmetrization.* Extremal properties of balls and thus uniqueness theorems characterizing balls can often by proved by means of symmetrization procedures. Sometimes symmetrization is also useful for obtaining stability properties of balls. In this way, Diskant (1971a) obtained Theorem 7.2.8, and the proof of Diskant (1968) for Theorem 7.2.11 also made use of symmetrizations. Earlier applications of such methods to similar results can be found in Fet (1963) and Diskant (1964, 1965). Symmetrization is also used in Schneider (1983) to obtain an estimate for convex hypersurfaces with unique projection to a hyperplane and satisfying an inequality $C_0 \geqq \alpha C_{n-1}$ with $\alpha > 0$.
4. *Characterizations of balls and stability.* Theorem 7.2.9, which characterizes tangential bodies of balls, appears in Schneider (1978c), and Theorem 7.2.10 was proved in Schneider (1979a). For the latter result, Kohlmann (1988) developed a new method of proof employing generalized Minkowskian integral formulae and permitting extensions to spaces of constant curvature.

   In the literature, there are several investigations, differential-geometric or convex-geometric, of the stability of the sphere within a class of convex hypersurfaces having some curvature restriction, such as almost constant Gauss–Kronecker curvature or mean curvature, or almost umbilical points. Besides the references in Note 3, we mention Diskant (1971b), Koutroufiotis (1971), Moore (1973), Treibergs (1985), Schneider (1990b), Pogorelov (1967), Rešetnjak (1968a), Guggenheimer (1969) and Vodop'yanov (1970).
5. *Uniqueness results involving curvature measures.* Aleksandrov's (1942a) proof of the case $j = 0$ of Theorem 7.2.12 proceeds by direct geometric reasoning (see also Busemann [12], p. 30); the proof for $j > 0$ by Schneider (1978a) has to use in addition the integral-geometric formula of Theorem 4.5.5. Using the latter method, the following more general version can be obtained. Let K, L, f be as in Theorem 7.2.12, and suppose that

$$\sum_{i=0}^{n-1} \alpha_i C_i(K, \beta) = \sum_{i=0}^{n-1} \alpha_i C_i(L, f(\beta))$$

   for each Borel set $\beta \subset \operatorname{bd} K$, where $\alpha_0, \ldots, \alpha_{n-1}$ are non-negative real constants with $\alpha_1 + \ldots + \alpha_{n-1} > 0$. Then $K = L$.

   In Schneider (1978a), the following result was also deduced. Let $K \in \mathcal{K}^n$ be a convex body with $o \in \operatorname{int} K$. Suppose that $C_j(K, H_{u,0}^+) = C_j(K, H_{u,0}^-)$ for some $j \in \{0, \ldots, n-1\}$ and all $u \in S^{n-1}$. Then K is centrally symmetric with respect to o.

The following characterization of the ball was proved for $n = 3$ by Blind (1977) and for $n \geqq 3$ by Schneider (1977e). Let $K \in \mathcal{K}_0^n$ be a smooth convex body with the property that every Jordan domain in the boundary of K that halves the curvature measure C_0 also halves the area C_{n-1}; then K is a ball.

6. *Characterizations involving inequalities*. Aleksandrov (1961) proved the following uniqueness theorem, which involves only inequalities. Suppose that K, $L \in \mathcal{K}_0^n$ are convex bodies for which
$$S_j(K, \cdot) \leqq S_j(L, \cdot) \quad \text{and } V_{j+1}(K) \geqq V_{j+1}(L)$$
for some $j \in \{1, \ldots, n-1\}$. Then K and L are translates of each other.
It would be very interesting to know whether a convex body $K \in \mathcal{K}_0^3$ satisfying the inequality
$$S_2(K, \cdot) - 2cS_1(K, \cdot) + c^2 S_0(K, \cdot) \leqq 0 \tag{7.2.23}$$
with some constant c is necessarily the sum of a ball and a segment (possibly degenerate). If true, this would imply that a regular closed convex surface of class C_+^2 in $\mathbb{E}^3$ whose principal curvatures k_1, k_2 satisfy $(k_1 - c)(k_2 - c) \leqq 0$ must be a ball. This is only proved for analytic surfaces and surfaces of revolution. Under the strong additional assumption that K admits some circular projection, the conjecture concerning (7.2.23) is easy to prove; see Schneider (1975b).
7. *Aleksandrov's theorem on projections*. The following theorem is due to Aleksandrov (1937b) (see also Chakerian 1967b).

Theorem. Let $i \in \{0, \ldots, n-2\}$ and let $K, L \in \mathcal{K}^n$ be centrally symmetric with respect to o and of dimension $\geqq n - i$. If
$$w_i(K|H_{u,0}) = w_i(L|H_{u,0}) \quad \text{for all } u \in S^{n-1},$$
where w_i denotes the ith quermassintegral in the $(n-1)$-dimensional subspace $H_{u,0}$, then $K = L$.

For the proof, one has to observe that the assumption implies $S_{n-1-i}(K, \cdot) = S_{n-1-i}(L, \cdot)$, by formula (5.3.32) and the uniqueness part of Theorem 3.5.3. The Aleksandrov–Fenchel–Jessen theorem (Corollary 7.2.5) then gives the assertion.

Stability results corresponding to the case $i = 0$ are due to Bourgain (1987), Bourgain & Lindenstrauss (1988a, b) and Campi (1988), and for the case $i = n - 2$ to Campi (1986) and Goodey & Groemer (1990). Further related references are given in Section 7.4, Note 1.

7.3. The difference-body inequality

In the final sections, we collect a number of inequalities for convex bodies, the proofs of which make use of the Brunn–Minkowski inequality or of inequalities for mixed volumes. The first of these inequalities concerns the difference body.

For a convex body $K \in \mathcal{K}^n$, the *difference body* DK is defined by
$$DK := K - K = \{x \in \mathbb{E}^n | K \cap (K + x) \neq \emptyset\}.$$

Theorem 7.3.1. *Let* $K \in \mathcal{K}_0^n$. *Then*
$$2^n V_n(K) \leqq V_n(DK) \leqq \binom{2n}{n} V_n(K). \tag{7.3.1}$$

Equality holds on the left precisely if K is centrally symmetric and on the right precisely if K is a simplex.

We note that the left-hand inequality of (7.3.1), together with the characterization of the equality case, is an immediate consequence of the Brunn–Minkowski theorem. For the right-hand inequality, which is due to Rogers & Shephard (1957), we first prove a lemma, following Chakerian (1967a).

Lemma 7.3.2. *Let $B \in \mathcal{K}_0^n$, $x_0 \in B$, and let $f: B \to \mathbb{R}$ be a non-negative concave function on B. If $h: \mathbb{R} \to \mathbb{R}$ is strictly increasing, then*

$$\int_B h(f(x))\,\mathrm{d}x \geqq nV_n(B)\int_0^1 h(tf(x_0))(1-t)^{n-1}\,\mathrm{d}t.$$

Equality holds if and only if, for each $y \in \operatorname{bd} B$, the function f is linear on $[x_0, y]$ and $f(y) = 0$.

Proof. We may assume that $x_0 = o$. For $x \in B\backslash\{o\}$, let $x \in [o, y] \subset B$ such that $[o, y]$ is maximal, and put $g(x) := f(o)(1 - |x|/|y|)$ and, further, $g(o) := f(o)$. Then $f \geqq g$, since f is concave and non-negative. It follows that

$$\begin{aligned}\int_B h(f(x))\,\mathrm{d}x &\geqq \int_B h(g(x))\,\mathrm{d}x\\ &= \int_{S^{n-1}}\int_0^{\rho(B,u)} h(f(o)(1 - r/\rho(B,u)))r^{n-1}\,\mathrm{d}r\,\mathrm{d}\mathcal{H}^{n-1}(u)\\ &= \int_{S^{n-1}}\rho(B,u)^n\int_0^1 h(tf(o))(1-t)^{n-1}\,\mathrm{d}t\,\mathrm{d}\mathcal{H}^{n-1}(u)\\ &= nV_n(B)\int_0^1 h(tf(o))(1-t)^{n-1}\,\mathrm{d}t.\end{aligned}$$

Equality holds if and only if $f = g$. ■

Proof of Theorem 7.3.1. Let $K \in \mathcal{K}_0^n$ and define $D(K, x) := K \cap (K + x)$ for $x \in DK$. Let $x_1, x_2 \in DK$, $\lambda \in [0, 1]$ and

$$a \in (1-\lambda)D(K, x_1) + \lambda D(K, x_2).$$

Then $a = (1-\lambda)a_1 + \lambda a_2$ with $a_i \in D(K, x_i)$, hence $a_i \in K$ and $a_i = b_i + x_i$ with $b_i \in K$ $(i = 1, 2)$. It follows that $a \in K$ and $a \in K + (1-\lambda)x_1 + \lambda x_2$, thus $a \in D(K, (1-\lambda)x_1 + \lambda x_2)$. We have proved that

$$(1-\lambda)D(K, x_1) + \lambda D(K, x_2) \subset D(K, (1-\lambda)x_1 + \lambda x_2).$$

From the Brunn–Minkowski theorem we conclude that the function defined by

$$f(x) := V_n(K \cap (K + x))^{1/n}, \quad x \in \mathrm{D}K,$$

is concave. Now Lemma 7.3.2, with $B = \mathrm{D}K$, $x_0 = o$, $h(\xi) = \xi^n$ and the function f, together with the identity

$$\int_{\mathrm{D}K} V_n(K \cap (K + x))\,\mathrm{d}x = \int_{\mathbb{E}^n}\int_{\mathbb{E}^n} \mathbf{1}_K(y)\mathbf{1}_K(y - x)\,\mathrm{d}y\,\mathrm{d}x$$
$$= V_n(K)^2,$$

where $\mathbf{1}$ denotes the characteristic function, yields the right-hand inequality of (7.3.1).

Equality holds if and only if f is linear on each segment joining o to a boundary point of $\mathrm{D}K$, and this is true if and only if $K \cap (K + x)$ is homothetic to K, for each $x \in \mathrm{D}K$. Simplices have this property, and it remains to show that they are characterized by it.

Assume, therefore, that $K \cap (K + x)$ is homothetic to K for each $x \in \mathrm{D}K$. Let $p \in \exp K$. Let $u \in S^{n-1}$ be such that the ray $\{p + \lambda u | \lambda \geqq 0\}$ meets K in a non-degenerate segment $[p, q]$. Let $\lambda \in (0, 1)$ and $K_\lambda := K \cap [K + \lambda(p - q)]$; then $K_\lambda = h_\lambda K$ with a homothety h_λ. Since $p \in \exp K$, there is a hyperplane H such that $K \cap H = \{p\}$. The point $h_\lambda p$ must be an exposed point of K_λ, lying in the supporting hyperplane $h_\lambda H$. It follows that $h_\lambda p = p$ and hence that p is the centre of homothety of h_λ.

Now we use the cones $P(K, p)$ and $S(K, p) = \mathrm{cl}\, P(K, p)$ defined in Section 2.2. Suppose that

$$x \in [P(K, p) + p] \cap [P(K, q) + q].$$

Then there are points $y, z \in K$ such that $y - p = \alpha(x - p)$ and $z - q = \beta(x - q)$ with $\alpha, \beta > 0$. For $\lambda < 1$ and sufficiently close to 1, the segments $[p, y]$ and $[q, z] + \lambda(p - q)$ will intersect in a point x_λ. Since $x_\lambda \in K_\lambda$ and, by similarity, $x_\lambda = h_\lambda x$, we deduce that $x \in K$. It follows that

$$[P(K, p) + p] \cap [P(K, q) + q] \subset K$$

and hence

$$K' := [S(K, p) + p] \cap [S(K, q) + q] \subset K.$$

Since $K \subset K'$ holds trivially, we conclude that $K = K'$. This implies that each two-dimensional halfplane bounded by $\mathrm{aff}\,\{p, q\}$ intersects K in a (possibly degenerate) triangle with side $[p, q]$. This holds for any point q for which $[p, q]$ is the (non-degenerate) intersection of K with a line. It is now easy to see that all possible points q make up a facet of K and that K is the convex hull of p and this facet. Since p was an arbitrary exposed point of K, induction with respect to the dimension now shows that K is a simplex. ■

Notes for Section 7.3

1. Special cases of the difference body inequality

$$V_n(\mathrm{D}K) \leqq \binom{2n}{n} V_n(K) \tag{7.3.2}$$

are treated in Section 53 of Bonnesen & Fenchel [8], where references to earlier work are given, and in the thesis of Godbersen (1938). The general case was settled by Rogers & Shephard (1957). Chakerian (1967a) slightly simplified and extended their argument; we have followed his approach. Lemma 7.3.2 appears in his paper, which also includes some generalizations of the difference body inequality. In Rogers & Shephard (1958a, b) one finds a variant of their proof and similar and related inequalities.

2. *Intersections of translates*. The characterization of simplices by equality in (7.3.2) depends on the following result: A convex body K with the property that each nonempty intersection $K \cap (K + x)$ is homothetic to K is necessarily a simplex. The proof given above is a modified version of a proof by Martini (1990). The original proof of Rogers and Shephard was rather long. A shorter proof for polytopes was given by Eggleston, Grünbaum & Klee (1964).

The Rogers–Shephard characterization of simplices can be considered as a special case of a rather general type of problem concerning characterizations of specific sets by intersection properties of their translates: Let $\mathcal{M}$ be a family of subsets of $\mathbb{E}^n$, and let σ be a binary relation on $\mathcal{M}$. The problem is to determine the subfamily $\mathcal{M}(\sigma) \subset \mathcal{M}$ defined by

$$\mathcal{M}(\sigma) := \{K \in \mathcal{M} \mid (K + x) \cap (K + y) \in \mathcal{M},\ x, y \in \mathbb{E}^n \Rightarrow (K + x) \cap (K + y)\ \sigma\ K\}.$$

One may also restrict the translation vectors x, y to smaller sets. Contributions to this problem, of various kinds, are due to Schneider (1967a, 1969b), Gruber (1969, 1970a, b, 1971), McMullen, Schneider & Shephard (1974) and Fourneau (1985) (the latter includes references to earlier work).

3. *A conjectured strengthening of the difference body inequality*. Godbersen (1938) and independently Makai jr. (1974) conjectured that

$$V(K[i], -K[n-i]) \leqq \binom{n}{i} V_n(K) \tag{7.3.3}$$

for $i \in \{1, \ldots, n-1\}$, with equality only for simplices. For $i = 1$ and $i = n - 1$, this follows from the fact that $-K \subset nK$ if K has centroid o (see Bonnesen & Fenchel [8], p. 53), but the general case is unknown. If (7.3.3) is true, it implies the inequality (7.3.2), by

$$V_n(\mathrm{D}K) = V_n(K - K) = \sum_{i=0}^{n} \binom{n}{i} V(K[i], -K[n-i])$$

$$\leqq \sum_{i=0}^{n} \binom{n}{i}^2 V_n(K) = \binom{2n}{n} V_n(K).$$

It would also imply the inequality

$$\int_0^1 V_n((1-\lambda)K + \lambda(-K))\,\mathrm{d}\lambda \leqq \frac{2^n}{n+1} V_n(K),$$

which was established in a different way by Rogers & Shephard (1958a).

4. *A generalization of the difference body*. For $K \in \mathcal{K}_0^n$ and a number $p \in \mathbb{N}$, define

$$\mathrm{D}_p K := \{(x_1, \ldots, x_p) \in (\mathbb{E}^n)^p \mid K \cap (K + x_1) \cap \ldots \cap (K + x_p) \neq \emptyset\},$$

so that $\mathrm{D}_1 K$ is the difference body of K. It was shown by Schneider (1970b) that the np-dimensional volume of $\mathrm{D}_p K$ satisfies

$$V_{np}(\mathrm{D}_p K) \leqq \binom{pn+n}{n} V_n^p(K),$$

with equality if and only if K is a simplex.

Write $\delta_p(K) := V_{np}(\mathrm{D}_p K) V_n(K)^{-p}$. For $n = 2$,

$$\delta_p(K) = \tfrac{1}{2}p(p+1)\delta_1(K) + 1 - p^2;$$

hence δ_p attains its minimum precisely for centrally symmetric convex bodies. This is not true for $n \geqq 3$ and $p \geqq 2$. For these cases, one may conjecture that the minimum of δ_p is attained by ellipsoids.

More generally than in Schneider (1970b), one can consider, for convex bodies $K_0, K_1, \ldots, K_p \in \mathcal{K}_0^n$, the translative integral-geometric quantity

$$\mathrm{D}(K_0, K_1, \ldots, K_p) := \int \ldots \int \chi(K_0 \cap (K_1 + x_1) \cap \ldots \cap (K_p + x_p))\,\mathrm{d}x_1 \ldots \mathrm{d}x_p.$$

Then an obvious extension of the argument used in Schneider (1970b) (essentially an application of Lemma 7.3.2) yields the inequality

$$\mathrm{D}(K_0, K_1, \ldots, K_p) \leqq \binom{pn+n}{n} \frac{V_n(K_0)V_n(K_1) \cdots V_n(K_p)}{V_n(K_0 \cap K_1 \cap \ldots \cap K_p)}.$$

Here each K_i may be replaced by any of its translates K_i'. The best estimate is obtained if $V_n(K_0 \cap K_1' \cap \ldots \cap K_p')$ is maximal.

7.4. Affinely associated bodies

The mixed volume $V(K_1, \ldots, K_n)$ remains unchanged if the same volume-preserving affine transformation of $\mathbb{E}^n$ is applied to each of the convex bodies $K_1, \ldots, K_n$. The general theory of mixed volumes thus belongs to the affine geometry of convex bodies. The Euclidean character of most of the related material treated in former sections is due to the fact that the mixed volume was specialized by taking balls for some of its arguments. However, several of the former results are of a purely affine nature, although metric tools, like support functions and mixed area measures, may have been used in their proofs.

In the present section, we collect a few more results and problems from the affine geometry of convex bodies. In particular, we consider some constructions leading to interesting affine-invariant functionals. More precisely, with a given convex body $K \in \mathcal{K}_0^n$ or $K \in \mathcal{K}_{00}^n$ we associate, in several ways, a convex body $\mathrm{T}K \in \mathcal{K}_{00}^n$ such that, for each $\alpha \in SL(n)$ (the special linear group of $\mathbb{E}^n$), either $\mathrm{T}\alpha K = \alpha \mathrm{T}K$ for all K or $\mathrm{T}\alpha K = \alpha^{-\mathrm{t}}\mathrm{T}K$ for all K. In the latter case, where $\alpha^{-\mathrm{t}}$ denotes the inverse of the adjoint linear map of α, the affine character is slightly obscured. This is a result of the fact that we have identified $\mathbb{E}^n$ with its dual space, using a fixed scalar product. The proper setting for the constructions under consideration would be the n-dimensional real vector space with a volume form, together with its dual space and the induced volume form. Then $\mathrm{T}K$, in the second case, should be defined as an object in the dual space, with the transformation rule $\mathrm{T}\alpha K = (\alpha^*)^{-1}\mathrm{T}K$, where α^* denotes the map dual to α. It is, however, sometimes convenient to use metric tools in definitions or proofs.

In the treatment of inequalities for the associated bodies, the Brunn–Minkowski theory is often useful, but symmetrization techniques also play an essential role. We give no proofs in this section, but rather refer to the original literature. Much more related material can be found in the survey article by Lutwak [53]. The affine geometry of convex bodies could fill a book in its own right; however, some central problems are still open.

First we recall equation (5.3.34). For $K \in \mathcal{K}_0^n$, the *projection body* ΠK of K was defined by

$$h(\Pi K, u) = V_{n-1}(K|H_{u,0}) = \frac{1}{2}\int_{S^{n-1}} |\langle u, v\rangle| \,\mathrm{d}S_{n-1}(K, v) \qquad (7.4.1)$$

for $u \in \mathbb{E}^n$. Thus ΠK is a zonoid with centre at the origin. Since K is n-dimensional, the origin is an interior point of ΠK, hence its polar body $(\Pi K)^*$ is defined. One abbreviates $(\Pi K)^*$ by $\Pi^* K$ and calls this the *polar projection body* of K.

It is not immediately obvious from the definition, but not difficult to show (see Petty 1967 or Bourgain & Lindenstrauss 1988b) that

$$\Pi \alpha K = \alpha^{-\mathrm{t}}\Pi K \quad \text{for } \alpha \in SL(n).$$

Since also

$$(\alpha K)^* = \alpha^{-\mathrm{t}} K^* \quad \text{for } \alpha \in SL(n),$$

this implies that

$$\Pi^* \alpha K = \alpha \Pi^* K \quad \text{for } \alpha \in SL(n),$$

and, further, that the second projection body $\Pi^2 K := \Pi\Pi K$ satisfies

$$\Pi^2 \alpha K = \alpha \Pi^2 K \quad \text{for } \alpha \in SL(n).$$

If K is an ellipsoid then both $\Pi^* K$ and $\Pi^2 K$ are homothetic to K. Weil (1971) determined the polytopes for which K and $\Pi^2 K$ are homothetic; these are the direct sums of centrally symmetric polygons or segments. Martini (1991) proved that an n-polytope P in $\mathbb{E}^n$, $n \geqq 3$, is a simplex if and only if its polar projection body $\Pi^* P$ is homothetic to its difference body $\mathrm{D}P$.

By suitably combining volumes of associated bodies as considered here with volumes of the original bodies, one can obtain affine-invariant functionals. They are continuous and hence attain their extreme values (see, e.g., Macbeath 1951). It is of considerable interest to determine these and the bodies characterized by the corresponding extremum properties. For centrally symmetric convex bodies $K \in \mathcal{K}_0^n$ it has been conjectured that

$$\kappa_{n-1}^n \kappa_n^{2-n} \leqq V_n(\Pi K) V_n(K)^{1-n} \leqq 2^n. \tag{7.4.2}$$

In the left-hand inequality, the restriction to centrally symmetric bodies is not essential. In fact, suppose that $K \in \mathcal{K}_0^n$ is a convex body for which $V_n(\Pi K) V_n(K)^{1-n}$ is minimal. For $L \in \mathcal{K}_0^n$ we have

$$V(\Pi K, L, \ldots, L) = V(\Pi L, K, \ldots, K),$$

by (5.1.18), (7.4.1) and Fubini's theorem. With $L := \Pi K$, Minkowski's inequality (6.2.2) yields

$$V_n(\Pi K) V_n(K)^{1-n} \geqq V_n(\Pi^2 K) V_n(\Pi K)^{1-n}.$$

By the assumed minimum property of K, the equality sign must hold here, hence $\Pi^2 K$ and K are homothetic. In particular, K must be a zonoid. Thus, the left-hand inequality of (7.4.2) need only be proved for zonoids. It was conjectured by Petty (1972) that this inequality holds and that equality characterizes ellipsoids. The right-hand inequality of (7.4.2), with equality precisely for direct sums of planar centrally symmetric convex bodies, was conjectured by Schneider (1982a). Both conjectures are open.

We remark that an integral representation for the volume of the projection body is obtained from (7.4.1) and Theorem 5.3.1, namely

$$V_n(\Pi K) = \frac{1}{n!} \int_{S^{n-1}} \cdots \int_{S^{n-1}} D_n(u_1, \ldots, u_n)\, \mathrm{d}S_{n-1}(K, u_1) \cdots \mathrm{d}S_{n-1}(K, u_n). \tag{7.4.3}$$

Concerning the volume of the polar projection body, which is given by

$$V_n(\Pi^* K) = \frac{1}{n} \int_{S^{n-1}} V_{n-1}(K | H_{u,0})^{-n}\, \mathrm{d}\sigma(u), \tag{7.4.4}$$

it is known that

$$n^{-n} \binom{2n}{n} \leqq V_n(\Pi^* K) V_n(K)^{n-1} \leqq (\kappa_n / \kappa_{n-1})^n \tag{7.4.5}$$

for $K \in \mathcal{K}_0^n$. The right-hand inequality, together with the equality condition characterizing ellipsoids, was proved by Petty (1972). This inequality is now known as the *Petty projection inequality*. The left-hand inequality in (7.4.5) had been conjectured by Ball (1990+a) and was proved by Zhang (1991); here equality holds if and only if K is a simplex.

For the definition of two other types of associated bodies, it is convenient to enlarge the domain of convex bodies slightly. A *star body* in $\mathbb{E}^n$ is a nonempty compact set K satisfying $[o, x] \subset K$ for all $x \in K$ and such that the *radial function* $\rho(K, \cdot)$, defined by

$$\rho(K, u) := \max\{\lambda \geqq 0 | \lambda u \in K\} \quad \text{for } u \in S^{n-1},$$

is positive and continuous. The set of star bodies in $\mathbb{E}^n$ is denoted by $\mathcal{S}_0^n$.

For $K \in \mathcal{S}_0^n$, there is a unique star body $\mathrm{I}K$ whose radial function satisfies

$$\rho(\mathrm{I}K, u) = V_{n-1}(K \cap H_{u,0}) \quad \text{for } u \in S^{n-1}$$

(where V_{n-1} is $(n-1)$-dimensional Lebesgue measure). It is called the *intersection body* of K. From the result of Busemann mentioned in Section 6.1, Note 7, it follows that $\mathrm{I}K$ is convex if K is convex and centrally symmetric with respect to the origin.

The intersection body plays an essential role in Busemann's (1949b, 1950) theory of area in Minkowski spaces. Let $B \in \mathcal{K}_0^n$ be a convex body that is centrally symmetric with respect to the origin. In the Minkowski space with unit ball B (i.e., $\mathbb{E}^n$ with the norm defined by the gauge function of B), Busemann introduced a notion of Minkowski surface area in such a way that, in particular, the surface area of a convex body $K \in \mathcal{K}_0^n$ is proportional to the mixed volume

$$V(K, \ldots, K, \mathrm{I}^*B),$$

where $\mathrm{I}^*B := (\mathrm{I}B)^*$, the polar of the intersection body of B. It follows from Minkowski's inequality that K solves the isoperimetric problem, that is, has maximal volume among all convex bodies with given Minkowski surface area, if and only if K is homothetic to the convex body I^*B. This body is, therefore, called the *isoperimetrix*.

A different, and in some sense 'dual', notion of surface area in the Minkowski space associated with B was proposed by Holmes & Thompson (1979). In their theory, the surface area of $K \in \mathcal{K}_0^n$ turns out to be proportional to the mixed volume

$$V(K, \ldots, K, \Pi B^*),$$

so that the isoperimetrix is now the zonoid ΠB^*. In both theories, interesting problems remain open. For example, which centrally symmetric convex bodies B yield the minimum or maximum for the affine-invariant functional $V(B, \dots, B, \mathrm{I}^*B)$ or for $V(B, \dots, B, \Pi B^*)$, the 'self-surface area' of the unit ball in the respective theory? And for which B is the isoperimetrix homothetic to B? More explicitly: If $\mathrm{I}B$ is homothetic to B^*, must B be an ellipsoid? If ΠB is homothetic to B^*, must B be an ellipsoid? These questions were posed by Busemann & Petty (1956) and Holmes & Thompson (1979), respectively. Further discussion and related problems can be found in the survey of Lutwak (1989+b).

Returning to the intersection body, we remark that its volume is given by

$$V_n(\mathrm{I}K) = \frac{1}{n} \int_{S^{n-1}} V_{n-1}(K \cap H_{u,0})^n \, \mathrm{d}\sigma(u). \tag{7.4.6}$$

Busemann (1953) proved that

$$V_n(\mathrm{I}K) V_n(K)^{1-n} \leqq \kappa_{n-1}^n \kappa_n^{2-n} \tag{7.4.7}$$

for $K \in \mathcal{K}_{00}^n$, with equality for $n \geqq 3$ if and only if K is an ellipsoid with centre at the origin. (7.4.7) is called the *Busemann intersection inequality*.

For the proof of (7.4.7), Busemann first established an integral-geometric identity of the Blaschke–Petkantschin type, namely

$$V_n(K_1) \cdots V_n(K_{n-1}) =$$
$$\frac{1}{2} \int_{S^{n-1}} \left[\int_{K_1 \cap H_{u,0}} \cdots \int_{K_{n-1} \cap H_{u,0}} D_{n-1}(x_1, \dots, x_{n-1}) \, \mathrm{d}x_1 \cdots \mathrm{d}x_{n-1} \right] \mathrm{d}\sigma(u), \tag{7.4.8}$$

where $\mathrm{d}x_i$ denotes the $(n-1)$-dimensional volume element at x_i in the hyperplane $H_{u,0}$. He then applied Steiner symmetrization to the right-hand side to obtain the inequality

$$V_n(K_1) \cdots V_n(K_{n-1}) \geqq \frac{1}{n} \frac{\kappa_n^{n-2}}{\kappa_{n-1}^n} \int_{S^{n-1}} \prod_{i=1}^{n-1} V_{n-1}(K_i \cap H_{u,0})^{n/(n-1)} \, \mathrm{d}\sigma(u). \tag{7.4.9}$$

Here for $n \geqq 3$ the equality sign holds if and only if the sets K_j are homothetic centred ellipsoids. Inequality (7.4.7) is a special case. Busemann assumed that $K_1, \dots, K_{n-1}$ are convex bodies, but this assumption can be relaxed.

Petty (1961a) reinterpreted (7.4.8) and (7.4.9) by introducing the so-called centroid bodies, which have since proved quite useful for obtaining other affine inequalities. The *centroid body* ΓK of a star body $K \in \mathcal{S}_0^n$ is defined by

$$h(\Gamma K, x) := \frac{1}{V_n(K)} \int_K |\langle x, y \rangle| \mathrm{d}y \quad \text{for } x \in \mathbb{E}^n,$$

where $\mathrm{d}y$ denotes the volume element at y. Thus

$$h(\Gamma K, x) = \frac{1}{(n+1)V_n(K)} \int_{S^{n-1}} |\langle x, u \rangle| \rho(K, u)^{n+1} \, d\sigma(u),$$

and ΓK is a zonoid with centre o. One has

$$\Gamma \alpha K = \alpha \Gamma K \quad \text{for } \alpha \in GL(n)$$

(see Lutwak (1990b), Section 7, for a systematic study of transformation rules for associated bodies). If $K \in \mathcal{K}_0^n$ is centrally symmetric with respect to o, then $\operatorname{bd} \Gamma K$ is the set of all centroids of the intersections $K \cap H_{u,0}^-$, $u \in S^{n-1}$. For this case (and $n = 3$), centroid bodies appear (though not under this name) as early as the paper of Blaschke (1917), §5, where they are attributed to Dupin. Blaschke also posed the problem solved by inequality (7.4.12).

Petty established the formula

$$V(\Gamma K_1, \ldots, \Gamma K_n) = \frac{2^n}{n!} \frac{1}{V_n(K_1) \cdots V_n(K_n)} \int_{K_1} \cdots \int_{K_n} D_n(x_1, \ldots, x_n) \, dx_1 \cdots dx_n, \tag{7.4.10}$$

which also follows from (5.3.35). In particular,

$$V_n(\Gamma K) = \frac{2^n}{n! V_n(K)^n} \int_K \cdots \int_K D_n(x_1, \ldots, x_n) \, dx_1 \cdots dx_n.$$

Together with Busemann's (1953) inequality

$$\int_K \cdots \int_K D_n(x_1, \ldots, x_n) \, dx_1 \cdots dx_n \geqq \frac{2}{n+1} \frac{\kappa_{n+1}^{n-1}}{\kappa_n^{n+1}} V_n(K)^{n+1} \tag{7.4.11}$$

this gives the *Busemann–Petty centroid inequality*

$$V_n(\Gamma K) \geqq \frac{2^{n+1}}{(n+1)!} \frac{\kappa_{n+1}^{n-1}}{\kappa_n^{n+1}} V_n(K) \tag{7.4.12}$$

(or

$$V_n(\Gamma K) \geqq V_n \left(\frac{2\kappa_{n-1}}{(n+1)\kappa_n} K \right), \tag{7.4.13}$$

if the identity $n! \kappa_n \kappa_{n-1} = 2^n \pi^{n-1}$ is used). Here, equality holds if and only if K is a centred ellipsoid.

Now (7.4.10), the inequality

$$V(\Gamma K_1, \ldots, \Gamma K_n)^n \geqq V_n(\Gamma K_1) \cdots V_n(\Gamma K_n), \tag{7.4.14}$$

which follows from the Aleksandrov–Fenchel inequality, and the inequality (7.4.12) together yield a generalization of (7.4.11), namely

$$\int_{K_1} \cdots \int_{K_n} D_n(x_1, \ldots, x_n) \, dx_1 \cdots dx_n \geqq \frac{2}{n+1} \frac{\kappa_{n+1}^{n-1}}{\kappa_n^{n+1}} V_n(K_1)^{(n+1)/n} \cdots V_n(K_n)^{(n+1)/n} \tag{7.4.15}$$

for $K_1, \dots, K_n \in \mathcal{S}_0^n$. The equality sign holds if and only if $K_1, \dots, K_n$ are homothetic centred ellipsoids.

Using the $(n-1)$-dimensional case of (7.4.15) in (7.4.8), Petty obtained (7.4.9) again.

Centroid bodies are also treated in Milman & Pajor (1989) and Lutwak (1990b).

A further type of associated convex body, to be considered now, may seem artificial at first sight, but is quite useful in establishing relations between different associated bodies and in proving affine inequalities. One says that the convex body $K \in \mathcal{K}^n$ has *curvature function* $f(K, \cdot): S^{n-1} \to \mathbb{R}$ if its surface area measure $S_{n-1}(K, \cdot)$ has $f(K, \cdot)$ as a density with respect to spherical Lebesgue measure or, equivalently, if

$$V(L, K, \dots, K) = \frac{1}{n} \int_{S^{n-1}} h(L, u) f(K, u) \, \mathrm{d}\sigma(u)$$

for all $L \in \mathcal{K}^n$. If K is of class C_+^2, then K has the curvature function $f = s_{n-1}$, the reciprocal Gauss curvature, viewed as a function of the unit normal vector. By $\mathcal{F}^n$ we denote the set of all convex bodies in $\mathcal{K}_0^n$ that have a positive continuous curvature function.

First we remark that one can define the *affine surface area* of $K \in \mathcal{F}^n$ by

$$\Omega(K) := \int_{S^{n-1}} f(K, u)^{n/(n+1)} \, \mathrm{d}\sigma(u). \tag{7.4.16}$$

This notion comes originally from affine differential geometry (see Blaschke 1923). Ω is invariant under volume-preserving affine transformations. The (extended) *affine isoperimetric inequality* says that

$$\Omega(K)^{n+1} \leqq n^{n+1} \kappa_n^2 V_n(K)^{n-1} \tag{7.4.17}$$

for $K \in \mathcal{F}^n$, with equality precisely if K is an ellipsoid. Under stronger differentiability assumptions, this is a classical result, due to Blaschke (for $n = 2, 3$ see Blaschke 1916a, 1923; it was extended to higher dimensions by Santaló (1949) and Deicke (1953)), which he obtained by means of Steiner symmetrization. The extension to $\mathcal{F}^n$, including the equality condition, is due to Petty (1985).

For the definition of the envisaged associated body, the curvature image, let $K \in \mathcal{K}_0^n$ be given. For $z \in \operatorname{int} K$, let

$$K^z := (K - z)^*.$$

As noted by Santaló (1949), there is a unique point $s \in \operatorname{int} K$ such that

$$V_n(K^s) \leqq V_n(K^z) \quad \text{for all } z \in \operatorname{int} K.$$

This point is called the *Santaló point* of K. The minimum property of s implies that

$$\int_{S^{n-1}} h(K - s, u)^{-(n+1)}\, u \, d\sigma(u) = o. \tag{7.4.18}$$

Since $\rho(K^s, u) = h(K - s, u)^{-1}$ for $u \in S^{n-1}$, equation (7.4.18) states that K^s has centroid o. Vice versa, if K^* has its centroid at o, then $K = K^{**}$ has its Santaló point at o.

Equation (7.4.18) says that the indefinite σ-integral of $h(K - s, \cdot)^{-(n+1)}$ satisfies the sufficient condition of Minkowski's existence theorem (Theorem 7.1.2). Hence, there exists a convex body $\mathrm{C}K \in \mathcal{K}_0^n$ with curvature function

$$f(\mathrm{C}K, \cdot) = h(K - s, \cdot)^{-(n+1)}. \tag{7.4.19}$$

$\mathrm{C}K$ is unique up to a translation, and to make it unique we may choose the translate that has its centroid at the origin. Clearly $\mathrm{C}K \in \mathcal{F}^n$. The body $\mathrm{C}K$ is called the *curvature image* of K. (Here we follow Lutwak (1990b) rather than [53], where the name 'curvature image' is reserved for the body ΛK, which in Lutwak (1991) was called the 'polar curvature image'. It is defined by

$$f(\Lambda K, \cdot) = \frac{\kappa_n}{V_n(K)} \rho(K, \cdot)^{n+1}, \tag{7.4.20}$$

where $K \in \mathcal{S}^n$ is a star body with centroid at the origin. For comparison, we note that

$$\mathrm{C}K = [V_n(K^s)/\kappa_n]^{1/(n-1)} \Lambda K^s \tag{7.4.21}$$

for $K \in \mathcal{K}_0^n$, up to a translation.) The curvature image satisfies

$$\mathrm{C}\alpha K = \alpha \mathrm{C}K \quad \text{for } \alpha \in SL(n)$$

(see Lutwak 1990b, (7.12)).

Now

$$\begin{aligned} V_n(K^s) &= \frac{1}{n} \int_{S^{n-1}} \rho(K^s, u)^n \, d\sigma(u) \\ &= \frac{1}{n} \int_{S^{n-1}} h(K - s, u)^{-n} \, d\sigma(u) \\ &= \frac{1}{n} \int_{S^{n-1}} f(\mathrm{C}K, u)^{n/(n+1)} \, d\sigma(u) \\ &= \frac{1}{n} \int_{S^{n-1}} h(K - s, u) f(\mathrm{C}K, u) \, d\sigma(u) \end{aligned}$$

and hence

$$nV_n(K^s) = \Omega(\mathrm{C}K) \tag{7.4.22}$$

and also

$$V_n(K^s) = V(K, \mathrm{C}K, \ldots, \mathrm{C}K). \tag{7.4.23}$$

Applying (7.4.23), Minkowski's inequality (6.2.2), the extended affine isoperimetric inequality (7.4.17) and (7.4.22), we arrive at the inequality

$$V_n(K)V_n(K^s) \leqq \kappa_n^2, \tag{7.4.24}$$

where equality holds if and only if K is an ellipsoid. The left-hand side of (7.4.24) is often called the *volume product* of the convex body K. Inequality (7.4.24) is known as the *Blaschke–Santaló inequality*. The approach sketched above is that of Santaló (1949), completed by Petty's (1985) treatment of the equality condition for the extended affine isoperimetric inequality.

For centrally symmetric convex bodies (where the Santaló point coincides with the centre), simpler proofs of the Blaschke–Santaló inequality, using Steiner symmetrization and the Brunn–Minkowski theorem, have been found by Saint Raymond (1981) and Meyer & Pajor (1990). The latter authors have obtained an improved version of (7.4.24) for general convex bodies (see Note 12 below).

There are many relations between the different associated bodies and the affine inequalities mentioned in this section. For a discussion and for generalizations we refer to Lutwak (1985a, 1986a, 1990b) and Lutwak's survey [53] and the literature quoted there.

Notes for Section 7.4

1. *Volumes and quermassintegrals of projections*. Extending definition (7.4.1), one may define a convex body $\Pi_i K$, for $K \in \mathcal{K}^n$ and $i = 1, \ldots, n-1$, by

$$h(\Pi_i K, u) = w_{n-i-1}(K^u) = \frac{1}{2}\int_{S^{n-1}} |\langle u, v\rangle| \, dS_i(K, v)$$

for $u \in S^{n-1}$, where $w_{n-i-1}(K^u)$ denotes the $(n-i-1)$th quermassintegral of $K^u = K|H_{u,0}$. Thus, the zonoid $\Pi_i K$ comprises all the information on the $(n-i-1)$th quermassintegrals of the orthogonal projections of K onto hyperplanes. If conclusions are to be drawn from partial information about volumes or other quermassintegrals of projections, this is equivalent to obtaining information of K from partial knowledge of $\Pi_i K$. Investigations of this kind, in several different directions, are found in Betke & McMullen (1983), Bourgain & Lindenstrauss (1988a, b), Campi (1986, 1988), Chakerian (1967b), Firey (1965a, 1970d), Goodey (1986), Goodey & Groemer (1990), Groemer (1991a), Schneider (1970c, 1977c) and Schneider & Weil (1970); see also the surveys on zonoids by Schneider & Weil [44], §10, and Goodey & Weil [47], Section 5.
2. *Mixed projection bodies*. The definition of the projection body can be extended as follows. For convex bodies $K_1, \ldots, K_{n-1} \in \mathcal{K}^n$, there is a convex body $\Pi(K_1, \ldots, K_{n-1})$ such that

$$h(\Pi(K_1, \ldots, K_{n-1}), u) = v(K_1^u, \ldots, K_{n-1}^u) \quad \text{for } u \in S^{n-1},$$

where $K_i^u = K_i|H_{u,0}$ and v denotes the $(n-1)$-dimensional mixed volume in $H_{u,0}$. $\Pi(K_1, \ldots, K_{n-1})$ is called the *mixed projection body* of $K_1, \ldots,$

K_{n-1}. It was introduced and thoroughly studied by Lutwak (1985b). The *mixed brightness* $\sigma(K_1, \ldots, K_{n-1}; u) = v(K_1^u, \ldots, K_{n-1}^u)$ also appears in an investigation of Chakerian (1967b). Lutwak (1985b, 1986b) obtained a number of inequalities for the volumes of polars of mixed projection bodies. Further, Lutwak (1989+a) proved inequalities for mixed projection bodies that are analogous to the Brunn–Minkowski, Minkowski and Aleksandrov–Fenchel inequalities. Quermassintegrals of (specialized) mixed projection bodies are treated by Lutwak (1990a).

3. *The volume of the projection body and related inequalities*. Petty's conjectured inequality

$$V_n(\Pi K)V_n(K)^{1-n} \leqq \kappa_{n-1}^n \kappa_n^{2-n}$$

can be put into a more symmetric form. Lutwak (1990c) showed that this inequality would follow from (and would imply) the following conjecture: If $K, L \in \mathcal{K}_0^n$, then

$$\int_{\mathrm{bd}\,K}\int_{\mathrm{bd}\,L} |\langle u, v\rangle| \mathrm{d}u\,\mathrm{d}v \geqq \frac{2n\kappa_{n-1}}{\kappa_n^{1-2/n}} [V_n(K)V_n(L)]^{(n-1)/n},$$

with equality if and only if K, L are homothetic to centred polar ellipsoids. Here $\mathrm{d}u$, $\mathrm{d}v$ denote the area elements on $\mathrm{bd}\,L$, $\mathrm{bd}\,K$ whose outer unit normal vectors are u, v and it is assumed that K, L are of class C_+^2.

Lutwak proved the following formal analogue of the above inequality. If $K, L \in \mathcal{K}_{00}^n$, then

$$\int_K\int_L |\langle x, y\rangle| \mathrm{d}x\,\mathrm{d}y \geqq \frac{2n\kappa_{n-1}}{(n+1)^2\kappa_n^{1+2/n}} [V_n(K)V_n(L)]^{(n+1)/n},$$

with equality if and only if K, L are dilates of centred polar ellipsoids. Here $\mathrm{d}x$, $\mathrm{d}y$ denote volume elements. Furthermore, Lutwak proved that

$$\int_K\int_L \langle x, y\rangle^2 \mathrm{d}x\,\mathrm{d}y \geqq \frac{n}{(n+2)^2\kappa_n^{4/n}} [V_n(K)V_n(L)]^{(n+2)/n}, \qquad (7.4.25)$$

with equality if and only if K, L are dilates of centred polar ellipsoids.

Let $K \in \mathcal{K}_0^n$ be centrally symmetric with respect to o. Inequality (7.4.25) with $L = K^*$ gives

$$\int_K\int_{K^*} \langle x, y\rangle^2 \mathrm{d}x\,\mathrm{d}y \geqq \frac{n}{(n+2)^2\kappa_n^{4/n}} [V_n(K)V_n(K^*)]^{(n+2)/n}, \qquad (7.4.26)$$

which was also proved by Ball (1988). Ball conjectured that

$$\frac{n\kappa_n^2}{(n+2)^2} \geqq \int_K\int_{K^*} \langle x, y\rangle^2 \mathrm{d}x\,\mathrm{d}y \qquad (7.4.27)$$

for centrally symmetric K, with equality only for centred ellipsoids. He proved this for K with the property that an affine image of K is symmetric with respect to the coordinate hyperplanes. In view of (7.4.26), inequality (7.4.27) would be stronger than the Blaschke–Santaló inequality (for symmetric bodies).

4. *Volume estimates from projections and sections*. Shephard (1964b) has posed the following question: If $K, L \in \mathcal{K}_0^n$ $(n \geqq 3)$ are centrally symmetric and if

$$V_{n-1}(K|H_{u,0}) > V_{n-1}(L|H_{u,0}) \quad \text{for all } u \in S^{n-1}, \qquad (7.4.28)$$

does it follow that

$$V_n(K) > V_n(L)? \qquad (7.4.29)$$

(Without the assumption of central symmetry, a non-spherical body of constant brightness and a suitable ball provide an immediate counter-example.) Petty (1967) and Schneider (1967c) independently showed that the answer is negative, but that (7.4.28) implies (7.4.29) if K is a zonoid and L is an arbitrary (not necessarily symmetric) convex body. Schneider moreover proved: If L is centrally symmetric, but not a zonoid, and is of class C_+^{n+3}, then there is a centrally symmetric convex body K such that (7.4.28) holds, but $V_n(K) < V_n(L)$.

The analogous question for sections is much harder. It was posed by Busemann & Petty (1956): If $K, L \in \mathcal{K}_0^n$ $(n \geqq 3)$ are centrally symmetric with respect to the origin and if

$$V_{n-1}(K \cap H_{u,0}) > V_{n-1}(L \cap H_{u,0}) \quad \text{for all } u \in S^{n-1}, \qquad (7.4.30)$$

does it follow that

$$V_n(K) > V_n(L)? \qquad (7.4.31)$$

The Busemann intersection inequality (7.4.7), together with its equality condition, implies an affirmative answer in the special case where L is an ellipsoid. Busemann (1960) gave a counterexample for the case where convexity is relaxed to starshapedness with respect to the centre. For three-dimensional coaxial bodies of revolution, affirmative answers were given by Hadwiger (1968c) and Giertz (1969), both of whom admitted also certain non-convex sets.

In analogy to the case of projections described above, Lutwak (1988b) proved that (7.4.30) implies (7.4.31) if L is an intersection body. He also showed that if K is sufficiently smooth and not an intersection body, then there exists a centred star body L such that (7.4.30) holds, but inequality (7.4.31) is reversed. The analogy is remarkable; in particular, the role that mixed volumes play in the treatment of projections is now played by Lutwak's dual mixed volumes.

M. Meyer (1990+) gave an affirmative answer for the case where L is a crosspolytope, and asked whether the result holds true if L is the polar of a zonoid.

A counterexample to the Busemann–Petty problem was first given by Larman & Rogers (1975). In their example, which works for dimensions $n \geqq 12$ only, the body K is a ball, while the body L is not found explicitly; its existence is established by means of a probabilistic argument. It came, therefore, as a considerable surprise when Ball (1988) observed that, for $n \geqq 10$, a suitable ball (for K) and a centred unit cube provide a counterexample. This result, however, depended on the non-trivial fact, established by Ball (1986), that every hyperplane section of the unit cube in $\mathbb{E}^n$ has $(n-1)$-dimensional volume at most $\sqrt{2}$. Replacing the cube by a suitable cylinder, Giannopoulos (1990) found counterexamples for $n \geqq 7$. Further contributions and partial results are due to Tanno (1987), Grinberg & Rivin (1990) and Bourgain (1991). The original question remains open for $3 \leqq n \leqq 6$.

Returning to projections, we mention the following interesting theorem due to Ball (1991): For every convex body $K \in \mathcal{K}_0^n$, there exists an affine image αK of volume 1 satisfying

$$V_{n-1}(\alpha K | H_{u,0}) \geqq 1 \quad \text{for all } u \in S^{n-1}.$$

The example of the cube shows that this result is best possible.

Ball (1990+b) also provided a strong negative answer to Shephard's question discussed above. He showed that (7.4.28) does imply

$$\tfrac{3}{2}\sqrt{n}\, V_n(K) \geqq V_n(L),$$

but that, up to the constant factor, no better estimate is possible for all n. More precisely, there is a constant $\delta > 0$ such that for each $n \in \mathbb{N}$ there is a symmetric body $K \in \mathcal{K}_0^n$ of volume 1 satisfying

$$V_{n-1}(K|H_{u,0}) \geqq \delta\sqrt{n} \quad \text{for all } u \in S^{n-1}.$$

Since for n-balls $B_n^{(1)}$ of volume 1, $V_{n-1}(B_n^{(1)}|H_{u,0})$ tends to a constant for $n \to \infty$, comparison with suitable balls verifies the claim.

For concentric sections, the situation seems to be different. It has been asked whether there exists a constant c, not depending on n, such that (7.4.30) for centrally symmetric convex bodies K, L implies $cV_n(K) \geqq V_n(L)$. The relations of this question to other problems on convex bodies are discussed in Section 5 of Milman & Pajor (1989).

5. *Projection and intersection bodies.* For $K \in \mathcal{K}_{00}^n$ $(n \geqq 3)$, Martini (1989) observed that $\mathrm{I}K \subset \Pi K$. Here equality holds if and only if K is a centred ellipsoid.
6. *Monotoneity for centroid bodies.* Two of the positive results mentioned in the preceding notes can be formulated as follows. Let $K, L \in \mathcal{K}_0^n$.
   (a) If $\Pi K \subset \Pi L$ and L is a projection body, then $V_n(K) \leqq V_n(L)$, with equality only if K and L are translates.
   (b) If $\mathrm{I}K \subset \mathrm{I}L$ and K is an intersection body, then $V_n(K) \leqq V_n(L)$, with equality only if $K = L$.

   Lutwak (1990b), Theorem (9.3), showed that there is a third result of a similar type, this time for centroid bodies.
   (c) If $\Gamma K \subset \Gamma L$ and L is a polar projection body, then $V_n(K) \leqq V_n(L)$, with equality only if $K = L$.
7. *Extensions of the affine surface area.* As mentioned, the notion of affine surface area from affine differential geometry has been extended to the class $\mathcal{F}^n$ of convex bodies with positive continuous curvature functions. Further extensions to arbitrary convex bodies were studied by Leichtweiß (1986, 1988, 1989), Lutwak (1991) and Schütt (1990+). In particular, Lutwak's theory includes an extension of the affine isoperimetric inequality, with the same equality conditions, to arbitrary convex bodies. Schütt (1991), Schütt & Werner (1990, 1990+) and Meyer & Reisner (1991) investigated convex floating bodies, which are closely related to an intuitive interpretation of affine surface area.

   Lutwak (1987b) defined the *mixed affine surface area* of $K_1, \ldots, K_n \in \mathcal{F}^n$ by
   $$\Omega(K_1, \ldots, K_n) := \int_{S^{n-1}} [f(K_1, u) \cdots f(K_n, u)]^{1/(n+1)} \,\mathrm{d}\sigma(u)$$
   and obtained a number of inequalities, for example,
   $$\Omega(K_1, \ldots, K_n)^{n+1} \leqq n^{n+1} \kappa_n^2 V(K_1, \ldots, K_n)^{n-1},$$
   with equality if and only if the K_i are homothetic ellipsoids.
8. *Petty's affine projection inequality.* If the left-hand inequality of conjecture (7.4.2) were true, then this inequality together with the affine isoperimetric inequality (7.4.17) would yield
   $$V_n(\Pi K) \geqq \frac{1}{n^{n+1}} \frac{\kappa_{n-1}^n}{\kappa_n^n} \,\Omega(K)^{n+1}$$
   for $K \in \mathcal{F}^n$. This inequality is known to be true; it was proved by Petty (1967). Equality holds if and only if K is an ellipsoid.
9. *The Petty projection inequality.* It should be observed that the Petty projection inequality (the right-hand inequality of (7.4.5)) provides an improvement of the classical isoperimetric inequality. Let $K \in \mathcal{K}_0^n$. Using the Cauchy surface area formula (see (5.3.27))
   $$S(K) = \frac{1}{\kappa_{n-1}} \int_{S^{n-1}} V_{n-1}(K|H_{u,0}) \,\mathrm{d}\sigma(u),$$

the Hölder (or Jensen) inequality and (7.4.4), one obtains the inequality

$$\frac{S(K)}{n\kappa_n^{1/n}V_n(K)^{(n-1)/n}} \geqq \frac{\kappa_n}{\kappa_{n-1}}\,[V_n(\Pi^*K)V_n(K)^{n-1}]^{-1/n}.$$

By (7.4.5), the right-hand side is $\geqq 1$, so that the isoperimetric inequality results.

Applications of the Petty projection inequality to slightly more remote topics appear in Kiener (1986) and Schneider (1987a).

10. *Geominimal surface area.* Petty (1972, 1974) introduced a new concept of affine surface area, the geominimal surface area, which serves, as he explained, as a connecting link between Minkowski geometry, relative differential geometry and affine differential geometry. For $K \in \mathcal{K}^n$, the *geominimal surface area* of K is defined by

$$G(K) := \inf\{nV(K, \ldots, K, Q^*) \,|\, Q \in \mathcal{K}_0^n,\ \ p_0(Q) = o,\ \ V_n(Q) = \kappa_n\},$$

where p_0 denotes the centroid (Petty's definition is slightly different, but equivalent). The geominimal surface area is invariant under volume-preserving affine transformations; it is homogeneous of degree $n-1$, monotone and continuous, and $G^{1/(n-1)}$ is concave under Minkowski addition. Besides many other results, Petty (1974) proved the inequalities

$$G(K)^n \leqq n^n\kappa_n V_n(K)^{n-1},$$

with equality if and only if K is an ellipsoid (see Petty (1985) for the equality case), and

$$\Omega(K)^{n+1} \leqq n\kappa_n G(K)^n$$

for $K \in \mathcal{F}^n$, where equality holds if and only if K is the curvature image of some convex body. Extensions of these results are proved by Lutwak (1991+b). Variants of the geominimal surface area are obtained if the infimum in its definition extends only over bodies Q restricted to an affine invariant subclass of $\mathcal{K}_0^n$ such as zonoids, polar zonoids or ellipsoids.

11. *A reverse isoperimetric inequality.* The case of ellipsoids mentioned in the preceding note amounts to considering, as a concept of affine surface area, the infimum of the areas of αK, taken over all volume-preserving affine transformations α. Petty (1961b) found the necessary and sufficient conditions for $K \in \mathcal{K}_0^n$ to have minimal surface area among its affine transforms of fixed volume. The analogous question for Minkowski surface area was treated by Clack (1990).

With this notion of affine surface area, a reverse isoperimetric inequality is meaningful. The following theorem was proved by Ball (1991a).

Theorem. If $K \in \mathcal{K}_0^n$ and if T denotes a regular n-simplex, there exists an affine image αK of K satisfying

$$V_n(\alpha K) = V_n(T) \quad \text{and} \quad S(\alpha K) \leqq S(T).$$

If K is centrally symmetric and if Q denotes the n-cube, there exists an affine image αK of K satisfying

$$V_n(\alpha K) = V_n(Q) \quad \text{and} \quad S(\alpha K) \leqq S(Q).$$

12. *The Blaschke–Santaló inequality.* Inequality (7.4.24) was first proved, for $n = 2$ and 3, by Blaschke (1917); see also Blaschke (1923). His approach, for $n = 3$, can be sketched as follows. One starts with a convex body K of class C_+^3. For its boundary surface bd K one defines, in equi-affine differential geometry, a certain Riemannian metric (a suitable multiple of the Euclidean second fundamental form) that is invariant under unimodular affine transformations. If Δ denotes the corresponding Laplace–Beltrami

operator, then $Y := \frac{1}{3}\Delta X$, where X is the position vector of $\operatorname{bd} K$, defines the affine normal vector Y. It traces out a surface, which Blaschke calls the curvature image (Krümmungsbild) of $\operatorname{bd} K$. Suppose (this is an additional assumption on K) that it is the boundary of a convex body $\bar{K}$, and let $\bar{K}^*$ be the polar body of $\bar{K}$. Then

$$3V(K, K, \bar{K}) = 3V_3(\bar{K}^*)\,\Omega,$$

where Ω is the affine surface area of K. Blaschke uses his affine isoperimetric inequality and Minkowski's inequality to deduce that

$$V_3(\bar{K})V_3(\bar{K}^*) \leqq \kappa_3^2.$$

Under his assumptions, equality holds only if $\bar{K}^*$ is an ellipsoid.

In Euclidean terms, one computes that

$$h(\bar{K}, \cdot) = f(K, \cdot)^{-1/4},$$

hence

$$\rho(\bar{K}^* \cdot) = f(K, \cdot)^{1/4}.$$

Furthermore, $\bar{K}^*$ has its centroid at the origin. It follows that the curvature image as defined by (7.4.19) leads from $\bar{K}$ to K (and hence is inverse to Blaschke's curvature image), thus K is a translate of $\mathrm{C}\bar{K}$. Comparison with (7.4.20) shows that K is a dilate of $\Lambda\bar{K}^*$. Blaschke himself pointed out that Minkowski's existence theorem can be used to construct K, starting from a star body $\bar{K}^*$ with centroid at the origin.

Santaló (1949), in his proof of the n-dimensional case of (7.4.24), used the curvature image $\mathrm{C}\bar{K}$ of a given convex body $\bar{K}$. The other ingredients of his proof are essentially the same as in Blaschke's proof, but extended to higher dimensions. Santaló had to apply the classical affine isoperimetric inequality, which requires differentiability properties, to a convex body $\mathrm{C}\bar{K}$ obtained as a solution of Minkowski's existence theorem. He did not mention the necessary assumptions and regularity results for the Minkowski problem that are necessary to justify such a procedure. The difficulties, in particular in connection with the characterization of the equality case, were overcome by Petty (1985).

A different proof for the two-dimensional case of the Blaschke–Santaló inequality was found by Heil (1967b).

Somewhat surprisingly, rather simple and direct proofs of the Blaschke–Santaló inequality (including identification of the equality case) have been found that use classical tools of convexity, namely Steiner symmetrization and the Brunn–Minkowski theorem. For centrally symmetric bodies, such proofs were given by Saint Raymond (1981) and Meyer & Pajor (1989, 1990). For arbitrary convex bodies, Meyer & Pajor (1989) obtained the following result, which is more general than the Blaschke–Santaló inequality and from which they deduced that equality in the latter holds only for ellipsoids.

Theorem. Let $K \in \mathcal{K}_0^n$ and let H be a hyperplane such that $H \cap \operatorname{int} K \neq \emptyset$. Then there exists a point $z \in H \cap \operatorname{int} K$ such that

$$V_n(K)V_n(K^z) \leqq \frac{\kappa_n^2}{4\lambda(1-\lambda)},$$

where $\lambda \in (0, 1)$ is defined by $V_n(K \cap H^+) = \lambda V_n(K)$.

Applications of the Blaschke–Santaló inequality to other problems from the geometry of convex bodies can be found in Lutwak (1983, 1989+c), Schneider (1969a, 1982b, 1987a) and Wieacker (1986).

13. *Lower bounds for the volume product.* It has repeatedly been conjectured that

$$\frac{(n+1)^{n+1}}{(n!)^2} \leqq V_n(K)V_n(K^*) \tag{7.4.32}$$

for $K \in \mathcal{K}_0^n$ with centroid at the origin, with equality precisely for simplices, and that

$$\frac{4^n}{n!} \leqq V_n(K)V_n(K^*) \tag{7.4.33}$$

for convex bodies K with centre o. Equality holds in (7.4.33) for affine transforms of cubes and crosspolytopes, but for $n \geqq 4$ not only for these. The latter fact was first pointed out by Saint Raymond (1981) and shows that the treatment of Guggenheimer (1973) is invalid. Conjectures (7.4.32) and (7.4.33) are still open.

For $n = 2$, inequalities (7.4.32) and (7.4.33) were proved by Mahler (1939b). The equality case, however, was settled by him only under the assumption that K is a polygon.

For the special case where K is a zonoid, inequality (7.4.33) was proved by Reisner (1985) (see also Saint Raymond 1986), and in Reisner (1986) he showed that equality holds only for parallelepipeds. A shorter proof was later given by Gordon, Meyer & Reisner (1988). The case $n = 2$ also settles the equality case in Mahler's inequality (7.4.33) for $n = 2$.

Another case where (7.4.33) has been proved is when K is the unit ball of a normed space with a 1-unconditional basis. Geometrically, this means that some affine transform of K is symmetric with respect to each of the coordinate hyperplanes. For such K, inequality (7.4.33) was proved by Saint Raymond (1981). The equality case was characterized independently by M. Meyer (1986) and Reisner (1987). Equality is attained precisely by bodies K of the following type: K can be obtained from n independent segments with centres at the origin by successively taking either the Minkowski sum or the convex hull of the union.

Non-sharp lower bounds for $V_n(K)V_n(K^*)$ in the case of centrally symmetric convex bodies were obtained by Mahler (1939a) and other authors. The strongest result of this kind is due to Bourgain & Milman (1985a, 1987, announced in 1985b), who proved that there exists a universal constant $c > 0$ such that

$$c^n \leqq \kappa_n^{-2} V_n(K)V_n(K^*).$$

Different proofs were given by Pisier (1989).

Appendix

Spherical harmonics

In several places in the geometry of convex bodies, spherical harmonics prove quite useful. In the present book, they were applied in Sections 3.3, 3.5, 4.3, 6.6 and 7.2 and some of the notes describe results that were obtained by means of this method. In this appendix, we briefly collect, without proofs, the basic facts about spherical harmonics, as they are needed in their applications to convexity. For introductory material, including proofs, we refer to Müller (1966), Seeley (1966) and the literature quoted there; for the connections with representations of the rotation group we refer also to Coifman & Weiss (1968).

A *spherical harmonic of degree m* on S^{n-1} is, by definition, the restriction to S^{n-1} of a homogeneous polynomial f (with respect to Cartesian coordinates) of degree m (or identically zero) on $\mathbb{E}^n$ that satisfies $\Delta f = 0$; here Δ denotes the Laplace operator. The set $\mathcal{S}^m$ of spherical harmonics of degree m is a vector subspace of $C(S^{n-1})$ of dimension

$$\dim \mathcal{S}^m = N(n, m) = \frac{(2m + n - 2)\Gamma(n + m - 2)}{\Gamma(m + 1)\Gamma(n - 1)}. \tag{A.1}$$

As a consequence of their definition, the spherical harmonics are eigenfunctions of the spherical Laplace operator Δ_S; each $Y_m \in \mathcal{S}_m$ satisfies

$$\Delta_S Y_m = -m(m + n - 2)Y_m. \tag{A.2}$$

By $L_2(S^{n-1})$ we denote the Hilbert space of square-integrable real functions on S^{n-1} (with the usual identification of functions coinciding almost everywhere) with scalar product $(\cdot, \cdot)$ defined by

$$(f, g) := \int_{S^{n-1}} fg \, d\sigma;$$

here σ denotes spherical Lebesgue measure on S^{n-1}. The induced L_2-norm is denoted by $\|\cdot\|_2$. Spherical harmonics of different degrees are orthogonal, that is,

$$(f, g) = 0 \quad \text{for } f \in \mathcal{S}^k, \quad g \in \mathcal{S}^j, \quad k \neq j.$$

The system of spherical harmonics is complete, thus

$$(f, Y) = 0 \quad \text{for } Y \in \mathcal{S}^m \text{ and all } m \in \mathbb{N}_0$$

for some $f \in L_2(S^{n-1})$ implies $f = 0$ (almost everywhere). Every continuous real function on S^{n-1} can be approximated uniformly by finite linear combinations of spherical harmonics. Consequently, if

$$\int_{S^{n-1}} Y \, \mathrm{d}\varphi = 0 \quad \text{for } Y \in \mathcal{S}^m \text{ and all } m \in \mathbb{N}_0$$

for some finite signed Borel measure φ on S^{n-1}, then $\varphi = 0$.

In each space $\mathcal{S}^m$ we choose an orthonormal basis $(Y_{m1}, \dots, Y_{mN(n,m)})$. Then $\{Y_{mj} | m \in \mathbb{N}_0, j = 1, \dots, N(n, m)\}$ is a complete orthonormal system in $L_2(S^{n-1})$. For $f \in L_2(S^{n-1})$ we write

$$\pi_m f := \sum_{j=1}^{N(n,m)} (f, Y_{mj}) Y_{mj}; \tag{A.3}$$

then $\pi_m f$ does not depend on the choice of the orthonormal basis, and the map $\pi_m: L_2(S^{n-1}) \to \mathcal{S}^m$ is the orthogonal projection. The Fourier series of $f \in L_2(S^{n-1})$ is given by

$$f \sim \sum_{m=0}^{\infty} \pi_m f;$$

it converges to f in the L_2-norm. The Parseval relation says that

$$\sum_{m=0}^{\infty} \|\pi_m f\|_2^2 = \|f\|_2^2, \tag{A.4}$$

or, more generally, that

$$\sum_{m=0}^{\infty} \sum_{j=1}^{N(n,m)} (f, Y_{mj})(g, Y_{mj}) = (f, g) \tag{A.5}$$

for $f, g \in L_2(S^{n-1})$.

We remark that the projections π_0 and π_1 have a simple meaning. Since $\mathcal{S}^0$ contains only constant functions,

$$\pi_0 f = \frac{1}{\omega_n} \int_{S^{n-1}} f \, \mathrm{d}\sigma$$

is the mean value of f. Let $(e_1, \dots, e_n)$ be an orthonormal basis of $\mathbb{E}^n$ and put

$$S_j(u) := \frac{1}{\sqrt{\kappa_n}} \langle e_j, u \rangle \quad \text{for } u \in S^{n-1}, \quad j = 1, \dots, n.$$

Using (1.7.5), we get

$$(S_i, S_j) = \frac{1}{\kappa_n} \int_{S^{n-1}} \langle e_i, u\rangle \langle e_j, u\rangle \, \mathrm{d}\sigma(u) = \langle e_i, e_j\rangle.$$

Since $\dim \mathcal{S}^1 = n$, we see that $(S_1, \ldots, S_n)$ is an orthonormal basis of $\mathcal{S}^1$ and hence that

$$\begin{aligned}(\pi_1 f)(v) &= \sum_{j=1}^{n} (f, S_j) S_j(v) = \sum_{j=1}^{n} \int_{S^{n-1}} f(u) \frac{1}{\sqrt{\kappa_n}} \langle e_j, u\rangle \, \mathrm{d}\sigma(u) \frac{1}{\sqrt{\kappa_n}} \langle e_j, v\rangle \\ &= \sum_{j=1}^{n} \langle e_j, s_f\rangle \langle e_j, v\rangle = \langle s_f, v\rangle\end{aligned}$$

with

$$s_f := \frac{1}{\kappa_n} \int_{S^{n-1}} f(u) u \, \mathrm{d}\sigma(u).$$

Thus

$$\pi_1 f = \langle s_f, \cdot\rangle.$$

In particular, for the support function h_K of a convex body $K \in \mathcal{K}^n$ we have

$$\pi_0 h_K = \tfrac{1}{2} b(K), \tag{A.6}$$

$$\pi_1 h_K = \langle s(K), \cdot\rangle, \tag{A.7}$$

in view of (1.7.2) and (1.7.3).

If we require spherical harmonics in the form $Y(u) = f(\langle e, u\rangle)$ with fixed $e \in S^{n-1}$, we are led to the Gegenbauer polynomials. Assuming that

$$\nu := \frac{n-2}{2} > 0,$$

the Gegenbauer polynomial C_m^ν of degree m and order ν can be defined by means of the generating function

$$(1 - tx + x^2)^{-\nu} = \sum_{m=0}^{\infty} C_m^\nu(t) x^m.$$

Explicitly,

$$C_m^\nu(t) = a_m^\nu (1 - t^2)^{-\nu+1/2} \left(\frac{\mathrm{d}}{\mathrm{d}t}\right)^m (1 - t^2)^{m+\nu-1/2}, \tag{A.8}$$

where

$$a_m^\nu = C_m^\nu(1) \left(-\frac{1}{2}\right)^m \frac{\Gamma\left(\frac{n-1}{2}\right)}{\Gamma\left(m + \frac{n-1}{2}\right)}, \tag{A.9}$$

$$C_m^\nu(1) = \frac{\Gamma(m + n - 2)}{\Gamma(n-2)\Gamma(m+1)}. \tag{A.10}$$

The function $C_m^\nu(\langle e, \cdot\rangle)$, where $e \in S^{n-1}$ is fixed, is a spherical harmonic of degree m on S^{n-1}.

The addition theorem says that

$$\sum_{j=1}^{N(n,m)} Y_{mj}(u) Y_{mj}(v) = \frac{N(n, m)}{\omega_n C_m^\nu(1)} C_m^\nu(\langle u, v\rangle) \tag{A.11}$$

for $u, v \in S^{n-1}$ and $m \in \mathbb{N}_0$. As a consequence, one obtains that the maximum norm $\|Y_m\|$ of $Y_m \in \mathcal{S}^m$ can be estimated from

$$\|Y_m\| \leqq \left[\frac{N(n, m)}{\omega_n}\right]^{1/2} \|Y_m\|_2. \tag{A.12}$$

The following result is particularly useful.

Funk–Hecke theorem. Let $f \in L_2([-1, 1])$ *be a function satisfying*

$$\int_{-1}^{1} |f(t)|(1 - t^2)^{(n-3)/2}\, \mathrm{d}t < \infty.$$

If $Y_m \in \mathcal{S}^m$ *is a spherical harmonic of degree m, then*

$$\int_{S^{n-1}} f(\langle u, v\rangle) Y_m(v)\, \mathrm{d}\sigma(v) = \lambda_m[f] Y_m(u) \tag{A.13}$$

for $u \in S^{n-1}$, *where*

$$\lambda_m[f] = \omega_{n-1}[C_m^\nu(1)]^{-1} \int_{-1}^{1} f(t) C_m^\nu(t)(1 - t^2)^{(n-3)/2}\, \mathrm{d}t. \tag{A.14}$$

We conclude with a remark on the development of sufficiently smooth support functions into series of spherical harmonics.

Let f be a real function of class C^∞ on S^{n-1} and let

$$f \sim \sum_{m=0}^{\infty} \sum_{j=1}^{N(n,m)} a_{mj} Y_{mj}, \tag{A.15}$$

with $a_{mj} = (f, Y_{mj})$, be its Fourier series with respect to the orthonormal system $\{Y_{mj}\}$. It follows from Theorems 4 and 5 of Seeley (1966) that

$$\mathrm{D}^\alpha f(x/|x|) = \sum_{m=0}^{\infty} \sum_{j=1}^{N(n,m)} a_{mj} \mathrm{D}^\alpha Y_{mj}(x/|x|), \tag{A.16}$$

where $\mathrm{D}^\alpha = \partial^{|\alpha|}/(\partial x_1)^{\alpha_1} \ldots (\partial x_n)^{\alpha_n}$, $|\alpha| = \alpha_1 + \ldots + \alpha_n$, and each of these series is uniformly convergent on S^{n-1}.

Let g be a function of class C^2 on S^{n-1}; let G be its positively homogeneous extension. By $\eta(g, x)$ we denote the smallest eigenvalue of the second differential of G at $x \in S^{n-1}$, restricted to the orthogonal complement of x; and we define

$$\eta(g) := \min\{\eta(g, x) | x \in S^{n-1}\}.$$

By Theorems 1.5.10 and 1.7.1, g is the restriction of a support function if and only if $\eta(g) \geqq 0$.

Now let f be the support function of a convex body K of class C^∞_+. Then $\eta(f) > 0$ by Corollary 2.5.2. Consider the partial sum

$$f_k := \sum_{m=0}^{k} \sum_{j=1}^{N(n,m)} a_{mj} Y_{mj}$$

of (A.15). Using (A.16) and uniform convergence, we may show that $\eta(f_k) \to \eta(f)$ for $k \to \infty$, hence $\eta(f_k) > 0$ for almost all k. It follows that the partial sum f_k is the restriction of a support function for all sufficiently large k.

Appendix Note

Applications to convex bodies. Spherical harmonics in space and Fourier series in the plane have been applied to a great variety of geometric problems on convex bodies. The first applications of this kind were Hurwitz's (1901) well-known proof of the isoperimetric inequality for closed (not necessarily convex) curves in the plane, the thorough investigation of Hurwitz (1902) in the plane and in three-space, and Minkowski's (1904) characterization of convex bodies of constant width. In the many different later contributions, two main principles of application are dominant. This is either the use of the Parseval relation to obtain quadratic inequalities, or the connection of spherical harmonics with irreducible representations of the rotation group. In particular, this often permits us to solve certain rotation invariant linear functional equations on the sphere by finding the spherical harmonics that satisfy the equation, and a number of geometric questions can be reduced to such equations.

Many authors have made geometric applications of Fourier series and spherical harmonics in the domain of convexity. Among them are Aleksandrov (1937b), Anikonov & Stepanov (1981), Berg (1969b), Blaschke (1914), [5], §23, (1916b, 1923), §90, (1936), Bol (1939), Bourgain (1987, 1991), Bourgain & Lindenstrauss (1988a, b, 1989), Bourgain, Lindenstrauss & Milman (1989), Campi (1981, 1984, 1986, 1988), Chernoff (1969), Cieślak & Góźdź (1987), Dinghas (1939, 1940a), Falconer (1983), Fillmore (1969), Fisher (1987), Focke (1969a), Fuglede (1986b, 1989), Fujiwara (1915, 1919), Fujiwara & Kakeya (1917), Geppert (1937), Gericke (1940a, b, 1941), Goodey & Groemer (1990), Görtler (1937a, b, 1938), Groemer (1990a, 1991a), Hayashi (1924, 1926a, b, c), Inzinger (1949), Kameneckii (1947), Knothe (1957b), Kubota (1918, 1920, 1922, 1925a, b), Matsumura (1930), Meissner (1909, 1918), Minoda (1941), Nakajima (1920), Oishi (1920), Petty (1961a), Rešetnjak (1968a), Sas (1939), Schneider (1967c, 1970a, d, 1971a, b, c, d, 1974a, 1975c, 1977c, 1984, 1989c), Shephard (1968b), Su (1927a, b). For details, see Groemer [49].

REFERENCES

A. Books on convexity

[1] Aleksandrov, A. D. 1955, *Die innere Geometrie der konvexen Flächen*. Akademie-Verlag, Berlin (Russian original: 1948).

[2] Aleksandrov, A. D. 1958, *Konvexe Polyeder*. Akademie-Verlag, Berlin, (Russian original: 1950).

[3] Bair, J. & Fourneau, R. 1980, *Etude géométrique des espaces vectoriels, II. Polyèdres et polytopes convexes. Lecture Notes in Math.* **802**, Springer, Berlin.

[4] Benson, R. V. 1966, *Euclidean Geometry and Convexity*. McGraw-Hill, New York.

[5] Blaschke, W. 1956, *Kreis and Kugel*, 2nd edn. W. de Gruyter, Berlin (1st edn: 1916).

[6] Boltjanski, V. G. & Soltan, P. S. 1978, *Combinatorial Geometry of Several Classes of Convex Sets* (in Russian). 'Stiinca', Kishinev.

[7] Bonnesen, T. 1929, *Les problèmes des isopérimètres et des isépiphanes*. Gauthiers-Villars, Paris.

[8] Bonnesen, T. & Fenchel, W. 1934, *Theorie der konvexen Körper*. Springer, Berlin. Reprint: Chelsea Publ. Co., New York, 1948. English translation: BCS Associates, Moscow, Idaho, 1987.

[9] Bourbaki, N. 1953, *Espaces vectoriels topologiques, Chap. I, II*. Hermann, Paris.

[10] Bourbaki, N. 1955, *Espaces vectoriels topologiques, Chap. III, IV, V*. Hermann, Paris.

[11] Brøndsted, A. 1983, *An Introduction to Convex Polytopes*. Springer, New York.

[12] Busemann, H. 1958, *Convex Surfaces*. Interscience, New York.

[13] Eggleston, H. G. 1958, *Convexity*. Cambridge Univ. Press.

[14] Fenchel, W. 1951, *Convex Cones, Sets and Functions*. Mimeographed lecture notes, Princeton Univ.

[15] Grünbaum, B. 1967, *Convex Polytopes*. Interscience, London.

[16] Guggenheimer, H. W. 1977, *Applicable Geometry. Global and Local Convexity*. R. E. Krieger, Huntington, NY.

[17] Hadwiger, H. 1955, *Altes und Neues über konvexe Körper*. Birkhäuser, Basel.

[18] Hadwiger, H. 1957, *Vorlesungen über Inhalt, Oberfläche und Isoperimetrie*. Springer, Berlin.

[19] Hadwiger, H., Debrunnner, H. E. & Klee V. 1964, *Combinatorial Geometry in the Plane*. Holt, Rinehart & Winston, New York.

[20] Jaglom, I. M. & Boltjanski W. G. 1956, *Konvexe Figuren*. Deutsch. Verl. d. Wiss., Berlin (Russian original: 1951).

[21] Kelly P. J. & Weiss, M. L. 1979, *Geometry and Convexity*. John Wiley, New York.
[22] Lay, S. R. 1982, *Convex Sets and Their Applications*. John Wiley, New York.
[23] Leichtweiß, K. 1980, *Konvexe Mengen*. Springer, Berlin; VEB Deutsch. Verl. d. Wiss., Berlin.
[24] Lyusternik, L. A. 1963, *Convex Figures and Polyhedra*. Dover, New York (Russian original: 1956).
[25] Marti, J. T. 1977, *Konvexe Analysis*. Birkhäuser Verlag, Basel.
[26] McMullen, P. & Shephard G. C. 1971, *Convex Polytopes and the Upper Bound Conjecture*. Cambridge Univ. Press.
[27] Pogorelov, A. V. 1973, *Extrinsic Geometry of Convex Surfaces. Translations of Mathematical Monographs* **35**, Amer. Math. Soc., Providence, RI (Russian original: 1969).
[28] Roberts, A. W. & Varberg, D. E. 1973, *Convex Functions*. Academic Press, New York.
[29] Rockafellar, R. T. 1970, *Convex Analysis*. Princeton Univ. Press, Princeton, NJ.
[30] Valentine, F. A. 1964, *Convex Sets*. McGraw-Hill, New York.

B. Some survey articles

[31] Chakerian, G. D. & Groemer, H. 1983, Convex bodies of constant width. In: *Convexity and its Applications* (eds P. M. Gruber & J. M. Wills), Birkhäuser, Basel, pp. 49–96.
[32] Danzer, L., Grünbaum, B. & Klee, V. 1963, Helly's theorem and its relatives. In: *Convexity, Proc. Symposia Pure Math.*, vol. VII, Amer. Math. Soc., Providence, RI, pp. 101–80.
[33] Gruber, P. M. 1983, Approximation of convex bodies. In: *Convexity and its Applications* (eds P. M. Gruber & J. M. Wills), Birkhäuser, Basel, pp. 131–62.
[34] Gruber, P. M. 1984, Aspects of convexity and its applications. *Expo. Math.* **2**, 47–83.
[35] Gruber, P. & Höbinger, J. 1976, Kennzeichnungen von Ellipsoiden mit Anwendungen. In: *Jahrbuch Überblicke Mathematik*, Bibliographisches Institut, Mannheim, pp. 9–29.
[36] Gruber, P. & Schneider, R. 1979, Problems in geometric convexity. In: *Contributions to Geometry, Proc. Geometry Symp.*, Siegen 1978 (eds J. Tölke & J. M. Wills), Birkhäuser, Basel, pp. 255–78.
[37] Grünbaum, B. 1963, Measures of symmetry for convex sets. In: *Convexity, Proc. Symposia Pure Math.*, vol. VII, Amer. Math. Soc., Providence, RI, pp. 233–70.
[38] Klee, V. 1971, What is a convex set? *Amer. Math. Monthy* **78**, 616–31.
[39] McMullen, P. 1979, Transforms, diagrams, and representations. In: *Contributions to Geometry, Proc. Geometry Symp.*, Siegen 1978 (eds J. Tölke & J. M. Wills), Birkhäuser, Basel, pp. 92–130.
[40] McMullen, P. & Schneider, R. 1983, Valuations on convex bodies. In: *Convexity and its Applications* (eds P. M. Gruber & J. M. Wills), Birkhäuser, Basel, pp. 170–247.
[41] Osserman, R. 1978, The isoperimetric inequality. *Bull. Amer. Math. Soc.* **84**, 1182–238.
[42] Petty, C. M. 1983, Ellipsoids. In: *Convexity and its Applications* (eds P. M. Gruber & J. M. Wills), Birkhäuser, Basel, pp. 264–76.
[43] Schneider, R. 1979, Boundary structure and curvature of convex bodies. In: *Contributions to Geometry, Proc. Geometry Symp.*, Siegen 1978 (eds J. Tölke & J. M. Wills), Birkhäuser, Basel, pp. 13–59.
[44] Schneider, R. & Weil, W. 1983, Zonoids and related topics. In: *Convexity and its Applications* (eds P. M. Gruber & J. M. Wills), Birkhäuser, Basel, pp. 296–317.

[45] Zamfirescu, T. 1991, Baire categories in convexity. *Atti Sem. Fis. Univ. Modena* **39**, 139–64.

The following articles in the *Handbook of Convex Geometry* (eds P. M. Gruber & J. M. Wills), North-Holland, Amsterdam (to appear 1993), are particularly relevant to the topic of this book.

[46] Florian, A., Extremum problems for convex discs and polyhedra.
[47] Goodey, P. & Weil, W., Zonoids and generalizations.
[48] Groemer, H., Stability of geometric inequalities.
[49] Groemer, H., Fourier series and spherical harmonics.
[50] Gruber, P. M., The space of convex bodies.
[51] Gruber, P. M., Baire categories in convexity.
[52] Leichtweiß, K., Convexity and differential geometry.
[53] Lutwak, E., Selected affine isoperimetric inequalities.
[54] Sangwine-Yager, J. R., Mixed volumes.
[55] Schneider, R., Convex surfaces, curvature and surface area measures.
[56] Schneider, R. & Wieacker, J. A., Integral geometry.

C. Articles and monographs

A date followed by a plus sign, e.g. '1990+', indicates that the article to which it refers appeared in preprint form in that year.

Ahrens, I. 1970, Über unendliche Reihen von Punktmengen bei Minkowskischer Addition und Subtraktion. Dissertation, Freie Universität Berlin.
1972, Über Folgen und Reihen von Punktmengen. *Arch. Math.* **23**, 92–103.
Aleksandrov, A. D., 1933, A theorem on convex polyhedra (in Russian). *Trudy Fiz.-Mat. Inst. Im. Steklova* **4**, 87.
1937a, Zur Theorie der gemischten Volumina von konvexen Körpern, I. Verallgemeinerung einiger Begriffe der Theorie der konvexen Körper (in Russian). *Mat. Sbornik* N.S. **2**, 947–72.
1937b, Zur Theorie der gemischten Volumina von konvexen Körpern, II. Neue Ungleichungen zwischen den gemischten Volumina und ihre Anwendungen (in Russian). *Mat. Sbornik* N.S. **2**, 1205–38.
1937c, Uber die Frage nach der Existenz eines konvexen Körpers, bei dem die Summe der Hauptkrümmungsradien eine gegebene positive Funktion ist, welche den Bedingungen der Geschlossenheit genügt. *C.R. (Doklady) Acad. Sci. URSS* **14**, 59–60.
1938a, Zur Theorie der gemischten Volumina von konvexen Körpern, III. Die Erweiterung zweier Lehrsätze Minkowskis über die konvexen Polyeder auf beliebige konvexe Flächen (in Russian). *Mat. Sbornik* N.S. **3**, 27–46.
1938b, Zur Theorie der gemischten Volumina von konvexen Körpern, IV. Die gemischten Diskriminanten und die gemischten Volumina (in Russian). *Mat. Sbornik* N.S. **3**, 227–51.
1939a, Anwendung des Satzes über die Invarianz des Gebietes auf Existenzbeweise (in Russian). *Izv. Akad. Nauk SSSR* **3**, 243–56.
1939b, Über die Oberflächenfunktion eines konvexen Körpers (in Russian). *Mat. Sbornik* N.S. **6**, 167–74.
1939c, Almost everywhere existence of the second differential of a convex function and some properties of convex surfaces connected with it (in Russian). *Uchenye Zapiski Leningrad. Gos. Univ., Math. Ser.* **6**, 3–35.
1942a, Existence and uniqueness of a convex surface with a given integral curvature. *C.R. (Doklady) Acad. Sci. URSS* **35**, 131–4.
1942b, Smoothness of the convex surface of bounded Gaussian curvature. *C. R. (Doklady) Acad. Sci. URSS* **36**, 195–9.

1956/8, Uniqueness theorems for surfaces in the large, I, II, III, IV, V (in Russian). *Vestnik Leningrad. Univ.* **11** (1956), No. 19, 5–17; **12** (1957), No. 7, 15–44; **13** (1958), No. 7, 14–26; **13** (1958), No. 13, 27–34; **13** (1958), No. 19, 5–8. English translation: *Amer. Math. Soc. Transl., Ser.* 2, **21** (1962), 341–416.

1961, A congruence condition for closed convex surfaces (in Russian). *Vestnik Leningrad. Univ.* **16**, 5–7.

1967, On mean values of support functions (in Russian) *Dokl. Akad. Nauk SSSR* **172**, 755–8. English translation: *Soviet Math. Dokl.* **8**, 149–53.

Alexander, R. 1988, Zonoid theory and Hilbert's fourth problem. *Geom. Dedicata* **28**, 199–211.

Alfsen, E. M. 1971, *Compact Convex Sets and Boundary Integrals*. Springer, Berlin.

Allendoerfer, C. B. 1948, Steiner's formula on a general S^{n+1}. *Bull. Amer. Math. Soc.* **54**, 128–35.

Ambartzumian, R. V. 1987, Combinatorial integral geometry, metrics, and zonoids. *Acta Appl. Math.* **9**, 3–27.

Anderson, R. D. & Klee, V. L. 1952, Convex functions and upper semi-continuous collections. *Duke Math. J.* **19**, 349–57.

Anikonov, Yu. E. & Stepanov, V. N. 1981, Uniqueness and stability of the solution of a problem of geometry in the large (in Russian). *Mat. Sbornik* **116** (1981), 539–46.

Arnold, R. 1989, Zur L^2-Bestapproximation eines konvexen Körpers durch einen bewegten konvexen Körper. *Monatsh. Math.* **108**, 277–93.

Arrow, K. J. & Hahn, F. H. 1971, *General Competitive Analysis*. Holden-Day, San Francisco.

Artstein, Z. 1974, On the calculus of closed set-valued functions. *Indiana Univ. Math. J.* **24**, 433–41.

1980, Discrete and continuous bang-bang and facial spaces or: Look for the extreme points. *SIAM Review* **22**, 172–85.

1987+ Convergence of sums of random sets. Preprint.

Artstein, Z. & Vitale, R. A. 1975, A strong law of large numbers for random compact sets. *Ann. Probab.* **3**, 879–82.

Ash, R. B. 1972, *Measure, Integration, and Functional Analysis*. Academic Press, New York.

Asplund, E. 1963, A k-extreme point is the limit of k-exposed points. *Israel J. Math.* **1**, 161–2.

1973, Differentiability of the metric projection in finite dimensional Euclidean space. *Proc. Amer. Math. Soc.* **38**, 218–19.

Assouad, P. 1980, Charactérisations de sous-espaces normés de L^1 de dimension finie. *Séminaire d'analyse fonctionnelle* (Ecole Polytechnique Palaiseau), 1979–80, exposé No. 19.

Baddeley, A. J. 1980, Absolute curvatures in integral geometry. *Math. Proc. Camb. Phil. Soc.* **88**, 45–58.

Baebler, F. 1957, Zum isoperimetrischen Problem. *Arch. Math.* **8**, 52–65.

Bair, J. 1975, Une mise au point sur la décomposition des convexes. *Bull. Soc. Roy. Sci. Liège* **44**, 698–705.

1976a, Extension du théorème de Straszewicz. *Bull. Soc. Roy. Sci. Liège* **45**, 166–8.

1976b, Une étude des sommands d'un polyèdre convexe. *Bull. Soc. Roy. Sci. Liège* **45**, 307–11.

1977a, Description des sommands de quelques ensembles convexes. *Rend. Ist. Mat. Univ. Trieste* **9**, 83–92.

1977b, Décompositions d'un convexe a l'aide de certains de ses dilatés. *Bull. Soc. Roy. Sci. Liège* **46**, 167–71.

1979, Sur la structure extrémale de la somme de deux convexes. *Canad. Math. Bull.* **22**, 1–7.

Bair, J., Fourneau, R. & Jongmans, F. 1977, Vers la domestication de l'extrémisme. *Bull.*

Soc. Roy. Sci. Liège **46**, 126–32.
Bair, J. & Jongmans, F. 1977, Sur les graves questions qui naissent quand la décomposition des ensembles atteint un stade avancé. *Bull. Soc. Roy. Sci. Liège* **46**, 12–26.
Ball, K. 1986, Cube slicing in R^n. *Proc. Amer. Math. Soc.* **97**, 465–73.
1988, Some remarks on the geometry of convex sets. In: *Geometric Aspects of Functional Analysis* (eds J. Lindenstrauss & V. D. Milman), *Lecture Notes in Math.* **1317**, Springer, Heidelberg, pp. 224–31.
1991a, Volume ratios and a reverse isoperimetric inequality. *J. London Math. Soc.* **44**, 351–9.
1991b, Shadows of convex bodies. *Trans. Amer. Math. Soc.* **327**, 891–901.
Banchoff, T. F. 1967, Critical points and curvature for embedded polyhedra. *J. Differential Geom.* **1**, 245–56.
1970, Critical points and curvature for embedded polyhedral surfaces. *Amer. Math. Monthly* **77**, 475–85.
Bandle, C. 1980, *Isoperimetric Inequalities and Applications*. Pitman, Boston.
Bandt, Ch. 1986, On the metric structure of hyperspaces with Hausdorff metric. *Math. Nachr.* **129**, 175–83.
1988, Estimates for the Hausdorff dimension of hyperspaces. In: *General Topology and its Relations to Modern Analysis and Algebra VI, Proc. 6th Prague Topological Symp. 1986*, Heldermann, Berlin, pp. 33–42.
Bandt, Ch. & Baraki, G. 1986, Metrically invariant measures on locally homogeneous spaces and hyperspaces. *Pacific J. Math,* **121**, 13–28.
Bangert, V. 1977, Konvexität in riemannschen Mannigfaltigkeiten. Dissertation, Univ. Dortmund.
1979, Analytische Eigenschaften konvexer Funktionen auf Riemannschen Mannigfaltigkeiten. *J. reine angew. Math.* **307/308**, 309–24.
Bantegnie, R. 1975, Convexité des hyperespaces. *Arch. Math.* **26**, 414–20.
Bárány, I. 1989, Intrinsic volumes and f-vectors of random polytopes. *Math. Ann.* **285**, 671–99.
Bárány, I. & Larman D. G. 1988, Convex bodies, economic cap coverings, random polytopes. *Mathematika* **35**, 274–91.
Bárány, I. & Zamfirescu, T. 1990, Diameters in typical convex bodies. *Canad. J. Math.* **42**, 50–61.
Barthel, W., 1959a, Zum Busemannschen und Brunn–Minkowskischen Satz. *Math. Z.* **70** (1958/59), 407–29.
1959b, Zur isodiametrischen und isoperimetrischen Ungleichung in der Relativgeometrie. *Comment. Math. Helvet.* **33**, 241–57.
Barthel, W & Bettinger, W. 1961, Die isoperimetrische Ungleichung für die innere Minkowskische Relativoberfläche. *Math. Ann.* **142** (1960/61), 322–7.
1963, Bemerkungen zum isoperimetrischen Problem. *Arch. Math.* **14**, 424–9.
Barthel, W & Franz, G. 1961, Eine Verallgemeinerung des Busemannschen Satzes vom Brunn–Minkowskischen Typ. *Math. Ann.* **144**, 183–98.
Batson, R. G. 1988, Necessary and sufficient conditions for boundedness of extreme points of unbounded convex sets. *J. Math. Anal. Appl.* **130**, 365–74.
Bauer, H. 1964, *Konvexität in topologischen Vektorräumen*. Vorlesungsausarbeitung, Hamburg.
1974, *Wahrscheinlichkeitstheorie und Grundzüge der Maßtheorie*. 2nd edn, W. de Gruyter, Berlin.
Baum, D. 1973a, Zur analytischen Darstellung des Randes konvexer Körper durch Stützhyperboloide, I. Halbachsenfunktion und Scheitelsätze. *Manuscripta Math.* **9**, 113–41.
1973b, Zur analytischen Darstellung des Randes konvexer Körper durch Stützhyperboloide, II. Die Randdarstellung durch die Halbachsenfunktion. *Manuscripta Math.* **9**, 307–22.

1987, Bemerkungen zur Halbachsenfunktion. *Analysis* **7**, 221–35.
1989, Randdarstellungen konvexer Körper. *Analysis* **9**, 185–94.
Beer, G. A. 1975, Starshaped sets and the Hausdorff metric. *Pacific J. Math.* **61**, 21–7.
1989, Support and distance functionals for convex sets. *Numer. Funct. Anal. Optim.* **10**, 15–36.
Bensimon, D. 1987, Sur l'espace topologique des points extrémaux d'un convexe compact. *C.R. Acad. Sci. Paris, Sér. I Math.* **304**, 391–2.
Benson, D. C. 1970, Sharpened forms of the plane isoperimetric inequality. *Amer. Math. Monthly* **77**, 29–34.
Berg, Ch. 1969a, Shephard's approximation theorem for convex bodies and the Milman theorem. *Math. Scand.* **25**, 19–24.
1969b, Corps convexes et potentiels sphériques. *Danske Vid. Selskab. Mat.-fys. Medd.* **37**, 6.
1971, Abstract Steiner points for convex bodies. *J. London Math. Soc.* (2) **4**, 176–80.
Bernštein, D. N. 1975, The number of roots of a system of equations. *Functional Anal. Appl.* **9**, 183–5.
Berwald, L. & Varga, O. 1937, Integralgeometrie 24. Über die Schiebungen im Raum. *Math. Z.* **42**, 710–36.
Besicovitch, A. S. 1963a, On singular points of convex surfaces. In: *Convexity, Proc. Symp. Pure Math.*, vol. 7, Amer. Math. Soc., Providence, RI, pp. 21–3.
1963b, On the set of directions of linear segments on a convex surface. In: *Convexity, Proc. Symp. Pure Math.*, vol. 7, Amer. Math. Soc., Providence, RI, pp. 24–5.
Betke, U. 1992, Mixed volumes of polytopes. *Arch. Math.* **58**, 388–91.
Betke, U. & Goodey, P. R. 1984, Continuous translation invariant valuations on convex bodies. *Abh. Math. Sem. Univ. Hamburg* **54**, 95–105.
Betke, U. & Gritzmann, P. 1986, An application of valuation theory to two problems in discrete geometry. *Discrete Math.* **58**, 81–5.
Betke, U. & Kneser, M. 1985, Zerlegungen und Bewertungen von Gitterpolytopen. *J. reine angew. Math.* **358**, 202–8.
Betke, U. & McMullen, P. 1983, Estimating the sizes of convex bodies from projections. *J. London Math. Soc.* **27**, 525–38.
Betke, U. & Weil, W. 1991, Isoperimetric inequalities for the mixed area of plane convex sets. *Arch. Math.* **57**, 501–7.
Björck, G. 1958, The set of extreme points of a compact convex set. *Ark. Mat.* **3**, 463–8.
Blackwell, D. 1951, The range of certain vector integrals. *Proc. Amer. Math. Soc.* **2**, 390–5.
Blaschke, W. 1914, Beweise zu Sätzen von Brunn und Minkowski über die Minimaleigenschaft des Kreises. *Jahresber. Deutsche Math.-Ver.* **23**, 210–34. *Ges. Werke* (1985b), pp. 106–30.
1915, Konvexe Bereiche gegebener konstanter Breite und kleinsten Inhalts. *Math. Ann.* **76**, 504–13. *Ges. Werke* (1985b), pp. 157–66.
1916a, Über affine Geometrie I: Isoperimetrische Eigenschaften von Ellipse und Ellipsoid. *Ber. Verh. Sächs. Akad. Leipzig, Math.-Phys. Kl.* **68**, 217–39. *Ges. Werke* (1985b), pp. 181–203.
1916b, Eine kennzeichnende Eigenschaft des Ellipsoids und eine Funktionalgleichung auf der Kugel. *Ber. Verh. Sächs. Akad. Leipzig, Math.-Phys. Kl.* **68**, 129–36. *Ges. Werke* (1985b), pp. 173–80.
1917, Affine Geometrie IX: Verschiedene Bemerkungen und Aufgaben. *Ber. Verh. Sächs. Akad. Leipzig, Math.-Phys. Kl.* **69**, 412–20. *Ges. Werke* (1985b), pp. 259–67.
1923, *Vorlesungen über Differentialgeometrie, II. Affine Differentialgeometrie.* Springer, Berlin.
1924, *Vorlesungen über Differentialgeometrie, I.* Springer, Berlin.
1933, Über die Schwerpunkte von Eibereichen. *Math. Z.* **36**, 166. *Ges. Werke* (1985b), p. 350.

1936, Integralgeometrie 10: Eine isoperimetrische Eigenschaft der Kugel. *Bull. Math. Soc. Roum. Sci.* **37**, 3–7. *Ges. Werke* (1985a), pp. 265–9.

1937a, *Vorlesungen über Integralgeometrie.* 3rd edn, VEB Deutsch. Verl. d. Wiss., Berlin, 1955 (1st edn: Part I, 1935; Part II, 1937).

1937b, Integralgeometrie 21, Über Schiebungen. *Math. Z.* **42**, 399–410. *Ges. Werke* (1985a), pp. 337–48.

1985a, *Gesammelte Werke* (eds W. Burau *et al.*), vol. 2, *Kinematik und Integralgeometrie.* Thales, Essen.

1985b, *Gesammelte Werke* (eds W. Burau *et al.*), vol. 3, *Konvexgeometrie.* Thales, Essen.

Blaschke, W. & Reidemeister, K. 1922, Über die Entwicklung der Affingeometrie. *Jahresber. Deutsche Math.-Ver.* **31**, 63–82.

Blind, R. 1977, Eine Charakterisierung der Sphäre im E^3. *Manuscripta Math.* **21**, 243–53.

Blumenthal, L. M. 1953, *Theory and Applications of Distance Geometry*. Clarendon Press, Oxford.

Boček, L. 1983, Eine Verschärfung der Poincaré-Ungleichung. *Casopis pěst. mat.* **108**, 78–81.

Böhm, J. & Hertel, E. 1980, *Polyedergeometrie in n-dimensionalen Räumen konstanter Krümmung.* VEB Deutsch. Verl. d. Wiss., Berlin.

Bokowski, J. 1973, Eine verschärfte Ungleichung zwischen Volumen, Oberfläche und Inkugelradius im R^n. *Elem. Math.* **28**, 43–4.

Bokowski, J., Hadwiger, H. & Wills, J. M. 1976, Eine Erweiterung der Croftonschen Formeln für konvexe Körper. *Mathematika* **23**, 212–19.

Bokowski, J. & Heil, E. 1986, Integral representations of quermassintegrals and Bonnesen-style inequalities. *Arch. Math.* **47**, 79–89.

Bokowski, J. & Mani-Levitska, P. 1984, Zur Approximation konvexer Körper im E^n. *Ann. Discrete Math.* **20**, 319–20.

1987, Approximation of convex bodies by polytopes with uniformly bounded valencies. *Monatsh. Math.* **104**, 261–4.

Bol, G. 1939, Zur Theorie der konvexen Körper. *Jahresber. Deutsche Math.-Ver.* **49**, 113–23.

1940, Ein isoperimetrische Problem. *Nieuw Arch. Wiskunde* (2) **20** (194), 171–5.

1941, Isoperimetrische Ungleichungen für Bereiche auf Flächen. *Jahresber. Deutsche Math.-Ver.* **51**, 219–57.

1942, Zur Theorie der Eikörper. *Jahresber. Deutsche Math.-Ver.* **52**, 250–66.

1943a, Einfache Isoperimetrie-Beweise für Kreis und Kugel. *Abh. Math. Sem. Univ. Hamburg* **15**, 27–36.

1943b, Beweis einer Vermutung von H. Minkowski. *Abh. Math. Sem. Univ. Hamburg* **15**, 37–56.

Bol, G. & Knothe, H. 1949, Über konvexe Körper mit Ecken und Kanten. *Arch. Math.* **1** (1948/1949), 427–31.

Bolker, E. D. 1969, A class of convex bodies. *Trans. Amer. Math. Soc.* **145**, 323–46.

1971, The zonoid problem (Research Problem). *Amer. Math. Monthly* **78**, 529–31.

Bonnesen, T. 1921, Über eine Verschärfung der isoperimetrischen Ungleichheit des Kreises in der Ebene und auf der Kugeloberfläche nebst einer Anwendung auf eine Minkowskische Ungleichheit für konvexe Körper. *Math. Ann.* **84**, 216–27.

1924, Über das isoperimetrische Defizit ebener Figuren. *Math. Ann.* **91**, 252–68.

1932, Beweis einer Minkowskischen Ungleichung. *Mat. Tidsskr.* B, 25–7.

Bonnice, W. & Klee, V. L. 1963, The generation of convex hulls. *Math. Ann.* **152**, 1–29.

Böröczky, K., Bárány, I., Makai, E. jr. & Pach, J. 1986, Maximal volume enclosed by plates and proof of the chessboard conjecture. *Discrete Math.* **60**, 101–20.

Borwein, J. M. & O'Brien, R. C. 1978, Cancellation characterizes convexity. *Nanta Math.* **11**, 100–2.

Bose, R. C. & Roy, S. N. 1935a, Some properties of the convex oval with reference to its perimeter centroid. *Bull. Calcutta Math. Soc.* **27**, 79–86.

1935b, A note on the area centroid of a closed convex oval. *Bull. Calcutta Math. Soc.* **27**, 111–18.

1935c, On the four centroids of a closed surface. *Bull. Calcutta Math. Soc.* **27**, 119–46.

Botts, T. 1942, Convex sets. *Amer. Math. Monthly* **49**, 527–35.

Bourgain, J. 1987, Remarques sur les zonoïdes (Projection bodies, etc.). In: *Sémin. d'analyse fonctionnelle*, Paris 1985/1986/1987, *Publ. Math. Univ. Paris VII* **28**, 171–86.

1991 On the Busemann–Petty problem for perturbations of the ball. *Geom. Functional Anal.* **1**, 1–13.

Bourgain, J. & Lindenstrauss, J. 1988a, Nouveaux résultats sur les zonoïdes et les corps de projection. *C.R. Acad. Sci. Paris* **306**, 377–80.

1988b, Projection bodies. In: *Geometric Aspects of Functional Analysis* (eds J. Lindenstrauss & V. D. Milman), *Lecture Notes in Math.* **1317**, Springer, Berlin, pp. 250–70.

1988c, Distribution of points on spheres and approximation by zonotopes. *Israel J. Math.* **64**, 25–31.

1989 Almost euclidean sections in spaces with a symmetric basis. In: *Geometric Aspects of Functional Analysis* (eds J. Lindenstrauss & V. D. Milman), *Lecture Notes in Math.* **1376**, Springer, Berlin, pp. 278–88.

Bourgain, J., Lindenstrauss, J. & Milman, V. D. 1986, Sur l'approximation de zonoïdes par des zonotopes. *C.R. Acad. Sci. Paris* **303**, 987–8.

1988, Minkowski sums and symmetrizations. In: *Geometric Aspects of Functional Analysis* (eds. J. Lindenstrauss & V. D. Milman), *Lecture Notes in Math.* **1317**, Springer, Berlin, pp. 44–66.

1989, Approximation of zonoids by zonotopes. *Acta Math.* **162**, 73–141.

Bourgain, J. & Milman, V. 1985a, On Mahler's conjecture on the volume of a convex symmetric body and its polar. Preprint IHES/M/85/17.

1985b, Sections euclidiennes et volume des corps symétriques convexes dans R^n. *C.R. Acad. Sci. Paris* **300**, 435–8.

1987, New volume ratio properties for convex symmetric bodies in R^n. *Invent. Math.* **88**, 319–40.

Brascamp, H. J. & Lieb, E. H. 1976, On extensions of the Brunn–Minkowski and Prékopa–Leindler theorems, including inequalities for log concave functions, and with an application to the diffusion equation. *J. Functional Anal.* **22**, 366–89.

Brehm, U. & Kühnel, W. 1982, Smooth approximation of polyhedral surfaces regarding curvatures. *Geom. Dedicata* **12**, 435–61.

Brøndsted, A. 1966, Convex sets and Chebyshev sets, II. *Math. Scand.* **18**, 5–15.

1977, The inner aperture of a convex set. *Pacific. J. Math.* **72**, 335–40.

1986, Continuous barycenter functions on convex polytopes. *Expo. Math.* **4**, 179–87.

Bronshtein, E. M. 1976, ε-entropy of convex sets and functions (in Russian). *Sibirskii Mat. Zh.* **17**, 508–14. English translation: *Siberian Math. J.* **17**, 393–8.

1978a, ε-entropy of the affine-equivalent convex bodies and Minkowski compacta (in Russian). *Optimizatsiya* **22**, 5–11.

1978b, Extremal convex functions (in Russian). *Sibirskii Mat. Zh.* **19**, 10–18. English translation: *Siberian Math J.* **19**, 6–12.

1978c, Extremal convex functions and sets (in Russian). *Optimizatcija* **22**, 12–23.

1979, Extremal H-convex bodies (in Russian). *Sibirskii Mat. Zh.* **20** (1979), 412–15. English translation: *Siberian Math. J.* **20**, 295–7.

1981, On extreme boundaries of finite-dimensional convex compacts (in Russian). *Optimizatsiya* **26** (43), 119–28.

Brooks, J. N. & Strantzen, J. B. 1989, Blaschke's rolling theorem in R^n. *Memoirs Amer. Math. Soc.* **80**, 101 pp.

Brown, A. L. 1980, Chebyshev sets and facial systems of convex sets in finite-dimensional spaces. *Proc. London Math. Soc.* **41**, 297–339.

Brunn, H. 1887, Über Ovale und Eiflächen. Dissertation, München.

1889, Über Curven ohne Wendepunkte. Habilitationsschrift, München.
1894, Referat über eine Arbeit: Exacte Grundlagen für eine Theorie der Ovale. *S.-B. Bayer. Akad. Wiss.*, 93–111.
Budach, L. 1989, Lipschitz–Killing curvatures of angular partially ordered sets. *Adv. Math.* **78**, 140–67.
Buldygin, V. V. & Kharazishvili, A. B. 1985, *The Brunn–Minkowski Inequality and its Applications* (in Russian), 'Naukova Dumka', Kiev.
Bunt, L. N. H. 1934, Bijdrage tot de theorie der convexe puntverzamelingen. Thesis, Univ. Groningen, Amsterdam.
Burago, Ju. D. & Zalgaller, V. A. 1978, Sufficient criteria of convexity. *J. Soviet Math.* **10**, 395–435.
1988, *Geometric Inequalities*. Springer, Berlin (Russian original: 1980).
Burckhardt, J. J. 1940, Über konvexe Körper mit Mittelpunkt. *Vierteljahresschr. Naturf. Ges. Zürich* **85**, 149–54.
Burton, G. R. 1976, On the sum of a zonotope and an ellipsoid. *Comment. Math. Helvet.* **51**, 369–87.
1979a, On the nearest-point map of a convex body. *J. London Math. Soc.* **19**, 147–52.
1979b, The measure of the s-skeleton of a convex body. *Mathematika* **26** (1979), 290–301.
1980a, Skeleta and sections of convex bodies. *Mathematika* **27**, 97–103.
1980b, Subspaces which touch a Borel subset of a convex surface. *J. London Math. Soc.* **21**, 167–70.
Burton, G. R. & Larman D. G. 1981, An inequality for skeleta of convex bodies. *Arch. Math.* **36**, 378–84.
Busemann, H. 1947, The isoperimetric problem in the Minkowski plane. *Amer. J. Math.* **69**, 863–71.
1949a, A theorem on convex bodies of the Brunn–Minkowski type. *Proc. Nat. Acad. Sci. USA.* **35**, 27–31.
1949b, The isoperimetric problem for Minkowski area. *Amer. J. Math.* **71**, 743–62.
1950, The foundations of Minkowskian geometry. *Comment. Math. Helvet.* **24**, 156–86.
1953, Volume in terms of concurrent cross-sections. *Pacific J. Math.* **3**, 1–12.
1959, Minkowski's and related problems for convex surfaces with boundaries. *Michigan Math. J.* **6**, 259–66.
1960, Volumes and areas of cross-sections. *Amer. Math. Monthly* **67**, 248–50. Correction: *ibid.*, p. 671.
Busemann, H. & Feller, W. 1935a, Bemerkungen zur Differentialgeometrie der konvexen Flächen, I. Kürzeste Linien auf differenzierbaren Flächen. *Mat. Tidsskr.* B, 25–36.
1935b, Bemerkungen zur Differentialgeometrie der konvexen Flächen, II. Über die Krümmungsindikatrizen. *Mat. Tidsskr.* B, 87–115.
1936a, Krümmungseigenschaften konvexer Flächen. *Acta Math.* **66**, 1–47.
1936b, Bemerkungen zur Differentialgeometrie der konvexen Flächen, II. Über die Gaussche Krümmung. *Mat. Tidsskr.* B, 41–70.
Busemann, H. & Petty, 1956, Problems on convex bodies. *Math. Scand.* **4**, 88–94.
Campi, S. 1981, On the reconstruction of a function on a sphere by its integrals over great circles. *Boll. Un. Mat. Ital.*, Ser. V, **18**, 195–215.
1984, On the reconstruction of a star-shaped body from its 'half-volumes'. *J. Austral. Math. Soc.*, Ser. A, 37, 243–57.
1986, Reconstructing a convex surface from certain measurements of its projections. *Boll. Un. Mat. Ital.* B (6) **5**, 945–59.
1988, Recovering a centred convex body from the areas of its shadows: A stability estimate. *Ann. Mat. Pura Appl.* **151**, 289–302.
Cassels, J. W. S. 1975, Measures of the non-convexity of sets and the Shapley–Folkman–Starr theorem. *Math. Proc. Camb. Phil. Soc.* **78**, 433–6.
Chakerian, G. D. 1966, Sets of constant width. *Pacific J. Math.* **19**, 13–21.

1967a, Inequalities for the difference body of a convex body. *Proc. Amer. Math. Soc.* **18**, 879–84.

1967b, Sets of constant relative width and constant relative brightness. *Trans. Amer. Math. Soc.* **129**, 26–37.

1971, Higher dimensional analogues of an isoperimetric inequality of Benson. *Math. Nachr.* **48**, 33–41.

1972, The mean volume of boxes and cylinders circumscribed about a convex body. *Israel J. Math.* **12**, 249–56.

1973, Isoperimetric inequalities for the mean width of a convex body. *Geom. Dedicata* **1**, 356–62.

Chakerian, G. D. & Sangwine-Yager, J. R. 1979, A generalization of Minkowski's inequality for plane convex sets. *Geom. Dedicata* **8**, 437–44.

Cheeger, J., Müller, W. & Schrader, R. 1984, On the curvature of piecewise flat spaces. *Commun. Math. Phys.* **92**, 405–454.

1986, Kinematic and tube formulas for piecewise linear spaces. *Indiana Univ. Math. J.* **35**, 737–54.

Cheng, S.-Y. & Yau, S.-T. 1976, On the regularity of the solution of the n-dimensional Minkowski problem. *Commun. Pure Appl. Math.* **29**, 495–516.

Chern, S. S. 1959, Integral formulas for hypersurfaces in euclidean space and their applications to uniqueness theorems. *J. Math. Mech.* **8**, 947–55.

Chern, S. S. & Lashof, R. K. 1958, On the total curvature of immersed manifolds, II. *Michigan Math. J.* **5**, 5–12.

Chernoff, P. R. 1969, An area–width inequality for convex curves. *Amer. Math. Monthly* **76**, 34–5.

Choquet, G. 1969a, *Lectures on Analysis*, vol. II. W. A. Benjamin, Reading MA.

1969b, *Lectures on Analysis*, vol. III, W. A. Benjamin, Reading, MA.

1969c, Mesures coniques, affines et cylindriques. In: *Proc. Symposia Math.*, INDAM, Rome 1968, vol. II, Academic Press, London, pp. 145–82.

Choquet, G., Corson, H. & Klee, V. 1966, Exposed points of convex sets. *Pacific J. Math.* **17**, 33–43.

Christoffel, E. B. 1865, Über die Bestimmung der Gestalt einer krummen Oberfläche durch lokale Messungen auf derselben. *J. reine angew. Math.* **64**, 193–209. *Ges. Math. Abh.*, vol. I, pp. 162–77. Teubner, Leipzig, 1910.

Cieślak, W. & Góźdź, S. 1987, On curves which bound special convex sets. *Serdica* **13**, 281–6.

Clack, R. 1990, Minkowski surface area under affine transformations. *Mathematika* **37**, 232–8.

Coifman, R. R. & Weiss, G. 1968, Representations of compact groups and spherical harmonics. *L'Einseignement math.* **14**, 121–73.

Collier, J. B. 1975, On the set of extreme points of a convex body. *Proc. Amer. Math. Soc.* **47** (1975), 184–6.

1976, On the facial structure of a convex body. *Proc. Amer. Math. Soc.* **61**, 367–70.

Corson, H. H. 1965, A compact convex set in E^3 whose exposed points are of the first category. *Proc. Amer. Math. Soc.* **16**, 1015–21.

Coxeter, H. S. M. 1962, The classification of zonohedra by means of projective diagrams. *J. Math. pures appl.* **41**, 137–56.

1963, *Regular Polytopes*, 2nd edn. Macmillan, New York.

Cressie, N. 1978, A strong limit theorem for random sets. *Suppl. Adv. Appl. Prob.* **10**, 36–46.

Curtis, D. W., Quinn, J. & Schori, R. M. 1977, The hyperspace of compact convex subsets of a polyhedral 2-cell. *Houston J. Math.* **3**, 7–15.

Czipszer, J. 1962, Über die Parallelbereiche nach innen von Eibereichen. *Publ. Math. Inst. Hungar. Acad. Sci.* **7**A, 197–202.

Dalla, L. 1985, The n-dimensional Hausdorff measure of the n-skeleton of a convex

w-compact set (body). *Math. Nachr.* **123**, 131–5.
1986, Increasing paths on the one-skeleton of a convex compact set in a normed space. *Pacific J. Math.* **124**, 289–94.
1987, On the measure of the one-skeleton of the sum of convex compact sets. *J. Austral. Math. Soc.* Ser. A **42**, 385–9.
Dalla, L. & Larman, D. G. 1980, Convex bodies with almost all *k*-dimensional sections polytopes. *Math. Proc. Camb. Phil. Soc.* **88**, 395–401.
Das Gupta, S. 1980, Brunn–Minkowski inequality and its after-math. *J. Multivariate Anal.* **10**, 296–318.
Davis, C. 1954, Theory of positive linear dependence. *Amer. J. Math.* **76**, 733–46.
Davy, P. J. 1976, Projected thick sections through multidimensional particle aggregates. *J. Appl. Prob.* **13**, 714–22. Correction: *J. Appl. Prob.* **15**, 456.
1978, Stereology – a statistical viewpoint. Thesis, Austral. Nat. Univ., Canberra.
1980a, The estimation of centroids in stereology. *Mikroskopie* (Wien) **37** (Suppl.), 94–100.
1980b, The stereology of location. *J. Appl. Prob.* **17**, 860–4.
1981, Interspersion of phases in a material. *J. Microscopy* **121**, 3–12.
Debrunner, H. 1955, Zu einem maßgeometrischen Satz über Körper konstanter Breite. *Math. Nachr.* **13**, 165–7.
Debs, G. 1978, Applications affines ouvertes et convexes compacts stables. *Bull. Sc. math.* (2) **102**, 401–14.
Deicke, A. 1953, Über die Finsler-Räume mit $A_i = 0$. *Arch. Math.* **4**, 45–51.
Delgado, J. A. 1979, Blaschke's theorem for convex hypersurfaces. *J. Differential Geom.* **14**, 489–96.
de Rham, G. 1957, Sur quelques courbes définies par des équations fonctionnelles. *Rend. Sem. Mat. Univ. Politecn. Torino* **16** (1956/57), 101–13.
Dierolf, P. 1970, Über den Auswahlsatz von Blaschke und ein Problem von Valentine. *Manuscripta Math.* **3**, 289–304.
Dinghas, A. 1939, Zur Theorie der konvexen Körper im *n*-dimensionalen Raum. *Abh. Preuß. Akad. Wiss., Phys.-Math. Kl.* **4**, 30 pp.
1940a, Geometrische Anwendungen der Kugelfunktionen. *Nachr. Ges. Wiss. Gött., Math.-Phys. Kl.*, **1**, 213–35.
1940b, Verallgemeinerung eine Blaschkeschen Satzes über konvexe Körper konstanter Breite. *Rev. Math. Union Interbalkan.* **3**, 17–20.
1940c, Verschärfung der isoperimetrischen Ungleichung für konvexe Körper mit Ecken. *Math. Z.* **47**, 669–75.
1942, Isoperimetrische Ungleichung für konvexe Bereiche mit Ecken. *Math. Z.* **48**, 428–40.
1943a, Über die lineare isoperimetrische Ungleichung für konvexe Polygone und Kurven mit Ecken. *Monatsh. Math. Phys.* **51**, 35–45.
1943b, Verschärfung der Minkowskischen Ungleichungen für konvexe Körper. *Monatsh. Math. Phys.* **51**, 46–56.
1944, Über einen geometrischen Satz von Wulff für die Gleichgewichtsform von Kristallen. *Zeitschr. Kristallographie* **105**, 304–14.
1948, Bemerkung zu einer Verschärfung der isoperimetrischen Ungleichung durch H. Hadwiger. *Math. Nachr.* **1**, 284–6.
1949a, Zur Theorie der konvexen Rotationskörper im *n*-dimensionalen Raum. *Math. Nachr.* **2**, 124–40.
1949b, Neuer Beweis einer verschärften Minkowskischen Ungleichung für konvexe Körper. *Math. Z.* **51**, 306–16.
1949c, Neuer Beweis einer isoperimetrischen Ungleichung von Bol. *Math. Z.* **51**, 469–73.
1949d, Über eine neue isoperimetrische Ungleichung für konvexe Polyeder. *Math. Ann.* **120**, 533–8.

1956, Zum Minkowskischen Integralbegriff abgeschlossener Mengen. *Math. Z.* **66**, 173–88.

1957a, Über das Verhalten der Entfernung zweier Punktmengen bei gleichzeitiger Symmetrisierung derselben. *Arch. Math.* **8**, 46–57.

1957b, Über eine Klasse superadditiver Mengenfunktionale von Brunn–Minkowski–Lusternikschem Typus. *Math. Z.* **68**, 111–25.

1961, *Minkowskische Summen und Integrale, superadditive Mengenfunktionale, isoperimetrische Ungleichungen.* Gauthier-Villars, Paris.

Diskant, V. I. 1964, Stability in Liebmann's theorem (in Russian). *Dokl. Akad. Nauk SSSR* **158**, 1257–59. English translation: *Soviet Math.* **5**, 1387–90.

1965, Theorems of stability for surfaces close to the sphere (in Russian). *Sibirskii Mat. Zh.* **6**, 1254–66.

1968, Stability of a sphere in the class of convex surfaces of bounded specific curvature (in Russian). *Sibirskii Mat. Zh.* **9**, 816–24. English translation: *Siberian Math. J.* **9**, 610–15.

1971a, Bounds for convex surfaces with bounded curvature functions (in Russian). *Sibirskii Mat. Zh.* **12**, 109–25. English translation: *Siberian Math. J.* **12**, 78–89.

1971b, Convex surfaces with bounded mean curvature (in Russian). *Sibirskii Mat. Zh.* **12**, 659–63. English translation: *Siberian Math. J.* **12**, 469–72.

1972, Bounds for the discrepancy between convex bodies in terms of the isoperimetric difference (in Russian). *Sibirskii Mat. Zh.* **13**, 767–72. English translation: *Siberian Math. J.* **13**, 529–32 (1973).

1973a, Stability of the solution of the Minkowski equation (in Russian). *Sibirskii Mat. Zh.* **14**, 669–73. English translation: *Siberian Math. J.* **14**, 466–69.

1973b, Strengthening of an isoperimetric inequality (in Russian). *Sibirskii Mat. Zh.* **14**, 873–77. English translation: *Siberian Math. J.* **14**, 608–11.

1973c, A generalization of Bonnesen's inequalities (in Russian). *Dokl. Akad. Nauk SSSR* **213**, 519–21. English translation: *Soviet Math. Dokl.* **14**, 1728–31.

1975, The stability of the solutions of the generalized Minkowski equations for a ball (in Russian). *Ukrain. Geom. Sb.* **18**, 53–9.

1976, Stability of a convex body under a change in the $(n-2)$nd curvature function (in Russian). *Ukrain. Geom. Sb.* **19**, 22–33.

1979, On the question of the order of the stability function in Minkowski's problem (in Russian). *Ukrain. Geom. Sb.* **22**, 45–7.

1982, Stability of the solution of the Minkowski equation for the surface area of convex bodies (in Russian). *Ukrain. Geom. Sb.* **25**, 43–51.

1984, A counterexample to an assertion of Bonnesen and Fenchel (in Russian). *Ukrain. Geom. Sb.* **27**, 31–3.

1985, Stability in Aleksandrov's problem for a convex body, one of whose projections is a ball (in Russian). *Ukrain. Geom. Sb.* **28**, 50–62, English translation: *J. Sov. Math.* **48** (1990), 41–9.

Dor, L. E. 1976, Potentials and isometric embeddings in L_1. *Israel. J. Math.* **24**, 260–8.

Drešević, M. 1970, A certain generalization of Blaschke's theorem to the class of m-convex sets (in Serbo-Croatian). *Mat. Vesnik* **7**, 223–6.

Dubins, L. E., 1962, On extreme points of convex sets. *J. Math. Anal. Appl.* **5**, 237–44.

Dubois, C. & Jongmans, F. 1982, Invitation à la gestion des stocks de cônes. *Bull. Soc. Roy. Sci. Liège* **51**, 107–20.

Dudley, R. 1974, Metric entropy of some classes of sets with differentiable boundaries. *J. Approximation Th.* **10**, 227–36. Corrigendum: *ibid.* **26**, 192–3.

1977, On second derivatives of convex functions. *Math. Scand.* **41**, 159–74.

Duporcq, E. 1896, Sur les centres de gravité des courbes parallèles. *Bull. Soc. Math. France* **24**, 192–4.

1897, Sur les centres de gravité des surfaces parallèles à une surface fermée. *C.R. Acad. Sci. Paris* **124**, 492–3.

Dziechcińska-Halamoda, Z. & Szwiec, W. 1985, On critical sets of convex polyhedra. *Arch. Math.* **44**, 461–6.

Eckhoff, J. 1979, Radon's theorem revisited. In: *Contributions to Geometry, Proc. Geometry Symp. Siegen 1978* (eds J. Tölke & J. M. Wills), Birkhäuser, Basel, pp. 164–85.

1980, Die Euler-Charakteristik von Vereinigungen konvexer Mengen im R^d. *Abh. Math. Sem. Univ. Hamburg* **50**, 133–44.

Edelstein, M. 1965, An example concerning exposed points. *Canad. Math. Bull.* **8**, 323–7.

Edelstein, M. & Fesmire, S. 1975, On the extremal structure and closure of sums of convex sets. *Bull. Soc. Roy. Sci. Liège* **44**, 590–9.

Efimow, N. W. 1957, *Flächenverbiegung im Großen.* Akademie-Verlag, Berlin.

Eggleston, H. G., Grünbaum, B. & Klee, V. 1964, Some semicontinuity theorems for convex polytopes and cell complexes. *Comment. Math. Helvet.* **39**, 165–88.

Eifler, L. Q. 1977, Semi-continuity of the face-function for a convex set. *Comment. Math. Helvet.* **52**, 325–8.

Ekeland, I. & Temam, R. 1976, *Convex Analysis and Variational Problems.* North-Holland, Amsterdam.

Ewald, G. 1964, Über die Schattengrenzen konvexer Körper. *Abh. Math. Sem. Univ. Hamburg* **27**, 167–70.

1965, Von Klassen konvexer Körper erzeugte Hilberträume. *Math. Ann.* **162**, 140–6.

1967, On Busemann's theorem of the Brunn–Minkowski type. In: *Proc. Colloquium Convexity*, Copenhagen 1965, Københavns Univ. Mat. Inst., pp. 69–71.

1988, On the equality case in Aleksandrov–Fenchel's inequality for convex bodies. *Geom. Dedicata* **28**, 213–20.

Ewald, G., Larman, D. G. & Rogers, C. A. 1970, The directions of the line segments and of the r-dimensional balls on the boundary of a convex body in Euclidean space. *Mathematika* **17**, 1–20.

Ewald, G. & Shephard, G. C. 1966, Normed vector spaces consisting of classes of convex sets. *Math. Z.* **91**, 1–19.

Falconer, K. J. 1983, Applications of a result on spherical integration to the theory of convex sets. *Amer. Math. Monthly* **90**, 690–3.

Faro Rivas, R. 1986, Aproximacion de cuerpos convexos simetricos. Thesis, Univ. de Extremadura, 1986, 147 pp.

Fáry, I. 1961, Functionals related to mixed volumes. *Illinois J. Math.* **5**, 425–30.

1983, Expectation of the isoperimetric deficit: Rényi's integral. *Z. Wahrscheinlichkeitsth. verw. Geb.* **63**, 289–95.

1986, The isoperimetric deficit as the integral of a positive function. *Bull. Sci. Math.* (2) **110**, 315–34. (This work is erroneous.)

Fáry, I. & Makai, E. jr. 1982, Isoperimetry in variable metric. *Studia Sci. Math. Hungar.* **17**, 143–58.

Favard, J. 1929, Problèmes d'extremums relatifs aux courbes convexes, I. *Ann. Sci. Ec. Norm. Sup.* **46**, 345–69.

1930, Sur les inegalités de Minkowski, *Mat. Tidsskr.* B, 33–40.

1933a, Sur les corps convexes. *J. Math. pures appl.* (9) **12**, 219–82.

1933b, Sur la détermination des surfaces convexes. *Bull. Acad. Roy. Belgique (Cl. Sci.)* **19**, 65–75.

1938, Sur la détermination des surfaces convexes. *C.R. Acad. Sci. Paris* **208**, 871–3.

Federer, H. 1959, Curvature measures. *Trans. Amer. Math. Soc.* **93**, 418–91.

1969, *Geometric Measure Theory*. Springer, Berlin.

Fedotov, V. P. 1978a, On the notions of faces of a convex compactum (in Russian). *Ukrain. Geom. Sb.* **21**, 131–41.

1978b, On the sum of pth surface functions (in Russian). *Ukrain. Geom. Sb.* **21**, 125–31.

1979a, Isolated faces of a convex compactum (in Russian). *Mat. Zametki* **25**, 139–47.

English translation: *Math. Notes* **25**, 73–7.
1979b, A counterexample to a hypothesis of Firey (in Russian). *Mat. Zametki* **26**, 269–75. English translation: *Math. Notes* **26**, (1980), 625–9.
1982, Polar representation of a convex compactum (in Russian). *Ukrain. Geom. Sb.* **25**, 137–8.
Fejes Tóth, L. 1948, The isepiphan problem for n-hedra. *Amer. J. Math.* **70**, 174–80.
1950, Elementarer Beweis enier isoperimetrischen Ungleichung. *Acta Math. Acad. Sci. Hungar.* **1**, 273–5.
Fenchel, W. 1936a, Inégalités quadratiques entre les volumes mixtes des corps convexes. *C.R. Acad. Sci. Paris* **203**, 647–50.
1936b, Généralisation du théorème de Brunn et Minkowski concernant les corps convexes. *C.R. Acad. Sci. Paris* **203**, 764–6.
1938, Über die neuere Entwicklung der Brunn–Minkowskischen Theorie der konvexen Körper. In: *Proc. 9th Congr. Math. Scand.*, Helsingfors 1938, pp. 249–72.
Fenchel, W. & Jessen, B. 1938, Mengenfunktionen und konvexe Körper. *Danske Vid. Selskab. Mat.-fys. Medd.* **16**, 3.
Ferrers, N. M. 1861, Note on a geometrical theorem of Mr. Steiner. *Quarterly J. Math.* **4**, 92–4.
Fet, A. I. 1963, Stability theorems for convex, almost spherical surfaces (in Russian). *Dokl. Akad Nauk SSSR* **153**, 537–9. English translation: *Soviet Math.* **4**, 1723–5.
Figiel, T., Lindenstrauss, J. & Milman, V. 1977, The dimension of almost spherical sections of convex bodies. *Acta Math.* **129**, 53–94.
Filliman, P. 1988, Extremum problems for zonotopes. *Geom. Dedicata* **27**, 251–62.
Fillmore, J. R. 1969, Symmetries of surfaces of constant width. *J. Differential Geom.* **3**, 103–10.
Firey, W. J. 1961a, Polar means of convex bodies and a dual to the Brunn–Minkowski theorem. *Canad. J. Math.* **13**, 444–53.
1961b, Mean cross-section measures of harmonic means of convex bodies. *Pacific J. Math.* **11**, 1263–6.
1962, p-means of convex bodies. *Math. Scand.* **10**, 17–24.
1963, The mixed area of a convex body and its polar reciprocal. *Israel J. Math.* **1**, 201–2.
1964, A generalization of an inequality of H. Groemer. *Monatsh. Math.* **68**, 393–5.
1965a, The brightness of convex bodies. Technical Report no. 19, Oregon State Univ.
1965b, Lower bound for volumes of convex bodies. *Arch. Math.* **16**, 69–74.
1966, Lower bounds for quermassintegrals of a convex body. *Portugaliae Math.* **25**, 141–6.
1967a, Blaschke sums of convex bodies and mixed bodies. In: *Proc. Colloquium Convexity*, Copenhagen 1965, Københavns Univ. Mat. Inst. 1967, pp. 94–101.
1967b, The determination of convex bodies from their mean radius of curvature functions. *Mathematika* **14**, 1–13.
1967c, Generalized convex bodies of revolution. *Canad. J. Math.* **14**, 972–96.
1968, Christoffel's problem for general convex bodies. *Mathematika* **15**, 7–21.
1969, An upper bound for volumes of convex bodies. *J. Austral. Math. Soc.* **9**, 503–10.
1970a, Local behaviour of area functions of convex bodies. *Pacific J. Math.* **35**, 345–57.
1970b, The determination of convex bodies by elementary symmetric functions of principal radii of curvature. Mimeographed manuscript.
1970c, Intermediate Christoffel–Minkowski problems for figures of revolution. *Israel J. Math.* **8**, 384–90.
1970d, Convex bodies of constant outer p-measure. *Mathematika* **17**, 21–7.
1972, An integral-geometric meaning for lower order area functions of convex bodies. *Mathematika* **19**, 205–12.
1973, Support flats to convex bodies. *Geom. Dedicata* **2**, 225–48.
1974a, Kinematic measures for sets of support figures. *Mathematika* **21**, 270–81.

1974b, Approximating convex bodies by algebraic ones. *Arch. Math.* **25**, 424–5.
1974c, Extension of some integral formulas of Chern. *Geom. Dedicata* **3**, 325–33.
1974d, Shapes of worn stones. *Mathematika* **21**, 1–11.
1975, Some open questions on convex surfaces. In: *Proc. Int. Congr. Math.*, Vancouver 1974 (ed. R. D. James), Canad. Mathematical Congress, pp. 479–84.
1976, A functional characterization of certain mixed volumes. *Israel J. Math.* **24**, 274–81.
1977, Addendum to R. E. Miles' paper on the fundamental formula of Blaschke in integral geometry. *Austral. J. Statist.* **19**, 155–6.
1979, Inner contact measures. *Mathematika* **26**, 106–12.
1981, Subsequent work on Christoffel's problem about determining a surface from local measurements. In: *E. B. Christoffel*, Aachen/Monschau 1979 (eds P. Butzer & F. Fehér), Birkhäuser, Basel, pp. 721–3.
Firey, W. J. & Grünbaum, B. 1964, Addition and decomposition of convex polytopes. *Israel J. Math.* **2**, 91–100.
Firey, W. J. & Schneider, R. 1979, The size of skeletons of convex bodies. *Geom. Dedicata* **8**, 99–103.
Fisher, J. C. 1987, Curves of constant width from a linear viewpoint. *Math. Mag.* **60**, 131–40.
Flaherty, F. J. 1973, Curvature measures for piecewise linear manifolds. *Bull. Amer. Math. Soc.* **79**, 100–2.
Flanders, H. 1966, The Steiner point of a closed hypersurface. *Mathematika* **13**, 181–8.
1968, A proof of Minkowski's inequality for convex curves. *Amer. Math. Monthly* **75**, 581–93.
Flatto, L. & Newman, D. J. 1977, Random coverings. *Acta Math.* **138**, 241–64.
Florian, A. 1986, Approximation of convex discs by polygons. *Discrete Comput. Geom.* **1** (1986), 241–63.
1989, On a metric for the class of compact convex sets. *Geom. Dedicata* **30**, 69–80.
Focke, J. 1969a, Symmetrische n-Orbiformen kleinsten Inhalts. *Acta Math. Acad. Sci. Hung.* **20**, 39–68.
1969b, Die beste Ausbohrung eines regulären n-Ecks. *Z. Angew. Math. Mech.* **49**, 235–48.
Focke, J. & Gensel, B. 1971, n-Orbiformen maximaler Randknickung. *Beitr. Anal.* **2**, 7–16.
Fourneau, R. 1979, Espaces métriques constitués de classes de polytopes convexes liés aux problèmes de décomposition. *Geom. Dedicata* **8**, 463–76.
1985, Choquet simplices in finite dimensions. In: *Discrete Geometry and Convexity* (eds J. E. Goodman, E. Lutwak, J. Malkevitch & R. Pollack), *Ann. New York Acad. Sci.* **440**, 147–62.
Fu, J. H. G. 1989, Curvature measures and generalized Morse theory. *J. Differential Geom.* **30**, 619–42.
1990, Kinematic formulas in integral geometry. *Indiana Univ. Math. J.* **39**, 1115–54.
1990+a, Curvature of singular spaces via the normal cycle. Preprint.
1990+b, Curvature measures of subanalytic sets. Preprint.
Fuglede, B. 1986a, Continuous selection in a convexity theorem of Minkowski. *Expo. Math.* **4**, 163–78.
1986b, Stability in the isoperimetric problem. *Bull. London Math. Soc.* **18**, 599–605.
1989, Stability in the isoperimetric problem for convex or nearly spherical domains in R^n. *Trans. Amer. Math. Soc.* **314**, 619–38.
1991, Bonnesen's inequality for the isoperimetric deficiency of closed curves in the plane. *Geom. Dedicata* **38**, 283–300.
Fujiwara, M., 1915, Über die einem Vielecke eingeschriebenen und umdrehbaren konvexen geschlossenen Kurven. *Sci. Rep. Tôhoku Univ.* **4**, 43–55.
1916, Über die Anzahl der Kantenlinien einer geschlossenen konvexen Fläche. *Tôhoku*

Math. J. **10**, 164–6.
1919, Über die innen-umdrehbare Kurve eines Vielecks. *Sci Rep. Tôhoku Univ.* **8**, 221–46.
Fujiwara, M. & Kakeya, S. 1917, On some problems of maxima and minima for the curve of constant breadth and the in-revolvable curve of the equilateral triangle. *Tôhoku Math. J.* **11**, 92–110.
Gage, M. E. 1990, Positive centers and Bonnesen's inequality. *Proc. Amer. Math. Soc.* **110**, 1041–8.
Gale, D. 1954, Irreducible convex sets. In: *Proc. Int. Congr. Math.*, Amsterdam 1954, vol. II, Noordhoff, Groningen; North-Holland, Amsterdam, pp. 217–18.
Gale, D. & Klee, V. 1959, Continuous convex sets. *Math. Scand.* **7**, 379–91.
1975, Unique reducibility of subsets of commutative topological groups and semigroups. *Math. Scand.* **36**, 174–98.
Gallivan, S. 1979, On the number of disjoint increasing paths in the one-skeleton of a convex body leading to a given exposed face. *Israel J. Math.* **32**, 282–8.
Gallivan, S. & Gardner, R. J. 1981, A counterexample to a 'simplex algorithm' for convex bodies. *Geom. Dedicata* **11**, 475–88.
Gallivan, S. & Larman, D. G. 1981, Further results on increasing paths in the one-skeleton of a convex body. *Geom. Dedicata* **11**, 19–29.
Geivaerts, M. 1972, Enkele eigenschappen van de relatie 'Homothetisch aanpasselijk' in de ruimte der konvexe lichamen. *Mededelingen Kon. Ac. voor Wetenschappen, Letteren en Schone kunsten van België* **34**, 6, 19 pp.
1974a, Classes of homothetically adaptable convex bodies. *Acad. Roy. Belgique, Bull. Classe Sci.* (5) **60**, 409–29.
1974b, Vector spaces consisting of classes of convex bodies. In: *Journée de Convexité*, Univ. de Liège, Inst. de Math.
1976, Vector spaces consisting of classes of convex bodies. *Geom. Dedicata* **5**, 175–87.
Geppert, H. 1937, Über den Brunn–Minkowskischen Satz. *Math. Z.* **42**, 238–54.
Gericke, H. 1937, Über eine Ungleichung für gemischte Volumina. Integralgeometrie 23. *Deutsche Math.* **2**, 61–7.
1940a, Über stützbare Flächen und ihre Entwicklung nach Kugelfunktionen. *Math. Z.* **46**, 55–61.
1940b, Stützbare Bereiche in komplexer Fourierdarstellung. *Deutsche Math.* **5**, 279–99.
1941, Zur Relativgeometrie ebener Kurven. *Math. Z.* **47**, 215–28.
Giannopoulos, A. A. 1990, A note on a problem of H. Busemann and C. M. Petty concerning sections of symmetric convex bodies. *Mathematika* **37**, 239–42.
Giertz, M. 1969, A note on a problem of Busemann, *Math. Scand.* **25**, 145–8.
Giné, E. & Hahn, M. G. 1985a, The Lévy-Hinčin representation for random compact convex subsets which are infinitely divisible under Minkowski addition. *Z. Wahrscheinlichkeitsth. verw. Geb.* **70**, 271–87.
1985b, M-infinitely divisible random compact convex sets. In: *Probability in Banach Spaces V, Proc. 5th Int. Conf.*, Medford MA 1984, *Lecture Notes in Math.* **1153**, Springer, pp. 226–48.
1985c, Characterization and domains of attraction of p-stable random compact sets. *Ann. Probab.* **13**, 447–68.
Giné, E., Hahn, M. G. & Zinn, J. 1983, Limit theorems for random sets: An application of probability in Banach space results. In: *Probability in Banach Spaces IV, Lecture Notes in Math.* **990**, Springer, pp. 112–35.
Gluck, H. 1972, The generalized Minkowski problem in differential geometry in the large. *Ann. Math.* **96**, 245–76.
1975, Manifolds with preassigned curvature – a survey. *Bull. Amer. Math. Soc.* **81**, 313–29.
Godbersen, C. 1938, Der Satz vom Vektorbereich in Räumen beliebiger Dimensionen. Dissertation, Göttingen.

Godet-Thobie, Ch. & The Laï, Ph. 1970, Sur le plongement de l'ensemble des convexes, fermés, bornés d'un espace vectoriel topologique localement convexe. *C.R. Acad. Sci. Paris, Sér.* A **271**, 84–7.
Goikhman, D. M. 1974, The differentiability of volume in Blaschke lattices (in Russian). *Sibirskii Mat. Zh.* **15**, 1406–8. English translation: *Siberian Math. J.* **15**, 997–9.
Goldberg, M. 1948, Circular-arc rotors in regular polygons. *Amer. Math. Monthly* **55**, 393–402.
1957, Trammel rotors in regular polygons. *Amer. Math. Monthly* **64**, 71–8.
1960, Rotors in polygons and polyhedra. *Math. Computation* **14**, 229–39.
1966, Rotors of variable regular polygons. *Elem. Math.* **21**, 25–7.
Goodey, P. R. 1977, Centrally symmetric convex sets and mixed volumes. *Mathematika* **24**, 193–8.
1981, Limits of intermediate surface area measures of convex bodies. *Proc. London Math. Soc.* (3) **43**, 151–68.
1982, Connectivity and freely rolling convex bodies. *Mathematika* **29**, 249–59.
1984a, Centrally symmetric convex bodies. *Mathematika* **31**, 305–22.
1984b, Degrees of symmetry of convex bodies. *Mitt. Math. Ges. Gießen* **166**, 39–48.
1986, Instability of projection bodies. *Geom. Dedicata* **20**, 295–305.
Goodey, P. R. & Groemer, H. 1990, Stability results for first order projection bodies. *Proc. Amer. Math. Soc.* **109**, 1103–14.
Goodey, P. R. & Howard, R. 1990, Processes of flats induced by higher dimensional processes. *Adv. Math.* **80**, 92–109.
Goodey, P. R. & Schneider, R. 1980, On the intermediate area functions of convex bodies. *Math. Z.* **173**, 185–94.
Goodey, P. R. & Weil, W. 1984, Distributions and valuations. *Proc. London Math. Soc.* (3) **49**, 504–16.
1987, Translative integral formulae for convex bodies. *Aequationes Math.* **34**, 64–77.
1990+, Centrally symmetric convex bodies and the spherical Radon transform. Preprint.
1991, Centrally symmetric convex bodies and Radon transforms on higher order Grassmannians. *Mathematika* **38**, 117–33.
1991+, The determination of convex bodies from the mean of random sections. Preprint.
Goodman, A. W. 1985, Convex curves of bounded type. *Int. J. Math. Math. Sci.* **8**, 625–33.
Gordon, Y. 1985, Some inequalities for Gaussian processes and applications. *Israel J. Math.* **50**, 265–89.
Gordon, Y., Meyer, M. & Reisner, S. 1988, Zonoids with minimal volume product – a new proof. *Proc. Amer. Math. Soc.* **104**, 273–6.
Görtler, H. 1937a, Zur Addition beweglicher ebener Eibereiche. *Math. Z.* **42**, 313–21.
1937b, Erzeugung stützbarer Bereiche I. *Deutsche Math.* **2**, 454–66.
1938, Erzeugung stützbarer Bereiche II. *Deutsche Math.* **3**, 189–200.
Gray, A. 1990. *Tubes*. Addison-Wesley, Redwood City, CA.
Gray, A. & Vanhecke, L. 1981, The volume of tubes in a Riemannian manifold. *Rend. Sem. Mat. Univ. Politecn. Torino* (3) **39**, 1–50.
Green, J. W. 1950, Sets subtending a constant angle on a circle. *Duke Math. J.* **17**, 263–7.
Grinberg, E. L. 1986, Isoperimetric inequalities for k-dimensional sections of convex bodies. Preprint IHES/M/89/28.
1991, Isoperimetric inequalities and identities for k-dimensional cross-sections of convex bodies. *Math. Ann.* **291**, 75–86.
Grinberg, E. L. & Rivin, I. 1990, Infinitesimal aspects of the Busemann–Petty problem. *Bull. London Math. Soc.* **22**, 478–84.
Gritzmann, P. 1988, A characterization of all loglinear inequalities for three quermass-integrals of convex bodies. *Proc. Amer. Math. Soc.* **104**, 563–70.
Gritzmann, P. & Sturmfels, B. 1990, Minkowski addition of polytopes: Computational

complexity and applications to Gröbner bases. Report No. 212, Inst. Math. Univ. Augsburg.

Gritzmann, P., Wills, J. M. & Wrase, D. 1987, A new isoperimetric inequality. *J. reine angew. Math.* **379**, 22–30.

Groemer, H. 1965, Eine neue Ungleichung für konvexe Körper. *Math. Z.* **86**, 361–4.

1972, Eulersche Charakteristik, Projektionen und Quermaßintegrale. *Math. Ann.* **198**, 23–56.

1973, Über einige Invarianzeigenschaften der Eulerschen Charakteristik. *Comment. Math. Helvet.* **48**, 87–99.

1974, On the Euler characteristic in spaces with a separability property. *Math. Ann.* **211**, 315–21.

1975, The Euler characteristic and related functionals on convex surfaces. *Geom. Dedicata* **4**, 91–104.

1977a, Minkowski addition and mixed volumes. *Geom. Dedicata* **6**, 141–63.

1977b, On translative integral geometry. *Arch. Math.* **29**, 324–30.

1978, On the extension of additive functionals on classes of convex sets. *Pacific J. Math.* **75**, 397–410.

1980a, The average measure of the intersection of two sets. *Z. Wahrscheinlichkeitsth. verw. Geb.* **54**, 15–20.

1980b, The average distance between two convex sets. *J. Appl. Prob.* **17**, 415–22.

1987a, Stability theorems for projections of convex sets. *Israel. J. Math.* **60**, 177–90.

1987b, On rings of sets associated with characteristic functions. *Arch. Math.* **49**, 91–3.

1988a, On the Brunn–Minkowski theorem. *Geom. Dedicata* **27**, 357–71.

1988b, Stability theorems for convex domains of constant width. *Canad. Math. Bull.* **31**, 328–37.

1990a, Stability properties of geometric inequalities. *Amer. Math. Monthly* **97**, 382–94.

1990b, On an inequality of Minkowski for mixed volumes. *Geom. Dedicata* **33**, 117–22.

1991a, Stability properties of Cauchy's surface area formula. *Monatsh. Math.* **112**, 43–60.

1991b, Stability theorems for projections and central symmetrization. *Arch. Math.* **51**, 394–9.

Groemer, H. & Schneider, R. 1991, Stability estimates for some geometric inequalities. *Bull. London Math. Soc.* **23**, 67–74.

Gromov, M. 1988, Convex sets and Kähler manifolds. Preprint IHES/M/88/10.

Gromov, M. & Milman, V. D. 1987, Generalization of the spherical isoperimetric inequality to uniformly convex Banach spaces. *Compositio Math.* **62**, 263–82.

Gruber, P. M. 1969, Zur Charakterisierung konvexer Körper. Über einen Satz von Rogers und Shephard I. *Math. Ann.* **181**, 189–200.

1970a, Zur Charakterisierung konvexer Körper. Über einen Satz von Rogers und Shephard II. *Math. Ann.* **184**, 79–105.

1970b, Über die Durchschnitte von translationsgleichen Polyedern. *Monatsh. Math.* **74**, 223–38.

1971, Über eine Kennzeichnung der Simplices des R^n. *Arch. Math.* **22**, 94–102.

1977, Die meisten konvexen Körper sind glatt, aber nicht zu glatt. *Math. Ann.* **229**, 259–66.

1978, Isometries of the space of convex bodies of E^d. *Mathematika* **25**, 270–8.

1980a, The space of compact subsets of E^d. *Geom. Dedicata* **9**, 87–90.

1980b, Isometrien des Konvexringes. *Colloq. Math.* **43**, 99–109.

1982, Isometries of the space of convex bodies contained in a Euclidean ball. *Israel J. Math.* **42**, 277–83.

1983, In most cases approximation is irregular. *Rend. Sem. Mat. Univ. Politecn. Torino* **41**, 19–33.

1984, Aspects of convexity and its applications. *Expo. Math.* **2**, 47–83.

1985, Results of Baire category type in convexity. In: *Discrete Geometry and Convexity* (eds J. E. Goodman, E. Lutwak, J. Malkevitch & R. Pollack), *Ann. New York Acad. Sci.* **440**, 163–9.

1988, Volume approximation of convex bodies by inscribed polytopes. *Math. Ann.* **281**, 229–45.

1989, Dimension and structure of typical compact sets, continua and curves. *Monatsh. Math.* **108**, 149–64.

1990+, Asymptotic best and step by step approximation of convex bodies. Preprint.

1991a, A typical convex surface contains no closed geodesic. *J. reine angew. Math.* **416**, 195–205.

1991b, Volume approximation of convex bodies by circumscribed polytopes. In: *Applied Geometry and Discrete Mathematics* (The Victor Klee Festschrift, eds P. Gritzmann & B. Sturmfels), DIMACS Series, vol. 4, Amer. Math. Soc., Providence, RI, pp. 309–17.

Gruber, P. M. & Kenderov, P. 1982, Approximation of convex bodies by polytopes. *Rend. Circ. Mat. Palermo* **31**, 195–225.

Gruber, P. M. & Lettl, G. 1979, Isometries of the space of compact subsets of E^d. *Studia Sci. Math. Hung.* **14**, 169–81.

1980, Isometries of the space of convex bodies in Euclidean space. *Bull. London Math. Soc.* **12**, 455–62.

Gruber, P. M. & Sorger, H. 1989, Shadow boundaries of typical convex bodies. Measure properties. *Mathematika* **36**, 142–52.

Gruber, P. M. & Tichy, R. 1982, Isometries of spaces of compact or compact convex subsets of metric manifolds. *Monatsh. Math.* **93**, 116–26.

Grzaślewicz, R. 1984, Extreme convex sets in R^2. *Arch. Math.* **43**, 377–80.

Guggenheimer, H. 1969, Nearly spherical surfaces. *Aequationes Math.* **3**, 186–93.

1973, Polar reciprocal convex bodies. *Israel J. Math.* **14**, 309–16. Correction: *ibid.* **29** (1978), 312 (Both works are erroneous.)

Guggenheimer, H. & Lutwak, E. 1976, A characterization of the n-dimensional parallelotope. *Amer. Math. Monthly* **83**, 475–8.

Hadamard, J. 1897, Sur certaines proprietés des trajectoires en dynamique. *J. Math. Pures Appl.* **3**, 331–87.

Hadwiger, H. 1941, Gegenseitige Bedeckbarkeit zweier Eibereiche und Isoperimetrie. *Vierteljahresschr. Naturf. Ges. Zürich* **86**, 152–6.

1944, Eine elementare Ableitung der isoperimetrischen Ungleichung für Polygone. *Comment. Math. Helvet.* **16** (1943/44), 305–9.

1945, Die erweiterten Steinerschen Formeln für ebene und sphärische Bereiche. *Comment. Math. Helvet.* **18**, 59–72.

1946a, Über die erweiterten Steinerschen Formeln für Parallelflächen. *Rev. Hisp.-Amer.* (4) **6**, 1–4.

1946b, Über das Volumen der Parallelmengen. *Mitt. Naturf. Ges. Bern.* **3**, 121–5.

1946c, Eine Erweiterung des Steiner–Minkowskischen Satzes für Polyeder. *Experientia* **2**, 2, 1–2.

1946d, Inhaltsungleichungen für innere und äußere Parallelmengen. *Experientia*, **2**, 12, 1–2.

1947, Über eine symbolisch-topologische Formel. *Elem. Math.* **2**, 35–41 (Portuguese translation: *Gazeta Mat.* **35**, 6–9).

1948, Kurzer Beweis der isoperimetrischen Ungleichung für konvexe Bereiche. *Elem. Math.* **3**, 111–12.

1949a, Ein Auswahlsatz für abgeschlossene Punktmengen. *Portugaliae Math.* **8**, 13–15.

1949b, Über konvexe Körper mit Flachstellen. *Math. Z.* **52**, 212–16.

1949c, Kurze Herleitung einer verschärften isoperimetrischen Ungleichung für konvexe Körper. *Revue Fac. Sci. Univ. Istanbul*, Sér. A **14**, 1–6.

1950a, Einige Anwendungen eines Funktionalsatzes für konvexe Körper in der räumlichen Integralgeometrie. *Monatsh. Math.* **54**, 345–53.
1950b, Minkowskische Addition und Subtraktion beliebiger Punktmengen und die Theoreme von Erhard Schmidt. *Math. Z.* **53**, 210–18.
1951, Beweis eines Funktionalsatzes für konvexe Körper. *Abh. Math. Sem. Univ. Hamburg* **17**, 69–76.
1952, Mittelpunktspolyeder und translative Zerlegungsgleichheit. *Math. Nachr.* **8**, 53–8.
1955a, Eulers Charakteristik und kombinatorische Geometrie. *J. reine angew. Math.* **194**, 101–10.
1955b, Konkave Eikörperfunktionale. *Monatsh. Math.* **59**, 230–7.
1956, Integralsätze im Konvexring. *Abh. Math. Sem. Univ. Hamburg* **20**, 136–54.
1959, Normale Körper im euklidischen Raum und ihre topologischen und metrischen Eigenschaften. *Math. Z.* **71**, 124–40.
1960, Zur Eulerschen Charakteristik euklidischer Polyeder. *Monatsh. Math.* **64**, 49–60.
1968a, Eine Schnittrekursion für die Eulersche Charakteristik euklidischer Polyeder mit Anwendungen innerhalb der kombinatorischen Geometrie. *Elem. Math.* **23**, 121–32.
1968b, Halbeikörper und Isoperimetrie. *Arch. Math.* **19**, 659–63.
1968c, Radialpotenzintegrale zentralsymmetrischer Rotationskörper und Ungleichheitsaussagen Busemannscher Art. *Math. Scand.* **23**, 193–200.
1969a, Zur axiomatischen Charakterisierung des Steinerpunktes konvexer Körper. *Israel J. Math.* **7**, 168–76. (This work is erroneous.)
1969b, Notiz zur Eulerschen Charakteristik offener und abgeschlossener Polyeder. *Studia Sci. Math. Hungar.* **4**, 385–7.
1969c, Eckenkrümmung beliebiger kompakter euklidischer Polyeder und Charakteristik von Euler–Poincaré. *L'Enseignement Math.* **15**, 147–51.
1971, Zur axiomatischen Charakterisierung des Steinerpunktes konvexer Körper; Berichtigung und Nachtrag. *Israel J. Math.* **9**, 466–72.
1972, Polytopes and translative equidecomposability. *Amer. Math. Monthly* **79**, 275–6.
1973, Erweiterter Polyedersatz und Euler–Shephardsche Additionstheoreme. *Abh. Math. Sem. Univ. Hamburg* **39**, 120–9.
1974, Begründung der Eulerschen Charakteristik innerhalb der ebenen Elementargeometrie. *L' Enseignement Math.* **20**, 33–43.
1975a, Eine Erweiterung der kinematischen Hauptformel der Integralgeometrie. *Abh. Math. Sem. Univ. Hamburg* **44**, 84–90.
1975b, Eikörperrichtungsfunktionale und kinematische Integralformeln. Studienvorlesung (Manuscript), Univ. Bern.
1975c, Das Wills'sche Funktional. *Monatsh. Math.* **79**, 213–21.
1979, Gitterpunktanzahl im Simplex und Wills'sche Vermutung. *Math. Ann.* **239**, 271–88.
Hadwiger, H. & Mani, P. 1972, On the Euler characteristic of spherical polyhedra and the Euler relation. *Mathematika* **19**, 139–43.
1974, On polyhedra with extremal Euler characteristic. *J. Combinat. Theory, Ser. A* **17**, 345–9.
Hadwiger, H. & Meier, Ch. 1973, Studien zur vektoriellen Integralgeometrie. *Math. Nachr.* **56**, 261–8.
Hadwiger, H. & Schneider, R. 1971, Vektorielle Integralgeometrie. *Elem. Math.* **26**, 49–57.
Halmos, P. R. 1948, The range of a vector measure. *Bull. Amer. Math. Soc.* **54**, 416–21.
Hammer, P. C. 1951, Convex bodies associated with a convex body. *Proc. Amer. Math. Soc.* **2**, 781–93.
1963, Approximation of convex surfaces by algebraic surfaces. *Mathematika* **10**, 64–71.
Haralick, R. M., Sternberg, S. R. & Zhuang, X. 1987, Image analysis using mathematical morphology. *IEEE Trans. Pattern Anal. Machine Intell.* **9**, 532–50.
Hardy, G. H., Littlewood, J. E. & Pólya, G. 1934, *Inequalities*. Cambridge Univ. Press.

Hartman. P. & Wintner, A. 1953, On pieces of convex surfaces. *Amer. J. Math.* **75**, 477–87.

Hausdorff, F. 1914, *Grundzüge der Mengenlehre.* Verlag von Veit, Leipzig.

1927, *Mengenlehre.* W. de Gruyter, Berlin.

Hayashi, T. 1924, On Steiner's curvature centroid. *Sci. Rep. Tôhoku Univ.* **13**, 109–32.

1926a, Some geometrical applications of the Fourier series. *Rend. Circ. Mat. Palermo* **50**, 96–102.

1926b, On in-revolvable and circum-revolvable curves of a regular polygon, I. *Sci. Rep. Tôhoku Univ.* **15**, 261–3.

1926c, On in-revolvable and circum-revolvable curves of a regular polygon, II. *Sci. Rep. Tôhoku Univ.* **15**, 499–502.

Heil, E. 1967a, Über die Auswahlsätze von Blaschke und Arzelà-Ascoli. *Math.-Phys. Semesterber.* **14**, 169–75.

1967b, Abschätzungen für einige Affininvarianten konvexer Kurven. *Monatsh. Math.* **71**, 405–23.

1976, Ungleichungen für die Quermaßintegrale polarer Körper. *Manuscripta Math.* **19**, 143–9.

1987, Extensions of an inequality of Bonnesen to D-dimensional space and curvature conditions for convex bodies. *Aequationes Math.* **34**, 35–60.

Heine, R. 1937, Der Wertevorrat der gemischten Inhalte von zwei, drei und vier ebenen Eibereichen. *Math. Ann.* **115**, 115–29.

Herglotz, G. 1943, Über die Steinersche Formel für Parallelflächen. *Abh. Math. Sem. Univ. Hamburg* **15**, 165–77.

Hilbert, D. 1910, Minkowskis Theorie von Volumen und Oberfläche. *Nachr. Ges. Wiss. Göttingen*, 388–406. Also in: *Grundzüge einer allgemeinen Theorie der linearen Integralgleichungen.* Teubner, Leipzig u. Berlin, 1912, Kap. 19.

Hildenbrand, W. 1974, *Core and Equilibria of a Large Economy*. Princeton Univ. Press.

1981, Short-run production functions based on microdata. *Econometrica* **49**, 1095–125.

Hildenbrand, W. & Neyman, A. 1982, Integrals of production sets with restricted substitution. *J. Math. Econom.* **9**, 71–82.

Hille, E. & Phillips, R. S. 1957, *Functional Analysis and Semi-groups. AMS Colloquium Publications* vol. 31, Amer. Math. Soc., Providence, RI.

Hiriart-Urruty, J.-B. 1986, A new set-valued second order derivative for convex functions. In: *Mathematics for Optimization* (ed. J.-B. Hiriart-Urruty), *Mathematical Studies Series* **129**, North-Holland, pp. 157–82.

Hiriart-Urruty, J.-B. & Seeger, A. 1989, The second order subdifferential and the Dupin indicatrices of a nondifferentiable convex function. *Proc. London Math. Soc.*, III. Ser., **58**, 351–65.

Hirose, T. 1965, On the convergence theorem for star-shaped sets in E^n. *Proc. Japan. Acad.* **41**, 209–11.

Holmes, R. B. 1975, *Geometric Functional Analysis and its Applications.* Springer, New York.

Holmes, R. D. & Thompson, A. C. 1979, N-dimensional area and content in Minkowski spaces. *Pacific J. Math.* **85**, 77–110.

Hörmander, L. 1955, Sur la fonction d'appui des ensembles convexes dans un espace localement convexe. *Arkiv Mat.* **3**, 181–6.

Horn, B. K. P. 1986, *Robot Vision.* MIT Press, Cambridge, MA; McGraw-Hill, New York.

Howe, R. 1982, Most convex functions are smooth. *J. Math. Econom.* **9**, 37–9.

Huck, H., Roitzsch, R., Simon, U., Vortisch, W., Walden, R., Wegner, B. & Wendland, W. 1973, *Beweismethoden der Differentialgeometrie im Großen. Lecture Notes in Math.* **335**, Springer, Berlin.

Hurwitz, A. 1901, Sur le problème des isopérimètres. *C.R. Acad. Sci. Paris* **132**, 401–3. *Math. Werke*, vol. I, pp. 490–1, Birkhäuser, Basel, 1932.

1902, Sur quelques applications géométriques des séries de Fourier. *Ann. Ecole Norm.* (3) **13**, 357–408. *Math. Werke*, vol. I, pp. 509–54, Birkhäuser, Basel, 1932.

Husain, T. & Tweddle, I. 1970, On the extreme points of the sum of two compact convex sets. *Math. Ann.* **188**, 113–22.

Inzinger, R. 1949, Stützbare Bereiche, trigonometrische Polynome und Defizite höherer Ordnung. *Monatsh. Math.* **53**, 302–23.

Ivanov, B. A. 1973, Über geradlinige Abschnitte auf der Berandung eines konvexen Körpers (in Russian). *Ukrain. Geom. Sbornik.* **13**, 69–71.

1976, Exceptional directions for a convex body (in Russian). *Mat. Zametki* **20**, 365–71. English translation: *Math. Notes* **20**, 763–6.

Jacobs, K. 1971, Extremalpunkte konvexer Mengen. In: *Selecta Mathematica* III, Springer, Berlin, pp. 90–118.

Jessen, B. 1929, Om konvekse Kurvers Krumning. *Mat. Tidsskr.* B, 50–62.

Johnson, K. & Thompson, A. C. 1987, On the isoperimetric mapping in Minkowski spaces. In: *Intuitive Geometry*, Siófok 1985, *Colloquia Math. Soc. János Bolyai* **48**, North-Holland, Amsterdam, pp. 273–87.

Jongmans, F. 1968, Théorème de Krein–Milman et programmation mathématique. *Bull. Soc. Roy. Sci. Liège* **37**, 261–70.

1973, Sur les complications d'une loi de simplification dans les espaces vectoriels. *Bull. Soc. Roy. Sci. Liège* **42**, 529–34.

1976, Réflections sur l'art de sauver la face. *Bull. Soc. Roy. Sci. Liège* **45**, 294–306.

1979, De l'art d'être a bonne distance des ensembles dont la décomposition atteint un stade avancé. *Bull. Soc. Roy. Sci. Liège* **48**, 237–61.

1981a, Contribution aux fondements de la calvitié mathématique. *Bull. Soc. Roy. Sci. Liège* **50**, 8–15.

1981b, Etude des cônes associés à un ensemble. In: *Séminaires dirigés par F. Jongmans et J. Varlet*, Univ. Liège 1980–1, 220 pp.

Jourlin, M. & Laget, B. 1987, Measure of asymmetry of plane convex bodies. *Acta Stereologica* **6**, 115–20.

1988, Convexity and symmetry, Part 1. In: *Image Analysis and Mathematical Morphology*, vol. 2 (ed. J. Serra), Academic Press, London, pp. 343–57.

Kahn, J. & Saks, M. 1984, Balancing poset extensions. *Order* **1**, 113–26.

Kallay, M. 1974, Reconstruction of a plane convex body from the curvature of its boundary. *Israel J. Math.* **17**, 149–61.

1975, The extreme bodies in the set of plane convex bodies with a given width function. *Israel J. Math.* **22**, 203–7.

1982, Indecomposable polytopes. *Israel. J. Math.* **41**, 235–43.

1984, Decomposability of polytopes is a projective invariant. In: *Convexity and Graph Theory*, Jerusalem 1981, *North-Holland Math. Stud.* **87**, North-Holland, Amsterdam, pp. 191–6.

Kalman, J. A. 1961, Continuity and convexity of projections and barycentric coordinates in convex polyhedra. *Pacific J. Math.* **11**, 1017–22.

Kameneckii, I. M. 1947, Die Lösung eines von Ljusternik gestellten geometrischen Problems (in Russian). *Uspechi mat. nauk* **2**, 199–202.

Karcher, H. 1968, Umkreise und Inkreise konvexer Kurven in der sphärischen und der hyperbolischen Geometrie. *Math. Ann.* **177**, 122–32.

Katsurada, Y. 1962, Generalized Minkowski formulas for closed hypersurfaces in Riemannian space. *Ann. Mat. pura appl.*, IV. Ser., **57**, 283–93.

Kellerer, A. M. 1983, On the number of clumps resulting from the overlap of randomly placed figures in a plane. *J. Appl. Prob.* **20**, 126–35.

1985, Counting figures in planar random configurations. *J. Appl. Prob.* **22**, 68–81.

Kellerer, H. G. 1984, Minkowski functionals of Poisson processes. *Z. Wahrscheinlichkeitsth. verw. Geb.* **67**, 63–84.

Khovanskii, A. G. 1984, Analogues of Aleksandrov–Fenchel inequalities for hyperbolic forms (in Russian). *Dokl. Akad. Nauk SSSR* **276**, 1332–4. English translation: *Soviet Math. Dokl.* **29**, 710–13.

Kiener, K. 1986, Extremalität von Ellipsoiden und die Faltungsungleichung von Sobolev. *Arch. Math.* **46**, 162–8.

Kincses, J. 1987, The classification of 3- and 4-Helly dimensional convex bodies. *Geom. Dedicata* **22**, 283–301.

Kiselman, C. O. 1986, How smooth is the shadow of a smooth convex body? *J. London Math. Soc.* **33**, 101–9; *Serdica* **12**, 189–95.

1987, Smoothness of vector sums of plane convex sets. *Math. Scand.* **60**, 239–52.

Klee, V. 1953, The critical set of a convex body. *Amer. J. Math.* **75**, 178–88.

1955, A note on extreme points. *Amer. Math. Monthly* **62**, 30–2.

1957a, Research problem No. 5. *Bull. Amer. Math. Soc.* **63**, 419.

1957b, Extremal structure of convex sets. *Arch. Math.* **8**, 234–40.

1958, Extremal structure of convex sets II. *Math. Z.* **69**, 90–104.

1959a, Some characterizations of convex polyhedra. *Acta Math.* **102**, 79–107.

1959b, Some new results on smoothness and rotundity in normed linear spaces. *Math. Ann.* **139**, 51–63.

1963a, On a theorem of Dubins. *J. Math. Anal. Appl.* **7**, 425–7.

1963b, The Euler characteristic in combinatorial geometry. *Amer. Math. Monthly* **70**, 119–27.

1968, Maximal separation theorems for convex sets. *Trans. Amer. Math. Soc.* **134**, 133–47.

1969a, Can the boundary of a d-dimensional convex body contain segments in all directions? *Amer. Math. Monthly* **76**, 408–10.

1969b, Separation and support properties of convex sets – a survey. In: *Control Theory and the Calculus of Variations* (ed. A. Balakrishnan), Academic Press, New York, pp. 235–304.

Klee, V. L. & Martin K. 1971, Semi-continuity of the face-function of a convex set. *Comment. Math. Helvet.* **46**, 1–12.

Klein, E. & Thompson, A. C. 1984, *Theory of Correspondences.* Wiley, New York.

Klima, V. & Netuka, I. 1981, Smoothness of a typical convex function. *Czech. Math. J.* **31**, 569–72.

Klötzler, R. 1975, Beweis einer Vermutung über n-Orbiformen kleinsten Inhalts. *Z. Angew. Math. Mech.* **55**, 557–70.

Kneser, H. 1970, Die Stützfunktion eines Durchschnitts konvexer Körper. *Arch. Math.* **21**, 221–4.

Kneser, H. & Süss, W. 1932, Die Volumina in linearen Scharen konvexer Körper. *Mat. Tidsskr.* B, 19–25.

Knothe, H. 1949, Über eine Vermutung H. Minkowskis. *Math. Nachr.* **2**, 380–5.

1957a, Contributions to the theory of convex bodies. *Michigan Math. J.* **4**, 39–52.

1957b, Inversion of two theorems of Archimedes. *Michigan Math. J.* **4**, 53–6.

Kohlmann, P. 1988, Krümmungsmaße, Minkowskigleichungen und Charakterisierungen metrischer Bälle in Raumformen. Dissertation, Univ. Dortmund.

Koutroufiotis, D. 1971, Ovaloids which are almost spheres. *Commun. Pure Appl. Math.* **24**, 289–300.

1972, On Blasche's rolling theorems. *Arch. Math.* **23**, 655–60.

Krein, M. & Milman, D. 1940, On extreme points of regularly convex sets. *Studia Math.* **9**, 133–8.

Kruskal, J. B. 1969, Two convex counterexamples: A discontinuous envelope function and a nondifferentiable nearest-point mapping. *Proc. Amer. Math. Soc.* **23**, 697–703.

Kubota, T. 1918, Über die Schwerpunkte der geschlossenen konvexen Kurven und Flächen. *Tôhoku Math. J.* **14**, 20–7.

1920, Einige Probleme über konvex-geschlossene Kurven und Flächen. *Tôhoku Math. J.* **17**, 351–62.

1922, Beweise einiger Sätze über Eiflächen. *Tôhoku Math. J.* **21**, 261–4.
1925a, Über die konvex-geschlossenen Mannigfaltigkeiten im n-dimensionalen Raume. *Sci. Rep. Tôhoku Univ.* **14**, 85–99.
1925b, Über die Eibereiche im n-dimensionalen Raume. *Sci. Rep. Tôhoku Univ.* **14**, 399–402.
Kuiper, N. H. 1971, Morse relations for curvature and tightness. In: *Proc. Liverpool Singularities Symp.* (ed. C. T. C. Wall), *Lecture Notes in Math.* **209**, Springer, Berlin, 1971, pp. 77–89.
Kuratowski, K. 1968, *Topology*, vol. II. Academic Press, New York (French original: 1961).
Kutateladze, S. S. 1973, Die Blaschkesche Struktur in der Programmierung isoperimetrischer Probleme (in Russian). *Mat. Zametki* **14**, 745–54. English translation: *Math. Notes* **14**, 985–9.
1976, Symmetry measures (in Russian). *Mat. Zametki* **19**, 615–22. English translation: *Math. Notes* **19**, 372–5.
Kutateladze, S. S. & Rubinov, A. M. 1969, Problems of isoperimetric type in a space of convex bodies (in Russian). *Optimalnoe Planirovanie* **14**, 61–79.
Lafontaine, J. 1987, Mesures de courbure des variétés lisses et des polyèdres (Séminaire Bourbaki, vol. 1985/86) *Astérisque* **145–6**, 241–56.
Lagarias, J. C. 1982, The van der Waerden conjecture: Two Soviet solutions. *Notices Amer. Math. Soc.* **29**, 130–3.
Laget, B. 1987, Sur l'ensemble critique d'un ensemble convexe. *Arch. Math.* **49**, 333–4.
Larman, D. G. 1971, On a conjecture of Klee and Martin for convex bodies. *Proc. London Math. Soc.* **23**, 668–82. Corrigendum: *ibid.* **36** (1978), 86.
1974, On the inner aperture and intersections of convex sets. *Pacific J. Math.* **55**, 219–32.
1977, On the one-skeleton of a compact convex set in a Banach space. *Proc. London Math. Soc.* (3) **34**, 117–44.
1980, The $d-2$ skeletons of polytopal approximations to a convex body in E^d. *Mathematika* **27**, 122–33.
Larman, D. G. & Mani, P. 1970, Gleichungen und Ungleichungen für die Gerüste von konvexen Polytopen und Zellenkomplexen. *Comment. Math. Helvet.* **45**, 199–218.
Larman, D. G. & Rogers, C. A. 1970, Paths in the one-skeleton of a convex body. *Mathematika* **17**, 293–314.
1971, Increasing paths on the one-skeleton of a convex body and the directions of line segments on the boundary of a convex body. *Proc. London Math. Soc.* **23**, 683–98.
1973, The finite dimensional skeletons of a compact convex set. *Bull. London Math. Soc.* **5**, 145–53.
1975, The existence of a centrally symmetric convex body with central sections that are unexpectedly small. *Mathematika* **22**, 164–75.
Lebesgue, H. 1914, Sur le problème des isopérimètres et sur les domaines de largeur constante. *Bull. Soc. Math. France C.R.*, 72–6.
Leichtweiß, K. 1965, Über eine analytische Darstellung des Randes konvexer Körper. *Arch. Math.* **16**, 300–19.
1981, Zum Beweis eines Eindeutigkeitssatzes von A. D. Aleksandrov. In: *E. B. Christoffel*, Aachen/Monschau 1979 (Eds P. Butzer & F. Fehér), Birkhäuser, Basel, pp. 636–52.
1986, Zur Affinoberfläche konvexer Körper. *Manuscripta Math.* **56**, 429–64.
1988, Über einige Eigenschaften der Affinoberfläche beliebiger konvexer Körper. *Results Math.* **13**, 255–82.
1989, Bemerkungen zur Definition einer erweiterten Affinoberfläche von E. Lutwak. *Manuscripta Math.* **65**, 181–97.
Leindler, L. 1972, On a certain converse of Hölder's inequality II. *Acta. Sci. Math. Szeged* **33**, 217–23.

Lenz, H. 1970, Mengenalgebra und Eulersche Charakteristik. *Abh. Math. Sem. Univ. Hamburg* **34**, 135–47.

Letac, G. 1983, Mesures sur le cercle et convexes du plan. *Ann. Sci. Univ. Clermont-Ferrand II, Prob. Appl.* **1**, 35–65.

Lettl, G. 1980, Isometrien des Raumes der konvexen Teilmengen der Sphäre. *Arch. Math.* **35**, 471–5.

Lévy, P. 1937, *Théorie de l'addition des variables aléatoires.* Gauthier-Villars, Paris.

Lewis, J. E. 1975, A Banach space whose elements are classes of sets of constant width. *Canad. Math. Bull.* **18**, 679–89.

Lewy, H. 1938, On differential geometry in the large, I (Minkowski's problem). *Trans. Amer. Math. Soc.* **43**, 258–70.

Liapounoff, A. A. 1940, On completely additive vector functions (in Russian). *Izv. Akad. Nauk SSSR* **4**, 465–78.

Lindelöf, L. 1869, Propriétés générales des polyèdres qui, sous une étendue superficielle donnée, renferment le plus grand volume. *Bull. Acad. Sci. St. Pétersbourg* **14**, 257–69. Extract: *Math. Ann.* **2** (1870), 150–9.

Lindenstrauss, J. 1966, A short proof of Liapounoff's convexity theorem. *J. Math. Mech.* **15**, 971–2.

Lindquist, N. F. 1975a, Approximation of convex bodies by sums of line segments. *Portugaliae Math.* **34**, 233–40.

1975b, Support functions of central convex bodies. *Portugaliae Math.* **34**, 241–52.

Linhart, J. 1986, Extremaleigenschaften der regulären 3-Zonotope. *Studia Sci. Math. Hungar.* **21**, 181–8.

1987, An upper bound for the intrinsic volumes of equilateral zonotopes. In: *Intuitive Geometry*, Siófok 1985, *Coll. Math. Soc. János Bolyai* **48**, North-Holland, Amsterdam, pp. 339–45.

1989, Approximation of a ball by zonotopes using uniform distribution on the sphere. *Arch. Math.* **53**, 82–6.

Linke, Ju. E. 1980, Application of Steiner's point for investigation of a class of sublinear operators (in Russian). *Dokl. Akad. Nauk SSSR* **254**, 1069–79. English translation: *Soviet Math. Dokl.* **22**, 522–5.

Little, J. J. 1983, An iterative method for reconstructing convex polyhedra from extended Gaussian images. In: *Proc. AAAI-83*, Washington, DC, pp. 247–50.

Ljašenko, N. N. 1979, Limit theorems for sums of independent compact random subsets of Euclidean space (in Russian). *Zap. Naučn. Sem. Leningrad. Otdel. Mat. Inst. Steklov* **85**, 113–28.

Lutwak, E. 1975a, Dual mixed volumes. *Pacific J. Math.* **58**, 531–8.

1975b, Width-integrals of convex bodies. *Proc. Amer. Math. Soc.* **53**, 435–9.

1975c, A general Bieberbach inequality. *Proc. Camb. Phil. Soc.* **78**, 493–6.

1975d, Dual cross-sectional measures. *Atti Accad. Naz. Lincei* **58**, 1–5.

1976, On cross-sectional measures of polar reciprocal convex bodies. *Geom. Dedicata* **5**, 79–80.

1977, Mixed width-integrals of convex bodies. *Israel J. Math.* **28**, 249–53.

1979a, Mean dual and harmonic cross-sectional measures. *Ann. Mat. Pura Appl.* **119**, 139–148.

1979b, On a complementary Minkowski inequality. *J. Math. Anal. Appl.* **72**, 70–4.

1983, A width–diameter inequality for convex bodies. *J. Math. Anal. Appl.* **93**, 290–5.

1984, A general isepiphanic inequality. *Proc. Amer. Math. Soc.* **90**, 415–21.

1985a, On the Blaschke–Santaló inequality. In: *Discrete Geometry and Convexity* (eds J. E. Goodman, E. Lutwak, J. Malkevitch & R. Pollack), *Ann. New York Acad. Sci.* **440**, pp. 106–12.

1985b, Mixed projection inequalities. *Trans. Amer. Math. Soc.* **287**, 91–106.

1986a, On some affine isoperimetric inequalities. *J. Differential Geom.* **23**, 1–13.

1986b, Volume of mixed bodies. *Trans. Amer. Math. Soc.* **294**, 487–500.

1987a, Rotation means of projections. *Israel J. Math.* **58**, 161–9.

1987b, Mixed affine surface area. *J. Math. Anal. Appl.* **125**, 351–60.

1988a, Inequalities for Hadwiger's harmonic quermassintegrals. *Math. Ann.* **280**, 165–75.

1988b, Intersection bodies and dual mixed volumes. *Adv. Math.* **71**, 232–61.

1989+a, Inequalities for mixed projection bodies. *Trans. Amer. Math. Soc.* (to appear).

1989+b, Larger objects with smaller projections. Preprint.

1989+c, A minimax inequality for inscribed cones. *J. Math. Anal. Appl.* (to appear).

1990a, On quermassintegrals of mixed projection bodies. *Geom. Dedicata* **33**, 51–8.

1990b, Centroid bodies and dual mixed volumes. *Proc. London Math. Soc.* (3) **60**, 365–91.

1990c, On a conjectured projection inequality of Petty. *Contemp. Math.* **113**, 171–82.

1991, Extended affine surface area. *Adv. Math.* **85**, 39–68.

1991+a, The Brunn–Minkowski–Firey theory, I. *J. Differential Geom.* (to appear).

1991+b, The Brunn–Minkowski–Firey theory, II. Preprint.

Macbeath, A. M. 1951, A compactness theorem for affine equivalence-classes of convex regions. *Canad. J. Math.* **3**, 54–61.

Maehara, H. 1984, Convex bodies forming pairs of constant width. *J. Geom.* **22**, 101–7.

Mahler, K. 1939a, Ein Übertragungsprinzip für konvexe Körper. *Casopis Pěst. Mat. Fys.* **68**, 93–102.

1939b, Ein Minimalproblem für konvexe Polygone. *Mathematica (Zutphen)* **B7**, 118–27.

Makai, E. 1959, Steiner type inequalities in plane geometry. *Period. Polytech. Elec. Engrg.* **3**, 345–355.

Makai, E. jr. 1974, Research Problem. *Periodica Math. Hungar.* **5**, 353–4.

Mani, P. (Mani-Levitska, P.) 1971, On angle sums and Steiner points of polyhedra. *Israel. J. Math.* **9**, 380–8.

1988, A simple proof of the kinematic formula. *Monatsh. Math.* **105**, 279–85.

Martini, H. 1987, Some results and problems around zonotopes. In: *Intuitive Geometry*, Siófok 1985 (eds K. Böröczky & G. Fejes Tóth), *Coll. Math. Soc. János Bolyai* **48**, pp. 383–418.

1989, On inner quermasses of convex bodies. *Arch. Math.* **52**, 402–6.

1990, A new view on some characterizations of simplices. *Arch. Math.* **55**, 389–93.

1991, Convex polytopes whose projection bodies and difference sets are polars. *Discrete Comput. Geom.* **6**, 83–91.

Mase, S. 1979, Random compact convex sets which are infinitely divisible with respect to Minkowski addition. *Adv. Appl. Prob.* **11**, 834–50.

Matheron, G. 1974a, Un théorème d'unicité pour les hyperplans poissoniens. *J. Appl. Prob.* **11**, 184–9.

1974b, Hyperplans poissoniens et compacts de Steiner. *Adv. Appl. Prob.* **6**, 563–79.

1975, *Random Sets and Integral Geometry*. Wiley, New York.

1978a, The infinitesimal erosions. In: *Geometrical Probability and Biological Structures: Buffon's 200th Anniversary* (eds R. E. Miles & J. Serra), *Lecture Notes in Biomath.* **23**, Springer, Berlin, pp. 251–69.

1978b, La formule de Steiner pour les érosions. *J. Appl. Prob.* **15**, 126–35.

Matsumura, S. 1930, Eiflächenpaare gleicher Breiten und gleicher Umfänge. *Japan. J. Math.* **7**, 225–6.

1932, Über Minkowskis gemischten Flächeninhalt. *Japan. J. Math.* **9**, 161–3.

Mazur, S. 1933, Über konvexe Mengen in linearen normierten Räumen. *Studia Math.* **4**, 70–84.

McKinney, R. L. 1962, Positive bases for linear spaces. *Trans. Amer. Math. Soc.* **103**, 131–48.

McMinn, T. J. 1960, On the line segments of a convex surface in E_3. *Pacific J. Math.* **10**, 943–6.

McMullen, P. 1970, Polytopes with centrally symmetric faces. *Israel J. Math.* **8**, 194–6.

1971, On zonotopes. *Trans. Amer. Math. Soc.* **159**, 91–109.

1973, Representations of polytopes and polyhedral sets. *Geom. Dedicata* **2**, 83–99.
1974a, A dice probability problem. *Mathematika* **21**, 193–8.
1974b, On the inner parallel body of a convex body. *Israel J. Math.* **19**, 217–19.
1975a, Non-linear angle-sum relations for polyhedral cones and polytopes. *Math. Proc. Camb. Phil. Soc.* **78**, 247–61.
1975b, Space tiling zonotopes. *Mathematika* **22**, 202–211.
1975c, Metrical and combinatorial properties of convex polytopes. In: *Proc. Int. Congr. Math.*, Vancouver 1974 (ed. R. D. James), Canadian Mathematical Congress, pp. 491–5.
1976a, On support functions of compact convex sets. *Elem. Math.* **31**, 117–19.
1976b, Polytopes with centrally symmetric facets. *Israel J. Math.* **23**, 337–8.
1977, Valuations and Euler type relations on certain classes of convex polytopes. *Proc. London Math. Soc.* **35**, 113–35.
1979, Transforms, diagrams and representations. In: *Contributions to Geometry, Proc. Geometry Symp.*, Siegen 1978 (eds J. Tölke & J. M. Wills), Birkhäuser, Basel, pp. 92–130.
1980, Continuous translation invariant valuations on the space of compact convex sets. *Arch. Math.* **34**, 377–84.
1983a, Weakly continuous valuations on convex polytopes. *Arch. Math.* **41**, 555–64.
1983b, Notes on Asplund's theorem (Private communication).
1984, The Hausdorff distance between compact convex sets. *Mathematika* **31**, 76–82.
1987, Indecomposable convex polytopes. *Israel J. Math.* **58**, 321–3.
1989, The polytope algebra. *Adv. Math.* **78**, 76–130.
1990, Monotone translation invariant valuations on convex bodies. *Arch. Math.* **55**, 595–8.

McMullen, P., Schneider, R. & Shephard, G. C. 1974, Monotypic polytopes and their intersection properties. *Geom. Dedicata* **3**, 99–129.

Mecke, J., Schneider, R., Stoyan, D. & Weil, W. 1990, *Stochastische Geometrie.* DMV Seminar 1989, vol. 16, Birkhäuser, Basel.

Meier, Ch. 1977, Multilinearität bei Polyederaddition. *Arch. Math.* **29**, 210–17.

Meissner, E. 1909, Über die Anwendung von Fourier-Reihen auf einige Aufgaben der Geometrie und Kinematik. *Vierteljahresschr. Naturf. Ges. Zürich* **54**, 309–29.
1918, Über die durch reguläre Polyeder nicht stützbaren Körper. *Vierteljahresschr. Naturf. Ges. Zürich* **63**, 544–51.

Melzak, Z. A. 1958, Limit sections and universal points of convex surfaces. *Proc. Amer. Math. Soc.* **9**, 729–34.

Menger, K. 1928, Untersuchungen über allgemeine Metrik. *Math. Ann.* **100**, 75–163.

Meschkowski, H. & Ahrens, I. 1974, *Theorie der Punktmengen.* Bibliographisches Institut, Mannheim.

Meyer, M. 1986, Une charactérisation volumique de certains espaces normés de dimension finie. *Israel J. Math.* **55**, 317–26.
1990+ On a problem of Busemann and Petty. Preprint.

Meyer, M. & Pajor, A. 1989, On Santaló's inequality. In: *Geometric Aspects of Functional Analysis* (eds J. Lindenstrauss & V. D. Milman), *Lecture Notes in Math.* **1376**, Springer, Berlin, pp. 261–3.
1990, On the Blaschke-Santaló inequality. *Arch. Math.* **55**, 82–93.

Meyer, M. & Reisner, S. 1991, A geometric property of the boundary of symmetric convex bodies and convexity of flotation surfaces. *Geom. Dedicata* **37**, 327–37.

Meyer, P. 1977, Eine Verallgemeinerung der Steinerschen Formeln. *Abh. Braunschweig. Wiss. Ges.* **28**, 119–23.

Meyer, W. 1969, Minkowski addition of convex sets. Thesis, Univ. of Wisconsin.
1970a, Characterization of the Steiner point. *Pacific J. Math.* **35**, 717–25.
1970b, Indecomposable polytopes. In: *Combinatorial Structures and Their Applications*, Calgary 1969 (eds R. Guy *et al.*), Gordon & Breach, New York, pp. 271–2.
1972, Decomposing plane convex bodies. *Arch. Math.* **23**, 534–6.

1974, Indecomposable polytopes. *Trans. Amer. Math. Soc.* **190**, 77–86.
Michael, E. 1951, Topologies on spaces of subsets. *Trans. Amer. Math. Soc.* **71**, 152–82.
Miles, R. E. 1974, The fundamental formula of Blaschke in integral geometry and geometric probability, and its iteration, for domains with fixed orientations. *Austral. J. Statist.* **16**, 111–18.
1976, Estimating aggregate and overall characteristics from thick sections by transmission microscopy. *J. Microscopy* **107**, 227–33.
Milka, A. D. 1973, Indecomposability of a convex surface (in Russian). *Ukrain. Geom. Sb.* **13**, 112–29. English translation: *Selected Transl. in Math., Statist. and Prob.* **15**, 129–43.
Milman, D. 1948, Isometry and extremal points. *Dokl. Akad. Nauk SSSR* **59**, 1241–4.
Milman, V. D. 1986, Inégalité de Brunn–Minkowski inverse et applications à la théorie locale des espaces normés. *C.R. Acad. Sci. Paris*, Sér. I, **302**, 25–8.
Milman, V. D. & Pajor A. 1989, Isotropic position and inertia ellipsoids and zonoids of the unit ball of a normed n-dimensional space. In: *Geometric Aspects of Functional Analysis* (eds J. Lindenstrauss & V. D. Milman), *Lecture Notes in Math.* **1376**, Springer, Berlin, pp. 64–104.
Minkowski, H. 1897, Allgemeine Lehrsätze über die konvexen Polyeder. *Nachr. Ges. Wiss. Göttingen*, 198–219. *Gesammelte Abhandlungen*, vol. II, Teubner, Leipzig, 1911, pp. 103–21.
1901a, Über die Begriffe Länge, Oberfläche und Volumen. *Jahresber. Deutsche Math.-Ver.* **9**, 115–21. *Gesammelte Abhandlungen*, vol. II, Teubner, Leipzig, 1911, pp. 122–27.
1901b, Sur les surfaces convexes fermées. *C.R. Acad. Sci. Paris* **132**, 21–24. *Gesammelte Abhandlungen*, vol. II, Teubner, Leipzig, 1911, pp. 128–30.
1903, Volumen und Oberfläche. *Math. Ann.* **57**, 447–95. *Gesammelte Abhandlungen*, vol. II, Teubner, Leipzig, 1911, pp. 230–76.
1904, Über die Körper konstanter Breite (in Russian). *Mat Sbornik* **25**, 505–8. German translation: *Gesammelte Abhandlungen*, vol. II, Teubner, Leipzig, 1911, pp. 277–9.
1910, *Geometrie der Zahlen*. Teubner, Leipzig.
1911, Theorie der konvexen Körper, insbesondere Begründung ihres Oberflächenbegriffs. *Gesammelte Abhandlungen*, vol. II, Teubner, Leipzig, pp. 131–229.
Minoda, T. 1941, On certain ovals. *Tôhoku Math. J.* **48**, 312–20.
Miranda, C. 1939, Su un problema di Minkowski. *Rend. Sem. Mat. Roma* **3**, 96–108.
Mityagin, B. S. 1969, Two inequalities for volumes of convex bodies (in Russian). *Mat. Zametki* **5**, 99–106. English translation: *Math. Notes* **5**, 61–5.
Molter, U. M. 1986, Tangential measure on the set of convex infinite cylinders. *J. Appl. Prob.* **23**, 961–72.
Moore, J. D. 1973, Almost spherical convex hypersurfaces. *Trans. Amer. Math. Soc.* **180**, 347–58.
Motzkin, T. S. 1935, Sur quelques propriétés caractéristiques des ensembles convexes. *Atti Real. Accad. Naz. Lincei, Rend. Cl. Sci. Fis., Mat., Natur.*, Serie VI, **21**, 562–7.
Müller, C. 1966, *Spherical Harmonics. Lecture Notes in Math.* **17**, Springer, Berlin.
Müller, H. R. 1953, Über Momente ersten und zweiten Grades in der Integralgeometrie. *Rend. Circ. Mat. Palermo,* II. Ser., **2**, 119–40.
Mürner, P. 1975, Translative Parkettierungspolyeder und Zerlegungsgleichheit. *Elem. Math.* **30**, 25–7.
1977, Translative Zerlegungsgleichheit von Polytopen. *Arch. Math.* **29**, 218–24.
Nádenîk, Z. 1965, Die Verschärfung einer Ungleichung von Frobenius für den gemischten Flächeninhalt der konvexen ebenen Bereiche. *Casopis Pěst. Mat.* **90**, 220–5.
1967, Die Ungleichungen für die Oberfläche, das Integral der mittleren Krümmung und die Breite der konvexen Körper. *Casopis Pěst. Mat.* **92**, 133–45.
1968, Erste Krümmungsfunktion der Rotationseiflächen. *Casopis Pěst. Mat.* **93**, 127–33.
Nadler, S. B. 1978, *Hyperspaces of Sets*. Marcel Dekker, New York.

Nadler, S., Quinn, J. & Stavrakas, N. M. 1975, Hyperspaces of compact convex sets I. *Bull. Acad. Polon. Sci.* **23**, 555–9.
1977, Hyperspaces of compact convex sets II. *Bull. Acad. Polon. Sci.* **25**, 381–5.
1979, Hyperspaces of compact convex sets. *Pacific J. Math.* **83**, 441–62.
Nakajima, S. 1920, On some characteristic properties of curves and surfaces. *Tôhoku Math. J.* **18**, 272–87.
1921, The circle and the straight line nearest to *n* given points, *n* given straight lines or a given curve. *Tôhoku Math. J.* **19**, 11–20.
Neveu, J. 1969, *Mathematische Grundlagen der Warscheinlichkeitstheorie.* Oldenbourg, München.
Nikliborc, W. 1932, Über die Lage des Schwerpunktes eines ebenen konvexen Bereiches und die Extrema des logarithmischen Flächenpotentials eines konvexen Bereiches. *Math. Z.* **36**, 161–5.
Nirenberg, L. 1953, The Weyl and Minkowski problems in differential geometry in the large. *Commun. Pure Appl. Math.* **6**, 337–94.
Oda, T. 1988, *Convex Bodies and Algebraic Geometry: An Introduction to the Theory of Toric Varieties.* Springer, Berlin.
Ohmann, D. 1952a, Eine Minkowskische Ungleichung für beliebige Mengen und ihre Anwendung auf Extremalprobleme. *Math. Z.* **55**, 299–307.
1952b, Ungleichungen zwischen den Quermaßintegralen beschränkter Punktmengen, I. *Math. Ann.* **124**, 265–76.
1953, Ein vollständiges Ungleichungssystem für Minkowskische Summe und Differenz. *Comment. Math. Helvet.* **27**, 151–6.
1954, Ungleichungen zwischen den Quermaßintegralen beschränkter Punktmengen, II. *Math. Ann.* **127**, 1–7.
1955a, Eine lineare Verschärfung des Brunn–Minkowskischen Satzes für abgeschlossene Mengen. *Arch. Math.* **6**, 33–5.
1955b, Eine Verallgemeinerung der Steinerschen Formel. *Math. Ann.* **129**, 209–12.
1956a, Ungleichungen für die Minkowskische Summe und Differenz konvexer Körper. *Comment. Math. Helvet.* **30**, 297–304.
1956b, Ungleichungen zwischen den Quermaßintegralen beschränkter Punktmengen, III. *Math. Ann.* **130**, 386–93.
1958, Ein allgemeines Extremalproblem für konvexe Körper. *Monatsh. Math.* **62**, 97–107.
Ohshio, S. 1955, Volume, surface area and total mean curvature. *Sci. Rep. Kanazawa Univ.* **4**, 21–8.
1958, Parallel series to a closed convex curve and surface and the differentiability of their quantities. *Sci. Rep. Kanazawa Univ.* **6**, 15–24.
1962, On an isoperimetric sequence. *Sci. Rep. Kanazawa Univ.* **8**, 13–23.
Oishi, K. 1920, A note on the closed convex surface. *Tôhoku Math. J.* **18**, 288–90.
Oliker, V. I. 1973, The uniqueness of the solution in Christoffel and Minkowski problems for open surfaces (in Russian). *Mat. Zametki* **13**, 41–9. English translation: *Math. Notes* **13**, 26–30.
1979a, Eigenvalues of the Laplacian and uniqueness in the Minkowski problem. *J. Differential Geom.* **14**, 93–8.
1979b, On certain elliptic differential equations on a hypersphere and their geometric applications. *Indiana Univ. Math. J.* **28**, 35–51.
1982, On the linearized Monge–Ampère equations related to the boundary value Minkowski problem and its generalizations. In: *Monge–Ampère Equations and Related Topics*, Florence 1980, Ist. Naz. Alta Mat. Francesco Severi, Roma, pp, 79–112.
1986, The problem of embedding S^n into R^{n+1} with prescribed Gauss curvature and its solution by variational methods. *Trans. Amer. Math. Soc.* **295**, 291–303.
Osserman, R. 1978, The isoperimetric inequality. *Bull. Amer. Math. Soc.* **84**, 1182–238.
1979, Bonnesen-style isoperimetric inequalities. *Amer. Math. Monthly* **86**, 1–29.

1987, A strong form of the isoperimetric inequality in R^n. *Complex Variables* **9**, 241–9.

Overhagen, T. 1975, Zur Gitterpunktanzahl konvexer Körper im 3-dimensionalen euklidischen Raum. *Math. Ann.* **216**, 217–24.

Panina, G. Y. 1988, Representation of an n-dimensional body in the form of a sum of $(n-1)$-dimensional bodies (in Russian). *Izv. Akad. Nauk Armyan. SSR*, Ser. Mat. **23**, 385–95. English translation: *Soviet J. Contemp. Math. Anal.* **23**, 91–103.

1989a, Convex bodies integral representations. In: *Geobild '89* (eds A. Hübler, W. Nagel, B. D. Ripley & G. Werner), Akademie-Verlag, Berlin, pp. 201–4.

1989b, Integral representations of convex bodies (in Russian). *Dokl. Akad. Nauk SSSR* **307**. English translation: *Soviet Math. Dokl.* **40** (1990), 116–18.

Panov, A. A. 1985a, On mixed discriminants connected with positive semidefinite quadratic forms (in Russian). *Doklady Akad. Nauk SSSR* **282**, 273–6. English translation: *Soviet Math. Dokl.* **31**, 465–7.

1985b, On some properties of mixed discriminants (in Russian). *Mat. Sbornik* **128**, 291–305. English translation: *Math. USSR (Sbornik)* **56**, 279–93.

Papaderou-Vogiatzaki, I. & Schneider, R. 1988, A collision probability problem. *J. Appl. Prob.* **25**, 617–23.

Papadopoulou, S. 1977, On the geometry of stable compact convex sets. *Math. Ann.* **229**, 193–200.

1982, Stabile konvexe Mengen. *Jahresber. Deutsche Math.-Ver.* **84**, 92–106.

Payá, R. & Yost, D. 1988, The two-ball property: Transitivity and examples. *Mathematika* **35**, 190–7.

Perles, M. A. & Sallee, G. T. 1970, Cell complexes, valuations, and the Euler relation. *Canad. J. Math.* **22**, 235–41.

Petermann, E. 1967, Some relations between breadths and mixed volumes of convex bodies in R^n. In: *Proc. Colloq. Convexity*, Copenhagen 1965, Københavns Univ. Mat. Inst., pp. 229–33.

Petty, C. M. 1961a, Centroid surfaces. *Pacific J. Math.* **11**, 1535–47.

1961b, Surface area of a convex body under affine transformations. *Proc. Amer. Math. Soc.* **12**, 824–8.

1967, Projection bodies. In: *Proc. Colloq. Convexity*, Copenhagen 1965, Københavns Univ. Mat. Inst., 234–41.

1972, Isoperimetric problems. In: *Proc. Conf. Convexity Combinat. Geom.*, Univ. Oklahoma 1971, pp. 26–41.

1974, Geominimal surface area. *Geom. Dedicata* **3**, 77–97.

1985, Affine isoperimetric problems. In: *Discrete Geometry and Convexity* (eds J. E. Goodman, E. Lutwak, J. Malkevitch & R. Pollack), *Ann. New York Acad. Sci.* **440**, 113–27.

Phelps, R. R. 1966, *Lectures on Choquet's Theorem.* Van Nostrand, Princeton.

1980, Integral representations for elements of convex sets. *Studies in Functional Analysis* (ed. R. G. Bartle), *MAA Studies in Math.* **21**, Math. Assoc. Amer., Washington DC, pp. 115–57.

Pisier, G. 1989, *The Volume of Convex Bodies and Banach Space Geometry.* Cambridge Univ. Press.

Pogorelov, A. V. 1953, On the question of the existence of a convex surface with a given sum of the principal radii of curvature (in Russian). *Uspehi Mat. Nauk* **8**, 127–30.

1967, Nearly spherical surfaces. *J. d'Analyse Math.* **19**, 313–21.

1975, *The Minkowski Multidimensional Problem.* Winston & Sons, Washington DC, 1978 (Russian original: 1975).

Pólya, G. & Szegö, G. 1951, *Isoperimetric Inequalities in Mathematical Physics.* Princeton Univ. Press, Princeton, NJ.

Pontrjagin, L. S. 1957, *Topologische Gruppen*, vol 1. Teubner, Leipzig (English translation of first Russian edn: 1946).

Positsel'skii, E. D. 1973, Characterization of Steiner points (in Russian). *Mat. Zametki* **14**,

243–7. English translation: *Math. Notes* **14**, 698–700.

Pranger, W. 1973, Extreme points of convex sets. *Math. Ann.* **205**, 299–302.

Prékopa, A. 1971, Logarithmic concave measures with applications to stochastic programming. *Acta. Sci. Math. Szeged* **32**, 301–16.

1973, On logarithmic concave measures and functions. *Acta Sci. Math. Szeged* **34**, 335–43.

Przesławski, K. 1985, Linear and Lipschitz continuous selectors for the family of convex sets in Euclidean vector spaces. *Bull. Polish. Acad. Sci. Math.* **33**, 31–3.

1988, Linear algebra of convex sets and the Euler characteristic. Report, Institute of Math., Phys. and Chem., Wyzsza Szkola Inzynierska, Zielona Góra, 65 pp.

Rademacher, H. 1921, Zur Theorie der Minkowskischen Stützebenenfunktion. *Sitzungsber. Berliner Math. Ges.* **20**, 14–19.

Radon, J. 1936, Annäherung konvexer Körper durch analytisch begrenzte. *Monatsh. Math. Phys.* **43**, 340–4. *Gesammelte Abhandlungen*, vol. 1 (eds P. M. Gruber *et al.*), Birkhäuser, Wien, 1987, pp. 370–4.

Rådström, H. 1952, An embedding theorem for spaces of convex sets. *Proc. Amer. Math. Soc.* **3**, 165–9.

Ratschek, H. & Schröder, G. 1977, Representation of semigroups as systems of compact convex sets. *Proc. Amer. Math. Soc.* **65**, 24–8.

Rauch, J. 1974, An inclusion theorem for ovaloids with comparable second fundamental forms. *J. Differential Geom.* **9**, 501–5.

Reay, J. R. 1965, Generalizations of a theorem of Carathéodory. *Amer. Math. Soc. Memoirs* **54**.

Reidemeister, K. 1921, Über die singulären Randpunkte eines konvexen Körpers. *Math. Ann.* **83**, 116–18.

Reisner, S. 1985, Random polytopes and the volume-product of symmetric convex bodies. *Math. Scand.* **57**, 386–92.

1986, Zonoids with minimal volume product. *Math. Zeitschr.* **192**, 339–46.

1987, Minimal volume-product in Banach spaces with a 1-unconditional basis. *J. London Math. Soc.* **36**, 126–36.

Reiter, H. B. & Stavrakas, N. M. 1977, On the compactness of the hyperspace of faces. *Pacific J. Math.* **73**, 193–6.

Rényi, A. 1946, Integral formulae in the theory of convex curves. *Acta Sci. Math.* **11**, 158–66.

Rešetnjak, Ju. G. 1968a, Some estimates for almost umbilical surfaces (in Russian). *Sibirskii Mat. Zh.* **9**, 903–17. English translation: *Siberian Math. J.* **9**, 671–82.

1968b, Generalized derivatives and differentiability almost everywhere (in Russian). *Mat. Sbornik* **75**, 323–34. English translation: *Math. USSR (Sbornik)* **4**, 293–302.

Reuleaux, F. 1875, *Theoretische Kinematik*, vol. I. Vieweg, Braunschweig.

Ricker, W. 1981, A new class of convex bodies. In: *Papers in Algebra, Analysis and Statistics*, Hobart 1981, *Contemp. Math.* **9**, Amer. Math. Soc., Providence, RI, pp. 333–40.

Rickert, N. W. 1967a, Measures whose range is a ball. *Pacific J. Math.* **23**, 361–71.

1967b, The range of a measure. *Bull. Amer. Math. Soc.* **73**, 560–3.

Rogers, C. A. 1970, *Hausdorff Measures*. Cambridge Univ. Press.

Rogers, C. A. & Shephard G. C. 1957, The difference body of a convex body. *Arch. Math.* **8**, 220–33.

1958a, Convex bodies associated with a given convex body. *J. London Math. Soc.* **33**, 270–81.

1958b, Some extremal problems for convex bodies. *Mathematika* **5**, 93–102.

Rota, G.-C. 1964, On the foundations of combinatorial theory, I. Theory of Möbius functions. *Z. Wahrscheinlichkeitsth. verw. Geb.* **2**, 340–68.

1971, On the combinatorics of the Euler characteristic. In: *Studies in Pure Math.* (Papers presented to Richard Rado), Academic Press, London. pp. 221–33.

Rother, W. & Zähle, M. 1990, A short proof of a principal kinematic formula and extensions. *Trans. Amer. Math. Soc.* **321**, 547–58.

1992, Absolute curvature measures II. *Geom. Dedicata* **41**, 229–40.

Roy, A. K. 1972, Facial structure of the sum of two compact convex sets. *Math. Ann.* **197**, 189–96.

Roy, N. M. 1987, Extreme points of convex sets in infinite dimensional spaces. *Amer. Math. Monthly* **94**, 409–22.

Roy, S. N. 1936, On the vector derivation of the invariants and centroid formulae for convex surfaces. *Bull. Calcutta Math. Soc.* **28**, 79–88.

Sacksteder, R. 1960, On hypersurfaces with no negative sectional curvatures. *Amer. J. Math.* **82**, 609–30.

Saint Pierre, J. 1985, Point de Steiner et sections lipschitziennes. In: *Sémin. Anal. Convexe*, Univ. Sci. Tech. Languedoc 15, Exp. No. 7, 42 pp.

Saint Raymond, J. 1981, Sur le volume des corps convexes symétriques. In: *Séminaire Initiation à l'Analyse*, Univ. Pierre et Marie Curie, Paris 1980/81, vol. 11, 25 pp.

1986, Nombre de sommets d'un polyèdre aléatoire. In: *Séminaire Initiation à l'Analyse*, Univ. Pierre et Marie Curie, Paris 1985/86, vol. 11, 13 pp.

Salinetti, G. & Wets, R. J.-B. 1979, On the convergence of sequences of convex sets in finite dimensions. *SIAM Rev.* **21**, 18–33.

Sallee, G. T. 1966, A valuation property of Steiner points. *Mathematika* **13**, 76–82.

1968, Polytopes, valuations, and the Euler relation. *Canad. J. Math.* **20**, 1412–24.

1971, A non-continuous 'Steiner point'. *Israel J. Math.* **10**, 1–5.

1972, Minkowski decomposition of convex sets. *Israel J. Math.* **12**, 266–76.

1974, On the indecomposability of the cone. *J. London Math. Soc.* **9**, 363–7.

1982, Euler's theorem and where it led. In: *Convexity and Related Combinatorial Geometry* (eds D. C. Kay & M. Breen), Marcel Dekker, New York, pp. 45–55.

1987, Pairs of sets of constant relative width. *J. Geom.* **29**, 1–11.

Sandgren, L. 1954, On convex cones. *Math. Scand.* **2**, 19–28.

Sangwine-Yager, J. R. 1980, The mean value of the area of polygons circumscribed about a plane convex body. *Israel J. Math.* **37**, 351–63.

1983, The mean quermassintegral of simplices circumscribed about a convex body. *Geom. Dedicata* **15**, 47–57.

1988a, A Bonnesen-style inradius inequality in 3-space. *Pacific J. Math.* **134**, 173–8.

1988b, Bonnesen-style inequalities for Minkowski relative geometry. *Trans. Amer. Math. Soc.* **307**, 373–82.

1989, The missing boundary of the Blaschke diagram. *Amer. Math. Monthly* **96**, 233–7.

1991, Stability for a cap-body inequality. *Geom. Dedicata* **38**, 347–56.

Santaló, L. A. 1946, Sobre los cuerpos convexos de anchura constante en E_n. *Portugaliae Math.* **5**, 195–201.

1949, Un invariante afin para los cuerpos convexos del espacio de n dimensiones. *Portugaliae Math.* **8**, 155–61.

1950, On parallel hypersurfaces in the elliptic and hyperbolic n-dimensional space. *Proc. Amer. Math. Soc.* **1**, 325–30.

1953, *Introduction to Integral Geometry*. Hermann & Cie., Paris.

1961, Sobre los sistemas completos de desigualdades entre los elementos de una figura convexa plana. *Math. Notae* **17**, 1959/61, 82–104.

1976, *Integral Geometry and Geometric Probability*. Addison-Wesley, Reading, MA.

Sas, E. 1939, Über eine Extremaleigenschaft der Ellipsen. *Compositio Math.* **6**, 309–29.

Saškin, Ju. A. 1973, Convex sets, extreme points, and simplexes (in Russian). *Itôgi Nauki i Tekhniki (Mat. Analiz)* **11**, 5–50. English translation: *J. Soviet Math.* **4**, 625–55.

Schaal, H. 1962, Prüfung einer Kreisform mit Hohlwinkel und Taster. *Elem. Math.* **17**, 33–8.

1963, Gleitkurven in regulären Polygonen. *Z. angew. Math. Mech.* **43**, 459–76.

Schmidt, K. D. 1986, Embedding theorems for classes of convex sets. *Acta Appl. Math.* **5**, 209–37. Acknowledgement of priority: *Acta Appl. Math.* **11**, 295.

Schmitt, K.-A. 1967, Hilbert spaces containing subspaces consisting of symmetry classes of convex bodies. In: *Proc. Colloq. Convexity*, Copenhagen 1965. Københavns Univ. Mat. Inst., pp. 278–80.

1968, Kennzeichnung des Steinerpunktes konvexer Körper. *Math. Z.* **105**, 387–92. (This work is erroneous.)

Schneider, R. 1966a, Ähnlichkeits- und Translationssätze für Eiflächen. *Arch. Math.* **17**, 267–73.

1966b, On A. D. Aleksandrov's inequalities for mixed discriminants. *J. Math. Mech.* **15**, 285–90. (This work is erroneous.)

1967a, Über die Durchschnitte translationsgleicher konvexer Körper und eine Klasse konvexer Polyeder. *Abh. Math. Sem. Univ. Hamburg* **30**, 118–28.

1967b, Eine allgemeine Extremaleigenschaft der Kugel. *Monatsh. Math.* **71**, 231–7.

1967c, Zu einem Problem von Shephard über die Projektionen konvexer Körper. *Math. Z.* **101**, 71–82.

1967d, Zur affinen Differentialgeometrie im Großen I. *Math. Z.* **101**, 375–406.

1969a, Über die Finslerräume mit $S_{ijkl} = 0$. *Arch. Math.* **19**, 656–8.

1969b, Characterization of certain polytopes by intersection properties of their translates. *Mathematika* **16**, 276–82.

1970a, Über eine Integralgleichung in der Theorie der konvexen Körper. *Math. Nachr.* **44**, 55–75.

1970b, Eine Verallgemeinerung des Differenzenkörpers. *Monatsh. Math.* **74**, 258–72.

1970c, On the projections of a convex polytope. *Pacific J. Math.* **32**, 799–803.

1970d, Functional equations connected with rotations and their geometric applications. *L'Enseignement Math.* **16**, 297–305.

1971a, On Steiner points of convex bodies. *Israel J. Math.* **9**, 241–9.

1971b, Gleitkörper in konvexen Polytopen. *J. reine angew. Math.* **248**, 193–220.

1971c, Zwei Extremalaufgaben für konvexe Bereiche. *Acta Math. Hungar.* **22**, 379–83.

1971d, The mean surface area of the boxes circumscribed about a convex body. *Ann. Polon. Math.* **25**, 325–8.

1972a, Krümmungsschwerpunkte konvexer Körper, I. *Abh. Math. Sem. Univ. Hamburg* **37**, 112–32.

1972b, Krümmungsschwerpunkte konvexer Körper, II. *Abh. Math. Sem. Univ. Hamburg* **37**, 204–17.

1973, Volumen und Schwerpunkt von Polyedern. *Elem. Math.* **28**, 137–41.

1974a, Equivariant endomorphisms of the space of convex bodies. *Trans. Amer. Math. Soc.* **194**, 53–78.

1974b, Summanden konvexer Körper. *Arch. Math.* **25**, 83–5.

1974c, Bewegungsäquivariante, additive und stetige Transformationen konvexer Bereiche. *Arch. Math.* **25**, 303–12.

1974d, Additive Transformationen konvexer Körper. *Geom. Dedicata* **3**, 221–8.

1974e, On asymmetry classes of convex bodies. *Mathematika* **21**, 12–18.

1975a, Isometrien des Raumes der konvexen Körper. *Colloquium Math.* **33**, 219–24.

1975b, Remark on a conjectured characterization of the sphere. *Ann. Polon. Math.* **31**, 187–90.

1975c, Zonoids whose polars are zonoids. *Proc. Amer. Math. Soc.* **50**, 365–8.

1975d, A measure of convexity for compact sets. *Pacific J. Math.* **58**, 617–26.

1975e, Kinematische Berührmaße für konvexe Körper. *Abh. Math. Sem. Univ. Hamburg* **44**, 12–23.

1975f, Kinematische Berührmaße für konvexe Körper und Integralrelationen für Oberflächenmaße. *Math. Ann.* **218**, 253–67.

1976, Bestimmung eines konvexen Körpers durch gewisse Berührmaße. *Arch. Math.* **27**, 99–105.

1977a, Das Christoffel-Problem für Polytope. *Geom. Dedicata* **6**, 81–5.

1977b, Eine kinematische Integralformel für konvexe Körper. *Arch. Math.* **28**, 217–20.

1977c, Rekonstruktion eines konvexen Körpers aus seinen Projektionen. *Math. Nachr.*

79, 325–9.
1977d, Kritische Punkte und Krümmung für die Mengen des Konvexringes. *L'Enseignement Math.* **23**, 1–6.
1977e, Eine Charakterisierung der Kugel. *Arch. Math.* **29**, 660–5.
1978a, Curvature measures of convex bodies. *Ann. Mat. Pura Appl.* **116**, 101–34.
1978b, On the skeletons of convex bodies. *Bull. London Math. Soc.* **10**, 84–5.
1978c, Über Tangentialkörper der Kugel. *Manuscripta Math.* **23**, 269–78.
1978d, Kinematic measures for sets of colliding convex bodies. *Mathematika* **25**, 1–12.
1979a, Bestimmung konvexer Körper durch Krümmungsmaße. *Comment Math. Helvet.* **54**, 42–60.
1979b, On the curvatures of convex bodies. *Math. Ann.* **240**, 177–81.
1980a, Parallelmengen mit Vielfachheit und Steiner-Formeln. *Geom. Dedicata* **9**, 111–27.
1980b, Curvature measures and integral geometry of convex bodies. *Rend. Sem. Mat. Univ. Politecn. Torino* **38**, 79–98.
1981a, Crofton's formula generalized to projected thick sections. *Rend. Circ. Mat. Palermo* **30**, 157–60.
1981b, A local formula of translative integral geometry. *Arch. Math.* **36**, 149–56.
1981c, Pairs of convex bodies with unique joining metric segments. *Bull. Soc. Roy. Sci. Liège* **50**, 5–7.
1982a, Random hyperplanes meeting a convex body. *Z. Wahrscheinlichkeitsth. verw. Geb.* **61**, 379–87.
1982b, Random polytopes generated by anisotropic hyperplanes. *Bull. London Math. Soc.* **14**, 549–53.
1983, Nonparametric convex hypersurfaces with a curvature restriction. *Ann. Polon. Math.* **51**, 57–61.
1984, Smooth approximation of convex bodies. *Rend. Circ. Mat. Palermo*, Ser. II **33**, 436–40.
1985a, On the Aleksandrov–Fenchel inequality. In: *Discrete Geometry and Convexity* (eds J. E. Goodman, E. Lutwak, J. Malkevitch & R. Pollack), *Ann. New York Acad. Sci.* **440**, 132–41.
1985b, Inequalities for random flats meeting a convex body. *J. Appl. Prob.* **22**, 710–16.
1986a, Affine-invariant approximation by convex polytopes. *Studia Sci. Math. Hungar.* **21**, 401–8.
1986b, Curvature measures and integral geometry of convex bodies II. *Rend. Sem. Mat. Univ. Politecn. Torino* **44**, 263–75.
1987a, Geometric inequalities for Poisson processes of convex bodies and cylinders. *Results Math.* **11**, 165–85.
1987b, Polyhedral approximation of smooth convex bodies. *J. Math. Anal. Appl.* **128**, 470–4.
1987c, Equidecomposable polyhedra. In: *Intuitive Geometry*, Siófok 1985, *Colloquia Math. Soc. János Bolyai* **48**, North-Holland, Amsterdam, pp. 481–501.
1988a, On the Aleksandrov–Fenchel inequality involving zonoids. *Geom. Dedicata* **27**, 113–26.
1988b, Closed convex hypersurfaces with curvature restrictions. *Proc. Amer. Math. Soc.* **103**, 1201–4.
1988c, Random approximation of convex sets. *J. Microscopy* **151**, 211–27.
1989a, On a morphological transformation for convex domains. *J. Geom.* **34**, 172–80.
1989b, Gemischte Volumina in Kanalscharen. *Geom. Dedicata* **30**, 223–34.
1989c, Stability in the Aleksandrov–Fenchel–Jessen theorem. *Mathematika* **36**, 50–9.
1990a, On the Aleksandrov–Fenchel inequality for convex bodies, I. *Results Math.* **17**, 287–95.
1990b, A stability estimate for the Aleksandrov–Fenchel inequality, with an application to mean curvature. *Manuscripta Math.* **69**, 291–300.

1990c, Curvature measures and integral geometry of convex bodies III. *Rend. Sem. Mat. Univ. Politecn. Torino* **46**, 111–23.
1991+, Equality in the Aleksandrov–Fenchel inequality – present state and new results. Preprint.
Schneider, R. & Weil, W. 1970, Über die Bestimmung eines konvexen Körpers durch die Inhalte seiner Projektionen. *Math. Z.* **116**, 338–48.
1986, Translative and kinematic integral formulae for curvature measures. *Math. Nachr.* **129**, 67–80.
Schneider, R. & Wieacker, J. A. 1981, Approximation of convex bodies by polytopes. *Bull. London Math. Soc.* **13**, 149–56.
1984, Random touching of convex bodies. In: *Proc. Conf. Stochastic Geom., Geom. Statist., Stereology*, Oberwolfach 1983 (eds R. V. Ambartzumian & W. Weil), Teubner, Leipzig, pp. 154–69.
Schürger, K. 1983, Ergodic theorems for subadditive superstationary families of convex compact random sets. *Z. Wahrscheinlichkeitsth. verw. Geb.* **62**, 125–35.
Schütt, C. 1990+, On the affine surface area. Preprint.
1991, The convex floating body and polyhedral approximation. *Israel J. Math.* **73**, 65–77.
Schütt, C. & Werner, E. 1990, The convex floating body. *Math. Scand.* **66**, 275–90.
1990+, The convex floating body of almost polygonal bodies. Preprint.
Schwarz, T. & Zamfirescu, T. 1987, Typical convex sets of convex sets. *J. Austral. Math. Soc* (A) **43**, 287–90.
Seeley, R. T. 1966, Spherical harmonics. *Amer. Math. Monthly* **73**, 115–21.
Sen'kin, E. P. 1966a, Stability of the width of a general closed convex surface depending on the change of the integral mean curvature (in Russian). *Ukrain. Geom. Sb.* **2**, 88–9.
1966b, Stability of the width of a general closed convex surface depending on the change of the integral mean curvature, II (in Russian). *Ukrain. Geom. Sb.* **3**, 93–4.
Serra, J. 1982, *Image Analysis and Mathematical Morphology*. Academic Press, London.
1988, (ed.) *Image Analysis and Mathematical Morphology. Vol. 2: Theoretical Advances.* Academic Press, London.
Shahin, J. K. 1968, Some integral formulas for closed hypersurfaces in Euclidean space. *Proc. Amer. Math. Soc.* **19**, 609–13.
Shephard, G. C. 1960, Inequalities between mixed volumes of convex sets. *Mathematika* **7**, 125–38.
1963, Decomposable convex polyhedra. *Mathematika* **10**, 89–95.
1964a, Approximation problems for convex polyhedra. *Mathematika* **11**, 9–18.
1964b, Shadow systems of convex bodies. *Israel J. Math.* **2**, 229–36.
1966a, The Steiner point of a convex polytope. *Canad. J. Math.* **18**, 1294–300.
1966b, Reducible convex sets. *Mathematika* **13**, 49–50.
1966c, A pre-Hilbert space consisting of classes of convex sets. *Israel J. Math.* **4**, 1–10.
1967, Polytopes with centrally symmetric faces. *Canad. J. Math.* **19**, 1206–13.
1968a, The mean width of a convex polytope. *J. London Math. Soc.* **43**, 207–9.
1968b, A uniqueness theorem for the Steiner point of a convex region. *J. London Math. Soc.* **43**, 439–44.
1968c, Euler-type relations for convex polytopes. *Proc. London Math. Soc.* (3) **18**, 597–606.
1971, Diagrams for positive bases. *J. London Math. Soc.* (2) **4**, 165–75.
1974a, Combinatorial properties of associated zonotopes. *Canad. J. Math.* **26**, 302–21.
1974b, Space-filling zonotopes. *Mathematika* **21**, 261–9.
Shephard, G. C. & Webster, R. J. 1965, Metrics for sets of convex bodies. *Mathematika* **12**, 73–88.
Silverman, R. 1973a, Decomposition of plane convex sets, Part I. *Pacific J. Math.* **47**, 521–30.

1973b, Decomposition of plane convex sets, Part II: Sets associated with a width function. *Pacific J. Math.* **48**, 497–502. (This work is erroneous.)

Simon, U. 1967, Minkowskische Integralformeln und ihre Anwendungen in der Differentialgeometrie im Großen. *Math. Ann.* **173**, 307–21.

Smilansky, Z. 1986, An indecomposable polytope all of whose facets are decomposable. *Mathematika* **33**, 192–6.

1987, Decomposability of polytopes and polyhedra. *Geom. Dedicata* **24**, 29–49.

Sorokin, V. A. 1968, Classes of convex sets as generalized metric spaces (in Russian). *Mat. Zametki* **4** (1968), 45–52. English translation: *Math. Notes* **4**, 517–21.

Spiegel, W. 1975, Ein Konvergenzsatz für eine gewisse Klasse kompakter Punktmengen. *J. reine angew. Math.* **277**, 218–20.

1976a, Zur Minkowski-Additivität bestimmter Eikörperabbildungen. *J. reine angew. Math.* **286/287**, 164–8.

1976b, Ein Zerlegungssatz für spezielle Eikörperabbildungen in den euklidischen Raum. *J. reine angew. Math.* **283/284**, 282–6.

1978, Ein Beitrag über additive, translationsinvariante, stetige Eikörperfunktionale. *Geom. Dedicata* **7**, 9–19.

1982, Nonnegative, motion-invariant valuations of convex polytopes. In: *Convexity and Related Combinatorial Geometry* (eds D. C. Kay & M. Breen), Marcel Dekker, New York, pp. 67–72.

Stachó, L. L. 1979, On curvature measures. *Acta. Sci. Math.* **41**, 191–207.

Stanley, R. P. 1981, Two combinatorial applications of the Aleksandrov–Fenchel inequalities. *J. Combinatorial Theory, Ser. A* **31**, 56–65.

1986, Two poset polytopes. *Discrete Comput. Geom.* **1**, 9–23.

Starr, R. 1969, Quasi-equilibria in markets with nonconvex preferences. *Econometrica* **37**, 25–38.

1981, Approximation of points of the convex hull of a sum of sets by points of the sum: An elementary approach. *J. Econom. Theory* **25**, 314–17.

Steenaerts, P. 1985, Mittlere Schattengrenzenlänge konvexer Körper. *Results Math.* **8**, 54–77.

Steiner, J. 1840a, Von dem Krümmungsschwerpunkte ebener Curven. *J. reine angew. Math.* **21**, 33–63. *Ges. Werke*, vol. 2, Reimer, Berlin, 1882, pp. 99–159.

1840b, Über parallele Flächen. *Monatsber. preuß. Akad. Wiss.*, Berlin, 114–18. *Ges. Werke*, vol. 2, Reimer, Berlin, 1882, pp. 171–6.

Stoka, M. I. 1968, *Géométrie intégrale.* Gauthier-Villars, Paris.

Stoker, J. J. 1936, Über die Gestalt der positiv gekrümmten offenen Flächen. *Compositio Math.* **3**, 55–88.

1950, On the uniqueness theorems for the embedding of convex surfaces in three-dimensional space. *Commun. Pure Appl. Math.* **3**, 231–57.

Stoyan, D., Kendall, W. S. & Mecke, J. 1987, *Stochastic Geometry and its Applications.* Akademie-Verlag, Berlin.

Straszewicz, S. 1935 Über exponierte Punkte abgeschlossener Punktmengen. *Fund. Math.* **24**, 139–43.

Streit, F. 1970, On multiple integral geometric integrals and their applications to probability theory. *Canad. J. Math.* **22**, 151–63.

1973, Mean-value formulae for a class of random sets. *J. R. Statist. Soc.* B **35**, 437–44.

1975, Results on the intersection of randomly located sets. *J. Appl. Prob.* **12**, 817–23.

Su, B. 1927a, On Steiner's curvature centroid. *Japan. J. Math.* **4**, 195–201.

1927b, On Steiner's curvature centroid, II. *Japan. J. Math.* **4**, 265–9.

Sung, C. H. & Tam, B. S. 1987, On the cone of a finite-dimensional compact convex set at a point. *Linear Algebra Appl.* **90**, 47–55.

Süss, W. 1929, Zur relativen Differentialgeometrie V: Über Eihyperflächen im R^{n+1}. *Tôhoku Math. J.* **30**, 90–5.

1931a, Bestimmung einer geschlossenen konvexen Fläche durch die Gaußsche Krümmung. *Sitzungsber. Preuß. Akad. Wiss.* 686–95.

1931b, Die Isoperimetrie der mehrdimensionalen Kugel. *Sitzungsber. Preuß. Akad. Wiss.* 342–4.

1932a, Zusammensetzung von Eikörpern und homothetische Eiflächen. *Tôhoku Math. J.* **35**, 47–50.

1932b, Eindeutigkeitssätze und ein Existenztheorem in der Theorie der Eiflächen im Großen. *Tôhoku Math. J.* **35**, 290–3.

1933, Bestimmung einer geschlossenen konvexen Fläche durch die Summe ihrer Hauptkrümmungsradien. *Math. Ann.* **108**, 143–8.

Sz.-Nagy, B. v. (Szökefalvi-Nagy, B.) 1940, Über ein geometrisches Extremalproblem. *Acta. Sci. Math.* **9**, 253–7.

1959, Über Parallelmengen nichtkonvexer ebener Bereiche. *Acta Sci. Math. Szeged* **20**, 36–47.

Sz.-Nagy, G. v. 1949, Schwerpunkt von konvexen Kurven und von konvexen Flächen. *Portugaliae Math.* **8**, 17–22.

Tanno, S. 1976, C^∞-approximation of continuous ovals of constant width. *J. Math. Soc. Japan* **28**, 384–95.

1987, Central sections of centrally symmetric convex bodies. *Kodai Math. J.* **10**, 343–61.

Teissier, B. 1979, Du théorème de l'index de Hodge aux inégalités isopérimétriques. *C.R. Acad. Sci. Paris* **288**, 287–9.

1981, Variétés toriques et polytopes. In: *Séminaire Bourbaki*, vol. 1980/81, *Lecture Notes in Math.* **901**, Springer, Berlin, pp. 71–84.

1982, Bonnesen-type inequalities in algebraic geometry I: Introduction to the problem. In: *Seminar on Differential Geometry* (ed. S. T. Yau), *Ann. Math. Studies* **102**, Princeton Univ. Press, pp. 85–105.

Tennison, R. L. 1967, An almost-measure of symmetry. *Amer. Math. Monthly* **74**, 820–3.

Thomas, C. 1984, Extremum properties of the intersection densities of stationary Poisson hyperplane processes. *Math. Operationsforsch. Statist. (Ser. Statist.)* **15**, 443–9.

Tolstonogov, A. A. 1977, Support functions of convex compacta (in Russian). *Mat. Zametki* **22**, 203–13. English translation: *Math. Notes* **22**, 604–9.

Treibergs, A. 1985, Existence and convexity for hyperspheres of prescribed mean curvature. *Ann. Scuola Norm. Sup. Pisa, Cl. Sci., Ser. IV*, **12**, 225–41.

1990, Bounds for hyperspheres of prescribed Gaussian curvature. *J. Differential Geom.* **31**, 913–26.

Tverberg, H. 1966, A generalization of Radon's theorem. *J. London Math. Soc.* **41**, 123–8.

1981, A generalization of Radon's theorem, II. *Bull. Austral. Math. Soc.* **24**, 321–5.

Uhrin, B. 1984, Sharpenings and extensions of Brunn–Minkowski–Lusternik inequality. Technical Report No. 203, Stanford Univ., Dept. of Statistics, Stanford, CA.

1987, Extensions and sharpenings of Brunn–Minkowski and Bonnesen inequalities. In: *Intuitive Geometry*, Siófok 1985, *Colloquia Math. Soc. János Bolyai* **48**, North-Holland, Amsterdam, pp. 551–71.

Urbański, R. 1976, A generalization of the Minkowski–Rådström–Hörmander theorem. *Bull. Acad. Polon. Sci., Sér. sci., math., astr., phys.* **24**, 709–15.

Valette, G. 1971, Couples de corps convexes dont la différence des largeurs est constante. *Bull. Soc. Math. Belg.* **23**, 493–9.

1974, Subadditive affine-invariant transformations of convex bodies. *Geom. Dedicata* **2**, 461–5.

van Heijenoort, J. 1952, On locally convex manifolds. *Commun. Pure Appl. Math.* **5**, 223–42.

Vidal Abascal, E. 1947, A generalization of Steiner's formulae. *Bull. Amer. Math. Soc.* **53**, 841–4.

Vincensini, P. 1935, Sur les corps convexes admettant un domaine vectoriel donné. *C. R. Acad. Sci. Paris* **201**, 761–3.

1936, Sur les domaines vectoriels des corps convexes. *J. Math. pures appl.* **15**, 373–83.

1937a, Sur le prolongement des séries linéaires de corps convexes. Applications. *Rend.*

Circ. Mat. Palermo **60**, 361–72.
1937b, Les domaines vectoriels et la théorie des corps convexes. *L'Enseignement Math.*, 69–80.
1937c, Sur une transformation des corps convexes et son application à la construction de l'ensemble des corps convexes de l'espace à n dimensions à partir de certains sous-ensembles bases. *Bull. Soc. Math. France* **65**, 175–89.
1938, *Corps convexes. Séries linéaires. Domaines vectoriels. Mem. des Sci. Math.* **44**, Gauthier-Villars, Paris.
1956, Sur l'application d'une méthode géométrique à l'étude de certains ensembles de corps convexes. In: *Colloque sur les questions de réalité en géométrie,* Liège 1955, Paris, pp. 77–94.
1957, Sur un invariant de partition de l'ensemble des corps convexes de l'espace euclidien à n dimensions. *C. R. Acad. Sci. Paris* **245**, 132–3.
1965, Questions liées à la notion de géométrie différentielle globale. *Rev. Roum. Math. Pures Appl.* **10**, 877–93.
Vincensini, P. & Zamfirescu, T. 1967, Sur une fibration de l'espace des corps convexes. *C. R. Acad. Sci. Paris* **264**, 510–11.
Vitale, R. A. 1979, Approximation of convex set-valued functions. *J. Approx. Theory* **26**, 301–16.
1984, On Gaussian random sets. In: *Proc. Oberwolfach Conf. Stochastic Geom., Geom. Statist., Stereology* (eds R. V. Ambartzumian & W. Weil), Teubner, Leipzig, pp. 222–4.
1985a, L_p metrics for compact, convex sets. *J. Approx. Theory* **45**, 280–7.
1985b, The Steiner point in infinite dimensions. *Israel J. Math.* **52**, 245–50.
1990, The Brunn–Minkowski inequality for random sets. *J. Multivariate Anal.* **33**, 286–93.
1991, The translative expectation of a random set. *J. Math. Anal. Appl.* **160**, 556–62.
Vlasov, L. P. 1973, Approximative properties of sets in normed linear spaces (in Russian). *Usp. Mat. Nauk* **28**, 3–66. English translation: *Russ. Math. Surveys* **28**, 1–66.
Vodop'yanov, S. K. 1970, Estimates of the deviation from a sphere of quasi-umbilical surfaces. *Sibirskii Mat. Zh.* **11**, 971–87. English translation: *Siberian Math. J.* **11**, 724–35.
Voiculescu, D. 1966, O ecuatie privind corpurile convexe si aplicatii la corpurile asociate unui corp convex. *Stud. Cerc. Mat.* **18**, 741–5.
Volkov, Yu. A. 1963, Stability of the solution of Minkowski's problem (in Russian). *Vestnik Leningrad. Univ., Ser. Mat. Meh. Astronom.* **18**, 33–43.
Volkov, Yu., A. & Oliker, V. I. 1970, Uniqueness of the solution of Christoffel's problem for nonclosed surfaces (in Russian). *Mat. Zametki* **8**, 251–7. English translation: *Math. Notes.* **8**, 611–14.
Volland, W. 1957, Ein Fortsetzungssatz für additive Eipolyederfunktionale im euklidischen Raum. *Arch. Math.* **8**, 144–9.
Waksman, Z. & Epelman, M. S. 1976, On point classification in convex sets. *Math. Scand.* **38**, 83–96.
Wallen, L. J. 1987, All the way with Wirtinger: A short proof of Bonnesen's inequality. *Amer. Math. Monthly* **94**, 440–2.
Webster, R. J. 1965, The space of affine-equivalence classes of compact subsets of Euclidean space. *J. London Math. Soc.* **40**, 425–432.
Wegmann, R. 1980, Einige Maßzahlen für nichtkonvexe Mengen. *Arch. Math.* **34**, 69–74.
Wegner, B. 1977, Analytic approximation of continuous ovals of constant width. *J. Math. Soc. Japan* **29**, 537–40.
Weil, W. 1971, Über die Projektionenkörper konvexer Polytope. *Arch. Math.* **22**, 664–72.
1973, Ein Approximationssatz für konvexe Körper. *Manuscripta Math.* **8**, 335–62.
1974a, Über den Vektorraum der Differenzen von Stützfunktionen konvexer Körper. *Math. Nachr.* **59**, 353–69.
1974b, Decomposition of convex bodies. *Mathematika* **21**, 19–25.

1975a, On mixed volumes of nonconvex sets. *Proc. Amer. Math. Soc.* **53**, 191–4.
1975b, Einschachtelung konvexer Körper. *Arch. Math.* **26**, 666–9.
1976a, Kontinuierliche Linearkombination von Strecken. *Math. Z.* **148**, 71–84.
1976b, Centrally symmetric convex bodies and distributions. *Israel J. Math.* **24**, 352–67.
1977, Blaschkes Problem der lokalen Charakterisierung von Zonoiden. *Arch. Math.* **29**, 655–9.
1979a, Centrally symmetric convex bodies and distributions, II. *Israel J. Math.* **32**, 173–82.
1979b, Berührwahrscheinlichkeiten für konvexe Körper. *Z. Wahrscheinlichkeitsth. verw. Geb.* **48**, 327–38.
1979c, Kinematic integral formulas for convex bodies. In: *Contributions to Geometry, Proc. Geometry Symp.*, Siegen 1978 (eds J. Tölke & J. M. Wills), Birkhäuser, Basel, pp. 60–76.
1980a, On surface area measures of convex bodies. *Geom. Dedicata* **9**, 299–306.
1980b, Eine Charakterisierung von Summanden konvexer Körper. *Arch. Math.* **34**, 283–8.
1981a, Zufällige Berührung konvexer Körper durch q-dimensionale Ebenen. *Resultate Math.* **4**, 84–101.
1981b, Das gemischte Volumen als Distribution. *Manuscripta Math.* **36**, 1–18.
1982a, Inner contact probabilities for convex bodies. *Adv. Appl. Prob.* **14**, 582–99.
1982b, An application of the central limit theorem for Banach space-valued random variables to the theory of random sets. *Z. Wahrscheinlichkeitsth. verw. Geb.* **60**, 203–8.
1982c, Zonoide und verwandte Klassen konvexer Körper. *Monatsh. Math.* **94**, 73–84.
1983a, Stereology: A survey for geometers. In: *Convexity and its Applications* (eds P. M. Gruber & J. M. Wills), Birkhäuser, Basel, pp. 360–412.
1983b, Stereological results for curvature measures. *Bull. Int. Statist. Inst.* **50**, 872–83.
1984, Densities of quermassintegrals for stationary random sets. In: *Proc. Conf. Stochastic Geom., Geom. Statist., Stereology*, Oberwolfach 1983 (eds R. V. Ambartzumian & W. Weil), Teubner, Leipzig, pp. 233–47.
1987, Point processes of cylinders, particles and flats. *Acta Applicandae Math.* **9**, 103–36.
1988, Expectation formulas and isoperimetric properties for non-isotropic Boolean models. *J. Microscopy* **151**, 235–45.
1989a, Collision probabilities for convex sets. *J. Appl. Prob.* **26**, 649–54.
1989b, Integral geometry, stochastic geometry, and stereology. *Acta Stereologica* **8**, 65–76.
1989c, Translative integral geometry. In: *Geobild 89* (eds A. Hübler *et al.*), *Math. Research* **51**, Akademie-Verlag, Berlin 1989.
1990a, Iterations of translative integral formulae and non-isotropic Poisson processes of particles. *Math. Z.* **205**, 531–49.
1990b, Lectures on translative integral geometry and stochastic geometry of anisotropic random geometric structures. *Rend. Sem. Mat. Messina* (Serie II), **13** (0), 79–97.
Weil, W. & Wieacker, J. A. 1984, Densities for stationary random sets and point processes. *Adv. Appl. Prob.* **16**, 324–46.
1988, A representation theorem for random sets. *Prob. Math. Statist.* **9**, 147–55.
Weiss, V. 1986, Relations between mean values for stationary random hyperplane mosaics of R^d. *Forschungserg. FSU* N/86/33.
Weiss, V. & Zähle, M. 1988, Geometric measures for random curved mosaics of R^d. *Math. Nachr.* **138**, 313–26.
Weissbach, B. 1972, Rotoren im regulären Dreieck. *Publ. Math. Debrecen* **19**, 21–7.
1977a, Zur Inhaltsschätzung von Eibereichen. *Beitr. Alg. Geom.* **6**, 27–35.
1977b, Ausbohrung von Rhomben. *Beitr. Alg. Geom.* **6**, 153–8.
Wets, R. J.-B. 1974, Über einen Satz von Klee und Strascewicz. *Oper. Res. Verf.* **19**, 185–9.

Weyl, H. 1917, Über die Starrheit der Eiflächen und konvexen Polyeder. *Sitzungsber. Preuß. Akad. Wiss. Berlin*, 250–66.

1939, On the volume of tubes. *Amer. J. Math.* **61**, 461–72.

Wieacker, J. A. 1982, *Translative stochastische Geometrie der konvexen Körper*. Dissertation, Univ. Freiburg.

1984, Translative Poincaré formulae for Hausdorff rectifiable sets. *Geom. Dedicata* **16**, 231–48.

1986, Intersections of random hypersurfaces and visibility. *Prob. Th. Rel. Fields* **71**, 405–33.

1988a, Helly-type decomposition theorems for convex sets. *Arch. Math.* **50**, 59–67.

1988b, The convex hull of a typical convex set. *Math. Ann.* **282**, 637–44.

1988c, Decomposition of closed convex sets. *Geom. Dedicata* **28**, 221–8.

1989, Geometric inequalities for random surfaces. *Math. Nachr.* **142**, 73–106.

Wiesler, H. 1964, On convexity for generalized barycentric coordinates (in Russian). *Stud. Cerc. Mat.* **15**, 369–73.

Wijsman, R. A. 1964, Convergence of sequences of convex sets, cones and functions. *Bull. Amer. Math. Soc.* **70**, 186–8.

1966, Convergence of sequences of convex sets, cones and functions II. *Trans. Amer. Math. Soc.* **123**, 32–45.

Wills, J. M. 1970, Zum Verhältnis von Volumen zu Oberfläche bei konvexen Körpern. *Arch. Math.* **21**, 557–60.

1973, Zur Gitterpunktanzahl konvexer Mengen. *Elem. Math.* **28**, 57–63.

1990a, Kugellagerungen und Konvexgeometrie. *Jahresber. Deutsche Math.-Ver.* **92**, 21–46.

1990b, Minkowski's successive minima and the zeros of a convexity function. *Monatsh. Math.* **109**, 157–64.

Willson, S. J. 1980, A semigroup on the space of compact convex bodies. *SIAM J. Math. Anal.* **11**, 448–57.

Wintgen, P. 1982, Normal cycle and integral curvature for polyhedra in Riemannian manifolds. In: *Differential Geometry*, Budapest 1979 (eds G. Soos & J. Szenthe), *Coll. Math. Soc. János Bolyai* **31**, North-Holland, Amsterdam, pp. 805–16.

Wintner, A. 1952, On parallel surfaces. *Amer. J. Math.* **74**, 365–76.

Witsenhausen, H. S. 1973, Metric inequalities and the zonoid problem. *Proc. Amer. Math. Soc.* **40**, 517–20.

1978, A support characterization of zonotopes. *Mathematika* **25**, 13–16.

Wu, H. 1974, The spherical images of convex hypersurfaces. *J. Differential Geom.* **9**, 279–90.

Wunderlich, W. 1939, Über eine Klasse zwangläufiger höherer Elementenpaare. *Z. Angew. Math. Mech.* **19**, 177–81.

Yano, K. & Tani, M. 1969, Integral formulas for closed hypersurfaces. *Kōdai Math. Sem. Rep.* **21**, 335–49.

Yost, D. 1991 Irreducible convex sets. *Mathematika* **38**, 134–55.

Zähle, M. 1982, Random processes of Hausdorff rectifiable sets. *Math. Nachr.* **108**, 49–72.

1984a, Curvature measures and random sets, I. *Math. Nachr.* **119**, 327–39.

1984b, Properties of signed curvature measures. In: *Proc. Conf. Stochastic Geom., Geom. Statist., Stereology*, Oberwolfach 1983 (eds R. V. Ambartzumian & W. Weil), Teubner, Leipzig, pp. 256–66.

1986a, Integral and current representation of Federer's curvature measures. *Arch. Math.* **46**, 557–67.

1986b, Curvature measures and random sets II. *Prob. Th. Rel. Fields* **71**, 37–58.

1987a, Curvatures and currents for unions of sets with positive reach. *Geom. Dedicata* **23**, 155–71.

1987b, Normal cycles and second order rectifiable sets. *Forschungsergebnisse FSU* N/87/40.

1987c, Polyhedron theorems for non-smooth cell complexes. *Math. Nachr.* **131**, 299–310.
1988, Random cell complexes and generalized sets. *Ann. Prob.* **16**, 1742–66.
1989, Absolute curvature measures. *Math. Nachr.* **140**, 83–90.
1990, Approximation and characterization of generalized Lipschitz–Killing curvatures. *Ann. Global. Anal. Geom.* **8**, 249–60.

Zajíček, L. 1979, On the differentiation of convex functions in finite and infinite dimensional spaces. *Czech. Math. J.* **29**, 340–8.

Zalgaller, V. A. 1967, Mixed volumes and the chance of a hit in a convex region under multidimensional normal distribution (in Russian). *Mat. Zametki* **2**, 97–104. English translation: *Math. Notes* **2**, 542–5.
1972, Über k-dimensionale Richtungen, die für einen konvexen Körper F in R^n singulär sind (in Russian). *Zapiski naučn. Sem. Leningrad. Otd. mat. Inst. Steklov* **27**, 67–72. English translation: *J. Soviet Math.* **3**, 437–41.

Zamfirescu, T. 1966a, Sur les corps associés à un corps convexe. *Rev. Roum. Math. Pures Appl.* **11**, 727–35.
1966b, Sur les séries linéaires de corps convexes à frontières non différentiables et applications à la réductibilité. *Rev. Roum. Math. Pures. Appl.* **11**, 1015–1022.
1966c, Réductibilité et séries linéaires de corps convexes. *L'Enseignement Math.* **12**, 57–67.
1967a, Sur la réductibilité des corps convexes. *Math. Z.* **95**, 20–33.
1967b, Sur quelques questions de continuité liées à la réductibilité des corps convexes. *Rev. Roum. Math. Pures Appl.* **12**, 989–98.
1967c, Reducibility of convex bodies. *Proc. London Math. Soc.* **17**, 653–68.
1967d, Conditions nécessaires et suffisantes pour la réductibilité des voisinages des corps convexes. *Rev. Roum. Math. Pures Appl.* **12**, 1523–7.
1973, Two characterizations of the reducible convex bodies. *Abh. Math. Sem. Univ. Hamburg* **39**, 69–75.
1975, Metric spaces consisting of classes of convex bodies. *Rend. Ist. Mat. Univ. Trieste* **7**, 128–36.
1980a, The curvature of most convex surfaces vanishes almost everywhere. *Math. Z.* **174**, 135–9.
1980b, Nonexistence of curvature in most points of most convex surfaces. *Math. Ann.* **252**, 217–19.
1982a, Most convex mirrors are magic. *Topology* **21**, 65–9.
1982b, Many endpoints and few interior points of geodesics. *Invent Math.* **69**, 253–7.
1984a, Intersecting diameters in convex bodies. *Ann. Discrete Math.* **20**, 311–16.
1984b, Points on infinitely many normals to convex surfaces. *J. reine angew. Math.* **350**, 183–7.
1985, Using Baire categories in geometry. *Rend. Sem. Mat. Univers. Politecn. Torino* **43**, 67–88.
1987a, Nearly all convex bodies are smooth and strictly convex. *Monatsh. Math.* **103**, 57–62.
1987b, Typical convex curves on convex surfaces. *Monatsh. Math.* **103**, 241–7.
1988a, Curvature properties of typical convex surfaces. *Pacific J. Math.* **131**, 191–207.
1988b, Too long shadow boundaries. *Proc. Amer. Math. Soc.* **103**, 587–90.
1989+, Long geodesics on convex surfaces. Preprint.
1990, Nondifferentiability properties of the nearest point mapping. *J. d'Analyse Math.* **54**, 90–8.
1991, On two conjectures of Franz Hering about convex surfaces. *Discrete Comput. Geom.* **6**, 171–80.

Zhang Gaoyong 1991, Restricted chord projection and affine inequalities. *Geom. Dedicata* **39**, 213–22.

Zivaljević, R. T. 1989, Extremal Minkowski additive selections of compact convex sets. *Proc. Amer. Math. Soc.* **105**, 697–700.

SYMBOLS

AUTHOR INDEX

Section references are given. Most of the references occur in the section notes.

SUBJECT INDEX